ORDER NUMBER

PO NYAATP08

BOOK NUMBER

2327677

BAY NUMBER

2748-11

AUTHOR	Ludeman, Lonnie C.
TITLE	Random Processes: Filtering, Estimation, and Detection

RANDOM PROCESSES

RANDOM PROCESSES
FILTERING, ESTIMATION, AND DETECTION

Lonnie C. Ludeman
New Mexico State University

WILEY-INTERSCIENCE

A JOHN WILEY & SONS PUBLICATION

Published by John Wiley & Sons, Inc., Hoboken, New Jersey.
Published simultaneously in Canada.

Library of Congress Cataloging-in-Publication Data is available.

ISBN 0-471-25975-6

10 9 8 7 6 5 4 3 2 1

To Miranda, Laurie,
and the memory of Robbie Ludeman

CONTENTS

PREFACE

Every day we are confronted with signals that we cannot model exactly by an analytical expression or in a deterministic manner. Examples of such signals are ordinary speech waveforms, selections of music, seismological signals, biological signals, passive sonar records, temperature histories, and communication signals. Signals representing the same spoken word by a number of people or by the same person at different times will have similarities and differences that we can't seem to handle with deterministic modeling. How do we model them? Such signals can be described as somehow random and can be mathematically characterized as random processes. The characterization takes on a different form than an analytical expression and has several levels of sophistication where precise statements about the signals as a collection can be made only in a probabilistic sense. For example, we could say that at a given time the probability that a given random process is greater than a certain value is 0.25.

Random processes are basic in the fields of electrical and computer engineering—especially in the communication theory, computer vision, and digital signal processing areas—as well as vibrational theory and stress analysis in mechanical engineering. The digital processing of random signals is critical to high-performance communication systems, and nonlinear signal processing is consistently used to perform this processing.

These random processes serve as inputs to deterministic linear and nonlinear systems. A speech signal passes through a microphone to an amplifier to a speaker. What can we say about the output of such systems? The outputs can be modeled also as random processes. So the logical question is: Knowing the statistical characterization of the input process, what are the statistical properties of the output process?

In many cases the random signals are contaminated by extraneous noise that must also be considered to be random. If both of these processes can be characterized by their statistical properties, we would like to find ways of extracting the signal process from the noise process or ways of estimating, filtering, or predicting the signal process. We have available many acceptable deterministic linear and nonlinear processing techniques, yet the measure of goodness for our estimates cannot be handled with deterministic arguments or modeling. How do we specify a measure of goodness, and how do we find optimum systems for estimation?

We have a sample that we know could come from one of several random processes and would like to tell from which process it comes. For example, we might be given a portion of a speech signal and would like to determine which of the five vowels it is from. How can we describe a procedure for obtaining the best way to decide which vowel, and what are reasonable performance measures for defining best?

It is the purpose of this book to join the fundamentals of random processes and the standard techniques of linear and nonlinear systems analysis and hypothesis testing to give signal estimation techniques, to specify certain optimum estimation procedures, to provide optimum decision rules for classification purposes, and to describe performance evaluation definitions and procedures for our resulting methods.

My main objective in writing this book was to present random variables and random processes, their interaction with linear and nonlinear systems, their estimation and properties of their estimators, and detection of such processes in noisy environments in a pedagogically sound, easy to follow format that would be suitable for the advanced undergraduate or first-year graduate student in electrical and computer, geophysical, or mechanical engineering.

To this effect the contents of the book are conveniently broken up into four main, yet interrelated topics: probability and characterizations of random variables and random processes, linear and nonlinear systems with random excitations, optimum estimation theory including the Wiener and Kalman filter, and detection theory for both discrete and continuous time measurements.

The book can be used at several educational levels depending on the selected course coverage. For example, a course in basic probability and introduction to random processes can be presented from the text at the advanced senior or first-year graduate level, whereas if the course contains random processes filtering, estimation, and detection, the material in the text can be presented at the first-year graduate level in electrical and computer engineering. The book can also be used as the basic text for an upper-level graduate course in detection and estimation theory. At the senior/graduate level, some understanding of probability and random variables is helpful although not necessary, while at the graduate level, a basic course in engineering probability and basic system theory would be needed as prerequisites. As there are many selected examples illustrating the main concepts the book can be used also in self-guided study.

Almost all parts of the book have been classroom tested over the past 15 years in courses I have taught in three different forms:

- An introduction to probability, random variables, and random processes using Chapters 1, 2, and 3, and selections from Chapters 4, 5, and 6.
- An introduction to random processes: filtering and estimation using Chapters 4, 5, 6, and parts of 7 and 8.
- An introduction to detection and estimation theory using Chapters 3, 7, 8, 9, and 10.

The current version of the book has evolved considerably from its initial stages, reflecting student comments regarding their understanding of the topics covered and the overall style of presentation. Versions of the text with videocassette presentations have been used in video courses presented at the graduate level to electrical and computer engineering, physics, and other engineering students at New Mexico State University, Holloman AFB, White Sands Missile Range, Kirtland AFB, Kwangju Institute of Science and Technology, South Dakota School of Mines and Technology, and RCA/Juarez, Mexico, as well as in standard classroom courses at the above-mentioned institutions and the Indian Institute of Science, Bangalore. Each presentation format has had an effect on the structure and form of the manuscript dictating changes to provide an understandable and useful text for a wide audience and a variety of presentation schemes.

Many extra homework problems are provided which emphasize the same concepts so that the text can be used from year to year with minimal overlapping of homework assignments, and for extra practice to strengthen the basic concepts. Most of the problems were developed for in-class and take-home examinations to specifically test the basic concepts emphasized in the text. A few more complex problems are included to allow the student to extend those concepts. Some such problems may require tenacity and occasionally endurance to complete their solutions. A number of problems resist a total analytical solution and thus require basic computing skills to solve. The ability to distinguish which problems will need a computer solution and which can be solved analytically is a step in the development and maturity of the student.

A great debt is owed to the many students and to the institutions above for allowing me to present the material in this book, to the students for enduring not so perfect rough drafts and for their valuable suggestions and criticisms, and to the institutions for fostering environments conducive to creative endeavors.

The book's overall style of presentation was greatly influenced by two important publications: first, the classic text *Probability, Random Variables, and Stochastic Processes* by Athanasios Papoulis, and second, the monumental treatise of Harry L. Van Trees in his *Detection, Estimation, and Modulation*, Part I. As far as I am concerned there is nothing that can compete directly with these two ageless texts for their level of presentation, thoroughness, and coverage of topics. My book was not written to be in direct competition with either of these texts as there is not the need. The path I have chosen is to present selected basic fundamental material from both, at a slightly lower mathematical level, with more detail and examples but not compromise mathematical integrity. My choice of topics, however, has been guided by these two books, and a great debt is owed to these exceptional authors, educators, and scholars.

Last, always, and most important, the author appreciates the encouragement and support of family and friends throughout the course of this project.

Lonnie C. Ludeman
Las Cruces, New Mexico

1 Experiments and Probability

1.1 DEFINITION OF AN EXPERIMENT

To fully appreciate the meaning of probability and acquire a strong mathematical foundation for analytical work, it is necessary to define precisely the concept of an experiment and sample space mathematically. These definitions provide consistent methods for the assignment of elementary probabilities in paradoxical situations, and thus allow for meaningful calculation of probabilities of events other than the elementary events. Although at the beginning this approach may seem stilted, it will lead to a concrete concept of probability and an interpretation of derived probabilities.

An **experiment** $\mathscr{E}$ is specified by the three tuple $(S, \mathscr{F}, \mathscr{P}(.))$, where S is a finite, countable, or noncountable set called the **sample space**, $\mathscr{F}$ is a Borel field specifying a set of events, and $\mathscr{P}(.)$ is a probability measure allowing calculation of probabilities of all events.

1.1.1 The Sample Space

The sample space S is a set of elements called **outcomes** of the experiment $\mathscr{E}$ and the number of elements could be finite, countable, or noncountable infinite. For example, S could be the set containing the six faces of a die, $S = \{f_1, f_2, f_3, f_4, f_5, f_6\}$, or the positive integers, $S = \{i : i = 1, 2, \ldots\}$, or the real values between zero and one, $S = \{x : 0 < x < 1\}$, respectively.

An **event** is defined as any subset of S. On a single trial of the experiment an outcome is obtained. If that outcome is a member of an event, it is said that the event has occurred. In this way many different events occur at each trial of the experiment. For example, if f_1 is the outcome of a single trial of the experiment then the events $\{f_1\}, \{f_1, f_2\}, \{f_1, f_3\}, \ldots, \{f_1, f_6\}, \ldots, \{f_1, f_3, f_5\}, \ldots,$ all occur. Events consisting of single elements, like f_i, are called **elementary events**. The **impossible event** corresponds to the empty set ϕ and never occurs, while the **certain event**, S, contains all outcomes and thus always occurs no matter what the outcome of the trial is. Events A and B are called **mutually exclusive** or disjoint if $A \cap B = \phi$, where ϕ is the **null set**.

Two events A and B are called **independent** if $P(A \cap B) = P(A) \cdot P(B)$. The events $A_1, A_2, \ldots, A_n$ are defined to be independent if the probabilities of all intersections two, three, ..., and n events can be written as products. This implies for all $i, j, k, \ldots,$ that the following conditions must be satisfied for independence

$$\begin{aligned} P(A_i \cap A_j) &= P(A_i)P(A_j) \\ P(A_i \cap A_j \cap A_k) &= P(A_i)P(A_j)P(A_k) \\ &\vdots \\ P(A_1 \cap A_2 \cap \ldots \cap A_n) &= P(A_1)P(A_2)\ldots P(A_n) \end{aligned} \tag{1.1}$$

1.1.2 The Borel Field

A **field** can be defined as a nonempty class of sets such that (1) if $a \in \mathscr{F}$, then the complement of $a \in \mathscr{F}$ and (2) if $a \in \mathscr{F}$ and $b \in \mathscr{F}$ then $a \cup b \in \mathscr{F}$. Thus a field contains all finite unions, and by virtue of compliments and DeMorgan's theorem, all intersections of the collection. If we further require that all infinite unions and intersections are present in the collection, a **Borel field** is defined.

The set of all events of our experiment that will have probabilities assigned to them (measurable events) must be a Borel field to have mathematical consistency. If A, a collection of events, has a finite number of elements, a Borel field can be formed as the set of those events plus all possible subsets obtained by unions and intersections of those events including the null set ϕ and entire set S.

If a set is noncountable, it is little harder to describe a Borel field. The most common Borel field, containing the real numbers, is the smallest Borel field containing the following intervals: $\{x : x \leq x_1\}$ for all $x_1 \in$ real numbers. This will contain all finite and infinite closed and open intervals of the form $[a, b]$, $[a, b)$, $(a, b]$, and (a, b), where a and b are real numbers and the intersections and unions of those intervals thereof.

1.1.3 The Probability Measure

The **probability measure**, $\mathscr{P}(.)$, must be a consistent assignment of probabilities such that the following conditions are satisfied

(1) For any event $A \in \mathscr{F}$, the probability of the event A, $P(A)$, is such that $P(A) \geq 0$.

(2) For the certain event, S, $P(S) = 1$.

(3) If A and B are any two events such that $A \cap B = \phi$ then $P(A \cup B) = P(A) + P(B)$.

(3a) If $A_i \in \mathscr{F}$ for $i = 1, 2, \ldots,$ and $A_i \cap A_j = \phi$ for all $i \neq j$, then
$$P(A_1 \cup A_2 \cup \ldots \cup A_i \cup \ldots) = P(A_1) + P(A_2) + \cdots + P(A_i) + \cdots.$$

With these conditions satisfied, the probability of any event $A \in \mathscr{F}$ can be calculated. How does one go about assigning probabilities such that we satisfy the conditions above?

For the case where S is a set with a finite number of elements the conditions above can be satisfied by assigning probabilities to all the events with only single outcomes, $\{\zeta_i\}$, where $\zeta_i \in S$, such that conditions (1) and (2) above are satisfied. This mapping from the sample space S to the positive reals is called the **distribution function** for the probability measure and equivalently specifies the probability measure.

When S is a set with a noncountably infinite number of elements the assignment above is not useful as the probabilities of most elementary events will be zero. In this case the assignment of probabilities is consistent if the probabilities of the events $\{x : x \leq x_1\}$ for all $x_1 \in$ real numbers are assigned such that

(1) $0 \leq P\{x : x \leq x_1\} \leq 1$ (**bounded**).
(2) $P\{x : x \leq x_2\} \geq P\{x : x \leq x_1\}$ for all $x_2 > x_1$ (**nondecreasing function** of x) and
(3) lim as $\varepsilon \to 0$ of $P\{x : x \leq x_1 + \varepsilon\}$ equals $P\{x : x \leq x_1\}$ (**continuous from the right side**).

This mapping: $x \to P\{x : x \leq x_1\}$, defined for all x, is called the **cumulative distribution function** and equivalently describes the probability measure for the noncountable case. From this distribution function we are able to calculate all probabilities of events that are members of the Borel field $\mathscr{F}$. The cumulative distribution function could have also been used to specify the probability measure for the case where S has a countable or finite number of elements.

A number of examples of experiments will now be presented. They will include a couple of coin-tossing experiments and a die-rolling experiment. The experiments will be described by specifying their sample space, Borel field, and probability measures.

EXAMPLE 1.1

This experiment consists of a single flipping of a coin that results in either a head or a tail showing. Give its description by specifying as $(S, \mathscr{F}, \mathscr{P}(.))$.

SOLUTION

The possible outcomes of the experiment are either a head or a tail. Thus the sample space can be described as the set $S = \{\text{head}, \text{tail}\}$.

The Borel field $\mathscr{F}$ consists of the elementary events $\{\text{head}\}$ and $\{\text{tail}\}$, the null set $\varnothing$, and S.

To complete the description of the experiment, a probability measure must be assigned. This particular assignment could be based on previous experience, careful experimentation, use of favorable to total alternatives, or any other interpretation of the concept of probability. There is no right or wrong assignment, but certain assignments (models) may be more appropriate in explaining the results of corresponding physical experiments. For the purpose of this example, we assume that this coin has been tampered with for more often a head comes up than a tail, which we specify by $P\{\text{head}\} = p$ and $P\{\text{tail}\} = 1 - p$. These two assignments comprise the distribution function and thus the probability measure $\mathscr{P}(.)$ for the experiment. ■

EXAMPLE 1.2

A more realistic assignment for the experiment of flipping a coin could be the experiment $(S, \mathscr{F}, \mathscr{P}(.)))$ defined as follows: $S = \{\text{head, tail, edge}\}$ with $\mathscr{P}(.)$ described by $P\{h\} = 0.49, P\{t\} = 0.49, P\{e\} = 0.02$. The Borel field $\mathscr{F}$ is defined as the power set of S.

(a) Identify all the events (elements) of the Borel field. (b) Calculate the probabilities of those events.

SOLUTION

(a) The Borel field $\mathscr{F}$ specifying the measurable events is the power set of S (all possible subsets of S) given by

$$\mathscr{F} = \{\varnothing, \{h\}, \{t\}, \{e\}, \{h, t\}, \{h, e\}, \{t, e\}, \{h, t, e\}\}$$

where $\varnothing$ is the impossible event, and $\{h, t, e\}$ is the certain event. The other events consist of all possible proper subsets of S, single elements, and combinations of two elements.

(b) By definition $P(\varnothing) = 0$ and the probabilities of the elementary events are given in the specification of the probability measure of the experiment as $P\{h\} = 0.49$, $P\{t\} = 0.49$, and $P\{e\} = 0.02$. The probabilities of the other events can be determined by repeated application of property (3) for the probability measure. For example the events $\{h\}$ and $\{e\}$ are mutually exclusive therefore $P\{h, e\}$ can be found as follows:

$$\begin{aligned} P\{h, e\} &= P(\{h\} \cup \{e\}) = P\{h\} + P\{e\} \\ &= 0.49 + 0.02 = 0.51 \end{aligned}$$

Similarly $P\{h, t\} = 0.98$, $P\{t, e\} = 0.51$, and $P\{h, t, e\} = 1$. ■

EXAMPLE 1.3

An experiment consists of the "random" selection of a point somewhere in the closed interval [0,1]. Let the sample space S be described as $S = \{x : 0 \leq x \leq 1\}$. Define the Borel field $\mathscr{F}$ to be the smallest field containing the events $\{x : 0 \leq x \leq x_1\}$ for all $x_1 \in$ real numbers. Define the probability measure $\mathscr{P}(.)$, for the experiment by the following: $P\{x : x \leq x_1\} = x_1$ for all $0 \leq x_1 \leq 1$.

(a) Give three or four outcomes of this experiment. (b) Give a couple examples of regions in the event space (Borel field). (c) Determine the probabilities of the following events:

$$\{0 \leq x \leq 0.5\}, \{0.25 < x \leq 0.75\}, \{x > 0.75\}, \{x < 0.5\}, \text{and } \{x = 0.5\}.$$

SOLUTION

(a) The outcomes of the experiment are any real numbers in the closed interval [0,1]. Some examples are 0.333, $\pi/4$, 0.00001, and 0.2346.

(b) Possible regions in the event space are (0.5, 0.75], [0.6235, 0.8976), [0.2, 0.3], (0.627, 0.8), and any union of these types of regions.

(c) The probabilities of events can be determined by using set operations, the probability measure, and property (3). By direct application of the probability measure given, $P\{0 \leq x \leq 0.5\}$ is seen as

$$P\{0 \leq x \leq 0.5\} = 0.5$$

$P\{0.25 < x \leq 0.75\}$ can be determined indirectly as follows: We know that

$$\{0 \leq x \leq 0.75\} = \{0 \leq x \leq 0.25\} \cup \{0.25 < x \leq 0.75\}$$

Since $\{0 \leq x \leq 0.25\} \cap \{0.25 < x \leq 0.75\} = \emptyset$, we have by property (3) that

$$P\{0 < x \leq 0.75\} = P\{0 \leq x \leq 0.25\} + P\{0.25 < x \leq 0.75\}$$

Then rearranging and using the definition for the probability measure, we have

$$\begin{aligned} P\{0.25 < x \leq 0.75\} &= P\{0 \leq x \leq 0.75\} - P\{0 \leq x \leq 0.25\} \\ &= 0.75 - 0.25 = 0.5 \end{aligned}$$

$P\{x > 0.75\}$ is found indirectly using $\{0 \leq x \leq 1\} = \{0 \leq x \leq 0.75\} \cup \{x > 0.75\}$. Since $\{0 \leq x \leq 0.75\}$ and $\{x > 0.75\}$ are mutually exclusive, we have

$$P\{0 \leq x \leq 1\} = P\{0 \leq x \leq 0.75\} + P\{x > 0.75\}$$

After rearranging, $P\{x > 0.75\}$ is easily seen to be

$$P\{x > 0.75\} = P\{0 \leq x \leq 1\} - P\{0 \leq x \leq 0.75\}$$
$$= 1 - 0.75 = 0.25$$

Using similar concepts, it can be shown that $P\{x < 0.5\} = 0.5$ and $P\{x = 0.5\} = 0$. ■

1.2 COMBINED EXPERIMENTS

Combined experiments play an important role in probability theory applications. There are many ways we can combine experiments, including cartesian products in which independent trials of the same or different experiments can be described. In some cases the probabilities of events will depend on the results of previous trials of experiments or random selection of different experiments. A number examples of combined experiments are now explored beginning with the classical case of sampling with replacement.

1.2.1 Cartesian Product of Two Experiments

Consider the case of having two separate experiments specified by the following: $\mathscr{E}_1 : (S_1, \mathscr{F}_1, \mathscr{P}_1(.))$ and $\mathscr{E}_2 : (S_2, \mathscr{F}_2, \mathscr{P}_2(.))$. The sample spaces S_1 and S_2 are usually different sets, for example, results of a coin toss and results of a die roll, but they could be the same sets representing separate trials of the same experiment as in repeated coin-tossing experiments. We can define a new combined experiment by using the cartesian product concept as $\mathscr{E} = \mathscr{E}_1 \circledast \mathscr{E}_2$, where the new sample space $S = S_1 \circledast S_2$ is the **cartesian product** of the two sample spaces expressible by the ordered pair of elements where the first element is from S_1 and the second is from S_2.

EXAMPLE 1.4

Let experiment $\mathscr{E}_1 : (S_1, \mathscr{F}_1, \mathscr{P}_1(.))$ and $\mathscr{E}_2 : (S_2, \mathscr{F}_2, \mathscr{P}_2(.))$ be defined as follows: $\mathscr{E}_1$ is the experiment of flipping a coin with outcomes head (h) and tail (t) with equal probability of occurrence. $\mathscr{E}_2$ is the experiment of random selection of a colored ball from a box with outcomes red (r), white (w), and blue (b) with replacement. Define the new experiment $\mathscr{E}$ as $\mathscr{E} = \mathscr{E}_1 \circledast \mathscr{E}_2$ with experiments $\mathscr{E}_1$ and $\mathscr{E}_2$ being performed independently of each other; that is, the outcome of experiment $\mathscr{E}_1$ in no way effects the outcome of $\mathscr{E}_2$, and vice versa. Set up a reasonable model for this new experiment.

SOLUTION

To specify the model it suffices to give $S, \mathscr{F}, \mathscr{P}(.)$ of the new experiment $\mathscr{E}$.

The new sample space S is the cartesian product of the two experiments and given by $S = S_1 \circledast S_2$. Elements of S are ordered pairs with the first element coming from $S_1 = \{h, t\}$ and the second from $S_2 = \{r, w, b\}$; therefore

$$S = \{(h, r), (h, w), (h, b), (t, r), (t, w), (t, b)\}$$

The new Borel field $\mathscr{F}$ is selected as the power set of S, that is all possible unions and intersections of S. This includes the null set, the entire set, and all possible pairs, triples, and so on, as shown below:

$$\mathscr{F} = \left\{ \begin{array}{l} \varnothing, \{(h,r)\}, \{(h,w)\}, \{(h,b)\}, \{(t,r)\}, \{(t,w)\}, \{(t,b)\} \\ \{(h,r),(h,w)\}, \{(h,r),(h,b)\}, \{(h,r),(t,r)\}, \{(h,r),(t,w)\}, \ldots \\ \{(h,r),(h,w),(h,b)\}, \{(h,r),(h,w),(t,r)\}, \{(h,r),(h,w),(t,w)\}, \ldots \\ \vdots \\ \{(h,r),(h,w),(h,b),(t,r),(t,w),(t,b)\} \end{array} \right\}$$

The probability measure $\mathscr{P}(.)$ can be described by specifying the probability of the elementary events. Once these are known, the probability of any event can be found by writing the event as a union of those events and using property (3). Since the experiments are independent, it is reasonable to assign probabilities of the elementary events as a product of the probabilities from each experiment. For example, $P\{(h, r)\} = P\{h\} \cdot P\{r\}$. If head and tail are equally probable in $\mathscr{E}_1$ then it is reasonable for $\mathscr{P}_1(.)$ to be described by $P\{h\} = P\{t\} = \frac{1}{2}$, and if we have reason to believe that red, white, and blue are not equally probable in $\mathscr{E}_2$, then $\mathscr{P}_2(.)$ could be given by $P\{r\} = 0.5, P\{w\} = 0.3, P\{b\} = 0.2$. Thus, for this example, the probability measure $\mathscr{P}(.)$ can be described by specifying the probabilities of the elementary events (the distribution function):

$$\begin{array}{ll} P\{(h,r)\} = P\{(h)\} \cdot P\{(r)\} = 0.25 & P\{(t,r)\} = P\{(t)\} \cdot P\{(r)\} = 0.25 \\ P\{(h,w)\} = P\{(h)\} \cdot P\{(w)\} = 0.15 & P\{(t,w)\} = P\{(t)\} \cdot P\{(w)\} = 0.15 \\ P\{(h,b)\} = P\{(h)\} \cdot P\{(b)\} = 0.1 & P\{(t,b)\} = P\{(t)\} \cdot P\{(b)\} = 0.1 \end{array}$$

The event of a head in the new experiment is $H = \{(h, r), (h, w), (h, b)\}$, and its probability of occurrence can be determined using property (3) as

$$\begin{aligned} P(H) &= P\{(h,r),(h,w),(h,b) = P(\{(h,r)\} \cup \{(h,w)\} \cup \{(h,b)\}\} \\ &= 0.25 + 0.15 + 0.1 = 0.5 \end{aligned}$$

■

1.2.2 Cartesian Product of *n* Experiments

Consider the case of having n separate experiments specified by $\mathscr{E}_k : (\mathrm{S}_k, \mathscr{F}_k, \mathscr{P}_k(.))$ for $k = 1, 2, \ldots, n$. Define a new combined experiment $\mathscr{E} : (\mathrm{S}, \mathscr{F}, \mathscr{P}(.))$ as a cartesian product: $\mathscr{E} = \mathscr{E}_1 \circledast \mathscr{E}_2 \circledast \cdots \circledast \mathscr{E}_n$ where the new sample space

$S = S_1 \circledast S_2 \circledast \cdots \circledast S_n$ is the cartesian product of the n spaces and expressible by the ordered n-tuples of elements whose first element is from S_1 and the second from $S_2 \ldots$, the nth from S_n. The $\mathscr{E}_k$ are, in general, different, but in many cases the experiment could be formed from independent trials of the same experiment. Also there is an important class of problems where the experiments are the same, yet they cannot be thought of as independent. A good example of this is the random selection of outcomes without replacement. Examples of each type are now presented.

Binomial Distribution. Consider the experiment $\mathscr{E}_1 : (S_1, \mathscr{F}_1, \mathscr{P}_1(.))$ where the outcomes of the experiment are either failure indicated by a 0 or a success indicated by a 1; therefore $S_1 = \{0, 1\}$. Assume that the probability measure $\mathscr{P}_1(.)$ for the experiment is given by $P\{\text{success}\} \triangleq P\{1\} = p$ and $P\{\text{failure}\} \triangleq P\{0\} = 1 - p$ and that the $\mathscr{F}_1$ is the set $\{\varnothing, \{0\}, \{1\}, \{0, 1\}\}$. Define a new experiment by $\mathscr{E} = \mathscr{E}_1 \circledast \mathscr{E}_1 \circledast \cdots \circledast \mathscr{E}_1$ where $\mathscr{E} : (S, \mathscr{F}, \mathscr{P}(.))$ describes the new experiment. Assume that this represents independent trials of the same experiment $\mathscr{E}_1$ where the probability of success or failure is the same for each trial.

The $S, \mathscr{F}, \mathscr{P}(.)$ are now described for this new experiment. The new S is the cartesian product $S = \mathrm{S}_1 \circledast \mathrm{S}_1 \circledast \ldots \circledast \mathrm{S}_1$ and consists of all possible n-tuples where the elements are either 0 or 1 as shown below:

$$S = \begin{Bmatrix} (0, 0, \ldots, 0, 0) \\ (0, 0, \ldots, 0, 1) \\ (0, 0, \ldots, 1, 0) \\ (0, 0, \ldots, 1, 1) \\ \vdots \\ (1, 1, \ldots, 1, 1) \end{Bmatrix} \tag{1.2}$$

The new probability measure is specified once the distribution function or probabilities of the elements of S are determined. By virtue of the independent experiment assumption, the probability of each elementary event of the new experiment is the product of the probabilities for the elementary events in the single trial. For example, the $P\{(0, 1, 1, 0, \ldots, 0, 1, 1)\}$ is given by

$$\begin{aligned} P\{(0, 1, 1, 0, \ldots, 0, 1, 1)\} &= P\{0\} \cdot P\{1\} \cdot P\{1\} \cdot P(0) \cdot \ldots \cdot P\{0\} \cdot P\{1\} \cdot P\{1\} \\ &= p^4(1-p)^{n-4} \end{aligned} \tag{1.3}$$

As a matter of fact every sequence that has only four ones (successes) will have this same probability. The total number of these sequences is the combination of n things taken four at a time, since we have n locations and we want only four of them with ones. The probability distribution function for the new experiment can then be determined as follows, where the first entry is the elementary sequence and after the arrow is the corresponding probability:

Outcome **Probability**

$$
\begin{aligned}
(0, 0, \ldots, 0, 0) &\rightarrow P(0, 0, \ldots, 0, 0) = P(0) \cdot P(0) \cdots P(0) \cdot P(0) = p^0(1-p)^n \\
(0, 0, \ldots, 0, 1) &\rightarrow P(0, 0, \ldots, 0, 1) = P(0) \cdot P(0) \cdots P(0) \cdot P(1) = p^1(1-p)^{n-1} \\
(0, 0, \ldots, 1, 0) &\rightarrow P(0, 0, \ldots, 1, 0) = P(0) \cdot P(0) \cdots P(1) \cdot P(0) = p^1(1-p)^{n-1} \\
(0, 0, \ldots, 1, 1) &\rightarrow P(0, 0, \ldots, 1, 1) = P(0) \cdot P(0) \cdots P(1) \cdot P(1) = p^2(1-p)^{n-2} \\
&\cdots\cdots\cdots\cdots\cdots\cdots\cdots\cdots\cdots\cdots \\
(1, 1, \ldots, 1, 1) &\rightarrow P(1, 1, \ldots, 1, 1) = P(1) \cdot P(1) \cdots P(1) \cdot P(1) = p^n(1-p)^0
\end{aligned}
\tag{1.4}
$$

The Borel field will be the power set associated with S and specifies all events for which probabilities are assigned.

Probabilities of different type of events for the above example can be determined by using the distribution function described. There are a wide number of applications as a success can mean all kinds of things. For example, a success could be obtaining an ace in drawing a card from a standard deck of cards with replacement, or obtaining successful reception of a binary symbol from a random communication channel.

The event of **exactly k successes out of n independent trials** appears frequently in physical situations and its probability will now be derived using the results above. Define the event A_1 as the event of exactly one success out of n trials. From the results above we see that the probability of exactly one success out of n trials is

$$
\begin{aligned}
P(A_1) &= P(\text{exactly one success out of } n \text{ trials}) \\
&= P(\{(0, 0, \ldots, 0, 1)\} \cup \{(0, 0, \ldots, 1, 0)\} \cup \ldots \cup \{(1, 0, \ldots, 0, 0)\} \\
&= P\{(0, 0, \ldots, 0, 1)\} + P\{(0, 0, \ldots, 1, 0)\} + \cdots + P\{(1, 0, \ldots, 0, 0)\} \\
&= \binom{n}{1} p^1(1-p)^{n-1}
\end{aligned}
\tag{1.5}
$$

Similarly the probability of exactly k successes out of n trials can be found to be

$$
\begin{aligned}
P(A_k) &= P(\text{exactly } k \text{ success out of } n \text{ trials}) \\
&= \binom{n}{k} p^k(1-p)^{n-k}
\end{aligned}
\tag{1.6}
$$

In some problems we may wish to know the probability that **the number of successes out of n trials is within a range of values**. The probability that the number of successes k is in the range $m \leq k \leq n$ can be obtained by adding the probabilities of exactly k successes for that range, since the events A_k and A_j are mutually exclusive events for all k and j such that $k \neq j$. Therefore

$$P\left(J \leq \begin{matrix}\text{number of}\\ \text{successes}\end{matrix} \leq K\right) = P\left(\begin{pmatrix}\text{exactly } J\\ \text{successes}\end{pmatrix} + \begin{pmatrix}\text{exactly } J+1\\ \text{successes}\end{pmatrix} + \cdots + \begin{pmatrix}\text{exactly } K\\ \text{successes}\end{pmatrix}\right)$$

$$= P\left(\bigcup_{k=J}^{K} A_k\right) = \sum_{k=J}^{K} \binom{n}{k} p^k (1-p)^{n-k} \tag{1.7}$$

EXAMPLE 1.5

Consider the experiment of tossing a fair coin with the sample space $S = \{h, t\}$ and probability distribution $P\{h\} = 0.6$ and $P\{t\} = 0.4$. Suppose the experiment is performed 10 times independently to define a new cartesian product experiment.

(a) Determine the probability that we get exactly 5 heads in the 10 trials.
(b) Determine the probability that we get greater than 7 heads in the 10 trials.
(c) Determine the probability that we get less than or equal 9 heads in the 10 trials.
(d) Determine the probability that the number of tails is greater than or equal to 4 and less than or equal 5.

SOLUTION

The desired probabilities are determined for Eqs. (1.6) and (1.7) as follows:

(a) $P(A_1) = P(\text{exactly 5 heads out of 10 trials})$

$$= \binom{10}{5}(0.6)^5(0.4)^{10-5} = 0.20066$$

(b) $$P\left(8 \leq \begin{matrix}\text{number}\\ \text{of heads}\end{matrix} \leq 10\right) = P\left(\begin{pmatrix}\text{exactly}\\ \text{8 heads}\end{pmatrix} + \begin{pmatrix}\text{exactly}\\ \text{9 heads}\end{pmatrix} + \begin{pmatrix}\text{exactly}\\ \text{10 heads}\end{pmatrix}\right)$$

$$= \binom{10}{8}(0.6)^8(0.4)^2 + \binom{10}{9}(0.6)^9(0.4)^1 + \binom{10}{10}(0.6)^{10}(0.4)^0$$

$$= 0.1269790$$

(c) $P(\text{less than or equal 9 heads out of 10 trials})$

$$= 1 - P(\text{exacly 10 heads out of 10 trials})$$

$$= 1 - \binom{10}{10}(0.6)^{10}(0.4)^0 = 0.99395$$

(d) $$P\left(4 \leq \begin{matrix}\text{number}\\ \text{of heads}\end{matrix} \leq 5\right) = P\left(\begin{pmatrix}\text{exactly}\\ \text{4 heads}\end{pmatrix} + \begin{pmatrix}\text{exactly}\\ \text{5 heads}\end{pmatrix}\right)$$

$$= \binom{10}{4}(0.6)^4(0.4)^6 + \binom{10}{5}(0.5)^5(0.4)^5 = 0.1315425$$ ■

Approximations for Binomial Probabilities. The probabilities of k successes in n trials of a Bernoulli experiment was found to be as in Eq. (1.6) and evaluating probabilities of ranges of successes is given in Eq. (1.7). In Figure 1.1 the $P(k$ successes out of n trials) is plotted for values of n equal to 5, 10, and 50 for all values of k, and a $P(\text{success}) = 0.6$. The graphs show a tendency toward a hill-shaped curve similar to a Gaussian function. These plots would be symmetrical around the point $0.5\,n$ if $p = 0.5$. However, when p does not equal 0.5, as for the cases given, the plot is not quite symmetrical. If the number of trials is large, the calculation load due to the factorials is considerable, and certain approximations to these probabilities become useful. Many of these approximations are good provided that the number of trials n, number of successes k, and probability of success p satisfy a given set of conditions. Of these approximations we will present the DeMoivre and Poisson approximations and the regions where the approximations are reliable.

DeMoivre-Laplace Approximation. For $np(1-p) \gg 1$ and $|k - np|$ of the order of $\sqrt{np(1-p)}$,

$$\binom{n}{k} p^k (1-p)^{n-k} \approx \frac{1}{\sqrt{2\pi np(1-p)}} \exp\left\{-\frac{(k-np)^2}{2np(1-p)}\right\} \tag{1.8}$$

Poisson Approximation. For $n \gg 1, p \ll 1$, and np of order 1,

$$\binom{n}{k} p^k (1-p)^{n-k} \approx e^{(-np)} \frac{(np)^k}{k!} \tag{1.9}$$

Approximation of Regions of Successes in Bernoulli Trials. Say that we are interested in approximating the probability that the number of successes k in n repeated trials of a Bernoulli experiment is in the range $k_1 \le k \le k_2$. If this range of successes

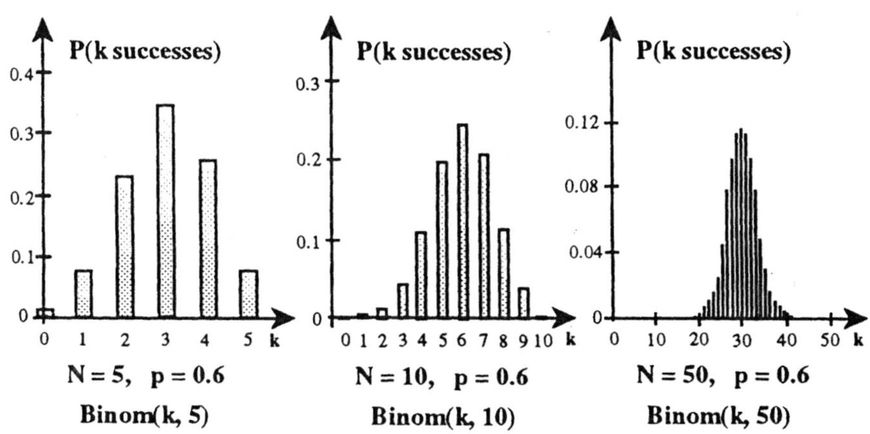

Figure 1.1 Plots of P (k successes out of n trials) for $p = 0.6$ and $n = 5, 10, 50$.

contains values that satisfy the DeMoivre-Laplace approximation, then the summation can be approximated by using the error function or the $\Phi(x)$ function as follows:

$$\sum_{k=k_1}^{k_2} \binom{n}{k} p^k (1-p)^{n-k} \approx \Phi\left(\frac{k_2 - np}{\sqrt{np(1-p)}}\right) - \Phi\left(\frac{k_1 - np}{\sqrt{np(1-p)}}\right) \tag{1.10}$$

$$\text{where } \Phi(x) = \int_{-\infty}^{x} \frac{1}{\sqrt{2\pi}} e^{-y^2/2} dy$$

The $\Phi(x)$ can only be determined by numerical integration as there is no antiderivative and it is convenient to use the table given in Appendix E.

EXAMPLE 1.6 (APPROXIMATIONS)

Consider the experiment of transmitting a single binary symbol across a communication channel that has an error probability $P(\text{error}) = 0.01$ (a pretty poor channel). Suppose that 1000 symbols are transmitted independently.

(a) Determine exactly the probability that exactly 10 errors occur in the 1000 trials.

(b) Are the conditions satisfied to use the DeMoivre-Laplace approximation? If so, evaluate the probability using the approximation and compare to the results in (a).

(c) Are the conditions satisfied to use the Poisson approximation? If not, calculate the probability using the approximation and compare to exact answer.

(d) Determine exactly the probability that the number of errors is greater than 7 but less than 16. Also use the DeMoivre approximation to determine the probability and compare it to the exact answer.

(e) Assuming 10,000 independent transmissions across a channel that has an error probability of 0.0001, determine the probability of exactly one error in the total transmission. Are the conditions satisfied to use the Poisson approximation? If so, determine the approximate results and compare to the exact results.

SOLUTION

(a) Using Eq. (1.6), the Bernoulli trials result, gives

$$P(\text{exactly 10 errors in 1000 trials}) = \binom{1000}{10} (0.01)^{10} (0.99)^{990} = 0.12574$$

(b) To see if the conditions are satisfied, it is necessary to calculate the following:

$$np(1-p) = 1000(0.01)(0.99) = 9.9$$
$$|k-np| = |10-1000(0.01)| = 0$$
$$\sqrt{np(1-p)} = \sqrt{1000(0.01)(0.99)} = 3.1464$$

Since $np(1-p) > 1$ and $|k-np|$ is of the same order of as $\sqrt{np(1-p)}$, the probability of exactly 10 errors can be approximated by the Gaussian function from Eq. (1.8) as follows:

$$\binom{1000}{10}0.01^{10}(0.99)^{990} \approx \frac{1}{\sqrt{2\pi}\,3.1464}\exp\left\{-\frac{(10-10))^2}{2(9.9)}\right\} = 0.126809$$

The exact result and approximate result differ in the third decimal. Although close, since $np(1-p)$ is only an order of magnitude rather than $\gg 1$, there is a loss in accuracy.

(c) To use the Poisson approximation the conditions $n \gg 1, p \ll 1$, and np of order 1 must be satisfied. At 1000, n is big enough, at 0.01 p is is small enough, but $np = 10$ is only marginally of order 1. Therefore we expect that the Poisson approximation will be marginal. Using the approximation given in Eq. (1.9) anyway gives

$$\binom{1000}{10}(0.01)^{10}(0.99)^{990} \approx e^{(-10)}\frac{(10)^{10}}{10!} = 0.12511$$

It is seen that this result is still pretty close to the exact value of 0.12574 determined in part (a).

(d) Since $|8-10|$ and $|15-10|$ are of order 3.1464 and $1000(0.01)(0.09) = 9.9$ is marginally $\gg 1$, then the approximation given in Eq. (1.10) can be used as follows:

$$\begin{aligned}\sum_{k=8}^{15}\binom{1000}{k}(0.01)^k(0.99)^{1000-k} &\approx \Phi\left(\frac{15-10}{3.1464}\right) - \Phi\left(\frac{8-10}{3.1464}\right)\\ &= \Phi(1.589) - \Phi(-0.635647)\\ &= 0.94408 - (1-0.73748) = 0.68156\end{aligned}$$

(e) The probability of exactly one error is given by the formula (1.6) as

$$\begin{aligned}P\left(\begin{matrix}\text{exactly 1 error}\\ \text{in 10,000 trials}\end{matrix}\right) &= \binom{10{,}000}{1}(0.0001)^1(0.9999)^{9999}\\ &= 0.367898\end{aligned}$$

To use the Poisson approximation the conditions $n \gg 1, p \ll 1$, and np of order 1 must be satisfied. At 10,000, n is big enough, at 0.0001, p is is small enough, and

$np = 1$ is definitely of order 1, so all conditions are satisfied and the approximation should be very good. The approximation from Eq. (1.9) is as follows:

$$\binom{10{,}000}{1} 0.0001^1 (0.9999)^{9999} \approx e^{(-1)} \frac{(1)^1}{1!} = 0.36788$$

It is seen that the approximation is accurate to the fourth decimal place. ■

Multinomial Distribution. A combined experiment that is an extension of the Bernoulli trial experiment just described, which resulted in the binomial distribution, is the case of multiple occurrences of several different events on multiple trials of the same experiment. Consider an experiment $\mathscr{E}_1 : (S_1, \mathscr{F}_1, \mathscr{P}_1(.))$. Define a set of events A_i, $i = 1$ to k, consisting of elements of S_1 such that their union is the certain event, they are pairwise disjoint, and their probabilities are given by $P(A_i) = p_i$ for $i = 1$ to k.

Define a new experiment $\mathscr{E}$ by $\mathscr{E} = \mathscr{E}_1 \circledast \mathscr{E}_1 \circledast \ldots \circledast \mathscr{E}_1$ described by $\mathscr{E} : (S, \mathscr{F}, \mathscr{P}(.))$. Assume that this represents n independent trials of the same experiment $\mathscr{E}_1$. Let n_i, $i = 1$ to k, be the number of times event A_i occurs in the new experiment, which is composed of n repeated trials. Also assume that $n_i \geq 0$ and $n_1 + n_2 + \cdots + n_k = n$.

It can be shown that the probability that A_i occurs exactly n_i times in the n trials is

$$P\begin{pmatrix} A_1 \text{ occurs exactly } n_1 \text{ times,} \\ A_2 \text{ occurs exactly } n_2 \text{ times,} \\ \vdots \\ A_k \text{ occurs exactly } n_k \text{ times} \end{pmatrix} = \frac{n!}{n_1!n_2!\ldots n_k!} p_1^{n_1} p_2^{n_2} \ldots p_k^{n_k} \tag{1.15}$$

EXAMPLE 1.7

An archer shoots at a target on a single trial of an experiment. Assume that the following events are defined: A_1 the arrow hits the bull's eye, A_2 the arrow hits the first ring of the target not on the bulls eye, and A_3 the arrow misses the target entirely. The probabilities of these events are given by

$$P(A_1) = 0.1, \quad P(A_2) = 0.7, \quad \text{and} \quad P(A_3) = 0.2$$

Assume that the archer shoots a number of arrows and that the results of each arrow shot are independent from the others. Find:

(a) The probability that on six arrows shot, the archer hit the bulls eye twice, hits the ring three times, and misses the target once.

(b) The probability that on five arrows shot, the archer hits the bulls eye three times and the ring twice.

SOLUTION

The desired probabilities are determined from Eq. (1-15) as follows

$$\text{(a)}\quad P\begin{pmatrix} A_1 \text{ occurs exactly 2 times,} \\ A_2 \text{ occurs exactly 3 times,} \\ A_3 \text{ occurs exactly 1 times} \end{pmatrix} = \frac{6!}{2!3!1!}(0.1)^2(0.7)^3(0.2)^1 = 0.04116$$

$$\text{(b)}\quad P\begin{pmatrix} A_1 \text{ occurs exactly 3 times,} \\ A_2 \text{ occurs exactly 2 times,} \\ A_3 \text{ occurs exactly 0 times} \end{pmatrix} = \frac{5!}{3!2!0!}(0.1)^3(0.7)^2(0.2)^0 = 0.049$$

Hypergeometric Distribution. In the previous compound experiments the trials of the experiment were considered to be independent. A very important class of experimentation problems is sampling done without replacement, for a finite number of elements, and without independent trials. For example, if a is the number of successful (s) elements and b is the number of failure (f) elements and n elements are drawn at random but not replaced, the corresponding outcomes of the experiment do not contain all possible n-tuples of s and f. So the n-tuple $\{s, s, \ldots, s\}$ and sequences containing more than a successes are not possible.

It can be shown for this experiment that the probability of k successes out of n trials can be determined as

$$P(k \text{ successes in } n \text{ trials}) = \frac{\binom{a}{k}\binom{b}{n-k}}{\binom{a+b}{n}}, \qquad k = 0, 1, \ldots \tag{1.16}$$

The next example is typical of the type of problems for which the hypergeometric formula above can be used.

EXAMPLE 1.8

Consider a standard deck of 52 playing cards with 4 aces, 4 kings, ..., and 4 twos. Say we draw 5 cards at random without replacement. What is the probability that our draw consists of (a) exactly 3 aces? (b) no aces? (c) more than two aces?

SOLUTION

(a) Using the formula (1.16) above, we find the probability of drawing exactly three aces as

$$P(3 \text{ successes in 5 trials}) = \frac{\binom{4}{3}\binom{48}{5-3}}{\binom{4+48}{5}} = 0.001736$$

(b) Using the formula (1.16) above, we find the probability of drawing exactly zero aces as

$$P(0 \text{ aces in } 5 \text{ trials}) = \frac{\binom{4}{0}\binom{48}{5-0}}{\binom{4+48}{5}} = 0.6588$$

(c) Using the formula (1.16) for two independent events, we find the probability of drawing more than two aces as

$$\begin{aligned} &P(\text{more than 2 aces in 5 draws}) \\ &= P((\text{exactly 3 aces in 5 draws}) \cup (\text{exactly 4 aces in 5 draws}) \\ &= \frac{\binom{4}{3}\binom{48}{5-3}}{\binom{48+4}{5}} + \frac{\binom{4}{4}\binom{48}{5-4}}{\binom{48+4}{5}} = 0.001754469 \end{aligned}$$

■

1.2.3 Counting Experiments

A special class of compound experiments deal with successive application of experiments where the total experiment may be stopped at any point depending on the results from the current experiment. Two common distributions result from these compound experiments.

Geometric Distribution. The geometric distribution is a result of another form of compound experiment. Define a countable number of identical experiments $\mathscr{E}_k : (\mathrm{S}_k, \mathscr{F}_k, \mathscr{P}_k(.))$ for $k = 1, 2, \ldots$. Each of these experiments will be of the Bernoulli type (two possible outcomes) previously discussed, with identical probabilities of success; the result of the experiment, if performed, is independent of previous or future experiments. The new experiment is described as follows: Perform experiment $\mathscr{E}_1$ if a success is obtained, then stop; if a success does not happen, then perform experiment $\mathscr{E}_2$; if a success occurs, then stop otherwise. Continue this process until a success occurs. Thus the number of trials is not the same each time the compound experiment is performed, and thus the sample space is not a cartesian product.

If p is the probability of success on each experiment, then the sample space, Borel field and probability measure can be described as follows: The elementary events are sequences of 1's (for success) and 0's for failures but with the property that they end on a 1 and only have 0's preceding the 1. Thus S can be described by

$$S = \{(1), (0, 1), (0, 0, 1), \ldots, (0, 0, \ldots, 0, 1), \ldots\} \tag{1.17}$$

Notice that (1, 0), (0, 1, 0), (1, 1), etc., are not elements of the sample space since the new experiment would stop at a 1 and not proceed to the next one. Events that are collections of outcomes are not of the same sequence length. For example, $\{(1), (0, 1)\}$ is an event where a success is obtained in less than or equal to 2 trials.

The probability measure $\mathscr{P}(.)$ can be specified by giving the distribution function for the elementary events. The $P\{(1)\}$ is the probability of getting a success on the very first trial. Since the probability of a success on the first experiment is p, the probability of getting $\{(1)\}$ is also p,

$$P\{(1)\} = p \tag{1.18}$$

The probability of getting $\{(0, 1)\}$ equals the probability of getting a failure on performing the first experiment and getting a success on the performance of the second experiment. Since results of each experiment are independent if performed, the $P\{(0, 1)\}$ can be written as a product:

$$P\{(0, 1)\} = (1 - p)p \tag{1.19}$$

Similarly the probability of the elementary event consisting of a sequence of $k - 1$ failures (0's) followed by a single success (1) is

$$\underset{(k-1)\text{zeros}}{P\{(0, 0, \ldots, 0, 1)\}} = (1 - p)^{k-1}p \tag{1.20}$$

These probabilities as $k \to \infty$ determine the distribution function, or equivalently the probability measure $\mathscr{P}(.)$. It can be shown that

$$\sum_{k=1}^{\infty}(1 - p)^{k-1}p = 1 \tag{1.21}$$

The Borel field is again defined as the power set of S, that is, all possible subsets of S. The probabilities of arbitrary events that are the subsets of S can then be found by adding up the probabilities of the elementary elements in the event. For example, define the event A to be the set of outcomes such that a success occurs on the performance of an odd number of experiments. Thus A can be written as

$$A = \{(1), (0, 0, 1), (0, 0, 0, 0, 1), \ldots, (0^{(2k)}, 1), \ldots\} \tag{1.22}$$

The probability of A can be determined by adding up the probabilities of the elementary events and using the properties of a geometric sequence as follows.

$$\begin{aligned} P(A) &= P(\{(1), (0, 0, 1), (0, 0, 0, 0, 1), \ldots, (0^{(2k)}, 1), \ldots\}) \\ &= \sum_{k=0}^{\infty}(1 - p)^{2k}p = p\left(\frac{1}{1 - (1 - p)^2}\right) = \frac{1}{2 - p} \end{aligned} \tag{1.23}$$

Negative Binomial Distribution. An extension of the geometric experiment is a compound experiment in which the number of trials necessary to obtain k successes is desired rather than the number of trials needed for a single success. The basic underlying experiment is to define a countable number of identical experiments $\mathscr{E}_k : (S_k, \mathscr{F}_k, \mathscr{P}_k(.))$ for $k = 1, 2, \ldots$. Each of these experiments will be of the Bernoulli type with identical probabilities of success, and the result of the experiment, if performed, is independent of previous or future experiments. The new experiment is described as follows: Perform experiment $\mathscr{E}_i$ if a success is obtained and it is the kth, then stop; if not, continue to the next experiment $\mathscr{E}_{i+1}$. Continue this process until the kth success occurs.

Let x be the number of trials in which the kth success occurs; then the probability of that event can be shown to be

$$P(\text{event that the kth success occurs on the } x\text{th trial}) = \binom{x-1}{k-1} p^k (1-p)^{x-k} \qquad x = k, k+1, k+2, \ldots \tag{1.24}$$

The following example illustrates the type of problems that will use the negative Binomial distribution given above.

EXAMPLE 1.9

Suppose the probability of getting a f_2 on the single toss of a die is 0.2. Find (a) the probability that the fourth f_2 occurs on the 6th trial; (b) that the fifth f_2 occurs before the 7th trial.

SOLUTION

(a) $P(\text{event that the 4th success occurs on the 6th trial})$

$$= \binom{6-1}{4-1} 0.2^4 (1-0.2)^{6-4} = 0.02048 \tag{1.25}$$

(b) $P(\text{event that the 5th success occurs before the 7th trial})$

$= P((\text{event that the 5th success occurs on the 5th trial})$

$\cup$ event that the 5th success occurs on the 6th trial))

$$= \binom{5-1}{5-1} 0.2^5 (1-0.2)^{5-5} + \binom{6-1}{5-1} 0.2^5 (1-0.2)^{6-5} \tag{1.26}$$

■

1.2.4 Selection Combined Experiment

A different type of combined experiment will now be considered that is a combination of three or more experiments. For purpose of illustration only three component experiments are presented. Let the three experiments be defined as

$$\mathscr{E}_0 : (S_0, \mathscr{F}_0, \mathscr{P}_0(.)), \quad \mathscr{E}_1 : (S_1, \mathscr{F}_1, \mathscr{P}_1(.)), \quad \mathscr{E}_2 : (S_2, \mathscr{F}_2, \mathscr{P}_2(.))$$

The sample space for $\mathscr{E}_0$ consists of only two outcomes (e_1 and e_2). These outcomes serve to help us select which one of the other experiments will be performed. If the outcome is e_1, then $\mathscr{E}_1$ is performed with outcome α_i; if the outcome is e_2 then $\mathscr{E}_2$ is performed with outcome β_i. Thus the combination experiment $\mathscr{E} : (S, \mathscr{F}, \mathscr{P}(.))$ can be described as follows: The sample space S will consist of ordered pairs of elements the first either e_1 or e_2 and the second either $\alpha_i \in S_1$ or $\beta_i \in S_2$. Thus a trial of the combined experiment results in an outcome $\zeta = (e, \delta)$, where $e \in \{e_1, e_2)$ and $\delta \in S_1 \cup S_2$.

The Borel field $\mathcal{F}$ will be defined to be the set of all possible subsets of S, and the associated probability measure can be described as

$$\begin{aligned} P(\{(e_i, \delta_j)\}) &= P_0(e_i) \cdot P(\delta_j) \quad \text{where} \\ P(\delta_j) &= P_1(\alpha_j) \quad \text{if } e_i = e_1 \quad \text{or} \\ P(\delta_j) &= P_2(\beta_j) \quad \text{if } e_i = e_2 \end{aligned} \tag{1.27}$$

The product of the probabilities is because of the assumption that the experiment $\mathcal{E}_0$ is independent of the experiments $\mathcal{E}_1$ and $\mathcal{E}_2$.

EXAMPLE 1.10

Define three experiments by $\mathcal{E}_0 : (S_0, \mathcal{F}_0, \mathcal{P}_0(.))$, $\mathcal{E}_1 : (S_1, \mathcal{F}_1, \mathcal{P}_1(.))$, and $\mathcal{E}_2 : (S_2, \mathcal{F}_2, \mathcal{P}_2(.))$. Experiment $\mathcal{E}_0$ is performed with $S_0 = \{e_1, e_2\}$. If the outcome is e_1 then box 1 is chosen and experiment $\mathcal{E}_1$ performed, while if e_2 is the outcome, $\mathcal{E}_2$ is performed. Experiment $\mathcal{E}_1$ is the random selection of a colored ball from a box with outcomes $S_1 = \{r, w, b\}$ where r means red, w means white, and b means blue ball chosen. The probability distribution for this experiment is given by

$$P_1(r) = 0.5, \quad P_1(w) = 0.3, \quad P_1(b) = 0.2$$

Similarly $\mathcal{E}_2$ is the random selection of a colored ball from box 2 having sample space with outcomes $S_2 = \{r, w, b\}$ and probability distribution given by

$$P_2(r) = 0.4, \quad P_2(w) = 0.1, \quad P_2(b) = 0.5$$

(a) Describe the sample space S for thẹ this new compound experiment.
(b) Assume $P_0(e_1) = 0.6$ and describe the probability measure $\mathcal{P}(.)$ for this new compound experiment.
(c) Show the event that the ball drawn is red.
(d) Show the event that box 1 is selected.

SOLUTION

(a) The new sample space S consists of ordered pairs

$$S = \{(e_1, r), (e_1, w), (e_1, b), (e_2, r), (e_2, w), (e_2, b)\}$$

This sample space is illustrated in Figure 1.2(a).

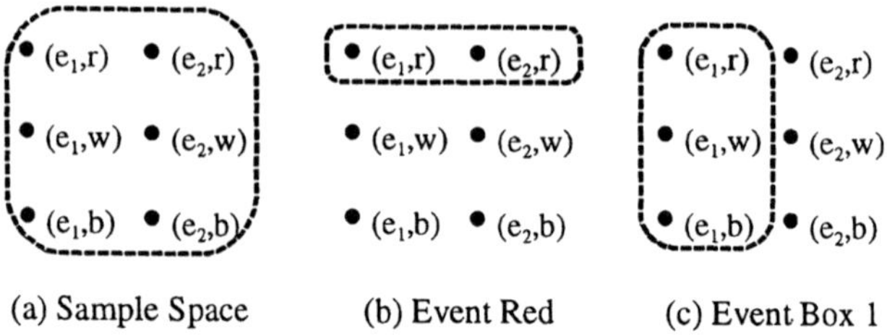

Figure 1.2 Set diagram for (a) Sample Space S, (b) Event red, (c) Event Box 1.

(b) The $\mathscr{P}(.)$ is specified by giving the distribution function. Using the results shown in the previous section because the experiments were given as independent, the distribution function is determined as

$$\begin{aligned}
&\text{Outcome} \rightarrow \text{Probability}\\
&(e_1, r) \rightarrow P\{(e_1, r)\} = P_0\{e_1\} \cdot P_1\{r\} = 0.6 \cdot 0.5 = 0.3\\
&(e_1, w) \rightarrow P\{(e_1, w)\} = P_0\{e_1\} \cdot P_1\{w\} = 0.6 \cdot 0.3 = 0.18\\
&(e_1, b) \rightarrow P\{(e_1, b)\} = P_0\{e_1\} \cdot P_1\{b\} = 0.6 \cdot 0.2 = 0.12\\
&(e_2, r) \rightarrow P\{(e_2, r)\} = P_0\{e_2\} \cdot P_2\{r\} = 0.4 \cdot 0.5 = 0.2\\
&(e_2, w) \rightarrow P\{(e_2, w)\} = P_0\{e_2\} \cdot P_2\{w\} = 0.4 \cdot 0.3 = 0.12\\
&(e_2, b) \rightarrow P\{(e_2, b)\} = P_0\{e_2\} \cdot P_2\{b\} = 0.4 \cdot 0.2 = 0.08
\end{aligned}$$

(c) The event the ball is red is shown in the Figure 1.2 and given by

$$\text{Ball is red} = \{(e_1, r), (e_2, r)\}$$

(d) The event box 1 is chosen, also shown in Figure 1.2, is as follows:

$$\text{Box 1 is chosen} = \{(e_1, r), (e_1, w), (e_1, b)\}$$ ■

1.3 CONDITIONAL PROBABILITY

The probability measure allows us to calculate the probabilities of all events that are members of the Borel field for the defined experiment. It will become useful to define the concept of conditional probability of events. Given a conditioning event C such that $P(C) > 0$, the **conditional probability of any event A assuming C** is defined as

$$P(A \mid C) \triangleq \frac{P(A \cap C)}{P(C)} \tag{1.28}$$

The following examples consider the calculation of conditional probabilities for discrete and continuous sample spaces. It is seen that the fundamental set operations and the probability measure for the underlying experiment are used to obtain conditional probabilities.

EXAMPLE 1.11

Consider an experiment defined as a single toss of a "crooked" die with sample space $S = \{f_1, f_2, f_3, f_4, f_5, f_6\}$. The probability measure, maybe based on past history, is known to be

$$P\{f_1\} = \tfrac{1}{2} \quad P\{f_2\} = \tfrac{1}{4} \quad P\{f_3\} = \tfrac{1}{8}$$
$$P\{f_4\} = \tfrac{1}{16} \quad P\{f_5\} = \tfrac{1}{32} \quad P\{f_6\} = \tfrac{1}{32}$$

Let the conditioning event C be given as $C = \{f_1, f_3, f_5\}$, the face of the die is odd, and calculate the $P(A \mid C)$ where $A = \{f_1, f_2\}$, the face is less than or equal 2.

SOLUTION

By the definition of conditional probability, Eq. (1.28), $P(A \mid C)$ is

$$P(A \mid C) = \frac{P(A \cap C)}{P(C)} = \frac{P(\{f_1, f_2\} \cap \{f_1, f_3, f_5\})}{P(\{f_1, f_3, f_5\})}$$
$$= \frac{P(\{f_1\})}{P(\{f_1, f_3, f_5\})} = \frac{1/2}{1/2 + 1/8 + 1/32} = \frac{16}{21} \quad \blacksquare$$

EXAMPLE 1.12

Consider the experiment of tossing two altered dice independently where the individual probabilities are as given in Example 1.11. Determine the probability of the event that one die is a six conditioned on the event that the sum of the two dice is 10.

SOLUTION

The sample space for the new experiment is the cartesian product of the single toss experiment, and thus the sample space S consists of the set of all ordered pairs with f_i as first element and f_j as second example:

$$S = \{(f_i, f_j) : i, j = 1, 2, \ldots, 6\}$$

The probability distribution for the sample space is the mapping from S to the reals, and by the independence condition, it is determined as the product

$$(f_i, f_j) \rightarrow P\{(f_i, f_j)\} = P\{f_i\} \cdot P\{f_j\} \qquad \text{for } i, j = 1, 2, \ldots, 6$$

The events under consideration are as follows:

$$\{\text{one die is 6}\} = \{(f_6, f_j) : j = 1, 2, \ldots, 5\} \cup \{(f_i, f_6) : i = 1, 2, \ldots, 5\} \cup \{(f_6, f_6)\}$$
$$\{\text{sum of dice equals 10}\} = \{(f_4, f_6), (f_6, f_4), (f_5, f_5)\}$$
$$\{\text{sum of dice equals 10}\} \cap \{\text{one die is 6}\} = \{(f_4, f_6), (f_6, f_4)\}$$

From the sets above the conditional probability desired is seen to be

$$\begin{aligned} P(\{\text{one die is a 6}\} \,|\, \{\text{sum is 10}\}) &= \frac{P(\{\text{one die is a 6}\} \cap \{\text{sum is 10}\})}{P(\{\text{sum is 10}\})} \\ &= \frac{P\{(f_4, f_6)\} + P\{(f_6, f_4)\}}{P\{(f_4, f_6)\} + P\{(f_4, f_6)\} + P\{(f_5, f_5)\}} \\ &= \frac{(1/16)(1/32) + (1/32)(1/16)}{(1/16)(1/32) + (1/32)(1/16) + (1/32)(1/32)} \\ &= \frac{4}{5} \end{aligned}$$

■

EXAMPLE 1.13

Define an experiment that has as outcomes, t, the set of real numbers greater than or equal to zero and that the outcome represents the time until a certain device fails. The probability measure for the experiment is described as

$$P(\text{failure time } t \leq x) = 1 - e^{-x} \qquad \text{for } x \geq 0$$

The Borel field is given by the smallest field containing the intervals $\{t : 0 \leq t \leq x\}$ for all $x \geq 0$. Find the $P(\text{failure time } t \leq 2 \,|\, \text{failure time } t > 1)$.

SOLUTION

From the definition of conditional probability the probability that the time to failure is less than or equal 2 for the device given that the failure time is greater than 1 is

$$P(\{\text{failure } t \leq 2\} \,|\, \{\text{failure } t > 1\}) = \frac{P(\{\text{failure } t \leq 2\} \cap \{\text{failure } t > 1\})}{P(\{\text{failure } t > 1\})}$$

Using set operations gives

$$\{\text{failure } t \leq 2\} \cap \{\text{failure } t > 1\} = \{\text{failure } 1 < t \leq 2\}$$

and the denominator is seen to be

$$\begin{aligned} P(\{\text{failure } t > 1\}) &= 1 - P(\{\text{failure } t \leq 1\}) \\ &= 1 - (1 - e^{-1}) = e^{-1} \end{aligned}$$

Substituting these two results into the first equation gives us the following result

$$P(\{\text{failure } t \leq 2\} \,|\, \{\text{failure } t > 1\}) = \frac{P(\{\text{failure} 1 < t \leq 2\})}{P(\{\text{failure } t > 1\})} = \frac{(1 - e^{-2}) - (1 - e^{-1})}{e^{-1}} = 1 - e^{-1} \quad \blacksquare$$

There are a number of special cases that deserve mentioning with regard to conditional probability. Usually A and C are overlapping but not necessarily included. The following special cases are for inclusion and nonoverlapping events:

$$\begin{aligned} &\text{If } A \subset C, \text{ then } A \cap C = A; \text{ thus } P(A \,|\, C) = \frac{P(A)}{P(C)} \\ &\text{If } A \supset C, \text{ then } A \cap C = C; \text{ thus } P(A \,|\, C) = \frac{P(C)}{P(C)} = 1 \\ &\text{If } A \cap C = \varnothing, \text{ then } P(A \,|\, C) = \frac{P(\varnothing)}{P(C)} = 0 \end{aligned} \tag{1.29}$$

If events A and C are independent, $P(A \cap C) = P(A)P(C)$, then conditional probabilities for the events can be written as

$$\begin{aligned} P(A \,|\, C) &= \frac{P(A \cap C)}{P(C)} = \frac{P(A)P(C)}{P(C)} = P(A) \\ P(C \,|\, A) &= \frac{P(C \cap A)}{P(A)} = \frac{P(C)P(A)}{P(A)} = P(C) \end{aligned} \tag{1.30}$$

Therefore, if events are independent, the conditional probabilities of the events are the same as the unconditional probabilities

1.3.1 Total Probability Theorem

Given n events $A_1, A_2, \ldots, A_n$ such that

$$\begin{aligned} &A_i \cap A_j = \varnothing \quad \text{for all } i \neq j \quad \text{(mutually exclusive)} \\ &\bigcup_{i=1}^{n} A_i = S \quad \text{(exhaustive)} \end{aligned} \tag{1.31}$$

Then it can be shown that the total probability of an arbitrary event B can be written in terms of the following conditional probabilities as

$$\begin{aligned} P(B) &= P(B \,|\, A_1)P(A_1) + P(B \,|\, A_2)P(A_2) + \cdots + P(B \,|\, A_n)P(A_n) \\ &= \sum_{i=1}^{n} P(B \,|\, A_i)P(A_i) \end{aligned} \tag{1.32}$$

1.3.2 Bayes's Theorem

A very important theorem that has many applications is Bayes's theorem which involves the determination of conditional probabilities $P(A_k \mid B)$ under the same framework as above. It is easily shown that

$$\begin{aligned} P(A_k \mid B) &= \frac{P(B \mid A_k)P(A_k)}{P(B \mid A_1)P(A_1) + P(B \mid A_2)P(A_2) + \cdots + P(B \mid A_n)P(A_n)} \\ &= \frac{P(B \mid A_k)P(A_k)}{\sum_{i=1}^{n} P(B \mid A_i)P(A_i)} \end{aligned} \tag{1.33}$$

An example is now presented that is representative of the type of problem that can be solved using Bayes's theorem and the total probability theorem.

EXAMPLE 1.14

Consider an experiment involving a random selection of one of three boxes. The random selection is of a single ball from the box chosen. The boxes contain red, white, and blue balls with specified probabilities of selection. Assume that

$$\begin{array}{rrr} P(\text{box1}) = 0.5 & P(\text{box2}) = 0.3 & P(\text{box3}) = 0.2 \\ P(\text{red} \mid \text{box1}) = 0.4 & P(\text{red} \mid \text{box2}) = 0.5 & P(\text{red} \mid \text{box3}) = 0.1 \\ P(\text{white} \mid \text{box1}) = 0.3 & P(\text{white} \mid \text{box2}) = 0.3 & P(\text{white} \mid \text{box3}) = 0.2 \\ P(\text{blue} \mid \text{box1}) = 0.3 & P(\text{blue} \mid \text{box2}) = 0.2 & P(\text{blue} \mid \text{box3}) = 0.7 \end{array}$$

(a) Find the probability of getting a red ball.

(b) Find the conditional probability, $P(A \mid B)$, where B is the event box 2 selected and A is the event a red ball is selected.

SOLUTION

(a) The total probability theorem can be used to obtain the probability of getting a red ball as follows:

$$\begin{aligned} P(\text{red}) &= P(\text{red} \mid \text{box1})P(\text{box1}) + P(\text{red} \mid \text{box2})P(\text{box2}) + P(\text{red} \mid \text{box3})P(\text{box3}) \\ &= 0.4 \cdot 0.5 + 0.3 \cdot 0.5 + 0.1 \cdot 0.2 = 0.37 \end{aligned}$$

(b) The probability of box 2 given that the ball is red can be determined by using Bayes's theorem and the results of part (a) as

$$\begin{aligned} P(\text{box 2} \mid \text{red}) &= \frac{P(\text{red} \mid \text{box2})P(\text{box2})}{p(\text{red})} \\ &= \frac{0.5 \cdot 0.3}{0.37} = 0.4054 \end{aligned}$$

■

1.4 RANDOM POINTS

The random placement of points in an interval is an important problem. Conceptually it is analogous to random arrival times used in basic inventory problems, and it is used as a basis for shot noise in communication theory. The random placement of the points can be uniformly or nonuniformly distributed on an interval as seen in the following sections.

1.4.1 Uniform Random Points in an Interval

Define an experiment $\mathscr{E} : (S, \mathscr{F}, \mathscr{P}(.))$ as the random placement of a point t somewhere in the closed interval $[0, T]$ as shown in Figure 1.3(*a*). The sample space is $S = \{t : 0 \le t \le T\}$, $\mathscr{F}$ is the smallest field containing the sets $\{t : t \le t_1\}$ for all $t_1 \in S$, and $\mathscr{P}(.)$ is defined by $P\{t : t \le t_1\} = t_1/T$, for all $t_1 \in S$. Thus it can be seen that the probability that the point selected will be in any given interval is the ratio of that interval's length to the total length T. Probabilities of other events that are unions of nonoverlapping intervals can be obtained by adding up the probabilities for each of the intervals.

The purpose of this section is to talk about the random placement of n points, not just one point, in an interval $[0, T]$ see Figure 1.3(*b*). A convenient way to design such an experiment is to form a new compound experiment $\mathscr{E}_n : (S_n, \mathscr{F}_n, \mathscr{P}_n(.))$ composed of ordered n-tuples of times, $(t_1, t_2, \ldots, t_n)$, obtained from repeating the experiment $\mathscr{E}$ defined above independently. Assume an independence of the trials so that the probability measure can be described as the product

$$P\{t_1 \le x_1, t_2 \le x_2, \ldots, t_n \le x_n\} = P\{t_1 \le x_1\}P\{t_2 \le x_1\} \ldots P\{t_n \le x_n\} \tag{1.34}$$

The Borel field is defined to be the smallest field containing the events $\{t_1 \le x_1, t_2 \le x_2, \ldots, t_n \le x_n\}$ for all $x_1, x_2, \ldots, x_n \in [0, T]$. We are interested in answering questions relating to calculating the probabilities that a certain number of points fall in a given interval or intervals.

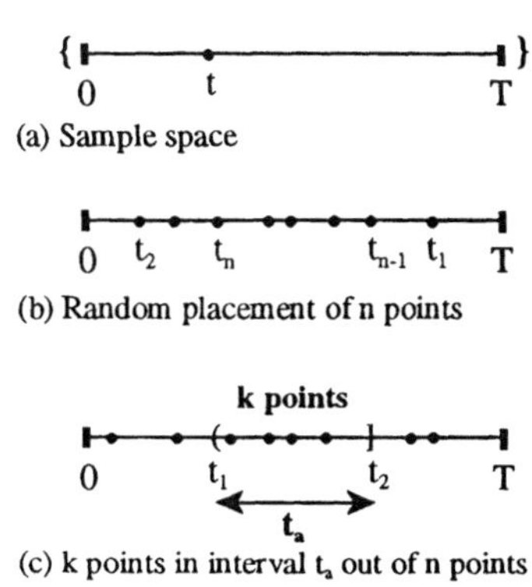

Figure 1.3 Random times in interval $[0, T]$.

Say that the probability that exactly k of the points fall in a given interval $(t_1, t_2]$ of length t_a as shown in Figure 1.3(c) is desired. Let A be the event that on a single trial the point selected at random falls in the given interval. Using the uniform probability measure $\mathscr{P}(.)$, we give the probability of A by $P(A) = (t_2 - t_1) = t_a$. Thus, on n independent trials, the probability of getting k points in the interval is binomially distributed as

$$P(k \text{ points in } (t_1, t_2) \text{ for } n \text{ trials}) = \binom{n}{k} p^k (1-p)^{n-k}$$
$$\text{where } p = \frac{t_a}{T} \tag{1.35}$$

Further assume that n, the number of trials, is very large, $n \gg 1$, and that the relative width of the interval is very small, $t_a/T \ll 1$, and k is of the order nt_a/T. The resulting Poisson approximation is written as

$$P\begin{pmatrix} \text{exactly } k \text{ points in } (t_1, t_2) \\ \text{out of } n \text{ trials} \end{pmatrix} \approx \exp\left(\frac{-nt_a}{T}\right) \frac{(nt_a/T)^k}{k!} \tag{1.36}$$
$$\text{where } t_a = t_2 - t_1$$

In the limiting case this result will give an interpretation in terms of an average number of points per unit interval. If $n \to \infty$, $T \to \infty$, and $n/T \to \lambda$, then it can be shown that

$$\lim_{t_a \to 0} \frac{P(\text{exactly 1point in } t_a)}{t_a} = \lambda \tag{1.37}$$

Another probability of interest is that of getting exactly k_a points in interval t_a and exactly k_b in interval t_b as indicated in Figure 1.4.

Let A, B, C equal the events that on a single trial of the experiment exactly one point falls in t_a, t_b, and not in t_a or t_b, respectively. Then the probability of getting exactly k_a points in interval t_a, exactly k_b points in interval t_b, and exactly $n - k_a - k_b$ points not in t_a or t_b out of n trials can be obtained from the multinomial result as

$$P\begin{pmatrix} k_a \text{ in } t_a \\ k_b \text{ in } t_b \\ n - k_a - k_b \\ \notin t_a \text{ or } t_b \end{pmatrix} = \frac{n!}{k_a! k_b! (n - k_a - k_b)!} \left(\frac{t_a}{T}\right)^{k_a} \left(\frac{t_b}{T}\right)^{k_b} \left(1 - \frac{t_a}{T} - \frac{t_b}{T}\right)^{n-k_a-k_b} \tag{1.38}$$

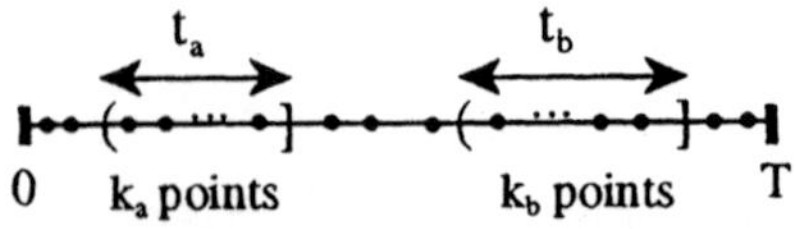

Figure 1.4 Exactly k_a points in interval t_a and exactly k_b in interval t_b.

The individual events exactly k_a points in t_a and exactly k_b points in t_b out of n trials have probabilities as follows:

$$\begin{aligned} P(k_a \text{ in } t_a) &= \binom{n}{k_a}\left(\frac{t_a}{T}\right)^{k_a}\left(1-\frac{t_a}{T}\right)^{n-k_a} \\ P(k_a \text{ in } t_a) &= \binom{n}{k_a}\left(\frac{t_b}{T}\right)^{k_b}\left(1-\frac{t_b}{T}\right)^{n-k_b} \end{aligned} \tag{1.39}$$

Since the product of the above two probabilities does not equal that of Eq. (1.38), the events exactly k_a points in t_a and exactly k_b points in t_b out of n trials are not independent events.

1.4.2 Nonuniform Random Points in an Interval

In certain problems the random points are not placed uniformly in the interval. A common way to describe a nonuniform rate is to assign the probability measure by using a weighting function $\alpha(t)$ that satisfies the following properties:

$$\begin{aligned} &\alpha(t) \geq 0 \quad \text{for all } t \in [0, T] \\ &\int_0^T \alpha(t)\, dt = 1 \end{aligned} \tag{1.40}$$

On a single trial of the experiment, the random placement of a single point in the interval $[0, T]$, the probability that the point selected is in $\{t_1, t_2\}$ is given by

$$P\{t \in (t_1, t_2]\} = \int_{t_1}^{t_2} \alpha(t)\, dt \tag{1.41}$$

Thus, if $\alpha(t)$ has a peak at $t = t_1$, it means that the point selected at random has a higher probability of being close to t_1 than other values of t. If n independent trials of this experiment are performed, the probability of exactly k points out of n trials being in the interval $\{t_1, t_2\}$ can be determined from Eq. (1.6) as

$$\begin{aligned} &P\begin{pmatrix} k \text{ points in } (t_1, t_2) \\ \text{out of } n \text{ trials} \end{pmatrix} = \binom{n}{k} p^k (1-p)^{n-k} \\ &\text{where } p = P\{t \in (t_1, t_2]\} = \int_{t_1}^{t_2} \alpha(t)\, dt \end{aligned} \tag{1.42}$$

For the case of a nonuniform rate with the assumption that n, the number of trials, is very large, $n \gg 1$, and that the relative width of the interval $t_a = (t_1, t_2)$ is very small, $t_a/T \ll 1$, and k of the order nt_a/T, the Poisson approximation results in the following probability where p is given as above:

$$P\begin{pmatrix} \text{exactly } k \text{ points in } (t_1, t_2) \\ \text{out of } n \text{ trials} \end{pmatrix} \approx \exp(-np)\frac{(np)^k}{k!} \tag{1.43}$$

1.5 SUMMARY

In this chapter the mathematical definitions of an experiment in terms of a sample space, a Borel field, and a probability measure were given. The concept of an experiment is basic to the understanding of random variables and random processes to be discussed in later chapters. The assignment of the probability measure for several experiments obtained by combining other experiments led to the binomial, multinormal, and hypergeometric distributions.

The important concept of conditional probability was defined, and several examples were presented to illustrate the relationship to the sample space and the probability measure. Using this definition a discussion of the total probability theorem and the Bayes theorem for obtaining the a posteriori probability followed. These concepts have a fundamental role in the detection and estimation of random variables and random processes as will be seen in Chapters 2, 8, and 9.

The chapter concluded with a short discussion on the random placement of points in a given interval. These experiments are important in analyzing problems involving random times of arrival and other related problems.

It was not the intent of this chapter to give an exhaustive presentation on experiments and their use but to provide background material that would be used in the remainder of the text. For those wishing a more thorough presentation there are many excellent texts, including Papoulis [1], Freund [3], Drake [4], and Parzen [7].

PROBLEMS

1.1 A regular 6 sided die is tossed twice. **(a)** What is the probability that the sum of the tosses is 6? **(b)** What is the probability that the die is a 1 on the first throw given that the sum is 6? **(c)** What assumptions did you make in obtaining your answers?

1.2 For a standard deck of 52 playing cards, what is

(a) the probability of being dealt at least 3 aces in a 5 card poker hand?

(b) the probability that a card drawn at random is a red card or an ace?

1.3 A small deck of cards contains four kings and four queens. Two cards are drawn without replacement. For this problem the suit is of no concern. Define an experiment, assuming that at the first draw each card is equally likely and that the second card drawn is equally likely from the remaining cards.

1.4 If the events A and B are mutually exclusive events and if A^c and B^c represent the compliments of A and B, respectively, prove or disprove that

(a) A^c and B are mutually exclusive events.

(b) A^c and B^c are mutually exclusive events.

(c) A and B are independent events.

(d) A^c and S are mutually exclusive events.

(e) A and $\varnothing$, the null event, are mutually exclusive events.

(f) A^c and B^c are independent events.

(g) A and B^c are independent events.

(h) A and S (the certain event) are independent events.

(i) A^c and $\varnothing$, the null event, are independent events.

1.5 A fisherman caught 10 fish, 3 of which were under six inches. Two of them are selected at random and measured. What is the probability that both fish selected are under six inches?

1.6 For three tosses of a fair coin, determine the probability of the following sequences:

(a) HHH.

(b) HTH.

(c) Two heads and one tail in any order.

(d) More heads than tails.

(e) More heads than tails given at least one tail.

(f) More heads than tails given less than two heads.

1.7 We have a box full of 50 balls. A trial of an experiment is the selection of a ball at random from a box. There are 5 balls of each of 10 different colors—call them $C_1, C_2, \ldots, C_{10}$.

(a) What is the probability of getting exactly 3 balls of color C_1 in 10 trials of the experiment *with* replacement?

(b) What is the probability of getting exactly 3 balls of color C_1 in 10 trials of the experiment *without* replacement?

(c) What is the probability of getting exactly 3 balls of C_1, 2 balls of C_2, and 3 balls of C_4 in 10 trials of the experiment with replacement.

1.8 We have a "crooked" coin, call it c, with the probability of getting a head being $\frac{3}{4}$ and a "fair" coin, call it f with probability of getting a head of $\frac{1}{2}$. An experiment is defined as selecting one of these coins at random, with probability of selecting the fair coin of 0.6, and flipping the coin selected. A single trial results in a head (h) or a tail (t).

(a) Determine the probability of getting a head on a single trial of the experiment.

(b) Given that on a single trial a head has occurred, determine the probability that the crooked coin was selected.

(c) Given two trials and that both outcomes were heads determine the probability that the crooked coin was selected.

1.9 Consider an urn with 3 red, 2 white, and 1 blue ball. An experiment consists in the drawing of an ordered pair of balls without replacement.

(a) Give the sample space for this experiment.

(b) What are the probabilities of all the elementary events. Explain any assumptions you made in assigning these probabilities.

(c) Calculate the probability of the event that a blue ball is drawn in a single trial of the experiment above.

1.10 Consider an urn with 3 red, 2 white, and 1 blue ball. An experiment is the drawing of an ordered triple of balls without replacement.

(a) Give the sample space for this experiment.

(b) What are the probabilities of all the elementary events? Explain any assumptions you made in assigning these probabilities.

(c) Calculate the probability of the event that none of the three balls drawn is blue in a single trial of the experiment above.

1.11 An urn contains two white balls and six red ones.

(a) What is the probability that one ball drawn at random is white?

(b) What assumptions did you make in arriving at your answer?

1.12 The probability that a rifleman hits the bull's-eye is 0.3, the probability that he hits the target but not the bull's-eye is 0.6 and the probability that he misses the target all together is 0.1. Determine the probability that in five shots at the target he will hit the bull's eye twice, the target but not the bull's eye twice and miss the target once.

1.13 In playing an opponent of equal ability, which is more probable: to win exactly three games out of four or to win exactly five games out of eight? Substantiate your answer.

1.14 A weighted coin is tossed ten times. On each trial the probability of a head is 0.8 and the probability a tail is 0.2. Determine the probabilities of the following events. **(a)** Exactly 5 heads. **(b)** At least 5 heads. **(c)** More than 5 heads. **(d)** Number of heads is greater than or equal to 5 and less than or equal to 7.

1.15 Given that a team is leading the world series 3 games to 2. What is the probability they will win the series if each team has an equal probability of winning at each game? (Remember the first team to win 4 games wins the series.)

1.16 We have 10 people each on a bus to Las Cruces which has four stops. Each person, acting independently from the others, has a probability of $\frac{1}{4}$ of getting off at each stop.

(a) What is the probability that on one trip to Las Cruces that exactly three people get off on stop 1, exactly two people get off at stop 2, exactly two people get off at stop 3, and exactly three people get off at stop 4?

(b) What is the probability that none get off at stop 1?

(c) What is the probability that less than 8 get off at stop 1?

1.17 The probability of getting a head on a single flip of a coin is p. The coin is tossed 10 times independently.

(a) Define the sample space for this experiment.

(b) Describe the probability assignment for the experiment.

(c) Give the event that an odd number of heads occur.

(d) Determine the probability of the event described in (c).

1.18 Given an honest standard deck of 52 playing cards a dealer draws one card at random from the deck and records it. He then slips the card back into the deck, shuffles, and is ready to draw another card.

(a) What is the probability that in 10 independent draws of the specified procedure above he will draw two or more aces?

(b) If the card was not replaced after each draw but kept out of the deck, what is the probability that of two cards drawn both are aces?

1.19 Say we have a hat with 10 letters, 3 a's, 5 b's, and 2 c's and a letter is drawn at random from the hat recorded and replaced. This experiment is repeated independently a number of times.

(a) What is the probability that on 10 draws that we get exactly k a's? Evaluate for $k = 3$.

(b) What is the probability that on 10 draws we get exactly 2 a's, 6 b's, and 2 c's?

(c) If we draw 10 times, what is the probability that we get more than 2 a's?

(d) If we draw 10 times, what is the probability that we get at least 2 a's?

(e) Determine the probability that we have 3 or 4 a's given that we have at least two a's.

(f) Suppose that on the first 10 trials we get 5 a's, 2 b's, and 3 c's. What is the probability that on the 11th trial that we get an a?

(g) If the letters are not replaced, find the probabilities for the following sequences: (a, a, a), (a, b, c), (c, c, c), and (a, a, b).

1.20 Given experiments $\mathscr{E}_1$ and $\mathscr{E}_2$ which are defined as follows:

$$\mathscr{E}_1 : (S_1, \mathscr{F}_1, \mathscr{P}_1(.)), S_1 = \{1, 2\}, \mathscr{F}_1 = \text{power set of } S_1,$$
$$\mathscr{P}_1(.) : P\{1\} = 2/3, P\{2\} = 1/3$$
$$\mathscr{E}_2 : (S_2, \mathscr{F}_2, \mathscr{P}_2(.)), S_2 = \{a, b\}, \mathscr{F}_2 = \text{power set of } S_2,$$
$$\mathscr{P}_2(.) : P\{a\} = 1/4, P\{b\} = 3/4$$

(a) Give all elements of $\mathscr{F}_2$.

(b) Considering $\mathscr{E}_1$ and $\mathscr{E}_2$ to be independent experiments, define a new experiment $\mathscr{E} = \mathscr{E}_1 \otimes \mathscr{E}_2$, the cartesian product, and specify its sample space, S, Borel field, $\mathscr{F}$, and probability measure, $\mathscr{P}(.)$. Be specific.

(c) Specify, for the combined experiment $\mathscr{E}$, the event that a 1 is obtained, and find the probability for this event.

1.21 Let $\mathscr{E}_1 : (S_1, \mathscr{F}_1, \mathscr{P}_1(.))$ and $\mathscr{E}_2 : (S_2, \mathscr{F}_2, \mathscr{P}_2(.))$ be independent experiments with

$$S_1 = \{a, b, c\}, \mathscr{F}_1 \text{ all subsets of } S_1, \text{ and}$$
$$\mathscr{P}_1 : (P\{a\} = 1/2, P\{b\} = 1/4, P\{c\} = 1/4)$$
$$S_2 = \{1, 2, 3\}, \mathscr{F}_2 \text{ all subsets of } S_2, \text{ and}$$
$$\mathscr{P}_1 : (P\{1\} = 1/3, P\{2\} = 1/3, P\{3\} = 1/3)$$

(a) List the elements of the cartesian product space $S = S_1 \otimes S_2$.

(b) What is the new $\mathscr{P}(.)$ for this combined experiment?

(c) For this new sample space, list the elements of the event that the number drawn is odd.

1.22 Say we have a big bag of red, blue, black, and white marbles. It is known that the probabilities of taking a red, blue, black, or white marble from the bag at random are, respectively, 0.3, 0.4, 0.2, and 0.1. Say we draw five marbles (with replacement) from the bag and the results at each trial are independent.

(a) Define the sample space for this new combined experiment.

(b) How many elements are there in this sample space?

(c) Specify the event R that at least one of the marbles is red.

(d) Specify the event Q that exactly two of the marbles are red.

(e) Give any elementary event for the new experiment and find its probability.

(f) Determine the probability that of the five marbles drawn exactly two of them are white and exactly one of them is red.

(g) What is the probability that the number of red marbles is greater than 1? Less than 2?

1.23 Evaluate the following integrals:

(a) $$\int_{-10}^{20} \frac{1}{\sqrt{2\pi}10} \exp\left\{-\frac{(x-10)^2}{200}\right\} dx$$

(b) $$\int_{-425}^{525} \frac{1}{\sqrt{2\pi}100} \exp\left\{-\frac{(x-500)^2}{20{,}000}\right\} dx$$

1.24 A fair coin, $P(\text{head}) = 0.5$, is tossed five times. Determine the probability that the number of heads occuring is greater than 2. Less than 2. Exactly equal to 2. Either 1 or 2.

1.25 An honest die is rolled 1000 times.

(a) Determine the probability that the number of 3's is between 120 and 150.

(b) Find the probability that the number of 3's is less than 20.

1.26 Consider the number of 3's that result from 600 tosses of a fair six-sided die. Find the probability that the number of 3's is between 60 and 140 inclusive.

1.27 An experiment of tossing a coin has the following outcomes: h, t, and e with given probabilities $P\{h\} = 0.5, P\{t\} = 0.495$, and $P\{e\} = 0.005$ (e means landed on edge).

(a) Give the exact expression for finding the probability that in 1000 independent trials that the number of heads is between and including 485 and 500.

(b) Give an approximate answer for the results in (a), if possible, and justify your approximation.

(c) In 1000 independent trials, what is the exact expression for the probability that the coin lands on edge less than two times.

(d) Give an approximate answer for your results in (c), if possible, and justify your approximation if you make one.

(e) Determine the probability that on 5 trials of the experiment we get exactly two h's, two t's and one e.

1.28 In a binary communication channel, the probability of making an error in the transmission of a single binary symbol is p. Assume that digits are sent along the channel independently.

(a) It is desired that the probability of 2 or fewer errors in the transmission of 100 symbols is 0.99. What is the largest value of p that satisfies this condition?

(b) Let $p = 0.001$. If less than 2 errors occur in 100 trials, determine the probability that there were no errors.

(c) If $p = 1$, "a worthless channel," determine the probability that the number of errors N out of 1000 trials is $400 \leq N \leq 600$.

1.29 In a binary communication channel the probability of making an error on a given transmission of a binary symbol is 0.01.

(a) In sending of 10,000 independent symbols, determine the probability that the number of errors is between and including 95 to 100 errors.

(b) What is the probability that the number of errors is 2 or less in the transmission of 100 symbols?

1.30 A player sits on the origin of a graph and flips a crooked coin where bias to heads is given by $P(H) = \frac{2}{3}$. If the result of the toss is a head, he moves one unit to the right, and if it is a tail, he moves one unit to the left. The coin is flipped again, and he moves from his new position in the same way. Repeating this procedure a total of 1200 times, what is the probability that the player ends up sitting on a number N where $350 \leq N \leq 450$?

1.31 If the probability that a person will catch a fish in 0.01 minutes is 0.02,

(a) Determine the probability that a person will not catch a single fish in one hour.

(b) Find the probability that in 2 hours he will catch at least 4 fish.

1.32 Suppose that there is a diagnostic test for measles. Define C as the event that a person tested has measles and A as the event that the test states that the person tested has the measles. Assume that $P(A \mid C) = 0.95$ and $P(A^c \mid C^c) = 0.95$. Say the probability that a person taking the test actually has measles is 0.005.

(a) Compute the probability that a person who tests positive actually does have measles.

(b) Compute the probability of A.

1.33 Consider a source that sends a 0 or a 1 with known probabilities

$$P(0 \text{ sent}) = p \text{ and } P(1 \text{ sent}) = 1 - p$$

The digit is sent over a noisy channel with the following transition probabilities:

$$P(0 \text{ received} \mid 1 \text{ sent}) = P(1 \text{ received} \mid 0 \text{ sent}) = 1/4$$
$$P(0 \text{ received} \mid 0 \text{ sent}) = P(1 \text{ received} \mid 1 \text{ sent}) = 3/4$$

(a) Find the probability that a 0 is received.

(b) What does p need to be for this probability to be $\frac{3}{8}$?

(c) What is the probability that a 1 was sent given that a 1 was received?

1.34 Given two boxes B_1 and B_2 where B_1 contains 8 red and 8 white balls and B_2 contains 5 red and 10 white balls. An experiment consists of selecting one of the boxes with $p(B_1) = 0.4$ and $P(B_2) = 0.6$, and then selecting a ball at random from the selected box.

(a) Find the probability that the ball drawn is red.

(b) If a red ball is chosen, determine the probability that it is from box B_1.

1.35 In a year's time if a tree is not watered, it will die with a probability of $\frac{1}{2}$, and if watered, it will die with probability of $\frac{1}{4}$. After a year, the tree is dead. Determine the probability that the tree was watered.

1.36 A crooked coin ($P\{\text{head}\} = 0.25$) is flipped twice with experimental outcome that at least one toss resulted in a head. Given this partial information, what is the conditional probability that both tosses resulted in heads.

1.37 From a master handwritten manuscript composed of the letters a, b, and c (the author has a limited vocabulary) with $P(a) = 0.5, P(b) = 0.2$, and $P(c) = 0.3$. A typist copies the manuscript making random errors when typing the letters a, b, and c with probabilities as follows $P(b|a) = 0.05$, $P(c|a) = 0.1$, $P(a|b) = 0.8$, $P(c|b) = 0.7$, $P(a|c) = 0.2$, $P(b|c) = 0.2$.

(a) What is the probability that a letter a typed was really a letter b in the original document?

(b) Given that a letter c was typed, what is the probability that there was a letter c in that position in the original manuscript?

REFERENCES

1. Papoulis, Athanasios, *Probability, Random Variables, and Stochastic Processes*, McGraw Hill, 1965.
2. Mood, Alexander M., Franklin A. Graybill, and Duane C. Boes, *Introduction to the Theory of Statistics*, 3rd ed. McGraw-Hill, 1974.
3. Freund, John E., *Mathematical Statistics*, Prentice-Hall, 1964.
4. Drake, Alvin W., *Fundamentals of Applied Probability Theory*, McGraw-Hill, 1967.
5. Davenport, Jr., William B., *Probability and Random Processes*, 1970.
6. Stark, Henry and John W. Woods, *Probability, Random Processes, and Estimation Theory for Engineers*, Prentice-Hall, 1986.
7. Parzen, Emanuel, *Modern Probability Theory and Its Applications*, Wiley, 1960.

2 Random Variables

2.1 DEFINITION OF A RANDOM VARIABLE

An **experiment** $\mathscr{E}$ is specified by the three tuple $(S, \mathscr{F}, \mathscr{P}(.))$ where S is a finite, countable, or noncountable set called the sample space, $\mathscr{F}$ is a Borel field specifying a set of events, and $\mathscr{P}(.)$ is a probability measure allowing calculation of probabilities of all events. Using an underlying experiment a **random variable** $X(\zeta)$ is defined as a real-valued function on S that satisfies the following: (a) $\{\zeta : X(\zeta) \leq x\}$ is a member of $\mathscr{F}$ for all x, which guarantees the existence of the cumulative distribution function, and (b) the probabilities of the events $\{\zeta : X(\zeta) = +\infty\}$ and $\{\zeta : X(\zeta) = -\infty\}$ are both zero. This means that the function is not allowed to be $+$ or $-$ infinite with a nonzero probability.

EXAMPLE 2.1

$\mathscr{E}$ is specified by $(S, \mathscr{F}, \mathscr{P}(.))$ where $S = \{a, b, c, d\}$, $\mathscr{F}$ is the power set of S, and $\mathscr{P}(.)$ is defined by $P\{a\} = 0.4,\ P\{b\} = 0.3,\ P\{c\} = 0.2$, and $P\{d\} = 0.1$. Define the following mapping $X(a) = 1,\ X(b) = 0,\ X(c) = 1,\ X(d) = 4$. Is the function $X(.)$ a random variable?

SOLUTION

To show that the function is a random variable, it must be shown that conditions (a) and (b), specified above, are satisfied. The sets $\{\zeta : X(\zeta) \leq x\}$ as x varies from $-\infty$ to ∞ are as follows:

$$\begin{aligned}
\text{For} \quad x < 0 \quad & \{\zeta : X(\zeta) \leq x\} = \varnothing \\
0 \leq x < 1 \quad & \{\zeta : X(\zeta) \leq x\} = \{b\} \\
1 \leq x < 4 \quad & \{\zeta : X(\zeta) \leq x\} = \{a, b, c\} \\
x \geq 4 \quad & \{\zeta : X(\zeta) \leq x\} = \{a, b, c, d\} = S
\end{aligned}$$

Since the sets described are subsets of S, they are members of the power set of S, and thus condition (a) is satisfied. If S has a finite number of elements and $\mathscr{F}$ is the power set of S, condition (a) will always be satisfied.

If the power set is not used for $\mathscr{F}$, it is possible to construct a function that is not a random variable. It is easily seen that

$$P\{\zeta : X(\zeta) = +\infty\} = P(\varnothing) = 0$$
$$P\{\zeta : X(\zeta) = -\infty\} = P(\varnothing) = 0$$

Thus condition (b) is satisfied and since (a) is also satisfied, $X(.)$ is a random variable. ■

EXAMPLE 2.2

An experiment $\mathscr{E}$ is specified by $(S, \mathscr{F}, \mathscr{P}(.))$ where $S = \{\Phi : 0 \le \Phi \le 2\pi\}$, $\mathscr{F}$ is the power set of S, and $\mathscr{P}(.)$ is defined by $P\{0 \le \Phi \le \phi\} = \phi/(2\pi)$. Define the following mapping $X(\phi) = \sin(\phi)$. Is this a random variable?

SOLUTION

The set $\{x : X(\phi) \le x\} = \{\phi : \sin(\phi) \le x\}$ is a subset of S and thus condition (a) is satisfied since $\mathscr{F}$, the power set of S, contains all subsets of S. Since $\sin(\phi)$ is bounded, the probabilities of the events $\{X = -\infty\}$ and $\{X = \infty\}$ are zero, thus satisfying condition (b) making $X(\phi)$ a random variable. ■

Most common functions defined on a given experiment are random variables, however, condition (a) can be easily violated if $\mathscr{F}$ is not the power set of S or condition (b) cannot be satisfied if X is defined to be ∞ for a set with finite probability.

Random variables are said to be **totally characterized** or described with relation to calculating probabilities of acceptable events (i.e., events that are a member of $\mathscr{F}$) by their cumulative distribution function or probability density function. Weaker characterizations, called "partial characterizations," would include specifying higher-order moments, variance, mean, and the like. Knowing just the mean of a random variable is certainly less information about the random variable than knowing the probability density function, yet it still provides some idea about values of the random variable. These characterizations are defined more precisely in the following sections.

2.1.1 Cumulative Distribution Function (CDF)

The **cumulative distribution function** $F_x(x)$ for a random variable $X(\zeta)$, or when convenient represented by just X, is defined for all x as

$$F_X(x) \triangleq P\{\zeta : X(\zeta) \le x\} = P\{X \le x\} \tag{2.1}$$

It is sufficient information to calculate the probabilities of all allowable events, and as a result is called a **total characterization**. Because of the properties of the probability measure $\mathscr{P}(.)$ described in the overall experiment, the cumulative distribution function (CDF) can be shown to have a number of important properties:

(1) $F_X(x)$ is **bounded** from above and below,

$$0 \leq F_X(x) \leq 1 \text{ for all } x \tag{2.2}$$

(2) $F_X(x)$ is a **nondecreasing** function of x,

$$F_X(x_2) \geq F_X(x_1) \qquad \text{for all } x_2 > x_1 \text{ and all } x_1 \tag{2.3}$$

(3) $F_X(x)$ is **continuous** from the right,

$$\lim_{a \to 0^+} F_X(x+a) = F_X(x) \tag{2.4}$$

(4) $F_X(x)$ can be used to **calculate probabilities** of events

$$\begin{aligned} P\{x_1 < X \leq x_2\} &= F_X(x_2) - F_X(x_1) \\ P\{x_1 < X \leq x_2\} &= F_X(x_2) - F_X(x_1^-) \\ P\{x_1 < X \leq x_2\} &= F_X(x_2^-) - F_X(x_1^-) \\ P\{x_1 < X \leq x_2\} &= F_X(x_2^-) - F_X(x_1) \\ \text{where } x^- &= \lim_{\varepsilon \to 0^+} F_X(x - \varepsilon) \qquad \text{(left-hand limit)} \end{aligned} \tag{2.5}$$

(5) Relation to the probability density function $f_X(x)$ (to be defined later) is written as

$$F_X(x) = \int_{-\infty}^{x} f_X(x)\,dx \tag{2.6}$$

EXAMPLE 2.3

For the random variable defined in Example 2.1 determine (a) the cumulative distribution function $F_X(x)$ and calculate the probabilities of the following events. Using this distribution function, determine (b) $P\{1 < X \leq 2\}$, (c) $P\{1 < X < 4\}$, (d) $P\{0 \leq x \leq 1\}$, (e) $P\{0 \leq x < 1\}$, (f) $P\{0 < x < 1\}$, (g) $P\{X > 1\}$, (h) $P\{X \leq 0\}$, (i) $P\{X = 1\}$, and (j) $P\{X = 3\}$.

SOLUTION

(a) The CDF $F_X(x)$ for the random variable X is obtained by determining the probabilities of the events $\{\zeta : X(\zeta) \leq x\}$ for all $x \in (-\infty, \infty)$. The results from

Example 2.1 help us determine $F_X(x)$ as follows for the following regions.

$$\begin{aligned}
x < 0 \quad & P\{\zeta : X(\zeta) \le x\} = P(\phi) = 0 \\
0 \le x < 1 \quad & P\{\zeta : X(\zeta) \le x\} = P\{b\} = 0.3 \\
1 \le x < 4 \quad & P\{\zeta : X(\zeta) \le x\} = P\{a, b, c\} \\
& \qquad = P\{a\} + P\{b\} + P\{c\} = 0.9 \\
x \ge 4 \quad & P\{\zeta : X(\zeta) \le x\} = P\{a, b, c, d\} = P(S) = 1
\end{aligned}$$

These results can be plotted as shown in Figure 2.1.

The CDF along with the probabilities of the intervals given in Eq. (2.5) will be used to determine the probabilities of the events listed

(a) $P\{1 < X \le 2\} = F_X(2) - F_X(1) = 0.9 - 0.9 = 0$

(b) $P\{1 < X < 4\} = F_X(4) - F_X(1) = 1 - 0.9 = 0.1$

(c) $P\{0 \le X \le 1\} = F_X(1) - F_X(0^-) = 0.9 - 0 = 0.9$

(d) $P\{0 \le X < 1\} = F_X(1^-) - F_X(0^-) = 0.3 - 0 = 0.3$

(e) $P\{0 < X < 1\} = F_X(1^-) - F_X(0) = 0.3 - 0.3 = 0$

(f) $P\{X > 1\} = F_X(\infty) - F_X(1) = 1 - 0.9 = 0.1$

(g) $P\{X \le 0\} = F_X(0) = 0.3$

(h) $P\{X = 1\} = F_X(1) - F_X(1^-) = 0.9 - 0.3 = 0.6$

(i) $P\{X = 3\} = F_X(3) - F_X(3^-) = 0.9 - 0.9 = 0$ ■

2.1.2 Probability Density Function (PDF)

The **probability density function** $f_X(x)$ for a random variable X is a total characterization and is defined as the derivative of the cumulative distribution function

$$f_X(x) \triangleq \frac{d}{dx} F_X(x) \tag{2.7}$$

If $F_X(x)$ has jump discontinuities, it is convenient to use delta functions so that a probability density function (PDF) will always be defined. Therefore a probability

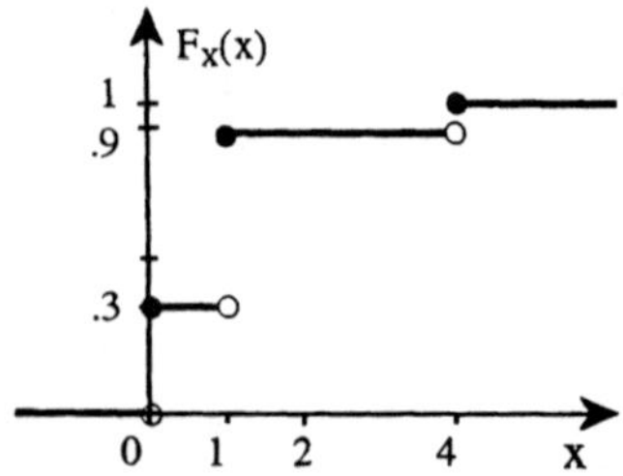

Figure 2.1 CDF $F_X(x)$ for Example 2.3.

density function will contain delta functions at the points of discontinuities of $F_X(x)$ with weights equal to the size of the jumps at those points. Important properties of the probability density function for a random variable X are as follows:

(1) **Positivity**

$$f_X(x) \geq 0 \qquad \text{for all } x \tag{2.8}$$

(2) Integral over all x (**unit area**),

$$\int_{-\infty}^{\infty} f_X(x)\,dx = 1 \tag{2.9}$$

(3) $f_X(x)$ used to **calculate probability of events**,

$$\begin{aligned}
p\{x_1 < X \leq x_2\} &= \int_{x_1+\epsilon}^{x_2+\epsilon} f_X(x)\,dx \triangleq \int_{x_1^+}^{x_2^+} f_X(x)\,dx \\
p\{x_1 \leq X < x_2\} &= \int_{x_1-\epsilon}^{x_2-\epsilon} f_X(x)\,dx \triangleq \int_{x_1^-}^{x_2^-} f_X(x)\,dx \\
p\{x_1 \leq X \leq x_2\} &= \int_{x_1-\epsilon}^{x_2+\epsilon} f_X(x)\,dx \triangleq \int_{x_1^-}^{x_2^+} f_X(x)\,dx \\
p\{x_1 < X < x_2\} &= \int_{x_1+\epsilon}^{x_2-\epsilon} f_X(x)\,dx \triangleq \int_{x_1^+}^{x_2^-} f_X(x)\,dx
\end{aligned} \tag{2.10}$$

(4) **Relationship to cumulative distribution function,**

$$f_X(x) = \frac{dF_X(x)}{dx} \tag{2.11}$$

A random variable is called a **discrete random variable** if its probability density function $f_X(x)$ is a sum of delta functions only, or correspondingly if its cumulative distribution function $F_X(x)$ is a staircase function. A random variable is called a **continuous random variable** if its cumulative distribution function has no finite discontinuities or equivalently its probability density function $f_X(x)$ has no delta functions. If a random variable is neither a continuous or discrete random variable we call it a **mixed random variable**.

Examples of these three types follow.

Discrete random variable,

$$f_X(x) = 0.2\delta(x+3) + 0.2\delta(x+2) + 0.4\delta(x-1) + 0.2\delta(x-3)$$

Continuous random variable,

$$f_X(x) = \tfrac{1}{2}e^{-|x|}$$

Mixed random variable,

$$f_X(x) = 0.5e^{-x}\mu(x) + 0.2\delta(x+2) + 0.1\delta(x-1) + 0.2\delta(x-2)$$

Probability density functions and cumulative distribution functions of these examples are shown in Figure 2.2.

EXAMPLE 2.4

Given a mixed random variable with probability density function $f_X(x)$ as

$$f_X(x) = \tfrac{1}{2}e^{-x}\mu(x) + \tfrac{1}{4}\delta(x+1) + \tfrac{1}{8}\delta(x) + \tfrac{1}{8}\delta(x-1)$$

Use this density function and properties of the delta function to calculate the the following:

(a) $P\{0 < X \leq 1\}$.
(b) $P\{0 \leq X < 1\}$.
(c) $P\{0 < X < 1\}$.
(d) $P\{0 \leq X \leq 1\}$.
(e) $P\{2 < X \leq 3\}$
(f) $P\{X = 1\}$.
(g) $P\{X = 2\}$.

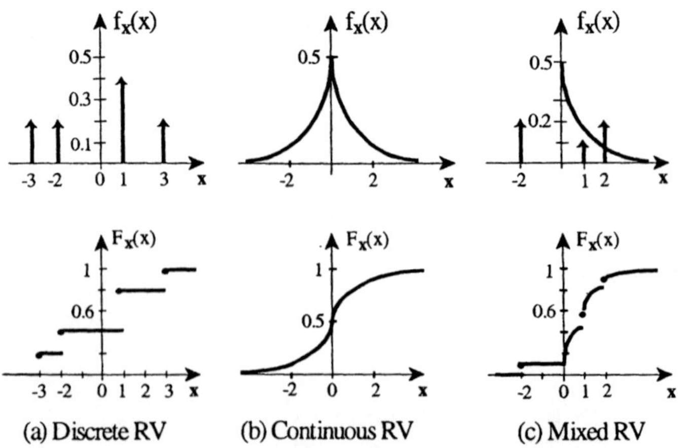

Figure 2.2 Examples of (a) discrete (b) continuous and (c) mixed random variables.

SOLUTION

The probabilities can be determined from the density function by using the proper integral in Eq. (2.10). The inequality signs for the event determine whether the delta functions will be included in the value of the integrals.

$$P\{0 < X \le 1\} = \int_{0+}^{1+} \tfrac{1}{2}e^{-x}\mu(x) + \tfrac{1}{4}\delta(x+1) + \tfrac{1}{8}\delta(x) + \tfrac{1}{8}\delta(x-1)\,dx$$
$$= \tfrac{1}{2}(1 - e^{-1}) + 0 + 0 + \tfrac{1}{8} = \left(\tfrac{5}{8} - \tfrac{1}{2}e^{-1}\right)$$

$$P\{0 \le X < 1\} = \int_{0-}^{1-} \tfrac{1}{2}e^{-x}\mu(x) + \tfrac{1}{4}\delta(x+1) + \tfrac{1}{8}\delta(x) + \tfrac{1}{8}\delta(x-1)\,dx$$
$$= \tfrac{1}{2}(1 - e^{-1}) + 0 + \tfrac{1}{8} + 0 = \left(\tfrac{5}{8} - \tfrac{1}{2}e^{-1}\right)$$

$$P\{0 \le X \le 1\} = \int_{0-}^{1+} \tfrac{1}{2}e^{-x}\mu(x) + \tfrac{1}{4}\delta(x+1) + \tfrac{1}{8}\delta(x) + \tfrac{1}{8}\delta(x-1)\,dx$$
$$= \tfrac{1}{2}(1 - e^{-1}) + 0 + \tfrac{1}{8} + \tfrac{1}{8} = \left(\tfrac{3}{4} - \tfrac{1}{2}e^{-1}\right)$$

$$P\{0 < X < 1\} = \int_{0+}^{1-} \tfrac{1}{2}e^{-x}\mu(x) + \tfrac{1}{4}\delta(x+1) + \tfrac{1}{8}\delta(x) + \tfrac{1}{8}\delta(x-1)\,dx$$
$$= \tfrac{1}{2}(1 - e^{-1}) + 0 + 0 + 0 = \tfrac{1}{2}\left(1 - e^{-1}\right)$$

$$P\{2 < X \le 3\} = \int_{2+}^{3+} \tfrac{1}{2}e^{-x}\mu(x) + \tfrac{1}{4}\delta(x+1) + \tfrac{1}{8}\delta(x) + \tfrac{1}{8}\delta(x-1)\,dx$$
$$= \tfrac{1}{2}(e^{-2} - e^{-3}) + 0 + 0 + 0 = \tfrac{1}{2}\left(e^{-2} - e^{-3}\right)$$

$$P\{X = 1\} = \int_{1-}^{1+} \tfrac{1}{2}e^{-x}\mu(x) + \tfrac{1}{4}\delta(x+1) + \tfrac{1}{8}\delta(x) + \tfrac{1}{8}\delta(x-1)\,dx$$
$$= 0 + 0 + 0 + \tfrac{1}{8} + 0 = \tfrac{1}{8}$$

$$P\{X = 3\} = \int_{3-}^{3+} \tfrac{1}{2}e^{-x}\mu(x) + \tfrac{1}{4}\delta(x+1) + \tfrac{1}{8}\delta(x) + \tfrac{1}{8}\delta(x-1)\,dx$$
$$= 0 + 0 + 0 + 0 = 0$$

■

EXAMPLE 2.5

Given a mixed random variable with probability density function

$$f_X(x) = \tfrac{1}{2}e^{-x}\mu(x) + \tfrac{1}{8}\delta(x+1) + \tfrac{1}{4}\delta(x) + \tfrac{1}{8}\delta(x-1)$$

Determine the cumulative distribution function $F_X(x)$.

SOLUTION

The cumulative distribution function $F_X(x)$ can be determined by the relationship given in (2.6).

$$\begin{aligned} F_X(x) &= \int_{-\infty}^{x} f_X(x)\,dx \\ &= \int_{-\infty}^{x} \tfrac{1}{2}e^{-x}\mu(x)\,dx + \int_{-\infty}^{x} [\tfrac{1}{8}\delta(x+1) + \tfrac{1}{4}\delta(x) + \tfrac{1}{8}\delta(x-1)]\,dx \\ &= [\tfrac{1}{2} - \tfrac{1}{2}e^{-x}]\mu(x) + \tfrac{1}{8}\mu(x+1) + \tfrac{1}{4}\mu(x) + \tfrac{1}{8}\mu(x-1) \end{aligned}$$

The integrals of the delta functions as x goes through the values gives unit steps, causing discontinuities or jumps, at the points $x = -1, 0, 1$ of sizes $\frac{1}{8}$, $\frac{1}{4}$, and $\frac{1}{8}$, respectively. A plot of the distribution function for X is shown in Figure 2.3. ■

2.1.3 Partial Characterizations

Important **partial characterizations** of random variables are the mean, variance, higher-order moments and central moments, and conditional distribution and density functions. Each provides information about the random variables but not necessarily enough information to calculate the probabilities of all events. Definitions for these partial characterizations follow.

The **mean** η_X of a random variable X or equivalently the expected value $E[X]$ of a random variable X is defined by

$$\eta_X = \int_{-\infty}^{\infty} x f_X(x)\,dx \triangleq E[X] \tag{2.12}$$

The mean of a random variable is also referred to as the average of the random variable as seen from the integral above. Knowing the mean of a random variable is not sufficient information to calculate probabilities of given events. It is just barely enough information to bound probabilities of some special events as seen in the

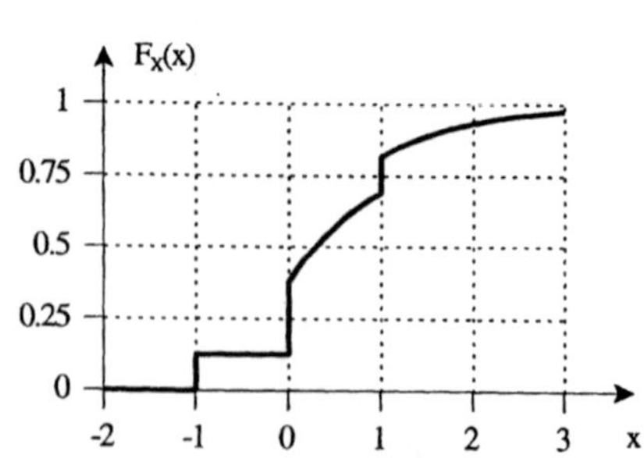

Figure 2.3 Plot of $F_X(x)$ for Example 2.5.

following inequality for the special case when the random variable takes on only positive values:

$$P\{X \geq a\} \leq \frac{E[X]}{a} \qquad \text{for } a > 0 \tag{2.13}$$

The **variance** σ_X^2 of a random variable X is defined by

$$\sigma_X^2 = \int_{-\infty}^{\infty} (x - n_X)^2 f_X(x)\, dx \triangleq E[(X - \eta_X)^2] \tag{2.14}$$

The **standard deviation** is defined as the square root of the variance and denoted as σ_X. The variance gives a sense of spread of the random variable as larger values would indicate a wider density function and as with the mean it is not sufficient information for calculation of probabilities of events. However, like the mean, bounds can be placed on the calculation of some special events.

Two of the most important bounds are described in the Tchebycheff inequalities for the events $\{|X - \eta_X| \geq k\sigma_X\}$ and $\{|X - \eta_X| < k\sigma_X\}$. These events are shown in Figure 2.4. Notice that these events represent the regions corresponding to the mean $\pm k$ standard deviations about the mean and the complement of that region.

The **Tchebycheff inequalities** corresponding to bounds on the probabilities of the tails and core of the density function, respectively, for any density function and $k \geq 1$ are

$$\begin{aligned} P\{|X - \eta_X| \geq k\sigma_X\} &\leq \frac{1}{k^2} \\ P\{|X - \eta_X| < k\sigma_X\} &\geq 1 - \frac{1}{k^2} \end{aligned} \tag{2.15}$$

These upper bounds are shown in Figure 2.5 along with the actual values determined for the Gaussian and other random variables. It is seen that the bound is rather loose for small values of k and, in particular, is useless for $k = 1$. Other random variables would yield exact probabilities that are below the bound shown.

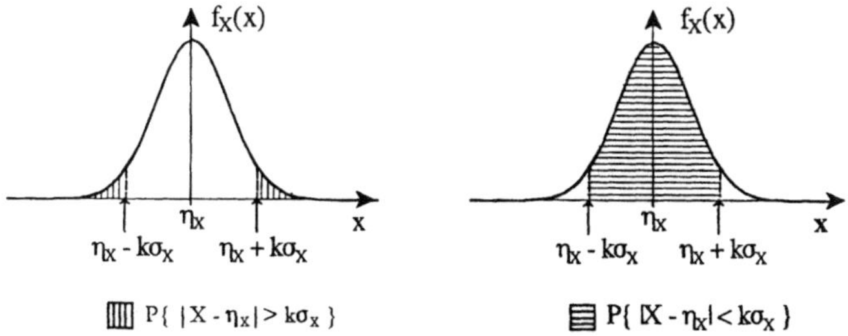

Figure 2.4 Regions for the Tchebycheff inequalities.

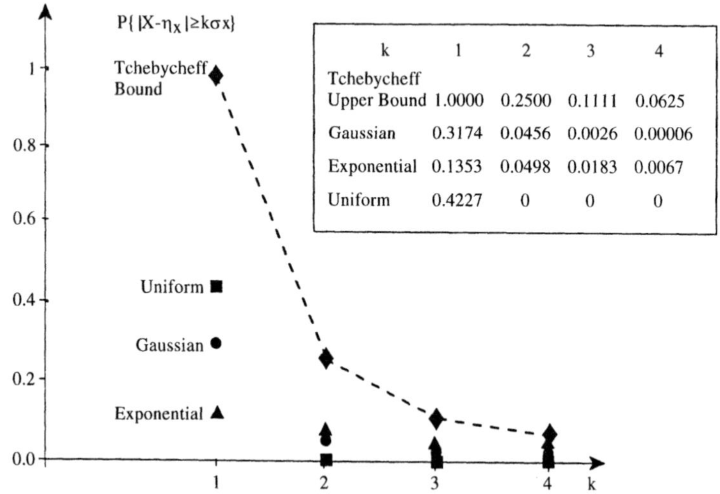

k	1	2	3	4
Tchebycheff Upper Bound	1.0000	0.2500	0.1111	0.0625
Gaussian	0.3174	0.0456	0.0026	0.00006
Exponential	0.1353	0.0498	0.0183	0.0067
Uniform	0.4227	0	0	0

Figure 2.5 Tchebycheff Bound on the $P\{|X - \eta_x| \geq k\sigma_X\}$ and actual values for several random variables.

For a given positive ε we can also bound the $P\{\eta_X - \varepsilon < X < \eta_X + \varepsilon\}$ which is the probability that X is in the region $\pm\varepsilon$ around the mean by

$$P\{\eta_X - \varepsilon < X < \eta_X + \varepsilon\} \geq 1 - \frac{\sigma_X^2}{\varepsilon^2} \tag{2.16}$$

Eq. (2.15) can be obtained from (2.14) by letting $k\sigma_X = \varepsilon$.

The **higher-order moments** m_k and **higher-order central moments** μ_k of a random variable X are partial characterizations and provide additional information about the statistical properties of random variables. They are defined as follows for $k > 2$:

$$m_k = \int_{-\infty}^{\infty} x^k f_X(x)\, dx \triangleq E[X^k] \tag{2.17}$$

$$\mu_k = \int_{-\infty}^{\infty} (x - \eta_X)^k f_X(x)\, dx \triangleq E[(X - \eta_X)^k] \tag{2.18}$$

A generalization of the Tchebycheff inequality called the **inequality of Bienayme** from Papoulis [1] gives a bound on the event $\{|X - a| \geq \varepsilon\}$ in higher-order moments about an arbitrary value a, not necessarily the mean, as

$$P\{|X - a| \geq \varepsilon\} \leq \frac{E\{|X - a|^n\}}{\varepsilon^n} \tag{2.19}$$

If $\varepsilon = k\varepsilon, a = \eta_X$, and $n = 2$, the inequality above reduces to the Tchebycheff inequality.

2.1.4 Conditional Cumulative Distribution Functions

In Chapter 1 the conditional probabilities of of events were defined and illustrated. Another partial characterization of a random variable X is the conditional distribution function defined for conditioning event C by

$$F_X(x \mid C) \triangleq P\{X \leq x \mid C\} \tag{2.20}$$

The determination of $F_X(x \mid C)$ for a given event C and a given $F_X(x)$ for a random variable X uses the definition of conditional probability as follows:

$$F_X(x \mid C) = \frac{P\{X \leq x, C\}}{P(C)} \tag{2.21}$$

Similarly the conditional probability density function for a random variable conditioned on the event C is defined in terms of the derivative of the conditional distribution function as

$$f_X(x \mid C) \triangleq \frac{dF_X(x \mid C)}{dx} \tag{2.22}$$

If C is given in terms of elements of the sample space S, then the $P\{X \leq x, C\}$ can be determined by working with the original sample space. In many problems the conditioning event is given in terms of the values of the random variable.

An example where the conditioning event is in terms of a range of x is now presented to illustrate the use of (2.21) and (2.22).

EXAMPLE 2.6

A random variable X is characterized by its cumulative distribution function $F_X(x) = (1 - e^{-x})\,\mu(x)$ and an event C is defined by $C = \{0.5 < X \leq 1\}$. Determine the conditional cumulative distribution function $F_X(x \mid C)$ and conditional probability density function $f_X(x \mid C)$ for the random variable X conditioned on C.

SOLUTION

By Eq. (2.21) the conditional distribution can be written as

$$F_X(x \mid C) = \frac{P\{X \leq x, 0.5 < X \leq 1\}}{P\{0.5 < X \leq 1\}}$$

The numerator will be different depending on where x is in the infinite interval as

for $x < 0.5$ $\quad P\{X \le x, 0.5 < X \le 1\} = P(\varnothing) = 0$

for $0.5 \le x \le 1$ $\quad P\{X \le x, 0.5 < X \le 1\} = P\{0.5 < X \le x\} = F_X(x) - F_x(0.5)$

$$= (1 - e^{-x}) - (1 - e^{-0.5}) = e^{-0.5} - e^{-x}$$

for $x > 1$ $\quad P\{X \le x, 0.5 < X \le 1\} = P\{0.5 < X \le 1\} = F_X(1) - F_x(0.5)$

$$= (1 - e^{-1}) - (1 - e^{-0.5}) = e^{-0.5} - e^{-1} = 0.2386512$$

But we know that the $P(C)$ is determined as

$$P(C) = P\{0.5 < X \le 1\} = F_X(1) - F_X(0.5)$$
$$= (1 - e^{-1}) - (1 - e^{-0.5}) = 0.2386512$$

Dividing the probabilities by $P(C)$ gives the final answer for the conditional distribution as

for $x < 0.5$ $\quad F_X(x|C) = 0$

for $0.5 \le x \le 1$ $\quad F_X(x|C) = (e^{-0.5} - e^{-x})/0.2386512$

for $x > 1$ $\quad F_X(x|C) = 1$

Plots of the $F_X(x)$ and $F_X(x|C)$ are shown for this example in Figure 2.6.

The conditional probability density function can be obtained by taking the derivative of the conditional distribution throughout the region of the conditioning event, and the result is

$$f_X(x|C) = \begin{cases} 4.1902157e^{-x} & 0.5 < x \le 1 \\ 0 & \text{elsewhere} \end{cases}$$

■

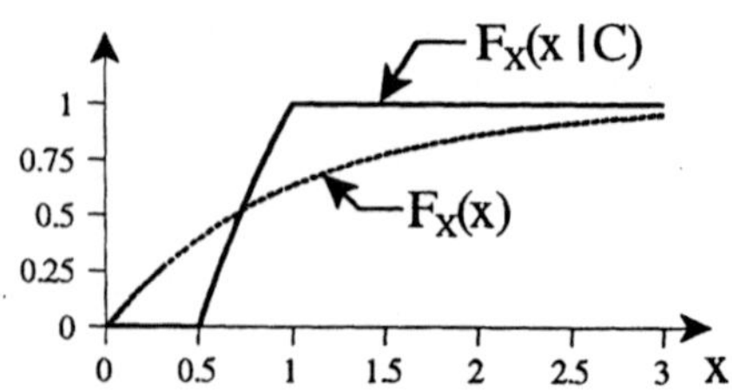

Figure 2.6 $F_X(x)$ and $F_X(x|C)$ for Example 2.6.

2.1.5 Characteristic Function

The characterization of a random variable can also be given by specifying its characteristic function $\Phi_X(\omega)$, which is the modified Fourier transform of $f_X(x)$ and defined as

$$\begin{aligned}\Phi_X(\omega) &\triangleq \int_{-\infty}^{\infty} f_X(x)e^{+j\omega x}\,dx = E[e^{j\omega X}] \\ &= \mathscr{F}(f_X(x))|_{\omega=-\omega}\end{aligned} \tag{2.23}$$

Specification of a characteristic function is equivalent to a total characterization, since its inverse transform yields the probability density function.

EXAMPLE 2.7

Determine the characteristic function for a Gaussian random variable with known mean n_X and variance σ_X^2.

SOLUTION

By the definition above, the characteristic function becomes

$$\Phi_X(\omega) = \int_{-\infty}^{\infty} \frac{1}{\sqrt{2\pi}\sigma_X} \exp\left\{-\frac{(x-\eta_X)^2}{2\sigma_X^2}\right\} e^{j\omega x}\,dx$$

After expanding out the term in the brackets and combining it with the other exponent over a common denominator, the characteristic function becomes

$$\Phi_X(\omega) = \int_{-\infty}^{\infty} \frac{1}{\sqrt{2\pi}\sigma_X} \exp\left\{-\frac{(x^2 - 2\eta_X x + \eta_X^2 - 2j\omega x\sigma_X^2)}{2\sigma_X^2}\right\} dx$$

Next complete the square in the brackets and multiply by the proper correction so that the characteristic function becomes

$$\begin{aligned}\Phi_X(\omega) = \int_{-\infty}^{\infty} &\frac{1}{\sqrt{2\pi}\sigma_X} \exp\left\{-\frac{(x - (\eta_X x + j\omega\sigma_X^2))^2}{2\sigma_X^2}\right\} \\ &\exp\left\{\frac{-\eta_X^2 + (n_X^2 + j\omega\sigma_X^2\eta_X)^2}{2\sigma_X^2}\right\} dx\end{aligned}$$

The exponential term on the right can be taken through the integral sign, since it is not a function of x and since the integral is of a Gaussian density it gives a value of 1. We can finally write, after simplifying the second exponent, the characteristic function for a Gaussian random variable as

$$\Phi_X(\omega) = \exp\left\{j\omega\eta_X - \frac{\omega^2\sigma_X^2}{2}\right\} \tag{2.24}$$

The higher-order moments can also be determined from the characteristic function as

$$E[X^k] = (-j)^k \frac{\partial^k \Phi_X(\omega)}{\partial^k \omega}\bigg|_{\omega=0} \tag{2.25}$$

Equation (2.25) can be used to determine the moments of any random variable by obtaining the characteristic function or modified Fourier transform and performing the differentiation and evaluation at zero as shown in the following example for an exponentially distributed random variable. The characteristic functions for several common random variables appear in Appendix D.

EXAMPLE 2.8

A random variable X has a probability density function $f_X(x)$ given by $f_X(x) = ae^{-ax}\mu(x)$. Determine the moments by using the characteristic function.

SOLUTION

The characteristic function can be obtained by taking the modified Fourier transform of the $f_X(x)$ as

$$\Phi_X(\omega) = \int_{-\infty}^{\infty} f_X(x) e^{+j\omega x} dx = \mathscr{F}(ae^{-ax}\mu(x))\bigg|_{\omega=-\omega}$$

$$= \frac{a}{j\omega + a}\bigg|_{\omega=-\omega} = \frac{a}{-j\omega + a}$$

By Eq. (2.25) the moments are determined as follows:

$$E[X^1] = (-j)^1 \frac{\partial \Phi_X(\omega)}{\partial \omega}\bigg|_{\omega=0} = (-j)^1 \frac{\partial[a/(-j\omega + a)]}{\partial \omega}\bigg|_{\omega=0}$$

$$= (-j)^1 a(-j)^1/(-j\omega + a)^2\Big|_{\omega=0} = \frac{1}{a}$$

$$E[X^2] = (-j)^2 \frac{\partial^2 \Phi_X(\omega)}{\partial^2 \omega}\bigg|_{\omega=0} = (-j)^2 \frac{\partial^2[a/(-j\omega + a)]}{\partial^2 \omega}\bigg|_{\omega=0}$$

$$= (-j)^2 a(-j)^2/(-j\omega + a)^3\Big|_{\omega=0} = \frac{1}{a^2}$$

On continuing this process, it is easy to see that

$$E[X^k] = (-j)^k \frac{\partial^k \Phi_X(\omega)}{\partial^k \omega}\bigg|_{\omega=0} = (-j)^k a \frac{\partial^k[1/(-j\omega + a)]}{\partial^k \omega}\bigg|_{\omega=0}$$

$$= (-j)^k a(-j)^k \frac{1}{(-j\omega + a)^{k+1}}\bigg|_{\omega=0} = \frac{1}{a^k}$$

■

Another function that can be used to determine the moments for a random variable X is the **moment-generating function** defined as follows:

$$M_X(t) \triangleq \int_{-\infty}^{\infty} f_X(x)e^{tx}dx = E[e^{tX}] \tag{2.26}$$

As the name implies, the moment-generating function can be used to calculate the moments for any random variable for which the derivatives exist and this relationship can be shown to be

$$E[X^k] = \left.\frac{\partial^k M_X(t)}{\partial^k t}\right|_{t=0} \tag{2.27}$$

The moment-generating function does not always exist, but when it does, it provides an alternative to using the characteristic function.

2.1.6 Higher-Order Moments for Gaussian Random Variables

Let X be a Gaussian random variable with known mean n_X and variance σ_X^2. The higher-order moments, $m_k = E[X^k]$, can be written in terms of the probability density function as

$$E[X^m] = \int_{-\infty}^{\infty} \frac{x^m}{\sqrt{2\pi}\sigma_X} \exp\left\{-\frac{(x-\eta_X)^2}{2\sigma_X^2}\right\}dx \tag{2.28}$$

If $x - n_X$ is replaced by y, the integral can be written as

$$E[X^m] = \int_{-\infty}^{\infty} \frac{(y+\eta_X)^m}{\sqrt{2\pi}\sigma_X} \exp\left\{-\frac{y^2}{2\sigma_X^2}\right\}dy \tag{2.29}$$

After expanding the $(y+\eta_X)^m$, the higher-order moments are written as

$$\begin{aligned} E[X^m] &= \int_{-\infty}^{\infty} \sum_{i=0}^{m} \binom{m}{i} y^{m-i}\eta_X^i \frac{1}{\sqrt{2\pi}\sigma_X} \exp\left\{-\frac{y^2}{2\sigma_X^2}\right\}dy \\ &= \sum_{i=0}^{m} \binom{m}{i} E[Y^{m-i}]\eta_X^i \end{aligned} \tag{2.30}$$

But Y, defined by $X - n_X$, is a Gaussian random variable with zero mean and variance equal to the variance of X. By symmetry of the zero mean Gaussian density, all higher-order odd moments of Y are zero and (2.30) can be rewritten as

$$E[X^m] = \begin{cases} \displaystyle\sum_{j=0}^{m/2-1} \binom{m}{2j} E[Y^{m-2j}]\eta_X^{2j} + \eta_X^m & m \text{ even} \\ \displaystyle\sum_{j=1}^{(m-1)/2} \binom{m}{2j-1} E[Y^{m-2j+1}]\eta_X^{2j-1} + \eta_X^m & m \text{ odd} \end{cases} \tag{2.31}$$

The higher-order moments up to order six are easily determined from (2.31) as follows:

$$
\begin{aligned}
E[X] &= \eta_X \\
E[X^2] &= \sigma_X^2 + \eta_X^2 \\
E[X^3] &= 3\eta_X\sigma_X^2 + \eta_X^3 \\
E[X^4] &= 3\sigma_X^4 + 6\sigma_X^2\eta_X^2 + \eta_X^4 \\
E[X^5] &= 15\sigma_X^4\eta_X + 10\sigma_X^2\eta_X^3 + \eta_X^5 \\
E[X^6] &= 15\sigma_X^6 + 45\sigma_X^4\eta_X^2 + 15\sigma_X^2\eta_X^4 + \eta_X^6
\end{aligned} \tag{2.32}
$$

If $\eta_x = 0$, then it is seen from the above that $E[X^{(2k-1)}] = 0$ for $k = 1, 2, \ldots,$ as expected.

2.2 COMMON CONTINUOUS RANDOM VARIABLES

Gaussian Random Variable. A random variable X is called a **Gaussian** or **normal** random variable if it has a probability density function $f_X(x)$ given by

$$
\begin{aligned}
f_X(x) &= \frac{1}{\sqrt{2\pi}\sigma_X}\exp\left\{-\frac{(x-\eta_X)^2}{2\sigma_X^2}\right\} \\
&\triangleq g(x; \eta_X, \sigma_X^2)
\end{aligned} \tag{2.33}
$$

It can be shown that η_X and σ_X^2 are the mean and variance of the Gaussian random variable. The **standard deviation** is defined as σ_X. The $g(x; \eta_X, \sigma_X^2)$ defined above is sometimes used as a convenience so that the entire exponential function will not have to be written.

Rather than write the expression (2.33), it is common practice to use the notation $X \sim N(\eta_X, \sigma_X^2)$, which means that X is a Gaussian random variable with mean η_X and variance σ_X^2.

The higher-order central moments can be determined for the Gaussian density in terms of the σ_x^2 as follows:

$$
E[(X-\eta_X)^n] = \begin{cases} 1 \cdot 3 \cdots (n-1)\sigma_X^n & \text{for } n \text{ even} \\ 0 & \text{for } n \text{ odd} \end{cases} \tag{2.34}
$$

Uniform Random Variable. A random variable is uniform if its probability density on an interval is given by

$$
f_X(x) = \begin{cases} \dfrac{1}{b-a} & a \le x \le b \\ 0 & \text{otherwise} \end{cases} \tag{2.35}
$$

The mean and variance of a uniformly distributed random variable can be shown to be

$$\eta_X = \frac{a+b}{2}, \quad \sigma_X^2 = \frac{(b-a)^2}{12} \tag{2.36}$$

Exponential Random Variable. An exponential random variable has the following probability density function for $a > 0$:

$$f_X(x) = \begin{cases} ae^{-ax} & x \geq 0 \\ 0 & \text{otherwise} \end{cases} \tag{2.37}$$

The mean and variance for an exponentially distributed random variable are

$$\eta_X = \frac{1}{a}, \quad \sigma_X^2 = \frac{1}{a^2} \tag{2.38}$$

Rayleigh Random Variable. A Rayleigh distributed random variable has for $b > 0$ the following probability density function:

$$f_X(x) = \begin{cases} \dfrac{x}{\alpha^2}\exp\left(-\dfrac{x^2}{2\alpha^2}\right), & x \geq 0 \\ 0, & \text{otherwise} \end{cases} \tag{2.39}$$

The mean and variance for a Rayleigh distributed random variable are

$$\eta_X = \sqrt{\pi/2}\alpha, \quad \sigma_X^2 = \left(2 - \frac{\pi}{2}\right)\alpha^2 \tag{2.40}$$

Gamma Random Variable. A gamma distributed random variable has for $\alpha > 0$ and $\beta > 0$ the following probability density function:

$$f_X(x) = \begin{cases} \dfrac{1}{\beta^\alpha \Gamma(\alpha)} x^{\alpha-1} e^{-x/\beta}, & x \geq 0 \\ 0, & \text{otherwise} \end{cases} \tag{2.41}$$

where $\Gamma(\alpha) = \int_0^\infty y^{\alpha-1} e^{-y} dy$

If α is a positive integer, then it is well known that $\Gamma(\alpha) = (\alpha - 1)!$, so it is not necessary to evaluate the integral.

The mean and variance for a Gamma distributed random variable are

$$\eta_X = \beta\alpha, \quad \sigma_X^2 = \beta^2\alpha \tag{2.42}$$

Cauchy Random Variable. A random variable is Cauchy distributed if its probability density takes the following form for $a > 0$:

$$f_X(x) = \frac{a}{\pi} \frac{1}{a^2 + x^2} \tag{2.43}$$

The mean and variance for a Cauchy distributed random variable are

$$\eta_X = \text{does not exist} \quad \sigma_X^2 = \text{does not exist} \tag{2.44}$$

Chi-square Random Variable. A random variable is chi-squared distributed with degree of freedom N if

$$f_X(x) = \frac{x^{(N/2)-1}}{2^{N/2}\Gamma(N/2)} \exp\left(-\frac{x}{2}\right)\mu(x) \tag{2.45}$$

The mean and variance for the chi-square random variable are

$$\eta_X = N, \quad \sigma_X^2 = 2N \tag{2.46}$$

Log Normal Random Variable. A random variable is log normally distributed if its probability density function is of the form

$$f_X(x) = \begin{cases} \dfrac{1}{\sqrt{2\pi}x\sigma}\exp\left\{-\dfrac{1}{2\sigma^2}(\log_e x - m)^2\right\}, & x > 0 \\ 0, & \text{otherwise} \end{cases} \tag{2.47}$$

The mean and variance in terms of the positive parameters n and σ are

$$n_X = \exp[\mu + \sigma^2/2], \quad \sigma_X^2 = \exp[2\mu + 2\sigma^2] - \exp[2\mu + \sigma^2] \tag{2.48}$$

Beta Random Variable. A beta-distributed random variable has for $\alpha > 0$ and $\beta > 0$ the probability density function

$$f_X(x) = \begin{cases} kx^{\alpha-1}(1-x)^{\beta-1}, & 0 < x < 1 \\ 0, & \text{otherwise} \end{cases}$$
$$k = \frac{\Gamma(\alpha+\beta)}{\Gamma(\alpha)\Gamma(\beta)} \quad \text{and} \quad \Gamma(\alpha) = \int_0^\infty y^{\alpha-1}e^{-y}dy \tag{2.49}$$

If α, β, and $\alpha + \beta$ are positive integers then $\Gamma(\alpha) = (\alpha - 1)!$, $\Gamma(\beta) = (\beta - 1)!$, and $\Gamma(\alpha + \beta) = (\alpha + \beta - 1)!$.

The mean and variance for a beta-distributed random variable can be shown to be

$$\eta_X = \frac{\alpha}{\alpha+\beta}, \quad \sigma_X^2 = \frac{\alpha\beta}{(\alpha+\beta)^2(\alpha+\beta+1)} \tag{2.50}$$

There are many other continuous random variables with various special uses that could be presented. The list above are among the most common for our purposes and are summarized in Appendix D.

2.3 COMMON DISCRETE RANDOM VARIABLES

A few discrete random variables that are commonly used in engineering applications are now presented. The delta function notation will be used rather than a probability mass function. The density functions for these discrete random variables follow, and they are summarized along with their moments in Appendix C.

Bernoulli Random Variable. A discrete random variable is Bernoulli if its probability density is given by

$$f_X(x) = p\delta(x) + (1-p)\delta(x-1) \tag{2.51}$$

The mean and variance of a Bernoulli distributed random variable can be shown to be

$$\eta_X = 1-p, \quad \sigma_X^2 = p - p^2 \tag{2.52}$$

Discrete Uniform Random Variable. A discrete random variable is uniform if its probability density on a range is given by

$$f_X(x) = \sum_{i=0}^{n} \frac{1}{n+1}\,\delta\left(x - \left(a + i\frac{(b-a)}{n}\right)\right) \tag{2.53}$$

The mean and variance of a uniformly distributed random variable can be shown to be

$$\eta_X = \frac{a+b}{2}, \quad \sigma_X^2 = \frac{(n+2)(b-a)^2}{12n} \tag{2.54}$$

Poisson Distribution. A random variable X is Poisson distributed if its probability density function is given by

$$f_X(x) = \sum_{k=0}^{\infty} \lambda^k \frac{e^{-\lambda}}{k!}\delta(x-k) \tag{2.55}$$

The mean and variance of a poisson distributed random variable can be shown to be

$$\eta_X = \lambda \quad \sigma_X^2 = \lambda \tag{2.56}$$

Binomial Random Variable. A random variable X is binomially distributed if its probability density function is given by

$$f_X(x) = \sum_{k=1}^{n} \binom{n}{k} p^k (1-p)^{n-k} \delta(x-k) \tag{2.57}$$

The mean and variance of a binomially distributed random variable can be shown to be

$$\eta_X = np, \quad \sigma_X^2 = np(1-p) \tag{2.58}$$

X could be interpreted as the number of successes in n repeated trials of the Bernoulli type.

Geometric Random Variable. A random variable X is geometrically distributed if its probability density function is given by

$$f_X(x) = \sum_{k=1}^{\infty} p(1-p)^{k-1}\delta(x-k) \tag{2.59}$$

The mean and variance of a geometric random variable can be shown to be

$$E[X] = \frac{1}{p}, \quad \sigma_X^2 = \frac{1-p}{p^2} \tag{2.60}$$

X could be interpreted as the number of the trial for which the first success occurs for repeated trials of the Bernoulli type.

Negative Binomial Random Variable. A random variable X is negative binomially distributed if its probability density function is given by

$$f_X(x) = \sum_{j=k}^{\infty} \binom{j-1}{k-1} p^k (1-p)^{n-k} \delta(x-j) \tag{2.61}$$

The mean and variance of a binomially distributed random variable can be shown to be

$$\eta_X = \frac{k}{p}, \quad \sigma_X^2 = \frac{k}{p}\left(\frac{1}{p} - 1\right) \tag{2.62}$$

X could be interpreted as the number of the trials on which the kth success occurs for repeated trials of the Bernoulli type.

2.4 TRANSFORMATIONS OF ONE RANDOM VARIABLE

A very important concept in the development of many problems in communication theory is that of a transformation or function of a random variable. Remember that a random variable can be partially characterized in many different ways but can always be thought of as a function defined on a sample space S that satisfies the properties described earlier. Let $X(\zeta)$ be a random variable defined on a sample space S. Define a real-valued function $g(x)$ on the real numbers. If $x(\zeta)$ for all ζ a member of S is the domain of $g(x)$, then the range becomes the set of all $Y(\zeta)$ such that $Y(\zeta) = g(X(\zeta))$. This can be described by

$$\zeta_i \Rightarrow X(\zeta_i) \Rightarrow g(X(\zeta_i)) \triangleq Y(\zeta_i) \quad \text{for all } \zeta_i \text{ member of } S \tag{2.63}$$

and illustrated in Figure 2.7. For most functions, $Y(\zeta)$ is a mapping from the sample space S to the real numbers that satisfies all the conditions of a random variable. As we usually drop the ζ from the random variable $X(\zeta)$ and use just X, it is expedient to drop the index from $Y(\zeta)$ and use just Y. Therefore Eq. (2.63) is usually represented as

$$Y = g(X) \tag{2.64}$$

The basic question is: Knowing the statistical characterizations of X, either partial or total, what are the resulting characterizations of Y? It will be seen that in many cases a partial characterization of X may not be sufficient to specify the corresponding partial characterization of Y.

In a similar fashion we can define a function of many random variables, several functions of several random variables, and several functions of a countable number of random variables. Furthermore the functions can be continuous, discontinuous, or flat over certain regions, while the random variables can be continuous, discrete, or mixed. In the next few sections we will explore many of these possibilities with the hope that the background and methodologies presented will allow you to solve a variety of problems. To reiterate, the basic problem explored is that of statistical characterization (partial or total) of transformed random variables and to answer the question of what statistical information is required to determine those characteristics? These questions will be answered by working on a set of problems that progress in difficulty.

2.4.1 Transformation of One Random Variable

Let X be a random variable, $g(x)$ be a real valued function, and Y be the resulting random variable described by $Y = g(X)$. It will seem incongruous that the problem of finding the simplest characterization of Y, that is, the mean $E[Y]$, cannot be solved knowing just the mean, $E[X]$, of the random variable X. In general, unless $g(x)$ is a linear function of x described by $g(x) = ax + b$, with a and b constants, or other special information regarding the type of probability and type of function are specified, the expected value of Y cannot be determined. In general, for $Y = g(X)$, the mean of Y is not the function $g(.)$ of the mean, that is,

$$E[Y] \neq g(E[X]) \tag{2.65}$$

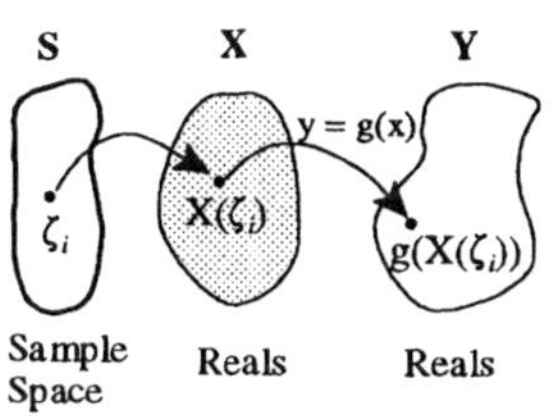

$\zeta \in S$	$X(\zeta)$	$Y=g(X(\zeta))$
ζ_1	$X(\zeta_1)$	$Y=g(X(\zeta_1))$
ζ_2	$X(\zeta_2)$	$Y=g(X(\zeta_2))$
$\vdots$	$\vdots$	$\vdots$
ζ_i	$X(\zeta_i)$	$Y=g(X(\zeta_i))$
$\vdots$	$\vdots$	$\vdots$

Figure 2.7 Transformation of random variables as a mapping and a tabular presentation.

Therefore our discussion begins with the problem of total characterization of the random variable $Y = g(X)$ from a total characterization of the random variable X. Each of the cases of discrete, continuous, and mixed random variables is presented.

Probability Density Function (Discrete Random Variable). If X is a real discrete random variable it takes on a finite or countable set $\mathscr{X}$ of possible values x_i. Therefore X is totally characterized by its probability density function $f_X(x)$ consisting of a sum of weighted impulse functions at the x_i as follows:

$$f_X(x) = \sum_{x_i \in \mathscr{X}} P(x_i)\delta(x - x_i) \tag{2.66}$$

In the expression above $P(x_i)$ is the probability that $X = x_i$ or more precisely is denoted by $P\{\zeta : X(\zeta) = x_i\}$. If X takes on only the values x_i, it is reasonable that Y takes on only the values $y_i = g(x_i)$. Thus Y is a discrete random variable whose probability density function $f_Y(y)$ can be written as

$$f_Y(y) = \sum_{x_i \in \mathscr{X}} P(x_i)\delta(y - g(x_i)) \tag{2.67}$$

In some cases the nonlinear function $g(x)$ is such that $g(x_i)$ of several different x_i give the same value y_j, thus allowing us to write the density as

$$f_Y(y) = \sum_{y_j \in \mathscr{Y}} P(y_j)\delta(y - y_j) \tag{2.68}$$

where $\mathscr{Y}$ is the set of all unique $g(x_i)$ and $P(y_j)$ is the sum of the probabilities. Consider the following example.

EXAMPLE 2.9

Let X be a discrete random variable characterized by its probability density function $f_X(x)$ as follows:

$$\begin{aligned} f_X(x) &= 0.1\delta(x+3) + 0.2\delta(x+1) + 0.25\delta(x) + 0.15\delta(x-1) \\ &\quad + 0.05\delta(x-2) + 0.3\delta(x-3) \end{aligned}$$

Define the function $y = g(x)$ as shown in Figure 2.8.

SOLUTION

Each point is taken through the transformation, and like values are collected to give

$$\begin{aligned} f_y(y) &= 0.1\delta(y-9) + 0.2\delta(y-1) + 0.25\delta(y) + 0.15\delta(y-1) \\ &\quad + 0.05\delta(y-4) + 0.3\delta(y-4) \\ &= 0.1\delta(y-9) + 0.35\delta(y-1) + 0.25\delta(y) + 0.35\delta(y-4) \end{aligned}$$

■

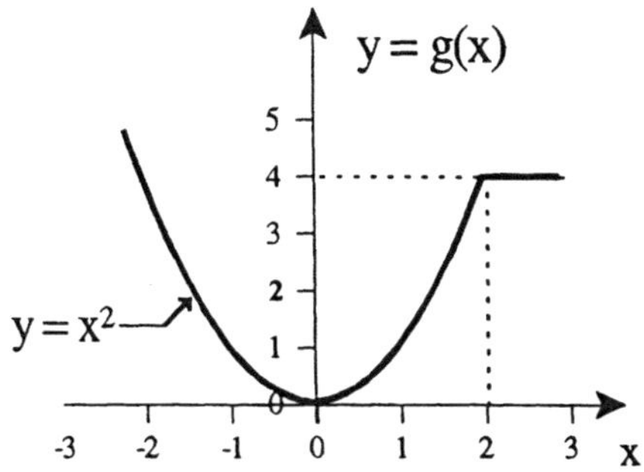

Figure 2.8 Function $y = g(x)$ for Example 2.9.

Probability Density Function (Continuous Random Variable). If X is a continuous random variable and $g(x)$ is a special type continuous function, then the probability density function for $Y = g(X)$ can be determined directly by using the transformation theorem that is called the "fundamental theorem" by Papoulis [1], or indirectly by first finding the distribution function and then differentiating, or using an incremental method.

The fundamental theorem is given below.

Theorem. If X is a continuous random variable characterized by its probability density function $f_X(x)$ and $g(x)$ is a continuous function with no flat spots (finite lengths of constant value), then the probability density function $f_Y(y)$, which characterizes the random variable $Y = g(X)$, is given for each value of y as follows

$$f_Y(y) = \sum_{x_i \in \mathscr{X}} \frac{f_X(x_i)}{|dg(x)/dx|\Big|_{x=x_i}} \tag{2.69}$$

where $\mathscr{X}$ is the set of all real solutions of $y = g(x)$

$f_y(y) = 0$ if no real solution of $y = g(x)$ exists

The theorem above can be used to find the probability density function $f_Y(y)$ for $Y = g(X)$, and it requires solving $y = g(x)$ for each value of y from $-\infty$ to ∞. A simple example illustrating the use of the theorem is now presented.

EXAMPLE 2.10

Given a random variable X with probability density function $f_X(x)$, find the probability density function $f_Y(y)$ for a random variable Y defined by

$$Y = X^2$$

as shown in Figure 2.9. Work the problem for the following cases:

(a) $f_X(x) = 0.5\ \exp\{-|x|\}$.
(b) $f_X(x) = \exp\{-x\}u(x)$.

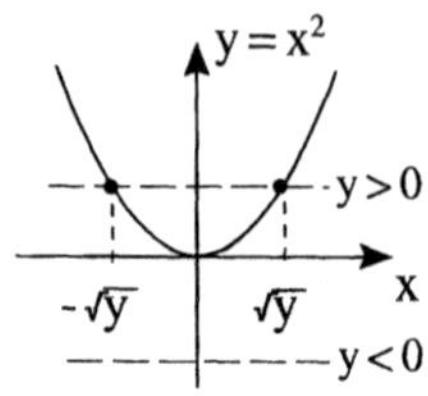

Figure 2.9 Transformation $y = x^2$ for Example 2.10.

SOLUTION

(a) Conditions are satisfied to use the fundamental theorem directly as $g(x)$ is a continuous function and X is a continuous random variable. For any $y < 0, y = x^2$ has no real solution (the dotted line for an arbitrary $y < 0$ doesn't intersect the curve $y = x^2$). Therefore $f_Y(y) = 0$ for those values of y.

If $y > 0$, then, $y = x^2$ has two real roots, $x_1 = +\sqrt{y}$ and $x_2 = -\sqrt{y}$. Using the fundamental theorem gives for $y > 0$,

$$f_Y(y) = \frac{f_X(x)}{|2x|}\bigg|_{x=x_1=\sqrt{y}} + \frac{f_X(x)}{|2x|}\bigg|_{x=x_2=-\sqrt{y}}$$

Using the two-sided exponential density for $f_X(x), f_Y(y)$ can be simplified to

$$f_Y(y) = \left(\frac{0.5e^{-|\sqrt{y}|}}{|2\sqrt{y}|} + \frac{0.5e^{-|-\sqrt{y}|}}{|-2\sqrt{y}|}\right)\mu(y) = \frac{1}{2\sqrt{y}}e^{-\sqrt{y}}\mu(y)$$

(b) Using $f_X(x) = \exp\{-x\}\mu(x)$, it is seen that for $y < 0$, $f_Y(y) = 0$, and that for $y > 0$,

$$f_Y(y) = \frac{e^{-x}\mu(x)}{|2x|}\bigg|_{x=\sqrt{y}} + \frac{e^{-x}\mu(x)}{|2x|}\bigg|_{x=-\sqrt{y}} = \frac{e^{-\sqrt{y}}}{2\sqrt{y}}\mu(y)$$ ∎

If X is a continuous random variable and $g(x)$ a continuous function except for a finite or countable number of flat spots, then the probability density for $Y = g(X)$ can be found by a slight modification of the fundamental theorem. The probability density function $f_Y(y)$ is composed of two parts. For all values of y such that there are no flat spots, find the density as given by the fundamental theorem, and then add to this a sum of delta functions, one for each flat spot, at y_j with weight equal to the probability that the random variable X produces y_j after passing through $g(x)$. In Figure 2.10 a typical continuous $g(x)$ is shown that contains flat spots at y_j for values

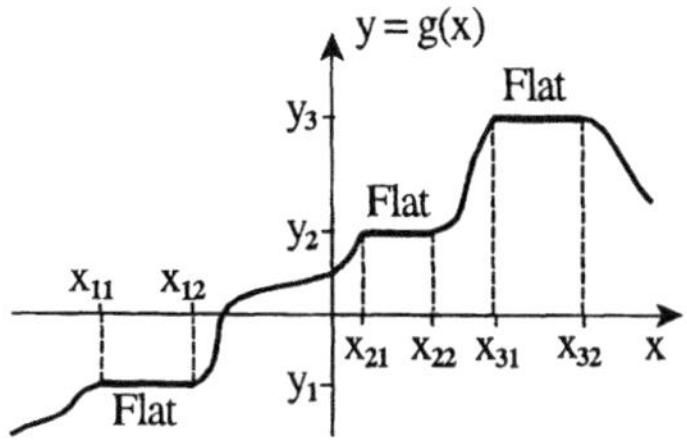

Figure 2.10 Function $y = g(x)$ that contains flat spots.

of x on the intervals $x_{j1} \leq x \leq x_{j2}$, respectively. In this notation the probability density function $f_Y(y)$ can be written as

$$f_Y(y) = \begin{cases} \sum_{x_i \in \mathscr{X}} \dfrac{f_X(x_i)}{|dg(x)/dx|}\Bigg|_{x_i} + \sum_j P(x_{j1} \leq x \leq x_{j2})\delta(y - y_j) \\ 0, \qquad \text{if no real solution of } y = g(x) \text{ exists} \end{cases} \tag{2.70}$$

The following example illustrates this modification.

EXAMPLE 2.11

Let X be a Gaussian random variable with zero mean and unity variance. The function $y = g(x)$ is defined in Figure 2.11, and the random variable Y is specified by the transformation $Y = g(X)$. Find the probability density function $f_Y(y)$ for Y.

SOLUTION

In this example the function $y = g(x)$ has two flat spots, one at $y = 9$ and the other at $y = 4$. From the graph of $g(x)$ it is seen that probability that $\{Y = 9\}$ is the probability that $\{X \leq -3\}$, while the probability that $\{Y = 4\}$ equals the probability that

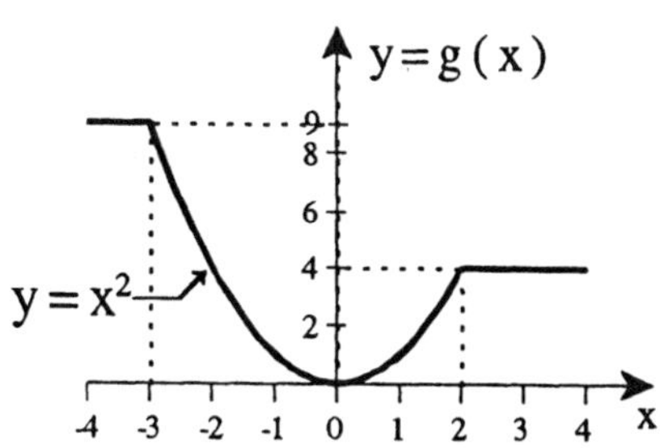

Figure 2.11 Transformation $y = g(x)$ for Example 2.11.

$\{X \geq 0.2\}$. Using the Gaussian density and the parameters given, these probabilities become

$$P(Y = 9) = P(X \leq -3) = \int_{-\infty}^{-3} \frac{1}{\sqrt{2\pi}} \exp\left(\frac{-x^2}{2}\right) dx = \Phi(-3) = 0.0013$$

$$P(Y = 4) = P(X \geq 2) = \int_{2}^{\infty} \frac{1}{\sqrt{2\pi}} \exp\left(\frac{-x^2}{2}\right) dx = 1 - \Phi(2) = 0.0228$$

where $\Phi(x) = \int_{-\infty}^{x} \frac{1}{\sqrt{2\pi}} \exp\left(\frac{-x^2}{2}\right) dx$

Treating the continuous part of the transformation separately it is seen that there are four different situations as y goes from $-\infty$ to ∞.

For $y < 0, y = g(x)$ has no real roots, therefore $f_Y(y) = 0$.

For $y > 9$, again no real roots so $f_Y(y) = 0$.

For $0 < y < 4$, there are two real roots at $+\sqrt{y}$ and $-\sqrt{y}$. Therefore $f_Y^{(c)}(y)$ is given by

$$\begin{aligned} f_Y^{(c)}(y) &= \frac{1}{\sqrt{2\pi}} \frac{\exp(-x^2/2)}{|2x|}\bigg|_{x=\sqrt{y}} + \frac{1}{\sqrt{2\pi}} \frac{\exp(-x^2/2)}{|2x|}\bigg|_{x=-\sqrt{y}} \\ &= \frac{\exp(-y/2)}{\sqrt{2\pi y}} \end{aligned}$$

For $4 < y < 9, y = g(x)$ has only one real root at $x = \sqrt{y}$, and by the fundamental theorem $f_Y^{(c)}(y)$ becomes

$$f_Y^{(c)}(y) = \frac{\exp(-y/2)}{2\sqrt{2\pi y}}$$

Using unit step functions, the continuous and discrete parts can be combined to give the final answer as

$$\begin{aligned} f_Y(y) = &\frac{\exp(-y/2)}{\sqrt{2\pi y}} [\mu(y) - \mu(y-4)] + \frac{\exp(-y/2)}{2\sqrt{2\pi y}} [\mu(y-4) - \mu(y-9)] \\ &+ 0.0228\ \delta(y-4) + 0.0013\ \delta(y-9) \end{aligned}$$

■

Probability Density Function (Mixed Random Variables). If X is a mixed random variable, its probability density function $f_X(X)$ can be written as

$$f_X(x) = f_X^{(c)}(X) + \sum_i P[X = x_i]\delta(x - x_i) \tag{2.71}$$

where $f_X^{(c)}(x)$ represents the density without delta functions. If $g(x)$ is a continuous function, then $f_y(y)$ can be obtained by adding the results from using the fundamental theorem on $f_X^{(c)}(x)$ to the results obtained by handling the impulses as previously described for transformation of discrete random variables. If $g(x)$ has flat spots, then the modification of the fundamental theorem can be used on $f_X^{(c)}(x)$ with the results added to the discrete part.

2.4.2 Cumulative Distribution Function

In many cases it is desirable to find the cumulative distribution of a transformed random variable. Previously we found the probability density function via the fundamental theorem, so indirectly we could determine the density and then integrate it to find the distribution. It will now be shown that finding the distribution function directly leads to an alternative method for determining the density function, since differentiation of the distribution gives the density.

Assume that a random variable X is characterized by its probability density function $f_X(x)$ and that a random variable Y is defined by $Y = g(X)$, where $g(x)$ is a real-valued function. It is desired to find the cumulative distribution function $F_Y(y)$ directly, and then use it to obtain the $f_Y(y)$. Using the basic definition of a cumulative distribution function gives

$$\begin{aligned} F_Y(y) &\triangleq P\{Y \leq y\} = P\{g(X) \leq y\} = P\{x : g(x) \leq y\} \\ &= \int_{I_y} f_X(x)\, dx \end{aligned} \tag{2.72}$$

where $I_y = \{x : g(x) \leq y\}$. This region is illustrated in Figure 2.12.

EXAMPLE 2.12

Assume that X is a random variable with probability density function $f_X(x)$ given by

$$f_X(x) = \frac{1}{\sqrt{2\pi}} \exp(-x^2/2)$$

Define the random variable Y by $Y = g(X)$, where $g(x)$ is shown in Figure 2.15. Find the cumulative distribution function $F_Y(y)$ for the random variable Y.

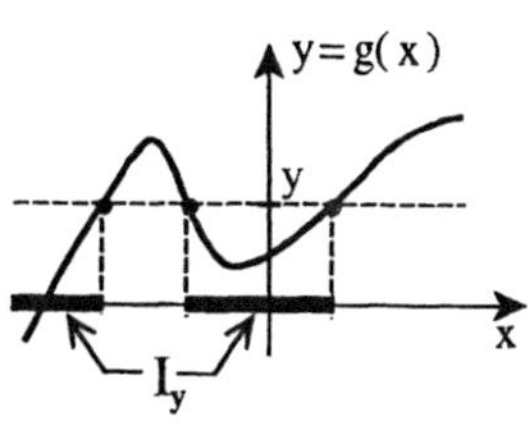

Figure 2.12 Region of integration I_y to obtain the cumulative distribution function.

SOLUTION

To find $F_Y(y)$ using Eq. (2.72) it is necessary to identify the regions I_y for all values of y. For this problem it turns out that there are four distinct regions for different ranges of y as shown in Figure 2.13.

Region 1. For $y < 0$, $\{x : g(x) \leq y\}$ is the null set ϕ which results in $F_Y(y) = 0$.

Region 2. For $0 < y < 4, I_y = \{x : g(x) < y\} = \{x : -\sqrt{y} \leq x \leq \sqrt{y}\}$. Therefore $F_Y(y)$ becomes

$$F_Y(y) = \int_{-\sqrt{y}}^{\sqrt{y}} \frac{1}{\sqrt{2\pi}} \exp\left(\frac{-x^2}{2}\right) dx = \Phi(\sqrt{y}) - \Phi(-\sqrt{y})$$

Region 3. For $4 < y < 9, I_y = \{x : x \geq -\sqrt{y}\}$ and $F_Y(y)$ is

$$F_Y(y) = \int_{-\sqrt{y}}^{\infty} \frac{1}{\sqrt{2\pi}} \exp\left(\frac{-x^2}{2}\right) dx = 1 - \Phi(-\sqrt{y})$$

Region 4. For $y > 9, I_y = \{\text{all } x\}$; therefore $F_Y(y) = \int_{-\infty}^{\infty} f_x(x)\, dx = 1$.

Summarizing the results for the four regions above the cumulative distribution function $F_Y(y)$ becomes

$$F_Y(y) = \begin{cases} 0, & y \leq 0 \\ \Phi(\sqrt{y}) - \Phi(-\sqrt{y}), & 0 < y < 4 \\ 1 - \Phi(-\sqrt{y}), & 4 < y < 9 \\ 1, & y \geq 9 \end{cases}$$

which is illustrated in Figure 2.14.

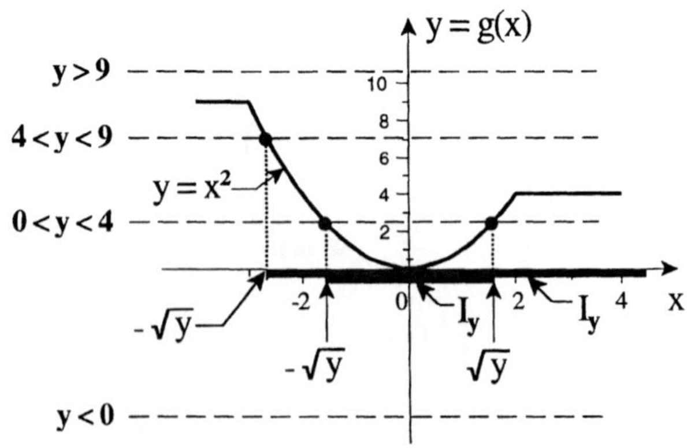

Figure 2.13 Regions for Example 2-12.

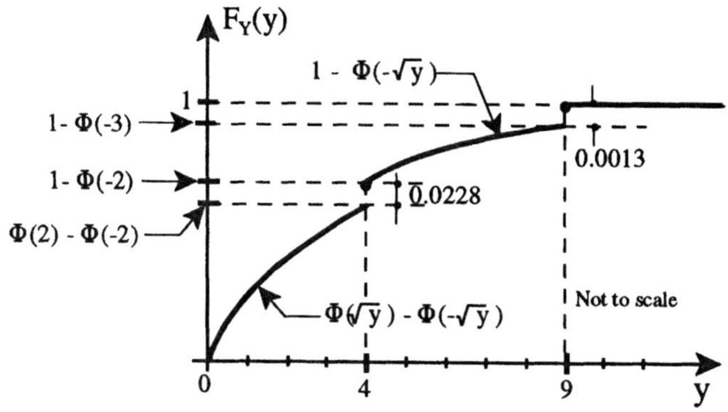

Figure 2.14 Cumulative distribution function for Example 2.12.

Notice in Figure 2.14 that there are discontinuities at $y = 4$ and $y = 9$. These discontinuities will give impulses in the probability density function. Taking the derivative of $F_Y(y)$ gives the probability density function $F_Y(y)$ for Y as

$$F_Y(y) = \frac{d}{dy}\left(\int_{-\sqrt{y}}^{\sqrt{y}} \frac{1}{\sqrt{2\pi}} \exp\left(\frac{-x^2}{2}\right) dx\right) (\mu(y) - \mu(y-4)) + 0.0228\delta(y-4)$$
$$+ \frac{d}{dy}\left(\int_{-\sqrt{y}}^{\infty} \frac{1}{\sqrt{2\pi}} \exp\left(\frac{-x^2}{2}\right) dx\right)(\mu(y-4) - \mu(y-9)) + 0.0013\delta(y-9)$$

In using this method the issue frequently comes up of taking the derivative of an integral with respect to a variable that appears in the limits of the integral. It is actually expedient to use the **Liebnitz rule**, which in its most general form is

$$\frac{d}{dx}\left[\int_{\alpha(x)}^{\beta(x)} f(x, \gamma)\, d\gamma\right] = f(x, \beta(x))\frac{d\beta(x)}{dx} - f(x, \alpha(x))\frac{d\alpha(x)}{dx} + \int_{\alpha(x)}^{\beta(x)} \frac{d}{dx} f(x, \gamma)\, d\gamma \tag{2.73}$$

Upon our applying this formula to the terms in parentheses, the $f_Y(y)$ becomes

$$f_y(y) = \left(\frac{1}{\sqrt{2\pi}} \exp\left(\frac{-y}{2}\right) y^{-\frac{1}{2}}/2 - \frac{1}{\sqrt{2\pi}} \exp\left(\frac{-y}{2}\right)(-1) y^{-\frac{1}{2}}/2\right)(\mu(y) - \mu(y-4)))$$
$$+ \left(\frac{-1}{\sqrt{2\pi}} \exp\left(\frac{-y}{2}\right)(-1) y^{-\frac{1}{2}}/2\right)(\mu(y-4) - \mu(y-9))$$
$$+ 0.0228\delta(y-4) + 0.0013\delta(y-9)$$

Further simplification yields the final answer for the transformed random variable probability density function as

$$f_y(y) = \left(\frac{1}{\sqrt{2\pi y}}\exp(-y/2)\right)(\mu(y) - \mu(y-4)) + 0.0228\delta(y-4)$$
$$+ \left(\frac{1}{2\sqrt{2\pi y}}\exp(-y/2)\right)(\mu(y-4)) - \mu(y-9)) + 0.0013\delta(y-9)$$

■

2.5 COMPUTATION OF EXPECTED VALUES

For a random variable X with probability density function $f_X(x)$ and a real-valued function $g(x)$, a random variable Y can be defined by $Y = g(X)$. In the previous sections methods for characterizing Y totally with its probability density function or its cumulative distribution have been presented. It is possible to describe with $f_X(x)$ and $f_Y(y)$ discrete and mixed random variables that use impulses in representing their densities. There are three common ways to compute the expected value of $Y = g(X)$.

Method 1. By using $f_Y(y)$, which yields

$$E[Y] = \int_{-\infty}^{\infty} y f_Y(y)\,dy \tag{2.74}$$

Method 2. By using $f_X(x)$, which yields

$$E[Y] = E[g(X)] = \int_{-\infty}^{\infty} g(x) f_X(x)\,dx \tag{2.75}$$

Method 3. By using the Monte Carlo technique, which is essentially synthetic sampling.

In method 1, one must first find the probability density function $f_Y(y)$ and then compute the integral, whereas in method 2, the $E[Y]$ can be calculated without going through the intermediate step of calculating the probability density function for Y.

Method 3 represents an alternative to methods 1 and 2 and it is not an analytical method. Monte Carlo Sampling is useful in problems where the integrals are of such a nature that analytical solutions are not easily obtained or do not exist. Roughly speaking, it is an experimental method using synthetic sampling from the density on X to generate samples x, computing samples $y = g(x)$, and then computing the average of those results to get the approximation for $E[Y]$. The method is not within the scope of this book but is mentioned here because of its overall usefulness. Interested readers could look at the presentation by Shanmugan and Breipohl [9] for a brief discussion of Monte Carlo techniques if they so desire.

The following example shows an application of the first two methods for computing the expected value of a function of one random variable.

EXAMPLE 2.13

In Example 2.10 a random variable Y was defined as $Y = g(X) = X^2$ and $f_X(x)$ was given by

$$f_X(x) = \tfrac{1}{2}\exp(-|x|)$$

For this problem compute the $E[Y]$ using methods 1 and 2.

SOLUTION

Method 1. The density $f_Y(y)$ determined in Example 2.9 is used to compute the integral

$$E[Y] = \int_{-\infty}^{\infty} y f_Y(y)\, dy = \int_0^{\infty} \frac{y e^{-\sqrt{y}}}{2\sqrt{y}}\, dy$$

By making a change of variables $x = (y)^{\frac{1}{2}}$, $E[Y]$ becomes

$$E[Y] = \int_0^{\infty} x^2 e^{-x} dx = \frac{e^{-x}}{-1}(x^2 + 2x + 2)\Big]_0^{\infty} = 2$$

Method 2. The original density for X and $g(x)$ is used:

$$E[Y] = E[g(X)] = \int_{-\infty}^{\infty} g(x) f_X(x)\, dx$$

$$= \int_{-\infty}^{\infty} x^2 \frac{e^{-|x|}}{2} dx = \int_0^{\infty} x^2 e^{-x} dx = 2 \qquad \blacksquare$$

As this example shows, the relationship between the two methods is a change of variables within the integration. Clearly, if the probability density function for Y is not needed, its calculation is an unnecessary step in the process of computing the expected value and the second method would be preferred.

2.6 TWO RANDOM VARIABLES

An experiment $\mathscr{E}$ is specified by the three-tuple $(S, \mathscr{F}, \mathscr{P}(.))$, where S is a finite, countable, or noncountable set called the sample space, $\mathscr{F}$ is a Borel field specifying a set of events, and $\mathscr{P}(.)$ is a probability measure allowing calculation of probabilities of all events. Based on this underlying experiment, two random variables $X(\zeta)$ and $Y(\zeta)$ are defined as real-valued functions on the same S that satisfies the following conditions:

(a) $\{\zeta : X(\zeta) \leq x \text{ and } Y(\zeta) \leq y\}$ is a member of $\mathscr{F}$ for all x and y. This guarantees the existence of the cumulative distribution function.

(b) The probabilities of the events $\{\zeta : X(\zeta) = -\infty\}$, $\{\zeta : Y(\zeta) = +\infty\}$, $\{\zeta : X(\zeta) = +\infty\}$, and $\{\zeta : Y(\zeta) = -\infty\}$ are all zero. This means that the function is not allowed to be + or − infinity with a nonzero probability.

The real-valued functions for random variables X and Y are defined on the sample space as shown in Figure 2.15. Random variables X and Y are said to be **totally characterized** or described with relation to calculating probabilities of acceptable events (i.e., events a member of $\mathscr{F}$) by their cumulative distribution function or probability density function. Weaker characterizations called partial characterizations would include specifying higher-order moments, variance, and mean. These characterizations are defined precisely in the following sections.

2.6.1 Joint Cumulative Distribution Function

The **joint cumulative distribution** function $F_{YX}(x, y)$ for random variables $X(\zeta)$ and $Y(\zeta)$, represented by X and Y is defined for all x and y as

$$F_{XY}(x, y) = P\{\zeta : X(\zeta) \leq x \cap Y(\zeta) \leq y\} \triangleq P\{X \leq x, Y \leq y\} \tag{2.76}$$

It is sufficient information to calculate the probabilities of all allowable events, and thus is called a total characterization. Because of the properties of the probability measure $\mathscr{P}(.)$ described in the overall experiment, the joint cumulative distribution function can be shown to have a number of important properties.

(1) $F_{XY}(x, y)$ is **bounded** from above and below,

$$0 \leq F_{XY}(x, y) \leq 1 \quad \text{for all } x \text{ and } y \tag{2.77}$$

(2) $F_{XY}(x, y)$ is a **nondecreasing** function of x and y,

$$\begin{aligned} F_{XY}(x_2, y) &\geq F_{XY}(x_1, y) \qquad \text{for all } x_2 > x_1, \text{ all } x_1, \text{ and all } y \\ F_{XY}(x, y_2) &\geq F_{XY}(x, y_1) \qquad \text{for all } y_2 > y_1, \text{ all } y_1, \text{ and all } x \end{aligned} \tag{2.78}$$

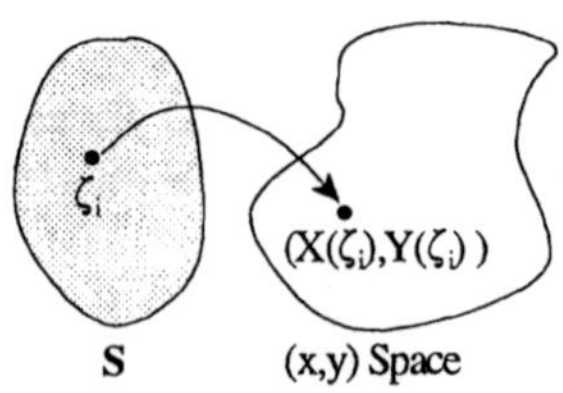

Figure 2.15 Random variables X and Y defined on the sample space **S**.

(3) $F_{XY}(x, y)$ is **continuous from the right** in both x and y,

$$\lim_{\substack{\varepsilon \to 0^+ \\ \delta \to 0^+}} F_{XY}(x+\varepsilon, y+\delta) = F_{XY}(x, y) \tag{2.79}$$

(4) $F_{XY}(x, y)$ can be used to **calculate probabilities** of rectangular events as

$$P\{x_1 < X \le x_2, y_1 < Y \le y_2\} = F_{XY}(x_2, y_2) - F_{XY}(x_1, y_1) \tag{2.80}$$

(5) $F_{XY}(x, y)$ is related **to the joint probability density function** by

$$F_{XY}(x, y) = \int_{-\infty}^{x} \int_{-\infty}^{y} f_{XY}(x, y)\, dx\, dy \tag{2.81}$$

An example of joint random variables and their corresponding distribution function is now presented.

EXAMPLE 2.14

Random variables X and Y are defined in Figure 2.16. In this example X and Y are both discrete random variables. Find the joint cumulative distribution function $F_X(x)$, $F_Y(x)$, and $F_{XY}(x, y)$ for X and Y.

SOLUTION

From the table it observed that X takes on values 0 and 1 with finite probabilities of 0.6 and 0.4, respectively, while Y takes on values 0, 1, and 2 with probabilities of 0.3, 0.4, and 0.3, respectively. Therefore $F_X(x)$ and $F_Y(y)$ are

$$F_X(x) = 0.6\mu(x) + 0.4\mu(x-1), \quad F_Y(y) = 0.3\mu(x) + 0.4\mu(x-1) + 0.3\mu(x-2)$$

To obtain the joint distribution it is necessary to compute probability of the event $\{X \le x, Y \le y\}$ as x and y are both varied from $-\infty$ to ∞. This will be done by fixing x in different regions and determining the probability of $\{X \le x, Y \le y\}$ as y varies

ζ	X(ζ)	Y(ζ)	P(ζ)
ζ_1	0	0	0.2
ζ_2	0	1	0.3
ζ_3	0	0	0.1
ζ_4	1	2	0.2
ζ_5	1	2	0.1
ζ_6	1	1	0.1

Figure 2.16 Random Variables X and Y for Example 2-14.

For $x < 0$ and **all** y the set $\{X \le x, Y \le y\} = \varnothing$ so $F_{XY}(x, y) = 0$.

For $0 \le x < 1$, there will be three different regions on the y-axis for which different values of the distribution are obtained as

$$\begin{aligned}
y < 0 \quad & F_{XY}(x, y) = P\{X \le x, Y \le y\} = P(\varnothing) = 0 \\
0 \le y < 1 \quad & F_{XY}(x, y) = P\{X \le x, Y \le y\} = P\{\zeta_1, \zeta_3\} = 0.2 + 0.1 = 0.3 \\
1 \le y \quad & F_{XY}(x, y) = P\{X \le x, Y \le y\} \\
& \qquad = P\{\zeta_1, \zeta_3, \zeta_2\} = 0.2 + 0.1 + 0.1 = 0.4
\end{aligned}$$

For $x \ge 1$, the results are similar to above but there are four different regions that give the joint cumulative distribution function as

$$\begin{aligned}
y < 0 \quad & F_{XY}(x, y) = P\{X \le x, Y \le y\} = P(\varnothing) = 0 \\
0 \le y < 1 \quad & F_{XY}(x, y) = P\{X \le x, Y \le y\} = P\{\zeta_1, \zeta_3\} = 0.2 + 0.1 = 0.3 \\
1 \le y < 2 \quad & F_{XY}(x, y) = P\{X \le x, Y \le y\} = P\{\zeta_1, \zeta_3, \zeta_2, \zeta_6\} = 0.2 + 0.1 + 0.3 \\
& \qquad + 0.1 = 0.7 \\
2 \le y \quad & F_{XY}(x, y) = P\{X \le x, Y \le y\} = P\{\zeta_1, \zeta_3, \zeta_2, \zeta_6, \zeta_4, \zeta_5\} = P(S) = 1
\end{aligned}$$

The joint cumulative distribution function is shown in Figure 2.17. ■

2.6.2 Joint Probability Density Function

The **joint probability density function** $f_{XY}(x, y)$ for the random variables X and Y is a total characterization and is defined as the derivative of the joint cumulative

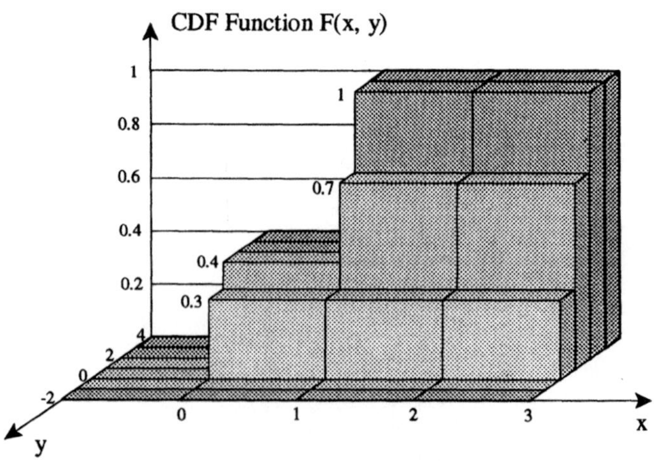

Figure 2.17 Cumulative distribution function $F_{XY}(x, y)$ for Example 2-14.

distribution function

$$f_{XY}(x, y) \triangleq \frac{\partial^2}{\partial x \partial y} F_{XY}(x, y) \tag{2.82}$$

If $F_{XY}(x, y)$ has jump discontinuities, it is convenient to use delta functions so that a joint probability density function will always be defined. Therefore a probability density function will contain delta functions at the points of discontinuities of $F_{XY}(x, y)$ with weights equal to the size of the jumps at those points. Important properties of the probability density function are as follow:

(1) **Positivity,**

$$f_{XY}(x, y) \geq 0 \quad \text{for all } x \text{ and } y \tag{2.83}$$

(2) **Integral over all x and y,**

$$\int_{-\infty}^{\infty} \int_{-\infty}^{\infty} f_{XY}(x, y)\, dx\, dy = 1 \tag{2.84}$$

(3) $f_{XY}(x, y)$ can be used to **calculate probability** of rectangular events as

$$P\{x_1 < X \leq x_2, y_1 < Y \leq y_2\} = \int_{x_1^+}^{x_2^+} \int_{y_1^+}^{y_2^+} f_{XY}(x, y)\, dx\, dy \tag{2.85}$$

where x_1^+, x_2^+, y_1^+ and y_2^+ are limits from the positive sides,
or any event A as

$$P(\{(X, Y) \in A\}) = \int\int_A f_{XY}(x, y)\, dx\, dy \tag{2.86}$$

(4) **Relationship to joint cumulative distribution function,**

$$\int_{-\infty}^{x^+} \int_{-\infty}^{y^+} f_{XY}(x, y)\, dx\, dy = F_{XY}(x, y) \tag{2.87}$$

Random variables are called **jointly discrete random variables** if their probability density function $f_{XY}(x, y)$ is a sum of two dimensional delta functions only, and correspondingly if its cumulative distribution function $F_{XY}(x, y)$ is a box staircase type function. Random variables are called **jointly continuous random variables** if their cumulative distribution function has no finite discontinuities or equivalently its probability density function $f_{XY}(x, y)$ has no delta functions. If random variables are neither jointly continuous or jointly discrete random variables, they are **jointly mixed random variables**.

EXAMPLE 2.15

The random variables described in Example 2.14 are jointly discrete random variables. Give the joint probability density function $f_{XY}(x, y)$ for these random variables X and Y.

SOLUTION

The joint density is seen to be

$$f_{XY}(x, y) = 0.3\delta(x, y) + 0.3\delta(x, y-1) + 0.3\delta(x-1, y-1) + 0.1\delta(x-1, y-2)$$

■

2.6.3 Partial Characterizations

Important partial characterizations for two random variables are the marginal densities, means, variances, covariances, higher-order joint moments, and joint central moments. Each provides information about the random variables but not enough information to calculate the probabilities of all events. Definitions follow.

For two random variables X and Y, the **marginal densities** for X and Y are defined as the densities of X and Y by themselves and will be denoted by $f_X(x)$ and $f_Y(y)$ as before. These marginal densities can be obtained from the joint probability density function for the two random variables as

$$\begin{aligned} f_X(x) &= \int_{-\infty}^{\infty} f_{XY}(x, y)\, dy \\ f_Y(y) &= \int_{-\infty}^{\infty} f_{XY}(x, y)\, dx \end{aligned} \tag{2.88}$$

The **conditional probability density functions** for X and Y are defined as

$$\begin{aligned} f_Y(y \mid x) &\triangleq \frac{f_{XY}(x, y)}{f_X(x)} \\ f_X(x \mid y) &\triangleq \frac{f_{X,Y}(x, y)}{f_Y(y)} \end{aligned} \tag{2.89}$$

The $f_Y(y \mid x)$ can be considered as the limit as $\Delta \to 0$ of the conditional density $f_Y(y \mid C)$, where C is the event $\{0 < X \leq x + \Delta\}$. The profile at $x = x_0$ is defined as $f_{XY}(x_0, y)$, which is a function of y. Figure 2.18 shows that the conditional density, $f_Y(y \mid x_0)$, is the scaled version of the joint density evaluated at x_0 with the scale factor $f_X(x_0)$.

The means η_X and η_Y for two random variables X and Y, or equivalently the expected values $E[X]$ and $E[Y]$ are as given earlier. Using the conditional densities, define the **conditional means** as

$$\begin{aligned} E[X \mid y] &= \int_{-\infty}^{\infty} x f_X(x \mid y)\, dx \triangleq E[X \mid Y = y] \\ E[Y \mid x] &= \int_{-\infty}^{\infty} y f_Y(y \mid x)\, dy \triangleq E[Y \mid X = x] \end{aligned} \tag{2.90}$$

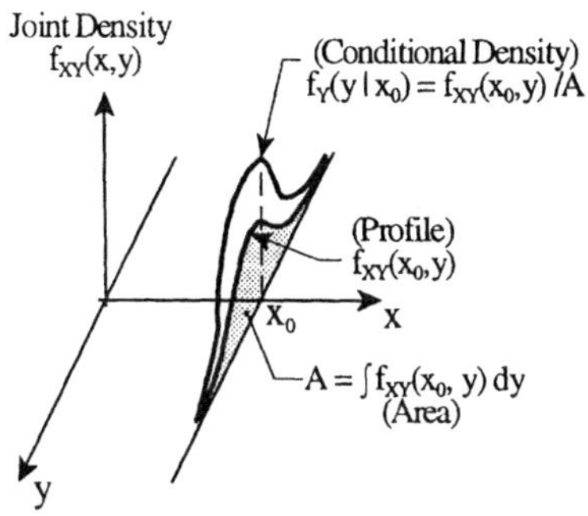

Figure 2.18 Conditional probability density functions.

The means and conditional means are partial characterizations for the random variables, and thus knowing just them is not sufficient information to calculate probabilities of given events.

By rearranging Eq. (2.89), we can rewrite the joint probability density function in terms of the conditional and marginal densities as

$$\begin{aligned} f_{XY}(x, y) &= f_Y(y \mid x)\, f_X(x) \\ f_{XY}(x, y) &= f_X(x \mid y)\, f_Y(y) \end{aligned} \tag{2.91}$$

By substituting (2.91) into (2.88), alternative formulas for determining the marginal densities for X and Y can be found as follows:

$$\begin{aligned} f_X(x) &= \int_{-\infty}^{\infty} f_X(x \mid y)\, f_Y(y)\, dy \\ f_Y(y) &= \int_{-\infty}^{\infty} f_Y(y \mid x)\, f_X(x)\, dx \end{aligned} \tag{2.92}$$

Examining (2.91), we see that it is also possible to write the relationships between the two conditional densities as

$$\begin{aligned} f_Y(y \mid x) &= f_X(x \mid y)\, f_Y(y) / f_X(x) \\ f_X(x \mid y) &= f_Y(y \mid x)\, f_X(x) / f_Y(y) \end{aligned} \tag{2.93}$$

These formulas are comparable to **Bayes's rule**, which expresses the relationship between conditional probabilities, except that they represent the relationships between the conditional probability density functions and not probabilities.

Should either X or Y be discrete random variables, these results would need to be re-written. For example, let X be a discrete random variable and Y a continuous random variable with known $f(y \mid \{X = x_i\})$ for all x_i and known $f_X(X)$:

$$f_X(x) = \sum_{i=1}^{N} P\{X = x_i\} \delta(x - x_i) \tag{2.94}$$

Then it is easily shown, for all $j = 1$ to N, that

$$P(\{X = x_j\}\,|\,y) = \frac{f(y\,|\,\{X = x_j\})P\{X = x_j\}}{\sum_{i=1}^{N} f(y\,|\,\{X = x_i\})P\{X = x_i\}} \tag{2.95}$$

Examples illustrating the mechanics of finding marginal and conditional densities and conditional means follow for the case of two continuous random variables with given joint probability density functions.

EXAMPLE 2.16

Given the joint probability density function $f_{XY}(x, y)$

$$f_{XY}(x, y) = \begin{cases} 8\,xy, & 0 < y < 1, 0 < x < y \\ 0, & \text{elsewhere} \end{cases}$$

Determine (a) $f_X(x)$, (b) $f_Y(y)$, and (c) $f_Y(y\,|\,x)$.

SOLUTION

(a) The $f_X(x)$ is obtained by integrating over all y at each value of x. For $x < 0$ and $x > 1$, $f_X(x) = 0$ because $f_{XY}(x, y)$ being zero leads to a zero integral. For $0 < x < 1$, $f_X(x)$ is determined as

$$f_X(x) = \int_{-\infty}^{\infty} f_{XY}(x, y)\,dy = \int_0^x 8xy\,dy = 4x^3$$

In summary $f_X(x)$ can be written as

$$f_X(x) = 4x^3[\mu(x) - \mu(x-1)]$$

(b) The $f_Y(y)$ is obtained by integrating over all x at each value of y. For $y < 0$ and $y > 1$, the $f_Y(y) = 0$ because $f_{XY}(x, y)$ is zero leads to a zero integral. For $0 < y < 1, f_Y(y)$ is determined as

$$f_Y(y) = \int_{-\infty}^{\infty} f_{XY}(x, y)\,dx = \int_0^y 8xy\;dx = 4y^3$$

Thus $f_Y(y)$ can be written as

$$f_Y(y) = 4y^3[\mu(y) - \mu(y-1)]$$

(c) The conditional density $f_Y(y\,|\,x)$ from (2.106) for $0 < y < x$ and $0 < x < 1$ is

$$f_Y(y\,|\,x) = \frac{f_{XY}(x, y)}{f_X(x)} = \frac{8xy}{4x^3} = \frac{2y}{x^2}$$

For $y > x$ the joint density is zero, and thus $f_Y(y|x)$ can be summarized as

$$f_Y(y|x) = \begin{cases} \dfrac{2y}{x^2}, & 0 < y < x \\ 0, & \text{elsewhere} \end{cases}$$

Several plots of $f_Y(y|x)$ for various values of x are shown in Figure 2.19. Notice the different domains for nonzero values. ■

EXAMPLE 2.17

Find $E[Y|x]$ for the random variables X and Y that have the joint probability density function given in Example 2.16.

SOLUTION

Using the conditional density determined in Example 2.16 and (2.90), the conditional expected value is determined for $0 < x < 1$ as

$$E[Y|x] = \int_{-\infty}^{\infty} y f_Y(y|x)\,dy = \int_0^x y\frac{2y}{x^2}\,dy$$
$$= \frac{2y^3}{3x^2}\Bigg]_0^x = \frac{2}{3}x$$

For x outside the interval [0, 1], the density is zero as is the expected value. ■

Two random variables X and Y are defined to be **statistically independent** if the joint probability density function can be written as a product of the two marginal densities for all x and y as follows:

$$f_{XY}(x, y) = f_X(x) \cdot f_Y(y) \tag{2.96}$$

The **correlation** between two random variables X and Y is defined as the expected value of their product:

$$R_{XY} = E[XY] \tag{2.97}$$

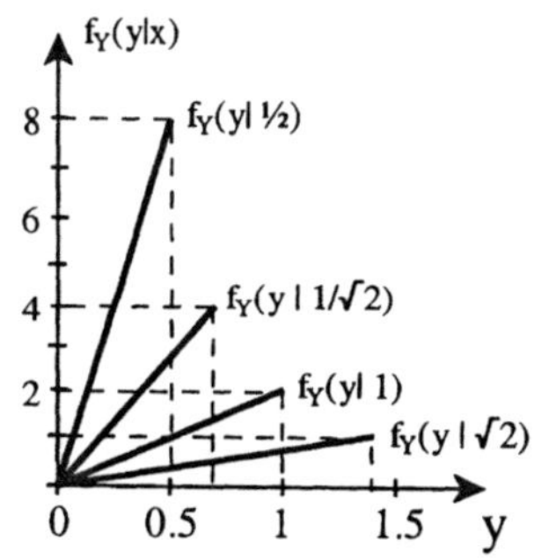

Figure 2.19 Conditional densities $f_Y(y|x)$ for Example 2.16 with $x = \frac{1}{4}, \frac{1}{2}, \frac{3}{4}$, and 1.

Two random variables are defined as **uncorrelated** if $R_{XY} = E[X]E[Y]$, which is the same as writing

$$E[XY] = E[X]E[Y] \tag{2.98}$$

Using Eq. (2.96), it is simple to show that if two random variables are statistically independent, then they are uncorrelated. The steps in the proof follow.

$$\begin{aligned} E[XY] &= \int_{-\infty}^{\infty}\int_{-\infty}^{\infty} xyf_{XY}(x,y)\,dx\,dy = \int_{-\infty}^{\infty}\int_{-\infty}^{\infty} xyf_X(x)\,f_Y(y)\,dx\,dy \\ &= \int_{-\infty}^{\infty} xf_X(x)\,dx \int_{-\infty}^{\infty} yf_Y(y)\,dy = E[X]E[Y] \end{aligned} \tag{2.99}$$

Two random variables X and Y are called **orthogonal** if $E[XY] = 0$. If the random variables X and Y are independent and both of the means are zero, or one of the means is zero and the other is finite, then they are orthogonal since by independence the $E[XY] = E[X] \cdot E[Y]$ and thus $E[X] \cdot E[Y] = 0$. These relationships between independence, orthogonality and uncorrelated random variables are shown in Figure 2.20. In the figure the single solid arrow indicates always true, whereas a dotted arrow indicates always true if the conditions written along side are true. The arrow back from uncorrelated to independence will be shown later to be true for jointly Gaussian random variables.

With two random variables the individual variances are as before for each random variable, but more information considering the relationship between X and Y is available through the **covariance** σ_{XY} defined as

$$\sigma_{XY} \triangleq E[(X - \eta_X)(Y - \eta_Y)] = \int_{-\infty}^{\infty}\int_{-\infty}^{\infty} (x - \eta_X)(y - \eta_Y)\,f_{XY}(x,y)\,dx\,dy \tag{2.100}$$

The **correlation coefficient** ρ_{XY} between the two random variables X and Y is defined as the normalized covariance by

$$\rho_{XY} \triangleq \frac{\sigma_{XY}}{\sigma_X \sigma_Y} \tag{2.101}$$

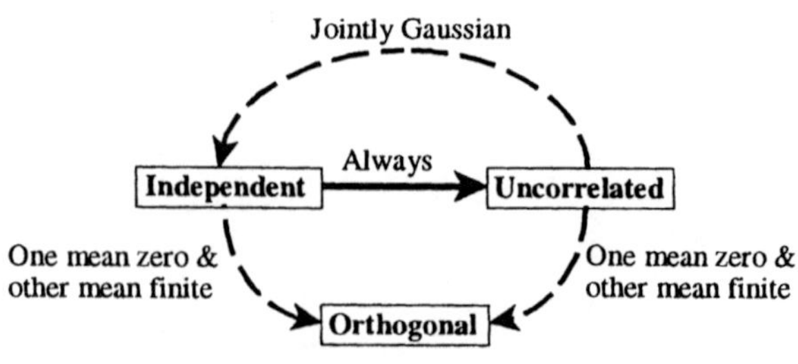

Figure 2.20 Relationship between the definitions of independent, uncorrelated, and orthogonal.

It provides a measure of linear relationship between the two random variables X and Y. The correlation coefficient for two random variables X and Y can be easily shown to be bounded by $+1$ and -1 as

$$-1 \leq |\rho_{XY}| \leq 1 \tag{2.102}$$

The closer ρ_{XY} is to -1 or $+1$ the more the random variables X and Y are said to be linearly related. In Figure 2.21, the joint probability density function for two random variables X and Y are shown for four different correlation coefficients. It is noticed that the closer ρ_{XY} is to $-$ or $+1$, the more ridgelike the probability density becomes, which indicates that one random variable can be written as almost a scaler multiple of the other.

The **higher-order moments** $\boldsymbol{m}_{jk}$ and the **central moments** $\boldsymbol{\mu}_{jk}$ for two random variables X and Y are defined by

$$\begin{aligned} m_{jk} &\triangleq E[X^j Y^k] \\ &= \int_{-\infty}^{\infty}\int_{-\infty}^{\infty} x^j y^k f_{XY}(x, y)\, dx\, dy \end{aligned} \tag{2.103}$$

$$\begin{aligned} \mu_{jk} &\triangleq E[(X - \eta_X)^j (Y - \eta_Y)^k] \\ &= \int_{-\infty}^{\infty}\int_{-\infty}^{\infty} (x - \eta_X)^j (y - \eta_Y)^k f_{XY}(x, y)\, dx\, dy \end{aligned} \tag{2.104}$$

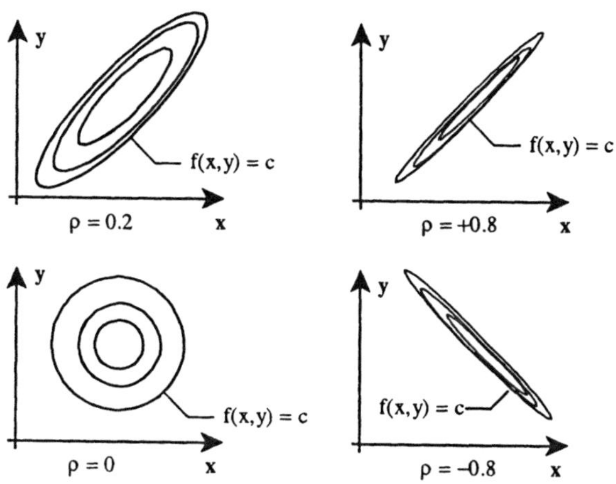

Figure 2.21 Contours of constant probability density functions for various values of correlation coefficients.

2.6.4 Jointly Normal Random Variables

The random variables X and Y are defined as **jointly Gaussian** or jointly normal if their probability density function $f_{XY}(x, y)$ has the following form:

$$f_{XY}(x,y) = \frac{1}{2\pi\sigma_X\sigma_Y r}\exp\left(-\frac{1}{2r}\left[\frac{(x-\eta_X)^2}{\sigma_X^2} - \frac{2\rho_{XY}(x-\eta_X)(y-\eta_Y)}{\sigma_X\sigma_Y} + \frac{(y-\eta_Y)^2}{\sigma_Y^2}\right]\right) \tag{2.105}$$

where $\eta_X, \eta_Y, \sigma_X^2, \sigma_Y^2$, and ρ_{XY} are, respectively, the mean of X, mean of Y, variance of X, variance of Y, and correlation coefficient of X and Y, and $r = \sqrt{1-\rho_{XY}^2}$. The marginal densities or densities of the random variables X and Y can be obtained by integrating out the proper variable to give

$$f_X(x) = \frac{1}{\sqrt{2\pi}\sigma_X}\exp\left\{-\frac{(x-\eta_X)^2}{2\sigma_X^2}\right\}, \quad f_Y(y) = \frac{1}{\sqrt{2\pi}\sigma_Y}\exp\left\{-\frac{(y-\eta_Y)^2}{2\sigma_Y^2}\right\} \tag{2.106}$$

By dividing the joint density by the marginal density and simplifying the conditional probability density functions for Y conditioned on $\{X = x\}$ and X conditioned on $\{Y = y\}$, the conditional probability density functions can be determined as

$$\begin{aligned}
f_Y(y|x) &= \frac{1}{\sqrt{2\pi}\sigma_{Y|x}}\exp\left\{-\frac{(y-\eta_{Y|x})^2}{2\sigma_{Y|x}^2}\right\}, \quad && \eta_{Y|x} = \eta_Y + \frac{\rho_{XY}\sigma_Y}{\sigma_X}(x-\eta_X) \\
& && \sigma_{Y|x}^2 = \sigma_Y^2(1-\rho_{XY}^2) \\
f_X(x|y) &= \frac{1}{\sqrt{2\pi}\sigma_{X|y}}\exp\left\{-\frac{(x-\eta_{X|y})^2}{2\sigma_{X|y}^2}\right\}, \quad && \eta_{X|y} = \eta_X + \frac{\rho_{XY}\sigma_X}{\sigma_Y}(y-\eta_Y) \\
& && \sigma_{X|y}^2 = \sigma_X^2(1-\rho_{XY}^2)
\end{aligned} \tag{2.107}$$

When the means are zero, the conditional means and variances are easily obtained from the above by setting $\eta_X = \eta_Y = 0$ to give

$$\begin{aligned}
f_Y(y|x) &= \frac{1}{\sqrt{2\pi}\sigma_{Y|x}}\exp\left\{-\frac{(y-\eta_{Y|x})^2}{2\sigma_{Y|x}^2}\right\}, \quad && \eta_{Y|x} = \frac{\rho_{XY}\sigma_Y}{\sigma_X}x \\
& && \sigma_{Y|x}^2 = \sigma_Y^2(1-\rho_{XY}^2) \\
f_X(x|y) &= \frac{1}{\sqrt{2\pi}\sigma_{X|y}}\exp\left\{-\frac{(x-\eta_{X|y})^2}{2\sigma_{X|y}^2}\right\}, \quad && \eta_{X|y} = \frac{\rho_{XY}\sigma_X}{\sigma_Y}y \\
& && \sigma_{X|y}^2 = \sigma_X^2(1-\rho_{XY}^2)
\end{aligned} \tag{2.108}$$

Note that although X and Y have zero means, the conditional means are not equal to zero.

A number of useful formulas for various moments of **jointly normal zero mean random variables** in terms of their given parameters σ_X, σ_Y, and ρ_{XY} follow without proof:

$$\begin{aligned}
E[XY] &= \rho_{XY}\sigma_X\sigma_Y \\
E[X^2Y^2] &= \sigma_X^2\sigma_Y^2 + 2\rho_{XY}\sigma_X^2 \\
E[X^2] &= \sigma_X^2, \quad E[Y^2] = \sigma_Y^2 \\
E[X^k] &= E[Y^k] = 0, \quad k \text{ odd} \\
E[X^4] &= 3\sigma_X^4, \qquad E[Y^4] = 3\sigma_Y^4 \\
E[|X||Y|] &= \frac{2\sigma_X\sigma_Y}{\pi}(\cos\alpha + \alpha\sin\alpha) \\
&\quad \text{where } \sin\alpha = \rho_{XY}, \ -\frac{\pi}{2} < \alpha < \frac{\pi}{2}.
\end{aligned} \tag{2.109}$$

2.7 TWO FUNCTIONS OF TWO RANDOM VARIABLES

The basic problem is to obtain the statistical characterizations of two random variables Z and W that are functions of two other random variables X and Y. The expressions for Z and W are specified by

$$W = g(X, Y) \quad \text{and} \quad Z = h(X, Y) \tag{2.110}$$

In particular, methods for obtaining the joint probability density function, joint cumulative distribution function, and expected values for W and Z are presented. As X and Y could be discrete, continuous, or mixed random variables, and $g(x, y)$ and $h(x, y)$ could be continuous functions, continuous functions with flat spots, or discontinuous functions of the x and y variables, there are quite a number of possible cases to investigate. Only the discrete case and the continuous functions of continuous random variables are discussed in the following sections.

2.7.1 Probability Density Function (Discrete Random Variables)

If X and Y are discrete random variables, their joint probability density function $f_{XY}(x, y)$ is composed of two-dimensional impulses:

$$f_{XY}(x, y) = \sum_{(x_i, y_i)} P(x_i, y_i)\delta(x - x_i, y - y_i) \tag{2.111}$$

Then the joint probability density function for $W = g(X, Y)$ and $Z = h(X, Y)$ will also contain only impulses with locations obtained by taking (x_i, y_i) through the functions $w = g(x, y)$ and $z = h(x, y)$, and with weights equal to the probabilities of each (x_i, y_i). The $f_{WZ}(w, z)$ is therefore seen to be

$$f_{WZ}(w, z) = \sum_{(x_i, y_i)} P(x_i, y_i)\delta\big(w - g(x_i, y_i), z - h(x_i, y_i)\big) \tag{2.112}$$

As before, this sum will be able to be simplified if there are different (x_i, y_i) that produce the same $g(x_i, y_i)$ and $h(x_i, y_i)$.

EXAMPLE 2.18

Given the following probability density function for the discrete random variables X and Y:

$$f_{XY}(x, y) = \tfrac{1}{4}\delta(x-1, y-1) + \tfrac{1}{2}\delta(x, y) + \tfrac{1}{6}\delta(x-1, y) + \tfrac{1}{12}\delta(x, y+1)$$

Consider two new random variables W and Z that are defined by the following transformations:

$$W = X^2 + Y^2, \quad Z = XY$$

Find the joint probability density function $f_{WZ}(w, z)$.

SOLUTION

Taking each delta function through the transformation gives the joint density $f_{WZ}(w, z)$ as

$$\begin{aligned} f_{WZ}(w, z) &= \tfrac{1}{4}\delta(w-2, z-1) + \tfrac{1}{2}\delta(w, z) + \tfrac{1}{6}\delta(w-1, z) + \tfrac{1}{12}\delta(w-1, z) \\ &= \tfrac{1}{4}\delta(w-2, z-1) + \tfrac{1}{2}\delta(w, z) + \tfrac{1}{4}\delta(w-1, z) \end{aligned}$$

■

2.7.2 Probability Density Function (Continuous Random Variables and Continuous Functions)

If X and Y are continuous random variables, and $g(x, y)$ and $h(x, y)$ are continuous functions of x and y with no flat spots, the counterpart to the one function of one random variable result can be written as follows for random variables $W = g(X, Y)$ and $Z = h(X, Y)$

$$f_{WZ}(w, z) = \sum_{(x_i, y_i)} \left. \frac{f_{XY}(x, y)}{|J(x, y)|} \right|_{(x,y)=(x_i, y_i)} \tag{2.113}$$

where (x_i, y_i) are all real solutions of $w = g(x, y)$ and $z = h(x, y)$. The $J(x, y)$ is the Jacobian of the transformation, which is expressed as the following determinant:

$$J(x, y) = \begin{vmatrix} \dfrac{\partial g(x, y)}{\partial x} & \dfrac{\partial g(x, y)}{\partial y} \\ \dfrac{\partial h(x, y)}{\partial x} & \dfrac{\partial h(x, y)}{\partial y} \end{vmatrix} \tag{2.114}$$

If no real solution of $w = g(x, y)$ and $z = h(x, y)$ exists for a given w and z, then the joint density $f_{WZ}(w, z) = 0$ for those values of w and z.

As before, if the functions $g(x, y)$ and $h(x, y)$ contain flat spots in either x or y or both, then the density will contain impulses. As in the one-dimensional case, if X and Y are mixed random variables, the continuous and discrete parts should be handled separately.

EXAMPLE 2.19

Let W and Z be two random variables defined by the following functions of two other random variables X and Y:

$$W = X^2 + Y^2 \quad \text{and} \quad Z = X^2$$

Let X and Y be independent Gaussian random variable described by $X \sim N(0, 1)$ and $Y \sim N(0, 1)$.

(a) Find the joint probability density function for the random variables X and Y.
(b) Are X and Y statistically independent?

SOLUTION

(a) The solution is obtained by using Eqs. (2.113) and (2.114). The Jacobian for the transformation defined is

$$J(x, y) = \begin{vmatrix} \dfrac{\partial(x^2 + y^2)}{\partial x} & \dfrac{\partial(x^2 + y^2)}{\partial y} \\ \dfrac{\partial(x^2)}{\partial x} & \dfrac{\partial(x^2)}{\partial y} \end{vmatrix} = \begin{vmatrix} 2x & 2y \\ 2x & 0 \end{vmatrix} = 4|xy|$$

We must find all real solutions (x, y) of the equations $w = x^2 + y^2$ and $z = x^2$ for all w and z.

For $w > 0$ and $z > 0$ the second equation can be solved for x in terms of z as $x = \pm\sqrt{z}$. Substituting this result into the first equation allows us to solve for y as $y = \pm\sqrt{(w - z)}$. Since y will not be real if $(w - z) < 0$, there is no real solution for $z > w$, so by the transformation theorem the joint density $f_{WZ}(w, z) = 0$.

Also there are no real solutions for any $w < 0$ or $z < 0$, since $w = x^2 + y^2$ and $z = x^2$ can never be negative, thus $f_{WZ}(w, z) = 0$ for those regions of w and z. In summary, there are only real roots for $w > 0, z > 0$, and $z < w$, and for that case there are four of them as combinations of the following: $x = \pm\sqrt{z}, y = \pm\sqrt{(w - z)}$.

By Eq. (2.113), the joint density for w and z becomes

$$f_{WZ}(w, z) = \left.\frac{f_{XY}(x, y)}{4|xy|}\right|_{\substack{x=\sqrt{z} \\ y=\sqrt{(w-z)}}} + \left.\frac{f_{XY}(x, y)}{4|xy|}\right|_{\substack{x=\sqrt{z} \\ y=-\sqrt{(w-z)}}}$$
$$+ \left.\frac{f_{XY}(x, y)}{4|xy|}\right|_{\substack{x=-\sqrt{z} \\ y=\sqrt{(w-z)}}} + \left.\frac{f_{XY}(x, y)}{4|xy|}\right|_{\substack{x=-\sqrt{z} \\ y=-\sqrt{(w-z)}}}$$

Substituting these roots into the following density function

$$f_{XY}(x, y) = \frac{1}{2\pi}\exp\left(-\frac{1}{2}(x^2 + y^2)\right)$$

and collecting like terms (they are all the same because of the squaring operation in the numerator and the Gaussian density), the $f_{WZ}(w, z)$ is easily seen to be

$$f_{WZ}(w, z) = \frac{e^{-w/2}}{2\pi\sqrt{z(w-z)}}\mu(w - Z)\mu(w)\mu(z)$$

(b) The random variables W and Z are not statistically independent because the joint density is not a product of the marginal densities. This can be easily seen by finding the marginal densities and multiplying them. However, it is easier to see that the product cannot be the same as the product would be nonzero for all $w > 0$ and $z > 0$ whereas the joint density is zero for $z > w$ and $w > 0$. ■

There are special cases where the transformation theorem cannot be used, though the functions are continuous and the random variables are continuous random variables. These cases arise in general when the Jacobian equals zero at x and y, the solutions of $w = g(x, y)$, and $z = h(x, y)$. Such cases imply a direct relationship between w and z as will be shown in the following examples. By using so-called line masses, the joint probability density function for W and Z can be determined, but it is usually better to use the joint distribution function to characterize the random variables W and Z.

EXAMPLE 2.20

Let Z and W be random variables defined by the following functions of two other random variables X and Y:

$$W = X + Y$$
$$Z = 2(X + Y)$$

Assume that $f_{XY}(x, y)$ is known and that it is required to find the joint density function $f_{WZ}(w, z)$.

SOLUTION

The transformational theorem cannot be used since the Jacobian is given by

$$J(x, y) = \begin{vmatrix} 1 & 1 \\ 2 & 2 \end{vmatrix} = 0$$

regardless of the roots of $w = x + y$ and $z = 2(x + y)$. Certainly W and Z are random variables, so they must have a probability density function. Notice that $Z = 2W$, that is, Z can be written as a linear function of W. Thus, as W takes on values w, Z takes on only the values $z = 2w$. In the w–z two-dimensional space, the joint density must be zero for (w, z), not on the line, and have meaning along the line $z = 2w$ as shown in Figure 2.22. This is what is meant by a line mass. Using the distribution function method, the derivative yields the joint density function as

$$f_{WZ}(w, z) = f_W(w)\delta(z - 2w)$$

where $f_W(w)$ is obtained by using the one function of two random variables approach. ■

EXAMPLE 2.21

Let $W = X + Y$ and $Z = \sin(X + Y)$ be random variables as functions of two other random variables, X and Y. Assume that $f_{XY}(x, y)$ is known. We want to find the joint density function $f_{WZ}(w, z)$.

SOLUTION

Let us proceed as if things are okay. The Jacobian becomes

$$f(x, y) = \begin{vmatrix} \dfrac{\partial g(x,y)}{\partial x} & \dfrac{\partial g(x,y)}{\partial y} \\ \dfrac{\partial h(x,y)}{\partial x} & \dfrac{\partial h(x,y)}{\partial y} \end{vmatrix} = \begin{vmatrix} 1 & 1 \\ \cos(x+y) & \cos(x+y) \end{vmatrix} = 0$$

As before the Jacobian is zero for all values of x and y. Again, it is possible to show that

$$f_{WZ}(w, z) = f_w(w)\delta(z - \sin w)$$

where $\delta(z - \sin w)$ is a line mass along the line shown in Figure 2.23. ■

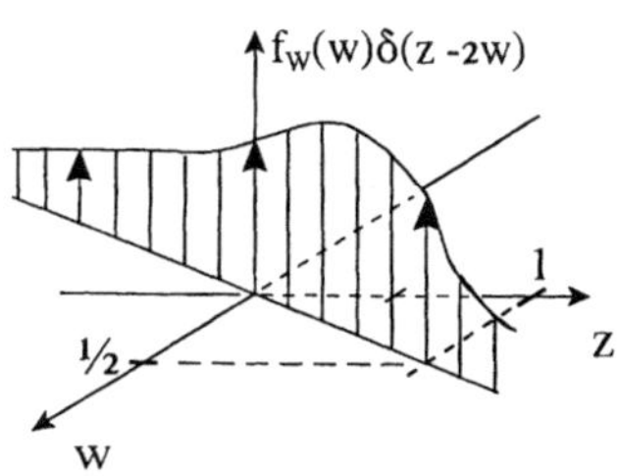

Figure 2.22 Line mass along $z = 2w$.

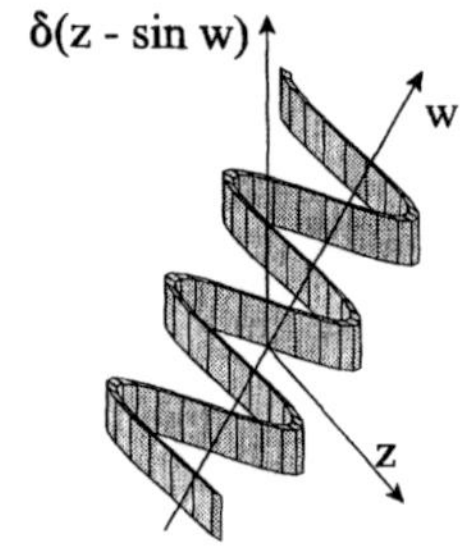

Figure 2.23 Line mass along $z = \sin w$.

From these two examples the general cases of linear and nonlinear dependence can be written as

$$\begin{aligned} W &= g(X, Y) \\ Z &= h(X, Y) = e(h(X, Y)) = e(W) \end{aligned} \tag{2.115}$$

and the joint density can be written as $f_{WZ}(w, z) = f_w(w)\delta(w - e(w))$.

Suppose that we wish to calculate the probability of a given event $\{(w, z) \in R\}$, where R is a closed region as shown in Figure 2.24. The probability can be obtained by integrating $f_Z(z)$ from z_1 to z_2, integrating $f_W(w)$ from w_1 to w_2, or integrating $f_{WZ}(w, z)$ over the region R.

$$\begin{aligned} P\{(W, Z) \in R\} &= \int_{z_1}^{z_2} f_Z(z)\, dz \text{ or} \\ &= \int_{w_1}^{w_2} f_W(w)\, dw \text{ or} \\ &= \iint_R f_{WZ}(w, z)\, dw\, dz \end{aligned} \tag{2.116}$$

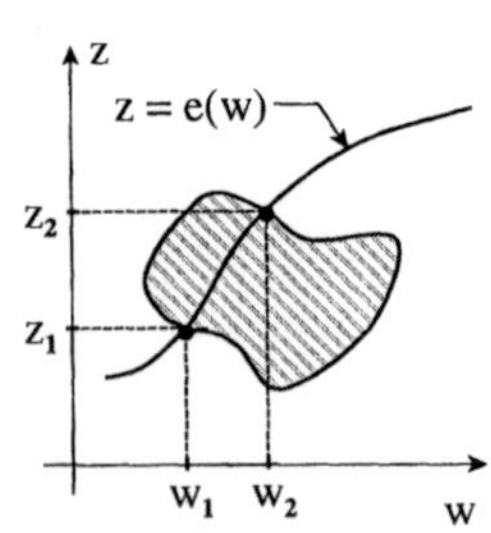

Figure 2.24 Region for obtaining the probability of an arbitrary event R in the w–z space.

EXAMPLE 2.22

Say that W and Z are defined by $W = 2(X + Y)$ and $Z = X + Y$ and that $f_{XY}(x, y)$ is given by

$$f_{XY} = 2e^{-(x+2y)}\mu(x)\mu(y)$$

Find the probability that $(W, Z) \in R$, where R is the set of values of w and z inside the unit circle given by

$$R = \{(w, z) : w^2 + z^2 \le 1\}$$

SOLUTION

Because this is a line mass problem the probability can be evaluated using any one of the three integrals shown in Eq. (2.116). As Z is a slightly simpler function of X and Y, we will evaluate the probability by integrating the $f_Z(z)$ from z_1 to z_2, where z_1 and z_2 can be obtained by solving for the intersections of the circle $w^2 + z^2 = 1$ and the line $w = 2z$ or $z = w/2$ as shown in Figure 2.25.

Solving for the intersections yields $z_1 = -\sqrt{\frac{1}{5}}$ and $z_2 = \sqrt{\frac{1}{5}}$. Since the integral uses $f_Z(z)$, it is necessary to obtain this density function which can be obtained by any method for one function of two random variables. For this problem it is noticed that X and Y are independent, and thus the density of their sum, which is Z, is the convolution of the marginal densities given by

$$\begin{aligned} f_Z(z) &= e^{-z}\mu(z) * 2e^{-2z}\mu(z) \\ &= 2(e^{-z} - 2e^{-2z})\mu(z) \end{aligned}$$

From this density function the desired probability can finally be obtained as

$$\begin{aligned} P\{(W, Z) \in R\} &= \int_{z_1}^{z_2} f_Z(z)\, dz = \int_{-\sqrt{1/5}}^{\sqrt{1/5}} (2e^{-z} - 2e^{-2z})\mu(z) \\ &= \left(-2e^{-z} + e^{-2z}\right)\Big|_0^{\sqrt{1/5}} = 0.13000 \end{aligned}$$

■

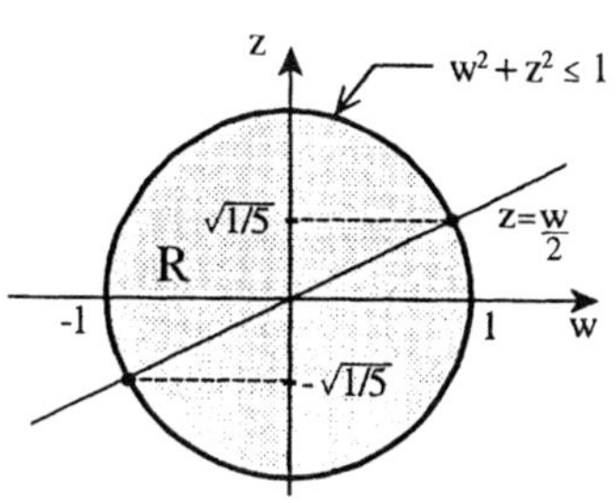

Figure 2.25 Limits for integral of Example 2.22.

2.7.3 Distribution Function (Continuous, Discrete, or Mixed)

The problem of finding the joint cumulative distribution function for the random variables W and Z defined by

$$W = g(X, Y) \quad Z = h(X, Y) \tag{2.117}$$

is similar to finding the cumulative distribution function of one function of one random variable. $F_{WZ}(w, z)$ can be written as

$$\begin{aligned} F_{WZ}(w, z) &= P\{W \leq w, Z \leq z\} \\ &= P\{g(X, Y) \leq w, h(X, Y) \leq z\} \end{aligned} \tag{2.118}$$

The $P\{h(x, y) \leq w, g(x, y) \leq z\}$ can be obtained by integrating the probability density function over the region I_{wz} defined in the x–y space as

$$I_{wz} = \{(x, y) : g(x, y) \leq w \text{ and } h(x, y) \leq z\} \tag{2.119}$$

Using this I_{WZ}, $F_{WZ}(w, z)$ can now be written as

$$F_{WZ}(w, z) = \iint_{I_{wz}} f_{XY}(x, y)\, dx\, dy \tag{2.120}$$

This approach can be used regardless whether the probability density function is from a discrete continuous or mixed random variable. If $f_{WZ}(w, z)$ is desired, it can be obtained by taking partial derivatives as follows:

$$f_{WZ}(w, z) = \frac{\partial}{\partial w}\frac{\partial}{\partial z} F_{wz}(w, z). \tag{2.121}$$

Special care is required if $F_{WZ}(w, z)$ is discontinuous, since various types of impulse functions will be generated.

The following example illustrates the procedure for finding the joint cumulative probability distribution function for two random variables defined as transformations of two other random variables.

EXAMPLE 2.23

Define two random variables W and Z by the following transformations of two random variables X and Y: $W = X^2 + Y^2$ and $Z = X^2$. Assume that X and Y are independent random variables characterized by their joint probability density function $f_{XY}(x, y)$, which is given by $f_{XY} = e^{-(x+y)}\mu(x)\mu(y)$. Find the joint cumulative distribution function $F_{WZ}(w, z)$ for W and Z, and then take partial derivitives to obtain the joint probability density function $f_{WZ}(w, z)$.

SOLUTION

Using Eq. (2.120) to obtain the cumulative distribution function depends on first obtaining the region I_{wz}, which for our problem is defined by

$$I_{wz} = \{(x, y) : x^2 + y^2 \leq w \text{ and } x^2 \leq z\}$$

and is shown in Figure 2.26.

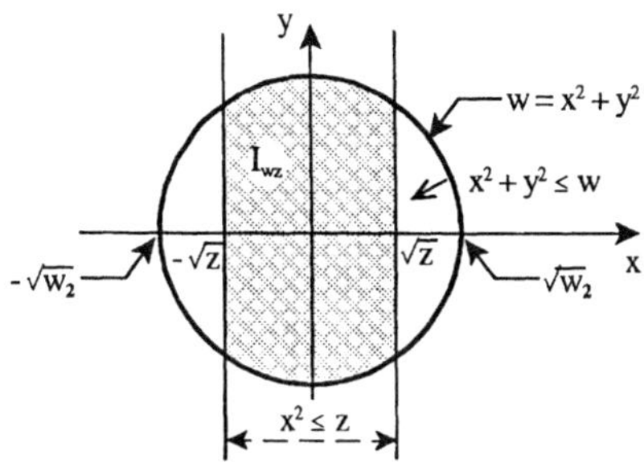

Figure 2.26 The region of integration I_{WZ} for Example 2.23.

For $w < 0$ and $z < 0$, $I_{WZ} = \emptyset$, the null set. Clearly, the integral over that region gives zero, and so $F_{WZ}(w, z) = 0$.

For $w \geq 0$ and $z \geq 0$, the region is shown as the shaded area in Figure 2.26. Thus the $F_{WZ}(w, z)$ can be written in terms of the integral of the joint probability density function for the random variables X and Y as follows:

$$F_{wz}(w, z) = \int_{-\sqrt{z}}^{\sqrt{z}} \left[\int_{-\sqrt{w-x^2}}^{\sqrt{w-x^2}} \exp -(x+y)\mu(x)\mu(y)\, dy \right] dx$$

$$= \int_0^{\sqrt{z}} e^{-x} \left[\int_0^{\sqrt{w-x^2}} e^{-y} dy \right] dx$$

This expression can be simplified to give

$$F_{wz}(w, z) = \int_0^{\sqrt{z}} e^{-x}(1 - e^{-\sqrt{w-x^2}})\, dx\, \mu(w)\mu(z)$$

The joint probability density function $f_{WZ}(w, z)$ is then obtained by taking the partial derivitives of $F_{WZ}(w, z)$ with respect to w and z as follows:

$$f_{WZ}(w, z) = \frac{\partial^2}{\partial w \partial z} F_{wz}(w, z) = \frac{\partial}{\partial w} \left(\frac{\partial}{\partial z} \int_0^{\sqrt{z}} e^{-x}(1 - e^{-\sqrt{w-x^2}})\, dx \right) \mu(w)\mu(z)$$

$$= \frac{\partial}{\partial w} (e^{-\sqrt{z}}(1 - e^{-\sqrt{w-z}}) \tfrac{1}{2}\, z^{-\frac{1}{2}})\, \mu(w)\mu(z)$$

In the equation above the partial with respect to z was obtained by Liebnitz's rule. Continuing and taking the partial with respect to w and simplifying gives

$$f_{WZ}(w, z) = \frac{e^{-\sqrt{w-z}} e^{-\sqrt{z}}}{4\sqrt{w-z}\sqrt{z}}\, \mu(w)\mu(z)$$

This result can be easily checked by using the transformation theorem for two functions of two random variables. ■

2.8 ONE FUNCTION OF TWO RANDOM VARIABLES

Define a random variable Z by $Z = g(X, Y)$, where X and Y are random variables. The basic problem is that of characterizing Z, knowing either partial or total characterizations of X and Y. This type of problem occurs in many different disciplines. For example, if the X and Y specify a random point in a two-dimensional space, then the distance from the origin to the point is given by $(X^2 + Y^2)^{\frac{1}{2}}$ and the angle with the x-axis is given by $\arctan(X, Y)$. If X represents a signal and Y a noise, then the output of a square law detector to the sum of the signal and noise is $(X + Y)^2$.

It was seen for one function of one random variable that a total characterization was required to determine the expected value of the transformed random variable. Similarly the determination of the mean of a random variable that is a function of two random variables usually requires a total characterization of each random variable and their joint relationship. This means that without other specified statistical information, it is necessary to have the joint probability density function $f_{XY}(x, y)$ to determine the $E[g(X, Y)]$, higher-order moments, or the probability density function for $Z = g(X, Y)$. Because X and Y could be discrete, continuous, or mixed random variables and $g(x, y)$ could be continuous, continuous with flat spots, or even discontinuous, a completely general approach seems almost impossible. The purpose of this section is to explore many of the most common cases and to describe procedures and methodologies that can be used to solve a variety of problems. The development parallels that for one function of one random variable as much as possible pointing out the similarities and the peculiarities of the two problems.

2.8.1 Probability Density Function (Discrete Random Variables)

If X and Y are discrete random variables, their joint probability density function $f_{XY}(x, y)$ is composed of two-dimensional delta functions as

$$f_{XY}(x, y) = \sum_{(x_i, y_i) \in S} P(x_i, y_i)\delta(x - x_i, y - y_i) \tag{2.122}$$

where $P(x_i, y_i) = P\{X = x_i,\ Y = y_i\}$ and S is the set of all pairs (x_i, y_i) considered. If $g(x, y)$ represents a function defined on real x and y, then the random variable $Z = g(X, Y)$ is a discrete random variable with probability density function $f_Z(z)$ given by

$$f_Z(z) = \sum_{(x_i, y_i) \in S} P(x_i, y_i)\delta(z - g(x_i, y_i)) \tag{2.123}$$

Some of the $g(x_i, y_i)$ will be the same, so adding those probabilities the $f_Z(z)$ can be written in terms of a reduced number of terms. If z_k is an element from the set of unique values of $g(x_i, y_i)$, where $(x_i, y_i) \in S$, then $f_Z(z)$ can be rewritten as

$$f_Z(z) = \sum_{z_k \in K} P(z_k)\delta(z - z_k) \tag{2.124}$$

where $P(z_k) = P\{(x_i, y_i) : g(x_i, y_i) = z_k\}$ and K is the set of all unique value of $g(x_i, y_i)$.

2.8.2 Probability Density Function (Continuous Random Variables)

Let X and Y be continuous random variables with known joint probability density function $f_{XY}(x, y)$ and $g(x, y)$ be a known continuous real-valued function of x and y. The basic problem is that of determining the probability density function $f_Z(z)$ for the random variable defined by $Z = g(X, Y)$. Four important basic methods for finding the density are (1) the cumulative distribution approach, (2) the auxiliary random variable approach, (3) the incremental probability approach and (4) the synthetic sampling or Monte Carlo approach. Of these only the synthetic sampling approach would be considered a direct approach.

Cumulative Distribution Approach. In this approach the cumulative distribution function $F_Z(z)$ is found and then differentiated to obtain $f_Z(z)$. The distribution function $F_Z(z)$ is determined as follows:

$$\begin{aligned} F_Z(z) &= P\{Z \leq z\} = P\{g(X, Y) \leq z\} \\ &= \iint_{I_z} f_{XY}(x, y)\, dx\, dy \end{aligned} \tag{2.125}$$

where $I_z = \{(x, y) : g(x, y) \leq z\}$. To obtain $F_Z(z)$ for z from $-\infty$ to ∞, $f_{XY}(x, y)$ must be integrated over changing regions of integration. The regions are composed of all (x, y) such that $g(x, y) \leq z$ with boundaries equal to the intersection of a plane at z and the function $g(x, y)$ as shown in Figure 2.27. The probability density function $f_Z(z)$ is then obtained by differentiating $F_Z(z)$ to give

$$f_Z(z) = \frac{d}{dz} F_Z(z) \tag{2.126}$$

For certain problems, whose regions can be expressed easily in the limits of the two-dimensional integral, Liebnitz's rule can be applied so that the integral may never need to be evaluated. However, for other problems the method is a tour de force using two-dimensional integration.

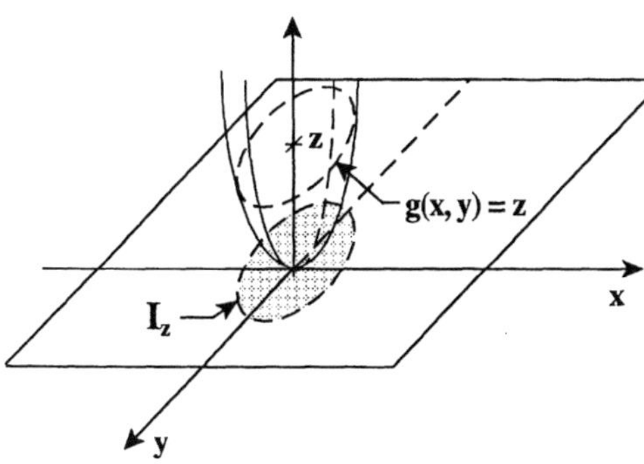

Figure 2.27 Obtaining the region of integration for the distribution method.

EXAMPLE 2.24

Let X and Y be a pair of random variables characterized by their joint probability density function $f_{XY}(x, y)$. Define a new random variable Z by $Z = X + Y$. Find the probability density function $f_Z(z)$ for Z by using the distribution method and determine the answer for the special case where X and Y are independent random variables.

SOLUTION

To find the distribution function $F_Z(z)$ the regions I_z must be identified for all z and are given by

$$I_Z = \{(x, y) : x + y \leq z\}$$

Thus the region is seen to be all points (x, y) below and including the boundary line $x + y = z$, or in standard form $y = z - x$ as seen in Figure 2.28.

Thus we can write $F_Z(z)$ for an arbitrary z as the following integral:

$$F_Z(z) = \int_{-\infty}^{\infty} \int_{-\infty}^{z-x} f_{XY}(x, y)\, dx\, dy$$

The probability density function is then given by

$$f_z(z) = \frac{d}{dz}\left[\int_{-\infty}^{\infty} \int_{-\infty}^{z-x} f_{XY}(x, y)\, dx\, dy\right]$$

Interchanging the first integral and the derivative and applying Liebnitz's rule to the second integral gives the answer as

$$f_Z(z) = \int_{-\infty}^{\infty} f_{XY}(x, z - x)\, dx \tag{2.127}$$

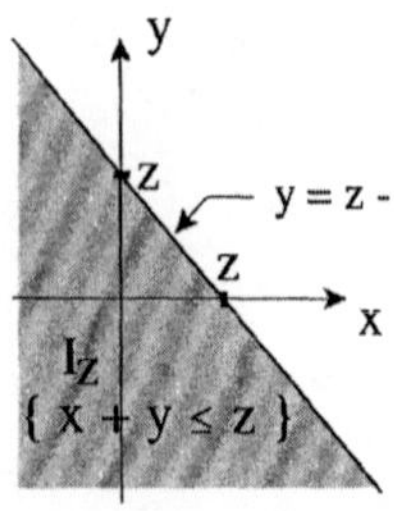

Figure 2.28 Region I_z for Example 2.24.

If X and Y are independent random variables, then $f(x, y) = f_X(x) f_Y(y)$. For this important case the $f_Z(z)$, for the random variable $Z = X + Y$, is determined from Eq. (2.127) to be the convolution integral as follows:

$$f_Z(z) = \int_{-\infty}^{\infty} f_X(x) f_Y(z - x)\, dx = f_X(z) * f_Y(z) \tag{2.128}$$

Auxiliary Random Variable Approach. The auxiliary random variable approach is a method for obtaining the probability density function $f_Z(z)$ for the random variable $Z = h(x, y)$. The approach consists of three basic steps: defining an auxiliary random variable, using the transformational theorem to obtain a joint probability density function, and an integration to obtain the marginal density for Z and thus the desired density.

In the first step a random variable W is defined by $W = g(X, Y)$ This is a fictitious random variable and selected as a vehicle for using the two-dimensional transformation theorem. Usually $g(X, Y)$ is selected as just X, just Y, or a simple function of X and Y that might be suggested by the form of the function $h(X, Y)$ which defines Z.

The second step is the application of the two-dimensional transformation theorem to the following:

$$\begin{aligned} W &= g(X, Y) \\ Z &= h(X, Y) \end{aligned} \tag{2.129}$$

to find the joint probability density $f_{WZ}(w, z)$ which is given by

$$f_{WZ}(w, z) = \sum_{(x_i, y_i)} \left. \frac{f_{XY}(x, y)}{|J(x, y)|} \right|_{(x,y)=(x_i,y_i)} \tag{2.130}$$

where (x_i, y_i) are all real solutions of $w = g(x, y)$ and $z = h(x, y)$.

The final step in the process of obtaining the probability density function $f_Z(z)$ is to integrate the joint density function $f_{ZW}(z, w)$ as follows:

$$f_Z(z) = \int_{-\infty}^{\infty} f_{WZ}(w, z)\, dw \tag{2.131}$$

EXAMPLE 2.25

Let X and Y be a pair of random variables characterized by their joint probability density function $f_{XY}(x, y)$. Define a new random variable Z by $Z = X + Y$ (same as in Example 2.24). Find the probability density function $f_Z(z)$ for Z by using the auxialary random variable method.

SOLUTION

Define the auxiliary random variable $W = X$, which now gives us two functions of two random variables $W = X$ and $Z = X + Y$. Using the two functions of two random variables theorem the joint probability density $f_{WZ}(w, z)$ is given by Eq. (2.130). There is only one solution to the set of equations, $w = x$ and $z = x + y$, which are easily seen to be $x = w$ and $y = z - w$. The Jacobian of the transformations is determined as

$$J(x, y) = \begin{vmatrix} \dfrac{\partial g(x, y)}{\partial x} & \dfrac{\partial g(x, y)}{\partial y} \\ \dfrac{\partial h(x, y)}{\partial x} & \dfrac{\partial h(x, y)}{\partial y} \end{vmatrix} = \begin{vmatrix} \dfrac{\partial x}{\partial x} & \dfrac{\partial x}{\partial y} \\ \dfrac{\partial (x + y)}{\partial x} & \dfrac{\partial (x + y)}{\partial y} \end{vmatrix} = \begin{vmatrix} 1 & 0 \\ 1 & 1 \end{vmatrix} = 1$$

Therefore the joint density can be written as

$$f_{wz}(w, z) = f_{XY}(x, y)\Big|_{\substack{x=w \\ y=z-w}} = f_{XY}(w, z - w)$$

The desired probability density function, $f_W(w)$, is obtained by integrating the joint density $f_{WZ}(w, z)$ to give the result

$$f_Z(z) = \int_{-\infty}^{\infty} f_{WZ}(w, z)\, dw = \int_{-\infty}^{\infty} f_{XY}(w, z - w)\, dw \qquad \blacksquare$$

Incremental Approach. The incremental approach can be used for certain problems of finding the probability density function $f_Z(z)$ for the random variable Z defined by $Z = h(X, Y)$. The method uses the fact that $f_z(z)\Delta z$ can be approximated by

$$f_z(z)\Delta z \approx P\{z < Z \le z + \Delta z\} \tag{1.132}$$

where $P\{z < Z \le z + \Delta z\}$ is equivalent to the probability that (x, y) is a member of $I_{\Delta z} = \{(x, y) : z < h(x, y) \le z + \Delta z\}$. Thus the approximation above can be written as

$$f_z(z)\Delta z \approx \iint_{I_{\Delta z}} f_{XY}(x, y)\, dx\, dy \tag{2.133}$$

The approach relies on the assumption that as $\Delta z \to 0$ the integral over $I_{\Delta z}$ becomes a linear function of Δz. When this happens, the Δz is canceled from both sides of the approximation thus leaving $f_Z(z)$ in the limit as $\Delta z \to 0$.

EXAMPLE 2.26

Suppose we are given random variables X and Y that are characterized by their joint probability density function $f_{XY}(x, y)$ and that $f_Y(y) = 0$ for $y < 0$. A new random variable Z is defined by $Z = XY$, the product of the two random variables. Find the probability density function $f_Z(z)$ for Z by the incremental method.

SOLUTION

Using the incremental method we must first determine the probability that Z is in the interval $(z, z + \Delta z)$, which is given by

$$p\{z < Z < z + \Delta z)\} = P\{z < XY < z + \Delta z)\} = P\left\{\frac{z}{Y} < X < \frac{(z + \Delta z)}{Y}\right\}$$

$$= \int_{-\infty}^{\infty} \int_{z/y}^{(z+\Delta z)/y} f_{XY}(x, y)\, dx\, dy \approx \int_{-\infty}^{\infty} f_{XY}\left(\frac{z}{y}, y\right) \frac{\Delta z}{y}\, dy$$

Using (2.132) gives

$$f_Z(z)\Delta z \approx \int_{-\infty}^{\infty} f_{XY}\left(\frac{z}{y}, y\right) \frac{\Delta z}{y}\, dy$$

Cancelling the Δz, the equation above in the limit reduces to the result

$$f_Z(z) = \int_{-\infty}^{\infty} \frac{1}{y} f_{XY}(z/y, y)\, dy$$

This proof can be modified to include the case where $f_Y(y) \neq 0$ for $y < 0$ to give the following result for the probability density $f_Z(z)$ for the product of any two random variables $Z = XY$:

$$f_Z(z) = \int_{-\infty}^{\infty} \frac{1}{|y|} f_{XY}(z/y, y)\, dy \tag{2.134}$$

■

2.9 COMPUTATION OF $E[h(X, Y)]$

As for the case of one random variable there are several basic methods for obtaining $E[Z]$, where $Z = h(X, Y)$. The first three methods described are comparable to those described for one function of one random variable. The fourth technique is very useful for functions of more than one random variable and does not have a counterpart in the function of one random variable case.

Method 1. Using $f_{XY}(x, y)$,

$$E[Z] = E[h(X, Y)] = \int_{-\infty}^{\infty} \int_{-\infty}^{\infty} h(x, y)\, f_{XY}(x, y)\, dx\, dy \tag{2.135}$$

Method 2. Using $f_z(z)$,

$$E[h(X, Y)] = E[Z] = \int_{-\infty}^{\infty} z f_Z(z)\, dz \tag{2.136}$$

Method 3. Using a Monte Carlo technique or synthetic sampling.

Method 4. Using iterated expected value,

$$E[Z] = E[h(X, Y)] = E\Big[E[h(x, Y)]\big|_{x=X}\Big] \\ = E\Big[E[h(X, y)]\big|_{y=Y}\Big] \tag{2.137}$$

Method 1 requires that a two-dimensional integral be evaluated and does not need the probability density $f_Z(z)$. Its main use is for cases where the $f_Z(z)$ is not required and the two-dimensional intergrals are conveniently evaluated.

Method 2, similar to the one function of one random variable, relies on being able to determine first the probability density function $f_Z(z)$ by using any of the techniques described earlier, and then performing the integral.

Method 3 uses a synthetic sampling procedure to obtain an approximation for the $E[Z]$. The procedure synthetically generates samples of x and y by using the given probability density for the random variables X and Y, computes $h(x, y)$ for each sample, and computes the average of those values. The method is particularly used to solve problems where the $h(x, y)$ and $f_{XY}(x, y)$ are very complex functions making the other analytical methods virtually impossible to solve and for problems where the analytical form of $h(x, y)$ is not given but only plots of $h(x, y)$ are known at discrete values of x and y.

Method 4 has no parallel for the case of one function of one random variable but can be a powerful method for computing expected values of functions of two random variables. The meaning of the **iterated expected value** operation described above is easily understood and represents conditional expected values as follows:

$$E[h(X, Y)] = \int_{-\infty}^{\infty}\int_{-\infty}^{\infty} f_{XY}(x, y)h(x, y)\,dx\,dy \tag{2.138}$$

Breaking the joint probability density function into the product of its conditional and marginal densities gives.

$$E[h(X, Y)] = \int_{-\infty}^{\infty}\int_{-\infty}^{\infty} f_Y(y|x)\,f_X(x)h(x, y)\,dx\,dy \tag{2.139}$$

Rearranging the integral on the right side of the equation above, and taking the $f_X(x)$ through the integral with respect to y (it is not a function of y), gives

$$E[h(X, Y)] = \int_{-\infty}^{\infty} f_X(x)\left[\int_{-\infty}^{\infty} h(x, y)\,f_Y(y|x)\,dy\right]dx \tag{2.140}$$

The term in brackets is recognized as the conditional expected value of $g(x, Y)$ or

$$E[h(x, Y)] = \int_{-\infty}^{\infty} h(x, y)\,f_Y(y|x)\,dy \tag{2.141}$$

Therefore $E[h(X, Y)]$ becomes

$$E[h(X, Y)] = \int_{-\infty}^{\infty} f_X(x)E[h(x, Y)]\,dx \tag{2.142}$$

But the integral on the right represents the $E[h(x, Y)]|_{x=X}$, which verifies the first equation specified in method 4, Eq. (2.137).

In a similar fashion $E[h(X, Y)]$ can be written in terms of the conditional expected value with respect to y as follows

$$E[Z] = E[h(X, Y)] = E[E[h(X, y)]|_{y=Y}] \tag{2.143}$$

EXAMPLE 1.27

Given $Z = \cos(X + Y)$ with X a uniform random variable on $(-\pi, \pi)$ and Y a Gaussian random variable with zero mean and variance equal to 1. Assume that X and Y are independent, and compute $E[Z]$ the expected value of Z using method 4.

SOLUTION

Using (2.137), we have the expected value of Z as

$$E[Z] = E[\cos(X + Y)] = E[E[\cos(X + y)]|_{y=Y}]$$

where

$$E[\cos(X + y)] = \int_{-\pi}^{\pi} \cos(x + y)\, f_X(x\,|\,y)\, dx$$

For this problem X and Y are independent, therefore the conditional density is the same as the marginal density, and the equation above can be written as

$$\begin{aligned} E[\cos(X + y)]\,|_{y=Y} &= \int_{-\pi}^{\pi} \cos(x + y) \frac{1}{2\pi}\, dx \Big|_{y=Y} \\ &= 0 \Big|_{y=Y} \end{aligned}$$

The expected value of Z can be found

$$E[Z] = E[0|_{y=Y}] = \int_{-\infty}^{\infty} 0 f_Y(y)\, dy = 0$$ ■

The following example calculates some higher-order moments for jointly Gaussian random variable also using the concept of iterated expected values.

EXAMPLE 2.28

Let X an Y be jointly normal random variables with parameters $\eta_X = \eta_Y = 0$ with σ_X, σ_Y, and ρ_{XY} known. Find $E[XY]$ and $E[X^2 Y^2]$.

SOLUTION

Using the iterated expected value formula just presented, the $E[XY]$ can be written as

$$E[XY] = \int_{-\infty}^{\infty} f_X(x)\left[\int_{-\infty}^{\infty} xyf_Y(y|x)\,dy\right]dx$$

where

$$f_Y(y|x) = \frac{1}{\sqrt{2\pi(1-\rho_{XY}^2)}\sigma_Y}\exp\left\{-\frac{(y-\rho_{XY}\sigma_Y\sigma_X^{-1}x)^2}{2\sigma_Y^2(1-\rho_{XY}^2)}\right\}$$

The term inside the square brackets can be simplified by taking x through the integral and using the result for the conditional mean Eq. (2.108) to give

$$\begin{aligned}\int_{-\infty}^{\infty} xyf_Y(y|x)\,dy &= x\int_{-\infty}^{\infty} yf_Y(y|x)\,dy \\ &= x\left(\rho_{XY}\sigma_Y\sigma_x^{-1}x\right)\end{aligned}$$

Substituting this result for the bracketed term and using the second moment for the random variable X gives the $E[XY]$ as

$$\begin{aligned}E[XY] &= \int_{-\infty}^{\infty} f_X(x)[x^2\rho_{XY}\sigma_Y\sigma_X^{-1}]\,dx \\ &= \rho_{XY}\sigma_Y\sigma_X^{-1}\int_{-\infty}^{\infty} f_X(x)x^2dx = \rho_{XY}\sigma_X\sigma_Y\end{aligned}$$

Similarly $E[X^2Y^2]$ can be found in the following fashion:

$$\begin{aligned}E[X^2Y^2] &= \int_{-\infty}^{\infty} f_X(x)\left(\int_{-\infty}^{\infty} x^2y^2f_Y(y|x)\,dy\right)dx \\ &= \int_{-\infty}^{\infty} f_X(x)\left(x^2\int_{-\infty}^{\infty} y^2f_Y(y|x)\,dy\right)dx\end{aligned}$$

The integral in the parentheses above represents the conditional second moment, and it can be written in terms of the conditional variance and the conditional mean from Eq. (2.107) as

$$\begin{aligned}E[X^2Y^2] &= \int_{-\infty}^{\infty} f_X(x)x^2[\sigma_{Y|x}^2 + \eta_{Y|x}^2]\,dx \\ &= \sigma_Y^2(1-\rho_{XY}^2)\int_{-\infty}^{\infty} f_X(x)x^2dx + \frac{\rho_{XY}^2\sigma_Y^2}{\sigma_X^2}\int_{-\infty}^{\infty} f_X(x)x^4dx\end{aligned}$$

The first integral is the variance of X, since X has a zero mean, and the second is the fourth-order moment for the one-dimensional Gaussian random variable, which is known from Eq. (2.32). Substituting these second-, and fourth-order moments gives

$$\begin{aligned} E[X^2Y^2] &= \sigma_Y^2(1-\rho_{XY}^2)[\sigma_X^2] + \frac{\rho_{XY}^2\sigma_Y^2}{\sigma_X^2}[3\sigma_X^4] \\ &= \sigma_X^2\sigma_Y^2 + 2\rho_{XY}^2\sigma_X^2\sigma_Y^2 \\ &= \sigma_X^2\sigma_Y^2(1+2\rho_{XY}^2) \end{aligned}$$

■

2.10 MULTIPLE RANDOM VARIABLES

There are many problems that involve more than two random variables. Using the framework set up earlier for two random variables, the partial and total characterizations as well as properties and partial characterizations for n random variables can be defined. This section gives just a brief summary of those definitions and properties for the total and partial characterizations. It also puts a little more emphasis on higher-order moments and Gaussian random variables.

2.10.1 Total Characterizations

The real-valued functions for random variables $X_1, X_2, \ldots, X_n$ and Y are defined on a sample space and are said to be **totally characterized** or described with relation to calculating probabilities of acceptable events (i.e., events a member of $\mathscr{F}$) by their cumulative distribution function or probability density function.

The **joint cumulative distribution function** $F_{X_1X_2\ldots X_n}(x_1, x_2, \ldots, x_n)$ for random variables $X_1, X_2, \ldots, X_n$ is defined by

$$F_{X_1X_2\ldots X_n}(x_1, x_2, \ldots, X_n) \triangleq P\{X_1 \le x_1, X_2 \le x_2, \ldots, X_n \le x_n\} \tag{2.144}$$

It satisfies the extensions of the bounded, nondecreasing, and continuity properties given in Section 2.6.1 and can be used to determine probabilities of given events. The **joint probability density function** $f_{X_1X_2\ldots X_n}(x_1, x_2, \ldots, x_n)$ is related to the joint cumulative distribution function through partial derivatives, and vice versa in terms of integrals, as

$$\begin{aligned} f_{X_1X_2\ldots X_n}(x_1, x_2, \ldots, x_n) &= \frac{\partial^n F_{X_1X_2\ldots X_n}(x_1, x_2, \ldots, x_n)}{\partial x_1 \partial x_2 \ldots \partial x_n} \\ F_{X_1X_2\ldots X_n}(x_1, x_2, \ldots, x_n) &= \int_{-\infty}^{x_1}\int_{-\infty}^{x_2} \cdots \int_{-\infty}^{x_n} f_{X_1X_2\ldots X_n}(x_1, x_2, \ldots, x_n)\, dx_1\, dx_2 \ldots dx_n \end{aligned} \tag{2.145}$$

The characterization function and moment-generating function, when they exist, are also total characterizations because of their direct relationship with the probability density function.

For a given vector **X** of random variables $X_1, X_2, \ldots, X_n$, the **characteristic function** $\Phi_{\mathbf{X}}(\omega_1, \omega_2, \ldots, \omega_n)$ is defined as the "modified" Fourier transform of the joint probability density function

$$\begin{aligned}\Phi_{\mathbf{X}}(\omega_1, \omega_2, \ldots, \omega_n) &= E[e^{j(\omega_1 X_1 + \omega_2 X_2 + \ldots + \omega_n X_n)}] \\ &= \int_{-\infty}^{\infty}\int_{-\infty}^{\infty} \cdots \\ &\quad \times \int_{-\infty}^{\infty} f_X(x_1, x_2, \ldots, x_n) e^{j(\omega_1 x_1 + \omega_2 x_2 + \ldots + \omega_n x_n)} dx_1 dx_2 \ldots dx_n \\ &= \mathscr{F}(f_X(x_1, x_2, \ldots, x_n)|_{\boldsymbol{\omega} \to -\boldsymbol{\omega}} \end{aligned} \tag{2.146}$$

where $\boldsymbol{\omega} = [\omega_1, \omega_2, \ldots, \omega_n]^T$. This integral does not necessarily exist for all probability density functions and has a given region of convergence.

The **moment-generating function** is defined as follows:

$$\begin{aligned}M_{\mathbf{X}}(t_1, t_2, \ldots, t_n) &= E[e^{(t_1 X_1 + t_2 X_2 + \ldots + t_n X_n)}] \\ &= \int_{-\infty}^{\infty}\int_{-\infty}^{\infty} \cdots \\ &\quad \times \int_{-\infty}^{\infty} f_X(x_1, x_2, \ldots, x_n) e^{(t_1 x_1 + t_2 x_2 + \ldots + t_n x_n)} dx_1 dx_2 \ldots dx_n \end{aligned} \tag{2.147}$$

2.10.2 Partial Characterizations

Partial characterizations include the various marginal and conditional densities and the higher-order moments. The higher-order moments can be determined from the definition of the moments. However, in many problems the characteristic function or moment-generating, function is useful for that purpose.

The **rth-order higher-order moments** for $k_1 + k_2 + \cdots + k_n = r$ are defined by

$$\begin{aligned}\mathrm{m}_{\mathrm{k}_1 \mathrm{k}_2 \ldots \mathrm{k}_n} &\triangleq E[X_1^{k_1} X_2^{k_2} \ldots X_n^{k_n}] \\ &= \int_{-\infty}^{\infty}\int_{-\infty}^{\infty} \cdots \int_{-\infty}^{\infty} x_1^{k_1} x_2^{k_2} \ldots x_n^{k_n} f(x_1, x_2, \ldots, x_n) dx_1 dx_2 \ldots dx_n \end{aligned} \tag{2.148}$$

where $k_1 + k_2 + \cdots + k_n = r$

A well-known theorem [10] gives the higher rth-order moments in terms of the characteristic function $\Phi_{\mathbf{X}}(\omega_1, \omega_2, \ldots, \omega_n)$ as follows:

$$\mathrm{m}_{k_1 k_2 \ldots k_n} \triangleq (-j)^r \left. \frac{\partial^r \Phi(\omega_1, \omega_2, \ldots, \omega_n)}{\partial^{k_1} \omega_1 \partial^{k_2} \omega_2 \ldots \partial^{k_n} \omega_n} \right]_{\omega_1 = 0, \omega_2 = 0, \ldots, \omega_n = 0}$$

where $k_1 + k_2 + \cdots + k_n = r$ and $r \leq n$

(2.149)

For example, all of the following are fourth order moments: $\mathrm{m}_{1111} = E[X_1^1 X_2^1 X_3^1 X_4^1]$, $\mathrm{m}_{4000} = E[X_1^4]$, $\mathrm{m}_{2020} = E[X_1^2 X_3^2]$, and $\mathrm{m}_{0130} = E[X_2^1 X_3^3]$.

Similarly the moments can be generated by using the **moment-generating function** through a corresponding theorem, from Davenport [5], that gives the higher-order moments in terms of the moment-generating function as follows:

$$\mathrm{m}_{k_1 k_2 \ldots k_n} = \left.\frac{\partial^r M_X(t_1, t_2, \ldots, t_n)}{\partial^{k_1} t_1 \partial^{k_2} t_2 \ldots \partial^{k_n} t_n}\right]_{t_1=0, t_2=0, \ldots, t_n=0} \tag{2.150}$$

where $k_1 + k_2 + \cdots + k_n = r$ and $r \leq n$

The moment-generating function is simpler to obtain in that it is determined as a real integral. However, the use of the existing Fourier transform table to obtain the characteristic function, and the fact the characteristic function exists for cases where the moment-generating function does not exist, makes the characteristic function more usable.

EXAMPLE 2.29

The random variables X_1, X_2, and X_3 are characterized by their joint probability density function $f_{X_1X_2X_3}(x_1, x_2, x_3)$, which is

$$f_{X_1X_2X_3}(x_1, x_2, x_3) = 6e^{-(x_1+2x_2+3x_3)}\mu(x_1)\mu(x_2)\mu(x_3)$$

Determine, by using Eq. (2.149), the following third-order moments: (a) m_{001}, (b) m_{111}, (c) m_{030}, (d) m_{210}.

SOLUTION

To obtain the moments from (2.149), we first need to calculate the characteristic function, which from (2.146) is as follows:

$$\begin{aligned}\Phi(\omega_1, \omega_2, \ldots, \omega_n) &= \int_0^\infty \int_0^\infty \int_0^\infty 6e^{-(x_1+2x_2+3x_3)} e^{j(\omega_1 x_1 + \omega_2 x_2 + \omega_3 x_3)} dx_1 dx_2 dx_3 \\ &= 6 \frac{1}{(-j\omega_1 + 1)} \frac{1}{(-j\omega_2 + 2)} \frac{1}{(-j\omega_1 + 3)}\end{aligned}$$

(a) The first-order moment m_{001} is determined from (2.149) as

$$\mathrm{m}_{001} = (-j)^1 \left.\frac{\partial\Phi}{\partial\omega_3}\right]_{\substack{\omega_1=0\\ \omega_2=0\\ \omega_3=0}} = \left.\frac{6(-j)(-1)(-j)}{(-j\omega_1 + 1)(-j\omega_2 + 2)(-j\omega_1 + 3)^2}\right]_{\substack{\omega_1=0\\ \omega_2=0\\ \omega_3=0}} = \frac{1}{3}$$

(b) The third-order moment m_{111} is determined from (2.149) as

$$\mathrm{m}_{111} = (-j)^3 \left.\frac{\partial^3\Phi}{\partial\omega_1\partial\omega_2\partial\omega_3}\right]_{\substack{\omega_1=0\\ \omega_2=0\\ \omega_3=0}} = \left.\frac{(-j)^3 6(-1)^3(-j)^3}{(-j\omega_1 + 1)^2(-j\omega_2 + 2)^2(-j\omega_1 + 3)^2}\right]_{\substack{\omega_1=0\\ \omega_2=0\\ \omega_3=0}} = \frac{1}{6}$$

(c) The third-order moment m_{030} is determined from (2.149) as

$$m_{030} = (-j)^3 \frac{\partial^3 \Phi}{\partial^3 \omega_3}\Bigg]_{\substack{\omega_1=0\\ \omega_2=0\\ \omega_3=0}} = \frac{(-j)^3 6(-1)(-2)(-3)(-j)^3}{(-j\omega_1+1)(-j\omega_2+2)^4(-j\omega_1+3)}\Bigg]_{\substack{\omega_1=0\\ \omega_2=0\\ \omega_3=0}} = \frac{1}{8}$$

(d) The third-order moment m_{210} is determined from (2.149) as

$$m_{210} = (-j)^3 \frac{\partial^3 \Phi}{\partial^2 \omega_1 \partial \omega_2}\Bigg]_{\substack{\omega_1=0\\ \omega_2=0\\ \omega_3=0}} = \frac{(-j)^3 6(-1)(-2)(-1)(-j)^3}{(-j\omega_1+1)^3(-j\omega_2+2)^2(-j\omega_1+3)}\Bigg]_{\substack{\omega_1=0\\ \omega_2=0\\ \omega_3=0}} = 1$$

Since the random variables for this problem are independent, it is easy to check these results by computing the moments shown as the product of various moments of the random variables taken by themselves. When the random variables are not independent, then the higher-order moments cannot be calculated as products of the individual moments, so the formula above becomes more useful for their evaluation. ■

Another useful characterization for random variables are the cumulants that will play a role in the analysis of nonlinear systems with random process excitations. The **cumulants** are defined in terms of the joint characteristic function as follows:

$$c_{k_1 k_2 \ldots k_n} = (-j)^r \frac{\partial^r \ln \Phi(\omega_1, \omega_2, \ldots, \omega_n)}{\partial^{k_1}\omega_1 \partial^{k_2}\omega_2 \ldots \partial^{k_n}\omega_n}\Bigg]_{\omega_1=0, \omega_2=0, \ldots, \omega_n=0}$$

where $k_1 + k_2 + \cdots + k_n = r$ and $r \leq n$

(2.151)

The only difference in the preceding formulas for moments and cumulants is that the natural log (ln) of the characteristic function is the object of the partial derivatives for the cumulants, whereas the characteristic function is used for determining the moments. The relationships between the cumulants and the moments depend on the joint probability density function and thus are different for each density. Characteristic functions that are exponential in character result in much easier calculation of the cumulants, as will be shown for the Gaussian case. Also note that if the argument of the exponential is a polynomial in the $\omega_1, \omega_2, \ldots, \omega_n$, there exist orders such that all higher-order cumulants are zero. This will be shown to be the case for the Gaussian density in the following section.

EXAMPLE 2.30

The random variables X_1 and X_2 are characterized by their joint probability density function $f_{X_1X_2}(x_1, x_2)$, which is

$$f_{X_1X_2}(x_1, x_2) = 2e^{-(x_1+2x_2)}\mu(x_1)\mu(x_2)$$

Determine, by using Eq. (2.151), the following cumulants: (a) c_{01}, (b) c_{10}, (c) c_{11}, and (d) the rth order cumulant c_{jk} for $j > 1, k > 1$, and $j + k = r$.

SOLUTION

To obtain the cumulants from (2.151), we first need to calculate the characteristic function, which from (2.146) is as follows

$$\Phi(\omega_1, \omega_2) = \int_0^{\infty}\int_0^{\infty} 2e^{-(x_1+2x_2)}e^{j(\omega_1 x_1+\omega_2 x_2)}dx_1 dx_2$$
$$= 2\frac{1}{(-j\omega_1 + 1)}\frac{1}{(-j\omega_2 + 2)}$$

(a) The first-order cumulant c_{01} is determined from (2.151) as

$$c_{10} = (-j)^1 \left.\frac{\partial(-\ln(-j\omega_1 + 1) - \ln(-j\omega_2 + 2) + \ln 2)}{\partial\omega_1}\right]_{\substack{\omega_1=0\\ \omega_2=0}}$$
$$= \left.\frac{-(-j)(-j)1}{(-j\omega_1 + 1)}\right]_{\substack{\omega_1=0\\ \omega_2=0}} = 1$$

(b) The first-order cumulant c_{01} is determined from (2.151) as

$$c_{01} = (-j)^1 \left.\frac{\partial(-\ln(-j\omega_1 + 1) - \ln(-j\omega_2 + 2) + \ln 2)}{\partial\omega_2}\right]_{\substack{\omega_1=0\\ \omega_2=0}}$$
$$= \left.\frac{-(-j)(-j)1}{(-j\omega_1 + 2)}\right]_{\substack{\omega_1=0\\ \omega_2=0}} = \frac{1}{2}$$

(c) The second-order cumulant c_{11} is determined from (2.151) as

$$c_{11} = (-j)^1 \left.\frac{\partial(-\ln(-j\omega_1 + 1) - \ln(-j\omega_2 + 2) + \ln 2)}{\partial\omega_1\partial\omega_2}\right]_{\substack{\omega_1=0\\ \omega_2=0}}$$
$$= \left.\frac{\partial}{\partial\omega_2}\left(\frac{-(-j)(-j)1}{(-j\omega_1 + 1)}\right)\right]_{\substack{\omega_1=0\\ \omega_2=0}} = 0$$

(d) The rth order cumulant c_{jk} for $j > 1, k > 1$, and $j + k = r$ is seen from the equation above to be 0. ■

A theorem presented by Nikias and Petropulu [10] originally presented by Brillinger gives the nth-order cumulant in terms of all the nth-order moments as follows:

Theorem. If $X_1, X_2, \ldots, X_n$ are jointly Gaussian random variables specified by their various order moments $m_{ij} = E[X_i X_j]$, then the nth order cumulant $c_{11\ldots1} = \text{cum}[X_1 X_2, \ldots X_n]$ can be written in terms of the $E[X_{i_1} X_{i_2}, \ldots, X_{i_n}]$ as follows:

$$c_{111\ldots1} = \sum_{\substack{\text{all}\\ \text{partitions}}} (-1)^{p-1}(p-1)! E[\Pi_{i\in s_1} X_i] \cdot E[\Pi_{i\in s_2} X_i] \cdot \ldots \cdot E[\Pi_{i\in s_p} X_i] \quad (2.152)$$

where the summation is taken over all partitions $\{s_1, s_2, \ldots, s_p\}, p = 1, 2, \ldots, n$ of the set of integers $\{1, 2, \ldots, n\}$.

EXAMPLE 2.31

Use the theorem above to determine the first, second, and third order cumulants in terms of various third-order moments (from Nikias and Petropulu [12]).

SOLUTION

The $c_1(X_1)$ and $c_{11}(X_1, X_2)$ are easily shown to be

$$\begin{aligned} c_1 &= c_1(X_1) = E[X_1] \\ c_{11} &= c_{11}(X_1, X_2) = E[X_1 X_2] - E[X_1]E[X_2] \end{aligned} \quad (2.153)$$

In order to determine the cumulant $c_{111} = c(X_1 X_2 X_3)$, it is necessary to determine all partitions of $\{1, 2, 3\}$. For $p = 1$, just one set we get $s_1 = \{1, 2, 3\}$. For $p = 2$, we have the following ways to partition: $s_1 = \{1\}$ and $s_2 = \{2, 3\}$, or $s_1 = \{2\}$ and $s_2 = \{1, 3\}$ or $s_1 = \{3\}$ and $s_2 = \{1, 2\}$. And finally for $p = 3$, there is only one way: $s_1 = \{1\}, s_2 = \{2\}$, and $s_3 = \{3\}$.

From (2.152) we can now write $c(X_1 X_2 X_3)$ as

$$\begin{aligned} c_{111} &\triangleq c_{111}(X_1, X_2, X_3) \\ &= (-1)^{1-1}(1-1)! E[X_1 X_2 X_3] \\ &\quad + (-1)^{2-1}(2-1)! \big(E[X_1]E[X_2 X_3] + E[X_2]E[X_1 X_3] + E[X_3]E[X_1 X_2]\big) \\ &\quad + (-1)^{3-1}(3-1)! E[X_1]E[X_2]E[X_3] \\ &= E[X_1 X_2 X_3] - (E[X_1]E[X_2 X_3] + E[X_2]E[X_1 X_3] + E[X_3]E[X_1 X_2]) \\ &\quad + 2E[X_1]E[X_2]E[X_3] \end{aligned} \quad (2.154)$$

Similarly it can be shown that c_{1111} is as follows:

$$
\begin{aligned}
c_{1111} &\triangleq c_{1111}(X_1, X_2, X_3, X_4) \\
&= E[X_1X_2X_3X_4] - E[X_1X_2]E[X_3X_4] - E[X_1X_3]E[X_2X_4] - E[X_1X_4]E[X_2X_3] \\
&\quad - E[X_1]E[X_2X_3X_4] - E[X_2]E[X_1X_3X_4] - E[X_3]E[X_1X_2X_4] - E[X_4]E[X_1X_2X_3] \\
&\quad + 2\big(E[X_1X_2]E[X_3]E[X_4] + E[X_1X_3]E[X_2]E[X_4] + E[X_1X_4]E[X_2]E[X_3] \\
&\quad + E[X_2X_3]E[X_1]E[X_4] + E[X_2X_4]E[X_1]E[X_3] + E[X_3X_4]E[X_1]E[X_2]\big) \\
&\quad - 6E[X_1]E[X_2]E[X_3]E[X_4] \qquad (2.155)
\end{aligned}
$$

■

For zero mean random variables the first, second, third, and fourth-order cumulants reduce to the following:

$$
\begin{aligned}
c_1(X_1) &= E[X_1] = 0 \\
c_{11}(X_1, X_2) &= E[X_1X_2] \\
c_{111}(X_1, X_2, X_3) &= E[X_1X_2X_3] \qquad (2.156) \\
c_{1111}(X_1, X_2, X_3, X_4) &= E[X_1X_2X_3X_4] - E[X_1X_2]E[X_3X_4] \\
&\quad - E[X_1X_3]E[X_2X_4] - E[X_1X_4]E[X_2X_3]
\end{aligned}
$$

It is thus seen that for zero mean random variables that the kth order cumulants are the same as the moments up to and including order three, but for all higher orders the cumulants are not just the moments but become increasingly more complicated functions of the lower-order moments.

2.10.3 Gaussian Random Vectors

When more than two jointly normal random variables are considered, it becomes convenient to use a vector formulation. Random variables $X_1, X_2, \ldots, X_N$ are **jointly Gaussian** if their joint probability density function can be written in the form

$$
f_{\mathbf{X}}(x_1, x_2, \ldots, x_N) = \frac{1}{(2\pi)^{N/2}|\mathbf{K}|^{1/2}} \exp\left[\frac{-(\mathbf{x}-\mathbf{m})^T\mathbf{K}^{-1}(\mathbf{x}-\mathbf{m})}{2}\right]
$$

where

$$
\begin{aligned}
\mathbf{X} &= [X_1X_2 \ldots X_N]^T \\
\mathbf{m} &= E[\mathbf{X}] = [m_1m_2 \ldots m_N]^T \\
\mathbf{K} &= E[(\mathbf{X}-\mathbf{m})(\mathbf{X}-\mathbf{m})^T] \qquad (2.157) \\
\mathbf{x} &= [x_1x_2 \ldots x_N]^T
\end{aligned}
$$

For the random vector **X**, **m** and **K** are the mean vector and covariance matrix, respectively. The $E[.]$ is the expected value operator. The vector **m**, called the **mean vector** of the random vector **X**, has the means of each random variable as its components. The argument of the exponential is a vector-matrix-vector product that yields a scaler result. The matrix **K** is called the **covariance matrix**. It can be easily shown that $\mathbf{K}$ is a square matrix with variances of the individual random variables down the main diagonal and covariances of the respective random variables off the main diagonal. It can also be shown that $\mathbf{K}$ is symmetric, positive definite, and has all its eigen values greater than 0. Instead of writing the long form given above, it is convenient to use a short-hand notation and write that $X \sim N(\mathbf{m}, \mathbf{K})$, which will mean that $\mathbf{X}$ is a Gaussian random vector with mean vector $\mathbf{m}$, covariance matrix $\mathbf{K}$, and have the joint density given in (2.157).

Moments and Cumulants for Gaussian Random Variables. Let $X_1, X_2, \ldots, X_n$ be jointly Gaussian random variables characterized by their mean vector $\mathbf{m}_X$ and covariance matrix $\mathbf{K}$. The cumulants and moments for the jointly Gaussian random variables will now be presented.

The joint characteristic function for a Gaussian random vector can be shown to be

$$\begin{aligned} \Phi(\omega_1, \omega_2, \ldots, \omega_n) &= \exp[j\boldsymbol{\omega}^T \mathbf{m}_X - \tfrac{1}{2}\boldsymbol{\omega}^T \mathbf{K}^{-1} \boldsymbol{\omega}] \\ \text{where } \boldsymbol{\omega}^T &= [\omega_1 \omega_2 \ldots \omega_n] \text{ and} \\ \mathbf{m}_X^T &= [E[X_1] \quad E[X_2] \quad \ldots \quad E[X_n]] \end{aligned} \tag{2.158}$$

Proofs of this result are available in many different texts; for example, see Stark and Woods [6] if interested.

The most direct way to find various moments is through the basic definition (2.148), or they can be found indirectly by (2.149) by using the characteristic function. The cumulants can also be found by using the characteristic function as in (2.151).

Using the characteristic function for jointly Gaussian random variables (2.158), it is straightforward to prove a number of useful results for jointly Gaussian random variables. The proofs are direct and involve taking partial derivatives, but they can be lengthy, and a test for patience or a zest for bookkeeping is required. One such example of this is the following theorem.

Theorem. If X_1, X_2, X_3, and X_4 are zero mean jointly Gaussian random variables specified by their second-order moments $\mathbf{m}_{ij} = E[X_i X_j]$, then the fourth-order moment $E[X_1 X_2 X_3 X_4]$ can be written in terms of the $E[X_i X_j]$ as follows:

$$E[X_1 X_2 X_3 X_4] = E[X_1 X_2]E[X_3 X_4] + E[X_1 X_3]E[X_2 X_4] + E[X_1 X_4]E[X_2 X_3] \tag{2.159}$$

2.11 *M* FUNCTIONS OF *N* RANDOM VARIABLES

In the previous sections the cases of one and two functions of one and two random variables have been presented. In general the techniques introduced in these discussions can be extended to one, two, and M functions of N random variables. The methods become increasingly more difficult to apply because of the complexity of the higher dimensions.

In general, we would wish to find the joint probability density function $f(y_1, y_2, \ldots, y_N)$ for random variables $Y_1, Y_2, \ldots, Y_N$ defined by

$$Y_i = g_i(X_1, X_2, \ldots, X_N) \quad i = 1, 2, \ldots, M$$

When M an N are equal, it is possible to find the joint density of the transformed variables by extending the transformation theorem given in Eq. (2.113). When M and N are not equal auxiliary random variables can be defined to make the orders equal then use the transformation theorem followed by integrating out with respect to the auxiliary variables.

The extension of the distribution function approach is possible, but usually the high-dimensional space makes it difficult to describe the regions over which we integrate, and the following partial derivatives of the distribution are certainly not a picnic either.

We are then led to obtaining the joint density by use of statistical sampling, where again the higher dimensions make the number of samples necessary to get a reasonable estimate of the probability density by a histogram becomes exceedingly large. This problem can be alleviated if the densities involved are of a certain parametric form.

2.12 SUMMARY

In this chapter the basic definitions for a single random variable and several joint random variables were presented along with the various forms of their characterizations, which included the mean, variance, covariance, and higher-order moments as partial characterizations. Total characterizations for random variables were given in terms of the probability and joint probability density functions and the corresponding probability distribution functions and joint probability distribution functions.

Common probability density functions for both continuous and discrete random variables were then discussed including the Gaussian, uniform, exponential Raleigh, chi-squared, beta, and Cauchy densities for the continuous case and uniform, Bernoulli, Poisson, binomial, geometric, and negative binomial densities for the discrete case. The mean, moments, and variance of these common random values were given to form a small catalog of densities and partial characterizations of the associated random variables.

The remainder of the chapter involved the presentation of techniques for obtaining total characterizations of random variables that are transformations of other random variables that have known probability density or distribution functions. Synthetic sampling, although not discussed, was offered as an alternative approach to be used when the functional forms are not known by a mathematical model or are just too complicated to lead to an analytical solution.

For a single function of a single random variable, the main techniques discussed for obtaining the density of the transformed random variable in terms of the given random variable were the transformation theorem and the distributional approach. The distributional approach involved first obtaining the distribution function and then obtaining the density by taking its derivative. These same basic techniques were used for a single function of multiple random variables by using an auxiliary random variable approach. In this way the n-dimensional transformation could be used first to obtain a joint probability density function followed by an integrating out of those auxiliary random variables from the joint density to give the required density.

It was also pointed out that these techniques involving the use of the transformation approach do not apply in cases where the Jacobian is zero. In these cases the density function will contain line or surface masses, and thus the use of the probability distribution function, as opposed to the probability density function, is usually a preferred alternative way of characterizing and determining probabilities of events.

The intent of this chapter was not to thoroughly explore the theory of random variables but to present methods useful for handling functions of random variables that are needed for the chapters to follow. The theory of random variables is extensively treated in texts by Papoulis [1], Parzen [7], Davenport [5], and Mood, Graybill, and Boes [2], to mention a few.

PROBLEMS

2.1 A sample space S is specified as follows: $S = \{f_1, f_2, f_3, f_4, f_5, f_6\}$. The probabilities of the elementary events and a mapping from the sample space to the real numbers are shown in Figure P2.1.

f_i	$P(f_i)$	$X(f_i)$
f_1	1/4	0
f_2	1/4	1
f_3	1/4	0
f_4	1/8	1
f_5	1/16	2
f_6	1/16	-1

P2.1

(a) What else do we need for this $X(f_i)$ mapping to be a random variable?

(b) Find the cumulative distribution function for the random variable X.

(c) Find the probability density function for the random variable X.

(d) What type of random variable is X? (Choices are discrete, continuous, or mixed.)

2.2 Consider an experiment $\mathscr{E} : (S, \mathscr{F}, \mathscr{P}(.))$, where $S = \{\zeta_i : i = 1, 2, \ldots, 6\}$, $\mathscr{F}$ is the set of all possible subsets of S, and $\mathscr{P}(.)$ is described by the $P(\{\zeta\})$ shown in the Figure P2.2. Define a random variable X, where $X(\zeta)$ is shown in third column.

ζ_i	$P(\zeta_i)$	$X(\zeta_i)$
ζ_1	1/4	0
ζ_2	1/8	1
ζ_3	1/4	0
ζ_4	1/16	2
ζ_5	1/4	0
ζ_6	1/16	3

P2.2

(a) How many possible events are there in $\mathscr{F}$?

(b) Given that ζ_1 was the outcome of the experiment. How many events have occurred?

Determine the following:

(c) $P\{X = \frac{1}{4}\}$.

(d) $P\{0 < X < 1\}$.

(e) $P\{X \leq 1\}$.

(f) $P\{X = 0\}$.

2.3 The table in Figure P2.3 specifies a random variable X.

(a) Find $F_X(x)$ and $f_X(x)$ for the random variable X.

(b) Specify $\{X \leq 1\}$ by giving set members.

(c) Specify $\{-1 < X \leq 1.5\}$ by giving set members.

(d) Specify $\{X < -2\}$ by giving set members

ζ_i	$P(\zeta_i)$	$X(\zeta_i)$
ζ_1	1/6	0
ζ_2	1/6	1
ζ_3	1/6	0
ζ_4	1/6	2
ζ_5	1/6	0
ζ_6	1/6	1

P2.3

2.4 A small deck of cards contains only the 2's and 3's of each suit—clubs, diamonds, hearts, and spades. Two cards are drawn from the set of eight cards at random with replacement. Define a random variable X such that the value for each fundamental outcome is the total of numbers obtained in the drawing. Find the probability distribution function for the random variable X.

2.5 Let $\mathscr{E}_1 : (S_1, \mathscr{F}_1, \mathscr{P}_1(.))$ and $\mathscr{E}_2 : (S_2, \mathscr{F}_2, \mathscr{P}_2(.))$ be independent experiments with $S_1 = \{a, b, c\}$, $\mathscr{F}_1$ all subjects of S_1, and $\mathscr{P}_1 : (P\{a\} = \frac{1}{2}$, $P\{b\} = \frac{1}{4}, P\{c\} = \frac{1}{4})$ $S_2 = \{1, 2, 3\}$, $\mathscr{F}_2$ all subjects of S_2, and $\mathscr{P}_1 : (P\{1\} = \frac{1}{3}P\{2\} = \frac{1}{3}$, $P\{3\} = \frac{1}{3})$

(a) List the elements of the Cartesian product space $S = S_1 \otimes S_2$.

(b) What is the new $\mathscr{P}(.)$ for this combined experiment?

(c) For this new sample space, list the elements of the event that the number drawn is odd.

(d) Define a random variable $X(\zeta)$ such that $X[(i, j)] = j$ for $i \in \{a, b, c\}$ and $j \in \{1, 2, 3\}$, and determine the cumulative distribution function $F_X(x)$ for this random variable.

2.6 A random variable X is defined by the following table:

ζ	$P(\zeta)$	$X(\zeta)$
ζ_1	0.1	−1
ζ_2	0.1	0
ζ_3	0.1	1
ζ_4	0.1	2
ζ_5	0.1	2
ζ_6	0.1	2
ζ_7	0.1	1
ζ_8	0.1	3
$[\zeta_a, \zeta_b]$	0.2	7

(a) Find and graph the probability distribution $F_X(x)$ for X.

(b) Is X a discrete, continuous, or mixed random variable? Why?

(c) Identify the following events

$$\{X < 2\} \quad \text{and} \quad \{0.1 \le X < 0.3\}$$

(d) Define the events $A = \{\zeta_4, \zeta_5, \zeta_6\}$ and $B = \{\zeta_3, \zeta_7\}$. Are A and B independent events? Why or why not?

(e) Are the events A and B defined above in (d) mutually exclusive events? Why or why not?

2.7 For a random variable X with cumulative distribution function $F_X(x)$ shown in Figure P2.7, find:

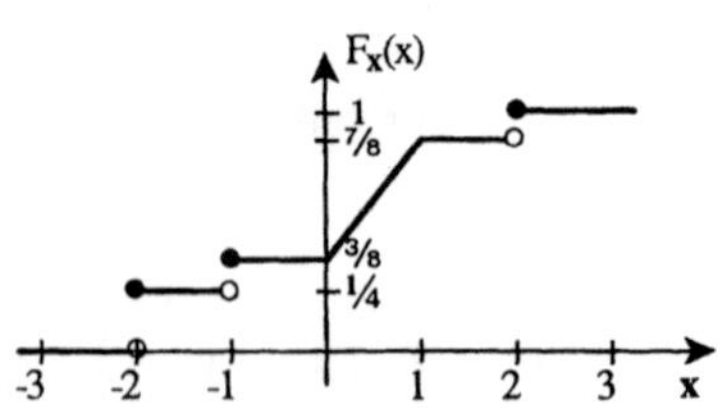

P2.7

(a) $P\{X \leq -1\}$.
(b) $P\{X < -1\}$.
(c) $P\{0 < X \leq 1.5\}$.
(d) $P\{X > 2\}$.
(e) $P\{X \geq 2\}$.
(f) $P\{-2 < X < 1\}$.
(f) $P\{X = 2\}$.
(g) $P(\{X \leq 1\} \mid \{-1 \leq X < 1\})$.
(h) $P(\{X \leq -1\} \mid \{X \leq -0.5\})$.
(i) $P\{X = 1\}$.
(j) The probability density function $f_X(x)$ and the cumulative distribution $F_X(x)$.

2.8 Suppose that X is a random variable with probability density $f_X(x)$ given by

$$f_X(x) = \frac{0.05}{\sqrt{2\pi}} \exp\left\{-\frac{(x-50)^2}{200}\right\} + \frac{3}{16}\delta(x-40) + \frac{5}{16}\delta(x)$$

(a) Show that $\int_{-\infty}^{\infty} f_X(x)\,dx = 1$. Calculate
(b) $P\{30 < X < 40\}$.
(c) $P\{X = 50\}$.
(d) $P\{X \geq 40\}$.
(e) $P\{X = 0\}$.
(f) $P(\{30 < X < 40\} \cup \{X = 0\})$.
(g) Roughly draw the cumulative distribution function $F_X(x)$ for the random variable X. Show specifically any jumps that occur.

2.9 A random variable X has a probability density function $f_X(x)$ given by

$$f_X(x) = \tfrac{1}{2}[\mu(x) - \mu(x-1)] + \tfrac{1}{8}\delta(x - \tfrac{1}{2}) + \tfrac{3}{8}\delta(x - 10)$$

Determine **(a)** $P\{X = \frac{1}{8}\}$. **(b)** $P\{X = \frac{1}{2}\}$. **(c)** $P\{X > 10\}$. **(d)** $P\{X \geq 10\}$. **(e)** $P\{0 < X < 10\}$. **(f)** $P\{\frac{1}{2} < X \leq 1\}$. **(g)** $P(\{x = \frac{1}{2}\} \cup \{X = 10\})$. **(h)** $F_X(x)$.

2.10 Given a function $f_X(x)$ as

$$f_X(x) = \tfrac{1}{2}\,\delta(x) + \tfrac{1}{4}\,\delta(x-1) + \tfrac{1}{4}\,\delta(x+1)$$

Define an experiment with a finite sample space S and a random variable X with $f_X(x)$ as its probability density function. Repeat the above for an experiment with a noncountable sample space.

2.11 Given that X is a Gaussian random variable with probability density function $f_X(x)$,

$$f_X(x) = \frac{1}{2\sqrt{2\pi}} \exp\left\{-\frac{(x-3)^2}{8}\right\}$$

Find: **(a)** $P\{X \le 9\}$. **(b)** $P\{X > 1\}$. **(c)** $P\{X = 3\}$. **(d)** $P\{1 < X \le 9\}$. **(e)** $P\{1 < X < 9\}$.

2.12 Given a random variable with density $f_X(x)$,

$$f_X(x) = \tfrac{1}{8} \sum_{k=1}^{4} \delta(x - 2k) + \tfrac{1}{4}\,[\mu(x+1) - \mu(x-1)]$$

and events A and B as $A = \{X \le 2\}$ and $B = \{0 < X \le 1\}$. Find: **(a)** $P(A)$. **(b)** $P(A|B)$. **(c)** $P(B|A)$. **(d)** $P\{X < 2\}$. **(e)** $P\{X = 1\}$. **(f)** $P\{X = 0\}$. **(g)** $P\{X = 2\}$. **(h)** $F_X(x)$.

2.13 If X is a uniform random variable on [0, 2], calculate $P\{|X - 2| < 1\}$.

2.14 For a random variable X with probability distribution function $F_X(x)$ shown in Figure P2.14 find:

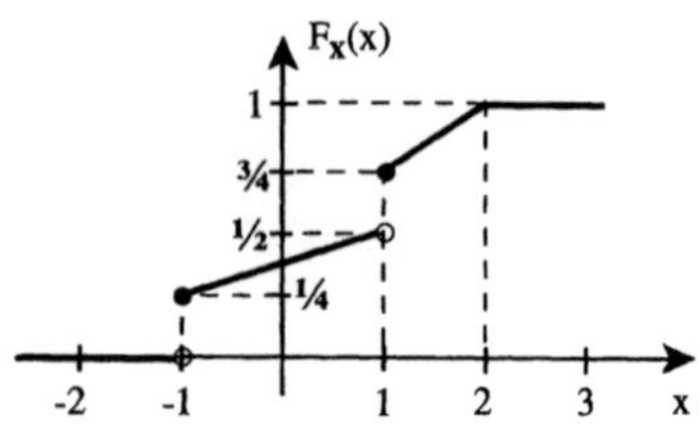

P2.14

(a) $P\{X \le 1\}$.
(b) $P\{-1 < X < 1\}$.
(c) $P\{-1 \le X \le 1\}$.
(d) $P\{-1 < X \le 2\}$.
(e) $P(\{-3 < X \le -1\} \cup \{-1 \le X < 2\})$
(f) $P(\{-1 < X < 1\} | \{-1 \le X \le 1\})$.
(g) $P\{X = 1\}$.
(h) $p\{X = 2\}$.

2.15 Given a random variable X with probability density function $f_X(x)$, where

$$f_X(x) = x\,[\mu(x) - \mu(x-1)] + \tfrac{1}{2}\,\delta(x - \tfrac{1}{4})$$

and the following events: $A = \{X \le 0.5\}$, $B = \{\frac{1}{8} \le X \le \frac{1}{4}\}$, $C = \{X = 1\}$, and $D = \{X = \frac{1}{2}\}$. Find:

(a) $P(B/A)$. **(b)** $P(A/B)$. **(c)** $P(C)$. **(d)** $P(D/A)$. **(e)** $F_X(x)$, the probability distribution function.

2.16 Let X be the life expectancy of a particular component in days with the probability density as shown in Figure P2.16. Determine the probability that the component will fail in the second day given that it has not failed in the first day?

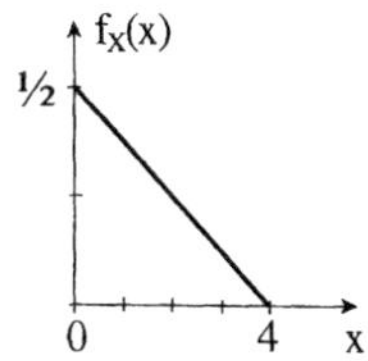

P2.16

2.17 The probability density function for a random variable X is given by

$$f_X(x) = \tfrac{1}{2}e^{-x}\,\mu(x) + \tfrac{1}{2}\delta(x-1)$$

(a) What is the probability of the event $\{X = 1\}$?

(b) What is the probability of the event $\{X = 3\}$?

(c) What is the probability of the event $\{-2 \le X \le 2\}$?

(d) Find $F_X(x)$ the probability distribution function for X.

2.18 Given a random variable X with probability density $f_X(x) = \lambda e^{-\lambda x}\mu(x)$.

(a) Let M be the event $\{X \ge T\}$. Find $f_X(x\,|\,M)$ and $E[X\,|\,M]$.

(b) Let C be the event $\{X \le b\}$. Find $f_X(x\,|\,C)$ and $E[X\,|\,C]$.

2.19 Given $f_X(x) = 0.25x[\mu(x) - \mu(x-2)] + 0.5\delta(x - \frac{1}{2})$. Find:
(a) $E[X]$. **(b)** $P\{X = 1\}$. **(c)** $P\{X = \frac{1}{2}\}$. **(d)** σ_X^2. **(e)** $F_X(x\,|\,A)$, where $A = \{1 \le X \le 2\}$. **(f)** $f_X(x\,|\,A)$, where $A = \{1 \le X \le 2\}$.

2.20 Given the following probability density function for X,

$$f_X(x) = \begin{cases} ae^{-ax} & \text{for } x \ge 0 \\ 0 & \text{elsewhere} \end{cases}$$

Find $f_X(x\,|\,M)$ and $F_X(x\,|\,M)$ where $M = \{X \le b\}$.

2.21 If X is a random variable with $E[X] = 3$ and $E[X^2] = 13$, find a lower bound on the $P\{-2 < X < 8\}$.

2.22 If X is a normal random variable with mean 1 and variance 2, what is the probability of $\{|X-1| \leq 3\}$?

2.23 Given a mixed random variable X with probability density function $f_X(x)$ as

$$f_X(x) = x\,[\mu(x) - \mu(x-1)] + \tfrac{1}{2}\,\delta(x - \tfrac{1}{4})$$

and the following events:

$$A = \{X \leq \tfrac{1}{2}\}, \quad B = \{\tfrac{1}{8} \leq X \leq \tfrac{1}{4}\}$$
$$C = \{X = 1\}, \quad D = \{X = \tfrac{1}{4}\}$$

Determine the following probabilities and distribution function:
(a) $P(B|A)$. **(b)** $P(A|B)$. **(c)** $P(C)$. **(d)** $P(D|A)$. **(e)** $F_X(x)$, the cumulative distribution function.

2.24 For a random variable X uniform on $[1, 3]$.
(a) Find $P\{|X-2| < 1\}$.
(b) Use the Tchebycheff inequality to bound $P\{|X-2| < 1\}$.
(c) Is this a tight or a loose bound for this case?

2.25 Given the random variable X and its associated characteristic function

$$\Phi_X(j\omega) = (1 - j\omega/\lambda)^{-r}$$

Find the mean and variance of X.

2.26 Given the following $\Phi(\omega_1, \omega_2) = e^{+ja_1\omega_1 + ja_2\omega_2}$. Is this a characteristic function? If your answer is yes, give the associated probability density function, whereas if your answer is no, tell why not. If your answer was I do not know, please try another answer.

2.27 For the $f_{XY}(x, y)$ given in Example 2.16, find the characteristic function and determine the $E[XY]$ using your characteristic function.

2.28 Given the probability density $f_X(x) = e^{-2x}\mu(x)$ for a random variable X.
(a) Determine $E[X^4]$ by using the characteristic function for X.
(b) Calculate the cumulant c_4 for X.

2.29 For two random variables X_1 and X_2 with joint probability density function $f_{X_1X_2}(x_1, x_2)$. Prove or disprove that the cumulants $c_{jk} = 0$ for all $j, k \geq 2$.

2.30 Define a random variable $Z = X_1 + X_2 + X_3$, where X_1, X_2, and X_3 are statistically independent identically distributed uniform random variables between 0 and 1. Find: **(a)** $E[Z]$. **(b)** σ_Z^2.

2.31 If two random variables X_1 and X_2 are jointly Gaussian, determine the following cumulants: **(a)** c_{11}. **(b)** c_{21}. **(c)** c_{k1} for $k > 2$.

2.32 If random variables $X_1, X_2, \ldots, X_n$ and are jointly Gaussian prove or disprove the following:
(a) $c_{11}(x_1, x_2) = E[X_1 X_2]$. **(b)** $c_{111}[X_1 X_2 X_3] = E[X_1 X_2 X_3]$.
(c) $c_{1111}[X_1 X_2 X_3 X_4] = E[X_1 X_2 X_3 X_4]$.

2.33 We know that $X_i, i = 1, 2, \ldots, n$ are independent identically distributed random variables with density functions $f_X(x) = \frac{1}{2}\delta(x_i + 1) + \frac{1}{2}\delta(x_i - 1)$. Given the random variable X defined by

$$X = \sum_{i=1}^{n} X_i$$

Determine the probability that $\{-3 \leq X \leq 3\}$.

2.34 Two random variables X and Y have joint probability density function $f_{XY}(x, y)$ given by

$$f_{XY}(x, y) = \begin{cases} x + y, & 0 < x < 1, 0 < y < 1 \\ 0, & \text{elsewhere} \end{cases}$$

Derive the marginal density $f_X(x)$ and find $f_Y(y \mid X = x)$.

2.35 Given the following probability density function:

$$f_{XY}(x, y) = xy \exp\left\{-\frac{x^2 + y^2}{2}\right\}\mu(x)\mu(y)$$

(a) Find $f_X(x), f_Y(y)$, and $f_X(x \mid y)$.
(b) Are X and Y independent? Why or why not?

2.36 Two random variables X and Y have joint probability density function $f_{XY}(x, y)$ given by

$$f_{XY}(x, y) = \begin{cases} x + y, & 0 < x < 1, 0 < y < 1 \\ 0, & \text{elsewhere} \end{cases}$$

Define the event C by $C = \{X \geq Y\}$. Determine $F_X(x \mid C)$ and $f_X(x \mid C)$.

2.37 Let X and Y denote random variables such that Y has a mean μ and variance σ_y^2. Let $E[Y \mid X = x] = g(x)$. Then form the random variable $Z = g(X)$. Find $E[Z]$ in terms of the given parameters.

2.38 Given $Y = \sin X$, where X is a random variable with probability density function $f_X(x)$. Assume that

$$f_X(x) = \begin{cases} 1/\pi, & |x| \leq \pi/2 \\ 0, & \text{elsewhere} \end{cases}$$

Find $E[Y]$ and σ_y^2.

2.39 Define a random variable H by $H = Q \log_2 Q + (1 - Q) \log_2 (1 - Q)$, where Q is a random variable specified by the density $f_Q(q)$ given by

$$f_Q(q) = \tfrac{1}{4}\, \delta(q - \tfrac{1}{2}) + \tfrac{3}{4}\, \delta(q - \tfrac{1}{4})$$

Find:
(a) $E[Q]$. **(b)** $E[H]$. **(c)** $E[H^2]$.

2.40 Assume that a certain nonlinear device has the transfer characteristic $y = (x + 1)/(x + 2)$. Let the input to the device be a random variable X with probability density function $f_X(x)$:

$$f_X(x) = \frac{1}{2} \frac{1}{\sqrt{2\pi}} \exp\left\{-\frac{x^2}{2}\right\} + \frac{1}{2}\delta(x)$$

(a) Find $f_Y(y)$. **(b)** Compute $E[X]$. **(c)** σ_X^2. **(d)** $E[Y]$. **(e)** σ_Y^2.

2.41 Given two random variables X and Y with joint density function $f_{XY}(x, y)$:

$$f_{XY}(x, y) = \begin{cases} 1, & -y < x < y, 0 < y < 1 \\ 0, & \text{elsewhere} \end{cases}$$

(a) Find $f_X(x)$ and $f_Y(y)$. Prove or disprove the following:
(b) X and Y are independent random variables.
(c) X and Y are uncorrelated random variables.
(d) X and Y are orthogonal random variables.

2.42 If $Y = a + bX + cX^2$ and $X \sim N(0, \sigma^2 = 2)$.
Calculate $E[Y]$ and σ_Y^2.

2.43 If X is a random variable with density

$$f_X(x) = \begin{cases} 1/\pi & 0 \leq x \leq \pi \\ 0 & \text{elsewhere} \end{cases}$$

Define a random variable Y by $Y = \sin^2 X$. Find $E[Y]$ and σ_y^2.

2.44 Define the random variable Y by $Y = (X - 3)(X - 1)$.

(a) If X is a normal random variable with zero mean and variance 1, find $f_Y(y)$ and $E[Y]$.

(b) If X is a uniform random variable between 2 and 3,

$$f_X(x) = \mu(x - 2) - \mu(x - 3)$$

find $f_Y(y)$.

(c) Use the results of (b) to calculate $E[Y]$ by two methods.

2.45 Given a random variable X with known probability density function

$$f_X(x) = e^{-x}\mu(x)$$

Define a new random variable $Y = g(X)$, where $g(x)$ is as shown in Figure P2.45. Find the cumulative distribution function $F_Y(y)$ and the probability density function $f_Y(y)$ for the random variable Y.

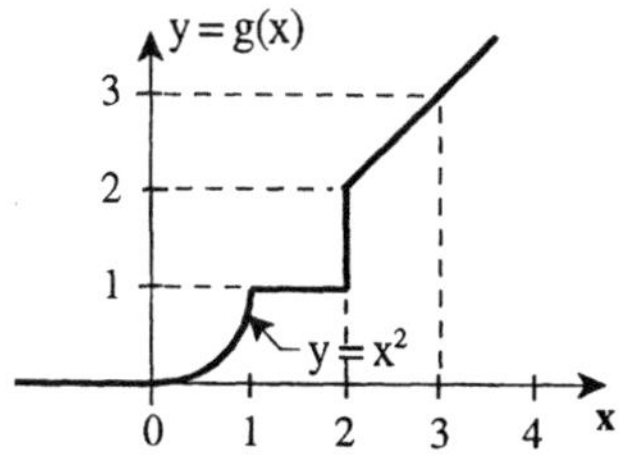

P2.45

2.46 Given a random variable X with known probability density function

$$f_X(x) = e^{-x}\mu(x)$$

Define a new random variable $Y = h(X)$, where $h(x)$ is shown in Figure P2.46. Find the cumulative distribution function $F_Y(y)$ and the probability density function $f_Y(y)$ for the random variable Y.

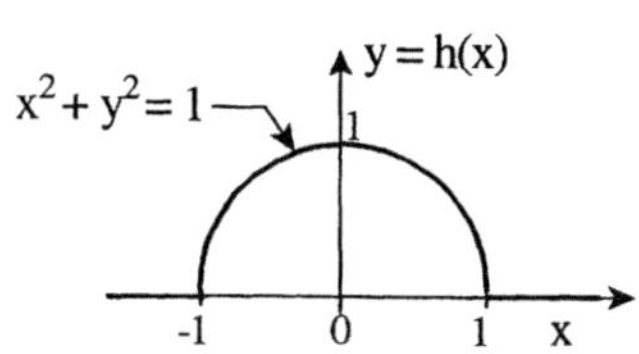

P2.46

2.47 Given a random variable X with uniform density on [0,1].
(a) Find the $f_Y(y)$ for $Y = (X - \frac{1}{2})^2$.
(b) Determine $E[Y]$.

2.48 Given $f_X(x)$ the density function for the random variable X and the function $y = g(x)$ shown in Figure P2.48. Find the cumulative distribution function $F_Y(y)$ and probability density function $f_Y(y)$ for the random variable $Y = g(X)$.

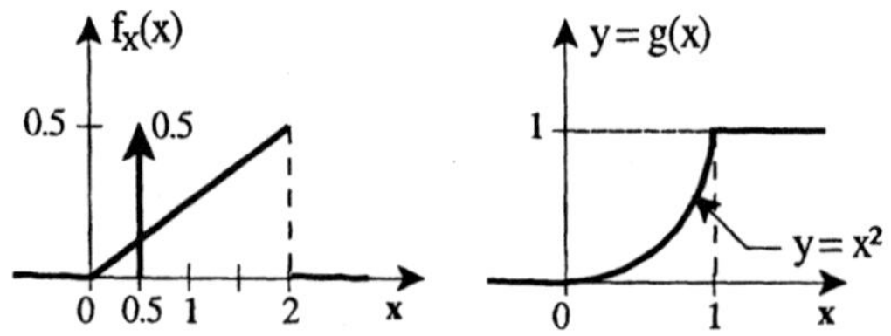

P2.48

2.49 Given X with probability density function $f_X(x)$ as

$$f_X(x) = \begin{cases} 0.1, & 0 \le x \le 10 \\ 0, & \text{elsewhere} \end{cases}$$

and $Y = -\ln X$. Find $f_Y(y)$.

2.50 A random variable X has probability density function $f_X(x)$ as

$$\begin{aligned} f_X(x) = {} & \frac{1}{2}e^{-x}\mu(x) + \frac{1}{4}\delta(x+1) + \frac{1}{8}\delta(x) \\ & + \frac{1}{16}\delta(x - \tfrac{1}{2}) + \frac{1}{32}\delta(x - 1.5) \\ & + \frac{1}{32}\delta(x-3) + \frac{1}{64}\delta(x-4) \end{aligned}$$

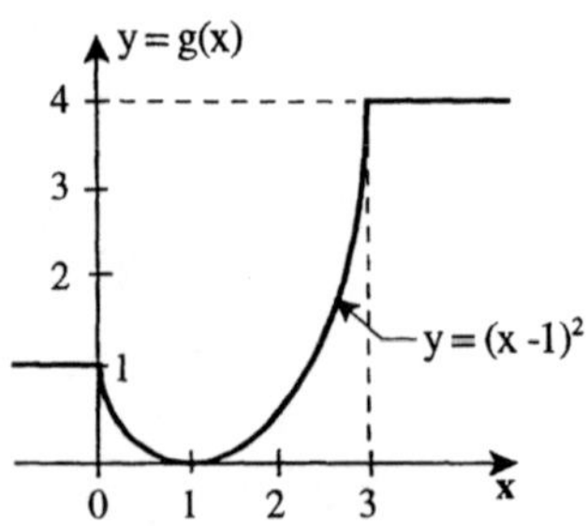

P2.50

Define a random variable $Y = g(X)$, where $y = g(x)$ is as shown in Figure P2.50.

(a) Find $f_Y(y)$ and $F_Y(y)$.

(b) Find $E[Y]]$ by two different methods.

2.51 Given a mixed random variable X with probability density function $f_X(x)$ as follows:

$$f_X(x) = 0.1\delta(x-2) + 0.4\delta(x-1) + \tfrac{1}{4}e^{-|x|}$$

Given the real-valued function $y = g(x)$ shown in Problem 2.48, define a random variable $Y = g(X)$ and find $f_Y(y)$.

2.52 Let Z be a random variable such that

$$f_Z(z) = \begin{cases} \frac{1}{2}\pi, & 0 \leq z \leq 2\pi \\ 0, & \text{otherwise} \end{cases}$$

and define the random variables X and Y by $X = \sin Z$ and $Y = \cos Z$.

(a) Are X and Y correlated random variables?

(b) Are X and Y orthogonal random variables?

(c) Are X and Y independent random variables?

2.53 The volt (x)–amp (y) characteristics of a particular device might be approximated by the curve shown in Figure P2.53. Assume that the voltage X is a random variable with density

$$f_X(x) = \tfrac{1}{2}e^{-|x|} \qquad \text{for all } x \in (-\infty, \infty)$$

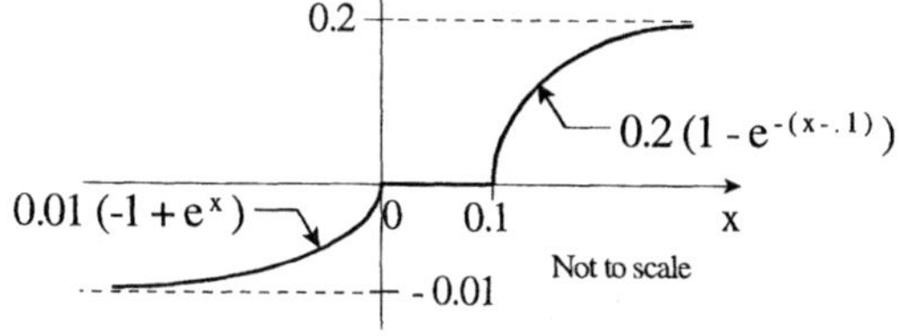

P2.53

(a) Find the density $f_Y(y)$ for the current Y.

(b) Give $E[Y]$ in an integral form and simplify if possible.

2.54 Let X and Y be identically distributed independent uniform random variables on $[0, 2\pi]$. Define a random variable Z by $Z = \cos(X + \beta \sin Y)$. Find **(a)** $E[Z]$. **(b)** σ_Z^2.

2.55 Suppose that X_1 and X_2 are jointly normal random variables with the following parameters: $\mu_1 = 3, \mu_2 = 1, \sigma_1^2 = 16, \sigma_2^2 = 25, \rho_{12} = \frac{3}{5}$. Find the probability density function for a random variable Z defined by $Z = X_1 + X_2$.

2.56 X and Y are statistically independent Gaussian random variables with zero means and unit variances. Define a random variable $Z = (X^2 + Y^2)^{3/2}$. Find $f_Z(z)$.

2.57 Find the density $f_Z(z)$ for $Z = XY$ by using the integration technique to find the distribution $F_Z(z)$ and then differentiating to get $f_Z(z)$. Assume that the joint probability density $f_{XY}(x, y)$ is known for the random variables X and Y. (Do not find by introducing an auxiliary random variable.)

2.58 A random variable Z is defined as the following function of two jointly normal random variables X and Y with parameters $(\eta_X, \eta_Y, \sigma_X^2, \sigma_Y^2, \rho_{XY})$:

$$Z = \int_0^X Y d\theta$$

Show that $E[Z] = \eta_X \eta_Y + \rho_{XY}\sigma_X\sigma_Y$

2.59 Suppose that X and Y are jointly normal uncorrelated random variables with $\sigma_X = 2, \sigma_Y = 4, \eta_X = 1,$ and $\eta_Y = 1$. Find the density of a new random variable Z defined by $Z = X + Y$.

2.60 Find $f_X(x)$, $E[X]$, $f_Y(y|x)$, $E[Y|X = x]$, $f_Y(y)$, and $E[Y]$ for the joint probability density functions shown below and the regions shown in Figure P2.60. Are X and Y independent random variables? Uncorrelated random variables? Orthogonal random variables? Support your answers.

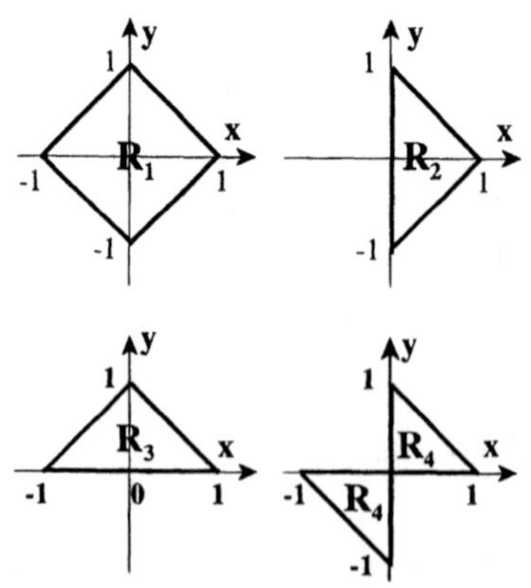

(a) $f(x, y) = \begin{cases} \dfrac{1}{\pi ab}, & \dfrac{x^2}{a^2} + \dfrac{y^2}{b^2} \le 1, a, b, > 0 \\ 0, & \text{elsewhere} \end{cases}$

(b) $f(x, y) = \begin{cases} \frac{1}{2}, & (x, y) \in R_1 \\ 0, & \text{elsewhere} \end{cases}$

(c) $f(x, y) = \begin{cases} 1, & (x, y) \in R_2 \\ 0, & \text{elsewhere} \end{cases}$

(d) $f(x, y) = \begin{cases} 1, & (x, y) \in R_3 \\ 0, & \text{elsewhere} \end{cases}$

(e) $f(x, y) = \begin{cases} 1, & (x, y) \in R_4 \\ 0, & \text{elsewhere} \end{cases}$

P2.60

2.61 Given $f_{XY}(x, y) = 16\,xy\,e^{-(x+y)/2}\,\mu(x)\mu(y)$.

(a) Find $f_X(x), f_Y(y)$, and $f_X(x\,|\,y)$.

(b) Are X and Y independent? Why or why not?

2.62 Given X and Y are statistically independent Gaussian random variables with $\eta_X = \eta_Y = 1$ and $\sigma_X^2 = \sigma_Y^2 = 2$. Define the random variables W and Z by $W = X^2 + Y^2$ and $Z = X^2 - Y^2$.

(a) Find $F_{WZ}(w, z)$.

(b) Determine $f_W(w)$.

(c) Are W and Z statistically independent? Why or why not?

(d) Are W and Z statistically uncorrelated? Why or why not?

2.63 Given X_1 and X_2 are statistically independent Gaussian random variables with $E[X_1] = E[X_2] = 0$ and $\sigma_1^2 = \sigma_2^2 = 1$. Define the random variables W and Z by $W = X_1^2 + X_2^2$ and $Z = X_2$. Find $f_{WZ}(w, z)$.

2.64 Say X and Y are random variables with joint probability density function $f_{XY}(x, y)$. Define two random variables Z and W by $W = \ln(X + Y)$ and $Z = X/Y$.

(a) Find the joint probability density function $f_{WZ}(z)$ in terms of $f_{XY}(x, y)$.

(b) One way to determine the probability distribution function for W and Z is to use the definition $F_{WZ}(w, z) = P\{W \leq w, Z \leq z\} = P\{(X, Y) \in D_{wz}\}$. For $w > 0$ and $z > 0$ illustrate the region D_{wz} specified by $D_{wz} = \{(x, y) : \ln(x + y) \leq w \text{ and } x/y \leq z\}$.

2.65 The random variables X and Y are jointly normal random variables with the following properties: $E[X] = 1$, $E[Y] = 2$, $E[XY] = 3$, $\sigma_X^2 = 1$, and $\sigma_Y^2 = 4$. Define two random variables W and Z by $W = X + Y$ and $Z = 2X + 2Y$.

(a) Is W a normal random variable? Why or why not?

(b) Find the probability density function for W.

(c) Is Z a normal random variable? Why or why not?

(d) Are W and Z jointly normal random variables? Why or why not?

(e) Find $f(z\,|\,W = w)$.

(f) Are W and Z statistically independent? Linearly independent? Explain.

2.66 X and Y are independent Gaussian random variables with $E[X] = E[Y] = 0$ and $\sigma_X^2 = \sigma_Y^2 = \sigma^2$. Define the random variable $Z = X^2 + Y^2$. Determine $E[Z]$.

2.67 A random variable Z is defined by $Z = X^2 + 2X + Y^2$ where the joint density for X and Y is known as $f_{XY}(x, y)$. We know that the cumulative distribution function $F_Z(z)$ can be written as $F_Z(z) = \int \int_{R_z} f_{XY}(x, y)\, dx\, dy$. Illustrate the region R_z for all z.

2.68 The random variables X and Y are statistically independent with known probability density functions $f_X(x) = \mu(x) - \mu(x-1)$ and $f_Y(y) = e^{-y}\mu(y)$. Define new random variables W and Z by $W = e^{-(X+Y)}$ and $Z = X$.

(a) Determine $f_{WZ}(w, z)$.

(b) Using part (a) find $f_W(w)$.

(c) The procedure given in parts (a) and (b) could be viewed as a technique for finding the density of W by the introduction of an auxiliary random variable Z. Give an easier way to determine $f_W(w)$ and do it.

2.69 Two independent random variables X and Y, with probability density functions $f_X(x) = \mu(x) - \mu(x-1)$ and $f_Y(y) = \mu(y) - \mu(y-1)$ are multiplied together to give a new random variable $Z = X \cdot Y$.

(a) Find $f_Z(z)$ by introduction of an auxiliary random variable and using the transformational theorem.

(b) Find $f_Z(z)$ by first finding $F_Z(z)$ using the basic definition of the distribution function and then differentiating.

2.70 The random variables X and Y are independent with probability density functions $f_X(x) = e^{-x}\mu(x)$ and $f_Y(y) = e^{-y}\mu(y)$. Find the joint probability density function density functions for the random variables Z and W defined by

$$W = \frac{X+Y}{X}, \quad Z = X + Y$$

2.71 Define the random variables W and Z by the transformation

$$W = X + Y, \quad Z = X^2 + 2XY + Y^2$$

If X and Y are statistically independent random variables with probability density functions given by $f_X(x) = e^{-x}\mu(x)$ and $f_Y(y) = e^{-y}\mu(y)$, find: **(a)** $f_{WZ}(w, z)$. **(b)** $P\{0 < W < 1, 0 < Z < 10\}$. **(c)** $F_{WZ}(w, z)$.

2.72 Given

$$f(x, y) = \begin{cases} 1, & 0 < x \le 1, 0 < y \le 1 \\ 0, & \text{elsewhere} \end{cases}$$

Define random variables R and S by $R = (X^2 + Y^2)^{\frac{1}{2}}$ and $S = \tan^{-1}(Y|X)$. Thus X and Y are related to R and S by $X = R\cos S$ and $Y = R\sin S$. Find $f_{RS}(r, s)$ and $f_S(s)$.

2.73 What are three common ways of determining the density of a random variable $Z = g(X, Y)$ knowing the joint density of X and Y?

2.74 We are given two random variables X and Y with joint probability density function

$$f_{XY}(x, y) = \begin{cases} 1, & 0 \le x \le 1, 0 \le y \le 1 \\ 0, & \text{elsewhere} \end{cases}$$

Define two new random variables $W = X - Y$ and $Z = X + Y$.
(a) Determine the joint probability density function $f_{WZ}(w, z)$.
(b) Illustrate the the region of the w–z space for which $f_{WZ}(w, z) \neq 0$.

2.75 Given

$$f_{X_1X_2X_3}(x_1, x_2, x_3) = \begin{cases} 8x_1x_2x_3, & 0 \le x_1 \le 1, 0 \le x_2 \le 1, 0 \le x_3 \le 1 \\ 0, & \text{elsewhere} \end{cases}$$

(a) Calculate $E[X_1X_2X_3]$. **(b)** Determine $E[X_1 | X_2 = x_2, X_3 = x_3]$.

2.76 X_1, X_2, and X_3 are independent identically distributed Gaussian random variables with zero means and variances σ^2. Define random variables Y_1, Y_2, and Y_3 by $Y_1 = X_1$, $Y_2 = (X_1 + X_2)/2$, and $Y_3 = (X_1 + X_2 + X_3)/3$.
(a) Find the joint density for Y_1, Y_2, and Y_3.
(b) Find the limit as $n \to \infty$ of the variance of Y_n, where Y_n is $Y_n = \sum_{i=1}^{n} X_i/n$.
(c) Determine $P\{\sigma^2 - 0.02\sigma^2 \le Y_n \le \sigma^2 + 0.02\sigma^2\}$.

2.77 Let X_i be identically distributed uniform random variables on [0, 1], and define Y by

$$Y = \sum_{i=1}^{20} X_i$$

(a) Determine the approximate density for Y.
(b) Using that approximate density, calculate the $P\{8 < Y < 12\}$.
(c) Compare the results of (b) with those of approximating the probability using the Tchebycheff inequality.

2.78 Let X_i be identically distributed uniform random variables on [0, 1], and define Y by $Y = \sum_{i=1}^{10} X_i^2$.
(a) Determine the approximate density for Y.
(b) Using that approximate density, calculate the $P\{6 < Y < 14\}$.

2.79 Let X_i be identically distributed independent Gaussian random variables with zero mean and variance σ^2. Define a new random variable Y by $Y = \sum_{i=1}^{10} X_i^2/10$.

(a) Compute $E[Y]$.

(b) Compute approximately the probability that $-1 \leq Y \leq 1$.

(c) Determine the probability density function for Y, and calculate the $P\{6 < Y < 14\}$.

2.80 Four random variables W, X, Y, and Z, are defined by the joint density

$$f(w, x, y, z) = \begin{cases} A\left(\dfrac{wx}{yz}\right)^2 \log\dfrac{w}{y}, & 1 \leq w \leq 2, 1 \leq x \leq 2, 1 \leq y \leq w \leq 2, \\ & 1 \leq z \leq 2 \\ 0, & \text{otherwise} \end{cases}$$

(a) Find $f(x \mid w, y, z)$.

(b) Comment on the results.

2.81 A random variable X, conditioned on another random variable A, has a Rician probability density function given by

$$f_X(x) = \begin{cases} \dfrac{x}{\sigma^2} \exp\left\{-\dfrac{(x^2 + a^2)}{2\sigma^2}\right\}, & x \geq 0 \\ 0, & \text{otherwise} \end{cases}$$

A is a Rayleigh distributed with probability density function $f^A(a)$ given by

$$f_A(a) = \frac{a}{\alpha^2} \exp\left(-\frac{a^2}{2\alpha^2}\right)\mu(a)$$

Find $f_X(x)$. (*Hint*: You may be able to use some properties of the Rician density functions and general density functions to evaluate the resulting integrals.)

2.82 Random variables X, Y, and Z are characterized by their joint probability density function $f_{XYZ}(x, y, z)$ given by

$$f_{XYZ}(x, y, z) = \begin{cases} (x + y)e^{-z}, & 0 < x < 1, 0 < y < 1, z > 0 \\ 0, & \text{elsewhere} \end{cases}$$

Determine the following:
(a) $f(x, y \mid z)$. **(b)** $f(y \mid x, z)$. **(c)** $f(x \mid z)$. **(d)** $E[Z]$.

2.83 The random variable X has the probability density function

$$f_X(x) = \tfrac{1}{2}\exp(-x)\mu(x) + \tfrac{1}{2}\delta(x - 1)$$

(a) What is the probability of the event $\{X = 1\}$?

(b) What is the probability of the event $\{X = 3\}$?

(c) What is the probability of the event $\{-2 < X \leq 2\}$?

(d) Find $F_X(x)$, the cumulative distribution function for the random variable X.

2.84 Given X and Y are random variables with joint probability density function $f_{XY}(x, y)$ as follows:

$$f(x, y) = \begin{cases} 1, & -y < x < y, 0 < y < 1 \\ 0, & \text{elsewhere} \end{cases}$$

(a) Are X and Y independent random variables? Why or why not?

(b) Are X and Y uncorrelated random variables? Why or why not?

(c) Are X and Y orthogonal random variables? Why or why not?

2.85 Suppose X and Y are jointly normal uncorrelated random variables with $\sigma_X = 2$, $\sigma_X = 4$, $\eta_X = 1$, and $\eta_y = 1$. Find the probability density function for a new random variable Z defined by $Z = X + Y$.

2.86 Define a random variable Z by $Z = X_1 + X_2 + X_3$, where X_1, X_2, and X_3 are statistically independent identically distributed uniform random variables on [0, 1].
Find: (a) $E[Z]$. (b) σ_Z^2. (c) $f_Z(z)$.

2.87 Let X and Y be statistically independent Gaussian random variables and both have a zero mean and unit variance. Define a random variable $Z = (X^2 + Y^2)^{3/2}$. Find $f_Z(z)$.

2.88 Define a random variable Y as $Y = e^{-|X|}$, where X is a normal random variable with zero mean and variance σ_X^2.

(a) Find $f_Y(y)$.

(b) Define by integrals the $E[Y]$ in two ways, and find the answer using one of them.

2.89 Given two independent random variables X and Y both with uniform probability densities on [0, 2].

(a) Define a random variable $Z = X + Y$.

(b) Determine $P\{1 < Z < 1.5\}$.

2.90 Define a random variable $Z = X^2 + Y^2$. Find $f_Z(z)$ if the joint density $f_{XY}(x, y)$ is

$$f_{XY}(x, y) = \begin{cases} x + y, & 0 < x < 1, \quad 0 < y < 1 \\ 0, & \text{elsewhere} \end{cases}$$

2.91 Random variables X and Y have the joint density

$$f_{XY} = 0.8(x + 3y)e^{-(x+2y)}\mu(x)\mu(y)$$

Find $f_Y(y|X = x)$, and calculate $E[Y|X = x]$.

2.92 Given two random variables with joint density $f_{XY}(x, y)$ as

$$f_{XY}(x, y) = \begin{cases} 2(x + y)/3, & 0 < x < 1, \quad 0 < y < 1 \\ 0, & \text{elsewhere} \end{cases}$$

Let $C = \{0 < Y \leq \frac{1}{2}\}$, and find:
(a) $F_X(x|C)$. **(b)** $f_X(x|C)$. **(c)** $f_Y(y|C)$. **(d)** $f_{XY}(x, y|C)$.

2.93 For the random variables given above determine:
(a) $f_X(x)$. **(b)** $f_Y(y|x)$. **(c)** $E[Y|X = x]$. **(d)** $E[Y|X]$. **(e)** $E[E[Y|X]]$. **(f)** Are X and Y independent random variables? Explain.

2.94 Define a random variable Z by $Z = Y + X^2$, where X and Y are random variables with joint density function $f_{XY}(x, y)$:

$$f_{XY}(x, y) = \begin{cases} x + y, & 0 < x < 1, \quad 0 < y < 1 \\ 0, & \text{elsewhere} \end{cases}$$

We desire to calculate the probability density function of Z by first finding $F_Z(z)$ and then differentiating.
(a) Illustrate the region D_z where $D_z = \{(x, y) : y + x^2 \leq z\}$.
(b) Give integral form for $F_Z(z)$ in terms of $f_{XY}(x, y)$.
(c) Take derivative of $F_Z(z)$ to obtain $f_Z(z)$.
(d) Find $f_Z(z)$ for the following $f_{XY}(x, y)$ using an auxiliary random variable.

2.95 Let X be a random variable with given probability density function $f_X(x)$. Define a new random variable Y by $Y = g(X)$, where $g(x) = F_x(x)$. Determine: **(a)** $f_Y(y)$. **(b)** $F_Y(y)$ in terms of $f_X(x)$.

2.96 Given a mixed random variable X with probability density function $f_X(x)$ as follows:

$$f_X(x) = 0.1\,\delta(x - 2) + 0.4\,\delta(x - 1) + \tfrac{1}{4}e^{-|x|}$$

An event B is defined as follows $B = \{0 < X \leq 1\}$. Determine
(a) $F_X(x|B)$. **(b)** $f_X(x|B)$ and $E[X|B]$.

2.97 A motorist stops at a stop sign (not a red light). The delay encountered can be considered a random variable X with probability distribution function $F_X(x)$ given by

$$F_X(x) = (1 - \tfrac{1}{4}\,e^{-x})\mu(x)$$

Let $C = \{0 < X \leq 1\}$. Find:
(a) $f_X(x)$. **(b)** $F_X(x\,|\,C)$. **(c)** $f_X(x\,|\,C)$. **(d)** $E[X]$. **(e)** σ_X^2.

2.98 X and Y are continuous random variables with joint density function $f_{XY}(x, y)$:

$$f_{XY}(x, y) = \begin{cases} \frac{1}{2}\,xy, & (x, y) \in R \\ 0, & \text{elsewhere} \end{cases}$$

where R is shown in Figure P2.98. Determine:
(a) $f_X(x)$. **(b)** $f_Y(y\,|\,x)$. **(c)** $E[Y\,|\,X = x]$. **(d)** $E[Y\,|\,X]$. **(e)** $E\{E[Y\,|\,X]\}$.

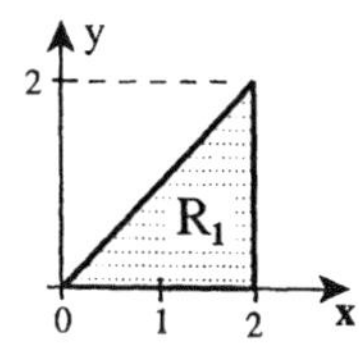

P2.98

2.99 A random variable X is defined by $X = \cos\Theta$, where Θ is a uniform random variable on $[0, 2\pi]$. Determine the variance of X.

2.100 Let X be a Gaussian random variable with zero mean and variance equal to 1. Assume that X is the input to a nonlinearity with input–output relationship $y = g(x)$ as shown in Figure P2.100.

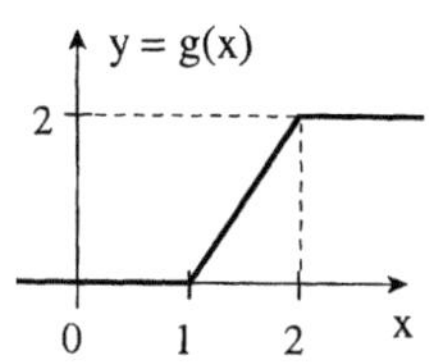

P2.100

(a) Find $E[g(X)]$ by using $f_X(x)$ directly.

(b) Find the probability density function $f_Y(y)$ for the random variable $Y = g(X)$.

2.101 The Tchebycheff polynomial of order 3 is given by $T_n(x) = 4x^3 - 3x$ as shown in Figure P2.101. If X is a Laplace random variable with

$$f_X(x) = \tfrac{1}{2}\, e^{-|x|}$$

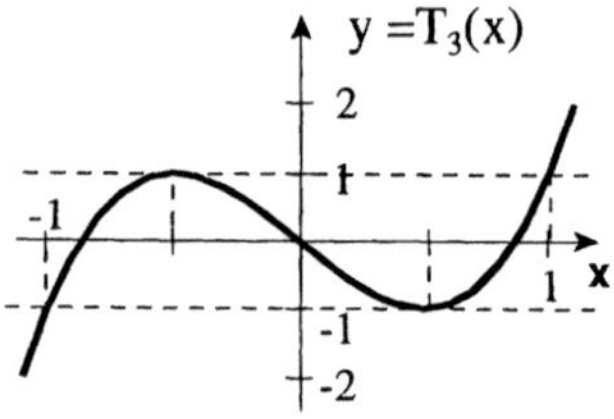

P2.101

Find $f_Y(y)$ for $Y = T_n(X)$ at the following values for y:
(a) $y = -26$.
(b) $y = 0$.
(c) $y = 26$.

2.102 Suppose that X is a uniform random variable on $[-1, 1]$. Find $f_Y(y)$ if $f_Y(y|X = x)$ is given by

$$f_Y(y|X = x) = \frac{1}{\sqrt{2\pi}\sigma} \exp\left\{-\frac{(y-x)^2}{2\sigma^2}\right\}$$

2.103 Given a quantizer with transfer function $y = g(x)$ shown in Figure P2.103.
(a) The input to this quantizer is a zero mean random variable X with variance σ^2. We say that an input value is outside the range of the quantizer if it is greater than 2 or less than -2. Find the value of σ^2 such that the probability of being out of range is less than 0.1.
(b) If the input to the quantizer is a Gaussian random variable with a zero mean and a variance of 0.5. Find $f_Y(y)$, the probability density for the output random variable $Y = g(X)$.
(c) Determine $E[Y]$.
(d) For this problem, does $E[Y] = g(E[X\})$?

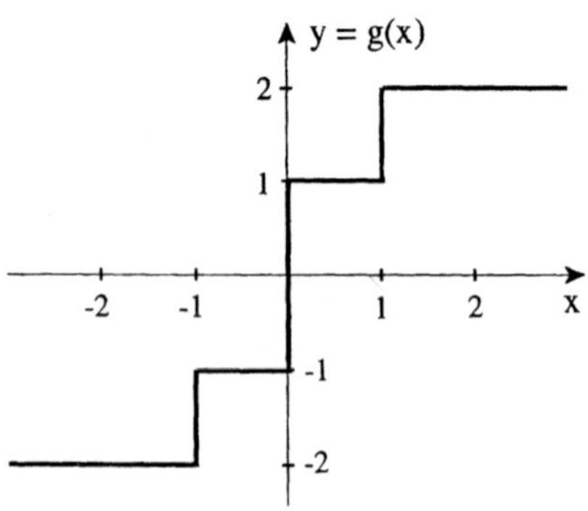

P2.103

2.104 Given a random variable X with $f_X(x)$ as

$$f_X(x) = \tfrac{1}{2}\delta(x - 0.3) + \tfrac{1}{4}\, e^{-|x|}$$

and a function $y = g(x)$ shown below in Figure P2.104.

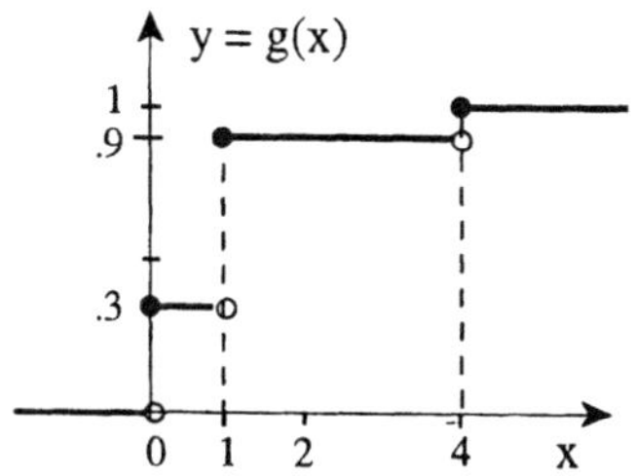

P2.104

(a) Find $f_Y(y)$ for the random variable $Y = g(X)$.

(b) Give two different integrals for evaluating $E[Y]$. Be specific and evaluate using one of them.

(c) Find $F_X(x \mid C)$, where $C = \{X \geq 0\}$.

2.105 X is a discrete random variable with probability density function $f_X(x)$ given by

$$f_X(x) = 0.3\delta(x - 1) + 0.5\delta(x - 2) + 0.2\delta(x - 3)$$

and Y is a continuous random variable with conditional density

$$f_Y(y|X = k) = \frac{1}{\sqrt{2\pi}} \exp\left\{ -\frac{(y - x)^2}{2} \right\}, \quad k = 1, 2, 3$$

(a) Find $f_Y(y)$.

(b) $P\{Y \leq 0\}$.

(c) $E[X]$.

(d) $E[X^2]$.

(e) $P\{Y = 5 \mid X = 2\}$.

(f) $P\{X = 2 \mid Y = 5\}$.

2.106 Given

$$f_{XY}(x, y) = \begin{cases} x + y, & 0 < x < 1, 0 < y < 1 \\ 0, & \text{elsewhere} \end{cases}$$

and the event $C = \{X \geq Y\}$. Find $F_X(x \mid C)$.

2.107 Given the random variables X and Y, with joint probability density function as

$$f_{XY}(x,y) = \begin{cases} \frac{1}{3}, & (x,y) \in R \\ 0, & \text{elsewhere} \end{cases}$$

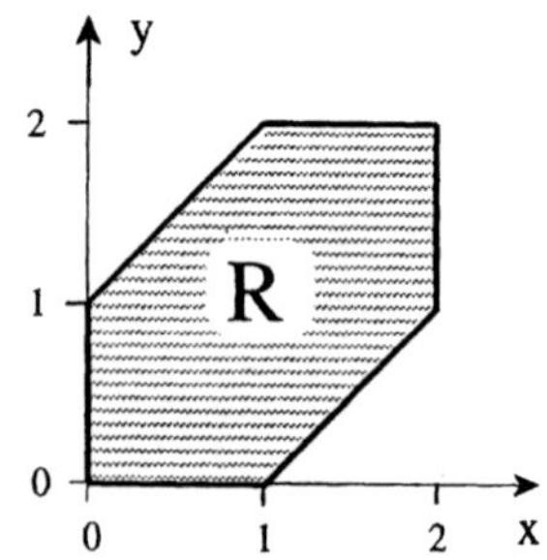

P2.107

where R is shown in Figure P2.107. Determine:

(a) $f_X(x)$.

(b) $f_Y(y|x)$.

(c) $E[Y|X = x]$.

(d) $f_{XY}(x, y)$.

(e) $P\{0 < X < 1, 1 < Y < 2\}$.

(f) Are X and Y independent? Why or why not?

(g) Find the normalized correlation coefficient r given by

$$r = \frac{E[(X - \eta_X)(Y - \eta_Y)]}{\sigma_X \sigma_Y}$$

(h) Are the random variables X and Y highly linearly related? Why or why not?

2.108 Let X be a Gaussian random variable, with $E[X] = 1$ and $\sigma_X^2 = 4$. Define a random variable Y by $Y = g(X)$, where $g(.)$ is shown in Figure P2.108.

(a) Determine $E[Y]$.

(b) Find $f_Y(y)$.

2.109 Let X be a Gaussian random variable with $E[X] = 1$ and $\sigma_X^2 = 4$. Define a random variable Y by $Y = g(X)$, where $g(\cdot)$ is shown in Figure P2.109.

(a) Determine $E[Y]$.

(b) Find $f_Y(y)$.

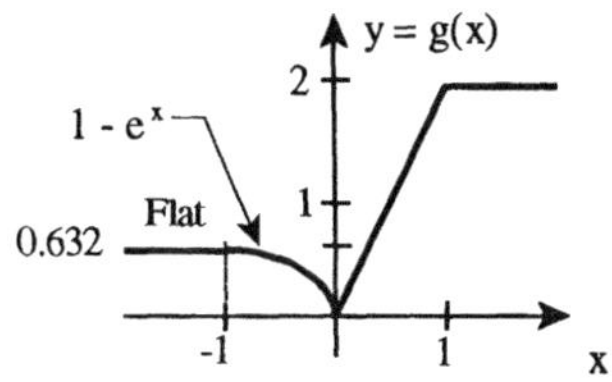

P2.108

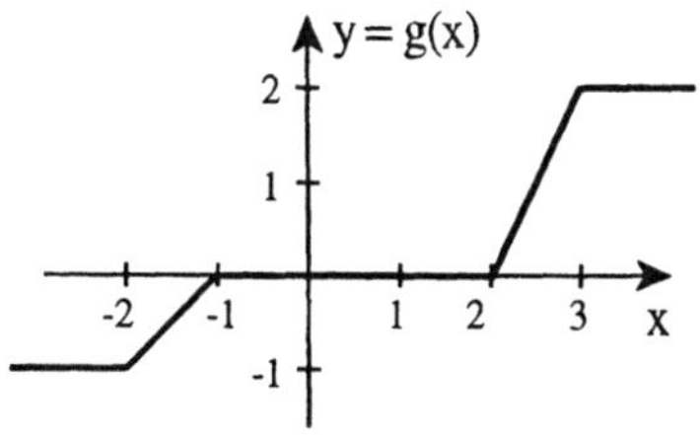

P2.109

2.110 Let X be a random variable with probability density function $f_X(x)$ given by

$$f_X(x) = \frac{1}{\sqrt{\pi}} \exp(-x^2)$$

Define a random variable Y by $Y = g(X)$, where $g(x)$ is defined in Figure P2.110.

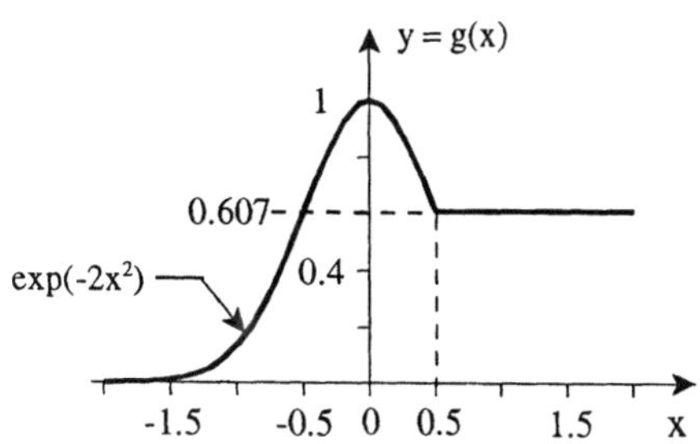

P2.110

2.111 Let X and Y be statistically independent Gaussian random variables with zero means and unity variances. Define the random variable $Z = (X^2 + Y^2)^{3/2}$. Find $f_Z(z)$. (Use previous results if you can.)

2.112 X and Y are jointly normal random variables with $\eta_X = 1$, $\eta_Y = 2$, $\sigma_X^2 = 1$, $\sigma_Y^2 = 4$, and $r_{XY} = 0.5$. Find $E[Y|X]$ and $E[E[Y|X]]$.

2.113 Let $X = \sin \Theta$ and $Y = \cos \Theta$, where Θ is a uniform random variable on $[0, 2\pi]$.
(a) Find $F_Y(y|X = \frac{1}{2})$. **(b)** Determine $P\{(X, Y) \in (X - \frac{1}{2})^2 + (Y - \frac{1}{2})^2 \leq \frac{1}{4}\}$.

2.114 Let $X = \sin \Phi$ and $Y = \cos \Phi$, where Φ is a uniform random variable on $[0, \pi/2]$. Find: **(a)** $E[X]$. **(b)** $E[XY]$. **(c)** $P\{XY > 0\}$.

2.115 Given $f_{XY}(x, y) = 1/(\pi A^2)$ for $(x, y) \in R$, where R is the interior of a circle of radius a and a center at (0, 0).
(a) Find $f_X(x)$ and $F_Y(y|X = x)$. **(b)** Determine $P\{a/2 < X \leq 3a/2, -a/2 < Y \leq a/2\}$.

2.116 Evaluate the following:

(a) $$I(m) = \frac{d}{dm}\left(\int_{-5}^{m^2} \frac{1}{\sqrt{2\pi}} \exp\left\{-\frac{(x-m)^2}{2}\right\} dx\right).$$

(b) $$\int_{-2}^{5} \frac{1}{\sqrt{2\pi}\,2} \exp\left\{-\frac{(x-2)^2}{8}\right\} dx.$$

2.117 Two resistors A and B are connected in parallel. The probability densities for the resistor values are as follows:

$$f_A(a) = \mu(a-1) - \mu(a-2), \quad f_B(b) = 0.5[\mu(b-1) - \mu(b-3)]$$

Find the probability density for the equivalent resistor Z of the parallel combination of A and B. (*Hint*: Find $Z = 1/Y$, where $Y = 1/A + 1/B$.)

2.118 Prove that the cumulants c_k for a Gaussian random variable with mean η and variance σ_2 are zero for $k > 2$.

2.119 Derive the general formula for the cumulant $c_{1111}(X_1, X_2, X_3, X_4)$ using (2.152) for $E[X_1] = E[X_2] = E[X_3] = E[X_4] = 0$.

REFERENCES

1. Papoulis, Athanasios, *Probability, Random Variables, and Stochastic Processes*, McGraw Hill, 1965.
2. Mood, Alexander M., Franklin A. Graybill, and Duane C. Boes, *Introduction to the Theory of Statistics*, 3rd Ed., McGraw-Hill, 1974.
3. Freund, John E., *Mathematical Statistics*, Prentice-Hall, 1964.
4. Drake, Alvin W., *Fundamentals of Applied Probability Theory*, McGraw-Hill, 1967.
5. Davenport, B., William Jr., *Probability and Random Processes*, 1970.

6. Stark, Henry, and John W. Woods, *Probability, Random Processes, and Estimation Theory for Engineers*, Prentice-Hall, 1986.
7. Parzen, Emanuel, *Modern Probability Theory and Its Applications*, Wiley, 1960.
8. Hogg, Robert V., and Allen T. Craig, *Introduction to Mathematical, Statistics*, 2nd ed., Macmillan, 1965.
9. Shanmugan, K. Sam, and Arthur M. Breipohl, *Random Signals: Detection, Estimation, and Data Analysis*, Wiley, 1988.
10. Nikias, Chrysostomos L., and Athina P. Petropuou, *Higher-Order Spectrum Analysis: A Nonlinear Processing Framework*, Prentice-Hall, 1993.
11. Nikias, Chrysostomas L., and Jerry M. Mendel, Signal processing with higher-order spectra, *IEEE Signal Processing*, July 1993, 10–35.

3 Estimation of Random Variables

3.1 ESTIMATION OF VARIABLES

In the sciences, especially engineering and physics, we are required to estimate variables that are not directly observable but are observed only through some other measurable variables—for example, estimating the amplitude, frequency, or phase of a known signal in noise. For some problems the variables can be modeled as random variables; others may not be realistically random just simply unknown deterministic variables. Naturally it is desirable to find the "best" estimates where best is in the sense of minimizing some assigned performance measure. The desired variables will be estimated through some function of the observations where the functional relationship is selected to minimize the performance measure. A logical question becomes: what are some special types of functions, and what are some realistic performance measures? Also of major importance is what statistical information is known and how this affects the procedures for finding the optimal estimators. These are the questions addressed in this chapter.

3.1.1 Basic Formulation for Estimation of Random Variables

Say we wish to estimate a random variable X by observing another random variable Z that is statistically related to X. An **estimator** for X, call it $\hat{X}$, is some function, $g(.)$, of the observable Z and is written as

$$\hat{X} = g(Z) \tag{3.1}$$

Thus $\hat{X}$ is itself a random variable. The function $g(Z)$ could be an unrestricted nonlinear form, or in special cases could be restricted to be of a particular form like a constant, linear, or polynomial function of the observed variable Z. The **estimation error** is defined as the difference between the true value X and the estimated value as follows:

$$\text{error} \triangleq \varepsilon = \begin{pmatrix} \text{true} \\ \text{value} \end{pmatrix} - \begin{pmatrix} \text{estimated} \\ \text{value} \end{pmatrix} = X - \hat{X} \tag{3.2}$$

This error is a random variable and is a measure of how well our estimator is performing. It seems reasonable that we would like this error to have a zero mean and a very small variance $E[(X - \hat{X})^2]$. This form of performance is a special case of the more general Bayes performance measure.

3.1.2 Bayes Performance Measure

A **performance measure** J, or "yardstick" by which we judge the **optimality** of our estimate, can take many functional forms. One of the most useful is described by the expected value of a **cost function** $C(.)$ that is a function of the error or difference between the true value and the estimated value. This is called the **Bayes performance measure**, and it can be expressed as

$$J \triangleq E[C(\varepsilon)] \text{ where } \varepsilon = X - \hat{X} \tag{3.3}$$

The **Bayes estimate** of X in terms of Z is given by $\hat{X} = g(Z)$, where $g(Z)$ is selected such that the J, given above, is minimized. If $C(\varepsilon)$ is selected as ε^2, then the performance measure is called the **mean squared error** and $g(Z)$ is the **minimum mean squared error estimate** of X. The mean squared error performance measure will be shown to be mathematically tractable for many situations involving estimating random variables in terms of other random variables.

Other common choices for $g(x)$ include the average absolute error and the average uniform error which are shown along with the average squared error in Figure 3.1. The mean squared error usually requires less statistical information and is more mathematically tractable than the others. Thus our study will concentrate in the beginning on using the mean squared error performance criterion. Estimates using the absolute error and the uniform cost functions will be presented in Section 3.5.

3.1.3 Statistical Characterizations of Data

Sufficient statistical information must be given to solve the optimum estimation problem. It will be shown that for linear estimation the second-order characterizations of the X and Z random variables, including means, correlations, and variances, are sufficient to find the minimum mean squared error estimates. Other cost func-

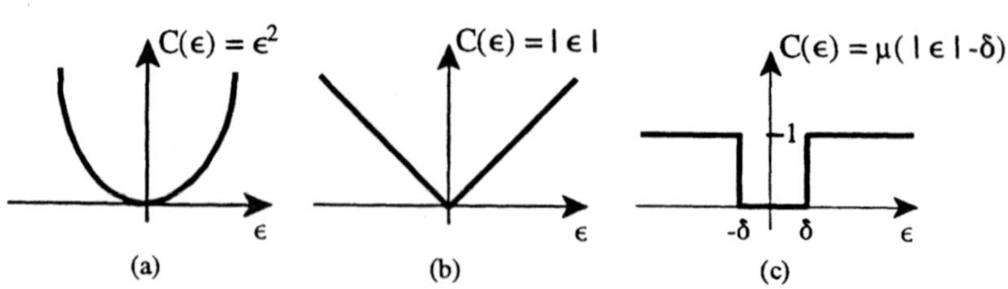

Figure 3.1 Common performance measures: (a) squared error, (b) absolute error, and (c) uniform error.

tions could require more statistical properties of X and Z, including higher-order moments and higher-order joint moments for X and Z. For the case of the absolute and uniform error cost functions, a total characterization of X and Z involving the joint and conditional density functions may be required. Thus, in general, more statistical information is usually required for obtaining optimum nonlinear estimators than for optimal linear estimators. Our discussion begins with finding optimal linear estimates followed by optimal nonlinear estimates.

3.2 LINEAR MMSE ESTIMATION

The mean squared error is a useful measure of performance for obtaining estimates of random variables. Its usefulness is based on mathematical convenience and tractability, and a subjective feeling about errors. In the next few sections optimum linear estimates will be found where optimality is in the minimum mean squared error sense. It will be shown that such estimates can be obtained in terms of the first- and second-order moments. Thus only a partial characterization of the random variables involved will be required.

3.2.1 Estimation of a Random Variable by a Constant

In many cases we wish to approximate a random variable by a deterministic constant. It may be desirable to choose this constant in such a way that a defined performance index is either maximized or minimized. One such performance index is the mean squared error.

Given a random variable Y that we desire to approximate by a constant a such that the mean squared error e_{ms} given by

$$e_{\text{ms}} = E[(Y - a)^2] = \int_{-\infty}^{\infty} (y - a)^2 f_Y(y)\, dy \tag{3.4}$$

is minimized. This selection of the optimum a corresponds to the best estimate of Y by a constant. A necessary condition for the minimization of e_{ms} can be obtained by taking the derivative of e_{ms} with respect to a and setting the result equal to zero to give

$$\frac{de}{da} = \int_{-\infty}^{\infty} 2(y - a)(-1) f_Y(y)\, dy = 0 \tag{3.5}$$

The interchange of the derivative and expected value, which is also an integral, is subject to the same rules for interchanging derivatives and integrals. Solving the equation above for a results in

$$a = \int_{-\infty}^{\infty} y f_Y(y)\, dy = E[y] \tag{3.6}$$

Clearly, it is seen that the constant a that minimizes the mean square error in approximating a random variable by a constant is the mean of the random variable.

The corresponding minimum mean squared error using this approximation can be obtained by substituting (3.6) into (3.4), and the variance of Y as follows:

$$e_{\text{mms}} = \int_{-\infty}^{\infty} (y - E[Y])^2 f_Y(y)\, dy = \sigma_Y^2 \tag{3.7}$$

EXAMPLE 3.1

Given a random variable Y with probability density function $f_Y(y)$ as $f_Y(y) = be^{-by}\mu(y)$. Find the best constant a that approximates the random variable Y, where best is in the sense that the mean squared error is minimized. Determine the minimum mean squared error for the approximation.

SOLUTION

From Eq. (3.6) the best estimate $\hat{Y} = a$, in the minimum mean squared error sense, is the mean or expected value of Y given by

$$a = \int_{-\infty}^{\infty} y\, be^{-by}\mu(y)\, dy = \frac{e^{-by}}{-b}\left(y + \frac{1}{b}\right)\Bigg|_0^{\infty} = \frac{1}{b}.$$

From Eq. (3.4) or (3.7) the minimum mean squared error is

$$\begin{aligned} e_{\text{mms}} &= \int_0^{\infty} \left(y - \frac{1}{b}\right)^2 be^{-by} dy \\ &= \sigma_Y^2 = \frac{1}{b^2} \end{aligned}$$

■

3.2.2 Linear Estimation of One Random Variable from Another

In other problems we may have access to one random variable X and desire to approximate or estimate another random variable Y that is somehow related to X. Assume that the two random variables X and Y have known means, standard deviations, and correlation specified by $\eta_X, \eta_Y, \sigma_X, \sigma_Y$, and R_{XY}, respectively. Let $\hat{Y}$ be the approximation of X and fix it as a linear function of X as

$$\hat{Y} = aX + b \tag{3.8}$$

It is desired to select the a and b such that the mean squared error e_{ms} given by

$$e_{\text{ms}} = E[(Y - \hat{Y})^2] = E[(Y - (aX + b))^2] \tag{3.9}$$

is minimized. Necessary conditions for a and b to minimize e_{ms} are

$$\frac{\partial e_{ms}}{\partial a} = E[2(Y - (aX + b))(-X)] = 0 \tag{3.10}$$

$$\frac{\partial e_{ms}}{\partial b} = E[2(Y - (aX + b))(-1)] = 0 \tag{3.11}$$

These equations can be simplified by taking expected values and rearranging to yield

$$\begin{aligned} E[XY] &= aE[X^2] + b\eta_X \\ \eta_Y &= a\eta_X + b \end{aligned} \tag{3.12}$$

Solving the second equation for b, substituting the result into the first equation, and solving for a gives a and b as

$$\begin{aligned} a &= \frac{E[XY] - \eta_X\eta_Y}{E[X^2] - \eta_X^2} = \frac{R_{XY} - \eta_X\eta_Y}{\sigma_X^2} = \frac{\mu_{XY}}{\sigma_X^2} \\ b &= \eta_Y - a\eta_X = \eta_Y - \frac{\mu_{XY}}{\sigma_X^2}\eta_X \end{aligned} \tag{3.13}$$

Since the matrix of the second derivatives, the Hessian, can be shown to be positive definite, this a and b will give the minimum mean square error. The corresponding minimum mean squared error is easily determined by substituting the a and b given in Eq. (3.13) into Eq. (3.9) and simplifying to give

$$e_{mms} = \sigma_Y^2 - \frac{\mu^2{}_{XY}}{\sigma_X^2} \tag{3.14}$$

Result (3.14) could also have been determined by expanding out (3.9) as follows:

$$e_{ms} = E[(Y - (aX + b))Y] - E[(Y - (aX + b))(aX + b)] \tag{3.15}$$

From Eqs. (3.10) and (3.11) the second expected value is zero, and after replacing the b in the first term by $\eta_Y - a\eta_X$ and writing the second Y as $Y - \eta_Y + \eta_Y$, we write the minimum mean squared error as

$$e_{ms} = E[(Y - \eta_Y - a(X - \eta_X))(Y - \eta_Y + \eta_Y)] \tag{3.16}$$

Using the a from Eq. (3.13) and taking the expected value, we easily see the equation above for e_{ms} to be the same as Eq. (3.14).

The minimum mean squared error is seen to be the original variance of Y, the random variable that is approximated, reduced by the ratio of the covariance between X and Y squared and the variance of X. If the covariance is zero, then $a = 0$, and the minimum mean squared error is equal to the variance of Y, which implies that we gain no information about Y from X by using a linear function.

Notice that the result depends only on the second-order moments of the involved random variables and that the joint density is not needed. This means that the results do not depend on the probability density functions, and thus all random variables that have the same second-order moments give the same results. If the second-order moments are not given but the joint probability density is the specified information, it is a trivial matter to compute the necessary moments, as shown in the following example.

EXAMPLE 3.2

Random variables X and Y are characterized by their joint probability density function $f_{XY}(x, y)$, written as

$$f_{XY}(x,y) = \begin{cases} 2, & \{(x,y): 0 \le y \le x, 0 \le x \le 1\} = R_1 \\ 0, & \text{elsewhere} \end{cases}$$

The region R_1 is shown in Figure 3.2. Find the best linear estimate of Y in terms of X, where best is in the minimum mean squared error sense and determine the corresponding minimum mean squared error.

SOLUTION

To obtain the best estimator of Y in terms of X, it is necessary to compute the second-order moments, which include all means, variances, and cross-correlation. These moments are calculated as follows: First, the means are determined as

$$E[X] = \int_0^1 \int_0^x x f_{XY}(x,y)\,dx\,dy = \int_0^1 (2xy]_0^x\,dx = \frac{2}{3}$$

$$E[Y] = \int_0^1 \int_0^x y f_{XY}(x,y)\,dx\,dy = \int_0^1 (y^2]_0^x\,dx = \frac{1}{3}$$

Next, the individual second-order moments are determined as

$$E[X^2] = \int_0^1 \int_0^x x^2 f_{XY}(x,y)\,dx\,dy = \int_0^1 (2x^2 y]_0^x\,dx = \frac{2}{3}$$

$$E[Y^2] = \int_0^1 \int_0^x y^2 f_{XY}(x,y)\,dx\,dy = \int_0^1 (y^3/3]_0^x\,dx = \frac{1}{12}$$

$$E[XY] = \int_0^1 \int_0^x xy f_{XY}(x,y)\,dx\,dy = \int_0^1 (xy^2]_0^x\,dx = \frac{1}{4}$$

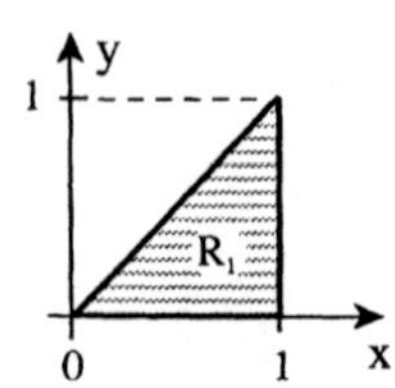

Figure 3.2 R_1 for Example 3.2.

From these moments the individual variances and the covariance are determined, by definition, as

$$\sigma_X^2 = E[X^2] - E^2[X] = \frac{1}{2} - \left(\frac{2}{3}\right)^2 = \frac{1}{18}$$

$$\sigma_Y^2 = E[Y^2] - E^2[Y] = \frac{1}{6} - \left(\frac{1}{3}\right)^2 = \frac{1}{18}$$

$$\mu_{XY} = E[XY] - E[X]E[Y] = \frac{1}{4} - \left(\frac{2}{3}\right)\left(\frac{1}{3}\right) = \frac{1}{36}$$

Using Eq. (3.13), the a and b for the minimum mean squared error linear estimate of Y in terms of X are determined from the various first- and second-order moments as

$$a = \frac{E[XY] - E[X]E[Y]}{E[X^2] - E^2[X]} = \frac{\frac{1}{4} - \left(\frac{2}{3}\frac{1}{3}\right)}{\frac{1}{18}} = \frac{1}{2}$$
$$b = E[Y] - aE[X] = \tfrac{1}{3} - \tfrac{1}{2}\tfrac{2}{3} = 0$$

Thus we have the best estimate of Y in terms of X as $\hat{Y} = \frac{1}{2}X$.

The minimum mean squared error can be determined from Eq. (3.14) as

$$e_{\text{mms}} = \sigma_Y^2 - \frac{\mu^2{}_{XY}}{\sigma_X^2} = \frac{1}{18} - \frac{\left(\frac{1}{36}\right)^2}{\frac{1}{18}}$$
$$= \frac{1}{18} - \frac{1}{72} = \frac{1}{24}$$

The first term $\frac{1}{18}$, being the variance, is the error for best estimate of Y in terms of a constant—namely the mean—while the second term is the reduction in the total mean squared error that results from the linear estimation thus giving the final minimum mean squared error for the linear estimate as the $\frac{1}{24}$. ■

3.2.3 Linear Estimation of a Random Variable from N Random Variables

The linear estimation problem can be easily extended to the case of estimating the random variable X_0 in terms of $X_1, X_2, \ldots, X_n$. It is assumed that the statistical relationships between the random variables are characterized by specifying the means and covariances (a second-order characterization):

$$\begin{aligned} E[X_i] &= \eta_i & i &= 0, 1, \ldots, n \\ E[(X_i - \eta_i)(X_j - \eta_j)] &= \mu_{ij}, & i,j &= 0, 1, \ldots, n \end{aligned} \tag{3.17}$$

The linear estimate of X_0 in terms of $X_1, X_2, \ldots, X_n$ and the mean squared error are given by

$$\hat{X}_0 = a_1X_1 + a_2X_2 + \cdots + a_nX_n + b \tag{3.18}$$

$$e_{\text{ms}} = E[(X_0 - \hat{X}_0)^2] = E[(X_0 - (a_1X_1 + a_2X_2 + \cdots + a_nX_n + b))^2] \tag{3.19}$$

The problem then is one of selecting $a_1, a_2, \ldots, a_n$, and b such that e_{ms} is minimized. Taking partial derivatives of e_{ms} with respect to the a_i and b, setting them equal to zero, and rearranging gives the following set of $n+1$ simultaneous linear equations in $n+1$ unknowns:

$$\begin{aligned}
E[X_0X_1] &= a_1E[X_1^2] + a_2E[X_1X_2] + \cdots + a_nE[X_1X_n] + bE[X_1] \\
E[X_0X_2] &= a_1E[X_2X_1] + a_2E[X_2^2] + \cdots + a_nE[X_2X_n] + bE[X_2] \\
&\vdots \\
E[X_0X_n] &= a_1E[X_nX_1] + a_2E[X_nX_2] + \cdots + a_nE[X_n^2] + bE[X_n] \\
E[X_0] &= a_1E[X_1] + a_2E[X_2] + \cdots + a_nE[X_n] + b
\end{aligned} \tag{3.20}$$

Solving the last equation for b, substituting the results into the other n equations, rearranging, and using $\mu_{ij} = E[X_iX_j] - \eta_i\eta_j$ results in the following set of n linear equations in n unknowns:

$$\begin{aligned}
\mu_{01} &= a_1\mu_{11} + a_2\mu_{12} + \cdots + a_n\mu_{1n} \\
\mu_{02} &= a_1\mu_{21} + a_2\mu_{22} + \cdots + a_n\mu_{2n} \\
&\vdots \\
\mu_{0n} &= a_1\mu_{1n} + a_2\mu_{2n} + \cdots + a_n\mu_{nn}
\end{aligned} \tag{3.21}$$

The solution of Eqs. (3.21) gives the a_i that minimize the mean square error and the last equation in (3.20) gives the corresponding b in terms of the a_i and the given means as

$$\begin{aligned}
b &= E[X_0] - a_1E[X_1] - a_2E[X_2] - \cdots - a_nE[X_n] \\
&= \eta_{X_0} - a_1\eta_{X_1} - a_2\eta_{X_2} - \cdots - a_n\eta_{X_n}
\end{aligned} \tag{3.22}$$

After substituting the optimal a_i from the solution of (3.21) and optimal b from (3.22) into the mean squared error of (3.19) and simplifying, the minimum mean squared error can be shown to be

$$e_{\text{mms}} = \sigma_{X_0}^2 - \sum_{i=1}^{n} \mu_{0i}a_i \tag{3.23}$$

The proof for this result will be presented after the introduction of the orthogonality principle by Eqs. (3.30) through (3.34). The following example illustrates the procedure for finding an estimate of one random variable in terms of three other random variables and the resulting performance indicated by the minimum mean squared error.

EXAMPLE 3.3

Given that the random variables X_0, X_1, X_2, and X_3 are jointly Gaussian with the following mean vector and covariance matrix:

$$\boldsymbol{\eta} = \begin{bmatrix} 0 \\ 0 \\ 0 \\ 0 \end{bmatrix}; \quad \boldsymbol{K} = \begin{bmatrix} 5 & 2 & 1 & 1 \\ 2 & 5 & 0 & 0 \\ 2 & 0 & 2 & -1 \\ 1 & 0 & -1 & 2 \end{bmatrix}$$

Find the best linear estimate, in the minimum mean squared error sense, of X_0 in terms of X_1, X_2, and X_3, and determine the minimum mean squared error for the estimate.

SOLUTION

For $\hat{X}_0 = a_1X_1 + a_2X_2 + a_3X_3 + b$, the optimum coefficients are the solution of (3.21) and (3.22). From (3.22) and the fact that the mean vector $\boldsymbol{\eta}$ is a zero vector, we see that

$$\begin{aligned} b &= \eta_{X_0} - a_1\eta_{X_1} - a_2\eta_{X_2} - a_3\eta_{X_3} \\ &= 0 - a_1 0 - a_2 0 - a_3 0 = 0 \end{aligned}$$

Using the entries from the given covariance matrix K and Eq. (3.17), we obtain a set of simultaneous linear equations in terms of the optimum coefficients a_1, a_2, and a_3 as

$$\begin{bmatrix} 2 \\ 1 \\ 1 \end{bmatrix} = \begin{bmatrix} 5 & 0 & 0 \\ 0 & 2 & -1 \\ 0 & -1 & 2 \end{bmatrix} \begin{bmatrix} a_1 \\ a_2 \\ a_3 \end{bmatrix}$$

Solving this matrix equation for the coefficients a_1, a_2, and a_3 gives $a_1 = 0.4$, $a_2 = 1$, and $a_3 = 1$, and since $b = 0$, the best linear estimate in the minimum mean squared error sense is

$$\hat{X}_0 = 0.4X_1 + X_2 + X_3$$

From Eq. (3.23) the minimum mean squared error resulting from using the optimum coefficients is determined as

$$\begin{aligned} e_{\text{mms}} &= \sigma_{X_0}^2 - \sum_{i=1}^{n} \mu_{0i}a_i \\ &= 5 - 2(0.4) - 1(1) - 1(1) = 2.2 \end{aligned}$$

Thus for this problem there is a significant reduction in the mean squared error when using the linear estimate as opposed to estimating X_0 by its mean only. ■

The Orthogonality Principle. In this section the problem of linear minimum mean squared error estimation will be viewed from a slightly different perspective. The orthogonality principle developed in this section provides an alternative starting place in approaching the linear MMSE problem. The assumptions are the same as given in the previous section; that is, we have $n+1$ random variables $X_0, X_1, X_2, \ldots, X_n$ whose means and second order moments are given by:

$$\begin{aligned} E[X_i] &= \eta_i, & i &= 0, 1, \ldots, n \\ E[(X_i - n_i)(X_j - n_j)] &= \mu_{ij}, & i,j &= 0, 1, \ldots, n \end{aligned} \tag{3.24}$$

The random variable X_0 is desired to be approximated by a linear combination of the so-called **data** random variables $X_1, X_2, \ldots, X_n$ as follows:

$$\hat{X}_0 = a_1X_1 + a_2X_2 + \cdots + a_nX_n + b \tag{3.25}$$

The error e involved with this approximation is given by

$$\mathscr{E} = X_0 - \hat{X}_0 = X_0 - (a_1X_1 + a_2X_2 + \cdots + a_nX_n + b) \tag{3.26}$$

As in the preceding section we desire the coefficients $a_1, a_2, \ldots, a_n$ and b such that the mean squared error given below is minimized:

$$e_{\text{ms}} = E[\mathscr{E}^2] = E[(X_0 - (a_1X_1 + a_2X_2 + \cdots + a_nX_n + b))^2] \tag{3.27}$$

By taking the partials of e_{ms} with respect to the a_i and b, it can be easily shown that

$$E[(X_0 - (a_1X_1 + a_2X_2 + \cdots + a_nX_n + b))X_i] = 0, \qquad i = 1, 2, \ldots, n \tag{3.28}$$

and

$$E[X_0] - a_1E[X_1] - a_2E[X_2] - \cdots - a_nE[X_n] - b = 0$$

or (3.29)

$$b = \eta_0 - (a_1\eta_1 + a_2\eta_2 + \cdots + a_n\eta_n)$$

Since the X_i, $i = 1, 2, \ldots, n$ are the **data** and $X_0 - (a_1X_1 + a_2X_2 + \cdots + a_nX_n + b)$ is the **error**, Eq. (3.28) can be interpreted as the **error is orthogonal to each member of the data**, and (3.29) gives the solution for the optimal b. This statement is the first part of the **orthogonality principle** for minimum mean squared error linear estimation. The second part of the orthogonality principle deals with the evaluation of the minimum mean squared error for this optimal estimater. The e_{mms} of Eq. (3.27), where the a_i are the solution to the simultaneous equations of (3.28), can be written in the expanded form as

$$\begin{aligned} e_{\text{mms}} &= E[(X_0 - (a_1X_1 + a_2X_2 + \cdots + a_nX_n + b))(X_0 - b)] \\ &\quad - E[(X_0 - (a_1X_1 + a_2X_2 + \cdots + a_nX_n + b))(a_1X_1 + a_2X_2 + \cdots + a_nX_n)] \end{aligned} \tag{3.30}$$

The second expected value of (3.30) is zero from (3.28), since it is the expected value of the error times a linear combination of the data. Thus the formula for the minimum mean squared error can be reduced to

$$e_{\text{mms}} = E[(X_0 - (a_1X_1 + a_2X_2 + \cdots + a_nX_n + b))(X_0 - b)] \tag{3.31}$$

Eq (3.31) is the **second part of the orthogonality principle**, and it gives a reduced expression for determining the minimum mean squared error. This result can be written in another form by substituting the b of Eq. (3.29) into (3.31) to get

$$\begin{aligned} e_{\text{mms}} = E\{&[(X_0 - \eta_0) - a_1(X_1 - \eta_1) - a_2(X_2 - \eta_2) - \cdots - a_n(X_n - \eta_n)] \\ &\cdot [(X_0 - \eta_0) - (a_1\eta_1 + a_2\eta_2 + \cdots + a_n\eta_n)]\} \end{aligned} \tag{3.32}$$

The equation above for e_{mms} can be simplified further by multiplying out the last terms in the two parts shown to give

$$\begin{aligned} e_{\text{mms}} &= E\{[(X_0 - \eta_0) - a_1(X_1 - \eta_1) - a_2(X_2 - \eta_2) - \cdots - a_n(X_n - \eta_n)] \cdot (X_0 - \eta_0)\} \\ &\quad - E\{[(X_0 - \eta_0) - a_1(X_1 - \eta_1) - a_2(X_2 - \eta_2) - \cdots - a_n(X_n - \eta_n)] \\ &\quad \cdot (a_1\eta_1 + a_2\eta_2 + \cdots + a_n\eta_n)]\} \end{aligned} \tag{3.33}$$

The first expected value gives a sum of covariances of X_0 and X_i, and the second expected value is zero, since it gives a weighted sum of central moments of the X_i, $E[X_i - \eta_i]$, which are zero by definition for all $i = 0, 1, \ldots, n$. Thus the expression above for the minimum mean squared error can be written in terms of just the covariances between X_0 and the data set X_i and the optimal coefficients as

$$e_{\text{mms}} = \sigma_{X_0}^2 - \sum_{i=1}^{n} \mu_{0i} a_i \tag{3.34}$$

Special Case. For the many problems where the expected values of the X_i, $i = 0, 1, \ldots, n$ are all zero the orthogonality principle can be simplified. From Eq. (3.29) it can be shown that $b = 0$ and then from (3.28) that the a_i are the solution of the set of equations

$$E[(X_0 - (a_1X_1 + a_2X_2 + \cdots + a_nX_n))X_i] = 0, \qquad i = 1, 2, \ldots, n \tag{3.35}$$

Setting $b = 0$ in (3.31) gives the resulting minimum mean squared error as the expected value of the product of the error and the variable X_0 that we are estimating

$$e_{\text{mms}} = E[(X_0 - (a_1X_1 + a_2X_2 + \cdots + a_nX_n))X_0] \tag{3.36}$$

Multiplying out the terms in (3.36) and taking expected values of the products allows the minimum mean squared error to be written in terms of the correlations between the data as follows:

$$e_{\text{mms}} = \sigma_{X_0}^2 - \sum_{i=1}^{n} a_i R_{0i} \tag{3.37}$$

The first term of this error, $\sigma_{X_0}^2$, is the error that would occur if X_0 were estimated by a constant with no data available, namely the mean of zero, whereas the terms in the sum represent the reduction in mean squared error due to the use of each member in the data set.

The following example presents an application of the orthogonality principle to the problem of estimating a "signal" random variable from a set of noisy measurements of that signal.

EXAMPLE 3.4

Suppose that we have the random variables $S, Z_1, Z_2, \ldots, Z_n$ where the Z_i are related to S by

$$Z_i = S + N_i, \qquad i = 1, 2, \ldots, n$$

Let S be a zero mean random variable with known variance σ_S^2 and the N_i, $i = 1, 2, \ldots, n$, be zero mean mutually orthogonal random variables each with variance σ_N^2 and each orthogonal to S. These assumptions can be written as using the Kronecker delta function δ_{ij} as

$$E[N_i N_j] = \sigma_N^2 \delta_{ij}, \qquad i, j = 1, 2, \ldots, n$$

$$E[SN_i] = 0 \qquad i = 1, 2, \ldots, n$$

From the Z_i we wish to approximate or estimate S by a linear combination of the Z_i given by

$$\hat{S} = \sum_{i=1}^{n} a_i Z_i + b$$

such that the mean squared error is minimized. The problem can be related to estimating a random constant value from a set of n noisy observations.

SOLUTION

The solution is based on a direct application of the orthogonality principle. From Eq. (3.29) the b is as follows:

$$b = E[S] - a_1 E[Z_1] - a_2 E[Z_2] - \cdots - a_n E[Z_n]$$

Since $E[Z_i] = E[S] + E[N_i] = 0$ for $i = 1, 2, \ldots, n$ and $E[S] = 0$, it is easily seen that $b = 0$. After substituting $b = 0$ into the first part of the orthogonality principle, Eq. (3.28), which states that the error is orthogonal to the data, gives

$$E[(S - (a_1 Z_1 + a_2 Z_2 + \cdots + a_n Z_n))Z_i] = 0, \qquad i = 1, 2, \ldots, n$$

This equation could have been obtained directly from the special case of zero means from (3.35). After taking the expected values, we can rearrange the set of simultaneous equations above to give

$$E[SZ_i] = a_1 E[Z_1 Z_i] + a_2 E[Z_2 Z_i] + \cdots + a_n E[Z_n Z_i], \qquad i = 1, 2, \ldots, n$$

To describe this set of equations, it is necessary to find the correlation of S and Z_i for all i and the correlation of Z_i and Z_j for all i and j. First the correlation of S and Z_i is determined as

$$E[SZ_i] = E[S(S + N_i)] = E[S^2] + E[SN_i] = \sigma_S^2, \qquad i = 1, 2, \ldots, n$$

The second term in the sum above, $E[SN_i]$, is zero by the original given properties on the expected value of SN_i. Similarly the $E[Z_iZ_j]$ can be found by using the given orthogonal properties of N_i and N_j as

$$\begin{aligned} E[Z_iZ_j] &= E[(S + N_i)(S + N_j)] = E[S^2] + E[N_iS] + E[SN_j] + E[N_iN_j] \\ &= \sigma_S^2 + \sigma_N^2 \delta_{ij} \end{aligned}$$

Substituting these results into the simultaneous equations given above results in the following set of n simultaneous equations in the n unknowns $a_1, a_2, \ldots, a_n$:

$$\begin{aligned} \sigma_S^2 &= (\sigma_S^2 + \sigma_N^2)a_1 + \sigma_S^2 a_2 + \cdots + \sigma_S^2 a_n \\ \sigma_S^2 &= \sigma_S^2 a_1 + (\sigma_S^2 + \sigma_N^2)a_2 + \cdots + \sigma_S^2 a_n \\ &\vdots \\ \sigma_S^2 &= \sigma_S^2 a_1 + \sigma_S^2 a_2 + \cdots + (\sigma_S^2 + \sigma_N^2)a_n \end{aligned}$$

By symmetry of the equations above, the a_i are equal, so the solution can be obtained from any of the equations above by equating the a_i as follows:

$$\sigma_S^2 = n\sigma_S^2 a_i + \sigma_N^2 a_i$$

Solving for a_i gives

$$a_i = \frac{\sigma_S^2}{n\sigma_S^2 + \sigma_N^2}, \qquad i = 1, 2, \ldots, n$$

With these coefficients the approximation of S can be written as

$$\hat{S} = \frac{\sigma_S^2}{n\sigma_S^2 + \sigma_n^2} \sum_{i=1}^{n} Z_i$$

As might be expected, the form of the estimator is a weighted sum of the measurements. This equation can be further rearranged to be written in terms of the average of the measurements in the following ways:

$$\hat{S} = \frac{n\sigma_S^2}{n\sigma_S^2 + \sigma_n^2} \left(\frac{1}{n} \sum_{i=1}^{n} Z_i \right) \quad \text{or}$$

$$\hat{S} = \frac{1}{1 + \sigma_n^2 n^{-1} \sigma_S^{-2}} \left(\frac{1}{n} \sum_{i=1}^{n} Z_i \right)$$

Using these a_i in the second part of the orthogonality principle equation (3.34), and identifying X_0 as S and X_i as Z_i, we can write the minimum mean squared error as

$$e_{\text{mms}} = \sigma_S^2 - \sum_{i=1}^{n} a_i E[SZ_i]$$

With the correlations of S and Z_i determined as above, e_{mms} becomes

$$\begin{aligned} e_{\text{mms}} &= \sigma_S^2 - \sum_{i=1}^{n} \frac{\sigma_S^2}{n\sigma_S^2 + \sigma_N^2} \sigma_S^2 \\ &= \frac{\sigma_N^2 \sigma_S^2}{n\sigma_S^2 + \sigma_N^2} \end{aligned}$$

For the special case where n, σ_S^2, and σ_N^2 are such that the effective signal to noise ratio, $n\sigma_S^2/\sigma_N^2$ is $\gg 1$, the coefficients for the linear approximation and the corresponding minimum mean squared error become

$$a_i \approx \frac{\sigma_S^2}{n\sigma_S^2} = \frac{1}{n}, \qquad e_{\text{mms}} \approx \sigma_N^2 \frac{\sigma_S^2}{n\sigma_S^2} = \frac{\sigma_N^2}{n}$$

The mean squared error is thus seen to be inversely proportional to the number of observations, and the coefficients make the estimate an average of the observations.

$$\hat{S} \approx \frac{1}{n} \sum_{i=0}^{n} Z_i$$

If n and σ_S^2 are such that $n\sigma_S^2 \ll \sigma_N^2$, low effective signal to noise ratio, the a_i and mean squared error become

$$a_i \approx \sigma_S^2/\sigma_N^2, \qquad e_{\text{mms}} \approx \sigma_S^2$$

This shows that in the low effective signal to noise case that the coefficients approach the ratio of the variances of S and N and the minimum mean squared error approaches the variance of S. Thus the form of the estimator is

$$\hat{S} \approx \frac{\sigma_S^2}{\sigma_N^2} \sum_{i=0}^{n} Z_i$$

Since the variance of S is the initial mean squared error without our taking any measurements, the results indicate that as σ_S^2/σ_N^2 approaches zero, the data do not significantly reduce the mean squared error. Therefore for this case the observations are almost not used to estimate S, and the mean of S, which for this problem is zero, is used as the best estimate. ■

EXAMPLE 3.5

Say we are given a random variable X that is exponentially distributed with probability density function given by

$$f_X(x) = 2e^{-2x}\mu(x)$$

Define a random variable Y related to X through $Y = e^{-X}$. We wish to find the best linear estimate, in the minimum mean squared error sense, of Y in terms of X. Let the optimal estimate be given by $\hat{Y} = aX + b$. Determine the a and b for this optimum estimate, and determine its minimum mean squared error.

SOLUTION

It is known that the optimal linear estimate can be found by using the orthogonality principle, which states that the error is orthogonal to the data. For our problem this relationship can be expressed as

$$E[(Y - aX - b)X] = 0$$

where the first term in the ( ) is the error and X is the data. Expanding out and taking the expected value gives an equation for a and b in terms of various first- and second-order moments of X and Y as

$$E[YX] = aE[X^2] + bE[X]$$

The other equation of the orthogonal principle from Eq. (3.29), recognizing that $X_0 = Y$ and $X_1 = X$, is

$$b = E[Y] - aE[X]$$

To solve these two equations for a and b we require the calculation of the various moments indicated. These expected values will now be determined

$$E[Y] = E[e^{-X}] = \int_0^\infty 2e^{-3x}dx = \frac{2}{3}$$

$$E[YX] = E[e^{-X}X] = \int_0^\infty 2xe^{-3x}dx = \frac{2}{9}$$

$$E[Y^2] = E[e^{-2X}] = \int_0^\infty 2e^{-4x}dx = \frac{1}{2}$$

$$E[X] = \int_0^\infty 2xe^{-2x}dx = \frac{1}{2}$$

$$E[X^2] = \int_0^\infty 2x^2e^{-2x}dx = \frac{1}{2}$$

Thus with these values the simultaneous equations become

$$\tfrac{2}{9} = \tfrac{1}{2}a + \tfrac{1}{2}b$$
$$b = \tfrac{1}{3} - \tfrac{1}{2}a$$

Solving for a and b yields $a = \frac{4}{9}$ and $b = \frac{8}{9}$. Therefore the best estimate is

$$\hat{Y} = -\tfrac{4}{9}X + \tfrac{8}{9}$$

From Eq. (3.23) the minimum mean squared error can be written as

$$\begin{aligned} e_{\text{mms}} &= \sigma_Y^2 - \mu_{YX}a \\ &= (\tfrac{1}{2} - \tfrac{4}{9}) - (\tfrac{2}{9} - \tfrac{1}{3})(-\tfrac{4}{9}) = \tfrac{1}{162} \end{aligned}$$

3.3 NONLINEAR MMSE ESTIMATION

In the previous sections, optimum linear estimates of one random variable in terms of one or more random variables were found in the minimum mean squared error sense. In many problems a further reduction in mean squared error can be obtained by using nonlinear estimation. This reduction is at the expense of requiring more statistical properties of the random variables. It will now be shown that nonlinear minimum mean squared error estimation can be performed if almost a total characterization of the random variables is specified. This is in sharp contrast with linear estimation where only first- and second-order moments are needed, that is, the means, variances, and covariances of the random variables involved.

3.3.1 Nonlinear Estimation of One Random Variable from Another

Given two random variables X and Y that are statistically related with joint probability density function $f_{XY}(x, y)$. By observing X, we desire to estimate Y by some nonlinear function $g(\cdot)$ of X, not necessarily a linear function, as

$$\hat{Y} = g(X) \tag{3.38}$$

Our estimate is to be the best in the sense that it gives the minimum mean squared error e as

$$e_{\text{ms}} = E[(Y - \hat{Y})^2] = E[(Y - g(X))^2] \tag{3.39}$$

The first step in minimizing e_{ms} is to write it in terms of $f_{XY}(x, y)$ as

$$e_{\text{ms}} = \int_{-\infty}^{\infty}\int_{-\infty}^{\infty} (y - g(x))^2 f_{XY}(x, y)\, dx\, dy \tag{3.40}$$

Using the fact that the joint density is a product of the conditional and marginal densities, $e_{\rm ms}$ can be rearranged to yield

$$e_{\rm ms} = \int_{-\infty}^{\infty} \left[\int_{-\infty}^{\infty} (y - g(x))^2 f_Y(y|x)\, dy \right] f_X(x)\, dx \tag{3.41}$$

Since the product of $f_Y(y|x)$ and $(y - g(x))^2$ is positive for all x and $f_X(x)$ is always positive, e can be minimized by making the term in brackets as small as possible for every x. That is, select $g(\cdot)$ in such a way that the term in brackets, $I(x)$, given by

$$I(x) = \int_{-\infty}^{\infty} (y - g(x))^2 f_Y(y|x)\, dy \tag{3.42}$$

is minimized. But for each x, $g(x)$ is just a constant, and we have shown previously (Section 3.2.1) that the integral is smallest when $g(x)$ is selected as the mean of the random variable, which for this case is in terms of the conditional probability density $f_Y(y|x)$. Therefore the optimum $g(x)$, in the minimum mean squared error sense, is given by

$$g(x) = E[Y|X = x] = \int_{-\infty}^{\infty} y f_Y(y|x)\, dy \tag{3.43}$$

The resulting $g(x)$ is sometimes called the **nonlinear regression curve**. The minimum mean squared error is obtained by substituting this optimal $g(x)$ into (3.39), the equation for the mean squared error, to give

$$\begin{aligned} e_{\rm mmse} &= \int_{-\infty}^{\infty} \left[\int_{-\infty}^{\infty} (y - E[Y|x])^2 f_Y(y|x)\, dy \right] f_X(x)\, dx \\ &= \int_{-\infty}^{\infty} \sigma_{Y|x}^2 f_X(x)\, dx \end{aligned} \tag{3.44}$$

Thus the minimum mean squared error is seen to be $E[(Y - E[Y|X])^2]$, where $E[Y|X]$ given by $E[Y|X = x]|_{x=X}$ is a random variable.

A series of nonlinear estimation examples are now presented. In the first a nonlinear estimate is found, in the second the best nonlinear estimate turns out to be a linear estimate, and the third shows the results for the case that the random variables are statistically independent.

EXAMPLE 3.6

The density for Y conditioned on $X = x$ and the marginal density $f_X(x)$ of two random variables X and Y are given by

$$f_Y(y|x) = xe^{-yx}\mu(y), \quad f_X(x) = \mu(x - 1) - \mu(x - 2)$$

Find the best nonlinear estimate of Y in terms of X, and calculate the corresponding minimum mean squared error for your estimate.

SOLUTION

The conditional density in y is seen as exponential with parameter $a = x$; thus we are trying to estimate this parameter from the given information and a sample of X. The $g(x)$ that gives the best nonlinear estimate in the minimum mean squared error sense is the conditional mean

$$g(x) = E[Y|X = x] = \int_{-\infty}^{\infty} y f_y(y|x)\, dy$$

$$= \int_0^{\infty} yxe^{-xy}\, dy = \left.\frac{xe^{-xy}(xy-1)}{x^2}\right]_0^{\infty} = \frac{1}{x}$$

Therefore $\hat{Y} = g(X) = 1/X$.

The minimum mean squared error for the optimal $g(x)$ can be determined from (3.41) as

$$e_{\text{mmse}} = \int_{-\infty}^{\infty} \left[\int_{-\infty}^{\infty} (y - 1/x)^2 f_Y(y|x)\, dy \right] f_X(x)\, dx$$

But the term in brackets is the conditional variance for the exponential density, which from Eq. 2.38 is just $1/x^2$. Therefore the mean squared error can be found as follows:

$$e_{\text{mmse}} = \int_{-\infty}^{\infty} \frac{1}{x^2} f_X(x)\, dx = \int_1^2 \frac{1}{x^2} 1\, dx = \left.\frac{-1}{x}\right]_1^2 = 1/2 \qquad \blacksquare$$

EXAMPLE 3.7

The random variables X and Y are statistically characterized by the following joint probability density function:

$$f(x, y) = \begin{cases} 2, & (x, y) \in R = \{(x, y) : 0 < y < x, 0 < x < 1\} \\ 0, & \text{elsewhere} \end{cases}$$

where R is shown in Figure 3.3.

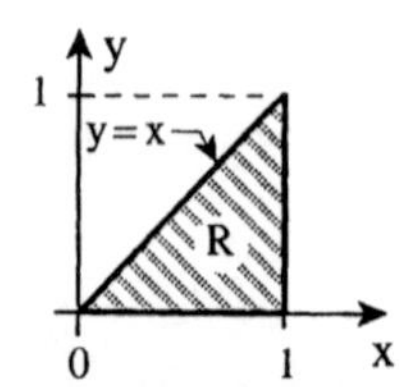

Figure 3.3 Region R for Example 3.7.

(a) Find the best nonlinear estimate of Y in terms of X.
(b) Find the best linear estimate of Y in terms of X.
(c) Comment on your results.

SOLUTION

(a) The best nonlinear estimate is the conditional expected value of Y given $X = x$. To find the conditional density, it is necessary to determine the marginal density $f_X(x)$. It is determined by integrating the joint density for each value of x:

$$f_X(x) = \int_{-\infty}^{\infty} f_{XY}(x, y)\, dy$$

For any x such that $0 < x < 1$ the $f_X(x)$ becomes

$$f_X(x) = \int_0^x 2dy = 2x$$

For x outside the interval (0, 1) the probability density function is zero, so the conditional density function can be summarized by using the unit step function as

$$f_Y(y/x) = \frac{f_{XY}(x, y)}{f_X(x)} = \frac{1}{x}(\mu(y) - \mu(y - x))$$

The function for the best nonlinear estimate of Y in terms of X for $0 < x < 1$ is the conditional mean as follows

$$\begin{aligned} g(x) = E[Y|X = x] &= \int_{-\infty}^{\infty} y f_Y(y|x)\, dy \\ &= \int_0^x y\frac{1}{x}\, dy = \left.\frac{y^2}{2x}\right]_0^x = \frac{x}{2} \end{aligned}$$

It is seen that for this example the best nonlinear estimate is a linear estimate.

(b) The best linear estimate of Y in terms of X by $ax + b$ was found in Example 3.2 to be $\hat{Y} = 1/2X$, so for this example the best linear estimate is the same as the best nonlinear estimate.

(c) It is rare that the best linear and nonlinear estimates are the same, but they will be shown in Section 3.3.3 to be the same for the special case where X and Y are jointly Gaussian. ■

EXAMPLE 3.8

The joint probability density function for two random variables X and Y is

$$f_{XY}(x, y) = e^{-(x+y)}\mu(x)\mu(y)$$

Find the best nonlinear estimate of Y in terms of X.

SOLUTION

Before the conditional density can be determined, the $f_X(x)$ must be found. For any x the marginal density for X is as follows:

$$f_X(x) = \int_{-\infty}^{\infty} f_{XY}(x, y)\, dy = \int_0^{\infty} e^{-(x+y)}\, dy\mu(x)$$
$$= e^{-x}\mu(x)$$

Using the marginal density, $f_X(x)$, we write the conditional density as

$$f_Y(y|x) = \frac{f_{XY}(x, y)}{f_X(x)} = e^{-y}\mu(y)$$

Finally the $g(x)$ is obtained as the conditional expected value

$$g(x) = E[Y|X = x] = \int_{-\infty}^{\infty} yf_Y(y|x)\, dy$$
$$= \int_0^{\infty} ye^{-y}\, dy = 1$$

This result states that the best nonlinear estimate of Y in terms of X is 1, not a function of x at all, and it equals the mean of Y. The reason for this result is that X and Y are statistically independent. So the estimation problem reduces to estimating the random variable Y by a constant for which we know that the best estimate is just the mean of Y, which is equal to 1 for this problem. The statistical independence of X and Y can easily be shown by computing the marginal density $f_Y(y)$ and seeing that it is the same as the conditional density for y. ■

EXAMPLE 3.9

Define two random variables X and Y characterized by their joint probability density function $f_{XY}(x, y)$ as

$$f_{XY}(x, y) = xe^{-x(y+1)}\mu(x)\mu(y)$$

Find the best nonlinear estimate in the minimum mean squared error sense for estimating Y in terms of X.

SOLUTION

To get the best nonlinear estimate, we must first determine the conditional density and then compute the conditional mean using that density. To get the conditional density, we first find the marginal density for the random variable X by integrating out the y variable of the joint density as follows:

$$
\begin{aligned}
f_X(x) &= \int_{-\infty}^{\infty} f_{XY}(x, y)\, dy = \int_0^{\infty} xe^{-x(y+1)}\mu(x)\, dy \\
&= xe^{-x}\int_0^{\infty} e^{-xy}\, dy\mu(x) = xe^{-x}\left(\frac{1}{-x}e^{-xy}\right]_0^{\infty}\mu(x) \\
&= e^{-x}\mu(x)
\end{aligned}
$$

Therefore the conditional density can be determined as

$$
\begin{aligned}
f_Y(y|x) &= \frac{f_{XY}(x, y)}{f_X(x)} = \frac{xe^{-x(y+1)}}{e^{-x}}\mu(x)\mu(y) \\
&= xe^{-xy}\mu(x)\mu(y)
\end{aligned}
$$

The function $g(x)$ for the nonlinear estimate of Y in terms of X is $E[Y|x]$, the conditional mean, which is derived as follows:

$$
\begin{aligned}
E[Y|x] &= \int_{-\infty}^{\infty} yf_Y(y|x)\, dy = \int_0^{\infty} yxe^{-xy} dy \\
&= \int_0^{\infty} \frac{1}{x} ze^{-z} dz = \frac{1}{x}
\end{aligned}
$$

Therefore the best nonlinear estimate of Y in terms of X is $\hat{Y} = 1/X$. ■

3.3.2 Nonlinear Estimation of One Random Variable from N Random Variables

The problem of estimating a random variable X_0 in terms of statistically related random variables $X_1, X_2, \ldots, X_n$, where the random variables are characterized by their joint probability density function $f_x(x_0, x_1, \ldots, x_n)$ is a common application in the sciences. In Section 3.2 formulas were given for finding the best linear estimate, while in this section the problem of nonlinear estimation will be examined.

We desire to approximate or estimate X_0 by a nonlinear function $g(X_0, X_1, \ldots, X_n)$ as

$$
\hat{X}_0 = g(X_1, X_2, \ldots, X_n) \tag{3.45}
$$

It is desired to find the $g(x_1, x_2, \ldots, x_n)$ that will minimize the mean squared error given by

$$
\begin{aligned}
e_{\text{ms}} &= E[(X_0 - \hat{X}_0)^2] \\
&= E[(X_0 - g(X_1, X_2, \ldots, X_n))^2]
\end{aligned} \tag{3.46}
$$

By extending the results given for the best nonlinear estimate of one random variable in terms of another, it can be easily shown that the optimal function $g(x_1, x_2, \ldots, x_n)$, in the minimum mean squared error sense, is the conditional mean given by

$$
\begin{aligned}
g(x_1, x_2, \ldots, x_n) &= E[X_0|X_1 = x_1, X_2 = x_2, \ldots, X_n = x_n] \\
&= \int_{-\infty}^{\infty} x_0 f(x_0|x_1, x_2, \ldots, x_n)\, dx_0
\end{aligned} \tag{3.47}
$$

It is worth noting that to obtain the minimum mean squared error nonlinear estimate of one random variable in terms of other random variables we require the knowledge of the conditional probability density function, $f(x_0|x_1, x_2, \ldots, x_n)$, which in many problems involves more statistical information than is available. It was shown in Section 3.2 that restricting $g(\cdot)$ to a linear function reduces the required information to the knowledge of the second-order properties, that is, the means, variances and covariances of the random variables involved.

EXAMPLE 3.10 (Random Variables)

Let X_0, X_1, and X_2 be random variables. The conditional density for X_0, given $X_1 = x_1$, and $X_2 = x_2$, is known to be the following:

$$
f_{X_0}(x_0|x_1, x_2) = kx_0^{(x_1-1)}(1 - x_0)^{x_2-1}, \qquad 0 < x_0 < 1
$$

$$
\text{where} \quad k = \frac{\Gamma(x_1)}{\Gamma(x_1)\Gamma(x_2)}
$$

Find the best nonlinear estimate of X_0 in terms of X_1 and X_2 where best is in the minimum mean squared error sense.

SOLUTION

From Eq. (3.47) we know that the optimal function is the conditional mean. For the conditional density function above the conditional mean can be shown from Eq. 2.50 to be

$$
\begin{aligned}
g(x_1, x_2) = E[X_0|x_1, x_2] &= \int_0^1 x_0 f_{X_0}(x_0|x_1, x_2)\, dx_0 \\
&= \frac{x_1}{x_1 + x_2}
\end{aligned}
$$

Thus the best nonlinear estimate of X_0 in terms of X_1 and X_2 is $\hat{X}_0 = X_1/(X_1 + X_2)$. ■

3.3.3 Nonlinear Estimation of Gaussian Random Variables

Given X and Y are jointly Gaussian random variables with known means, η_X, η_Y, variances, σ_X^2, σ_Y^2, and covariance σ_{XY}. The best nonlinear estimate of Y in terms of

X is desired where best is in the sense of minimizing the mean squared error. The solution for $g(x)$ has already been shown to be the conditional mean

$$g(x) = E[Y|X = x] = \int_{-\infty}^{\infty} y f_Y(y|x)\, dy \tag{3.48}$$

From Section 2.6.4, the conditional density for jointly Gaussian random variables is known to be Gaussian with mean and variance given by

$$\begin{aligned} \eta_{Y|x} &= \eta_Y + \frac{\rho_{XY}\sigma_Y(x - \eta_X)}{\sigma_X} \\ \sigma^2_{Y|x} &= \sigma^2_Y(1 - \rho^2_{XY}) \end{aligned} \tag{3.49}$$

Thus the optimal $g(x)$ for this Gaussian case is the conditional mean

$$g(x) = \eta_X + \frac{\rho_{XY}\sigma_Y}{\sigma_X}(x - \eta_X) \tag{3.50}$$

By the definition for the correlation coefficient, Eq. (2.101), where ρ_{XY} is

$$\rho_{XY} = \frac{\sigma_{XY}}{\sigma_X \sigma_Y} \tag{3.51}$$

the optimal $g(x)$ can be rewritten in terms of the various variances and covariances as

$$\begin{aligned} g(x) &= \eta_X + \frac{\sigma_{XY}}{\sigma^2_X}(x - \eta_X) \\ &= \frac{\sigma_{XY}}{\sigma^2_X} x + \eta_x\left(1 - \frac{\sigma_{XY}}{\sigma^2_X}\right) \end{aligned} \tag{3.52}$$

Thus the **best nonlinear estimate** for the jointly **Gaussian** case is seen to be a **linear estimate** given by

$$\begin{aligned} &\hat{Y} = g(X) = aX + b \\ \text{where} \quad &a = \frac{\sigma_{XY}}{\sigma^2_X}, \quad b = \eta_x\left(1 - \frac{\sigma_{XY}}{\sigma^2_X}\right) \end{aligned} \tag{3.53}$$

This fact gives more power to the linear estimate for the case of Gaussian random variables and releases us from needing to look for a best nonlinear estimate for that case. However, it is not possible to say that if the best nonlinear estimate is a linear estimate, the random variables involved are jointly Gaussian. A counterexample has already been shown in Example 3.7.

For the special case of zero mean Gaussian random variables, the a and b from Eq. (3.53) reduce to

$$a = \frac{\sigma_{XY}}{\sigma^2_X} \quad \text{and} \quad b = 0 \tag{3.54}$$

The extension of the result above to $n + 1$ jointly random variables is possible and the best nonlinear estimate of the Gaussian random variable X_0 in terms of the other jointly Gaussian random variables $X_1, X_2, \ldots, X_n$ can be shown to be a linear function of the random variables $X_1, X_2, \ldots, X_n$. This is expressed as follows:

$$\begin{aligned} g(x_1, x_2, \ldots, x_n) &= E[X_0|X_1 = x_1, X_2 = x_2, \ldots, X_n = x_n] \triangleq \eta_{X_0|x_1,x_2,\ldots,x_n} \\ &= \int_{-\infty}^{\infty}\int_{-\infty}^{\infty}\cdots\int_{-\infty}^{\infty} x_0 f(x_1, x_2, \ldots, x_n|x_0)\, dx_0 \\ &= a_1x_1 + a_2x_2 + \cdots + a_nx_n + b \end{aligned} \tag{3.55}$$

where the $\mathbf{a} = [a_1, a_2, \ldots, a_n]^T$ and b can be determined from (3.21) and (3.22). If the random vector $\mathbf{X}_{n+1} = [X_0, X_1, \ldots, X_n]^T = [X_0 \vdots \mathbf{X}_n]^T$ is jointly Gaussian with given covariance matrix $\mathbf{K}_{n+1}$ and mean vector $\mathbf{m}_{n+1} = [E[X_0], E[X_1], \ldots, E[X_n]]^T = [E[X_0] \vdots \mathbf{m}_n]^T$ then the best nonlinear estimate of X_0 in terms of $X_1, X_2, \ldots, X_n$ can be written as

$$\hat{X}_0 = [a_1, a_2, \ldots, a_n]\begin{bmatrix} X_1 \\ X_2 \\ \vdots \\ X_n \end{bmatrix} = \mathbf{a}\mathbf{X}_n + b \tag{3.56}$$

To obtain the **a** vector and b in terms of the given information, the covariance matrix $\mathbf{K}_{n+1}$ is first partitioned as

$$\mathbf{K}_{n+1} = \begin{bmatrix} \sigma_{X_0}^2 & \mu_{X_0}^T \\ \mu_{X_0} & \mathbf{K}_n \end{bmatrix} \tag{3.57}$$

$$\text{where} \quad \boldsymbol{\mu}_{X_0}^T = [\mu_{X_0X_1}, \mu_{X_0X_2}, \ldots, \mu_{X_0X_n}]$$

The desired **a** and b can now be obtained by using the linear estimate formulas given in (3.21) and (3.22) as follows

$$\boldsymbol{\mu}_{X_0}^T = \mathbf{K}_n\mathbf{a}, \quad b = E[X_0] - \mathbf{a}^T\mathbf{m}_n \tag{3.58}$$

where the solution for the coefficient vector **a** is easily obtained from the equation above as

$$\mathbf{a} = \mathbf{K}_n^{-1}\boldsymbol{\mu}_{X_0}^T \tag{3.59}$$

3.4 PROPERTIES OF ESTIMATORS OF RANDOM VARIABLES

There are a number of properties of estimators of random variables, whether they be linear or nonlinear functions of the observations, that are useful to define. Assume that X is a random variable that is statistically related to another random variable Z. Let $\hat{X}(Z) = g(Z)$ be an estimator for X. Assume that mean $E[X]$ is known or can be determined.

An estimator of a random variable X is **unbiased** if its expected value is equal to the expected value of the random variable being estimated, which can be written as

$$E[\hat{X}(Z)] = E[X] \tag{3.60}$$

If (3.60) is not satisfied, the estimator is **biased**.

Suppose that we are estimating a random variable X from a vector of observations $\mathbf{Z} = [Z_1, Z_2, \ldots, Z_N]^T$. The **Cramer-Rao bound**, from Van Trees [5], is the most important lower bound on the variance for estimating random variable X and is given below without proof:

$$E[(X - \hat{X}(\mathbf{Z}))^2] \geq \frac{1}{E\left[\left(\dfrac{\partial \ln p(\mathbf{z}|x)}{\partial x}\right)^2\right]} = \frac{1}{-E\left[\dfrac{\partial^2 \ln p(\mathbf{z}|x)}{\partial^2 x}\right]} \tag{3.61}$$

Any estimator satisfying this bound is called an **efficient estimator**.

3.5 BAYES ESTIMATION

The **performance measure** J_{BAYES}, or "yardstick" by which we judge the **optimality** for a Bayes estimate, is the expected value of a **cost function** $C(\cdot)$ that is a function of the error or difference between the true value and the estimated value.

$$\begin{aligned} J_{\text{BAYES}} &\triangleq E[C(\varepsilon)] \quad \text{where} \\ \varepsilon &= X - \hat{X} \end{aligned} \tag{3.62}$$

The **Bayes estimator** of X in terms of Z is given by $\hat{X} = g(Z)$ where $g(Z)$ is selected such that the J_{BAYES}, given above, is minimized. Common choices for $C(\varepsilon)$, shown in Figure 3.1, are the squared error, absolute error, and the uniform error, which are defined as

$$\begin{aligned} C_{\text{MS}}(\varepsilon) &= \varepsilon^2 \\ C_{\text{ABS}}(\varepsilon) &= |\varepsilon| \\ C_{\text{UN}}(\varepsilon) &= \begin{cases} 0, & |\varepsilon| \leq \delta \\ 1, & |\varepsilon| > \delta \end{cases} \end{aligned} \tag{3.63}$$

The choice of which cost function to use is subjective. It is seen that the $C_{\text{MS}}(\varepsilon)$ penalizes large errors severely and rewards errors less than 1, whereas the $C_{\text{ABS}}(\varepsilon)$ treats errors on a linear basis. $C_{\text{ABS}}(\varepsilon)$ does not penalize if the absolute value of the error is within the bound of δ but penalizes all errors bigger than δ in magnitude by the same value. Bayes estimates for each of these three cost functions are now presented.

3.5.1 Bayes Estimates

Squared Error Cost. Bayes estimates for random variables X in terms of a measured random variable Z for the squared error cost function have been thoroughly explored in the previous two sections. It was shown that the $g(z)$ for the optimal nonlinear estimate of X was obtained as the conditional mean of X given $Z = z$ and could be written as

$$\begin{aligned} g(z) &= E[X|Z = z] = \int_{-\infty}^{\infty} xp(x|z)\,dx \\ \hat{X} &= g(Z) \end{aligned} \tag{3.64}$$

Thus the Bayes estimate for the squared error cost is seen to be the **mean of the a posteriori density** $p(x|z)$ given by

$$\hat{x}_{MS} = \int_{-\infty}^{\infty} xp(x|z)\,dx \tag{3.65}$$

Unfortunately, what we usually have available, or can easily obtain, is $p(z|x)$. Thus $p(x|z)$ must be found by using Bayes's rule as follows:

$$p(x|z) = \frac{p(z|x)p(x)}{\displaystyle\int_{-\infty}^{\infty} p(z|x)p(x)\,dx} \tag{3.66}$$

Absolute Error Cost. To obtain the optimal estimate of X in terms of Z for the absolute error cost, we must select $g(z)$ such that J_{ABS} given below is minimized:

$$\begin{aligned} J_{ABS} &\triangleq E[C_{ABS}(\varepsilon)] = E[|X - \hat{X}|] = E[|X - g(Z)|] \\ &= \int_{-\infty}^{\infty}\int_{-\infty}^{\infty} |x - g(z)|\,p(x, z)\,dx\,dz \end{aligned} \tag{3.67}$$

Equation (3.67) can be rearranged by using the relationship for the joint and conditional densities as follows:

$$J_{ABS} = \int_{-\infty}^{\infty} p(z)\left[\int_{-\infty}^{\infty} |x - g(z)|\,p(x|z)\,dx\right]dz \tag{3.68}$$

To minimize J_{ABS}, it is sufficient to minimize the term in brackets for every value of z. This is true since $p(z)$ is always positive as is the term in brackets, I, which can be written as

$$I = \int_{-\infty}^{g(z)} (-x + g(z))p(x|z)\,dx + \int_{g(z)}^{\infty} (x - g(z))p(x|z)\,dx \tag{3.69}$$

For a fixed $z, g(z)$ is a constant setting the partial derivitive of the integral with respect to $g(z)$ equal to zero, and using Liebnitz's rule, we obtain a necessary condition for minimizing I as follows:

$$\begin{aligned}\frac{\partial I}{\partial g(z)} &= \int_{-\infty}^{g(z)} (-x+g(z))p(x|z)\,dx + \int_{g(z)}^{\infty} (x-g(z))p(x|z)\,dx \\ &= (-g(z)+g(z))p(g(z)|z) + \int_{-\infty}^{g(z)} p(x|z)\,dx \\ &\quad - (g(z)-g(z))p(g(z)|z) - \int_{g(z)}^{-\infty} p(x|z)\,dx = 0\end{aligned} \tag{3.70}$$

Recognizing the zero terms and rearranging the $g(z)$ for each z satisfies the equation

$$\int_{-\infty}^{g(z)} p(x|z)\,dx = \int_{g(z)}^{-\infty} p(x|z)\,dx \tag{3.71}$$

In this expression it is seen that the optimum $g(z)$, for each z is the **median of the a posteriori density** $p(x|z)$. Thus the optimum estimator in the minimum absolute error sense is seen to be the $\hat{x}_{\mathrm{ABS}}(z)$, which satisfies

$$\int_{-\infty}^{\hat{x}_{\mathrm{ABS}}(z)} p(x|z)\,dx = \int_{\hat{x}_{\mathrm{ABS}}(z)}^{-\infty} p(x|z)\,dx \tag{3.72}$$

As with the MS estimate, what we usually have available or can easily obtain is $p(z|x)$. Thus $p(x|z)$ must be found by using Bayes's rule as follows:

$$p(x|z) = \frac{p(z|x)p(x)}{\int_{-\infty}^{\infty} p(z|x)p(x)\,dx} \tag{3.73}$$

Uniform Cost. To obtain the optimal estimator of X in terms of Z for the uniform error cost function, the $g(z)$ is selected such that the J_{UN} given below is minimized:

$$\begin{aligned}J_{\mathrm{UN}} &= \int_{-\infty}^{\infty}\int_{-\infty}^{\infty} C_{\mathrm{UN}}(x-g(z))p(x,z)\,dx\,dz \\ &= \int_{-\infty}^{\infty} p(z)\left[\int_{-\infty}^{\infty} C_{\mathrm{UN}}(x-g(z))p(x|z)\,dx\right]dz\end{aligned} \tag{3.74}$$

The optimal estimator is as before, where the $g(z)$ minimizes the term in brackets given by

$$I(z) = \int_{-\infty}^{\infty} C_{\mathrm{UN}}(x-g(z))p(x|z)\,dx = 1 - \int_{|x-g(z)|<\delta} p(x|z)\,dx \tag{3.75}$$

Minimizing $I(z)$ for each z is equivalent to maximizing the integral on the right side of Eq. (3.75). If δ is very small, then that integral is maximized by selecting $g(z)$ to be the maximum of $p(x|z)$ viewed as a function of x for each z. So the estimator in the uniform cost sense is written as

$$\hat{x}_{\text{UN}}(z) = \max_{x} p(x|z) \triangleq \hat{x}_{\text{MAP}}(z) \tag{3.76}$$

and it is the **maximum of the a posteriori density** $p(x|z)$.

To obtain the estimates, it is necessary to determine the a posteriori density $p(x|z)$, but what is usually available is the $p(z|x)$. The a posteriori density can be written in terms of $p(z|x)$ as

$$p(x|z) = \frac{p(z|x)p(x)}{p(z)} \tag{3.77}$$

For each z the $p(z)$ is nothing more than a scale factor, so maximizing $p(x|z)$ is equivalent to **maximizing the numerator** $p(z|x)p(x)$

A necessary condition for the maximum a posteriori estimate is that it satisfy

$$\left.\frac{\partial(p(z|x)p(x))}{\partial x}\right|_{x=\hat{x}(z)} = 0 \tag{3.78}$$

In many problems where the conditional and marginal densities are exponential in form, a more expedient necessary condition can be obtained by taking the ln of the product of densites before the partial derivative is performed. This will not change the location of the maximum value, and thus it leads to the alternative necessary condition for the MAP estimate of X in terms of Z:

$$\left.\frac{\partial(\ln p(z|x))}{\partial x}\right|_{x=\hat{x}_{\text{MAP}}(z)} + \left.\frac{\partial(\ln p(x))}{\partial x}\right|_{x=\hat{x}_{\text{MAP}}(z)} = 0 \tag{3.79}$$

3.5.2 Examples of Bayes Estimators

It has been shown that Bayes estimators for the squared error, absolute error, and uniform error are the mean, median, and max or mode of the a posteriori density $p(x|z)$. In general, all of these estimators are different. This is illustrated in Figure 3.4 for a Raleigh a posteriori density.

A number of examples are now presented that illustrate the procedure for determining Bayes estimators.

EXAMPLE 3.11

We are given that $Z = \tanh(X) + N$ and that $X \sim N(0, \sigma_X^2)$ and $N \sim N(0, \sigma_N^2)$. Let X and N be independent random variables, and find the Bayes estimate of X from Z for the uniform cost function.

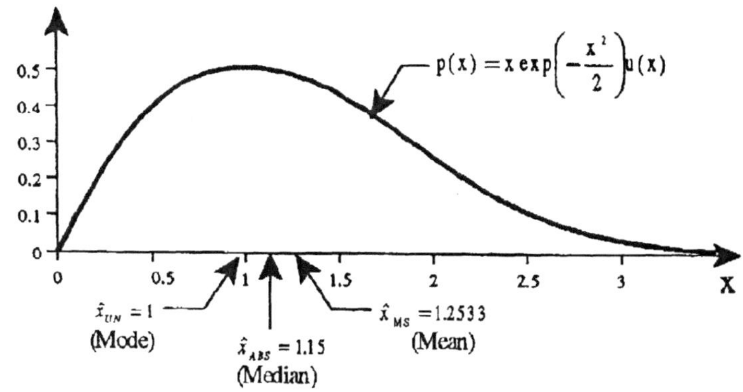

Figure 3.4 Bayes estimates for Uniform, Absolute, and squared error cost functions for a Raleigh a posteriori density $\alpha^2 = 1$.

SOLUTION

The uniform cost Bayes estimate is the maximum of the product of $p(z|x)$ and $p(x)$. The conditional density $p(z|x)$ is seen for the given function with x fixed as the translated version of the probability density for N which gives $N(\tanh(x), \sigma_N^2)$. Using this density and the given density for X, we obtain the desired product as

$$p(z|x)p(x) = \frac{1}{2\pi\sigma_N\sigma_X}\exp\left\{-\frac{(z-\tanh(x))^2}{2\sigma_N^2} - \frac{x^2}{\sigma_X^2}\right\}$$

We treat this as a function of x for a given z. Then we obtain the estimator form by finding the maximum with respect to x. Taking the derivative with respect to x gives us the following necessary condition for the maximization:

$$\frac{dp(z|x)p(x)}{dx} = \frac{1}{2\pi\sigma_N\sigma_X}\exp\left\{-\frac{(z-\tanh(x))^2}{2\sigma_N^2} - \frac{x^2}{2\sigma_X^2}\right\}\frac{d}{dx}\left\{-\frac{(z-\tanh(x))^2}{2\sigma_N^2} - \frac{x^2}{2\sigma_X^2}\right\}$$

Setting this equal to zero is equivalent to setting the derivative term equal to zero, since the exp term is nonzero and the standard deviations and π are constants. Thus the uniform cost estimator is found as the solution of

$$\frac{d}{dx}\left\{-\frac{(z-\tanh(x))^2}{2\sigma_N^2} - \frac{x^2}{\sigma_X^2}\right\} = 0$$

Taking the derivative with respect to x and rearranging, we simplify the equation above to

$$\frac{z\sec h^2(x)}{\sigma_N^2} - \frac{\sec h^2(x)\tanh(x)}{\sigma_N^2} - \frac{x}{\sigma_N^2} = 0$$

Factoring out the $\sec h^2(x)$ and discarding it, since it is always positive and cannot be zero, we are left with the equation

$$\frac{z}{\sigma_N^2} = \frac{\tanh(x)}{\sigma_N^2} - \frac{x}{\sigma_X^2 \sec h^2(x)} = 0$$

Rearranging this equation, we see that the solution for a given value of z is the Bayes estimate for that z and given by

$$z = \frac{\tanh(\hat{x}_{\mathrm{UN}})}{\sigma_N^2} + \frac{\hat{x}_{\mathrm{UN}}}{\sigma_X^2(1 - \tanh^2(\hat{x}_{\mathrm{UN}}))}$$

As an analytical solution for $\hat{x}_{\mathrm{UN}}$ in terms of the variable z was not available, a numerical solution could be obtained. For example, if we let $\sigma_X^2 = \sigma_N^2 = 1$ and vary z from -5.9 to 5.9, the graphical form $g(z)$ of the estimator is shown in Figure 3.5. Thus our estimator is $\hat{X}_{\mathrm{UN}}(Z) = g(Z)$. ■

EXAMPLE 3.12

A and X are two random variables. The conditional probability density function for X and the probability density function for A are given as

$$f_X(x|a) = ae^{-ax}\mu(x) \quad \text{and} \quad f_A(a) = \tfrac{1}{2}[\mu(a-1) - \mu(a-3)]$$

(a) Find the maximum a posteriori estimator for A.
(b) Compute $E[\hat{A}_{\mathrm{MAP}}(X)]$, and say whether the estimator is biased or unbiased.
(c) Compute the variance for your estimator.

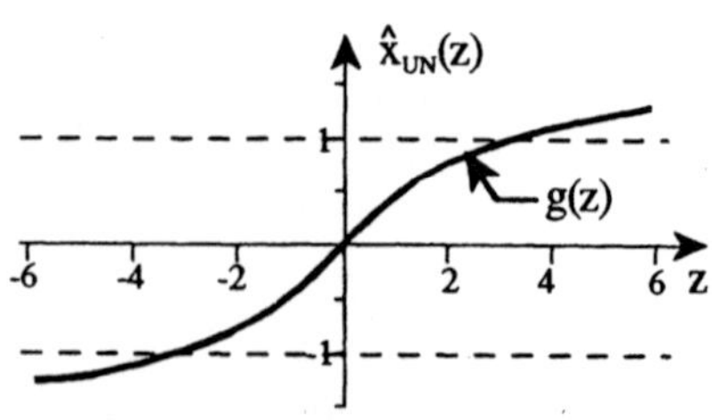

Figure 3.5 Graphical form for Bayes estimator for Example 3-10 with $\sigma_X^2 = \sigma_{N^2} = 1$.

SOLUTION

(a) To obtain the estimator, it is necessary to find the a posteriori density $f_A(a|x)$, which can be determined as

$$f(a|x) = \frac{f_X(x|a)\,f_A(a)}{f_X(x)}$$
$$= \frac{ae^{-ax}\mu(x)[\mu(a-1) - \mu(a-3)]}{f_X(x)}$$

Since the $f_X(x)$ is just a scale factor and does not depend on a, we need to select the a that maximizes the numerator with respect to a. Figure 3.6 shows for three separate regions for x the plot of $f(a|x)$ as a function of a. The maximum is located at different values of a depending on whether the maximum of the ae^{-xa}, which occurs at $1/x$, is in the interval [1, 3] or on the boundary.

From Figure 3.6 it is seen that if $1/x < 1$, the maximum occurs at the boundary, $a = 1$; if $1 < 1/x < 3$, the maximum occurs at $a = 1/x$; and if $1/x > 3$, then the maximum occurs at the boundary, $a = 3$. Thus we can write the maximum a posteriori estimate as follows:

$$\hat{A}_{\text{MAP}}(x) = \begin{cases} 1, & 0 < 1/x < 1 \\ 1/x, & 1 < 1/x < 3 \\ 3, & 1/x > 3 \end{cases}$$

Since $x > 0$, the inequalities can be reversed and rearranging gives the $\hat{A}_{\text{MAP}}(x)$ as

$$\hat{A}_{\text{MAP}}(x) = \begin{cases} 3, & 0 < x < \frac{1}{3} \\ 1/x, & \frac{1}{3} < x < 1 \\ 1, & x > 1 \end{cases}$$

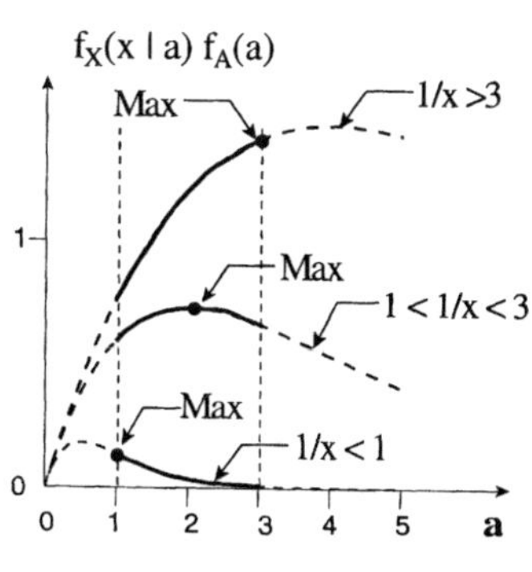

Figure 3.6 Plot of $f_X(x \mid a)f_A(a)$ versus a for Example 3-11.

(b) The expected value of the estimator can be written as

$$\begin{aligned} E[\hat{A}_{\mathrm{MAP}}(X)] &= \int_{-\infty}^{\infty} \hat{A}_{\mathrm{MAP}}(x)\, f_X(x)\, dx \\ &= \int_0^{1/3} 3 f_X(x)\, dx + \int_{1/3}^{1} (1/x)\, f_X(x)\, dx + \int_1^{\infty} 1 f_X(x)\, dx \end{aligned}$$

where the $f_X(x)$ is obtained by integrating the product of the conditional density by the marginal as

$$\begin{aligned} f_X(x) &= \int_{-\infty}^{\infty} f_X(x|a) f_A(a)\, da \\ &= \int_1^3 a e^{-ax} \mu(x) \frac{1}{2}\, da = \frac{-e^{-ax}}{2x}\left(a + \frac{1}{x}\right)\Bigg]_1^3 \\ &= \frac{1}{2}\left(-\frac{3}{x} e^{-3x} - \frac{1}{x^2} e^{-3x} + \frac{1}{x} e^{-x} - \frac{1}{x^2} e^{-x}\right)\mu(x) \end{aligned}$$

Using this $f_X(x)$ in the expected value above, and some creative integration, we obtain the expected value of our estimator as $E[\hat{A}_{\mathrm{MAP}}(X)] = 1.635584$. Since $E[A]$ is easily seen to be 2, this estimator is biased.

(c) The variance of our estimator is easier written down than evaluated as

$$\begin{aligned} \sigma^2_{\hat{A}_{\mathrm{MAP}}(X)} &= E[(A - \hat{A}_{\mathrm{MAP}}(X))^2 \\ &= \int_1^3 \int_0^{1/3} (a-3)^2 f_X(x|a) f_A(a)\, dx\, da \\ &\quad + \int_1^3 \int_{1/3}^{1} (a - 1/x)^2 f_X(x|a) f_A(a)\, dx\, da \\ &\quad + \int_1^3 \int_1^{\infty} (a-1)^2 f_X(x|a) f_A(a)\, dx\, da \end{aligned}$$

where $f_X(x|a)$ and $f_A(a)$ are as given in the problem assignment. ■

3.6 ESTIMATION OF NONRANDOM PARAMETERS

In the previous sections we were concerned with estimating one or more random variables in terms of other random variables. However, there are many problems where the variables we wish to estimate are not random variables but deterministic values or parameters. The techniques of the previous section cannot be applied, since there would be no probability density function for the variable to be estimated. The question then becomes how can we determine reliable estimates for nonrandom variables or parameters, and what do we mean by reliability?

In this section the maximum likelihood procedure will be presented and used to determine such estimates and establish a measure of performance. This measure of performance will be described in terms of the two qualities of mean and variance of the estimates. Suppose we are trying to estimate the parameter a as a function, $g(Z)$, of some observation Z which we can express as $\hat{a}(Z) = g(Z)$. The Z is random and somehow related to a. The $\hat{a}(Z)$ is a random variable because it is a function of a random variable Z. An example of its probability density function is shown in Figure 3.7. Intuitively it seems that for $\hat{a}(Z)$ to be a good estimate of the parameter a, the density for $\hat{a}(Z)$ should have a small spread or variance and be centered around the variable being estimated. Such a property can be expressed in terms of the $E[\hat{a}(Z)]$ and the variance $\sigma_{\hat{a}}^2 = E[\hat{a}(Z) - E[\hat{a}(Z)]]^2$.

An estimator of a **nonrandom variable** a is said to be an **unbiased estimator** of a if $E[\hat{a}(Z)] = a$. Therefore we usually seek estimates that are unbiased and have the smallest possible variance. Say that we are estimating a nonrandom parameter a from a vector of random observations **Z** and that the variance of our estimator is defined by $E[(a - \hat{a}(Z)^2]$. For even simple problems this variance may be complicated to obtain, and so bounds on estimates are important. The **Cramer-Rao bound** from Van Trees [5] is the most important lower bound on the variance. It is given by

$$E[(a - \hat{a}(Z))^2] \geq \frac{1}{E^2[\partial \ln p(z;\, a)/\partial a]} \tag{3.80}$$

Any estimator whose variance satisfies the lower bound given in (3.80) is called an **efficient estimator** of a. The maximum likelihood procedure to follow describes an estimator but says nothing about whether the estimator is unbiased or has a small variance. Nevertheless, if the maximum likelihood estimator is unbiased it will have the smallest variance of all estimators, which affirms our using the maximum likelihood procedure and estimators.

3.6.1 Maximum Likelihood Estimation

A random variable Z is observed that is a function of some unknown but nonrandom parameter a. For example, a constant unknown variable a is observed in additive fashion as $Z = a + X$, where the random variable X has a known statistical characterization and an estimate of a is desired as a function of Z. The maximum likelihood estimation procedure suggests one way to solve for an estimate but not in the sense of an optimality.

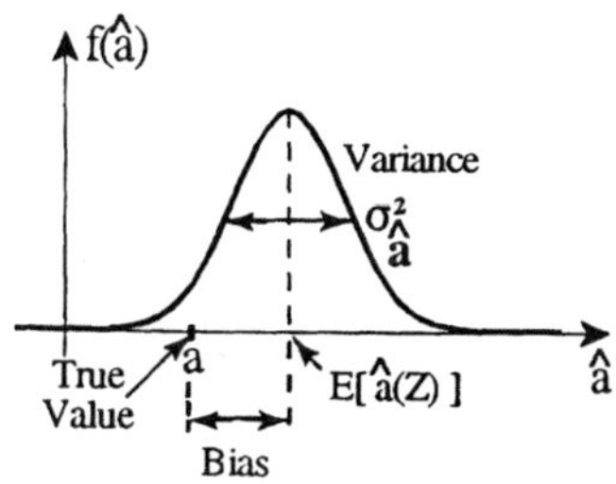

Figure 3.7 Probability density function for $\hat{a}(Z)$.

The **maximum likelihood estimator** $\hat{a}_{\text{ml}}(Z)$ of a nonrandom parameter a in terms of an observed random variable Z is defined to be the a that maximizes the **likelihood function** $f(z; a)$, which is defined as the probability density function for the observation Z in terms of a nonrandom parameter a that could be a vector. Mathematically the functional form for the estimator is expressed in terms of $f(z; a)$, which is viewed as a function of a and can be written as

$$\hat{a}_{\text{ml}}(z) = \max_{a} f(z; a) \tag{3.81}$$

Extending this concept to multiple observations, the maximum likelihood estimate of a nonrandom parameter in terms of observations $Z_1, Z_2, \ldots, Z_n$ is defined as the a that maximizes the likelihood function $f(z_1, z_2, \ldots, z_n; a)$ and given by

$$\hat{a}_{\text{ml}}(z_1, z_2, \ldots, z_n) = \max_{a} f(z_1, z_2, \ldots, z_n; a) \tag{3.82}$$

Furthermore, if we have multiple observations and are estimating a parameter vector, $\mathbf{a} = [a_1, a_2, \ldots, a_M]$, rather than a single variable or parameter, the maximum likelihood estimate of $\mathbf{a}$ is given by

$$\hat{\mathbf{a}}_{ml}(z_1, z_2, \ldots, z_n) = \max_{\mathbf{a}} f(z_1, z_2, \ldots, z_n; \mathbf{a}) \tag{3.83}$$

3.6.2 Maximum Likelihood Estimation Examples

Several examples are now presented to illustrate the procedure involved in obtaining maximum likelihood estimators. Their performance will be expressed in terms of both the mean and variance for the estimates.

EXAMPLE 3.13

The probability density function for an observation Z_i in terms of a nonrandom parameter a is given by

$$f(z_i; a) = ae^{-az_i}\mu(z_i)$$

Say there are n independent observations $Z_1, Z_2, \ldots, Z_n$.

(a) Find the maximum likelihood estimate $\hat{a}_{\text{ml}}$ of a.
(b) Determine the $E[\hat{a}_{\text{ml}}]$.

SOLUTION

(a) The likelihood function is a product of the likelihood for each measurement because of the independence of observations. It is determined as follows:

$$
\begin{aligned}
f(z_1, z_2, \ldots, z_n; a) &= \prod_{i=1}^{n} f(z_i; a) \\
&= a^n \exp\left(-a \sum_{i=1}^{n} z_i\right) \mu(z_1)\mu(z_2)\cdots\mu(z_n)
\end{aligned}
$$

A necessary condition for the maximum of the likelihood function is that its partial derivative with respect to a be set equal to zero, which is

$$
\begin{aligned}
\frac{\partial f(z_1, z_2, \ldots, z_n; a)}{\partial a} &= \frac{\partial}{\partial a}\left(a^n \exp\left(-a \sum_{i=1}^{n} z_i\right)\mu(z_1)\mu(z_2)\cdots\mu(z_n)\right) \\
&= n(a^{n-1}) \exp\left(-a \sum_{i=1}^{n} z_i\right) - \left(a^n \exp\left(-a \sum_{i=1}^{n} z_i\right)\right) \sum_{i=1}^{n} z_i \\
&= a^{n-1} \exp\left(-a \sum_{i=1}^{n} z_i\right)\left[n - a \sum_{i=1}^{n} z_i\right] = 0
\end{aligned}
$$

Thus we have relative extremum at $a = 0$ due to the a^{n-1} term and at $a = n/\sum_{i=1}^{n} Z_i$ because of the term in square brackets. But $a = 0$ is a minimum rather than a maximum, and a maximum occurs for the bracketed term. Therefore the maximum likelihood estimator becomes

$$
\hat{a}_{\mathrm{ML}}(Z_1, Z_2, \ldots, Z_n) = n \bigg/ \sum_{i=1}^{n} Z_i
$$

(b) The expected value of $\hat{a}_{\mathrm{ML}}(Z_1, Z_2, \ldots, Z_n)$ is given by:

$$
E[\hat{a}_{\mathrm{ML}}(Z_1, Z_2, \ldots, Z_n)] = E\left[n \bigg/ \sum_{i=1}^{n} Z_i\right] = nE\left[1 \bigg/ \sum_{i=1}^{n} Z_i\right]
$$

The expected value of the term in brackets is *not* the ratio of expected values. Its calculation is facilitated by defining a new random variable Z as

$$
Z = 1 \bigg/ \sum_{i=1}^{n} Z_i
$$

The general approach to determining the required expected value is first to find $f_Z(z)$ by using the Laplace transform (since the density will be a convolution), then to use the transformation theorem to obtain $f_W(w)$, where $W = 1/Z$, and finally to compute the expected value of W. It is left to the reader to show that

$$
E[\hat{a}_{\mathrm{ML}}(Z_1, Z_2, \ldots, Z_n)] = a
$$

The estimator is unbiased because its expected value is the unknown parameter or variable being estimated. ■

EXAMPLE 3.14

Let a_1 and a_2 be unknown nonrandom parameters observed in terms of

$$Z_1 = a_1 + a_2 + N_1$$
$$Z_2 = a_1 - a_2 + N_2$$

where N_1 and N_2 are independent Gaussian random variables with zero mean and equal variances of σ^2. Find the maximum likelihood estimates of a_1 and a_2 in terms of Z_1 and Z_2, and determine the expected value of your estimates.

SOLUTION

The likelihood function $f(z_1, z_2; a_1, a_2)$ is given as

$$f(z_1, z_2; a_1, a_2) = \frac{1}{2\pi\sigma^2}\exp\left\{\frac{-\frac{1}{2}[z_1 - (a_1 + a_2)]^2}{\sigma^2} - \frac{\frac{1}{2}[z_2 - (a_1 - a_2)]^2}{\sigma^2}\right\}$$

Since the $f(z_1, z_2; a_1, a_2)$ is exponential, it is convenient to first take the ln before performing the maximization. The location of the maximum will not be changed if any monotonic function is used. This leads to two necessary conditions for the extremum:

$$\frac{\partial}{\partial a_1}\ln f(z_1, z_2; a_1, a_2) = \frac{\partial}{\partial a_1}\left\{\frac{-\frac{1}{2}[z_1 - (a_1 + a_2)]^2}{\sigma^2} - \frac{\frac{1}{2}[z_2 - (a_1 - a_2)]^2}{\sigma^2}\right\}$$
$$= [z_2 - (a_1 + a_2)] + [z_2 - (a_1 - a_2)] = 0$$

$$\frac{\partial}{\partial a_2}\ln f(z_1, z_2; a_1, a_2) = \frac{\partial}{\partial a_1}\left\{\frac{-\frac{1}{2}[z_1 - (a_1 + a_2)]^2}{\sigma^2} - \frac{\frac{1}{2}[z_2 - (a_1 - a_2)]^2}{\sigma^2}\right\}$$
$$= [z_2 - (a_1 + a_2)] - [z_2 - (a_1 - a_2)] = 0$$

Simplifying the equations above gives simultaneous equations as

$$z_1 + z_2 - 2a_1 = 0$$
$$-z_1 + z_2 + 2a_2 = 0$$

These equations are easily solved to get

$$\hat{a}_1(z_1, z_2) = \tfrac{1}{2}(z_1 + z_2), \quad \hat{a}_2(z_1, z_2) = \tfrac{1}{2}(z_1 - z_2)$$

Since $E[N_1] = E[N_2\} = 0$, the expected values of our estimates can be determined as

$$
\begin{aligned}
E[\hat{a}_1(Z_1, Z_2)] &= E[\tfrac{1}{2}(Z_1 + Z_2)] \\
&= E[a_1 + N_1 + N_2] = a_1 \\
E[\hat{a}_2(Z_1, Z_2)] &= E[\tfrac{1}{2}(Z_1 - Z_2)] \\
&= E[a_2 + N_1 - N_2] = a_2
\end{aligned}
$$

Thus we have shown that the estimates are unbiased for this problem, since there expected values equal the parameters to be estimated. The variances of our estimates for this problem are easily found as

$$
\begin{aligned}
\sigma^2_{\hat{a}_1} &= E[(\hat{a}_1(Z_1, Z_2) - a_1)^2] = E[(\tfrac{1}{2}(Z_1 + Z_2) - a_1)^2] \\
&= E[\tfrac{1}{2}(N_1 + N_2)^2] = \sigma^2 \\
\sigma^2_{\hat{a}_2} &= E[(\hat{a}_2(Z_1, Z_2) - a_2)^2] = E[(\tfrac{1}{2}(Z_1 - Z_2) - a_2)^2] \\
&= E[\tfrac{1}{2}(N_1 - N_2)^2] = \sigma^2
\end{aligned}
$$

■

3.7 SUMMARY

In this chapter basic procedures were presented for optimally estimating random variables from other random variables where optimality was with respect to a given performance measure. Emphasis was placed on the minimum mean squared error performance measure, not only because of its tractability but because many practical problems do not have a full characterization of the involved random variables.

It was shown that for minimum mean squared error estimation of a random variable by a linear combination of other random variables, the only characterizations of the random variables required are the means, aucorrelations, and cross-correlations. The orthogonality principle provided another interpretation and the equations necessary to obtain optimum estimates in the MMSE sense.

For nonlinear MMSE estimation, it was shown that the mean of the conditional density was the needed function, which involved almost a total characterization of the random variables. For the special case of Gaussian random variables, it was shown that the optimal linear estimate is the same as the optimum nonlinear estimate when finding MMSE estimates. This result was shown by Meditch [7] to be true as well for a wider class of performance measures and a wider class of probability densities.

Examples were presented using the Bayes performance measure for special cases of squared error, absolute error, and uniform error cost functions. It was shown that these estimates are correspondingly the mean, median, and mode of the involved a posteriori probability density, which involves almost a total characterization.

The topic of estimating nonrandom variables was introduced as there are many problems in which the estimate parameters or variables to be estimated are not random. The one method presented was the maximum likelihood procedure, and several examples were given. The concept of variance and bias in describing the performance of estimators of nonrandom parameters or variables was explored, with the result being that we would like our estimates to have a small variance and be unbiased.

The concepts discussed in this chapter are basic background information for the estimation of random variables, which has many applications and is an important area in many disciplines. The concepts developed in this chapter will also enable the reader to better understand the optimum filters of Chapter 6, the Kalman filter of Chapter 7, and some related issues in Chapter 8 and 9 involving detection theory.

A rigorous and complete discussion of estimation theory cannot be accomplished in one chapter. The intent was to provide enough basics that would smooth the transition for further advanced topics and make subsequent chapters easier to understand. For the reader desiring more detailed discussions, a few good starting places are the excellent texts on estimation theory by Nahi [6], Meditch [7], Mendel [4], Poor [13], Van Trees [5], and Sharf [11].

PROBLEMS

3.1 The random variable S has the following conditional density function

$$f_S(s|\, P = p) = pe^{-ps}\mu(s) \quad \text{for } 0 \le p \le 1$$

P is uniformly distributed random variable on [0, 1]. Using the minimum mean squared error as a performance measure, find:

(a) The best linear estimate of S in terms of P.

(b) The best nonlinear estimate of S in terms of P.

3.2 Given random variables X and Y with joint probability density function $f_{xy}(x, y)$, where $E[Y] = E[X] = 0$. We desire the best quadratic estimate of Y in terms of X, where best is in the sense of minimizing the mean squared error.

(a) Find a, b, and c such that the mean squared error e given by

$$e = E[(Y - (a + bX + cX^2))^2] \quad \text{is minimized.}$$

(b) Repeat (a) if X and Y are jointly Gaussian with zero means and given variances, and covariances.

3.3 Given

$$f(x, y) = \begin{cases} 2, & 0 < x < y, 0 < y < 1 \\ 0, & \text{elsewhere} \end{cases}$$

(a) Find the best linear estimate of Y in terms of X.

(b) Find the best nonlinear estimate of Y in terms of X.

(c) Comment on your results.

3.4 Using the $f_{XY}(x, y)$ of Problem 2.36, find the best nonlinear estimate of Y in terms of X where best is in the sense of minimizing the mean squared error. Also evaluate the resulting minimum mean squared error.

3.5 The random variables X and Y are characterized by their conditional and marginal density as

$$f(y|x) = \frac{x}{\pi(x^2 + y^2)}, \quad f_X(x) = \exp(-x)\mu(x)$$

(a) Find the best nonlinear estimate of Y in terms of X.

(b) Find the best linear estimate of Y in terms of X.

3.6 Random variables X and Y are statistically characterized by the following joint probability density function:

$$f(y|x) = \frac{1}{\pi(1 + (x - y))^2} \quad f_X(x) = \exp(-x)\mu(x)$$

(a) Find the best linear estimate of Y in terms of X.

(b) Find the best nonlinear estimate of Y in terms of X.

3.7 The random variables X and Y have the joint probability density function

$$f(x, y) = \begin{cases} x + y, & 0 < x < 1, 0 < y < 1 \\ 0, & \text{elsewhere} \end{cases}$$

Use the minimum mean squared error as a performance measure.

(a) Find the best linear estimate of Y in terms of X.

(b) Find the best nonlinear estimate of Y in terms of X.

(c) Comment on your results.

3.8 The random variables X and Y have the joint probability density function

$$f(x, y) = \begin{cases} \frac{2}{3}(x + 2y), & 0 < x < 1, \ 0 < y < 1 \\ 0, & \text{elsewhere} \end{cases}$$

Use the minimum mean squared error performance measure.

(a) Find the best linear estimate of Y in terms of X.

(b) Find the best nonlinear estimate of Y in terms of X.

(c) Comment on your results.

3.9 The random variable X is defined by $X = e^{-(S+N)}$, where S and N are independent random variables with probability density functions $f_S(s) = e^{-s}\mu(s)$ and $f_N(n) = 2e^{-n}\mu(n)$.

We wish to estimate the signal S by $\hat{S} = aX + b$, where a and b are such that $e_{\text{ms}} = E[(S - \hat{S}^2]$ is minimized.

(a) Determine a and b.

(b) Evaluate the minimum mean squared error for your estimate.

(c) Find $f_X(x|N = n)$.

3.10 Consider two independent Gaussian random variables X and N with known means and variances as follows η_S, η_N, σ_S^2, and σ_N^2. Define a random variable R (a received signal in multiplicative noise) as $R = S \cdot N$. We desire the best estimate of S by a linear function of R, where best is in the minimum mean squared error sense and the estimate is given by $\hat{S} = aR + b$.

(a) Find a and b.

(b) Determine the minimum mean squared error for this case.

3.11 The joint probability density function for two random variables X and Y is

$$f(x, y) = xye^{-(x+y)}\mu(x)\mu(y)$$

(a) Find the best nonlinear estimate of Y in terms of X.

(b) Find the best linear estimate of Y in terms of X.

3.12 Given two random variables X and Y with the joint probability density function

$$f(x, y) = e^{x(y-1)}\mu(x)\mu(y)$$

(a) Find the best linear estimate of Y in terms of X.

(b) Find the best nonlinear estimate of Y in terms of X.

(c) Comment on your results.

3.13 Given X and Y are random variables with joint probability density function $f_{XY}(x, y)$ as follows:

$$f(x, y) = \begin{cases} 2, & 0 < x < y, 0 < y < 1 \\ 0, & \text{elsewhere} \end{cases}$$

Use the minimum mean squared error as a performance measure.

(a) Find the best linear estimate of Y in terms of X.

(b) Find the best nonlinear estimate of Y in terms of X.

(c) Comment on your results.

3.14 Given $f_{XY}(x, y)$ as

$$f_{XY}(x, y) = \begin{cases} \dfrac{1}{6\pi}, & 5x^2 + 4xy + 8y^2 \leq 36 \\ 0, & \text{elsewhere} \end{cases}$$

(a) Find $f_X(x)$.

(b) Find $f_Y(y|x)$.

(c) Find the best nonlinear estimator of Y in terms of X, where best is in the minimum mean squared error sense.

(d) Find the minimum mean squared error for your estimator.

3.15 X and Y are random variables that are related nonlinearly as $Y = X^2$. Let X be a normal random variable with mean η_X and variance σ_X^2. For some unknown reason we want to approximate Y by a linear function of X, which can be expressed as $\hat{Y} = aX + b$.

(a) Find a and b such that the mean squared error $E[(Y - \hat{Y})^2]$ is minimized, and determine the resulting mean squared error for your approximation.

(b) Repeat (a) if Y is related to X by $Y = X^3$.

(c) Simplify your answers for (a) and (b) if $\eta_X = 0$, and explain your results.

3.16 Random variables Z and Y related to the random variable X by the function $Z = XY$. Assume that Z is observed and X is to be estimated.

(a) Assume that X and Y are independent Gaussian random variables with known means η_X and η_Y and known variances σ_X^2 and σ_Y^2. Find a and b such that $E[(X - (aZ + b))^2]$ is minimized.

(b) Use the assumptions in (a) to find the best nonlinear estimator of X in the minimum mean squared error sense, that is, find $g(x)$ such that $E[(X - g(Z)^2]$ is minimized.

(c) Assume that X and Y are zero mean random variables, not necessarily Gaussian, and that they are statistically related by their joint probability density function $f_{XY}(x, y)$. Find the best linear and nonlinear estimators of X in terms of Z, where best is in the sense of minimizing the mean squared error.

3.17 Given

$$f_{X_1X_2X_3}(x_1, x_2, x_3) = \begin{cases} 8x_1x_2x_3, & 0 \le x_1 \le 1, 0 \le x_2 \le 1, 0 \le x_3 \le 1 \\ 0, & \text{elsewhere} \end{cases}$$

(a) In the minimum mean squared error sense, find the best linear estimate of X_1 in terms of X_2 and X_3.

(b) In the minimum mean squared error sense, find the best nonlinear estimate of X_1 in terms of X_2 and X_3.

3.18 The random variables X and Y have a probability density function $f_{XY}(x, y)$ given by

$$f_{XY}(x, y) = \begin{cases} 21x^2y^3, & 0 < x < y < 1 \\ 0, & \text{elsewhere} \end{cases}$$

(a) Illustrate the region in the x–y plane for which $f_{XY}(x, y) \neq 0$.

(b) In the minimum mean squared error sense, find the best linear estimate of Y in terms of X.

(c) In the minimum mean squared error sense, find the best nonlinear estimate of Y in terms of X.

3.19 Given $Y_1 = C \cdot S + N$ and $Y_2 = D \cdot S + N$, where C and D are group independent of S and N. Assume that C and D are independent random variables with uniform densities on $[-1, 1]$ and that S and N are orthogonal random variables with zero means and known variances σ_S^2 and σ_N^2, respectively.

We desire the best linear estimate of S in terms of Y_1 and Y_2, where best is in the minimum mean squared error sense.

(a) Find a_1 and a_2 for the linear estimate $\hat{S} = a_1 Y_1 + a_2 Y_2$.

(b) Determine the mean squared error for the results in (a).

(c) Is there sufficient information to determine the best nonlinear estimate of S in terms of Y_1 and Y_2. If so, determine it; if not, say why it cannot be found.

3.20 Given $R_1 = S + N_1$ and $R_2 = S + N_2$. The following second-order properties for the random variables are known:

$$
\begin{array}{lll}
E[S] = 0 & E[S^2] = 1 & E[SN_1] = 0.5 \\
E[N_1] = 0 & E[N_1^2] = 3 & E[SN_2] = 0.5 \\
E[N_2] = 0 & E[N_2^2] = 3 & E[N_1 N_2] = 0
\end{array}
$$

We desire the best linear estimate of S in terms of R_1 and R_2, where best is in the minimum mean squared error sense.

(a) Find a_1 and a_2 for the best linear estimate $\hat{S} = a_1 R_1 + a_2 R_2$.

(b) Determine the mean squared error for the results in (a).

(c) Is there sufficient information to determine the best nonlinear estimate of S in terms of R_1 and R_2. If so, determine it; if not, say why it cannot be found.

3.21 The random variables X, Y, and Z have the conditional density function

$$
f(x, y|z) = \begin{cases} z + (1 - z)(x + y), & 0 < x < 1, 0 < y < 1, 0 < z < 2 \\ 0, & \text{elsewhere} \end{cases}
$$

where Z has the density function $f_Z(z) = \frac{1}{2}[\mu(z) - \mu(z-2)]$. Use the minimum mean squared error as a performance measure to find:

(a) The best nonlinear estimate of Z in terms of X and Y.

(b) $f_{XY}(x, y)$.

(c) The best linear estimate of Z in terms of X and Y.

3.22 Four random variables W, X, Y and Z, are defined by the following joint density:

$$f(w,x,y,z) = \begin{cases} A\left(\dfrac{wx}{yz}\right)^2 \log\dfrac{w}{y}, & 1 \le w \le 2, 1 \le x \le 2, 1 \le y \le w \le 2, \\ & 1 \le z \le 2 \\ 0, & \text{otherwise} \end{cases}$$

Calculate the best nonlinear estimate of X in terms of W, Y, and Z.

3.23 Random variables X, Y, and Z are characterized by their joint probabilty density function $f_{XYZ}(x,y,z)$ given by

$$f_{XYZ}(x,y,z) = \begin{cases} (x+y)e^{-z}, & 0 < x < 1, 0 < y < 1, z > 0 \\ 0, & \text{elsewhere} \end{cases}$$

Find the best nonlinear estimate of Y in terms of X and Z, where best is in the minimum mean squared error sense.

3.24 The following problem is the estimation of the parameter for a Rayleigh distributed random variable. The density for Y conditioned on $X = x$ and the marginal density $f_X(x)$ of the two random variables X and Y are given by

$$f_Y(y|x) = x^2 y e^{-y^2x^2/2}\mu(y), \quad f_X(x) = \mu(x-1) - \mu(x-2)$$

(a) Find the best nonlinear estimate of Y in terms of X, and calculate the corresponding minimum mean squared error for your estimate.

(b) Determine the minimum mean squared error for the same problem except for the fact that $f_X(x) = e^{-x}\mu(x)$, and comment on your unexpected result.

3.25 Given random variables Z, X, and N with X and N independent Gaussian random variables characterized by $X \sim N(0, \sigma_X^2)$ and $N \sim N(0, \sigma_N^2)$, and Z defined by $Z = \cos(X) + N$. Find the form of the Bayes estimator of X from Z where the uniform cost function is used.

3.26 Given random variables Z, X, and N with X and N independent random variables characterized by $f_X(x) = ae^{-ax}\mu(x)$ and $f_N(n) = be^{-bn}\mu(n)$, and Z defined by $Z = X + N$. If $a > b$, find the Bayes estimator of X from Z where the uniform cost function is used. Repeat for $a < b$.

3.27 Given random variables Z, X, and N with X and N independent random variables characterized by $f_X(x) = ae^{-ax}\mu(x)$, and $f_N(n) = be^{-bn}\mu(n)$, and Z defined by $Z = X^2 + N$. Find the Bayes estimator of X from Z where the uniform cost function is used.

3.28 Given random variables Z, X, and N with X and N independent random variables characterized by $X \sim N(0, \sigma_X^2)$ and $N \sim N(0, \sigma_N^2)$, and Z defined by $Z = X^2 + N$. Find the Bayes estimator of X from Z where the uniform cost function is used.

3.29 Given random variables Z, X, and N with X and N independent random variables characterized by $f_X(x) = ae^{-ax}\mu(x)$ and $f_N(n) = be^{-bn}\mu(n)$, and Z defined by $Z = XN$. Find the Bayes estimator of X from Z where the uniform cost function is used.

3.30 Given random variables Z, X, and N with X and N independent random variables characterized by $X \sim N(0, \sigma_X^2)$, $N \sim$ uniform on [1, 2], and Z defined by $Z = X + N$. Find the Bayes estimator of X from Z where the uniform cost function is used.

3.31 Given random variables Z, X, and N with X and N independent random variables characterized by $X \sim N(0, 2\sigma^2)$ and $N \sim N(1, \sigma^2)$ and Z defined by $Z = X + N$.

(a) Find the Bayes estimator of X from Z where the absolute error cost function is used.

(b) Find the Bayes estimator of X from Z where the squared error cost function is used.

3.32 Given random variables Z, X, and N with X and N independent random variables characterized by $X \sim N(0, 2\sigma^2)$ and $N \sim N(1, \sigma^2)$, and Z defined by $Z = XN$.

(a) Find the Bayes estimator of X from Z where the absolute error cost function is used.

(b) Find the Bayes estimator of X from Z where the squared error cost function is used.

3.33 Given random variables Z, X, and N with X and N independent random variables characterized by X uniform on $[-2, 2]$ and N uniform on $[1, 3]$, and Z defined by $Z = X + N$. Find the Bayes estimator of X from Z where the uniform cost function is used.

3.34 Assume that A and X are two random variables with the conditional probability density function for X and the probability density function A given as

$$f_X(x|a) = ae^{-ax}\mu(x) \quad \text{and} \quad f_A(a) = [\mu(a-1) - \mu(a-2)]$$

(a) Find the maximum a posteriori estimator for A.

(b) Compute $E[\hat{A}_{\mathrm{MAP}}(X)]$, and say whether the estimate is biased or unbiased.

(c) Give a formula for the variance for your estimator.

3.35 The random variable Z is the sum of two independent random variables X and N and writtten as $Z = X + N$. Let $N \sim N(0, \sigma_N^2)$ and $f_X(x) = \frac{1}{2}\,\delta(x+1) + \frac{1}{2}\,\delta(x-1)$.

(a) Find the MAP estimator for X in terms of Z, and determine the expected value of your estimator.

(b) Is your estimator unbiased? Why or why not.

(c) Determine the variance of your estimator.

3.36 Repeat Problem 3.35, but with $f_X(x) = \frac{1}{3}\,\delta(x+1)\frac{2}{3}\,\delta(x-1)$.

3.37 Repeat Problem 3.35, but with $f_X(x) = \frac{1}{2}(\mu(x+1) - \mu(x-1))$

3.38 The random variable Z is the sum of two independent random variables X and N and writtten as $Z = X + N$. Let $N \sim N(0, \sigma_N^2)$ and $f_X(x) = \frac{1}{2}\,\delta(x+1) + \frac{1}{2}\,\delta(x-1)$.

(a) Find the MMSE estimator for X in terms of Z, and determine the expected value of your estimator.

(b) Is your estimator unbiased? Why or why not.

(c) Determine the variance of your estimator.

(d) Compare your result to that of Problem 3.35.

3.39 Given $Z = BX + N$, where X and N are independent Gaussian random variables characterized as $X \sim N(1, 1)$ and $N \sim N(0, 1)$. Assume that B is a uniform random variable on the interval $[1, 3]$ independent of both X and N. We wish to estimate X from Z. Find:

(a) $\hat{X}_{\mathrm{MAP}}(Z)$.

(b) $\hat{X}_{\mathrm{MMSE}}(Z)$.

3.40 Given $Z = a + N$, where N is uniform random variable on the interval $[1, 2]$. Find the maximum likelihood estimator for a in terms of Z, and compute the expected value and variance for your estimator. Is your estimator unbiased?

3.41 Given $Z = aN$, where N is a Gaussian random variable with a mean of 1 and a variance of σ^2. Find the maximum likelihood estimator for a in terms of Z, and compute the expected value and variance for your estimator. Is your estimator unbiased?

3.42 Given $Z_i = \sin(2\pi f_0 t) + N_i$ for $t_i = -1, 0, 1$. Find the maximum likelihood estimator for f_0, and determine the expected value for your estimator. Is it unbiased?

3.43 Determine the $E[\hat{a}_{\text{ML}}(Z_1, Z_2, \ldots, Z_n)]$ for the estimator determined in Example 3.12.

REFERENCES

1. Papoulis, Athanasios, *Probability, Random Variables, and Stochastic Processes*, McGraw-Hill, 1965.
2. Drake, Alvin W., *Fundamentals of Applied Probability Theory*, McGraw-Hill, 1967.
3. Srinath, M. D., P. K. Rajasekaran, and R. Viswanathan, *Statistical Signal Processing with Applications*, Prentice-Hall, 1996.
4. Mendel, Jerry M., *Discrete Techniques of Parameter Estimation*, Dekker, 1973.
5. Van Trees, H. L., *Detection, Estimation and Modulation Theory*, Vol. 1, Wiley, 1968.
6. Nahi, Nasser E., *Estimation Theory and Applications*, Wiley, 1969.
7. Meditch, J. S, *Stochastic Optimal Linear Estimation and Control*, McGraw-Hill, 1969.
8. McDonough, Robert N., and Anthony D. Whalen, *Detection of Signal in Noise*, Academic Press, 1995.
9. Davenport, Jr., William B., *Probability and Random Processes*, 1970.
10. Mendel, Jerry M., *Lessons in Digital Estimation Theory*, Prentice-Hall, 1987.
11. Scharf, Louis L., *Statistical Signal Processing*, Addison-Wesley, 1991.
12. Kailath, T., *Lectures on Linear Least-Squares Estimation*, Springer, 1976.
13. Poor, H. Vincent, *An Introduction to Signal Detection and Estimation*, 2nd Ed., Springer, 1994.
14. Parzen, Emanuel, *Modern Probability Theory and Its Applications*, Wiley, 1960.
15. Sage, Andrew P., and James L. Melsa, *Estimation Theory with Applications to Communications and Control*, McGraw-Hill, 1971.

4 Random Processes

4.1 DEFINITION OF A RANDOM PROCESS

From Chapters 2 and 3 we learned that an experiment is specified by the three tuple $(S, BF, P(\cdot))$, where S is a countable or noncountable set representing the **outcomes** of the experiment, BF is a **Borel field** specifying the set of events for which consistent probabilities exist, and $P(\cdot)$ is a **probability measure** that allows the calculation of the probability of all the events in the Borel field. A **real random variable** $X(\xi)$ was defined as a real-valued function of S subject to (a) the event specified by $\{\xi : X(\xi) \leq x\}$ is a member of the BF for all x, thus guaranteeing the existence of the cumulative distribution, and (b) $P\{\xi : X(\xi) = \infty\} = 0$ or $P\{\xi : x(\xi) = -\infty\} = 0$, or both.

If, instead of assigning a real value to each ξ, a time function $X(t, \xi)$ is defined for each ξ, we say a **random process** is specified. Roughly speaking, a random process is a family of time functions together with a probability measure. For a finite sample space S we can visualize the random process as in Figure 4.1, that of a mapping from the sample space to a space of time waveforms.

A random process can also be viewed as shown in Figure 4.2. For a particular outcome ξ_i, with probability P_i, the time waveform shown $X(t, \xi_i)$ occurs. The n time signals represent an ensemble of time waveforms.

If we evaluate these time waveforms at t_0, the values on the second column from the right are obtained. Coming down that column of values, we have a mapping from the outcomes on the far left, and this mapping describes a particular random variable $X(t_0)$. Similarly a random variable $X(t_1)$ is defined for the mapping shown at time t_1 as in the last column. A simple example of random process will now be given.

EXAMPLE 4.1

Consider the sample space $S = \{f_1, f_2, \ldots, f_6\}$, where f_i are the faces on a four-sided die or outcomes of tossing a die. Here the Borel field consists of the power set of S, that is, the set of all possible subsets of S. The probability measure is defined by giving the probability of the elementary events $\{f_i\}$. Assuming an honest die, we would have $p\{f_i\} = \frac{1}{6}$ for $i = 1, 2, \ldots, 6$. Consider the following assignment of waveforms to each outcome: $f_i \to X(t, f_i) = \sin(2\pi i t)$.

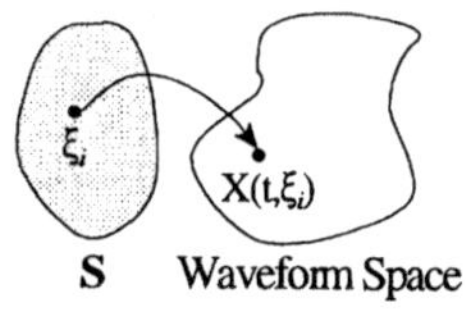

Figure 4.1 A Random process viewed as a functional mapping.

On a single trial of the experiment a sine wave is specified with a frequency equal to the number on the face of the die. The collection of waveforms together with the outcomes and the probability measure thus specify the random process $X(t, \xi)$. ■

It should be noted that a **random process** $X(t, \xi)$ is a **function** of two variables, t and ξ, where t is the time variable and ξ is the outcome variable. For a particular time $t = t_0$, $X(t_0, \xi)$ specifies a **random variable**, since it is a function of the sample space. If $\xi = \xi_i$, or the outcome is fixed, then $X(t, \xi_i)$ is a particular **time function** called a sample function or realization of the random process. With both t and ξ fixed, say at t_0 and ξ_i, $X(t_0, \xi_i)$ becomes a **real number** for a real random process or a complex value for a complex random process.

As the notation for a random variable is usually a capital letter X, Y, or Z rather than the more formal $X(\xi)$, $Y(\xi)$, $Z(\xi)$, a random process is usually denoted as $X(t)$ rather than $X(t, \xi)$. Also it is common to use a small x, or $x(t)$, to indicate a particular value of the random variable X or a particular time function of a

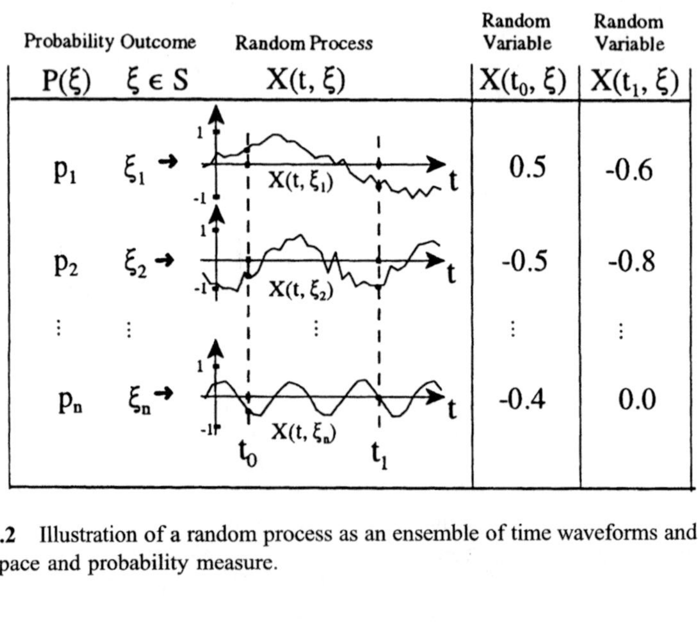

Figure 4.2 Illustration of a random process as an ensemble of time waveforms and a related sample space and probability measure.

random process $X(t)$, respectively. Such an $x(t)$, or $X(t, \xi_i)$, is called a **sample function** or a **realization** of the process.

Also, as $X(t, \xi)$ is a particular random variable for each t, a common alternative definition of a random process is an indexed set of random variables where $t \in \Upsilon$ the indexing set. If the indexing set Υ can be put into a one to one correspondence with the set of integers, the random process is called a random sequence or a **discrete time random process** if t is time. While, if the indexing set takes on a continuum of values, we will call the resulting process a **continuous random process** or continuous time random process.

4.2 CHARACTERIZATIONS OF A RANDOM PROCESS

We saw in Chapter 2 that partial characterizations for a random variable include its mean, variance, moments, and the like, and similarly, useful partial characterizations for a random process include the mean, variance, and moments. However, all will, in general, be functions of time. For random processes new types of characterizations like the nth-order densities, autocorrelation function, and spectral density will be defined that will be useful in analyzing random processes. These characterizations give different statistical properties of the process and different amounts of information concerning the process. A hierarchical relationship between the various types of characterizations exists and is shown in Figure 4.3. Each of the characterizations will now be defined and their relationships explained.

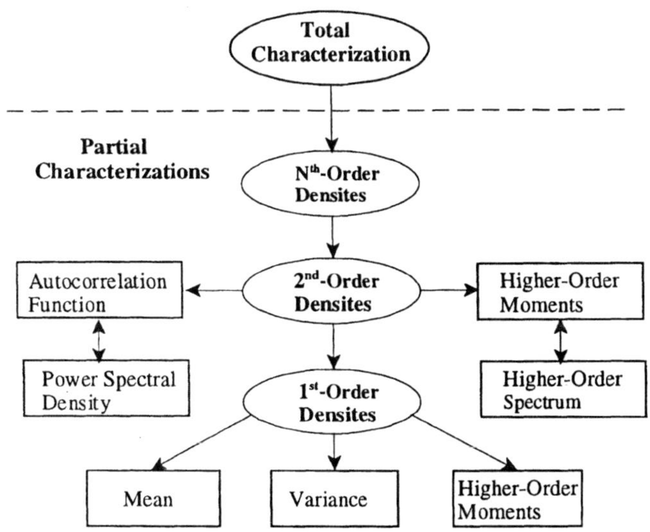

Figure 4.3 Relationships between various characterizations for random processes.

4.2.1 Total Characterization of a Random Process

Remember that for every $t, X(t)$ is a random variable. This gives us a countably infinite or infinite number of random variables described for the random process. A random process is defined to be completely or **totally characterized** if the joint densities for the random variables $X(t_1), X(t_2), \ldots, X(t_n)$ are known for all times $t_1, t_2, \ldots, t_n$ and all n. In general, a complete characterization is a luxury, and except in rare cases, such a characterization is not realistically possible. As a result it is useful to define and work with various partial characterizations. It will be shown later that for particular classes of systems, a partial characterization suffices to specify the output characterization from the input characterization.

4.2.2 First-Order Densities of a Random Process

Cutting across the ensemble of waveforms representing a random process at a specific time t specifies a random variable. As we do not expect the random variable at one time to be the same as that at another time, the probability densities could be different. The **first-order probability density functions** of the random variables $X(t) \triangleq X_t$ defined for all time t will be denoted by $f_{X_t}(x_t)$ or $f(x; t)$. If they are all the same, then $f(x; t)$ does not depend on t and we call the resulting density the first-order density of the process; otherwise, we have a family of first-order densities. An example of such a family of first-order densities is shown in Fig. 4.4.

The first-order densities are only a partial characterization as they do not contain information that specifies the joint densities of the random variables defined at two or more different times.

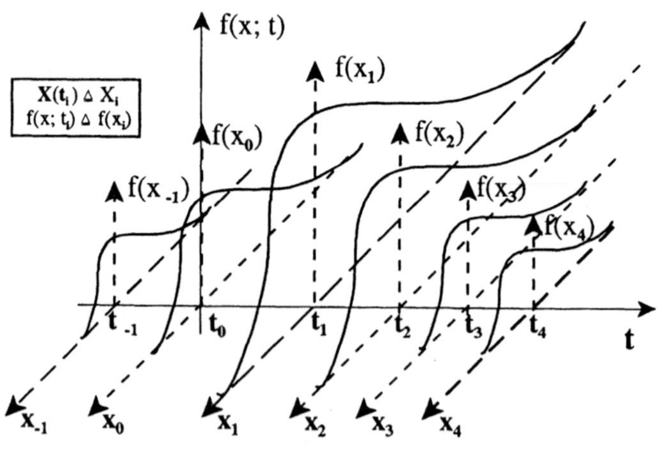

Figure 4.4 Family of first order densities for different times t_i.

4.2.3 Mean of a Random Process

The first-order density of the process, $f(x; t)$, gives the probability density of the random variables X_t defined for all time. The **mean** of a random process, $\eta_X(t)$, is thus a function of time specified by

$$\eta_X(t) = E[X(t)] = E[X_t] = \int_{-\infty}^{\infty} x_t f_{X_t}(x_t)\, dx_t \tag{4.1}$$

For the case where the mean of X_t does not depend on t, we use the notation $E[X(t)] = \eta_X$, where η_X is a constant.

4.2.4 Variance of a Random Process

The **variance of the process** $X(t)$ is similarly defined by

$$\sigma_X(t) = E[(X(t) - \eta_X(t))^2] = E[X_t^2] - \eta_X^2(t) \tag{4.2}$$

Just knowing the mean and variance of the process, along with the Tchebycheff inequality, allows us an interpretation of the region of the space where a sample function from the process will most likely be. The shaded area between $\eta_X(t) - 3\sigma_X(t)$ and $\eta_X(t) + 3\sigma_X(t)$ shown in Figure 4.5 represents such a region.

4.2.5 Second-Order Densities of a Random Process

For any two times t_1 and t_2, the random variables $X(t_1) = X_1$ and $X(t_2) = X_2$ are defined. The second-order densities of the random process $X(t)$ are the family of densities specified for all t_1 and t_2, where t_1 and t_2 are not equal. These densities can be denoted as $f(x_1, x_2)$ or $f(x_1, x_2;\ t_1, t_2)$. In general, these two random variables will not be independent, and therefore the second-order density cannot be determined from just the first-order density functions. For any other pair of times we do not necessarily expect to have this same form of density. Thus the second-order density of a process is not known until the joint densities of every pair of times is specified.

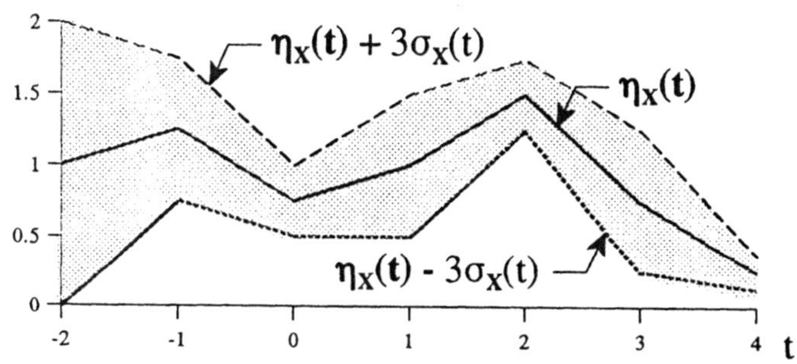

Figure 4.5 Shaded region representing highly probable region for realizations of the random process.

4.2.6 Autocorrelation and Autocovariance Functions of Random Processes

Given the two random variables $X(t_1)$ and $X(t_2)$ we know that a measure of linear relationship between them is specified by the correlation $E[X(t_1)X(t_2)]$. As t_1 and t_2 go through all possible values, this correlation can change and is thus a function of t_1 and t_2. The **autocorrelation function** of a real process is, in general, a two-dimensional function of the variables t_1 and t_2 defined by

$$R_{XX}(t_1, t_2) = E[X(t_1)X(t_2)] \tag{4.3}$$

Since $X(t_1)X(t_2) = X(t_2)X(t_1)$, $R_{XX}(t_1, t_2)$ is seen to be symmetrical in t_1 and t_2 as

$$R_{XX}(t_1, t_2) = R_{XX}(t_2, t_1) \tag{4.4}$$

In certain cases, what we will define latter as wide sense stationarity, the autocorrelation function is seen to depend only on the time difference $\tau = t_1 - t_2$. For this case we define a one-dimensional autocorrelation function $R_{XX}(\tau)$ by

$$R_{XX}(\tau) = E[X(t+\tau)X(t)] \tag{4.5}$$

Since the $X(t+\tau)$ can be interchanged with $X(t)$, we see that $R_{XX}(\tau)$ is an even function of τ, which is written as

$$R_{XX}(-\tau) = R_{XX}(\tau) \tag{4.6}$$

An **autocovariance function,** $C_{XX}(t_1, t_2)$, is defined by

$$\begin{aligned} C_{XX}(t_1, t_2) &= E[(X(t_1) - \eta_X(t_1))(X(t_2) - \eta_X(t_2))] \\ &= R_{XX}(t_1, t_2) - \eta_X(t_1)\eta_X(t_2). \end{aligned} \tag{4.7}$$

If $C_{XX}(t_1, t_2)$ is normalized, we obtain the **normalized autocovariance** function defined by

$$\rho_{XX}(t_1, t_2) = \frac{C_{XX}(t_1, t_2)}{(C_{XX}(t_1, t_1)C_{XX}(t_2, t_2))^{\frac{1}{2}}} \tag{4.8}$$

In general, the **autocorrelation function** for a **complex random process** $X(t)$ is defined by

$$R_{XX}(t_1, t_2) \triangleq E[X(t_1)X^*(t_2)] \tag{4.9}$$

where * stands for complex conjugate. By writing $X(t)$ as $X_R(t) + jX_I(t)$, we can easily show the autocorrelation function to have the properties

$$\begin{aligned} R_{XX}(t_1, t_2) &= R^*_{XX}(t_2, t_1) \\ R_{XX}(-\tau) &= R^*_{XX}(\tau) \end{aligned} \tag{4.10}$$

An **autocovariance function** $C_{XX}(t_1, t_2)$ for a complex process $X(t)$ is defined by

$$\begin{aligned} C_{XX}(t_1, t_2) &= E[(X(t_1) - \eta_X(t_1))(X(t_2) - \eta_X(t_2))^*] \\ &= R_{XX}(t_1, t_2) - \eta_X(t_1)\eta^*_X(t_2) \end{aligned} \tag{4.11}$$

If $C_{XX}(t_1, t_2)$ for a complex process is normalized, the **normalized autocovariance** function is defined by

$$\rho_{XX}(t_1, t_2) = \frac{C_{XX}(t_1, t_2)}{(C_{XX}(t_1, t_1)C^*_{XX}(t_2, t_2))^{1/2}} \tag{4.12}$$

4.2.7 Power Spectral Density of a Random Process

A very important class of random processes are those for which the autocorrelation function $R_{XX}(t_1, t_2)$ can be written as a function of the time difference $t_1 - t_2 = \tau$ as $R_{XX}(\tau)$. This type of process will be described later as being stationary in autocorrelation. For these processes another form of characterization is available that is described in the frequency domain. This characterization is called a **power spectral density** $S_{XX}(\omega)$ and defined as the Fourier transform of the autocorrelation function as follows:

$$S_{XX}(\omega) = \int_{-\infty}^{\infty} R_{XX}(\tau)e^{-j\omega\tau}\, d\tau \tag{4.13}$$

This formulation roughly describes the region in the frequency domain where power of the process exists and the relative proportion of power at each frequency.

The power spectral density of a given real random process has the following properties:

$$\begin{aligned}
&S_{XX}(-\omega) = S_{XX}(\omega) \qquad \text{Even function of } \omega \\
&S_{XX}(\omega) \geq 0 \qquad \text{Real and nonnegative} \\
&S_{XX}(\omega) = F(R_{XX}(\tau)) \\
&R_{XX}(\tau) = F^{-1}(S_{XX}(\omega)) \qquad \text{Fourier Transform pairs} \\
&\frac{1}{2\pi}\int_{-\infty}^{\infty} S_{XX}(\omega)\, d\omega = R_{XX}(0) \qquad \text{Total average power in the process} \\
&\frac{1}{\pi}\int_{\omega_1}^{\omega_2} S_{XX}(\omega)\, d\omega = \frac{1}{2\pi}\int_{-\omega_2}^{-\omega_1} S_{XX}(\omega)\, d\omega + \frac{1}{2\pi}\int_{\omega_1}^{\omega_2} S_{XX}(\omega)\, d\omega \\
&\text{Average power in frequency band } (\omega_1, \omega_2)
\end{aligned} \tag{4.14}$$

4.2.8 Higher-Order Moments

For the n random variables $X(t_1), X(t_2), \ldots, X(t_n)$ the **higher-order moments** include $E[X_1X_2X_3]$, $E[X_1X_2X_3X_4]$, $E[X_1^2, X_2^2]$, and so on. In linear systems we will find that these higher-order characterizations are not necessary and that the mean and autocorrelation are sufficient to discuss input-output characteristics. On the other hand, for nonlinear systems the order of densities and moments required depends entirely on the type of system considered.

4.2.9 Higher-Order Spectra

For certain random processes, especially those generated as the output of a nonlinear system, it becomes convenient to define **higher-order spectra**. Precise definitions and properties of higher-order spectra are presented in Chapter 6.

4.2.10 *N*th-Order Densities

The ***n*th order density functions** for the random process $X(t)$ at times $t_1, t_2, \ldots, t_n$, are given by $f(x_1, x_2, \ldots, x_n)$ or $f(x_1, x_2, \ldots, x_n;\ t_1, t_2, \ldots, t_n)$ to emphasize the times.

Each of the characterizations of a random process contains statistical information concerning the process, and the various relationships shown in Figure 4.3 provides a way to view the many different statistical characterizations for random processes. Relationships between forms of statistical stationarity exist and will be found useful for describing further statistical properties of random variables.

4.3 STATIONARITY OF RANDOM PROCESSES

In analyzing random processes, various forms of statistical regularity between the random variables defined at different times prove to be useful.

A random process $X(t)$ is called strong sense or **strict sense stationary** if the sets of random variables $X(t_1), X(t_2), \ldots, X(t_n)$ and $X(t_1+\varepsilon), X(t_2+\varepsilon), \ldots, X(t_n+\varepsilon)$ have the same probability density function for all t_i, all n and all ε. That is, for all $t_1, t_2, \ldots, t_n, \varepsilon$, and n, we have

$$f(x_1, x_2, \ldots, x_n;\ t_1, t_2, \ldots, t_n) = f(x_1, x_2, \ldots, x_n;\ t_1+\varepsilon, t_2+\varepsilon, \ldots, t_n+\varepsilon) \tag{4.15}$$

This form of stationarity is quite uncommon and except for some Gaussian processes is hardly ever seen.

A slightly weaker form of stationarity is **stationarity of order** ***N*** where the conditions above (4.15) are true for all $n \leq N$ for N a fixed integer. If the conditions are true for $n = N$, the joint density may be integrated to give one less variable, making the conditions true for $n = N - 1$. Continuing this process, we see that the conditions above are true for all n less than N. Therefore a useful definition for stationarity of order N is that Eqs. (4.15) are satisfied for $n = N$. Stationarity of order N does not preclude that the process could be stationary of an order greater than N. However, if it is not stationary of order N, it will be not stationary of any order greater than N.

4.3.1 Wide Sense Stationary Random Processes

A much weaker form of stationarity, even weaker than stationarity of order two, that plays an important role in analysis of linear systems with random inputs is weak sense or wide sense stationarity. A process $X(t)$ is weak sense or **wide sense stationary** if

(1) the expected value of the process $E[X(t)]$ is a constant, thus independent of the time variable, and

(2) the autocorrelation function $R_{XX}(t_1, t_2)$ is a function of time difference $t_2 - t_1$ only.

When $R_{XX}(t_1, t_2)$ can be written as a function of time difference, $\tau = t_2 - t_1$ only, we will denote the autocorrelation of the process as $R_{XX}(\tau)$ or sometimes as just $R_X(\tau)$.

If condition (2) above is satisfied, the process is said to be **stationary in autocorrelation**. If condition (1) is satisfied, the process has the lowest form of stationarity called stationarity in mean. A random process $X(t)$ is **stationary in mean** if $E[X(t)] = \eta_X$, a constant for all time. If a process is stationary of order one, then it must be stationary in mean. However, a random process can be stationary in mean without being stationary of order one or stationary in autocorrelation. In Figure 4.6 the relationships between the various forms of stationarity are shown. A solid arrow means implication, no arrow means no implication, while a dotted arrow with the further information given by the arrow means implication with that information.

A random process is called **asymptotically stationary** of a particular sense if conditions for that sense are satisfied in the limit as all times approach infinite. Some processes are in a sense stationary for particular choices of the time variables. We say that a process is **cyclo stationary** of a particular sense if the sense is satisfied for times that are periodically displaced by an amount T.

It has been shown that if a random process is Gaussian and weak sense stationary, it is strict sense stationary. This is truly a remarkable result as it means that the autocorrelation function and the mean of a weak sense stationary process must contain all the information necessary to determine the nth-order density function. This property will be examined further in Section 4.7 on Gaussian random processes.

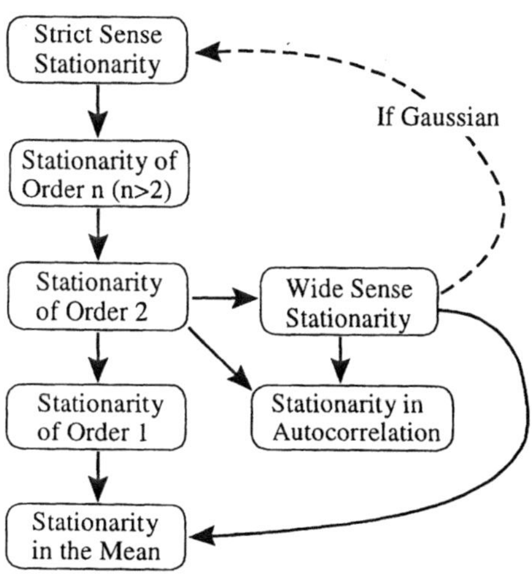

Figure 4.6 Conceptual relationship between some common forms of stationarity.

4.3.2 Properties for Wide Sense Stationary Random Processes

The autocorrelation function for a wide sense stationary process is a function of the time difference and can thus be written in terms of that variable as $R_{XX}(\tau)$. For a real process the following properties can be shown to be true:

$$
\begin{aligned}
&R_{XX}(-\tau) = R_{XX}(\tau) \qquad \text{even function of } \tau \\
&|R_{XX}(\tau)| \le R_{XX}(0) \qquad \text{bounded fuction of } \tau \\
&R_{XX}(\tau) = F^{-1}S_{XX}(\omega), \quad S_{XX}(\omega) = F(R_{XX}(\tau)) \qquad \text{Fourier transform pairs} \\
&F(R_{XX}(\tau)) \ge 0 \text{ for all } \omega \qquad \text{real and positive} \\
&R_{XX}(0) = \frac{1}{2\pi}\int_{\infty}^{\infty} S_{XX}(\omega)\,d\omega \qquad \text{average power in the process} \\
&R_{XX}(\tau) = C_{XX}(\tau) + \eta_X^2 \qquad \text{relationship to autocovariance function of process}
\end{aligned}
\tag{4.16}
$$

4.4 EXAMPLES OF RANDOM PROCESSES

In this section a number of random processes will be described. Some of the processes will be seen to be closely related to actual signals seen frequently in communications, while others are presented to illustrate the procedures in describing and characterizing random processes.

4.4.1 Straight Line Process

Given a random process $X(t)$ defined as

$$X(t) = P + Qt \tag{4.17}$$

This expression is an abbreviation for

$$X(t, \xi) = P(\xi) + Q(\xi)t \tag{4.18}$$

For every ξ_i, $X(t, \xi_i)$ is a straight line with intercept $P(\xi_i)$ and slope $Q(\xi_i)$. Some realizations of the process are presented in Figure 4.7.

EXAMPLE 4.2

We are asked to find the following for the straight line process:

(a) First-order densities of the process.
(b) Second-order densities $f(x_1, x_2; t_1, t_2)$ of the process.
(c) Mean $\eta_X(t)$ of the process.
(d) Autocorrelation function $R_{XX}(t_1, t_2)$ of the process.
(e) The different types of stationarity for this process.

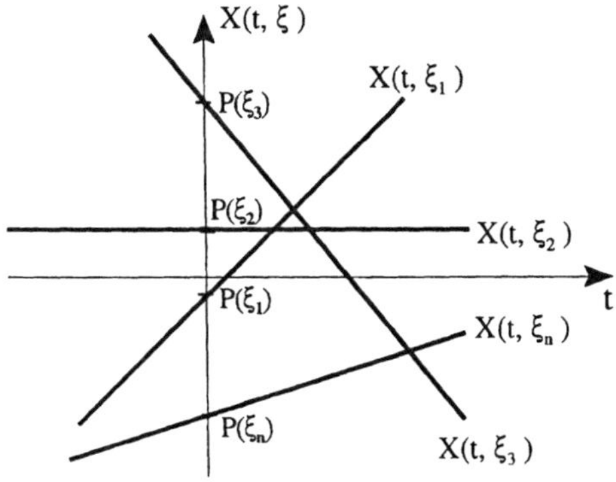

Figure 4.7 Four Realizations of the straight line process.

SOLUTION

(a) The first-order density $f_X(x_t)$ of the process can be obtained from the transformational theorem as X_t can be written as $X_t = P + Qt$, the sum of the two random variables P and Qt. By defining an auxiliary random variable $Y_t = P$, the joint density for X_t and Y_t becomes

$$f_{X_t Y_t}(x_t, y_t) = \sum_{(p_i, q_i)} \frac{f_{PQ}(p, q)}{|J(p, q)|}\bigg|_{(p_i, q_i)} \tag{4.19}$$

where p_i and q_i are the roots of $x_t = p + qt$ and $y_t = p$. The only solution to these equations is

$$p = y_t, \quad q = \frac{x_t - y_t}{t} \tag{4.20}$$

The Jacobian $J(p, q)$ is easily calculated as

$$J(p, q) = \begin{bmatrix} 1 & t \\ 1 & 0 \end{bmatrix} = -t \tag{4.21}$$

Since the desired marginal density is obtained by integrating out the auxiliary variable, we have

$$f_{X_t}(x_t) = \int_{-\infty}^{\infty} \frac{f_{PQ}(y_t, (x_t - y_t/t))}{|t|} dy_t \tag{4.22}$$

Notice that if P and Q are statistically independent, $f_p(p, q)$ can be written as a product and $f_X(x_t)$ can be determined by the convolution of the two densities as X_t is a sum of two independent random variables.

(b) The second-order densities of the process can be obtained by using the transformational theorem for two functions of two random variables, since $X(t_1) = X_1$ and $X(t_2) = X_2$ are given by

$$X_1 = P + Qt_1, \quad X_2 = P + Qt_2 \tag{4.23}$$

The set of equations

$$x_1 = p + qt_1, \quad x_2 = p + qt_2 \tag{4.24}$$

has only one solution given by

$$p = \frac{(t_2 x_1 - t_1 x_2)}{t_2 - t_1}, \quad q = \frac{x_2 - x_1}{t_2 - t_1} \tag{4.25}$$

The Jacobian is easily calculated as $(t_2 - t_1)$, so the joint probability density function for X_1 and X_2 can be shown to be

$$f_{X_1 X_2}(x_1, x_2) = \frac{f_{PQ}((t_2 x_1 - t_1 x_2)/(t_2 - t_1), (-x_1 + x_2)/(t_2 - t_1))}{|t_2 - t_1|} \tag{4.26}$$

(c) The mean of the process, $\eta_X(t)$, is computed by finding $E[X(t)]$. One way to get the mean is to use the first-order density of $X(t)$ determined in part (a), that is,

$$\eta_X(t) = \int_{-\infty}^{\infty} x_t f_{X_t}(x_t)\, dx_t \tag{4.27}$$

An alternative approach is to notice that the random variable X_t is really a function of the two random variables P and Q. In this way $n_X(t)$ can be written as

$$\eta_X(t) = E[X_t] = E[P + Qt] = E[P] + E[Q]t \tag{4.28}$$

Therefore, the mean of the process is a straight line that has intercept of $E[P]$ and a slope of $E[Q]$. It is not necessary that the mean be one of the realizations or sample functions of the process.

(d) To obtain the autocorrelation function for the random process, it is necessary to determine

$$R_{XX}(t_1, t_2) = E[X(t_1)X(t_2)]. \tag{4.29}$$

Again, this function can be found in several different ways. First, it can be found by using the joint density of the random variables $X(t_1) = X_1$ and $X(t_2) = X_2$ determined in part (b) as follows:

$$R_{XX}(t_1, t_2) = \int_{-\infty}^{\infty} \int_{-\infty}^{\infty} x_1 x_2 f_{X_1 X_2}(x_1, x_2)\, dx_1\, dx_2 \tag{4.30}$$

If you do not have the second-order density, you must determine it before the integral above can be evaluated. An alternative approach is to notice that the product $X(t_1)X(t_2)$ is a function of the two random variables P and Q and times t_1 and t_2, from which the expected value can be calculated as

$$\begin{aligned} R_{XX}(t_1, t_2) &= E[(P + Qt_1)(P + Qt_2)] \\ &= E[P^2] + E[PQ](t_1 + t_2) + E[Q^2]t_1 t_2 \end{aligned} \tag{4.31}$$

Therefore, once $f_{PQ}(p, q)$ is specified, the moments and joint moments given above can be calculated and $R_{xx}(t_1, t_2)$ can be determined.

(e) The mean of the process is given by $E[X(t)] = E[P] + E[Q]t$. If the $E[Q] = 0$, the mean is not a function of time and the random process $X(t)$ is stationary in the mean. Otherwise, it will not be stationary in the mean. The first-order density given is inherently a function of time. Therefore the process $X(t)$ is not stationary of order 1, and consequently not stationary of any higher-order densities.

For the process to be stationary in autcorrelation $R_{XX}(t_1, t_2)$ must be a function of time difference only. Except for the degenerative case of $E[PQ] = 0$ and $E[Q^2] = 0$, the process will not be stationary in autocorrelation, and thus it cannot be wide sense stationary as well. ■

4.4.2 Semirandom Binary Transmission Process

The problem of transmission of random binary information across a channel is a basic one in communication theory. A source generates a sequence of two symbols, say h and t. The h and t could represent the outcome of flipping an honest coin. It will be assumed that the probability of getting a head on a single trial is $\frac{1}{2}$ and that tosses at different times are independent. The sample space consists of all possible doubly infinite sequences of heads and tails (a cartesian product of the outcome of a single toss):

$$S = \{\zeta : \zeta = (\ldots, q_{-1}, q_0, q_1, q_2, \ldots)\, for\ q_i \in \{h, t\}\} \tag{4.32}$$

With the above sample space, probability measure, and the Borel field equal to the power set, the semirandom binary transmission process, $X(t)$, using 1 and -1 as values, is described on each subinterval of the time axis as

$$X(t, \zeta) = \begin{cases} 1 & \text{if } q_i = h \\ -1 & \text{if } q_i = t \end{cases} \qquad \text{for } iT \leq t < (i+1)T \tag{4.33}$$

A realization of the process $X(t, \zeta_a)$ is shown in Figure 4.8 for the particular outcome ζ_a of the experiment described by $\zeta_a = (\ldots, h, t, t, h, h, h, t, h, t, t, h, h, h, t, h, h, \ldots)$.

The defined process $X(t)$ is a collection or ensemble of time waveforms that change from 1 to -1, change from -1 to 1, or remain the same at possible transition time points that are equally spaced a distance T apart along the time axis. We are unable for this process to write a single analytical expression in terms of random variables for the process. We can, however, characterize the process by specifying

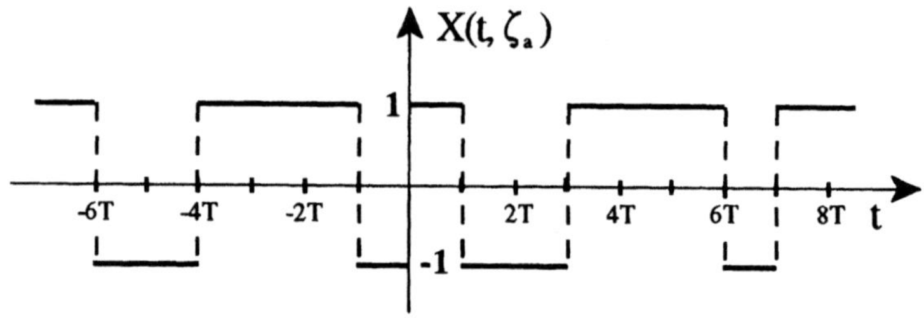

$\zeta_a = (\ldots, h, t, t, h, h, h, t, h, t, t, h, h, h, t, h, h, \ldots)$

Figure 4.8 A Realization of the Semirandom Binary Transmission Process.

certain statistical properties. In the following paragraphs the first-order density, the mean, the second-order density function, and the autocorrelation function of the process will be found.

First-Order Density. By cutting across the process at an arbitrary time t, a random variable is defined. Since the values of $X(t)$ for all t are either 1 or -1, this random variable is discrete for each t in the interval iT to $(i+1)T$. Also during each interval the random variables are the same. From the honest coin assumption the probability of getting either a head or a tail (1 or -1 for the process during each interval) is $\frac{1}{2}$. Therefore the first-order density is as follows for all t:

$$f_{X_t}(x_t) = \tfrac{1}{2}\delta(x_t + 1) + \tfrac{1}{2}\delta(x_t - 1) \tag{4.34}$$

Mean. The mean, $E[X(t)]$, of the process can be found directly from the first-order density by

$$\begin{aligned} \eta_X(t) = E[X(t)] &= \int_{-\infty}^{\infty} x_t f_{X_t}(x_t)\, dx_t \\ &= \int_{-\infty}^{\infty} x_t[\tfrac{1}{2}\delta(x_t + 1) + \tfrac{1}{2}\delta(x_t - 1)]\, dx_t \\ &= \tfrac{1}{2}(-1) + \tfrac{1}{2}(1) = 0 \end{aligned} \tag{4.35}$$

Second-Order Density. X_1 and X_2 representing $X(t_1)$ and $X(t_2)$ for arbitrary times t_1 and t_2 are discrete random variables which take on values of $+1$ and -1. Their joint density function is given by

$$\begin{aligned} f(x_1, x_2) = {} & P\{X_1 = 1, X_2 = 1\}\delta(x_1 - 1, x_2 - 1) \\ & + P\{X_1 = 1, X_2 = -1\}\delta(x_1 - 1, x_2 + 1) \\ & + P\{X_1 = -1, X_2 = 1\}\delta(x_1 + 1, x_2 - 1) \\ & + P\{X_1 = -1, X_2 = -1\}\delta(x_1 + 1, x_2 + 1) \end{aligned} \tag{4.36}$$

If t_1 and t_2 are in different intervals, X_1 and X_2 are independent random variables as the tosses were independent. If t_1 and t_2 are in the same intervals, they are the same random variable. The second-order density thus depends on whether t_1 and t_2 are in the same or different intervals.

For t_1 and t_2 in different intervals,

$$P\{X_1 = 1, X_2 = 1\} = P\{X_1 = 1\} \cdot P\{X_2 = 1\} = \tfrac{1}{2} \cdot \tfrac{1}{2} = \tfrac{1}{4} \tag{4.37}$$

Similarly $P\{X_1 = 1, X_2 = -1\}$, $P\{X_1 = -1, X_2 = 1\}$, and $P\{X_1 = -1, X_2 = -1\}$ are seen to be equal to $\frac{1}{4}$. Thus the second-order density from equation (4.36) becomes

$$\begin{aligned} f(x_1, x_2) = \tfrac{1}{4}[&\delta(x_1 - 1, x_2 - 1) + \delta(x_1 - 1, x_2 + 1) + \delta(x_1 + 1, x_2 - 1) \\ &+ \delta(x_1 + 1, x_2 + 1)] \end{aligned} \tag{4.38}$$

For t_1 and t_2 in the same interval, the probabilities needed can be obtained by using the conditional probabilities as follows:

$$P\{X_1 = 1, X_2 = 1\} = P\{X_2 = 1 | X_1 = 1\} P\{X_1 = 1\} \tag{4.39}$$

In the same interval $P\{X_2 = 1 | X_1\} = 1$. Therefore

$$P\{X_1 = 1, X_2 = 1\} = P\{X_1 = 1\} = \tfrac{1}{2} \tag{4.40}$$

Similarly $P\{X_1 = -1, X_2 = -1\} = \frac{1}{2}$. Since X_1 and X_2 in the same interval cannot change sign, $P\{X_2 = 1 | X_1 = -1\}$ and $P\{X_2 = -1 | X_1 = 1\}$ are both zero. Thus

$$P\{X_1 = 1, X_2 = -1\} = P\{X_2 = -1, X_1 = 1\} = 0 \tag{4.41}$$

Using these results in Eq. (4.36) gives the second-order density of the random process for times t_1 and t_2 in the same interval as

$$f(x_1, x_2) = \tfrac{1}{2}\delta(x_1 - 1, x_2 - 1) + \tfrac{1}{2}\delta(x_1 + 1, x_2 + 1) \tag{4.42}$$

Autocorrelation Function. The autocorrelation function $R_{xx}(t_1, t_2)$ is found by computing the expected value of the product of $X(t_1)$ and $X(t_2)$ as

$$R_{XX}(t_1, t_2) = E[X(t_1)X(t_2)] = E[X_1 X_2] = \int_{-\infty}^{\infty}\int_{-\infty}^{\infty} x_1 x_2 f_{X_1 X_2}(x_1, x_2)\, dx_1\, dx_2 \tag{4.43}$$

As the density $f(x_1, x_2)$ is different for the cases of t_1 and t_2 in the same and different intervals, the autocorrelation function is computed as follows:

For t_1 and t_2 in different intervals,

$$\begin{aligned} R_{XX}(t_1, t_2) &= \int_{-\infty}^{\infty}\int_{-\infty}^{\infty} x_1 x_2 \tfrac{1}{4}[\delta(x_1 - 1, x_2 - 1) + \delta(x_1 - 1, x_2 + 1) \\ &\quad + \delta(x_1 + 1, x_2 - 1) + \delta(x_1 + 1, x_2 + 1)]\, dx_1\, dx_2 \\ &= \tfrac{1}{4}[(1)(1) + (1)(-1) + (-1)(1) + (-1)(-1)] = 0 \end{aligned} \tag{4.44}$$

For t_1 and t_2 in the same interval,

$$\begin{aligned} R_{XX}(t_1, t_2) &= \int_{-\infty}^{\infty}\int_{-\infty}^{\infty} [\tfrac{1}{2}\delta(x_1 - 1, x_2 - 1)] + \tfrac{1}{2}\delta(x_1 + 1, x_2 + 1)]\, dx_1\, dx_2 \\ &= (1)(1)\tfrac{1}{2} + (-1)(-1)\tfrac{1}{2} = 1 \end{aligned} \tag{4.45}$$

Therefore $R_{xx}(t_1, t_2)$ is the two-dimensional "snake of blocks" function along the line $t_1 = t_2$ shown in Figure 4.9.

Stationarity. The semirandom binary transmission process is stationary in the mean since $E[X(t)] = 0$, which is not a function of time t, and stationary of order 1 since its first-order densities are not a function of time. However, the semirandom process is not stationary of order 2, and thus all higher orders, because the second-order densities are not the same for shifted versions of time differing by a set amount. Similarly the process is seen to be not stationary in autocorrelation and thus not wide sense stationary also.

4.4.3 Random Binary Transmission Process

In the previous example we considered a semirandom binary transmission signal. It was called semirandom since we knew the transition times were synchronous. A more realistic case is where we do not have the realizations synchronized. In this example the random binary transmission process will be defined. Let D be a random variable that is independent of the semirandom binary process $X(t)$ described in the previous section. A random variable is independent of a process if it is group independent of every set of random variables defined across the process for all

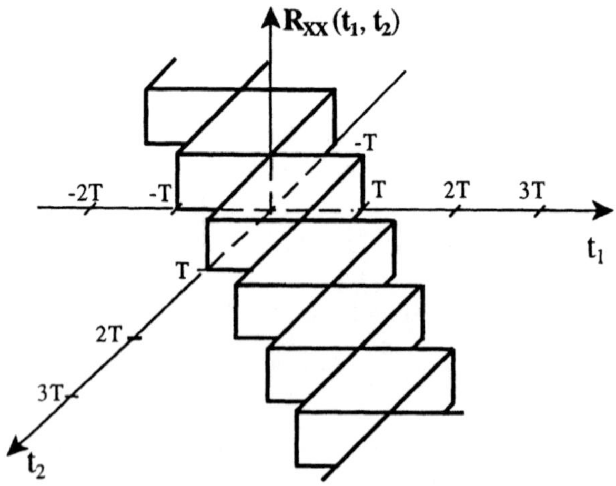

Figure 4.9 The autocorrelation function $R_{XX}(t_1, t_2)$ for the semirandom binary transmission process.

times in the indexing set. Assume that D has a uniform probability density function, $f_D(d)$, on $[0, T]$ that allows total uncertainty with respect to the time origin up to one transition duration.

The **random binary transmission process**, $Y(t)$, is defined by

$$Y(t) = X(t - D) \tag{4.46}$$

where $X(t)$ is the semirandom binary transmission process and D is a uniform random variable on $[0, T]$ that is independent of $X(t)$.

First Order Densities of Random Binary Transmission Process. The first-order density for this process can be obtained by first finding the first-order density function for $X(t - d)$ conditioned on a particular value of D and then integrating this conditional density with the probability density function for D. The first-order density function for $X(t - d)$ is obtained similarly to (4.34), and since d plays no part, the conditional density is the same. Since the result is not a function of d, the integration gives the first-order density as the same as that for the semirandom binary process:

$$f(x_t) = \frac{1}{2}\delta(x_t - 1) + \frac{1}{2}\delta(x_t + 1) \tag{4.47}$$

Mean of Random Binary Transmission Process. Since the first-order density is the same for the random binary transmission process, the expected value will also be the same, that is, zero. Alternatively, we can find the mean by using the iterated expected value formula as follows:

$$E[Y(t)] = E[E[Y(t - d)]|_{d=D}] = 0 \tag{4.48}$$

Second Order Densities of Random Binary Transmission Process. To obtain the second-order densities, it will be convenient to use the conditional probability density functions for $X(t - d)$, where d is any value between 0 and T. Thus the second-order density conditioned on d is as follows: For values of t_1 and t_2 inside and outside the doubly shaded area shown in Figure 4.10, the density becomes:

inside:
$$f(x_1, x_2|D = d) = \tfrac{1}{2}\delta(x_1 - 1, x_2 - 1) + \tfrac{1}{2}\delta(x_1 + 1, x_2 + 1) \tag{4.49a}$$

outside:
$$\begin{aligned} f(x_1, x_2|D = d) = {} & \tfrac{1}{4}\delta(x_1 - 1, x_2 - 1) + \tfrac{1}{4}\delta(x_1 - 1, x_2 + 1) \\ & + \tfrac{1}{4}\delta(x_1 + 1, x_2 - 1) + \tfrac{1}{4}\delta(x_1 + 1, x_2 + 1) \end{aligned} \tag{4.49b}$$

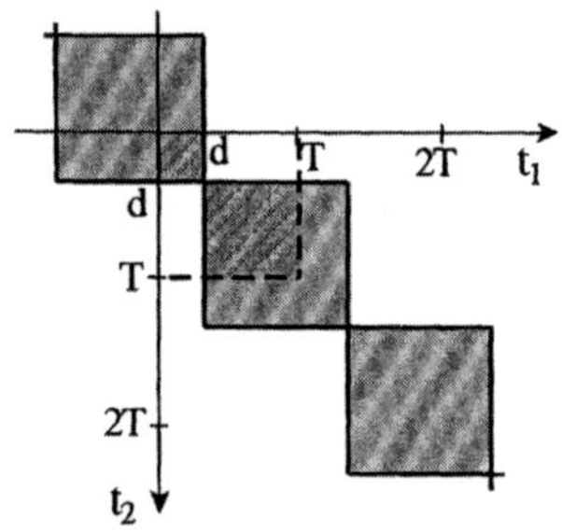

Figure 4.10 Regions for the conditional densities.

For $t_1 > t_2$ multiplying (4.49a) and (4.49b) by the appropriate probabilities and adding gives

$$f(x_1, x_2) = \left[\frac{1}{2} - \frac{(t_1 - t_2)}{4T}\right](\delta(x_1 - 1_1x_2 - 1) + \delta(x_1 + 1_1x_2 + 1)) \\ + \frac{(t_1 - t_2)}{4T}[\delta(x_1 + 1_1x_2 - 1) + \delta(x_1 - 1_1x_2 + 1)] \tag{4.50}$$

Autocorrelation Function of Random Binary Transmission Process. The autocorrelation function for the random binary transmission process can be found by using the iterated expected value approach as follows:

$$\begin{aligned} R_{YY}(t_1, t_2) &= E[Y(t_1)Y(t_2)] = E[X(t_1 - D)X(t_2 - D)] \\ &= E[R_{XX}(t_1 - d, t_2 - d)|_{d=D}] \\ &= \frac{1}{T}\int_0^T R_{XX}(t_1 - d, t_2 - d)\, dd \end{aligned} \tag{4.51}$$

where $R_{XX}(t_1, t_2)$, is known to be 1 for (t_1, t_2) in the same box as shown in Figure 4.10 and zero outside that region. So the $R_{XX}(t_1 - d, t_2 - d)$ will be one provided that the point $(t_1 - d, t_2 - d)$ falls in one of the intervals. Assume that $t_1 > t_2$, that $t_1 - t_2 = \tau$, and that $0 < \tau \leq T$. Then the point $(t_1 - d, t_2 - d)$ will appear as in Figure 4.11. It will be in a proper interval provided that $t_1 - d < T$ and $t_2 - d > 0$.

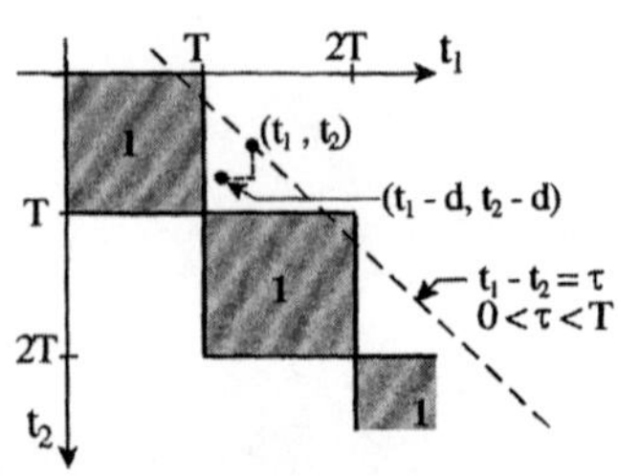

Figure 4.11 Regions for the random binary transmission process.

Rearranging these conditions to be put as a region for d gives $d > t_1 - T$ and $d < t_2$ to be in the proper interval for the value of the $R_{XX}(t_1 - d, t_2 - d)$ to be one. Thus from (4.51) we have

$$\begin{aligned} R_{YY}(t_1, t_2) &= \frac{1}{T}\int_{t_1 - T}^{t_2} 1 dd = \frac{t_2 - (t_1 - T)}{T} \\ &= 1 - \frac{t_1 - t_2}{T} \end{aligned} \tag{4.52}$$

Repeating the steps above for $t_1 < t_2$, we find that $R_{YY}(t_1, t_2) = 1 - (t_1 - t_2)/T$. If $t_1 - t_2 = \tau$ where $\tau > T$, the point will never fall in an interval because d is always $\leq T$, so $R_{YY}(t_1, t_2) = 0$ for those values of t_1 and t_2. Thus the autocorrelation function for the random binary transmission process can be expressed as

$$R_{YY}(t_1, t_2) = \begin{cases} 1 - |t_2 - t_1|/T, & -T \leq (t_1 - t_2) \leq T \\ 0, & \text{otherwise} \end{cases} \tag{4.53}$$

Stationarity of the Random Binary Transmission Process. The random binary random transmission wave is thus seen to be stationary of mean, stationary in autocorrelation, stationary of order one, and wide sense stationary.

4.4.4 Semirandom Telegraph Waves Process

Consider the experiment of a random selection of times at a uniform rate λ per unit time such that the probability of k points in the interval $[0, t]$ is given by

$$P(k \text{ points} \in [0, t]) = \frac{(\lambda t)^k e^{-\lambda t}}{k!} \tag{4.54}$$

The outcome of a single trial of the experiment results in a sequence of times

$$\zeta_i = [\ldots, t_{-2}, t_{-1}, t_0, t_1, t_2, \ldots] \tag{4.55}$$

Define the **semirandom telegraph process** as follows:

$$X(t_1\zeta_i) \triangleq \begin{cases} 1 \text{ at time } t \text{ if } \zeta_i \text{ has an even number of points in } [0, t] \\ 1 \text{ at time } t \text{ if } \zeta_i \text{ has an odd number of points in } [0, t] \end{cases} \tag{4.56}$$

An example of a realization of the semirandom telegraph process is shown in Figure 4.12.

We would like to characterize this process by finding its first-order densities, second-order densities, mean, and autocorrelation function.

First-Order Density. At any given time t, $X(t)$ is a discrete random variable denoted by X_t which takes on only the values 1 or -1. The probability density function for X_t can be written as

$$f_{X_t}(x_t) = P\{X_t = 1\}\delta(x_t - 1) + P\{X_t = -1\}\delta(x_t + 1) \tag{4.57}$$

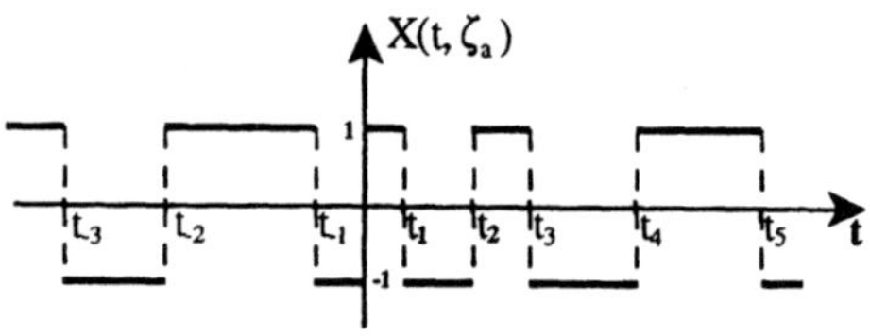

Figure 4.12 Realization of the semirandom telegraph process.

where

$$\begin{aligned} P\{X_t = 1\} &= P\{\text{the number of points} \in [0, t] \text{ is even}\} \\ &= P\{\text{exactly 0 points} \in [0, t] \\ &\qquad \text{or exactly 2 points} \in [0, t] \\ &\qquad\qquad \ldots \text{or exactly } 2k \text{ points} \in [0, t] \\ &\qquad\qquad\qquad \ldots or \ldots\} \end{aligned} \tag{4.58}$$

Since the events exactly $2k$ and exactly $2j$ points are mutually exclusive for $k \neq j$, the probability of the union of all the events can be written as the sum of probabilities. Substituting the probability of exactly $2k$ points from Eq. (4.54) into (4.58) gives the following for $t > 0$:

$$\begin{aligned} P\{X_t = 1\} &= \sum_{k=0}^{\infty} P\{2k \text{ points in } [0, t]\} \\ &= \sum_{k=0}^{\infty} \frac{e^{-\lambda t}(\lambda t)^{2k}}{(2k)!} \end{aligned} \tag{4.59}$$

Factoring out the $e^{-\lambda t}$ and recognizing the sum as $\cosh(\lambda t)$, $P\{X_t = 1\}$ simplifies to

$$P\{X_t = 1\} = e^{-\lambda t} \cosh(\lambda t) \tag{4.60}$$

In a similar fashion the $P\{X_t = -1\}$ can be written as

$$\begin{aligned} p\{X_t = -1\} &= \sum_{k=0}^{\infty} P\{2k + 1 \text{ points in } [0, t]\} \\ &= \sum_{k=0}^{\infty} \frac{e^{-\lambda t}(\lambda t)^{2k+1}}{(2k + 1)!} = e^{-\lambda t} \sinh(\lambda t) \end{aligned} \tag{4.61}$$

From these results the first-order density of the semirandom telegraph wave becomes

$$f_{X_t}(x_t) = e^{-\lambda t} \cosh(\lambda t)\delta(x_t - 1) + e^{-\lambda t} \sinh(\lambda t)\delta(x_t + 1) \tag{4.62}$$

For $t < 0$, the density is obtained by replacing λt by $-\lambda t$.

Mean of $X(t)$. Using the above obtained first-order density function for the process $X(t)$, we calculate the mean $E[X(t)]$ as

$$\begin{aligned} E[X(t)] &= \int_{-\infty}^{\infty} x_t f_{X_t}(x_t)\, dx_t \\ &= \int_{-\infty}^{\infty} x_t[e^{-\lambda t}\cosh(\lambda t)\delta(x_t - 1) + e^{-\lambda t}\sinh(\lambda t)\delta(x_t + 1)]\, dx_t \end{aligned} \tag{4.63}$$

The sampling property of the delta function can be used to simplify $E[X(t)]$ as follows:

$$E[X(t)] = e^{-\lambda t}\cosh(\lambda t) - e^{-\lambda t}\sinh(\lambda t) = e^{-2\lambda t} \tag{4.64}$$

For $t < 0$, $E[X(t)]$ becomes $e^{2\lambda}$, and thus $E[X(t)]$ can finally be written as

$$E[X(t)] = e^{-2\lambda|t|} \tag{4.65}$$

A plot of the mean $E[X(t)]$ is shown in Figure 4.13. It is noticed that for positive time it starts at 1 and goes to zero as $t \to \infty$. Close to the origin it is almost 1, since each realization of the process starts at 1.

Second-Order Density. The random variables $X(t_1)$ and $X(t_2)$ denoted by X_1 and X_2, respectively, are discrete random variables taking on only values 1 and -1 with joint probability density function $f_{X_1X_2}(x_1, x_2)$ as

$$\begin{aligned} f_{X_1X_2}(x_1, x_2) = P\{X_1 &= +1, X_2 = +1\}\,\delta(x_1 - 1, x_2 - 1) \\ + P\{X_1 &= +1, X_2 = -1\}\,\delta(x_1 - 1, x_2 + 1) \\ + P\{X_1 &= -1, X_2 = +1\}\,\delta(x_1 + 1, x_2 - 1) \\ + P\{X_1 &= -1, X_2 = -1\}\,\delta(x_1 + 1, x_2 + 1) \end{aligned} \tag{4.66}$$

The weights of the delta functions are found by writing the random variables joint probability in terms of a conditional probability and marginal probability as follows:

$$P\{X_1 = +1, X_2 = -1\} = P\{X_2 = -1|X_1 = +1\}P\{X_1 = 1\} \tag{4.67}$$

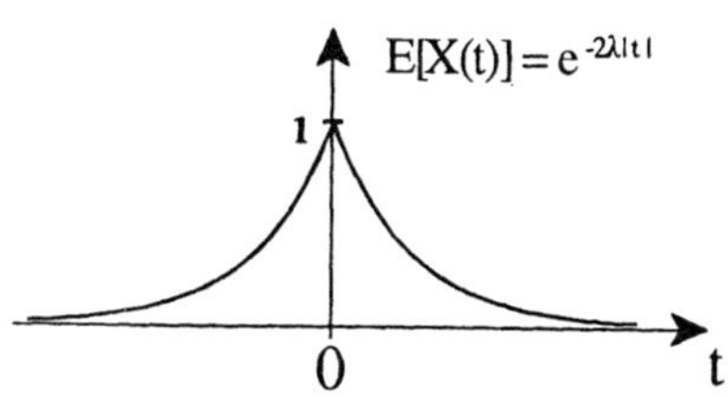

Figure 4.13 Mean of the semirandom telegraph wave process.

The conditional probability that an odd number of points occur in interval $[t_1, t_2]$ is for $t_1 > t_2$ the following

$$P\{X_2 = -1 | X_1 = +1\} = e^{\lambda(t_2 - t_1)} \sinh(\lambda(t_2 - t_1)) \tag{4.68}$$

The $P\{X_1 = +1\}$ was found earlier as

$$P\{X_1 = +1\} = e^{-\lambda t_1} \cosh(\lambda t_1) \tag{4.69}$$

Using the conditional and marginal probabilities just determined gives the joint probability as

$$P\{X_1 = +1, X_2 = -1\} = e^{-\lambda t_2} \cosh(\lambda t_1) \sinh(\lambda(t_2 - t_1)) \tag{4.70}$$

In a similar way the other joint probabilities can be determined, resulting in the joint probability density function

$$\begin{aligned} f_{X_1X_2}(x_1, x_2) = e^{-\lambda t_2} &\cosh(\lambda t_1) \cosh(\lambda(t_2 - t_1))\, \delta(x_1 - 1, x_2 - 1) \\ + e^{-\lambda t_2} &\cosh(\lambda t_1) \sinh(\lambda(t_2 - t_1))\, \delta(x_1 - 1, x_2 + 1) \\ + e^{-\lambda t_2} &\sinh(\lambda t_1) \sinh(\lambda(t_2 - t_1))\, \delta(x_1 + 1, x_2 - 1) \\ + e^{-\lambda t_2} &\sinh(\lambda t_1) \cosh(\lambda(t_2 - t_1))\, \delta(x_1 + 1, x_2 + 1) \end{aligned} \tag{4.71}$$

Autocorrelation Function. From the $f_{X_1X_2}(x_1, x_2)$ above the autocorrelation function $R_{XX}(t_1, t_2)$ is obtained as

$$R_{XX}(t_1, t_2) = \int_{-\infty}^{\infty} \int_{-\infty}^{\infty} x_1 x_2\, f_{X_1X_2}(x_1, x_2)\, dx_1\, dx_2 \tag{4.72}$$

Carrying out the integration and using the sampling property of the two-dimensional delta function, $R_{XX}(t_1, t_2)$, can be simplified to give

$$R_{XX}(t_1, t_2) = e^{-2\lambda(t_2 - t_1)} \qquad \text{for } t_2 > t_1 \tag{4.73}$$

By reversing the roles of t_2 and t_1 and repeating the procedure described above, $R_{XX}(t_1, t_2)$ can be shown to be

$$R_{XX}(t_1, t_2) = e^{-2\lambda|t_2 - t_1|} \qquad \text{for all } t_1 \text{ and } t_2 \tag{4.74}$$

A plot of $R_{XX}(t_1, t_2)$ as a function of t_1 and t_2 is shown in Figure 4.14a and is a "trough" of absolute exponentials. In Figure 4.14b, $R_{XX}(t_1, t_2)$ is plotted as a function of the time difference. Since $R_{XX}(t_1, t_2)$ is a function of time difference only, it can be written as $R_{XX}(\tau)$ where $\tau = t_1 - t_2$. Thus the semirandom telegraph wave process is stationary in autocorrelation. However, it is not wide sense stationary since $E[X(t)]$ is a function of t as shown in Figure 4.13.

4.4.5 Random Telegraph Process

Let $X(t)$ be the semirandom telegraph random process. It is not stationary in the mean as its mean is exponential, not constant with time, since the telegraph process

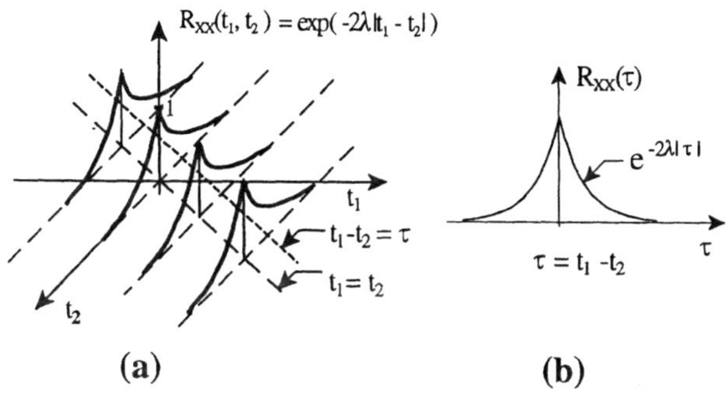

Figure 4.14 Autocorrelation function for the semirandom telegraph wave process. (a) plot of $R_{XX}(t_1, t_2)$ as function of t_1 and t_2, (b) plot of $R_{XX}(\tau)$.

always starts at the origin in a positive fashion. The **random telegraph wave** can be defined as

$$Y(t) = AX(t) \tag{4.75}$$

where A is discrete random variable with probability density as

$$f_A(a) = \tfrac{1}{2}\delta(a+1) + \tfrac{1}{2}\delta(a-1) \tag{4.76}$$

A is assumed to be independent of the random process $X(t)$. A random variable is independent of a random process if it is group independent of all random variables defined across the process.

The expected value or mean of the process $Y(t)$ can be determined as follows since the expected value of A is zero:

$$E[Y(t)] = E[AX(t)] = E[A]E[X(t)] = 0 \tag{4.77}$$

The autocorrelation function for $Y(t)$ can be found as follows:

$$\begin{aligned} R_{YY}(t_1, t_2) &= E[AX(t_1)AX(t_2)] = E[A^2]E[X(t_1)X(t_2)] \\ &= 1 \cdot R_{XX}(t_1, t_2) = e^{-2\lambda|t_2 - t_1|} \end{aligned} \tag{4.78}$$

The expected value of the product above can be written as the product of the expected values because A^2 is also independent of the process (functions of independent processes are independent), and since $E[A^2] = 1$, the autocorrelation function can be written in terms of the time difference τ as

$$R_{YY}(\tau) = e^{-2\lambda|\tau|} \tag{4.79}$$

Since the mean is a finite constant, (zero), and the autocorrelation function is a function of time difference only, we have shown that the random telegraph wave is a wide sense stationary random process.

A logical question at this time is: Is the random telegraph wave stationary of any other variety? For example, is it stationary of order 1? In order to answer that question the first-order density for the semirandom binary transmission must be found. One way of obtaining the density is to integrate out the conditional density for Y_t:

$$f_{Y_t}(y_t) = \int_{-\infty}^{\infty} f(y_t|a) f_A(a)\, da \tag{4.80}$$

where $f_A(a)$ is given by Eq. (4.76) and $f_{Y_t}(y_t)$ can be determined by modifying Eq. (4.62), which for the semirandom telegraph first-order density, at amplitude $A = a$, is

$$f(y_t|a) = e^{-\lambda t}\cosh(\lambda t)\delta(y_t - a) + e^{-\lambda t}\sinh(\lambda t)\delta(y_t + a) \tag{4.81}$$

Substituting the two densities into the equation for $f_{Y_t}(y_t)$ and integrating using the sampling property of the delta function gives, after simplification, the density as

$$f_{Y_t}(y_t) = 0.5\delta(y_t + 1) + 0.5\delta(y_t - 1) \tag{4.82}$$

The first-order density is thus shown to be independent of time t, and hence the random telegraph wave is stationary of order one.

The answer to the question of stationarity of order two is saved for the reader's homework.

4.4.6 Random Sinusoidal Signals

Sinusoidal signals are ubiquitous in physics, electrical engineering, and the sciences. In many situations collections of these signals have different radian frequencies, different phases, and different amplitudes. If these parameters can be modeled as random variables, then the **random sinusoidal signal random process** can be defined by

$$x(t) = A\ \cos(\Omega t + \Phi) \tag{4.83}$$

where A, Ω and Φ are random variables. By making different assumptions about the characterizations of $A, \Omega,$ and Φ, various partial or even total characterizations of $X(t)$ can be obtained. For example, if $A, \Omega,$ and Φ are characterized by their joint probability density function, we can determine the mean, autocorrelation function, first-order density, and so forth, of the process $X(t)$. However, if all we know is the mean and variances of the random variables $A, \Omega,$ and Φ, we will be, in general, unable to determine even the lowest-order characterization, the mean of the random process. The following example will explore these relationships for a useful and common random process, a sinusoidal process.

EXAMPLE 4.3

Define a random process $X(t)$ as $X(t) = A\cos(\Omega_0 t + \Phi)$, where A and Φ are independent random variables with known probability density functions $f_A(a)$ and $f_\Phi(\phi)$ and where Ω_0 is deterministic and known. This is the case of a known sinusoidal frequency signal with random amplitude and phase. Further it is assumed that A is a

Rayleigh random variable and Φ is a uniform random variable with the following probability density functions:

$$f_A(a) = ae^{-a^2/2}\mu(a) \quad f_\Phi(\phi) = \begin{cases} \dfrac{1}{2\pi}, & 0 < \phi < 2\pi \\ 0, & \text{elsewhere} \end{cases} \tag{4.84}$$

For this defined random process determine (a) the mean of the process, (b) the autocorrelation function of the process, (c) the first-order densities of the process, (d) the second order density functions, and (e) the types of stationarity of the process.

SOLUTION

(a) The mean of the process, $E[X(t)]$, is

$$E[X(t)] = E[A\cos(\Omega_0 t + \Phi)] = E[A]E[\cos(\Omega_0 t + \Phi)] \tag{4.85}$$

The second step can be made since the expected value of a product of independent random variables is the product of the expected values. Since Φ is uniform on $[0, 2\pi]$, the second expected value above is zero, thus giving for all t,

$$E[X(t)] = 0 \tag{4.86}$$

(b) The autocorrelation function $R_{XX}(t_1, t_2)$ is

$$\begin{aligned} R_{XX}(t_1, t_2) &= E[A\ \cos(\Omega_0 t_1 + \Phi)A\cos(\Omega_0 t_2 + \Phi)] \\ &= E[A^2]E[\cos(\Omega_0 t_1 + \Phi)\cos(\Omega_0 t_2 + \Phi)] \end{aligned} \tag{4.87}$$

Again the second step can be done since functions of independent random variables are independent. The first term is easily obtained from the density $f_A(a)$, whereas the second term takes slightly more work. Using the product identity for cosines the second term becomes

$$\begin{aligned} &E[\cos(\Omega_0 t_1 + \Phi)\cos(\Omega_0 t_2 + \Phi)] \\ &\quad = E[\tfrac{1}{2}\cos(\Omega_0(t_1 + t_2) + 2\Phi)] + E[\tfrac{1}{2}\cos(\Omega_0(t_1 - t_2))] \end{aligned} \tag{4.88}$$

The first expected value on the right side of the equation above is zero since the integration of a cosine over two periods is zero, while the second expected value gives the bracketed term since it is nonrandom. Therefore the autocorrelation function becomes

$$R_{XX}(t_1, t_2) = \tfrac{1}{2}E[A^2]\cos(\Omega_0(t_1 - t_2)) \tag{4.89}$$

(c) The first-order densities $f(x; t)$ can be obtained in several different ways for the random variable X_t defined by

$$X_t = A\cos(\Omega_0 t + \Phi) \tag{4.90}$$

From the equation above, X_t is one function of the two random variables A and Φ. Defining a random variable B by $B = \cos(\Omega_0 t + \Phi)$, X_t is seen as the product of A and B. The density for the product X_t has been shown to be

$$f_{X_t}(x_t) = \int_{-\infty}^{\infty} f_{X_t}(x_t|b) f_B(b)\, db \tag{4.91}$$

where $f_B(b)$ is known to be

$$f_B(b) = \begin{cases} \dfrac{1}{\pi\sqrt{1-b^2}}, & |b| < 1 \\ 0, & \text{elsewhere} \end{cases} \tag{4.92}$$

and $f_A(a)$ is given in the problem statement. Strange as it may seem, the result of performing the integration gives $f(x_t)$ as

$$f_{X_t}(x_t) = \frac{1}{\sqrt{2\pi}\sigma_{X_t}} \exp\left\{\frac{-x_t^2}{2\sigma_{X_t}^2}\right\} \tag{4.93}$$

Therefore the first order density for the process is Gaussian with variance given by evaluating Eq. (4.89) for $t_1 = t_2 = t$ as

$$\sigma_{X_t}^2 = \frac{1}{2}E[A^2] = \frac{1}{2}\left(2 - \frac{\pi}{2}\right) = 0.2146 \tag{4.94}$$

(d) The second-order density for $X(t_1)$ and $X(t_2)$ defined by X_1 and X_2 as functions of the random variables A and Φ, respectively, can be found using the two functions of two random variables transformation theorem on the following transformation:

$$\begin{aligned} X(t_1) &\triangleq X_1 = A\ \cos(\Omega_0 t_1 + \Phi) \\ X(t_2) &\triangleq X_2 = A\ \cos(\Omega_0 t_2 + \Phi) \end{aligned} \tag{4.95}$$

Another way to obtain the density uses the result obtained in Chapter 2, that $X(t_1)$ and $X(t_2)$ are Gaussian random variables since Φ is a Rayleigh distributed random variable independent of the random variable A which is uniformly distributed. By this result, the problem of characterizing the two Gaussian random variables X_1 and X_2 is equivalent to finding the means, variances, and correlation for those variables. It becomes expedient in finding those moments that we expand out the cosine functions from above to give

$$\begin{aligned} X_1 &= A\cos\Phi\cos(\Omega_0 t_1) + A\sin\Phi\sin\ (\Omega_0 t_1) \\ X_2 &= A\cos\Phi\cos(\Omega_0 t_2) + A\sin\Phi\sin\ (\Omega_0 t_2) \end{aligned} \tag{4.96}$$

since $A\cos\Phi$ and $A\sin\Phi$ are independent and Gaussian, the random variables X_1 and X_2 are jointly Gaussian as they are linear functions of Gaussian random variables. By taking expected values, we can show the moments to be

$$\begin{aligned} &E[X_1]=0, \quad E[X_1^2]=\tfrac{1}{2}E[A^2] \quad E[X_1X_2]=E[A^2]R_{XX}(t_1,t_2) \\ &E[X_2]=0 \quad E[X_2^2]=\tfrac{1}{2}E[A^2], \end{aligned} \tag{4.97}$$

These five parameters determine the second-order Gaussian density function for the random sinusoidal signal described in this example.

(e) Since the mean of $X(t)$ is zero for all time, the process is stationary in mean. As the autocorrelation function is a function of time difference only, the process is stationary in autocorrelation, and since it is stationary in mean, it is also wide sense stationary. Further, since $X(t)$ is a Gaussian random process and wide sense stationary, it is strict sense stationary.

With different assumptions on A, Ω, and Φ we obtain different results for the mean and autocorrelation function. Some of these appear in the Problems section for this chapter. ■

4.4.7 Random Walk Process

Consider the experiment of tossing a coin countably infinite number of times. The outcome of the experiment is a cartesian product of single tosses that are either heads or tails. The sample space for the random process is then the set of all countably infinite sequences of heads and tails. The Borel field is defined as the power set of the sample space that is the set of all possible subsets of this sample space. Assume that the probability of getting a head at each toss is equal to p and that each toss is "independent" of the others.

The **random walk process** is defined along uniform segments of the time axis for each outcome $\xi_a = (\ldots h, h, t, t, \ldots)$ as follows:

$$\begin{aligned} &X(t,\zeta_a)=cS, \quad (n-1)T \le t < nT, \qquad n=1,2,\ldots \\ &c = \text{number of heads} - \text{number of tails in first } n \text{ elements of } \zeta_a \end{aligned} \tag{4.98}$$

The T and S are time duration and step size, respectively, and fixed. If there are k heads in the first n elements, then the number of tails is $(n-k)$ and the number of heads minus the number of tails is $k-(n-k)$ or $2k-n$. Therefore the process can be defined as follows for each outcome:

$$X(t,\zeta)=(2K-n)S, \quad (n-1)T \le t < nT, \qquad n=1,2,\ldots \tag{4.99}$$

where K is a random variable representing the number of heads. A few realizations of the process are shown in Figure 4.15.

The first-order density, mean, second-order density, and autocorrelation function for the random walk are now presented.

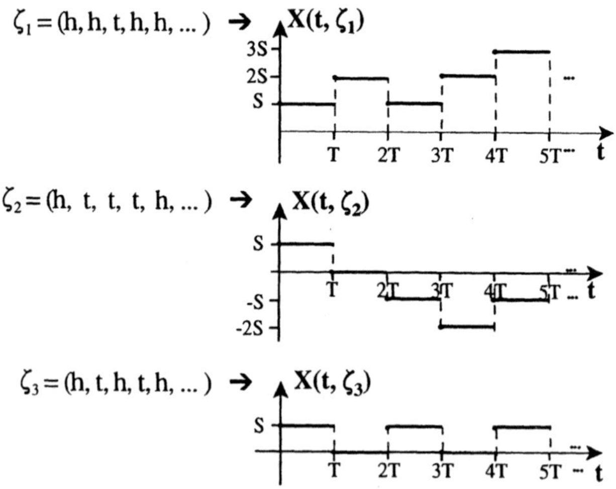

Figure 4.15 Realizations of the random walk process.

First Order Density. By the independent assumption, K has a binomial density as follows:

$$f_K(k) = \sum_{j=0}^{n} \binom{n}{j} p^j (1-p)^{n-j} \delta(k-j) \tag{4.100}$$

where p is the probability of a head on each toss and $(1-p)$ is the probability of a tail. Since $X_t = (2K - n)S$ or a transformation of K, the probability density of X_t for $(n-1)T < t < nT$ becomes

$$f_{X_t}(x_t) = \sum_{k=0}^{n} \binom{n}{k} p^k (1-p)^{nk} \delta(x_t - (2k-n)S) \tag{4.101}$$

The result above was obtained by using the procedure for transformations of a discrete random variable. If we fix t in the first interval $0 \le t < T$, the density becomes

$$f_{X_t}(x_t) = (1-p)\delta(x_t + S) + p\delta(x_t - S) \tag{4.102}$$

and for any $1 \le t < 2$, the second interval, we have

$$f_{X_t}(x_t) = (1-p)^2\delta(x_t + S) + 2p(1-p)\delta(x_t) + p^2\delta(x_t - S) \tag{4.103}$$

Representative probability density functions for several t's are shown in Figure 4.16. As t becomes large, the envelope of the density functions approaches a Gaussian function centered at np and with a variance parameter of $np(p-1)$.

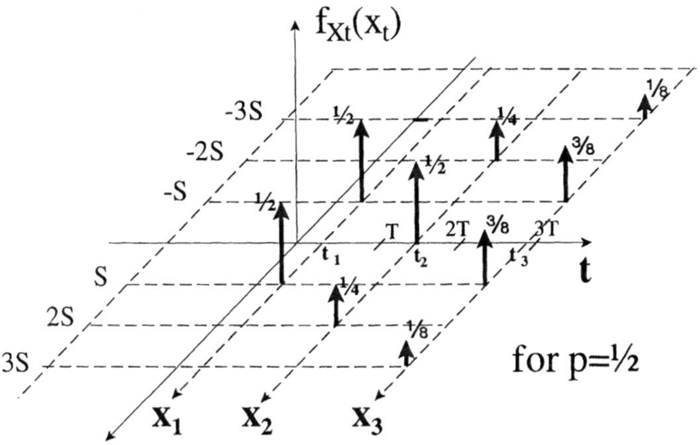

Figure 4.16 Representative first order probability density functions for the random walk process.

Mean. Using the density function for X_t given in Eq. (4.101), we can write the mean of $X(t)$ as

$$\begin{aligned} E[X(t)] = E[X_t] &= \int_{-\infty}^{\infty} x_t f_{X_t}(x_t)\,dx_t \\ &= \int_{-\infty}^{\infty} x_t \left(\sum_{k=0}^{n} \binom{n}{k} p^k (1-p)^{n-k} \delta(x_t - (2k-n)S) \right) dx_t \end{aligned} \tag{4.104}$$

Using the sampling property of the delta function, we can write the expected value of the random walk process for $(n-1)T \leq t < nT$ as

$$E[X(t)] = \sum_{k=0}^{n} (2k-n)S \binom{n}{k} p^k (1-p)^{n-k} \tag{4.105}$$

The first part of the sum is just the average value for the binomial density times 2 and the second is just n times 1, since it is a sum over the binomial density. Therefore $E[X(t)]$, for $(n-1)T \leq t < nT$, is

$$E[X(t)] = (np)2S - (n)S = n(2p-1)S \tag{4.106}$$

A plot of the expected value is shown in Figure 4.17 for $p > \frac{1}{2}$, $p = \frac{1}{2}$, and $p < \frac{1}{2}$. For all p other than $\frac{1}{2}$ the line $E[X(t)]$ as $t \to \infty$ is either $+\infty$ or $-\infty$, whereas for $p = \frac{1}{2}$, the $E[X(t)]$ is zero for all t. Thus, unless $p = \frac{1}{2}$, the random walk process is not stationary in the mean.

Autocorrelation Function. One approach that can be used to find the autocorrelation is to first find the second-order density and then compute the expected value using that density.

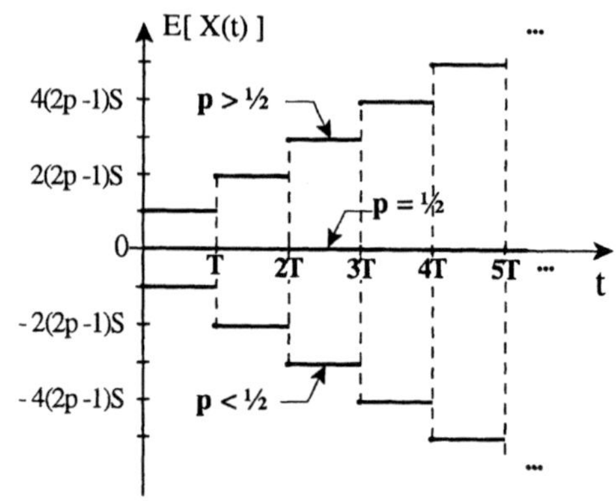

Figure 4.17 The mean of the random walk process for various values of p, the probability of a heads.

An alternative approach, which for this particular problem is easier, is to compute the expected value directly as a function of other random variables. To do this, we have to view the random process as the sum of independent identically distributed random variables:

$$X(t) = \sum_{i=1}^{n} X_i, \quad (n-1)T \leq t < nT, \quad n = 1, 2, \ldots \tag{4.107}$$

where X_i are discrete independent random variables with identical probability density functions

$$f_{X_i}(x_i) = (1-p)\delta(x_i + S) + p\delta(x_i - S) \tag{4.108}$$

Reminiscent of the random telegraph wave development, the autocorrelation function $R_{xx}(t_1, t_2)$ will take on different values depending on whether t_1 and t_2 are in the same or different intervals along the t axis.

Assuming $t_1 \geq t_2$, where t_1 is in the mth interval ($(m-1)T \leq t_1 < mT$), and t_2 is in the nth interval, $(n-1)T \leq t < nT$, the autocorrelation function becomes for $m > n$,

$$\begin{aligned} R_{XX}(t_1, t_2) &= E[X(t_1)X(t_2)] = E\left[\sum_{i=1}^{m} X_i \sum_{j=1}^{n} X_j\right] \\ &= E\left[\left(\sum_{i=1}^{n} X_i + \sum_{i=n+1}^{m} X_i\right)\sum_{j=1}^{n} X_j\right] \\ &= E\left[\left(\sum_{i=1}^{n} X_i\right)\left(\sum_{j=1}^{n} X_j\right)\right] + E\left[\left(\sum_{i=n+1}^{m} X_i\right)\left(\sum_{j=1}^{n} X_j\right)\right] \end{aligned} \tag{4.109}$$

Because the X_i are independent and identically distributed, the first expected value can be written as $nE[X_i^2] + (n^2 - n)E^2[X_i]$ and the second becomes

$(m-n)nE^2[X_i]$, since products contain all different random variables. Therefore $R_{XX}(t_1, t_2)$ becomes

$$R_{XX}(t_1, t_2) = nE[X_i^2] + (n^2 - n)E^2[X_i] + (m-n)nE^2[X_i] \tag{4.110}$$

Since the X_i are identically distributed with densities, Eq. (4.108), their first- and second-order moments are

$$\begin{aligned} E[X_i] &= (1-p)(-S) + pS = (2p-1)S \\ E[X_1^2] &= (1-p)(-S)^2 + p(S)^2 = S^2 \end{aligned} \tag{4.111}$$

Substituting these moments into Eq. (4.110) and simplifying, gives $R_{XX}(t_1, t_2)$, for $m > n$, as follows:

$$R_{XX}(t_1, t_2) = nS^2 + n(m-1)(2p-1)^2 S^2 \tag{4.112}$$

Interchanging m and n gives us the following for $m < n$:

$$R_{XX}(t_1, t_2) = mS^2 + m(n-1)(2p-1)^2 S^2 \tag{4.113}$$

If we set $p = \frac{1}{2}$, then $R_{XX}(t_1, t_2)$ becomes as shown in Figure 4.18.

4.5 DEFINITE INTEGRALS OF RANDOM PROCESSES

In many practical problems the integral of random processes appear naturally. For instance, the output of linear time invariant system to a random input can be expressed as a convolution integral of the input with the impulse response. For a given random process $X(t, \zeta)$ and a given ζ, say ζ_1, we have the definite integral of that particular waveform by

$$I(\zeta_1) = \int_a^b X(t, \zeta_1)\, dt \tag{4.114}$$

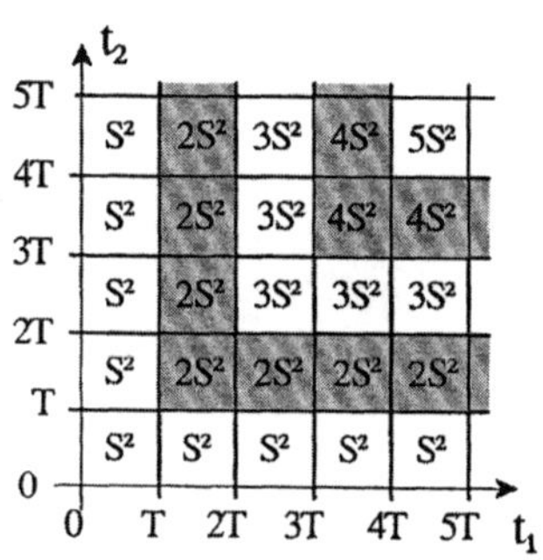

Figure 4.18 Autocorrelation function for the random walk when $p = 0.5$.

As ζ goes over the sample space, $I(\zeta)$ specifies a random variable. For simplicity, we will write that random variable $I(\zeta)$ and denote it by I:

$$I = \int_a^b X(t)\,dt \tag{4.115}$$

It is logical at this point to ask what is the mean and the variance of such a random variable and what information about the random process do we need to find them. The mean of I is given as

$$E[I] = E\left[\int_a^b X(t)\,dt\right] \tag{4.116}$$

It can be shown in most cases that the expected value and the integral can be interchanged to give

$$E[I] = \int_a^b E[X(t)]\,dt = \int_a^b \eta_X(t)\,dt \tag{4.117}$$

In a similar fashion the second moment of I can be found as

$$\begin{aligned} E[I^2] &= E\left(\int_a^b X(t)\,dt \int_a^b X(u)\,du\right) = \int_a^b \int_a^b E[X(t)X(u)]\,dt\,du \\ &= \int_a^b \int_a^b R_{XX}(t, u)\,dt\,du \end{aligned} \tag{4.118}$$

Given $E[I]$ and $E[I^2]$, the variance of I is easily seen to be

$$\sigma_I^2 = E[I^2] - E^2[I] = \int_a^b \int_a^b C_{XX}(t, u)\,dt\,du \tag{4.119}$$

Therefore the mean and variance of I are determined by the second-order characterization of the process, namely its mean $\eta_X(t)$ and autocovariance function $C_X(t, u)$. Higher-order moments require higher-order characterizations of the process $X(t)$ itself. In order to obtain the probability density function for I, a total characterization of the process $X(t)$ on the interval of integration is required.

If the process $X(t)$ is a wide sense stationary process with mean η_X and autocorrelation function $R_{XX}(\tau)$ the formulas for $E[I]$ and σ_I^2 can be simplified. The expected value of I from equation then becomes

$$E[I] = E[\int_a^b X(t)\,dt] = \int_a^b \eta_X(t)\,dt = (b - a)\eta_X \tag{4.120}$$

Since the autocorrelation function $R_{XX}(\tau)$ is a function of the time difference $\tau = t - u$, the two-dimensional integral given in the equation can be changed to a

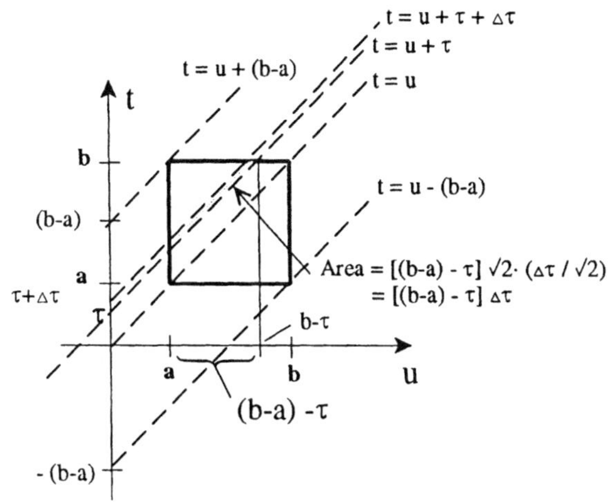

Figure 4.19 Incremental area for the change of variables $t - u = \tau$.

one-dimensional integral. The incremental area required in this change is shown in Figure 4.19, and the resulting integral for the variance becomes

$$\sigma_I^2 = \int_{-(b-a)}^{b-a} [(b-a) - |\tau|][R_{XX}(\tau) - \eta_X^2]\, d\tau$$

or (4.121)

$$\sigma_I^2 = \int_{-(b-a)}^{b-a} [(b-a) - |\tau|] C_{XX}(\tau)\, d\tau$$

An example illustrating the procedure for finding the mean and variance of a definite integral of a wide sense stationary process is now presented.

EXAMPLE 4.4

A wide sense stationary random process $X(t)$ is characterized by its mean and autocorrelation function as follows: $E[X(t)] = 2$ and $R_{XX}(\tau) = 4 + e^{-|\tau|}$. Define a random variable I as $I = \int_1^3 X(t)\, dt$. Determine $E[I]$ and σ_I^2.

SOLUTION

The expected value of the random variable I is easily obtained from (4.120) as

$$\begin{aligned} E[I] &= (b-a)\eta_X \\ &= (3-1)2 = 4 \end{aligned}$$

From the given mean and autocorrelation function, the variance of I can be determined using (4.121) as follows:

$$\begin{aligned}\sigma_I^2 &= \int_{-(b-a)}^{b-a} [(b-a) - |\tau|][R_{XX}(\tau) - \eta_X^2]\, d\tau \\ &= \int_{-2}^{2} [2 - |\tau|][4 + e^{-|\tau|} - 4]\, d\tau \\ &= 2\int_0^2 (2-\tau)e^{-\tau}\, d\tau = 4e^{-2}\end{aligned}$$

■

4.6 JOINT CHARACTERIZATIONS OF RANDOM PROCESSES

In most situations there are many random processes to consider. For example, if the input to a linear or nonlinear system is a random process, then the output will also be a random process. For these type of problems it is desired to characterize the processes jointly. Definitions that prove useful in describing the joint statistical properties include joint probability densities, cross correlations, cross-spectral densities, stationarity, and independence. Such joint characterizations will now be defined.

Two random processes $X(t)$ and $Y(t)$ are defined to be **independent** if any set of random variables defined across the random process $X(t)$ is group independent of any set of random variables defined across the other process $Y(t)$. Two random processes are called **uncorrelated** if their cross-correlation function can be written as the product

$$\begin{aligned}E[X(t_1)Y(t_2)] &= E[X(t_1)]E[Y(t_2)] \qquad \text{for all } t_1, t_2 \\ &= \eta_X(t_1)\eta_X(t_2)\end{aligned} \tag{4.122}$$

which implies that any random variable defined across the process $X(t)$ at time t_1 is uncorrelated with any random variable defined across the process $Y(t)$ at time t_2.

4.6.1 First-Order Joint Densities

Define the random variables $X(t_1)$ and $Y(t_2)$, notation X_1 and Y_2 respectively. The second-order joint density is given as the joint probability density function for these two random variables as $f(x_1, y_2;\ t_1, t_2)$. The notation $f(x_1, y_2)$ is commonly used. The first-order joint densities consist of all joint densities resulting from all choices of times t_1 and t_2.

4.6.2 Cross-correlation Function

The **cross-correlation function**, denoted as $R_{XY}(t_1, t_2)$, is defined as

$$R_{XY}(t_1, t_2) \triangleq E[X(t_1)Y(t_2)] \tag{4.123}$$

If the random processes are complex processes, a useful definition for the cross-correlation function can be obtained by taking the complex conjugate of the second process as

$$R_{XY}(t_1, t_2) = E[X(t_1)Y^*(t_2)] \tag{4.124}$$

4.6.3 Cross-covariance Function

Similar to the auto correlation function except now defined across two processes the **cross-covariance function** $C_{XY}(t_1, t_2)$ is defined by

$$\begin{aligned} C_{XY}(t_1, t_2) &\triangleq E[(X(t_1) - E[X(t_1)])(Y(t_2) - E[Y(t_2)])] \\ &= E[(X(t_1) - \eta_X(t_1))(Y(t_2) - \eta_Y(t_2))] \end{aligned} \tag{4.125}$$

If the processes are complex, then the following definition is used for the cross correlation function:

$$C_{XY}(t_1, t_2) = E[(X(t_1) - \eta_X(t_1))(Y(t_2) - \eta_Y(t_2))^*] \tag{4.126}$$

The cross covariance function and cross correlation functions are related by

$$C_{XY}(t_1, t_2) = R_{XY}(t_1, t_2) - \eta_X(t_1)\eta_Y^*(t_2) \tag{4.127}$$

where $\eta_X(t_1)$ and $\eta_Y(t_2)$ represent the means of processes $X(t)$ and $Y(t)$ as functions of t_1 and t_2, respectively.

4.6.4 Joint Stationarity

Various forms of stationarity can be defined similar to those for just one process. The first and most important is jointly wide sense stationarity. The random processes are **jointly wide sense stationary** if each process $X(t)$ and $Y(t)$ are wide sense stationary themselves and the cross-covariance function is only a function of time difference and $E[X(t_1)Y(t_2)]$ is a constant and not a function of t_1 or t_2. When this is true, the cross-correlation function and cross-covariance function are written in terms of the time difference τ only:

$$C_{XY}(\tau) = R_{XY}(\tau) - \eta_X\eta_Y \tag{4.128}$$

Joint strict sense stationarity would imply that for every time translate of amount ε for any n, the joint probability densities are the same. This can be expressed as

$$\begin{aligned} &f(x_1, x_2, \ldots, x_n, y_1, y_2, \ldots, y_m;\ t_1 + \varepsilon, t_2 + \varepsilon, \ldots, t_n + \varepsilon, t_{n+1} + \varepsilon, t_{n+2} + \varepsilon, \ldots, \\ &t_{n+m} + \varepsilon) = f(x_1, x_2, \ldots, x_n, x_{n+1}, x_{n+2}, \ldots, x_{n+m}; t_1, t_2, \ldots, t_n, t_{n+1}, t_{n+2}, \ldots, t_{n+m}) \end{aligned} \tag{4.129}$$

4.6.5 Cross-spectral Density

When the random processes $X(t)$ and $Y(t)$ are jointly wide sense stationary, it is convenient to define a cross-spectral density that gives spectral content information to the joint process. The **cross-spectral density** for processes $X(t)$ and $Y(t)$ is defined as the Fourier transform of the cross-correlation function $R_{XY}(\tau)$ as

$$
\begin{aligned}
S_{XY}(\omega) &= \int_{-\infty}^{\infty} R_{XY}(\tau)e^{-j\omega\tau}\,d\tau \\
S_{YX}(\omega) &= \int_{-\infty}^{\infty} R_{YX}(\tau)e^{-j\omega\tau}\,d\tau
\end{aligned}
\tag{4.130}
$$

4.7 GAUSSIAN RANDOM PROCESSES

A random process $X(t)$ is a **Gaussian random process** if the random variables $X(t_1), X(t_2), \ldots, X(t_n)$ are jointly Gaussian for all choices of $t_1, t_2, \ldots, t_n$ and for all n. There are a number of properties and theorems related to Gaussian random processes. Three of the most useful follow.

(1) A random process that is wide sense stationary and Gaussian is strict sense stationary.

(2) Integrals of Gaussian random processes over a given time interval are Gaussian random variables and random variables over many different intervals are jointly Gaussian random variables.

(3) The process defined by the integral of a Gaussian random process over the interval $-\infty$ to t, as t varies from $-\infty$ to ∞, is a Gaussian random process, which can be written as

$$
Y(t) = \int_{-\infty}^{t} X(t)\,dt \qquad \text{for } -\infty < t < \infty \tag{4.131}
$$

4.7.1 First-Order Densities for Gausssian Random Processes

If a Gaussian random process is characterized by its mean $\eta_X(t)$ and autocorrelation function $R_{XX}(t_1, t_2)$, its first- and second-order density functions can be written in terms of the mean and autocovariance function $C_{XX}(t_1, t_2)$. The first-order density is

$$
f(x; t) = \frac{1}{\sqrt{2\pi}C_{XX}^{\frac{1}{2}}(t, t)} \exp\left(-\frac{(x - \eta_X(t))^2}{2C_{XX}(t, t)}\right) \tag{4.132}
$$

4.7.2 Second-Order Densities for Gausssian Random Processes

The second-order density for the random variables $X(t_1)$ and $X(t_2)$ defined, respectively, as X_1 and X_2 is

$$f(x_1, x_2;\ t_1, t_2) = \frac{1}{2\pi|K|^{\frac{1}{2}}} \exp\left\{\frac{1}{2}(\mathbf{x}-\mathbf{m})^T\mathbf{K}^{-1}(\mathbf{x}-\mathbf{m})\right\}$$

where (4.133)

$$\mathbf{x} = \begin{bmatrix} x_1 \\ x_2 \end{bmatrix}, \quad \mathbf{m} = \begin{bmatrix} \eta_X(t_1) \\ \eta_X(t_2) \end{bmatrix}, \quad \mathbf{K} = \begin{bmatrix} C_{XX}(t_1, t_1) & C_{XX}(t_1, t_2) \\ C_{XX}(t_2, t_1) & C_{XX}(t_2, t_2) \end{bmatrix}$$

When the Gaussian random process is wide sense stationary, the first- and second-order densities can be written in terms of the mean η_X and autocovariance function $C_{XX}(\tau)$ as

$$f(x;t) = \frac{1}{\sqrt{2\pi}C_{XX}^{\frac{1}{2}}(0)} \exp\left(-\frac{(x-\eta_X)^2}{2C_{XX}(0)}\right) \tag{4.134}$$

and

$$f(x_1, x_2;\ t_1, t_2) = \frac{1}{2\pi|K|^{\frac{1}{2}}} \exp\left\{-\frac{1}{2}(\mathbf{x}-\mathbf{m})^T\mathbf{K}^{-1}(\mathbf{x}-\mathbf{m})\right\}$$

where (4.135)

$$\mathbf{x} = \begin{bmatrix} x_1 \\ x_2 \end{bmatrix}, \quad \mathbf{m} = \begin{bmatrix} \eta_X \\ \eta_X \end{bmatrix}, \quad \mathbf{K} = \begin{bmatrix} C_{XX}(0) & C_{XX}(t_1 - t_2) \\ C_{XX}(t_2 - t_1) & C_{XX}(0) \end{bmatrix}$$

4.8 WHITE RANDOM PROCESSES

For ease in analysis, it is convenient to define the white noise process. The name comes from the fact that the spectral density is assumed to be flat and contain all frequencies. A useful definition is that a **white noise random process** is characterized by its power spectral density $S_{XX}(w)$ as follows:

$$S_{XX}(\omega) = \tfrac{1}{2}N_0 \qquad \text{for } -\infty < \omega < \infty \tag{4.136}$$

The autocorrelation function for the white noise process is the inverse Fourier transform of the power spectral density, which becomes

$$R_{XX}(\tau) = \tfrac{1}{2}N_0\,\delta(\tau) \tag{4.137}$$

The autocorrelation function and power spectral density for the white noise process are given in Figure 4.20. This particular definition implies the following:

(1) $E[X(t)] = 0$.

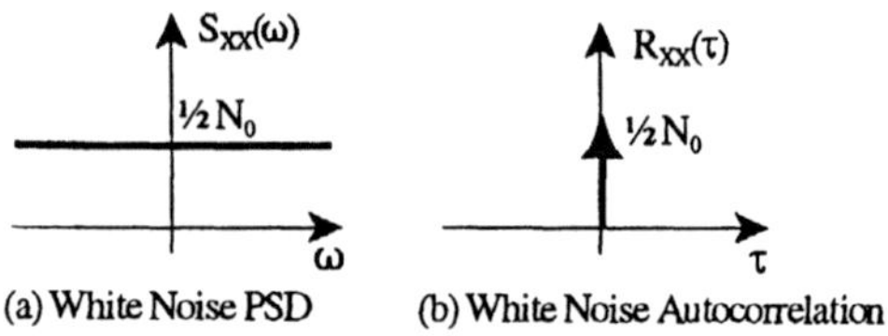

Figure 4.20 Power spectral density and autocorrelation function for the white noise random process.

(2) $E[X(t_1)X(t_2)] = 0$ for t_1 not equal t_2. This means that the random variables defined at two nonequal times are orthogonal and because of the zero mean, they are also uncorrelated. Furthermore, if the white noise process is Gaussian, they are independent random variables.

(3) The average power is infinite. Thus the white noise process is not physically realizable, but it can be used as a basis for generating realizable processes by passing white noise through a filter. White noise should not be thought of as an observable process; only its output after filtering is observable. Nonetheless, it is convenient to make white noise assumptions to simplify analysis.

No other characterizations are given in the definition, so it is possible to have any first-order and higher-order probability density functions. In many problems it is convenient to use densities that are Gaussian, and thus the words **Gaussian white noise** (GWN) are commonly used to identify a white noise random process with the same autocorrelation as given in Eq. (4.137) and the nth order densities being Gaussian.

In some problems the power is not constant but varies with time. A white noise process is defined as a **nonstationary white random process** if its autocorrelation function is written as

$$R_{XX}(t, u) = \tfrac{1}{2}N_0(t)\delta(t - u) \tag{4.138}$$

4.9 ARMA RANDOM PROCESSES

In working with continuous linear time-varying and time-invariant systems, a useful system model is a differential equation in terms of the input and output. For time-invariant systems the coefficients of the differential equation are constant, and the Laplace transform is convenient and expedient for analysis. For discrete time shift invariant linear systems, a difference equation in terms of input and output is a useful model, and the Z-transform plays a significant role in analyzing such systems. Further discussions on these relationships will be presented in Chapter 5, while in this section we investigate a special class of random processes resulting from driving a discrete time system with a white noise process.

If $w(n)$ represents the input to a linear time-invariant causal system and $x[n]$ the output, the input–output relationship can be written as

$$x[n] = -\sum_{k=1}^{p} a_k x[n-k] + \sum_{k=0}^{q} b_k w[n-k] \tag{4.139}$$

It is usually assumed that $p > q$ and that the equation above is defined for either all n or for $n \geq 0$.

If $w[n]$ is a white sequence with $E[w[n]] = 0$ and $E[w[j]w[k]] = \sigma^2\delta_{jk}$ for all j and k, the resulting process $x[n]$ is called an **autoregressive moving average** (ARMA) random process of order p, q, which is written as ARMA(p, q). Such processes are common in the modeling and analysis of random sequences or data in a wide variety of applications for system identification.

If all the $a_k = 0$ for all $k = 1, 2, \ldots, p$, the resulting random process $x[n]$ is called a **moving average (MA)** process of order q abbreviated as MA(q) and governed by

$$x[n] = \sum_{k=0}^{q} b_k w[n-k] \tag{4.140}$$

If $b_k = 0$ for all $k = 1, 2, \ldots, q$, the resulting random process $x[n]$ is called an **auto regressive (AR)** process of order p, written as AR(p) and governed by

$$x[n] = -\sum_{k=1}^{p} a_k x[n-k] + b_0 w[n] \tag{4.141}$$

The word "autoregressive" comes from the statistical community and implies that the $x[n]$ regresses on itself, which means that it can be written as a linear function of its own previous values.

Depending on the statistical properties assumed for $w[k]$, the ARMA processes defined by (4.139) have varying properties with respect to characterizations like mean, variance, autocorrelation, first-order densities, and so on. Also different results are obtained if the $w[n]$ is defined for all time indexes or just for indexes greater than or equal zero. If $w[n]$ and $x[n]$ are defined for all n, the system will have reached a steady state output. However, if they are defined for only time indexes $n \geq 0$, the beginning samples will contain a transient portion reaching steady state only for large values of the time index n.

4.9.1 Moving Average Processes, MA(q)

For this discussion let the governing equation for a moving average process of order q, MA(q) in Eq. (4.140), be defined as

$$x[n] = \sum_{k=0}^{q} b_k w[n-k], \quad n \geq 0 \tag{4.142}$$

where $w[n]$ is a white stationary process, not necessarily Gaussian, with zero mean, $E\{w[n]\} = 0$, and variance given by $E\{w^2[n]\} = \sigma^2$ for all $n \geq 0$. We will characterize the resulting process by finding its mean, variance, and autocorrelation function.

Mean of MA(q) Process. The mean of the moving average process of order q can be found by taking the expected value of Eq. (4.142) as

$$E\{x[n]\} = E\left\{\sum_{k=0}^{q} b_k w[n-k]\right\}, \qquad n \geq 0$$
$$= \sum_{k=0}^{q} b_k E\{w[n-k]\} \tag{4.143}$$

Since $E\{w[n-k]\} = 0$ for $k = 0$ to q, the mean of the MA(q) process is given by

$$E\{x[n]\} = 0, \qquad n \geq 0 \tag{4.144}$$

Variance of the Moving Average Process, MA(q). The variance of the MA(q) process is given by

$$\sigma_X^2[n] = E\{(x[n] - E\{x[n]\})^2\}, \qquad n \geq 0 \tag{4.145}$$

The variance $\sigma_X^2[n]$ is zero by definition for all $n < 0$. From (4.144) we know that $E\{x[n]\} = 0$ for all $n \geq 0$, so the variance is just the second moment. After substituting the $x[n]$ of Eq. (4.142) into Eq. (4.145), the variance becomes

$$\sigma_X^2[n] = E\{x^2[n]\} = E\left\{\left(\sum_{k=0}^{q} b_k w[n-k]\right)^2\right\}$$
$$= E\left\{\sum_{k=0}^{q} b_k w[n-k] \sum_{j=0}^{q} b_j w[n-j]\right\} \tag{4.146}$$

Multiplying out the terms in the brackets and taking the expected value through the summations gives for $n \geq q$,

$$\sigma_X^2[n] = \sum_{k=0}^{q}\sum_{j=0}^{q} b_k b_j E\{w[n-k]w[n-j]\}$$
$$= \sigma^2 \sum_{k=0}^{q} b_k^2 \qquad n \geq q \tag{4.147}$$

The last step is possible because the only terms where $E\{w[n-k]x[n-j]\}$ are nonzero are those for which $j = k$. Notice that this variance is thus a constant for $n \geq q$.

If $0 \leq n < q$, the result above is modified to include only those $w(n-k)$ and $w(n-j)$ that have arguments ≥ 0, and is thus a function of n:

$$\sigma_X^2[n] = \sigma^2 \sum_{k=0}^{n} b_k^2, \quad 0 \leq n < q \tag{4.148}$$

The variance $\sigma_X^2[n]$ is thus seen to be a function of n and a part of the initial transient period until n reaches q yielding a steady state after that.

Autocorrelation Function for the Moving Average Process. MA(q). For k and j less than zero the autocorrelation function $R_{XX}(j, k)$ for the MA(q) process is defined to be zero, while in general, it will depend on two time indexes $k + m$ and k, where k and m are positive integers and can be written as

$$E\{x[k+m]x[k]\} = E\left\{\sum_{i=0}^{q} b_k w[k+m-i] \sum_{j=0}^{q} b_j w[k-j]\right\}, \quad k, m \geq 0 \qquad (4.149)$$

Let $k > q$ to begin with and $0 \leq m \leq q$. For this range of m, Eq. (4.149) can be written as

$$\begin{aligned} R_{XX}(k+m, k) = E\{&(b_0 w[k+m] + b_1 w[k+m-1] + \cdots + b_q w[k+m-q]) \\ &\cdot (b_0 w[k] + b_1 w[k-1] + \cdots + b_q w[k-q])\} \end{aligned} \qquad (4.150)$$

Looking at all cross product terms, we see that the only ones that give a nonzero value of σ^2 are those with equal indexes, so Eq. (4.150) reduces to

$$\begin{aligned} R_{XX}[k+m, k] &= \sum_{r=0}^{q-m} b_r b_{r+m} E\{w^2[k-r]\} \\ &= \sigma^2 \sum_{r=0}^{q-m} b_r b_{r+m}, \qquad 0 \leq m \leq q \end{aligned} \qquad (4.151)$$

For $m > q$ there are no overlaps in the products, so we have

$$R_{XX}[k+m, k] = 0. \qquad (4.152)$$

Clearly, the autocorrelation function for the MA(q) process is independent of the starting time k and a function of the time difference m, provided that $k > q$. Thus after the time q, the process becomes wide sense stationary since the mean was previously shown to be a constant, namely 0, and the autocorrelation has been shown to be a function of time difference only.

EXAMPLE 4.5

Find the mean, autocorrelation function, and variance for the MA(1), MA(2), and MA(3) random processes assuming that a steady state has been reached, $k > q$, for the models given and with the $w[n]$ as a zero mean white random process.

SOLUTION

For the MA(q) random process the means are all zero because of the assumption that $w[n]$ is a white random process; $E\{x[n]\} = 0$ for $n \geq 0$, and defined to be zero for $n < 0$.

MA(1) For this process we have the defining equation

$$x[n] = b_0 w[n] + b_1 w[n-1] \qquad \text{for } n \geq 0 \qquad (4.153)$$

The autocorrelation function evaluated at lags of 0,1, and 2 after steady state are given from (4.151) as

$$
\begin{aligned}
R_{XX}[0] &= \sigma^2 \sum_{r=0}^{1} b_r b_r = \sigma^2 (b_0^2 + b_1^2) \\
R_{XX}[1] &= \sigma^2 \sum_{r=0}^{0} b_r b_{r-1} = \sigma^2 b_0 b_1 \\
R_{XX}[k] &= 0, \qquad k > 1
\end{aligned}
\tag{4.154}
$$

The steady state variance $\sigma_x^2[k]$ is given by $\sigma_x^2[k] = R_{XX}(0)$, since the mean is zero. The autocorrelation function for negative time difference is obtained by symmetry $R_{XX}[-k] = R_{XX}[k]$.

In a similar fashion we obtain the autocorrelation functions for the MA(2) and MA(3) as follows:

MA(2)

$$
x[n] = b_0 w[n] + b_1 w[n-1] + b_2 w[n-2] \qquad \text{for } n \geq 0 \tag{4.155}
$$

The autocorrelation function after steady state is easily found from (4.151) to be

$$
\begin{aligned}
R_{XX}[0] &= \sigma^2 (b_0^2 + b_1^2 + b_2^2) \\
R_{XX}[1] &= \sigma^2 (b_0 b_1 + b_1 b_2) \\
R_{XX}[2] &= \sigma^2 b_0 b_2 \\
R_{XX}[k] &= 0, \qquad k > 2 \\
R_{XX}[-k] &= R_{XX}[k]
\end{aligned}
\tag{4.156}
$$

The steady state variance $\sigma_X^2[k]$ is given by $\sigma_X^2[k] = R_{XX}(0)$, since the mean is zero.

MA(3)

$$
x[n] = b_0 w[n] + b_1 w[n-1] + b_2 w[n-2] + b_3 w[n-3] \qquad \text{for } n \geq 0 \tag{4.157}
$$

The autocorrelation function after steady state is easily determined from (4.151) as

$$
\begin{aligned}
R_{XX}[0] &= \sigma^2 (b_0^2 + b_1^2 + b_2^2 + b_3^2) \\
R_{XX}[1] &= \sigma^2 (b_0 b_1 + b_1 b_2 + b_2 b_3) \\
R_{XX}[2] &= \sigma^2 (b_0 b_2 + b_1 b_3) \\
R_{XX}[3] &= \sigma^2 b_0 b_3 \\
R_{XX}[k] &= 0, \qquad k > 3 \\
R_{XX}[-k] &= R_{XX}[k]
\end{aligned}
\tag{4.158}
$$

The steady state variance $\sigma_X^2[k]$ is given by $\sigma_X^2[k] = R_{XX}(0)$, since the mean is zero. ■

4.9.2 Autoregressive Processes, AR(p)

An autoregressive process, AR(p), is a special case of the ARMA(p, q) process. For this discussion it is assumed that AR(p) is governed for $n \geq 0$ by

$$x[n] = -\sum_{k=1}^{p} a_k x[n-k] + b_0 w[n], \quad n \geq 0 \tag{4.159}$$

The excitation process $w[n]$, for $n \geq 0$, is a white stationary random process with $E\{w[j]w[k]\} = \sigma^2 \delta_{jk}$. We would like to determine the mean, variance, and autocorrelation for this AR(p) random process.

Mean of AR(p) Process. The mean of the AR(p) process is obtained by taking the expected value of the governing Eq. (4.159) to yield the following difference equation:

$$E\{x[n]\} = -\sum_{k=1}^{p} a_k E\{x[n-k]\} + b_0 E\{w[n]\}, \quad n \geq 0 \tag{4.160}$$

Since $E\{w[n]\} = 0$, and $x[n] = 0$, for $n < 0$, the solution of the difference equation yields

$$E\{x[n]\} = 0, \quad n \geq 0 \tag{4.161}$$

Variance of AR(p) Process. The variance of the defined AR(p) process can be found directly from the definition of the variance and the process $x[n]$ given in Eq. (4.159). Using the zero mean assumption, we write the variance as

$$\begin{aligned} \sigma_X^2[n] &= E\{x^2[n]\} \\ &= E\left\{\left(-\sum_{k=0}^{p} a_k x[n-k] + b_0 w[n]\right) \cdot \left(-\sum_{j=0}^{p} a_j x[n-j] + b_0 w[n]\right)\right\} \end{aligned} \tag{4.162}$$

The expected value of all cross products of $w[n]$ with the summations of the $x(n-k)$, given by

$$E\left\{\left(-\sum_{k=0}^{p} a_k x[n-k]\right) w[n]\right\} = 0 \tag{4.163}$$

are all zero, since $x[n-k]$ are not functions of $w[n]$ as they are from a time preceding $w[n]$. Thus the variance reduces to

$$\begin{aligned} \sigma_X^2[n] &= E\{x^2[n]\} \\ &= E\left\{\left(-\sum_{k=0}^{p} a_k x[n-k] w[n]\right) \cdot \left(-\sum_{j=0}^{p} a_j x[n-j]\right)\right\} + b_0^2 \sigma^2 \end{aligned} \tag{4.164}$$

Multiplying the product terms out, taking the expected value and rearranging, we find the variance from Eq. (4.164) to be

$$\sigma_X^2[n] = \sum_{k=1}^{p} \sum_{j=1}^{p} a_j a_k R_{XX}[n-k, n-j] + b_0^2 \sigma^2 \tag{4.165}$$

The variance at time index n is thus seen to be a function of the variances at the p previous times, plus a function of various preceding autocorrelation functions at various time indexes, plus a weighted sum of the white sequence variance.

An alternative form for the variance can be obtained by multiplying the $x[n]$ of Eq. (4.159) by $x[n]$ and taking the expected value as

$$\begin{aligned} \sigma_{XX}^2[n] &= E\{x^2[n]\} \\ &= E\left\{\left(-\sum_{k=1}^{p} a_k x[n-k] + b_0 w[n]\right) x[n]\right\} \end{aligned} \tag{4.166}$$

Multiplying out and taking the expected value of the terms results in the following equation for the variance which, because of the zero mean, is the autocorrelation function at n and n:

$$R_{XX}[n, n] = -\sum_{k=1}^{p} a_k R_{XX}[n, n-k] + b_0^2 \sigma^2 \tag{4.167}$$

In this form the variance is written in terms of the autocorrelation functions for indexes $n, n-k$ as k goes from 1 to p. This same approach will now be used to determine the autocorrelation function for the autoregressive process at other lags.

Autocorrelation of AR(p) Process. The autocorrelation function, in general, goes through a transient period since the signal starts at zero with zero initial conditions. To derive the autocorrelation function, it is assumed that $n > p$. The derivation begins with the single step difference by taking the expected value of $x[n]x[n-1]$ as follows:

$$E\{x[n]x[n-1]\} = E\left\{\left(-\sum_{k=1}^{p} x[n-k] + b_0 w[n]\right) x[n-1]\right\} \tag{4.168}$$

Multiplying out, taking the expected value through the sum, and using the fact that $E\{w[n]x[n-1]\} = 0$, since $x\{n-1\}$ is not a function of $w[n]$ and $w[n]$ is a white process, yields the equation

$$R_{XX}[n, n-1] = -\sum_{k=1}^{p} a_k R_{XX}[n-1, n-k] \tag{4.169}$$

Similarly multiplying $x[n]$ of Eq. (4.159) by $x[n-j]$ for $j = 2$ to p, taking the expected value, and simplifying leads to

$$R_{XX}[n, n-j] = -\sum_{k=1}^{p} a_k R_{XX}[n-j, n-k], \qquad j = 2, 3, \ldots, p \tag{4.170}$$

Thus Eqs. (4.167), (4.169), and (4.170) give a set of equations that determines the autocorrelation functions at lags 0 to p in terms of various other autocorrelation functions, and these are the key equations in solving for the steady state. If $j > p$, Eq. (4.170) holds as well.

If we assume that a steady state has been reached, the $R_{XX}(r, s)$ are functions of time difference $r - s$ only. So we can rewrite Eqs. (4.169) and (4.170) as

$$R_{XX}[j] = -\sum_{k=1}^{p} a_k R_{XX}[k-j], \qquad j = 1, 2, \ldots, p \tag{4.171}$$

By the symmetry property of the autocorrelation function for a stationary random process, $R_{XX}(-j) = R_{XX}(j)$, these equations can be put into the following matrix form:

$$\begin{bmatrix} R_{XX}[0] & R_{XX}[1] & \ldots & R_{XX}[p-1] \\ R_{XX}[1] & R_{XX}[0] & \ldots & R_{XX}[p-2] \\ \vdots & \vdots & \vdots & \vdots \\ R_{XX}[p-1] & R_{XX}[p-2] & \ldots & R_{XX}[0] \end{bmatrix} \begin{bmatrix} a_1 \\ a_2 \\ \vdots \\ a_p \end{bmatrix} = - \begin{bmatrix} R_{XX}[1] \\ R_{XX}[2] \\ \vdots \\ R_{XX}[p] \end{bmatrix} \tag{4.172}$$

The equations represented in this matrix form are called the **Yule-Walker equations** for the autoregressive process defined in Eq. (4.159). If we know the $R_{XX}(0)$ through $R_{XX}(p)$, or their estimates, these equations may be solved for the AR parameters $a_1, a_2, \ldots, a_p$ and thus be used for system identification. The Levinson recursion algorithm [11] can be used to expedite the solution, since the coefficient matrix is a Toeplitz matrix.

However, if we have the $a_1, a_2, \ldots, a_p$ values, meaning the system is known, and we want to solve for the autocorrelation function, we must use Eq. (4.167) for $R_{XX}(0)$ and rearrange the equations of (4.172) so that the $R_{XX}[k]$ are the unknowns. This way we obtain the following equation in terms of the unknown autocorrelations at the lags 0 through p:

$$\begin{bmatrix} 1 & a_1 & a_2 & \ldots & a_{p-2} & a_{p-1} & a_p \\ a_1 & 1+a_2 & a_3 & \ldots & a_{p-1} & a_p & 0 \\ a_2 & a_1+a_3 & 1+a_4 & \ldots & a_p & 0 & 0 \\ \vdots & \vdots & \vdots & \vdots & \vdots & \vdots & \vdots \\ 0 & a_p & a_{p-1} & \ldots & a_3 & 1+a_2 & a_1 \\ a_p & a_{p-1} & a_{p-2} & \ldots & a_2 & a_1 & 1 \end{bmatrix} \begin{bmatrix} R_{XX}[0] \\ R_{XX}[1] \\ R_{XX}[2] \\ \vdots \\ R_{XX}[p-1] \\ R_{XX}(p) \end{bmatrix} = - \begin{bmatrix} b_0\sigma^2 \\ 0 \\ 0 \\ \vdots \\ 0 \\ 0 \end{bmatrix} \tag{4.173}$$

Examples 4.6 and 4.7 illustrate this procedure for finding the autocorrelation function for AR(1) and AR(2).

EXAMPLE 4.6

Find the mean, autocorrelation, and variance for the autoregressive process AR(1) random process defined by

$$x[n] = -a_1 x[n-1] + b_0 w[n], \qquad n \geq 0 \tag{4.174}$$

where a_1 is such that a stable system is represented.

SOLUTION

From (4.167) and (4.169) the autocorrelation function at 0 and 1 must satisfy the two equations

$$\begin{aligned} R_{XX}[n,n] &= -a_1 R_{XX}[n, n-1] + b_0^2\sigma^2 \\ R_{XX}[n, n-1] &= -a_1 R_{XX}[n-1, n-1] \end{aligned} \tag{4.175}$$

If we assume that a steady state has been reached, then the autocorrelation function can be written in terms of time difference only and with the symmetry property of the autocorrelation functions (4.175) becomes

$$\begin{aligned} R_{XX}[0] &= -a_1 R_{XX}[1] + b_0^2\sigma^2 \\ R_{XX}[1] &= -a_1 R_{XX}[0] \end{aligned} \tag{4.176}$$

Solving these two equations for $R_{XX}[0]$ and $R_{XX}[1]$ gives

$$R_{XX}[0] = \frac{b_0^2\sigma^2}{1-a_1^2}, \qquad R_{XX}[1] = \frac{-a_1 b_0^2\sigma^2}{1-a_1^2} \tag{4.177}$$

To obtain the autocorrelation function at lags greater that 1, Eq. (4.170) and $j > 1$, we have

$$R_{XX}[n, n-j] = -a_1 R_{XX}[n-j, n-1] \tag{4.178}$$

and in the steady state we obtain

$$\begin{aligned} R_{XX}[j] &= -a_1 R_{XX}(j-1) = (-a_1)^{j-1} R_{XX}(1) \\ &= \frac{b_0^2\sigma^2}{1-a_1^2}(-a_1)^j, \qquad j > 1 \end{aligned} \tag{4.179}$$

■

EXAMPLE 4.7

Find the mean and autocorrelation function for a second-order autoregressive process AR(2) with the assumptions given in Section 4.9.2 and

$$x[n] = -a_1 x[n-1] - a_2 x[n-2] + b_0 w[n], \qquad n \geq 0 \tag{4.180}$$

It is further assumed that the parameters a_1, a_2 are such that (4.180) gives a stable causal system.

SOLUTION

The $E\{x[k]\}$ has been shown to be zero for all $n \geq 0$. The autocorrelation function can be solved using the results of (4.167), (4.169), and (4.170). These equations are applicable to any autoregressive process with the assumption of zero initial conditions and $w[n]$ a white random process. They are repeated here for AR(2) as

$$\begin{aligned} R_{XX}[n,n] &= -\sum_{k=1}^{2} a_k R_{XX}[n,n-k] + b_0^2\sigma^2 \\ R_{XX}[n,n-1] &= -\sum_{k=1}^{2} a_k R_{XX}[n-1,n-k] \\ R_{XX}[n,n-2] &= -\sum_{k=1}^{2} a_k R_{XX}[n-2,n-k] \end{aligned} \tag{4.181}$$

Assuming that n is sufficiently large enough so that a steady state has been reached and using the symmetry property of the autocorrelation function, $R_{XX}[-k] = R_{XX}[k]$, the autocorrrelation functions of (4.181) can be written in terms of time difference only as

$$\begin{aligned} R_{XX}[0] &= -a_1 R_{XX}[1] - a_2 R_{XX}[2] + b_0^2\sigma^2 \\ R_{XX}[1] &= -a_1 R_{XX}[0] - a_2 R_{XX}[1] \\ R_{XX}[2] &= -a_1 R_{XX}[1] - a_2 R_{XX}[0] \end{aligned} \tag{4.182}$$

The three equations can now be solved for the unknowns: $R_{XX}[0], R_{XX}[1]$, and $R_{XX}[2]$. For this problem it is convenient to solve the middle equation for $R_{XX}[0]$ in terms of $R_{XX}[1]$, and substitute this result into the third equation to solve the third equation for $R_{XX}[2]$ in terms of $R_{XX}[1]$, which gives

$$\begin{aligned} R_{XX}[0] &= \frac{-(1+a_2)}{a_1} R_{XX}[1] \\ R_{XX}[2] &= \frac{-a_1^2 + a_2 + a_2^2}{a_1} R_{XX}[1] \end{aligned} \tag{4.183}$$

Substituting both of these results into the first equation, the $R_{XX}[1]$, can be shown to be

$$R_{XX}[1] = \frac{a_1 b_0^2 \sigma^2}{(-1 - a_2 + a_1^2 - a_1^2 a_2 + a_2^2 + a_2^3)} \tag{4.184}$$

Using the results of (4.184), we determine the $R_{XX}[0]$ and $R_{XX}[1]$ in (4.183) as follows:

$$\begin{aligned} R_{XX}[0] &= \frac{-(1+a_2)b_0^2\sigma^2}{(-1-a_2+a_1^2-a_1^2a_2+a_2^2+a_2^3)} \\ R_{XX}[2] &= \frac{(-a_1^2+a_2+a_2^2)b_0^2\sigma^2}{(-1-a_2+a_1^2-a_1^2+a_2^2+a_2^3)} \end{aligned} \quad (4.185)$$

For lags greater than two, we can begin with Eq. (4.170), which is

$$R_{XX}[n, n-j] = -\sum_{k=1}^{2} a_k R_{XX}[n-k, n-j] \quad (4.186)$$

In the steady state the equation above can be written as

$$R_{XX}[j] = -\sum_{k=1}^{2} a_k R_{XX}[j-k] \quad (4.187)$$

If we set $j = 3$ in (4.187) it reduces to

$$R_{XX}[3] = -\sum_{k=1}^{2} a_k R_{XX}[3-k] = -a_1 R_{XX}[2] - a_2 R_{XX}[1] \quad (4.188)$$

Knowing $R_{XX}[2]$ and $R_{XX}[1]$ from (4.184) and (4.185), the $R_{XX}[3]$ can be determined from (4.188). Thus (4.187) gives a recursive formula for the generation of autocorrelation function at all lags $j > p$. ■

4.9.3 Autoregressive Moving Average Processes, ARMA (p, q)

For this discussion it will be assumed that the governing equation for a moving average process of order q, ARMA (p, q) given in Eq. (4.139) is defined as

$$x[n] = -\sum_{k=1}^{p} a_k x[n-k] + \sum_{k=0}^{q} b_k w[n-k], \qquad n \geq 0 \quad (4.189)$$

for positive time index n from 0 to ∞ and that $w[n]$ is a white stationary process, not necessarily Gaussian, with $E\{w[n]\} = 0$ for all $n \geq 0$, and variance given by $E\{w^2[n]\} = \sigma^2$ for all $n \geq 0$. We would like to partially characterize the resulting ARMA (p, q) process by finding its mean, variance, and autocorrelation function.

Mean of ARMA (p, q). The mean of the ARMA (p, q) process is easily found by taking the expected value of Eq. (4.189) to give

$$E\{x[n]\} = -\sum_{k=1}^{p} a_k E\{x[n-k]\} + \sum_{k=0}^{q} b_k E\{w[n-k]\}, \qquad n \geq 0 \quad (4.190)$$

Since $E\{w[n-k]\} = 0$ for all $k = 0, 1, \ldots, q$, Eq. (4.190) can be written as a difference equation for $E\{x[n]\}$ with a zero driving function:

$$E\{x[n]\} = -\sum_{k=1}^{p} a_k E\{x[n-k]\}, \qquad n \geq 0 \tag{4.191}$$

Since all initial conditions are assumed equal to zero, the mean $E\{x[n]\} = 0$ for $n \geq 0$.

Variance of ARMA (p, q). Since the mean is zero for all n, the variance can be written in terms of the autocorrelation function evaluated at time indexes n and n as

$$\begin{aligned} \sigma^2[n] &= E\{x^2[n]\} = R_{XX}[n, n] \\ &= E\left\{\left(-\sum_{k=1}^{p} a_k x[n-k] + b_0 \sum_{k=0}^{q} w[n-k]\right) x[n]\right\} \end{aligned} \tag{4.192}$$

On taking the expected value in Eq. (4.192), we can show the variance to be

$$\sigma^2[n] = -\sum_{k=1}^{p} a_k R_{XX}[n, n-k] + \sum_{k=0}^{q} b_k E\{w[n-k]x[n]\} \tag{4.193}$$

This result can be rewritten as

$$\sigma^2[n] = -\sum_{k=1}^{p} a_k R_{XX}[n, n-k] + \sum_{k=0}^{q} b_k R_{XW}[n, n-k] \tag{4.194}$$

Thus the variance is written in terms of the weighted sum of the autocorrelation function evaluated at different times and the weighted sum of various cross correlations at different times. These terms will be determined in the following section on the determination of the autocorrelation function for a general ARMA process.

Autocorrelation for ARMA (p, q). The autocorrelation function, in general, goes through a transient period since the signal starts at zero with zero initial conditions. To derive the autocorrelation function, it is assumed that $n > p$. The derivation begins with the expected value of the product of $x[n]$ and $x[n-1]$, where $x[n]$ is from (4.189), as follows:

$$E\{x[n]x[n-1]\} = E\left\{\left(-\sum_{k=1}^{p} x[n-k] + \sum_{k=0}^{q} b_k w[n-k]\right) x[n-1]\right\} \tag{4.195}$$

Multiplying out and taking the expected value through the sum yields the equation

$$R_{XX}[n, n-1] = -\sum_{k=1}^{p} a_k R_{XX}[n-1, n-k] + \sum_{k=0}^{q} b_k E\{w[n-k]x[n-1]\} \tag{4.196}$$

Similarly multiplying $x[n]$ of Eq. (4.189) by $x[n-j]$ for $j = 2$ to p, taking the expected value, and simplifying leads to

$$R_{XX}[n, n-j] = -\sum_{k=1}^{p} a_k R_{XX}[n-j, n-k] + \sum_{k=0}^{q} b_k E\{w[n-k]x[n-j]\},$$
$$j = 2, 3, \ldots, p \quad (4.197)$$

Thus Eqs. (4.196) and (4.197) give a set of equations that determine the autocorrelation functions at lags 1 to p in terms of various other autocorrelation functions. If we assume that a steady state has been reached, the $R_{XX}(r, s)$ are functions of time difference $r - s$ only. So we can rewrite Eqs. (4.196) and (4.197) as

$$R_{XX}[j] = -\sum_{k=1}^{p} a_k R_{XX}[k-j] + f_j(\mathbf{a}, \mathbf{b}) \qquad j = 1, 2, \ldots, p \quad (4.198)$$

where the $f_j(\mathbf{a}, \mathbf{b})$ are the second sums shown in Eqs. (4.196) and (4.197). These $f_j(\mathbf{a}, \mathbf{b})$ in the simultaneous equations above are rather complex nonlinear function of the ARMA(p, q) model parameter vectors $\mathbf{a}$ and $\mathbf{b}$. Using the symmetry property of the autocorrelation function for a stationary random process, $R_{XX}(-j) = R_{XX}(j)$, allows the equations of (4.198) to be put into the following matrix form:

$$\begin{bmatrix} R_{XX}[0] & R_{XX}[1] & \cdots & R_{XX}[p-1] \\ R_{XX}[1] & R_{XX}[0] & \cdots & R_{XX}[p-2] \\ \vdots & \vdots & \vdots & \vdots \\ R_{XX}[p-1] & R_{XX}[p-2] & \cdots & R_{XX}[0] \end{bmatrix} \begin{bmatrix} a_1 \\ a_2 \\ \vdots \\ a_p \end{bmatrix} + \begin{bmatrix} f_1(\mathbf{a}, \mathbf{b}) \\ f_2(\mathbf{a}, \mathbf{b}) \\ \vdots \\ f_p(\mathbf{a}, \mathbf{b}) \end{bmatrix}$$
$$= -\begin{bmatrix} R_{XX}[1] \\ R_{XX}[2] \\ \vdots \\ R_{XX}[p] \end{bmatrix} \quad (4.199)$$

Thus the solution for the autocorrelation function $R_{XX}(k)$ at all lags is no longer a solution of simultaneous linear equations but a solution of a set of simultaneous nonlinear equations. The following example illustrates the overall procedure in finding the mean, variance, and autocorrelation function of the simplest ARMA process, namely the ARMA(1, 1) process.

EXAMPLE 4.8

Find the (a) mean, (b) autocorrrelation function, and (c) variance for the ARMA(1, 1) random process given by the difference equation

$$x[n] = -a_1 x[n-1] + b_0 w[n] + b_1 w[n-1], \qquad n \geq 0 \quad (4.200)$$

Assume that $w[n]$ is a zero mean white random process with variance $E[w(j)w(k)] = \sigma^2\delta_{jk}$. Assume all initial conditions for negative time index to be zero and that a steady state has been obtained.

SOLUTION

(a) It was shown in Eq. (4.191) that the mean of ARMA process is zero. Therefore $E\{x[n]\} = 0$ for all $n \geq 0$.

(b) The variance for n is the same as the autocorrelation function at a zero lag since the mean is zero. Therefore we have using (4.193) that

$$\begin{aligned}\sigma^2[n] &= E\{x^2[n]\} = R_{XX}[n, n] \\ &= E\{x[n](-a_1x[n-1] + b_0w[n] + b_1w[n-1])\} \\ &= -a_1R_{XX}[n, n-1] + b_0E\{x[n]w[n]\} + b_1E\{x[n]w[n-1]\}\end{aligned} \tag{4.201}$$

The last two terms in the expression above will be evaluated separately. Substituting the difference equation expression for $x[n]$, multiplying out, and taking the expected value gives

$$\begin{aligned}E\{x[n]w[n-1]\} &= E\{(-a_1x[n-1] + b_0w[n] + b_1w[n-1])w[n-1]\} \\ &= -a_1E\{x[n-1]w[n-1]\} + b_1\sigma^2 \\ &= -a_1E\{(-a_1x[n-2] + b_0w[n-1] \\ &\quad + b_1w[n-2])w[n-1]\} + b_1\sigma^2 \\ &= -a_1b_0\sigma^2 + b_1\sigma^2\end{aligned} \tag{4.202}$$

The first expected value in the expression above is zero because $x(n-1)$ is not a function of $w[n]$, and the last term is zero because the $w[n]$ is a white sequence.

The other expected value is given as

$$\begin{aligned}E\{x[n]w[n]\} &= E\{(-a_1x[n-1] + b_0w[n] + b_1w[n-1])w[n]\} \\ &= b_0\sigma^2\end{aligned} \tag{4.203}$$

Substituting the results from (4.202) and (4.203) into (4.201), we finally obtain the variance as

$$\sigma^2[n] = -a_1R_{XX}[n, n-1] + b_0^2\sigma^2 - a_1b_1b_0\sigma^2 + b_1^2\sigma^2 \tag{4.204}$$

The $R_{XX}[n, n-1]$ can be calculated by substituting the definition of the ARMA(1, 1) process into the expected value:

$$\begin{aligned} R_{XX}(n, n-1) &= E\{x[n]x[n-1]\} = E\{(-a_1x[n-1] \\ &\quad + b_0w[n] + b_1x[n-1])x[n-1]\} \\ &= -a_1R_{XX}[n-1, n-1] + b_0E\{w[n]x[n-1]\} \\ &\quad + b_1E\{w[n-1]x[n-1]\} \end{aligned} \tag{4.205}$$

The $E\{w[n]x[n-1]$ term is zero because $x[n-1]$ is not a function of $w[n]$. The third term above $E\{w[n-1]x[n-1]\}$ is determined by replacing $x[n-1]$ by its equivalent from the defining equation:

$$\begin{aligned} E\{w[n-1]x[n-1]\} &= E\{w[n-1](-a_1x[n-2] \\ &\quad + b_0w[n-1] + b_1w[n-2])\} \\ &= b_0\sigma^2 \end{aligned} \tag{4.206}$$

Thus, after substituting (4.206) into (4.205), the single lag autocorrelation function of the ARMA(1, 1) process becomes

$$R_{XX}\ [n, n-1] = -a_1R_{XX}\ [n-1, n-1] + b_0\ b_1\sigma^2 \tag{4.207}$$

To obtain the steady state solution, we assume that the autocorrelation function can be written in terms of time difference only. Thus (4.204) and (4.207) are written as

$$\begin{aligned} R_{XX}[0] &= -a_1R_{XX}[1] + b_0^2\sigma^2 - a_1b_1b_0\sigma^2 + b_1^2\sigma^2 \\ R_{XX}[1] &= -a_1R_{XX}[0] + b_0b_1\sigma^2 \end{aligned} \tag{4.208}$$

Solving these two equations for $R_{XX}[0]$ and $R_{XX}[1]$ yields the following values for the autocorrelation function at 0 and 1:

$$\begin{aligned} R_{XX}[0] &= \frac{(b_0^2 + b_1^2 - 2a_1b_0b_1)\sigma^2}{1 - a_1^2} \\ R_{XX}[1] &= \frac{(a_1^2b_0b_1 - a_1b_0^2 - a_1b_1^2 + b_0b_1)\sigma^2}{1 - a_1^2} \end{aligned} \tag{4.209}$$

To find the autocorrelation function for higher lags, $j = 2$, we have that $R_{XX}[n, n-2]$ with $E\{w[n]x[n-2]\} = 0$ and $E\{w[n-1]x[n-2]\} = 0$ is determined as

$$\begin{aligned} R_{XX}[n, n-2] &= E\{x[n]x[n-2]\} \\ &= E\{(-a_1x[n-1] + b_0w[n] + b_1w[n-1])x[n-2]\} \\ &= -a_1R_{XX}[n-1, n-2] \end{aligned} \tag{4.210}$$

In a similar fashion it is easy to show that $R_{XX}[n, n-j] = -a_1R_{XX}[n, n-j+1]$ for lags $j > 2$. This property is general for ARMA(p, q) processes whenever j is greater than q. Evaluating this result in the steady state yields the autocorrelation function for lags greater than one for the ARMA(1, 1) process as follows:

$$R_{XX}[j] = (-a_1)^{j-1} R_{XX}[1], \qquad j \geq 2 \tag{4.211}$$

(c) The steady state variance is just the $R_{XX}(0)$ given in (4.209), since the mean is zero. Thus

$$\sigma_{XX}^2(n) = R_{XX}[0] = \frac{(b_0^2 + b_1^2 - 2a_1 b_0 b_1)\sigma^2}{1 - a_1^2} \tag{4.212}$$

It becomes increasingly more difficult to obtain closed form solutions for the autocorrelation function and variance as the order p and q of the ARMA(p, q) process is increased. ■

4.10 PERIODIC RANDOM PROCESSES

A wide sense stationary random process is defined as **periodic** if its autocorrelation function is periodic. The random sinusoidal process $X(t) = A \sin(\Omega_0 t + \Phi)$ (defined in Example 4.3) has an autocorrelation function from (4.89):

$$R_{XX}(\tau) = K \cos(\Omega_0 \tau) \tag{4.213}$$

This process is a periodic random process since its autocorrelation function is periodic.

In a similar fashion a random process $X(t)$ defined by a Fourier series with independent zero mean random variables A_k and B_k as amplitudes given by

$$X(t) = A_0 + \sum_{k=0}^{\infty} A_k \cos(k\Omega_0 t) + \sum_{k=0}^{\infty} B_k \sin(k\Omega_0 t) \tag{4.214}$$

can be shown to have a periodic autocorrelation function and thus be periodic.

4.11 SAMPLING OF CONTINUOUS RANDOM PROCESSES

Since the processing of random signals is moving toward digital processing, it is important to examine sampled continuous random processes and determine their characterizations in terms of the continuous time process characterizations. Let $X(t)$ be a random process that is characterized by its mean $\eta_X(t)$, autocorrelation function $R_{XX}(t_1, t_2)$ and various first- and higher-order density functions $f(x_1, \ldots, x_2, x_n; t_1, t_2, \ldots, t_n)$. It is assumed that the continuous process $X(t)$ is not necessarily wide sense stationary.

The purpose of this section is to determine the statistical properties of a discrete time random process $X[k]$ that is generated by sampling uniformly, with spacing T, a continuous time random process and thus defined by $X[k] = X(kT)$.

The mean of $X[k]$ can be easily seen to be

$$E[X[n]] = E[X(nT)] = \eta_X(nT) \tag{4.215}$$

The autocorrelation function $R_{XX}[k_1, k_2]$ is easily determined as

$$R_{XX}[k_1, k_2] = E[X(k_1T)X(k_2T)] = R_{XX}(k_1T, k_2T) \tag{4.216}$$

Therefore the mean and autocorrelation function of a process generated by sampling a continuous process are just the sampled versions of the mean and autocorrelation function of the continuous process.

It is also easy to see that if $X(t)$ is a wide sense stationary continuous time process, the random sequence generated by uniform sampling is a wide sense stationary discrete time sequence, that is, its mean is independent of n and its autocorrelation function, depends only on the time difference, $k_1 - k_2$. The converse of this statement is not necessarily true; that is, if a discrete time process that is a sampled version of a continuous time process is wide sense stationary, the continuous time process is not necessarily wide sense stationary. Similarly, if the sampled version of a random process is in some sense stationary, this does not imply that the continuous time process is the same sense stationary.

The first- and higher-order densities are easily seen to be just a sampled version of the respective densities, which can be expressed for $X(t_1), X(t_2), \ldots, X(t_n)$, as

$$f[x_1, x_2, \ldots, x_n; k_1, k_2, \ldots, k_n] = f(x_1, x_2, \ldots, x_n; k_1T, k_2T, \ldots, k_nT) \tag{4.217}$$

where $t_1 = k_1T, t_2 = k_2T, \ldots, t_n = k_nT$.

4.12 ERGODIC RANDOM PROCESSES

Say we are given a random process $X(t)$ with a known mean $E[X(t)]$. A realization of the process $X(t, \zeta_i,)$ is a deterministic waveform and has a time average

$$N_T(\zeta_i) = \frac{1}{2T}\int_{-T}^{T} X(t, \zeta_i)\,dt$$

The time average is a real value and possibly different for each outcome ζ_i, thus defining a random variable N_T. We will define N as the limit of N_T as $T \to \infty$ as

$$N = \lim_{T\to\infty} N_T$$

As $T \to \infty$ the random variable N_T may converge in some sense to another random variable. The definitions of convergence for the deterministic or nonrandom case must be modified to be able to discuss convergence of random variables as $T \to \infty$. There are many different forms of convergence that have been defined, including mean squared convergence, convergence in the mean, convergence in probability, and convergence with probability 1. These forms of convergence and theorems relating to ergodicity are introduced in Papoulis [1], and a thorough treatment of them would take us far afield of our intent of introducing basic concepts of random processes.

A random process $X(t))$ is defined as **ergodic in the mean** if the N_T converges to a degenerate random variable with deterministic value equal to the statistical average given by $E[X(t)] = \eta_X$. Notice that this definition implies that a random process $X(t)$ that is not stationary in the mean cannot be ergodic in the mean.

Similarly, if a random variable $R_T(\tau)$ is defined by

$$R_T(\tau) = \frac{1}{2T}\int_{-T}^{T} X(t+\tau)X(\tau)\,dt$$

and we define $R(\tau)$ as the limit of $R_T(\tau)$ as $T \to \infty$, we can define a process to be **ergodic in autocorrelation** if $R_T(\tau)$ for all τ converges to a degenerate random variable with deterministic value equal to the statistical average given by $E[X(t+\tau)X(t)]$, which is the autocorrelation function evaluated at τ. Again, we see that if the process $X(t)$ is not wide sense stationary, then it cannot possibly be ergodic in autocorrelation.

The statistical means and autocorrelation functions for ergodic processes are thus interchangable with their time determined counterparts.

4.13 SUMMARY

The main objective of this chapter was to present the basic background material in random processes including process formulation and analysis. The definition of a random process in terms of an indexed set of random variables where the indexing set could be continuous or discrete links the concepts of random variables and random processes. A random process could also be viewed as a mapping from the sample space of a defined experiment to a set of output functions. Evaluating a random process at a particular time gives a random variable, at a particular outcome gives a function of time, and evaluation of a random variable at a time and an outcome gives a real value.

Although random processes can always be described in terms of an underlying experiment that gives a total characterization with respect to the calculation of probabilities of events, various partial characterizations prove to be useful. These include the mean, variance, autocorrelation function, first- and second-order densities, as well as higher-order densities and moments. In later chapters it will be shown that partial characterizations may provide all the information needed for many problems involving Gaussian statistics and linear systems.

The partial characterizations of the mean and autocorrelation function for random processes are particularly important in optimum linear estimation and the analysis of linear system interaction with random processes.

Various forms of stationarity of random process were defined including stationarity in the mean, variance, autocorrelation function, first- and second-order densities, and higher-order moments and densities. Perhaps most important here is the concept of wide sense stationarity. A random process is wide sense stationary provided that its mean is a constant, independent of time, and its autocorrelation function is a function of time difference only and does not depend on the particular times. The partial characterization of mean and autocorrelation functions for wide sense stationary random processes will be shown to be enough information about the process to perform the various optimum filtering operations explored in Chapters 7 and 8.

Next a series of important examples of random processes were presented, including the straight line process, semirandom and random transmission processes, semirandom and random telegraph processes, the random sinusoidal process, and the random walk process. The main emphasis was on presenting techniques for obtaining the partial characterizations of mean, variance, autocorrelation, first- and second-order densities, and power spectral densities of those processes, and we explored as well their stationarity properties.

The concept of one random process was then extended to two and multiple random processes, and new partial characterizations of joint probability densities and cross-correlation functions were defined along with joint forms of stationarity.

A random process was defined as Gaussian if its density functions of any order were Gaussian. Knowing the mean and autocorrelation function of a Gaussian random process was sufficient information to determine its densities of all orders.

Important special cases of a white noise random process were discussed and processes generated by passing white noise through a discrete time linear system were shown to lead to MA, AR, and ARMA random processes. The mean and autocorrelation of these type of processes were determined in the steady state. These processes will be analyzed further using the Z-transform in Chapter 5 to complete the discussion. They play a particularly important role in system identification, modeling of random processes, and spectral analysis. Further information on these random processes is provided in many excellent texts, including Kay [11] and Bendat [10,12].

In sum, this chapter establishes the framework for analyzing the interaction of random processes and linear systems in Chapter 5 and nonlinear systems in Chapter 6. There we will be primarily concerned with obtaining first- and second-order densities, means, autocorrelation functions, and cross-correlation functions for the output process in terms of the input process characteristics.

PROBLEMS

4.1 Given an experiment $\mathscr{E} : (S, P(\cdot), \mathscr{F})$, where $\mathscr{F}$ is all possible subsets of S and $S = \{\zeta_{kj} : k = 1, 2, j = 0, 1, 2\}$ with $P\{\zeta_{kj}\} = P(k = k)P(j = j)$ and $P(k = 1) = 0.75$, $P(k = 2) = 0.25$, $P(j = 0) = 0.5$, $P(j = 1) = 0.25$, $P(j = 2) = 0.25$. Define a random process $X(t)$ by the mapping:

$$\zeta_{kj} - k\cos(j\pi t)$$

(a) Illustrate all the sample functions.

(b) Find the first order density function $f(x, t)$ for $t = 0$ and $t = 0.5$.

(c) Find $E\{X(t)\}$ for $t = 0$ and $t = 0.5$.

(d) Find $f(x_1, x_2, t_1, t_2)$ for $t_1 = 0$ and $t_2 = 0.5$.

(e) Find $R_{XX}(t_1, t_2) = E[X(t_1)X(t_2)]$ for $t_1 = 0$ and $t_2 = 0.5$.

(f) What form of stationarity do we have if any?

4.2 Given the following experiment $\mathscr{E} = \{S, \mathscr{F}, P(\cdot)\}$ defined by $S = \{\zeta_1, \zeta_2, \zeta_3, \zeta_4\}$; $\mathscr{F}$ = set of all possible subsets of S; and $P\{\zeta_1\} = \frac{1}{4}$, $P\{\zeta_2\} = \frac{1}{3}$, $P\{\zeta_3\} = \frac{1}{6}$, $P\{\zeta_4\} = \frac{1}{4}$. Define a random process $X(t)$ by the mapping

$$X(t, \zeta_1) = 2\cos 2\pi t, \quad X(t, \zeta_2) = \sin 2\pi t$$
$$X(t, \zeta_3) = t, \quad X(t, \zeta_4) = e^{-t}$$

(a) Evaluate $X(2, \zeta_4)$.

(b) Give several realizations of the random process $X(t)$, and tell how many realizations the process has.

(c) Find the first order density $f(x; t)$ for $t = 1$.

(d) Compute the mean, $\eta_X(t)$, of the random process $X(t)$.

(e) Is this process stationary in mean? Stationary in autocorrelation?

(f) Find the second-order density for $X(0)$ and $X(0.5)$.

4.3 Consider the experiment of the random selection of points on a line using a uniform density with $\lambda = 1$. Define a random process $X(t)$ in the following way:

$X(t) = 1$	If the total number of points between 0 and t is $0, 3, 6, \ldots$
$X(t) = 0$	If the total number of points between 0 and t is $1, 4, 7, \ldots$
$X(t) = -1$	If the total number of points between 0 and t is $2, 5, 8, \ldots$

A realization of the process is shown in Figure P4.3.

(a) Find the first order density $f(x; t)$ for this process.

(b) Find $E[X(t)]$.

(c) Find the second order density $f(x_1, x_2;\ t_1, t_2)$.

(d) Find the autocorrelation function $R_{XX}(t_1, t_2)$.

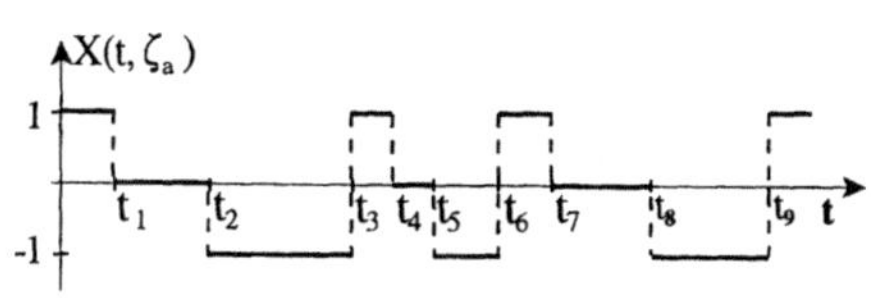

$\zeta_a = (\ldots, t_{-1}, t_1, t_2, t_3, t_4, t_5, t_6, t_7, t_8, \ldots)$

P4.3

(e) Illustrate a few realizations (sample functions) of the process.

(f) Comment on the stationarity of this process.

4.4 Define a random process $X(t)$ in the following manner:

$$X(t) = A_i \qquad iT \leq t \leq (i+1)T \qquad \text{for } i = -\infty, \ldots, \infty$$

The A_i are independent identically distributed random variables with probability density functions $f(a_i)$ given by

$$f_{A_i}(a_i) = \tfrac{1}{4}\delta(a_i + 1) + \tfrac{1}{2}\delta(a_i) + \tfrac{1}{4}\delta(a - 1)$$

(a) Draw a couple of the realizations of the process $X(t)$.

(b) Find $f(x; t)$, the first-order density of the process.

(c) Determine $E[X(t)]$, the mean of the process. Is the mean of this process a realization of the process?

(d) Is this process stationary of order 1? Why or why not?

(e) Find $f(x_1, x_2;\ t_1, t_2)$, the second-order density function of the process.

(f) Find $R_{XX}(t_1, t_2)$, the autocorrelation function for the process.

(g) Is the process wide sense stationary? Why or why not? Narrow sense stationary? Why or why not?

(h) Is the process stationary in autocorrelation? Explain.

4.5 A random signal is defined as in Figure P4.5, such that the amplitude level changes every b seconds, where b is a fixed constant. The amplitude after any given shift is assumed statistically independent of the amplitude after any other shift and is uniformly distributed on [0, 1]. Also assume that the phase of the signal is completely random, that is, E has a uniform probability density function on $[0, b]$.

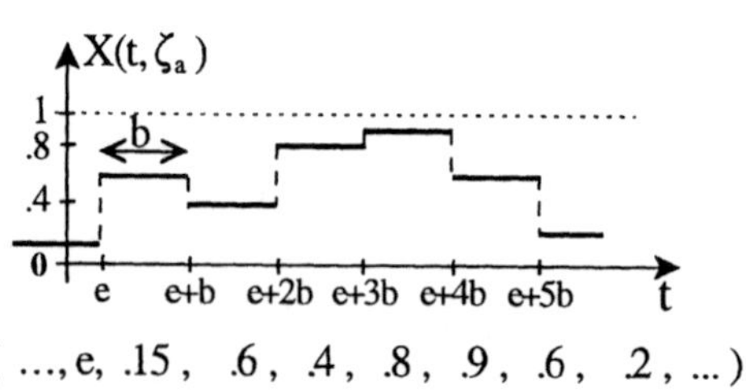

P4.5

(a) Find the mean of the defined random process.

(b) Find the autocorrelation function $R_X(t_1, t_2)$ for the defined process.

4.6 Repeat Problem 4.5 where the amplitudes are identically distributed statistically independent Gaussian random variables with zero means and unity variances.

4.7 On–off signaling is one prevalent method for sending binary streams of data. For each bit interval a known deterministic signal $s(t)$ is sent if the binary information value is 1, whereas no signal is sent if the binary value is 0. For example, a realization for the binary sequence $\zeta_i = [\ldots, 1, 0, 1, 1, 0, 1, \ldots]$ would result in the signal $x(t, \zeta_i)$ shown in the figure. If the A_i are discrete random variables independent of all other $A_j, j \neq i$, the process $X(t)$ can be written as

$$X(t) = \sum_{i=-\infty}^{\infty} A_i s(t - iT)$$

If the probability density functions for the A_i are given by

$$f(a_i) = \tfrac{1}{2}\delta(a_i) + \tfrac{1}{2}\delta(a_i - 1)$$

Find:

(a) $E[X(t)]$.

(b) $f(x; t)$.

(c) $R_{XX}(t_1, t_2)$

(d) Is the process wide sense stationary? If the process is nonstationary, how could we change the problem statement to make the process wide sense stationary?

(e) The power spectral density $S_{XX}(\omega)$ for the wide sense stationary process.

4.8 Repeat Problem 4.5 where the amplitudes are identically distributed independent discrete random variables with probability mass function as follows: $P\{A_i = 0\} = \frac{1}{4}$ and $P\{A_i = 1\} = \frac{3}{4}$.

4.9 Consider the combination of the random walk problem and the random telegraph signal. The experiment is the random selection of times t_i and a step up or down of height s at each t_i. Let $P(\text{step up}) = P(\text{step down}) = \frac{1}{2}$, and call the resulting process $Z(t)$.

(a) Draw a few realizations or sample functions of the process.

(b) Find $f(z; t)$.

(c) Compute $E[Z(t)]$.

(d) Find $R_{ZZ}(t_1, t_2)$—not easy.

4.10 Define a random process $X(t)$ by

$$X(t) = A_i, \quad (i - 1) \leq t < i, \quad i = 1, 2, \ldots$$

The A_i are defined as follows:

$$A_i = \sum_{j=1}^{i} (0.5)^{j-1} X_{i-j}, \quad i = 1, 2, \ldots$$

Let X_i be statistically independent normal random variables with zero means and unit variances.

(a) Is $X(t)$ a continuous or discrete random process?

(b) Draw a few realizations of the process.

(c) Find $R_{XX}(t_1, t_2)$ where $t_1 = 1.5$ and $t_2 = 2.4$.

(d) Find $f(x; t)$ for $t = 2.5$.

4.11 A random variable X_0 is normally distributed with zero mean and variance σ_0^2. Let $U_i \sim N(0, \sigma_u^2)$ be identically distributed independent Gaussian random variables that are group independent of X_0. Define a random process $X(t)$ in the following manner:

$$\begin{aligned} X(t) &= X_0 && 0 \le t < 1 \\ &= \tfrac{1}{2}X_0 + U_1 = X_1 && 1 \le t < 2 \\ &= \tfrac{1}{2}X_1 + U_2 = X_2 && 2 \le t < 3 \\ &\vdots && \vdots \\ &= \tfrac{1}{2}X_{k-1} + U_k = X_k && k \le t < k+1 \end{aligned}$$

etc.

(a) Illustrate a few realizations of the process.

(b) Find the first-order density function $f(x; t)$ at $t = 1.5, 1.75, 2.5$.

(c) Determine $f(x_1, x_2;\ t_1, t_2)$ where t_1 and t_2 are 0.5 and 1.5, respectively.

(d) Is this process wide sense stationary? Asymptotically wide sense stationary?

4.12 Define a random process $X(t)$ as a cosine function with random frequency and phase:

$$X(t) = \cos(\Omega\, t + \Phi)$$

The phase Φ is uniformly distributed on $[0, 2\pi]$, while the frequency Ω, which is independent of Φ, is a normally distributed with a mean of m and a variance of σ^2.

(a) Give expressions for the following probability density functions (should be able to write these out quickly): **(i)** $f_{\Omega\Phi}(\omega, \phi)$. **(ii)** $f_\Omega(\omega|\phi)$. **(iii)** $f(x; t|\omega)$. **(iv)** $f(x; t)$.

(b) Compute $E[X(t)]$ in two different ways.

(c) Compute the autocorrelation function $R_{XX}(t + \tau, t)$ for the process $X(t)$. *Hint*: A simple approach to this part of the problem involves computing $E[X(t + \tau)X(t)|\Omega = \omega]$ and using characteristic functions.

(d) Is this $X(t)$ a wide sense stationary process? Why or why not?

4.13 Consider the experiment of tossing a four-sided die with faces f_1, f_2, f_3, f_4. The outcome of a new experiment is defined as repeated independent trials of this experiment expressible as an infinite cartesian product. The sample space for a random process is thus the set of all countably infinite sequences of faces. The Borel field is defined as the power set of the sample space, which is the set of all possible subsets of this sample space. Assume that the probabilities of getting each face are equal on each trial.

A random process similar to a random walk but with variable step size is defined in the following manner: Start at the origin with interval size of 1. In the first interval $0 \le t < 1, X(t) = 1$ if f_1 is the first outcome, $X(t) = 2$ if the first outcome is $f_2, X(t) = -1$ if f_3 is the first outcome, $X(t) = -2$ if the first outcome is f_4. On subsequent intervals the value of $X(t)$ is set to the sum of all the values associated with the previous outcomes up to that time.

(a) Illustrate a few realizations of the process.

(b) Find the first-order density function $f(x; t)$ at $t = 1.5, 2.5$.

(c) Determine $f(x_1, x_2;\ t_1, t_2)$ where t_1 and t_2 are given as 0.5 and 1.5, respectively.

(d) Find $E[X(t)]$.

(e) Is the process stationary in the mean? Why or why not?

(f) Is this process wide sense stationary? Asymptotically wide sense stationary?

4.14 A random process $X(t)$ is defined in terms of the unit step function $\mu(t)$ as follows:

$$X(t) = \mu(t) - \mu(t - A)]$$

The A is a random variable with the following probability density function:

$$f_A(a) = \mu(a) - \mu(a - 1)$$

(a) Find $E[X(t)]$.

(b) Is $X(t)$ stationary in mean?

(c) Determine $R_{XX}(t_1, t_2)$ for all t_1 and t_2.

(d) Is $X(t)$ stationary in autocorrelation? Wide sense stationary?

4.15 A random process $X(t)$ is defined in terms of the unit step function $\mu(t)$ as follows:

$$X(t) = \mu(t - A) - \mu(t - T - A)$$

The A is a random variable with the following probability density function

$$f_A(a) = 1/T[\mu(a) - \mu(a - T)]$$

(a) Find $E[X(t)]$.

(b) Is $X(t)$ stationary in mean?

(c) Determine $R_{XX}(t_1, t_2)$ for all t_1 and t_2.

(d) Is $X(t)$ stationary in autocorrelation? Wide sense stationary?

4.16 A random process $X(t)$ is defined in terms of the unit step function $\mu(t)$ as follows:

$$X(t) = B[\mu\ (t - A) - \mu(t - T - A)]$$

The A and B are independent random variables with the following probability density functions:

$$f_A(a) = \frac{1}{T\ [\mu(a) - \mu(a - T)]}, \quad f_B(b) = 0.5\ [\mu(b + 1) - \mu(b - 1)]$$

(a) Find $E[X(t)]$.

(b) Is $X(t)$ stationary in mean?

(c) Determine $R_{XX}(t_1, t_2)$ for all t_1 and t_2.

(d) Is $X(t)$ stationary in autocorrelation? Wide sense stationary?

4.17 Rework Problem 4.14 where $f_A(a) = e^{-a}\mu(a)$.

4.18 Define a random process $Y(t)$ by the following:

$$Y(t) = e^{-At}\cos(\Omega t + \Theta)\mu(t)$$

A, Ω, and Θ are independent random variables with probability density functions given by

$$f_A(a) = \mu(a) - \mu(a - 1)$$
$$f_\Omega(\omega) = 0.5[\mu(\omega) - \mu(\omega - 2)]$$
$$f_\Theta(\theta) = \frac{\mu(\theta) - \mu(\theta - 2\pi)}{2\pi}$$

(a) Illustrate a few realizations of the process.

(b) Find $E[Y(t)]$.

(c) Find $R_{YY}(t_1, t_2)$ for all t_1 and t_2.

(d) Is the process wide sense stationary? Support your answer.

(e) Discuss the procedure (in detail) for finding the first-order density of the process $Y(t)$.

4.19 Given a random process $X(t)$ defined as

$$X(t) = Ae^{-Bt}\mu(t)$$

where A and B are independent uniform random variables on $[-1, 1]$ and $[0, 1]$, respectively.

(a) Illustrate a few realizations of the process.

(b) Find $f(x; t)$ the first-order density for the process.

(c) Compute $E[X(t)]$ the mean of the process.

(d) Determine $R_{XX}(t, u)$, the autocorrelation function for the process.

(e) Is $X(t)$ wide sense stationary? Strict sense stationary? Stationary in mean? Stationary in autocorrelation? Stationary of order two? Justify your answers.

4.20 Given a random process $X(t)$ specified by

$$X(t) = a\ \cos(\omega_0 t + \Theta)$$

where a and ω_0 are deterministic values and Θ is a uniform random variable on $[0, 2\pi]$.

(a) Find $E[X(t)]$.

(b) Find $f(x; t)$.

(c) Is $X(t)$ a mean square periodic process?

(d) If so, what is its periodic expansion on the interval $-T/2$ to $T/2$, where $T = 1/\omega$. If not, why isn't it periodic?

(e) Prove or disprove the following: The process $X(t)$ is ergodic in the mean.

4.21 Let $X(t) = A\ \sin(\Omega t + \Phi)$ be a random process, where A, Ω, and Φ are independent random variables. Assume that A and Ω have known probability density functions $f_A(a)$ and $f_\Omega(\omega)$ and that Φ is a uniform random variable on $[0, 2\pi]$.

(a) Find $E[X(t)]$.

(b) Discuss the procedure in detail for finding the first-order density for $X(t)$.

(c) Is $X(t)$ a wide sense stationary process? Support your answer.

(d) If Ω and Φ are uniform random variables on $[0, \pi]$ and $f_A(a) = \frac{1}{2}\ \delta(a-1) + \frac{1}{2}\ \delta(a+1)$, find $f(x_1, t_1)$, where $t_1 = 1$.

4.22 Let $X(t) = \sin(\Omega t + \Phi)$ be a random process, where Ω, and Φ are independent random variables. Assume that Ω is Gaussian with known mean m and variance σ^2 and that Φ is a uniform random variable on $[0, 2\pi]$. X_t is the random variable defined by $X(t)$. Determine:

(a) $f_{\Omega\Phi}(\omega, \phi)$.

(b) $f_\Omega(\omega|\phi)$.

(c) $f_{X_t}(x_t|\omega)$.

(d) $f_{X_t}(x_t)$

(e) $E[X_t]$

4.23 Define a random process $Y(t)$ as follows:

$$Y(t) = A + e^{-B|t|}$$

where A and B are independent uniform random variables on $[-1, 1]$ and $[0, 2]$ respectively.

(a) Draw a few sample functions of the process.

(b) Find $E[Y(t)]$.

(c) Find $R_{YY}(t_1, t_2)$.

(d) Give the forms of stationarity that apply if any.

(e) Find the first-order density $f(y; t)$ for the process.

4.24 Define a random process $Y(t)$ by

$$Y(t) = A + e^{-B|t|}$$

where A and B are independent random variable with probability density functions $f_A(a)$ and $f_B(b)$ as follows:

$$f_A(a) = 0.25\delta(a) + 0.5\delta(a-1) + 0.25\delta(a+2)$$
$$f_B(b) = 0.3\delta(b-3) + 0.7\delta(b-5)$$

(a) How many different sample functions does $Y(t)$ have?

(b) Illustrate at least three of them.

(c) Find $E[Y(t)]$.

(d) Calculate $E[Y(0)Y(1)]$.

(e) Is the process stationary in mean? Why or why not?

(f) Is the process wide sense stationary? Why or why not?

(g) Find the first-order density $f(y; t)$.

4.25 A random process is partially characterized by its mean and autocorrelation function as follows:

$$\eta_X(t) = 0, \quad R_{XX}(\tau) = \frac{\exp(-\tau/\sigma^2)}{2\pi\sigma}$$

Is $X(t)$ a Gaussian random process? Justify your answer.

4.26 Let $X(t)$ be a Gaussian random process characterized by $E[X(t)] = 0$ and $R_{XX}(\tau) = e^{-|\tau|}$. Define the random variable A as follows:

$$A = \int_0^1 X(t)\,dt$$

(a) Find $E[A]$.

(b) Compute σ_A^2.

(c) Determine the probability density function $f_A(a)$.

4.27 Let $X(t)$ be a Gaussian random process characterized by $E[X(t)] = 0$ and $R_{XX}(\tau) = e^{-|\tau|}$. Define the random variable A in terms of a random variable B and the random process $X(t)$ as follows:

$$A = \int_0^B X(t)\,dt$$

where B is a Gaussian random variable with a mean of 1 and a variance equal to $\frac{1}{4}$ that is independent of the random process $X(t)$.

(a) Find $E[A]$.

(b) Compute σ_A^2.

(c) Determine the probability density function $f_A(a)$.

4.28 Let $X(t)$ be a random process defined as follows:

$$X(t) = A\cos(2\pi t + \Phi)$$

where A is a Gaussian random variable with $\eta_A = 0$ and $\sigma_A^2 = 2$ and Φ is a uniform random variable on $[-\pi, \pi]$ statistically independent of A. Define the random variable S by

$$S = \int_0^1 X(t)\,dt$$

(a) Compute $E[S]$.

(b) Determine the variance of S.

(c) Was your answer expected after further examining the process? Why or why not?

(d) Is the process $X(t)$ a Gaussian random process? Defend your answer.

4.29 Given a zero mean Gaussian random process $X(t)$ with known autocorrelation function $R_X(\tau)$. Define a random variable Z as

$$Z = \frac{1}{T}\int_0^T X^2(t)\,dt$$

Z may be considered to be an estimate of the average power of the process $X(t)$. Let the autocorrelation function $R_X(\tau)$ be described as follows:

$$R_X(\tau) = (1 - |\tau|)\,[\mu(\tau + 1) - \mu(\tau - 1)]$$

(a) Find $E[Z]$.

(b) Is Z an unbiased estimator of the power of the process? (An estimator is unbiased if its expected value is the quantity we wish to estimate.)

(c) Compute the variance σ_Z^2.

(d) For Z to be a good estimator what do we want to have for σ_Z^2?

(e) Is Z a Gaussian random variable?

(f) How large of a T would we need for Z to be within 10% of the actual power with probability greater than 0.95?

4.30 Repeat Problem 4.29 for $R_{XX}(\tau) = \exp\{-2|\tau|\}$.

4.31 A random process $X(t)$ is defined by $X(t) = a + N(t)$, where $N(t)$ is a zero mean random process with autocorrelation function $R_N(\tau) = 0.5\delta(\tau)$. We desire to estimate the deterministic constant a by

$$\hat{a} = \frac{1}{T}\int_0^T X(t)\,dt$$

(a) Does $E[\hat{a}]$ equal a? Is it an unbiased estimate of a?

(b) Determine the variance of $\hat{a}$.

(c) Assuming $X(t)$ is a Gaussian random process, how long must the integration length T be for the estimate variance to be less than 0.1? If $X(t)$ is not a Gaussian process, answer the same question.

4.32 Assume that $X(t)$ is a zero mean Gaussian random process with autocorrelation function $R_{XX}(\tau) = \exp\{-2|\tau|\}$. A is a uniform random variable on [0,1]. Define the random variable Y as the definite integral

$$Y = \int_0^A X^2(t)\,dt$$

(a) Find $E[Y]$.

(b) Calculate σ_Y^2.

(c) If you have enough information, find $f_Y(y)$. If you do not have enough information, tell what else you need.

4.33 Let $X(t)$ be a wide sense stationary random process with known mean $\eta_X(t) = 1$ and known autocorrelation function $R_{XX}(\tau) = e^{-2|\tau|}$.

(a) Evaluate $E[X(t_1)X(t_2)]$ for $t_1 = 1$ and $t_2 = 0.5$.

(b) If $X(t)$ is a Gaussian random process, determine the second-order density function $f(x_1, x_2;\ t_1, t_2)$.

(c) If $X(t)$ is a Gaussian random process, find the third-order density function for the process. Use $X(t_1) = X_1, X(t_2) = X_2, X(t_3) = X_3$.

4.34 $X(t)$ is a wide sense stationary Gaussian random process characterized by its mean $\eta_X(t) = 0$ and autocorrelation function $R_{XX}(\tau) = N_0\delta(\tau)/2$. Define a random variable S by

$$S = \frac{1}{T}\int_0^T X(t)\,dt$$

Find:

(a) $E[S]$.

(b) Variance of S.

(c) Probability density function for S.

(d) The limit of the variance of S as $T \to \infty$.

(e) Discuss ergodicity for the process $X(t)$.

4.35 $X(t)$ is a wide sense stationary random binary waveform process characterized by its mean $\eta_X(t) = 0$ and autocorrelation function $R_{XX}(\tau) = \exp(-2|\tau|)$. Define a random variable S by

$$S = \int_0^1 t X(t)\, dt$$

Find:
(a) $E[S]$.
(b) Variance of S.
(c) Probability density function for S.

4.36 Given a random process $X(t)$ with mean $E[X(t)] = 1$ and autocorrelation function $R_{XX}(\tau) = \cos(2\pi\tau)$. Define the random variables A_i by the integral

$$A_i = \int_0^1 X(t) \cos(2\pi\, it)\, dt$$

(a) Determine $E[A_i]$ for $i = 0, 1, 2, \ldots$.
(b) Find $E[A_i, A_j]$ for all i and j.
(c) If $X(t)$ is a Gaussian random process, are A_i and A_j independent random variables? Why or why not?
(d) Find the variances of A_i for $i = 0, 1, \ldots$.

4.37 Given the autoregressive process AR(3), $n \geq 0$.
(a) Find the mean for all $n \geq 0$.
(b) Give the Yule-Walker equation that the autocorrelation must satisfy.
(c) Give the matrix form that the autocorrelation function must satisfy in terms of the coefficients of the AR(3) model, and give a formula for finding the autocorrelation function at all lags.

4.38 For an ARMA(2, 1) process:
(a) Find the equations that determine the values of the autocorrelation function at lags 0, 1, and 2 in terms of the coefficients of the model and put them in a matrix form.
(b) Give a formula for determining the autocorrelation values for lags greater than 2.

4.39 Find the autocorrelation function for all lag indexes, both positive and negative, for a MA(4) process in terms of the coefficients $b_0, b_1, \ldots, b_4$. If $E\{w^2[k]\} = 2$ for $k \geq 0$, evaluate for the steady state the autocorrelation function for the following MA(4) process:

$$x[n] = \tfrac{1}{2} w[n] - \tfrac{1}{4} w[n-1] + \tfrac{1}{8} w[n-2] - \tfrac{1}{16} w[n-3] + \tfrac{1}{32} w[n-4]$$

REFERENCES

1. Papoulis, Athanasios, *Probability, Random Variables, and Stochastic Processes*, McGraw-Hill, 1965.
2. Davenport, Jr., William B., *Probability and Random Processes*, McGraw-Hill, 1970.
3. Shanmugan, K. Sam, and Arthur M. Breipohl, *Random Signals: Detection, Estimation, and Data Analysis*, Wiley, 1988.
4. Lathi, B. P., *An Introduction to Random Signals and Communication Theory*, International Textbook Company, 1968.
5. Srinath, M. D., and P. K. Rajasekaran, *An Introduction to Statistical Signal Processing with Applications*, Wiley, 1979.
6. Thomas, John B., *An Introduction to Statistical Communication Theory*, Wiley, 1969.
7. O'Flynn, Michael, *Probabilities, Random Variables, and Random Processes*, Harper Row, 1982.
8. Brown, Robert Grover, and Patrick Y. C. Hwang, *Introduction to Random Signals and Applied Kalman Filtering*, 2nd Ed., Wiley, 1992.
9. Stark, Henry, and John W. Woods, *Probability, Random Processes, and Estimation Theory for Engineers*, Prentice-Hall, 1986.
10. Bendat, Julius S., and Allan G. Piersol, *Engineering Applications of Correlation and Spectral Analysis*, Wiley, 1980.
11. Kay, Steven M., *Modern Spectral Estimation-Theory and Application*, Prentice-Hall, 1988.
12. Bendat, Julius S., and Allan G. Piersol, *Random Data: Analysis and Measurement Procedures*, Wiley Interscience, 1971.

5 Linear Systems: Random Processes

5.1 INTRODUCTION

In many scientific disciplines the description of input–output relationships for various types of systems is paramount to understanding the system. The inputs could be many different things. However, they are many times considered some form of excitation to the system, with the output representing the response of the system. System description can take many forms. For example, the input and output could be coupled through a differential equation or transfer function or impulse response. When the input to a system can be thought of as a realization from a particular random process, the output signal can be determined by using the system's definition. Each realization of the process in turn generates an output signal. In this way there exists a mapping through the system from a set of outcomes governing the input random process and thus defining an output random process.

In this chapter we will explore the relationships that exist between input process characterizations and output process characterizations for a special class of systems called linear systems, while in the next chapter, we will explore the same problem for nonlinear systems. For example, the following question is the main theme: Knowing the mean, autocorrelation function, first-order density, and stationarity of the input process, what are the mean, autocorrelation function, first-order density, stationarity of the output process? It will be seen that a particular partial characterization of one type may not be sufficient for determining corresponding partial characterization of the same type for the output process. The presentation begins with definitions of signals and systems pertinent to this development and is followed by deeper exploration of the partial characterizations of the output process for linear time-varying and time-invariant systems of both discrete time and continuous time representations.

5.2 CLASSIFICATION OF SYSTEMS

A system can be thought of as a mapping from a closed set, S_I, called the input signal set to a closed set, S_O, called the output signal set. If the input and output signals are continuous time signals, then the system represented by $T[\cdot]$ is called a **continuous time system**. Similarly, if the inputs and output signals are discrete time signals, the system is called a **discrete time system**. It is conventional to indicate continuous time input and output signals as $x(t)$ and $y(t)$, respectively, whereas discrete time input and output signals are indicated by $x[n]$ and $y[n]$ respectively. The input and output functional relationships for both continuous time and discrete time systems will be indicated by $T[\cdot]$ as shown in Figure 5.1. Also shown are two other types of mixed continuous time and discrete time formulations, specifically discrete-continuous time and continuous-discrete time systems. Examples are digital to analog and analog to digital convertors.

Let us consider some special characteristics that these systems possess, for example linearity, causality, and time invariance. Letting $T[\cdot]$ represent the mapping of a system and $x(t)$ and $y(t)$, the input and output, respectively, a continuous time system can be represented as follows:

$$y(t) = T[x(t)] \tag{5.1}$$

Although the following presentation uses continuous time system formality, the concepts and definitions that follow will apply to any of the four types of systems previously presented in Figure 5.1.

A system is called a **linear** if for every $x_1(t)$ and $x_2(t) \in S_I$, the input set, and all constants a_1 and a_2,

$$T[a_1x_1(t) + a_2x_2(t)] = a_1T[x_1(t)] + a_2T[x_2(t)] \tag{5.2}$$

In words, the response of the system to a weighted sum of input signals is simply the sum of the weighted responses of each of the inputs responses as if administered separately. This is commonly expressed by saying superposition holds. If a system is not linear, it is called a **nonlinear** system. Further descriptions and classifications of nonlinear systems are given in Chapter 6.

Let $x(t)$ and $y(t)$ represent the input and output of the system. A system is described as being **time invariant** if the response to a time-translated version of that input signal results in a time-translated version of the output signal for all choices of input signal. For all $x(t) \in S_1$,

$$\text{if } y(t) = T[x(t)], \text{ then } T[x(t-\alpha)] = y(t-\alpha) \tag{5.3}$$

A system that is not time invariant will be called **time varying**.

A system is called **causal** if its output at time t_0, $y(t_0)$, resulting from input $x(t)$ does not depend on $x(t)$ for any time $t \geq t_0$, for all t_0 and all inputs $x(t)$. This means that the output of a causal filter cannot precede or anticipate the input signal to

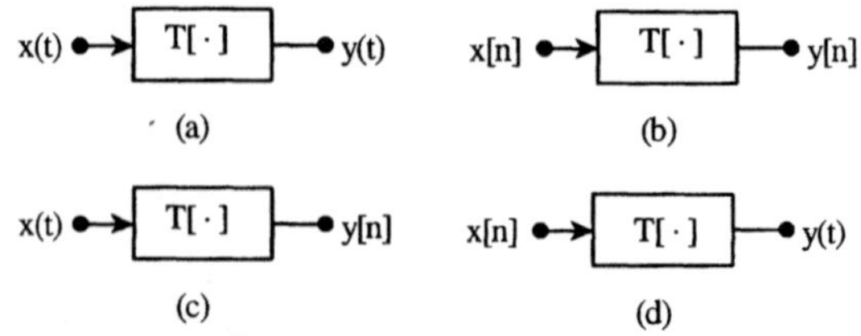

Figure 5.1 Systems: (a) Continuous time, (b) Discrete-time, (c) Continuous-discrete time and (d) Discrete-continuous time.

produce the output. Therefore other words are used to describe causality, including nonanticipatory. If the system is not causal it is called **noncausal** or anticipatory.

A system is called a **random system** if the output depends not only on the input but on the outcome of an underlying random experiment. In many cases its response can be seen to be a function of one or more random variables. For example, a system's output could depend on a random toss of a coin, or its impulsive response could be given by $h(t, A)$ or $h(t, \tau, A)$, where A is a random variable. If a system is not random, it is called a **deterministic system**. Unless specified otherwise, all filters discussed will be deterministic, and they may have either deterministic or random signal inputs.

5.2.1 Linear Time-Invariant Systems

A linear continuous time-invariant filter with input $x(t)$ and output $y(t)$ can be characterized by the following rule of correspondence (the convolution integral):

$$\begin{aligned} y(t) &= T[x(t)] \\ &= \int_{-\infty}^{\infty} h(t-\tau)x(\tau)\,d\tau \end{aligned} \tag{5.4}$$

By making a change of variables the convolution integral can also be written as

$$y(t) = \int_{-\infty}^{\infty} h(\tau)x(t-\tau)\,d\tau \tag{5.5}$$

If the input $x(t)$ is an impulse function $\delta(t)$, then the output $y(t)$, after using the sampling property for $\delta(t)$, equals $h(t)$. Therefore $h(t)$ is referred to as the impulse response of the filter.

The linear time-invariant system specified by (5.4) above will be causal if $h(t) = 0$ for all $t < 0$. For a causal filter the input output equations are easily seen to effect the limits for the integrals above and can be written as

$$y(t) = \int_{-\infty}^{t} h(t-\tau)x(\tau)\,d\tau = \int_{0}^{\infty} h(\tau)x(t-\tau)\,d\tau \tag{5.6}$$

5.2.2 Linear Time-Varying Systems

A certain class of linear time-varying filters can be characterized by an $h(t, \tau)$ which represents the response to an impulse at time τ. The response of a system to an impulse at time τ_0 is not a translated version of the response of the system to an impulse at 0. An example of an $h(t, \tau)$ is shown in Figure 5.2. It is seen that the response to an impulse at time τ_2 is longer and lower than the response of the system to an impulse at time τ_1, which could represent a system that becomes sluggish as

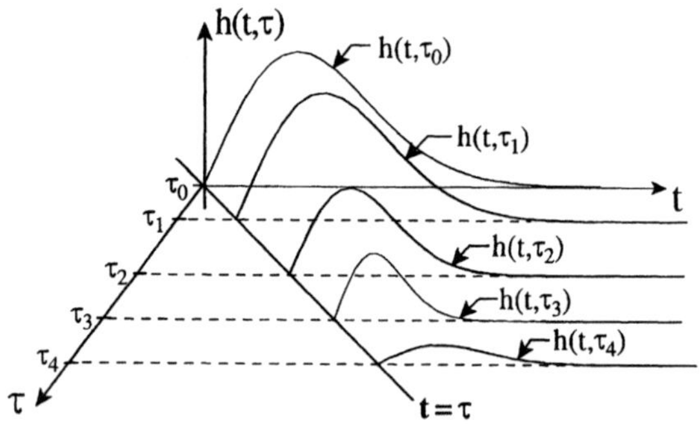

Figure 5.2 An example of the impulse response function $h(t, \tau)$ for a linear time varying system.

time increases. The $h(t, \tau_1)$ and $h(t, \tau_2)$ are just two of the profiles that make up $h(t, \tau)$. The input-output relationship for such filters is given by

$$y(t) = \int_{-\infty}^{\infty} h(t, \tau)x(\tau)\,d\tau \tag{5.7}$$

The linear time varying filter will be defined to be **causal** if $h(t, \tau) = 0$ for $t < \tau$.

The input–output statistical characterizations are now presented for continuous and discrete linear time-varying and time-invariant systems.

5.3 CONTINUOUS LINEAR TIME-INVARIANT SYSTEMS (RANDOM INPUTS)

Consider a random process $X(t, \xi)$ to be the input to a nonrandom linear filter with impulsive response $h(t)$. For every ξ_i an input waveform $X(t, \xi_i)$ is specified, and an output waveform $Y(t, \xi_i)$ can be given by the convolution integral of the realization with the impulse response as

$$Y(t, \xi_i) = \int_{-\infty}^{\infty} X(t - \tau, \xi_i)h(\tau)\,d\tau \tag{5.8}$$

Thus $Y(t, \xi)$ is a mapping from S, the sample space of the underlying experiment for $X(t)$, to a set of time functions and is, therefore, a random process $Y(t, \xi)$. For notation purposes the expression above can be written as a function of the input random process $X(t)$ in the following way:

$$Y(t) = \int_{-\infty}^{\infty} X(t - \tau)h(\tau)\,d\tau \tag{5.9}$$

The problem of prime importance is the statistical description of the output process $Y(t)$ in terms of the given characterizations of the input process $X(t)$. For example, we might ask what is the mean $\eta_y(t)$ of the output process, and is the knowing of the mean $\eta_x(t)$ of the input process sufficient to determine it? Similar questions may be asked about autocorrelation functions, first-order densities, and nth-order densities. These questions are answered in the following sections.

5.3.1 Mean in–Mean out (Linear Time-Invariant Filters)

The mean of the output process for a linear time invariant system will now be found taking special care to discuss every step in the derivation as many of the sections that follow will go through similar steps. Taking the expected value of both sides of Eq. (5.9) gives the mean of the output random process as

$$\eta_Y(t) = E[Y(t)] = E[\int_{-\infty}^{\infty} X(t-\tau)h(\tau)\,d\tau] \tag{5.10}$$

If $E[X(t-\tau)h(\tau)]$ is finitely integrable, it is possible to interchange the expected value operator, which itself is an integral operator, with the integral for the convolution. Interchanging the expected value and the integration gives

$$\eta_Y(t) = E[Y(t)] = \int_{-\infty}^{\infty} E[X(t-\tau)h(\tau)]\,d\tau \tag{5.11}$$

Assuming that the linear filter is not random we can take $h(\tau)$ outside the expected value to obtain

$$\eta_Y(t) = \int_{-\infty}^{\infty} E[X(t-\tau)]h(\tau)\,d\tau \tag{5.12}$$

Since $E[X(t-\tau)] = \eta_x(t-\tau)$, the following very important result is obtained:

$$\eta_Y(t) = \int_{-\infty}^{\infty} \eta_X(t-\tau)h(\tau)\,d\tau = \eta_X(t) * h(t) \tag{5.13}$$

or in the Laplace domain

$$\eta_Y(t) = \mathscr{L}^{-1}[N_X(p)H(p)]$$

where $N_X(p)$ and $H(p)$ are the Laplace transforms[‡] of $\eta_X(t)$ and $h(t)$. Thus the mean of the output process is seen to be the convolution of the mean of the input process and the impulse response of the system.

When the mean of the input process is a constant, the mean of the output process will be constant as well:

$$\eta_Y(t) = \eta_X \int_{-\infty}^{\infty} h(\tau)\,d\tau \triangleq \eta_Y \tag{5.14}$$

[‡] For Laplace transform variable we will use p rather than s to eliminate ambiguities with the notation of power spectral densities and the use of $s(t)$ for a general time signal.

If the system transfer function is defined as the Fourier transform of the impulse response, the mean of the output process can written as

$$\eta_Y = \eta_X H(j0)$$
$$\text{where} \qquad H(j\omega) = \int_{-\infty}^{\infty} h(t)e^{-j\omega t}\,dt \tag{5.15}$$

Thus it is seen that the mean of the output process $Y(t)$ can be obtained by multiplying the input mean η_X by the dc gain $H(j0)$.

5.3.2 Autocorrelation in–Autocorrelation out (Linear Time-invariant Filters)

If the input process has an autocorrelation function $R_{XX}(t, u)$, a procedure paralleling the development given in the previous section is used to obtain the output correlation function $R_{YY}(t, u)$ and thus the relationship between the input autocorrelation and the output autocorrelation. Using the definition of the autocorrelation of the output process and (5.9) representing the linear filter, we have

$$\begin{aligned} R_{YY}(t, u) &= E[Y(t)Y(u)] \\ &= E\left[\int_{-\infty}^{\infty} X(t-\alpha)h(\alpha)\,d\alpha \int_{-\infty}^{\infty} X(u-\beta)h(\beta)\,d\beta\right] \end{aligned} \tag{5.16}$$

Interchanging integrals and expected values, and assuming a deterministic filter, we rewrite (5.16) as

$$\begin{aligned} R_{YY}(t, u) &= \int_{-\infty}^{\infty}\int_{-\infty}^{\infty} E[X(t-\alpha)X(u-\beta)]h(\alpha)h(\beta)\,d\alpha\,d\beta \\ &= \int_{-\infty}^{\infty}\int_{-\infty}^{\infty} R_{XX}(t-\alpha, u-\beta)h(\alpha)h(\beta)\,d\alpha\,d\beta \end{aligned} \tag{5.17}$$

If the input process is wide sense stationary, its autocorrelation function $R_{XX}(\tau)$ is a function of the time difference only. By letting $t - u = \tau$ and using the fact that $X(t)$ is a wide sense stationary process, we can treat $R_{YY}(t, u)$ as a function of the time difference $t - u$ alone. This we denote by $R_{YY}(\tau)$:

$$\begin{aligned} R_{YY}(\tau) &= \int_{-\infty}^{\infty}\int_{-\infty}^{\infty} R_{XX}(\tau+\beta-\alpha)h(\alpha)h(\beta)\,d\alpha\,d\beta \\ &= h(\tau) * R_{XX}(\tau) * h(-\tau) \\ \text{or } R_{YY}(\tau) &= \mathscr{L}_{\beta}^{-1}[H(p)\Phi_{XX}(p)H(-p)] \end{aligned} \tag{5.18}$$

Equation (5.18) expresses the output autocorrelation function $R_{YY}(\tau)$ in terms of the convolution of $h(\tau)$, $R_{XX}(\tau)$ and the time reversed impulse function $h(-\tau)$. By taking the Fourier transform of both sides of (5.18), the relationship between in the input and output power spectral densities can be obtained as

$$\begin{aligned} S_{YY}(\omega) &= H(j\omega)S_{XY}(\omega)H^*(j\omega) \\ &= |H(j\omega)|^2 S_{XX}(\omega) \end{aligned} \tag{5.19}$$

EXAMPLE 5.1

Let the random process $X(t)$ be the input to a linear system represented by its impulse response $h(t)$ resulting in the output process $Y(t)$. The input was applied at minus infinite. We have $h(t)$ given by $h(t) = 2e^{-t}\mu(t)$, and that $X(t)$ is a random telegraph process with autocorrelation function $R_{XX}(\tau) = e^{-2|\tau|}$. Find the power spectral density $S_{YY}(\omega)$ for the output process.

SOLUTION

The output power spectral density depends upon the input power spectral density and the magnitude squared of the transfer function. The power spectral density of the input process is determined as

$$\begin{aligned} S_{XX}(\omega) &= \mathscr{F}[R_{XX}(\tau)] \\ &= \int_{-\infty}^{\infty} e^{-2|\tau|} e^{-j\omega\tau} d\tau = \frac{4}{\omega^2 + 4} \end{aligned}$$

The transfer function $H(j\omega)$ is obtained from the impulse response as

$$\begin{aligned} H(j\omega) &= \mathscr{F}[h(t)] \\ &= \int_{-\infty}^{\infty} e^{-\tau}\mu(t) e^{-j\omega\tau}\, dt = \frac{1}{j\omega + 1} \end{aligned}$$

Therefore from (5.19) the power spectral density of the output can be written as

$$\begin{aligned} S_{YY}(\omega) &= |H(j\omega)|^2 S_{XX}(\omega) \\ &= \left| \frac{1}{j\omega + 1} \right|^2 \frac{4}{\omega^2 + 4} = \frac{4}{(\omega^2 + 1)(\omega^2 + 4)} \end{aligned}$$

The power spectral density could have also been obtained in terms of the Laplace transform as follows:

$$S_{XX}(\omega) = [H(p)\Phi_{XX}(p)H(-p)]|_{p=j\omega}$$

■

5.3.3 Cross Correlation of the Input and Output

For an input random process $X(t)$ a linear time invariant filter produces an output process $Y(t)$. It is natural to expect that these two processes are in some way statistically related. Such a relationship can take the form of joint density functions of various orders for the processes. A particularly useful joint characterization for two random processes is the cross-correlation function. It provides a statistical measure of the correlation that exists between the processes.

If $X(t)$ represents the input process to a linear time-invariant filter characterized by an impulse response $h(t)$, the output $Y(t)$ can be expressed by the convolution integral given in (5.9). The cross-correlation function for $X(t)$ and $Y(t)$ is computed by taking the expected value of the product of $X(t)$ and $Y(u)$, where $Y(u)$ is expressed by the convolution integral. This results in

$$\begin{aligned} R_{XY}(t, u) &= E[X(t)Y(u)] \\ &= E[X(t)\int_{-\infty}^{\infty} h(\beta)X(u-\beta)\,d\beta] \\ &= \int_{-\infty}^{\infty} h(\beta)E[X(t)X(u-\beta)]\,d\beta \end{aligned} \tag{5.20}$$

Recognizing the expected value to be the autocorrelation function for the input process $X(t)$, the cross-correlation function between the input and output becomes

$$R_{XY}(t, u) = \int_{-\infty}^{\infty} h(\beta)R_{XX}(t, u-\beta)\,d\beta \tag{5.21}$$

Thus the cross-correlation function between the input and output process can be determined by knowing the input process autocorrelation and the impulse response defining the system.

If the input process is wide sense stationary, the autocorrelation function can be written in terms of the time difference only as

$$R_{XY}(t, u) = \int_{-\infty}^{\infty} h(\beta)R_{XX}(t-u+\beta)\,d\beta \tag{5.22}$$

Thus the cross-correlation function between input and output processes for a wide sense stationary input process is a function of time difference $\tau = t - u$ only, and it can be written as

$$\begin{aligned} R_{XY}(\tau) &= \int_{-\infty}^{\infty} h(\beta)R_{XX}(\tau+\beta)\,d\beta \\ &= \int_{-\infty}^{\infty} R_{XX}(\tau-\beta)h(-\beta)\,d\beta \end{aligned} \tag{5.23}$$

The $R_{XY}(\tau)$ is recognized as a convolution integral of the autocorrelation function of the input process and a time-reversed version of the impulse response of the system:

$$\begin{aligned} R_{XY}(\tau) &= R_{XX}(\tau) * h(-\tau) \\ \text{or } R_{XY}(\tau) &= \mathscr{L}_{\beta}^{-1}[\Phi_{XX}(p)H(-p)] \end{aligned} \tag{5.24}$$

Following a similar development for the cross-correlation between $Y(t)$ and $X(t)$, it can be shown that $R_{YX}(\tau)$ is

$$\begin{aligned} R_{YX}(\tau) &= h(\tau) * R_{XX}(\tau) \\ \text{or } R_{YX}(\tau) &= \mathscr{L}_{\beta}^{-1}[H(p)\Phi_{XX}(p)] \end{aligned} \tag{5.25}$$

This result can also be obtained by replacing τ by $-\tau$ in (5.24) and using the fact that $R_{XX}(\tau)$ is an even function of τ.

Since the cross-correlation function has been shown to be a function of time difference only, the input autocorrelation function is a function of time difference only, and the output autocorrelation function is a function of time difference only, the input and output process $X(t)$ and $Y(t)$ are jointly wide sense stationary.

As a nemonic notice that if the right variable is changed from X to Y passing through a system with impulsive response $h(\tau)$ that $R_{XX}(\tau)$ is convolved with a time-reversed impulsive response $h(-\tau)$ or if a left variable is changed from X to Y passing through a system with impulse response $h(t)$, that $h(\tau)$ is convolved with $R_{XX}(\tau)$. This relationship is shown in Figure 5.3.

Notice that (5.18) can be easily confirmed by using this nemonic as both right and left variables will be changed. This approach can be used for ease in calculating cross-correlation and autocorrelation functions for outputs of linear systems with multiple inputs and multiple outputs.

EXAMPLE 5.2

A given linear time-invariant causal filter has impulse response $h(t) = e^{-t}\mu(t)$, and its input $X(t)$ is a zero mean wide sense stationary process characterized by its autocorrelation function $R_{XX}(\tau) = 1/2e^{-2|\tau|}$. Suppose that the input was applied at time $t = -\infty$, resulting in the output process $Y(t)$. Find the following:

(a) The mean $\eta_Y(t)$ of the output process $Y(t)$.
(b) The cross-correlation function $R_{YX}(\tau)$.
(c) The cross-correlation function $R_{XY}(\tau)$.
(d) The output power spectral density $S_{YY}(\omega)$.
(e) The output autocorrelation function $R_{YY}(\tau)$.

SOLUTION

(a) The mean of the output process can be found by using (5.13) and is determined as

$$\eta_X(t) = \eta_X(t) * h(t) = 0 * e^{-\tau}\mu(t) = 0$$

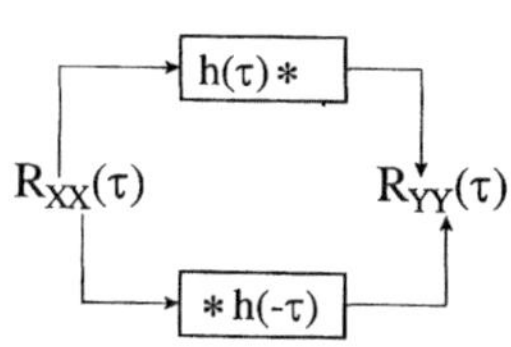

Figure 5.3 Relationships between input, output, and cross correlation functions for a linear time invariant system.

(b) The cross-correlation function $R_{YX}(\tau)$ between output process $Y(t)$ and input process $X(t)$ can be determined from (5.25) as

$$\begin{aligned}R_{YX}(\tau) &= \mathscr{L}_\beta^{-1}[H(p)\Phi_{XX}(p)] = \mathscr{L}_\beta^{-1}\{\mathscr{L}_u[e^{-t}\mu(t)]\cdot\mathscr{L}_\beta(\tfrac{1}{2}e^{-2|\tau|})\}\\ &= \mathscr{L}_\beta^{-1}\left[\frac{1}{p+1}\cdot\frac{1}{2}\left(\frac{1}{p+2}+\frac{1}{-p+2}\right)\right]\\ &= \mathscr{L}_\beta^{-1}\left[\frac{2}{(p+1)(p+2)(-p+2)}\right]\end{aligned}$$

To obtain the inverse transform, a partial fraction expansion for the term in the brackets is performed followed by the inverse bilateral Laplace transform to get

$$\begin{aligned}R_{YX}(\tau)] &= \mathscr{L}_\beta^{-1}\left[\frac{-\frac{2}{3}}{(p+1)}+\frac{-\frac{1}{2}}{(p+2)}+\frac{\frac{1}{6}}{(-p+2)}\right]\\ &= \left(-\frac{2}{3}e^{-\tau}-\frac{1}{2}e^{-2\tau}\right)\mu(\tau)+\frac{1}{6}e^{2\tau}\mu(-\tau)\end{aligned}$$

(c) To find $R_{XY}(\tau)$, we can use the fact that $R_{XY}(\tau) = R_{YX}(-\tau)$ to obtain

$$R_{XY}(\tau) = R_{YX}(-\tau) = \left(-\frac{2}{3}e^{\tau}-\frac{1}{2}e^{2\tau}\right)\mu(-\tau)+\frac{1}{6}e^{-2\tau}\mu(\tau)$$

(d) To find $S_{YY}(\omega)$, the power spectral density $S_{XX}(\omega)$ of the input must first be found. From the definition of power spectral density, we have

$$S_{XX}(\omega) = \mathscr{F}[R_{XX}(\tau)] = \mathscr{F}[\tfrac{1}{2}e^{-2|\tau|}] = \frac{2}{\omega^2+4}$$

From (5.19) the power spectral density $S_{YY}(\omega)$ of the output process $Y(t)$ in terms of the input power spectral density is

$$\begin{aligned}S_{YY}(\omega) &= |H(j\omega)|^2 S_{XX}(\omega) = \left|\frac{1}{j\omega+1}\right|^2\cdot\frac{2}{(\omega^2+4)}\\ &= \frac{2}{(\omega^2+1)(\omega^2+4)}\end{aligned}$$

(e) The autocorrelation function $R_{YY}(\tau)$ can be found from Eq. (5.18) as the convolution or as the inverse transform. The easiest way is to use the Bilateral inverse as follows:

$$\begin{aligned}R_{YY}(\tau) &= \mathscr{L}_\beta^{-1}[H(p)\Phi_{XX}(p)H(-p)]\\ &= \mathscr{L}_\beta^{-1}\left\{\frac{1}{(p+1)}\,\frac{2}{(p+2)(-p+2)}\,\frac{1}{(-p+1)}\right\}\end{aligned}$$

A partial fraction expansion is performed for the function in { }, and then an inverse bilateral Laplace transform to give the output autocorrelation function as

$$R_{YY}(\tau)] = \mathscr{L}_{\beta}^{-1}\left[\frac{\frac{1}{3}}{(p+1)} + \frac{\frac{1}{3}}{(-p+1)} + \frac{-\frac{1}{6}}{(p+2)} + \frac{-\frac{1}{6}}{(-p+2)}\right]$$

$$= \frac{1}{3}e^{-\tau}\mu(\tau) + \frac{1}{3}e^{\tau}\mu(-\tau) - \frac{1}{6}e^{-2\tau}\mu(\tau) - \frac{1}{6}e^{2\tau}\mu(-\tau)$$

$$= \frac{1}{3}e^{-|\tau|} - \frac{1}{6}e^{-2|\tau|}$$

■

5.3.4 *n*th-Order Densities in–*n*th-Order Densities Out

To arrive at the output densities, it is worthwhile to discuss the first-order density of the output. Rewriting the input–output relationship (5.4) as

$$Y(t) = \int_{-\infty}^{\infty} X(\tau)h(t-\tau)\,d\tau \tag{5.26}$$

it is easily seen that the output random variable $Y(t) = Y_t$ defined at any time t not only depends on the input $X(t)$ at time t but on $X(t)$ for all $-\infty < t < \infty$. Therefore, in general, the determination of the output first-order density requires a complete characterization of the input process. Although the second-order density is just as complicated, it should be noted that the mean and autocorrelation of the output can be determined, (5.13) and (5.18), without knowing the first- or second-order densities of the input process provided the mean and autocorrelation of the input process are known. This is an important result, and it demonstrates that for linear systems the mean and autocorrelation of the input process are sufficient for the same partial characterizations of the output. Later we will see for the special case where the input process is Gaussian that the first, second, and *n*th order densities of the output process $Y(t)$ can be easily determined.

5.3.5 Stationarity of the Output Process

For a wide sense stationary process as input to a linear time-invariant system it has been shown, (5.14) and (5.18), that the mean of the output process is a constant and not a function of time and that the autocorrelation function depends only on the time difference, $t - u$ or τ. These results allow the following statement to be made:

> If the input to a linear time-invariant filter is a wide sense stationary random process, the output is also a wide sense stationary random process.

We also might ask if the input process is stationary of order one or two, what can we say about the output? From the argument given in the previous section, there is no reason to believe that the first-order density would be independent of time or that the second-order densities would be equal for each pair of times t_i and t_j. Thus an output process is not necessarily stationary of order two just because the input is.

5.4 CONTINUOUS TIME-VARYING SYSTEMS WITH RANDOM INPUT

If the input to a linear time-varying system, characterized by its $h(t, \tau)$, is a random process $X(t)$ with known partial characterizations, the characterizations of the output process are desired. The partial characteristics of the output process consist of mean, autocorrelation, first- and higher-order densities along with various stationarity properties.

5.4.1 Mean in–Mean out (Linear Time-Varying Filter)

Taking the expected value of both sides of (5.7), the input-output equation for a linear time varying filter with impulse response $h(t, \tau)$, the mean of the output random process becomes

$$\eta_Y(t) = E[Y(t)] = E[\int_{-\infty}^{\infty} h(t, \tau)X(\tau)\, d\tau] \tag{5.27}$$

Interchanging the expected value and the integration operations give

$$\eta_Y(t) = E[Y(t)] = \int_{-\infty}^{\infty} E[h(t, \tau)X(\tau)]\, d\tau \tag{5.28}$$

We assume that the linear filter is not random, so $h(t, \tau)$ can be taken outside the expected value to yield

$$\eta_Y(t) = \int_{-\infty}^{\infty} h(t, \tau)E[X(\tau)]\, d\tau \tag{5.29}$$

Now, recognizing that $E[X(\tau)] = \eta_X(\tau)$, we write the mean of the output process as

$$\eta_Y(t) = \int_{-\infty}^{\infty} h(t, \tau)\eta_X(\tau)\, d\tau \tag{5.30}$$

The mean, $\eta_X(t)$, of the output process is thus functionally related to the mean of the input process and system impulse response $h(t, \tau)$ through the integral above.

When the input process is stationary in mean, $\eta_X(t)$ is constant, the mean of the output process is

$$\eta_Y(t) = \eta_X \int_{-\infty}^{\infty} h(t, \tau)\, d\tau \tag{5.31}$$

From this result it is seen that the mean $\eta_Y(t)$ of the output process $Y(t)$ will be a function of time, since the integral of $h(t, \tau)$ with respect to τ is a function of time. Thus the output process is not stationary in the mean even if the input process is stationary in mean.

5.4.2 Autocorrelation in–Autocorrelation out (Linear Time-Varying Filter)

Assume that the input process has an autocorrelation function $R_{XX}(t, u)$, and the output a correlation function $R_{YY}(t, u)$; the relationship between the input autocorrelation and the output autocorrelation is now determined. Using the definition of the autocorrelation of the output process and (5.7), representing the linear filter, we write the output autocorrelation as

$$\begin{aligned} R_{YY}(t, u) &= E[Y(t)Y(u)] \\ &= E[\int_{-\infty}^{\infty} h(t, \alpha)X(\alpha)\, d\alpha \int_{-\infty}^{\infty} h(u, \beta)X(\beta)\, d\beta] \end{aligned} \tag{5.32}$$

After interchanging the integrals and expected values, and assuming a deterministic filter, we rewrite (5.32) as

$$\begin{aligned} R_{YY}(t, u) &= \int_{-\infty}^{\infty}\int_{-\infty}^{\infty} h(t, \alpha)h(u, \beta)E[X(\alpha)X(\beta)]\, d\alpha\, d\beta \\ &= \int_{-\infty}^{\infty}\int_{-\infty}^{\infty} h(t, \alpha)h(u, \beta)R_{XX}(\alpha, \beta)\, d\alpha\, d\beta \end{aligned} \tag{5.33}$$

If the input process is wide sense stationary, its autocorrelation function $R_{XX}(\tau)$ is a function of the time difference only. Thus the output process autocorrelation function, $R_{YY}(t, u)$, can be written as

$$R_{YY}(t, u) = \int_{-\infty}^{\infty}\int_{-\infty}^{\infty} h(t, \alpha)h(u, \beta)R_{XX}(\alpha - \beta)\, d\alpha\, d\beta \tag{5.34}$$

The equation above expresses the output autocorrelation function $R_{YY}(t, u)$ in terms of the autocorrelation function of the input process, $R_{XX}(\tau)$, but since $h(t, \alpha)h(u, \beta)$ is not necessarily a function of $t - u$, the double integral will in general not be a function of $t - u$. Thus the output process will not be stationary in autocorrelation even if the input process is stationary in autocorrelation.

5.4.3 Cross Correlation of the Input and Output (Linear Time-Varying Filter)

For an input random process $X(t)$ a linear time-varying filter produces an output process $Y(t)$. These two processes are in some ways statistically related. The cross-correlation function provides a statistical measure of the correlation that exists between the processes. For a time-varying linear filter characterized by its $h(t, \tau)$, the output $Y(t)$ is given by (5.7). Thus the cross correlation can be written as

$$\begin{aligned} R_{XY}(t, u) &= E[X(t)Y(u)] = E[X(t) \int_{-\infty}^{\infty} h(u, \beta)X(\beta)\, d\beta] \\ &= \int_{-\infty}^{\infty} h(u, \beta)E[X(t)X(\beta)]\, d\beta \end{aligned} \tag{5.35}$$

Recognizing the autocorrelation function for the input process $X(t)$, the cross-correlation function between input and output becomes

$$R_{XY}(t, u) = \int_{-\infty}^{\infty} h(u, \beta) R_{XX}(t, \beta)\, d\beta \tag{5.36}$$

Thus the cross-correlation function between input and output can be determined by knowing only the input autocorrelation function $R_{XX}(t, u)$ and the system's impulse response $h(t, \tau)$.

If $X(t)$, the input process, is wide sense stationary, the $R_{XX}(t, \beta)$ can be written in terms of the time difference. The cross-correlation function of (5.36) becomes

$$R_{XY}(t, u) = \int_{-\infty}^{\infty} h(u, \beta) R_{XX}(t - \beta)\, d\beta \tag{5.37}$$

The cross-correlation function can be rewritten in a slightly different form by making a change of variables $t - \beta = \alpha$ in (5.37), resulting in

$$R_{XY}(t, u) = \int_{-\infty}^{\infty} h(u, t - \alpha) R_{XX}(\alpha)\, d\alpha \tag{5.38}$$

The cross-correlation function is thus not necessarily a function of time difference $\tau = t - u$ only.

The cross correlation between $Y(t)$ and $X(u)$ can be similarly found to be

$$\begin{aligned} R_{YX}(t, u) &= E[Y(t)X(u)] \\ &= \int_{-\infty}^{\infty} h(t, \beta) R_{XX}(u, \beta)\, d\beta \end{aligned} \tag{5.39}$$

Clearly, even if the input process is wide sense stationary, the input and output processes are not jointly wide sense stationary for a linear time-varying system.

5.4.4 *n*th-Order Densities in–*n*th-Order Densities out (Linear Time-Varying Filter)

By a similar argument to that given in Section 5.3.4 for a linear time-invariant system, the first- and higher-order densities of the output of a linear time-varying filter cannot be determined knowing only the same order characterizations of the input. In general, a total characterization of the input process would be required to even get the first-order densities.

5.4.5 Stationarity of the Output Process (Linear Time-Varying Filter)

From (5.31), it is seen that the mean of the output process will be a function of time even if the input process is stationary in its mean, and from (5.34), that the autocorrelation function does not depend only on the time difference $t - u$ even when the input process is stationary in autocorrelation. From these results the following statement can be made:

If the input to a linear time-varying filter is a wide sense stationary random process, the output random process will not necessarily be stationary in any sense.

5.5 DISCRETE TIME-INVARIANT LINEAR SYSTEMS WITH RANDOM INPUTS

In the study of linear continuous time systems with random inputs, relationships were obtained for the output mean and the autocorrelation function in terms of the input autocorrelation and mean. Similar formulas will now be developed for discrete time linear shift-invariant systems characterized by the following input–output relationships (discrete convolution sum).

$$\begin{aligned} y[n] &= \sum_{k=-\infty}^{\infty} h[k]x[n-k] \quad \text{or} \\ &= \sum_{k=-\infty}^{\infty} h[n-k]x[k] \end{aligned} \tag{5.40}$$

The derivations assume that the input process was applied at time $-\infty$, and results are interpreted as being in steady state. In (5.40) the $h[n]$ is the unit-sample response. If $h[n]$ is zero for all $n < 0$, the discrete system is causal, and Eq. (5.40) can be written as

$$y[n] = \sum_{k=0}^{\infty} h[k]x[n-k] \tag{5.41}$$

5.5.1 Mean in–Mean Out

The mean of the output random sequence is seen from (5.40) for a nonrandom $h[n]$ to be

$$\begin{aligned} E[Y[n]] &= \sum_{k=-\infty}^{\infty} h[k]E[x[k-n]] \\ &= h[n] * \eta_X[n] \end{aligned} \tag{5.42}$$

In other words, the output mean is the convolution of the input mean with the unit-sample response $h[n]$, similar to the result for continuous time systems.

If the input process is wide sense stationary with mean $E[X[n]] = \eta_X$, for all n, then $E[Y[n]]$ can be written as

$$\begin{aligned} E[Y[n]] &= \eta_X \sum_{k=-\infty}^{\infty} h[k] \\ &= \eta_X Z(h[n])|_{z=1} \equiv \eta_X H(1) \end{aligned} \tag{5.43}$$

where $H(1) = H(z)|_{z=1}$ and $H(z)$ is the Z-transform of $h[n]$.

This result compares with that for the continuous case, the difference being $z = 1$ in $H(z)$ instead of $p = 0$ in the $H(p)$.

5.5.2 Autocorrelation in–Autocorrelation Function Out

In a straightforward manner, the output autocorrelation function $R_{YY}[k_1, k_2]$ defined by

$$R_{YY}[k_1, k_2] = E[Y[k_1]Y[k_2]] \tag{5.44}$$

can be found using (5.40). Assuming that the linear filter is not random, we can take $h[n]$ and $h[m]$ outside the expected value to yield

$$R_{YY}[k_1, k_2] = \sum_{m=-\infty}^{\infty} \sum_{n=-\infty}^{\infty} h[n]h[m]R_{XX}[k_1 - n, k_2 - m] \tag{5.45}$$

For a wide sense stationary input sequence $X[n]$, the output autocorrelation function $R_{YY}[k_1, k_2]$ can be written as a function of $k_1 - k_2 = k$:

$$R_{YY}[k] = \sum_{m=-\infty}^{\infty} \sum_{n=-\infty}^{\infty} h[n]h[m]R_{XX}[k - n + m] \tag{5.46}$$

Equation (5.46) is the discrete form of a double convolution sum and can be written more compactly as

$$\begin{aligned} R_{YY}[k] &= h[k] * R_{XX}[k] * h[-k] \\ &= Z^{-1}[H(z)\Psi_{XX}(z)H(z^{-1})] \end{aligned} \tag{5.47}$$

and the Z-transform of (5.47) can be expressed as

$$\Psi_{YY}(z) = H(z)\Psi_{XX}(z)H(z^{-1}) \tag{5.48}$$

where $\Psi_{XX}(z)$, $\Psi_{YY}(z)$, and $H(z)$ represent the Z-transforms of $R_{XX}[k]$, $R_{YY}[k]$, and $h[k]$, respectively.

The **power spectral density** for a random sequence $X[n]$ is defined as

$$\psi_{XX}(\omega) \underline{\Delta} \Psi_{XX}(z)|_{z=e^{j\omega}} \tag{5.49}$$

Therefore from (5.48) and (5.49) it is possible to write the power spectral density of the output process in terms of the power spectral density of the input process as

$$\begin{aligned} \psi_{YY}(\omega) &= H(z)\Psi_{XX}(z)H(z^{-1})|_{z=e^{j\omega}} \\ &= |H(e^{j\omega})|^2 \psi_{XX}(\omega) \end{aligned} \tag{5.50}$$

The results above can be compared to those determined for a continuous random input process $X(t)$ and a continuous system characterized by $h(t)$ with output process $Y(t)$. The formulas are analogous, with the discrete time convolution replacing the continuous time convolution, the Z-transform replacing the Laplace transform, and the $e^{j\omega}$ replacing the $j\omega$.

5.5.3 Cross-correlation Functions

As in the continuous time case the corresponding cross-correlation functions of input and output random processes for discrete time can be determined as convolutions:

$$\begin{aligned} R_{YX}[k] &= h[k] * R_{XX}[k] \quad \text{or} \quad R_{YX}(k) = Z^{-1}[H(z)\Psi_{XX}(z)] \\ R_{XY}[k] &= R_{XX}[k] * h[-k] \quad \text{or} \quad R_{XY}(k) = Z^{-1}[\Psi_{XX}(z)H(z^{-1})] \end{aligned} \tag{5.51}$$

Their Z-transform equivalents are determined as

$$\begin{aligned} \Psi_{YX}(z) &= Z(R_{YX}[n]) = H(z)\Psi_{XX}(z) \\ \Psi_{XY}(z) &= Z(R_{XY}[k]) = \Psi_{XX}(z)H(z^{-1}) \end{aligned} \tag{5.52}$$

5.5.4 *n*th-Order Densities

By arguments similar to those for the continuous case the first-, second-, and nth-order densities can be examined. Since the output $Y[n]$ at any given n is a function of the input process $X[k]$ random variables for all k form $-\infty$ to ∞, the first-order density cannot be determined without knowing the nth-order joint densities for all input random variables $X[k]$. Thus a first-order characterization of the input process is insufficient information to obtain the first-order densities for the output process. Similarly the second- and nth-order densities cannot be determined unless the input process is totally characterized. However, if the input process is a Gaussian random process and characterized by its mean and autocorrelation function, it is totally characterized. Thus the corresponding first-, second-, and all higher-order densities can be determined, totally characterizing the output.

5.5.5 Stationarity

If the input process is stationary in mean, stationary in autocorrelation, or wide sense stationary the output process is correspondingly stationary in mean, stationary in autocorrelation, and wide sense stationary. This is verified by the functional relationships for the output mean in terms of the input mean given in (5.43) and the output autocorrelation in terms of the input autocorrelation function given in (5.47). For a discrete time-invariant linear system specified by its impulse response $h[n]$, the mean and autocorrelation of the input are sufficient to determine the mean and output autocorrelation function, so only a partial characterization of the input process is required. However, input stationarity of order 1 does not imply output stationarity of order 1, nor does input stationarity of order n imply output stationarity of order n.

5.5.6 MA, AR, and ARMA Random Processes

A number of examples will now be presented that use the results in the previous three sections on the determination of the mean, autocorrrelation function, cross-correlation function, and spectral density for random processes generated by passing a given random process through a linear discrete time-invariant linear system. We will treat a special class of random processes that are generated by passing a discrete white noise process through a system characterized by a rational transfer function. Results for the transient and steady state responses of the MA, AR, and ARMA processes were presented in Chapter 4 using the time domain, whereas in this chapter we examine only the steady state characterizations using the Z-transform and assuming that the excitation and difference equation are defined from $-\infty$ to $+\infty$.

Given a discrete time linear system with input $x[n]$ and output $y[n]$ characterized by a difference equation as

$$y[n] = -\sum_{k=1}^{p} a_k y[n-k] + \sum_{k=0}^{q} b_k x[n-k] \qquad \text{for all } n \tag{5.53}$$

The system function or transfer function can be obtained by first taking an inverse Z-transform of both sides of Eq. (5.53), ignoring initial conditions to obtain the following:

$$Y(z) = -\sum_{k=1}^{p} a_k z^{-k} Y(z) + \sum_{k=0}^{q} b_k z^{-k} X(z) \tag{5.54}$$

We then rearrange (5.54) to obtain $H(z)$ as $Y(z)$ over $X(z)$:

$$H(z) = \frac{Y(z)}{X(z)} = \frac{\sum_{k=0}^{q} b_k z^{-k}}{1 + \sum_{k=1}^{p} a_k z^{-k}} \tag{5.55}$$

Thus the transfer function $H(z)$ is a rational expression in z^{-1}, where q is assumed less than p. The impulse response of such a system is obtained by taking the inverse Z-transform of $H(z)$ which is written as $h[n] = Z^{-1}(H(z))$. If the input process is a white random sequence, then the output process is an ARMA(p, q) process.

If all the $b_k = 0$ for $k = 1, 2, \ldots, q$, then the transfer function can be written in terms of an **all pole** model as

$$H(z) = \frac{b_0}{1 + \sum_{k=1}^{p} a_k z^{-k}} \tag{5.56}$$

If the input to a system represented by the $H(z)$ above is a white random sequence, then the output process is an autoregressive process of order p, AR(p).

If all the $a_k = 0$ for $k = 1, 2, \ldots, p$, then the transfer function can be written in terms of the **all zero** model as

$$H(z) = \sum_{k=0}^{q} b_k z^{-k} \tag{5.57}$$

and if the input process is a white random sequence, the output process is said to be a moving average process, MA(q). The examples now presented are special cases of

MA, AR, and ARMA models, so the resulting autocorrelation functions can be compared to those determined in Chapter 4.

EXAMPLE 5.3

Find the steady state mean, autocorrelation and power spectral density for the MA(1), processes where the input process $X[n]$ has a zero mean for all n and an autocorrelation function given by $E\{X[m+k]x[m]\} = \sigma^2\delta[k]$ for all m.

SOLUTION

The mean of the output process is obtained from Eq. (5.42) as the convolution of the impulse response of the system with the mean of the input process. Since the mean of the input process is zero for all n, the convolution is zero for all n, and thus $E\{Y[n]\} = 0$ for all n.

From Eq. (5.47) the output autocorrelation function of the output process $Y[n]$ is a convolution of the impulse response, the input autocorrelation function, and the time-reversed impulse response. It can also be determined as the inverse Z-transform of the product of their transforms

$$R_{YY}[k] = h[k] * R_{XX}[k] * h[-k] = Z^{-1}[H(z)\Psi_{XX}(z)H(z^{-1})] \tag{5.58}$$

where $H(z)$ from (5.57) is $H(z) = b_0 + b_1 z^{-1}$ and $\Psi_{XX}[z] = Z(\sigma^2\delta[m]) = \sigma^2$. The autocorrelation function from (5.58) is seen to be

$$\begin{aligned} R_{YY}[k] &= Z^{-1}\{(b_0 + b_1 z^{-1})\sigma^2(b_0 + b_1 z)\} \\ &= \sigma^2 Z^{-1}\{b_0 b_1 z + (b_0^2 + b_1^2) + b_1 b_0 z^{-1}\} \\ &= \sigma^2 b_0 b_1 \delta[k+1] + \sigma^2(b_0^2 + b_1^2)\delta[k] + \sigma^2 b_1 b_0 \delta[k-1] \end{aligned} \tag{5.59}$$

The values for $k = 0$ and $k = 1$ check with the results determined in Example 4.6, as do the zero values for $k > 1$ and $k < -1$.

The power spectral density $\phi_{YY}(\omega)$ is obtained from $\Psi_{YY}(z)$ using (5.50) as follows:

$$\begin{aligned} \psi_{YY}(\omega) &= \Psi_{YY}(z)|_{z=e^{j\omega}} = H(z)\Psi_{XX}(z)H(z^{-1})|_{z=e^{j\omega}} \\ &= \{(b_0 + b_1 z^{-1})\sigma^2(b_0 + b_1 z)\}|_{z=e^{j\omega}} \\ &= (b_0^2 + b_1^2 + 2b_0 b_1 \cos(\omega))\sigma^2 \end{aligned} \tag{5.60}$$

Similarly the autocorrelation function and power spectral density for the MA(q) process are easily determined using the proper transfer functions in (5.57) and (5.50). The following example finds the power spectral density for the MA(q) process: ■

EXAMPLE 5.4

Find the power spectral density for the MA(q) process.

SOLUTION

The power spectral density for the MA(q) process uses the transfer function given in (5.57) and is determined as follows:

$$\begin{aligned}\psi_{YY}(\omega) &= \Psi_{YY}(z)|_{z=e^{j\omega}} = H(z)\Psi_{XX}(z)H(z^{-1})|_{z=e^{j\omega}} \\ &= \left\{\sum_{k=0}^{q} b_k z^{-k}(\sigma^2)\sum_{k=0}^{q} b_k z^k\right\}|_{z=e^{j\omega}} \\ &= \left|\sum_{k=0}^{q} b_k e^{-jk\omega}\right|^2 \sigma^2\end{aligned} \tag{5.61}$$

■

EXAMPLE 5.5

Find the steady state mean, autocorrelation, and power spectral density for the AR(1), processes where the input process $X[n]$ has a zero mean for all n and an autocorrelation function given by $E\{X[m+k]x[m]\} = \sigma^2\delta[k]$ for all m.

SOLUTION

The mean of the output process is given from Eq. (5.42) as the convolution of the impulse response of the system with the mean of the input process. Since the mean of the input process is zero for all n, the convolution is zero for all n, and thus $E\{X[n]\} = 0$ for all n.

Using Eq. (5.47) where $H(z)$ from (5.56) is $H(z) = b_0/(1 + a_1 z^{-1})$ and $\Psi_{XX}[z] = Z(\sigma^2\delta[m]) = \sigma^2$, the autocorrelation function $R_{YY}[k]$ is seen to be

$$\begin{aligned}R_{YY}[k] &= Z^{-1}\{H(z)\Psi_{XX}(z)H(z^{-1})\} \\ &= Z^{-1}\left\{\frac{b_0}{(1+a_1 z^{-1})}\sigma^2\frac{b_0}{(1+a_1 z^1)}\right\} \\ &= \frac{b_0^2\sigma^2}{a_1} Z^{-1}\left\{\frac{z}{(z+a_1)(z+1/a_1)}\right\}\end{aligned} \tag{5.62}$$

Expanding the rational expression in terms of a partial fraction expansion gives

$$R_{YY}[k] = \frac{b_0^2\sigma^2}{a_1} Z^{-1}\left\{z\left(\frac{A_1}{(z+a_1)} - \frac{A_1}{(z+1/a_1)}\right)\right\}$$
$$\text{where} \quad A_1 = \frac{a_1}{1-a_1^2} \tag{5.63}$$

After taking the inverse Z-transform and recognizing the first term as a positive time sequence and the second as a negative time sequence, we find the desired autocorrelation function $R_{YY}[k]$ to be

$$R_{YY}[k] = \frac{b_0^2\sigma^2}{1-a_1^2}\left\{(-a_1)^{-k}u[k] + \left(\frac{-1}{a_1}\right)^{-k} u[-k-1]\right\} \tag{5.64}$$

Evaluating $R_{YY}[k]$ for $k \geq 0$ gives

$$R_{YY}[k] = \frac{b_0^2\sigma^2}{1-a_1^2}(-a_1)^k \tag{5.65}$$

which checks with the result determined in (4.179) by a time domain method.

The power spectral density can be determined directly from this autocorrelation by taking the Z-transform and evaluating at $z = e^{j\omega}$, or directly from (5.50) which gives $\psi_{YY}(\omega)$ as

$$\begin{aligned}\psi_{YY}(\omega) &= \Psi_{YY}(z)|_{z=e^{j\omega}} = H(z)\Psi_{XX}(z)H(z^{-1})|_{z=e^{j\omega}} \\ &= \left\{\frac{b_0}{(1+a_1z^{-1})}\sigma^2\frac{b_0}{(1+a_1z^{-1})}\right\}|_{z=e^{j\omega}} \\ &= b_0^2\sigma^2\left\{\frac{1}{(1+a_1e^{-j\omega})(1+a_1e^{j\omega})}\right\}\end{aligned} \tag{5.66}$$

Multiplying the terms in the denominator and simplifying the power spectral density reduces to

$$\psi_{YY}(\omega) = \frac{b_0^2\sigma^2}{1+a_1^2+2a_1\cos(\omega)} \tag{5.67}$$

■

EXAMPLE 5.6

Find the steady state power spectral density for the AR(p) process where the input process $X[n]$ has a zero mean for all n and an autocorrelation function given by $E\{X[m+k]X[m]\} = \sigma^2\delta[k]$ for all m.

SOLUTION

The transfer function for the general AR(p) process from Eq. (5.56) is

$$H(z) = \frac{b_0}{1 + \sum_{k=1}^{p} a_k z^{-k}} \tag{5.68}$$

Thus the power spectral density can be determined as

$$\begin{aligned}\psi_{YY}(\omega) &= \Psi_{YY}(z)|_{z=e^{j\omega}} = H(z)\Psi_{XX}(z)H(z^{-1})|_{z=e^{j\omega}} \\ &= \left\{\frac{b_0}{1 + \sum_{k=1}^{p} a_k z^{-k}} \sigma^2 \frac{b_0}{1 + \sum_{k=1}^{p} a_k z^{k}}\right\}\Bigg|_{z=e^{j\omega}} \\ &= \frac{b_0^2\sigma^2}{|1 + \sum_{k=1}^{p} a_k e^{-jk\omega}|^2}\end{aligned} \tag{5.69}$$

■

EXAMPLE 5.7

Find the steady state mean, autocorrelation, and power spectral density for the ARMA(1, 1), processes where the input process $X[n]$ has a zero mean for all n and an autocorrelation function given by $E\{X[m+k]x[m]\} = \sigma^2\delta[k]$ for all m.

SOLUTION

Since the mean of the input process is zero the mean of the output process will also be zero. The transfer function for the ARMA(1, 1) process from Eq. (5.55) is

$$H(z) = \frac{b_0 + b_1 z^{-1}}{1 + a_1 z^{-1}} \tag{5.70}$$

Thus the autocorrelation function in the steady state can be found from Eq. (5.64) by using the inverse Z-transform as follows:

$$\begin{aligned}R_{YY}[k] &= Z^{-1}\{H(z)\Psi_{XX}(z)H(z^{-1})\} \\ &= Z^{-1}\left\{\frac{b_0 + b_1 z^{-1}}{1 + a_1 z^{-1}} \sigma^2 \frac{b_0 + b_1 z}{1 + a_1 z}\right\} \\ &= \frac{\sigma^2}{a_1} Z^{-1}\left\{\frac{b_0 b_1 z^2 + (b_0^2 + b_1^2)z + b_1 b_0}{(z + a_1)(z + 1/a_1)}\right\}\end{aligned} \tag{5.71}$$

Expanding the term in brackets, divided by z, and expanding into partial fractions gives

$$R_{YY}[k] = \frac{\sigma^2}{a_1} Z^{-1}\left\{ z\left[\frac{A}{z} + \frac{B}{z + a_1} + \frac{C}{z + 1/a_1} \right] \right\} \tag{5.72}$$

Where A, B, and C are determined by partial fraction expansion method to yield

$$\begin{aligned} A &= b_0 b_1 \\ B &= \frac{b_0 b_1 a_1^2 - a_1 b_0^2 - a_1 b_1^2 + b_0 b_1}{a_1^2 - 1} \\ C &= -\frac{b_0 b_1 a_1^2 - a_1 b_0^2 - a_1 b_1^2 + b_0 b_1}{a_1^2 - 1} \end{aligned} \tag{5.73}$$

Multiplying out the z and taking the inverse Z-transform shown in Eq. (5.71), we have the autocorrelation function $R_{YY}[k]$ for the ARMA(1, 1) process as

$$R_{YY}[k] = \frac{\sigma^2}{a_1}\{A\delta[k] + B(-a_1)^k \mu[k] - C(-a_1^{-1})^{-k-1}\mu[-k-1]\} \tag{5.74}$$

Thus the $\delta[k]$ term gives a contribution at $k = 0$ only, while away from the origin the autocorrelation function is a power of $(-a_1)$ for positive k with coefficient $\sigma A/a_1$ as indicated by the second term and the third term is the autocorrelation function for time index less than zero. If we evaluate the autocorrelation at zero, we must add contributions of both of the first terms:

$$R_{YY}[0] = \frac{\sigma^2(A + B)}{a_1} \tag{5.75}$$

Substituting the A and B from (5.73) and simplifying gives the $R_{YY}[0]$ as

$$R_{YY}[0] = \frac{(b_0^2 + b_1^2 - 2b_0 b_1 a_1)\sigma^2}{1 - a_1^2} \tag{5.76}$$

The formula above for $R_{YY}[0]$ verifies the result determined in (4.209). After evaluating (5.74) for $k > 0$ using the B of (5.73), the $R_{YY}[k]$ can be simplified to give

$$\begin{aligned} R_{YY}[k] &= \frac{\sigma^2}{a_1}\{B\,(-a_1)^k \mu[k]\} \\ &= \frac{\sigma^2(a_1^2\, b_0\, b_1\; - a_1 b_0^2 - a_1 b_1^2 + b_0 b_1)}{1 - a_1^2}(-a_1)^{k-1} \end{aligned} \tag{5.77}$$

This result verifies that determined in the steady state development given in Chapter 4, Eq. (4.209) for $k = 1$ and (4.211) for $k > 1$. It is also easily shown that $R_{YY}[-k] = R_{YY}[k]$ by evaluating (5.74) for negative k.

The power spectral density can be determined directly from this autocorrelation by taking the Z-transform and evaluating at $z = e^{j\omega}$ or determined directly from (5.50), which gives $\phi_{YY}(\omega)$ as

$$\begin{aligned}\psi_{YY}(\omega) &= \Psi_{YY}(z)|_{z=e^{j\omega}} = H(z)\Psi_{XX}(z)H(z^{-1})|_{z=e^{j\omega}} \\ &= \left\{\frac{b_0 + b_1 z^{-1}}{1 + a_1 z^{-1}}\,\sigma^2\,\frac{(b_0 + b_1 z^1)}{1 + a_1 z}\right\}\Bigg|_{z=e^{j\omega}} \\ &= b_0^2\,\sigma^2\left\{\frac{(b_0 + b_1 e^{-j\omega})\,(b_0 + b_1 e^{j\omega})}{(1 + a_1 e^{-j\omega})\,(1 + a_1 e^{j\omega})}\right\}\end{aligned} \tag{5.78}$$

Multiplying out the terms in the numerator and denominator and simplifying the power spectral density reduces to

$$\psi_{YY}(\omega) = \frac{b_0^2 + b_1^2 + 2b_0 b_1 \cos(\omega)}{1 + a_1^2 + 2a_1 \cos(\omega)}\,b_0^2\sigma^2 \tag{5.79}$$

■

EXAMPLE 5.8

Find the steady state power spectral density for the ARMA(p, q) process where the input process $X[n]$ has a zero mean for all n and an autocorrelation function given by $E\{X[m+k]x[m]\} = \sigma^2\delta[k]$ for all m.

SOLUTION

The power spectral density for the ARMA(p, q) is easily found by substituting the transfer function for the system $H(z)$ given in Eq (5.55) into Eq. (5.50) to get

$$\begin{aligned}\psi_{YY}(\omega) &= \Psi_{YY}(z)|_{z=e^{j\omega}} = H(z)\psi_{XY}(z)H(z^{-1})|_{z=e^{j\omega}} \\ &= \left\{\frac{\sum_{k=0}^{q} b_k z^{-k}}{1 + \sum_{k=1}^{p} a_k z^{-k}}\,\sigma^2\,\frac{\sum_{k=0}^{q} b_k z^{k}}{1 + \sum_{k=1}^{p} a_k z^{k}}\right\}\Bigg|_{z=e^{j\omega}} \\ &= \frac{|\sum_{k=0}^{q} b_k e^{-jk\omega}|^2}{\left|1 + \sum_{k=1}^{p} a_k e^{-jk\omega}\right|^2}\,b_0^2\,\sigma^2\end{aligned} \tag{5.80}$$

■

5.6 DISCRETE TIME-VARYING LINEAR SYSTEMS WITH RANDOM INPUTS

In the study of linear continuous time systems with random inputs, relationships were obtained for the output mean and the autocorrelation function in terms of the input autocorrelation and mean. Similar formulas will now be developed for discrete time-varying linear systems that can be characterized by the following input–output relationship:

$$y[n] = \sum_{m=-\infty}^{\infty} h[n, m]x[m] \tag{5.81}$$

where $h[n, m]$ is the response of the system to unit sample impulse at time m. If $h[n, m]$ is zero for all $m > n$, the discrete time-varying system is causal, and Eq. (5.81) can be rewritten as

$$y[n] = \sum_{m=-\infty}^{n} h[n, m]x[m] \tag{5.82}$$

5.6.1 Mean in–Mean out (Time-Varying Discrete Time System)

The mean of the output random sequence is seen from (5.81) for a nonrandom $h[m, n]$ to be

$$\begin{aligned} E[Y[n]] &= \sum_{m=-\infty}^{\infty} h[n, m]E[X[m]] \\ &= \sum_{m=-\infty}^{\infty} h[n, m]\eta_X[m] \end{aligned} \tag{5.83}$$

In other words, the output mean is the result of taking the input mean through the system similar to the result for continuous time-varying systems.

If the input process is stationary in the mean, $E[X[m]] = \eta_X$, for all m, then $E[Y[n]]$, from (5.83) can be written as

$$\begin{aligned} E[Y[n]] &= \eta_X \sum_{m=-\infty}^{\infty} h[n, m] \\ &= \eta_X Z_m(h[n, m])\,|_{z=1} \equiv \eta_X H(n, 1) \end{aligned} \tag{5.84}$$

This result can be compared with that determined for the continuous time-varying case.

5.6.2 Autocorrelation in–Autocorrelation Function out (Time-Varying Discrete Time System)

The definition of the output autocorrelation function $R_{YY}[k_1, k_2]$ is given by

$$R_{YY}[k_1, k_2] = E[Y[k_1]Y[k_2]] \tag{5.85}$$

We substitute $Y(k_1)$ and $Y(k_2)$ from (5.81) into (5.85), and assuming that the linear filter is not random, we now have

$$R_{YY}[k_1, k_2] = \sum_{n=-\infty}^{\infty} \sum_{m=-\infty}^{\infty} h[k_1, m]h[k_2, n]R_{XX}[m, n] \tag{5.86}$$

For a wide sense stationary input random sequence $X[n]$, the output autocorrelation function $R_{YY}[m, n]$ can be written as

$$R_{YY}[k_1, k_2] = \sum_{m=-\infty}^{\infty} \sum_{n=-\infty}^{\infty} h[k_1, n]h[k_2, m]R_{XX}[n - m] \tag{5.87}$$

There is no reason to believe that this double sum will be just a function of the difference between k_1 and k_2. Even though the input process is wide sense stationary, the output of a linear discrete time-varying system will not necessarily be wide sense stationary.

5.6.3 Cross-correlation Functions (Time-Varying Discrete Time System)

Using $Y(k_1)$ and $Y(k_2)$ from (5.81) in the definitions of the cross-correlation functions between the input and output processes, we can show the $R_{YX}[k_1, k_2]$ and $R_{XY}[k_1, k_2]$ to be

$$\begin{aligned} R_{YX}[k_1, k_2] &= E[Y(k_1)X(k_2)] = \sum_{m=\infty}^{\infty} h[k_1, m]R_{XX}[m, k_2] \\ R_{XY}[k_1, k_2] &= E[X(k_1)Y(k_2)] = \sum_{m=-\infty}^{\infty} h[k_2, m]R_{XX}[k_1, m] \end{aligned} \tag{5.88}$$

5.6.4 *n*th-Order Densities

Since the output process at any time n is a function of the input process random variables at all times, a total characterization is required to determine the first-order densities of the output random process. Thus a first-order characterization of the input process is insufficient information to determine the first-order densities of the output process. Similarly the second and nth-order densities cannot be determined unless the input process is totally characterized. If the input process is a Gaussian random process with known mean and autocorrelation function, the output process will be Gaussian and thus totally characterized by its mean and autocorrelation functions given in (5.83) and (5.86).

5.6.5 Stationarity

For a discrete linear time-varying system characterized by its unit-sample response, the output process is not necessarily correspondingly stationary in mean, stationary in autocorrelation, or stationary in a wide sense if the input process is stationary in mean, stationary in autocorrelation, or stationary in a wide sense. For a discrete time-varying linear system, specified by an $h(n, m)$, knowing the second-order information of the input process is, however, sufficient information to determine the second-order output information with the relationships given in (5.83) and (5.86).

5.7 LINEAR SYSTEM IDENTIFICATION

If the input $X(t)$ to a linear time-invariant filter, specified by its impulse response $h(t)$, is a white process with autocorrelation function

$$R_{XX}(\tau) = \frac{N_0}{2}\delta(\tau) \tag{5.89}$$

the cross-spectral density $S_{YX}(jw)$ can be obtained from (5.25) as

$$S_{YX}(\omega) = S_{XX}(\omega)H(j\omega) \tag{5.90}$$

Dividing through by $S_{XX}(w)$ which is $N_0/2$ gives the transfer function $H(j\omega)$ as

$$H(j\omega) = \frac{S_{YX}(\omega)}{S_{XX}(\omega)} = \frac{2}{N_0} S_{YX}(\omega) \tag{5.91}$$

That is, by measuring the cross-spectral density between output and input when white noise is the input, the transfer function of an unknown linear system can be determined.

5.8 DERIVATIVES OF RANDOM PROCESSES

Let $X(t)$ be a random process with known characterization in terms of mean and autocorrelation function. If a realization of this process is the input to a differentiator an output signal is obtained, thus establishing a mapping from the sample space to the output and specifying the output as a random process. The derivative process is denoted as

$$Y(t) = \frac{dX(t)}{dt} \tag{5.92}$$

The logical question is: What are the statistical characterizations of this derivative process? The derivative operation can be viewed as a linear system with transfer function $H(p) = p$, so the formulas given earlier can be used to find the autocorrelation and cross-correlation functions for the input and output. The cross-correlation function between input and output can be expressed as

$$\begin{aligned} R_{YX}(\tau) = h(\tau) * R_{XX}(\tau) &= \frac{d\delta(\tau)}{d\tau} * R_{XX}(\tau) \\ &= \frac{d}{d\tau} R_{XX}(\tau) \end{aligned} \tag{5.93}$$

The autocorrelation function of the output is determined as

$$\begin{aligned} R_{YY}(\tau) &= \mathscr{L}^{-1}[H(p)\Phi_{XX}(p)H(-p)] \\ &= \mathscr{L}^{-1}[-p^2\Phi_{XX}(p)] \end{aligned} \tag{5.94}$$

A property of the inverse Laplace transform is that

$$\mathscr{L}^{-1}[p^n F(p)] = \frac{d^n f(t)}{d^n t} \tag{5.95}$$

Using (5.95) in (5.94), we obtain the autocorrelation function of the output process as

$$R_{YY}(\tau) = -\frac{d^2}{d^2\tau} R_{XX}(\tau) \tag{5.96}$$

In a similar fashion the autocorrelation function of $Z(t)$ can be found, where $Z(t)$ is defined as the nth-order derivative of a random process $X(t)$ and denoted by

$$Z(t) = \frac{d^n X(t)}{d^n t} \tag{5.97}$$

The autocorrrelation function for $Z(t)$ can be shown to be

$$R_{ZZ}(\tau) = (-1)^n \frac{d^{2n}}{d^2\tau} R_{XX}(\tau) \tag{5.98}$$

The derivatives of wide sense stationary random processes are statistically related and it is easily shown, similar to the previous derivations, that their cross-correlation functions are given as

$$E\left[\frac{d^m X(t+\tau)}{d^m t} \frac{d^n X(t)}{d^n t}\right] = (-1)^n \frac{d^{(m+n)} R_{XX}(\tau)}{d^{(m+n)}\tau} \tag{5.99}$$

A word of caution on using these formulas is necessary as all derivatives of random processes do not necessarily exist, but if they do, the formulas are meaningful.

5.9 MULTI-INPUT, MULTI-OUTPUT LINEAR SYSTEMS

Multiple-input and multiple-output linear systems are becoming increasingly significant as systems are becoming more complicated. An example of a two-input, two-output system is shown in Figure 5.4. The input signals are $x_1(t)$ and $x_2(t)$, and the output signals are $y_1(t)$ and $y_2(t)$. Each output is the sum of two signals. For example,

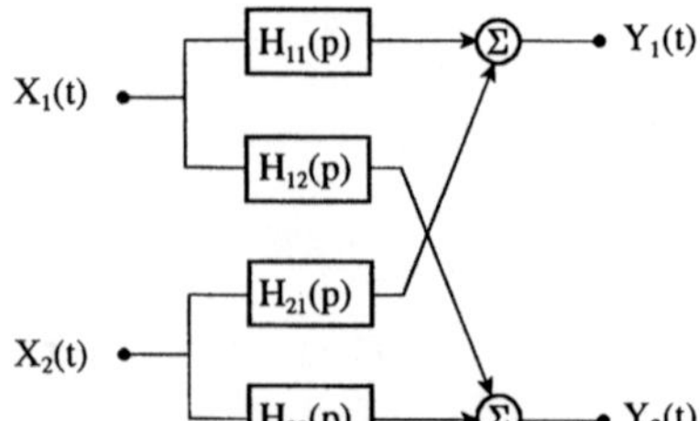

Figure 5.4 A general linear two-input two output system denoted by MIMO(2, 2).

$y_1(t)$ is the sum of one signal resulting from passing $x_1(t)$ through a linear system with transfer function $H_{11}(p)$ and the other from passing $x_2(t)$ through a linear system with transfer function $H_{21}(p)$. Similarly $y_2(t)$ is the sum of a two responses one due to passing $x_1(t)$ through $H_{12}(p)$ and the other from passing $x_2(t)$ through $H_{22}(p)$. In general, $H_{12}(p)$ and $H_{21}(p)$ are not the same. Thus the two-input, two-output system is specified by a 2 by 2 matrix of transfer functions. The basic problem is knowing the means of the inputs, the autocorrelation functions of the inputs, and the cross-correlation between inputs determine the corresponding characterizations of the outputs.

5.9.1 Output Means for MIMO(2, 2)

If the inputs to the two-input, two-output linear system shown in Figure 5.4 are random processes $X_1(t)$ and $X_2(t)$, the output processes $Y_1(t)$ and $Y_2(t)$ are easily seen to be the sum of two convolutions:

$$\begin{aligned} Y_1(t) &= \int_{-\infty}^{\infty} X_1(t-\alpha)h_{11}(\alpha)\,d\alpha + \int_{-\infty}^{\infty} X_2(t-\beta)h_{21}(\beta)\,d\beta \\ Y_2(t) &= \int_{-\infty}^{\infty} X_1(t-\alpha)h_{12}(\alpha)\,d\alpha + \int_{-\infty}^{\infty} X_2(t-\beta)h_{22}(\beta)\,d\beta \end{aligned} \tag{5.100}$$

The means are determined by taking the expected value of the equations in (5.100) and interchanging the expected value operator and the integral sign to get

$$\begin{aligned} E[Y_1(t)] &= \int_{-\infty}^{\infty} E[X_1(t-\alpha)]h_{11}(\alpha)\,d\alpha + \int_{-\infty}^{\infty} E[X_2(t-\beta)h_{21}(\beta)\,d\beta \\ E[Y_2(t)] &= \int_{-\infty}^{\infty} E[X_1(t-\alpha)]h_{12}(\alpha)\,d\alpha + \int_{-\infty}^{\infty} E[X_2(t-\beta)h_{22}(\beta)\,d\beta \end{aligned} \tag{5.101}$$

Recognizing the integrals as convolutions permits us to write the means in terms of convolutions as

$$\begin{aligned} E[Y_1(t)] &= E[X_1(t)] * h_{11}(\alpha) + E[X_2(t)] * h_{21}(\beta) \\ E[Y_2(t)] &= E[X_1(t)] * h_{12}(\alpha) + E[X_2(t)] * h_{22}(\beta) \end{aligned} \tag{5.102}$$

When the input processes are jointly wide sense stationary, the input means are constant, and the output means can be written in terms of the various transfer functions evaluated at zero:

$$\begin{aligned} \eta_{Y_1} &= \eta_{X_1} H_{11}(0) + \eta_{X_2} H_{21}(0) \\ \eta_{Y_2} &= \eta_{X_1} H_{12}(0) + \eta_{X_2} H_{22}(0) \end{aligned} \tag{5.103}$$

5.9.2 Cross-correlation Functions for MIMO(2, 2) Linear Systems

There are four cross-correlation functions that need to be examined for the two-input, two-output linear system given by $R_{Y_1X_1}(t, u), R_{Y_1X_2}(t, u), R_{Y_2X_1}(t, u)$ and $R_{Y_2X_2}(t, u)$. Consider the first one of these, which can be written as

$$\begin{aligned} R_{Y_1X_1}(t, u) &= E[Y_1(t)X_1(u)] \\ &= E\left[\left(\int_{-\infty}^{\infty} h_{11}(\alpha)X_1(t-\alpha)\,d\alpha + \int_{-\infty}^{\infty} h_{21}(\beta)X_2(t-\beta)\,d\beta\right)X_1(u)\right] \end{aligned} \tag{5.104}$$

After we multiply out and interchange the integral and expected value operations, (5.104) becomes

$$R_{Y_1X_1}(t, u) = \int_{-\infty}^{\infty} h_{11}(\alpha)R_{X_1\,X_1}(t-\alpha, u)\,d\alpha + \int_{-\infty}^{\infty} h_{21}(\beta)R_{X_2\,X_1}(t-\beta, u)\,d\beta \tag{5.105}$$

Let us assume that $X_1(t)$ and $X_2(t)$ are jointly wide sense stationary. Then this cross-correlation function can be written in terms of $\tau = t - u$ as follows:

$$R_{Y_1X_1}(\tau) = h_{11}(\tau) * R_{X_1X_1}(\tau) + h_{21}(\tau) * R_{X_2\,X_1}(\tau) \tag{5.106}$$

The other input–output cross-correlation functions for jointly wide sense stationary inputs can be similarly shown to be

$$\begin{aligned} R_{Y_1X_2}(\tau) &= h_{11}(\tau) * R_{X_1X_2}(\tau) + h_{21}(\tau) * R_{X_2\,X_2}(\tau) \\ R_{Y_2X_1}(\tau) &= h_{12}(\tau) * R_{X_1X_2}(\tau) + h_{22}(\tau) * R_{X_2\,X_2}(\tau) \\ R_{Y_2X_2}(\tau) &= h_{12}(\tau) * R_{X_1X_2}(\tau) + h_{22}(\tau) * R_{X_2\,X_2}(\tau) \end{aligned} \tag{5.107}$$

As was done in previous sections, these convolutions may be conveniently calculated by using inverse Laplace transforms of various products of Laplace transforms of the system impulse responses and various autocorrelation and cross-correlation functions of the input processes.

5.9.3 Autocorrelation Functions for Outputs of MIMO(2, 2) Linear Systems

The autocorrelation function for the output $Y_1(t)$ can be written as

$$\begin{aligned} R_{Y_1Y_1}(t, u) &= E[Y_1(t)Y_1(u)] \\ &= E\left[\left(\int_{-\infty}^{\infty} h_{11}(\alpha)X_1(t-\alpha)\,d\alpha + \int_{-\infty}^{\infty} h_{21}(\beta)X_2(t-\beta)\,d\beta\right)\right. \\ &\quad \left.\cdot\left(\int_{-\infty}^{\infty} h_{11}(\gamma)X_1(u-\gamma)\,d\gamma + \int_{-\infty}^{\infty} h_{21}(\theta)X_2(u-\theta)\,d\theta\right)\right] \end{aligned} \tag{5.108}$$

Multiplying out the terms in parentheses to get four terms, taking the expected value through the integral, and evaluating the expected value gives the autocorrelation function $R_{Y_1Y_1}(t, u)$ as

$$\begin{aligned} R_{Y_1Y_1}(t, u) &= \int_{-\infty}^{\infty} h_{11}(\alpha)h_{11}(\gamma)R_{X_1X_1}(t-\alpha, u-\gamma)\,d\alpha\,d\gamma \\ &+ \int_{-\infty}^{\infty} h_{11}(\alpha)h_{21}(\theta)R_{X_1X_2}(t-\alpha, u-\theta)\,d\alpha\,d\theta \\ &+ \int_{-\infty}^{\infty} h_{21}(\beta)h_{11}(\gamma)R_{X_2X_1}(t-\beta, u-\gamma)\,d\beta\,d\gamma \\ &+ \int_{-\infty}^{\infty} h_{21}(\beta)h_{21}(\theta)R_{X_2X_2}(t-\beta, u-\theta)\,d\beta\,d\theta \end{aligned} \tag{5.109}$$

If the input processes are jointly wide sense stationary, then the equation above can be rewritten as

$$\begin{aligned} R_{Y_1Y_1}(t, u) &= \int_{-\infty}^{\infty}\int_{-\infty}^{\infty} h_{11}(\alpha)h_{11}(\gamma)R_{X_1X_1}(\tau-(\alpha-\gamma))\,d\alpha\,d\gamma \\ &+ \int_{-\infty}^{\infty}\int_{-\infty}^{\infty} h_{11}(\alpha)h_{21}(\theta)R_{X_1X_2}(\tau-(\alpha-\theta))\,d\alpha\,d\theta \\ &+ \int_{-\infty}^{\infty}\int_{-\infty}^{\infty} h_{21}(\beta)h_{11}(\gamma)R_{X_2X_1}(\tau-(\beta-\gamma))\,d\beta\,d\gamma \\ &+ \int_{-\infty}^{\infty}\int_{-\infty}^{\infty} h_{21}(\beta)h_{21}(\theta)R_{X_2X_2}(\tau-(\beta-\theta))\,d\beta\,d\theta \end{aligned} \tag{5.110}$$

As before, for the single-input, single-output linear system, the integrals above are recognized as convolutions, and thus the autocorrelation function becomes

$$\begin{aligned} R_{Y_1Y_1}(\tau) &= h_{11}(\alpha) * R_{X_1X_1}(\tau) * h_{11}(-\tau) + h_{11}(\tau) * R_{X_1X_2}(\tau) * h_{21}(-\tau) \\ &+ h_{21}(\tau) * R_{X_2X_1}(\tau) * h_{11}(-\tau) + h_{21}(\tau)R_{X_2X_2}(\tau) * h_{21}(-\tau) \end{aligned} \tag{5.111}$$

In a similar development the autocorrelation function for the other output $Y_2(t)$ can be shown to be

$$\begin{aligned} R_{Y_2Y_2}(\tau) &= h_{12}(\tau) * R_{X_1X_1}(\tau) * h_{12}(-\tau) + h_{12}(\tau) * R_{X_1X_2}(\tau) * h_{22}(-\tau) \\ &+ h_{22}(\tau) * R_{X_2X_1}(\tau) * h_{12}(-\tau) + h_{22}(\tau)R_{X_2X_2}(\tau) * h_{22}(-\tau) \end{aligned} \tag{5.112}$$

As before, these cross-correlation functions are more conveniently determined in the Laplace domain.

5.9.4 Cross-correlation Functions for Outputs of MIMO(2, 2) Linear Systems

The cross-correlation function for the two outputs $Y_1(t)$ and $Y_2(t)$ is determined in a similar fashion to that for the autocorrelation functions:

$$\begin{aligned} R_{Y_1Y_2}(t,u) &= E[Y_1(t)Y_2(u)] \\ &= E\left[\left(\int_{-\infty}^{\infty} h_{11}(\alpha)X_1(t-\alpha)\,d\alpha + \int_{-\infty}^{\infty} h_{21}(\beta)X_2(t-\beta)\,d\beta\right)\right. \\ &\quad \left.\cdot\left(\int_{-\infty}^{\infty} h_{12}(\gamma)X_1(u-\gamma)\,d\gamma + \int_{-\infty}^{\infty} h_{22}(\theta)X_2(u-\theta)\,d\theta\right)\right] \end{aligned} \tag{5.113}$$

After we multiply out the integrals, assuming jointly wide sense stationary inputs, the cross correlation between outputs can be written as

$$\begin{aligned} R_{Y_1Y_2}(\tau) &= h_{11}(\alpha) * R_{X_1X_1}(\tau) * h_{12}(-\tau) + h_{11}(\tau) * R_{X_1X_2}(\tau) * h_{22}(-\tau) \\ &\quad + h_{21}(\tau) * R_{X_2X_1}(\tau) * h_{12}(-\tau) + h_{21}(\tau)R_{X_2X_2}(\tau) * h_{22}(-\tau) \end{aligned} \tag{5.114}$$

As the orders of the MIMO(n, n) system increase, the autocorrelation and cross-correlation functions become very complex, simplifying only when the input processes are all uncorrelated.

5.10 TRANSIENTS IN LINEAR SYSTEMS

Let $X(t)$ be the input and $y(t)$ the output of a linear system that is governed by the following linear constant coefficient differential equation:

$$y(t) = -\sum_{k=1}^{n} a_k \frac{d^k y(t)}{dt^k} + x(t), \qquad t > 0 \tag{5.115}$$

If the input is a random process, then the output is also a random process described from the mapping of each realization of the input process to a realization of the output process. Assume that $y(t) = 0$ for all $t < 0$ and that the initial conditions are

$$\left.\frac{d^k y(t)}{dt^k}\right|_{t=0} = 0 \qquad \text{for } k = 0 \text{ to } n-1 \tag{5.116}$$

This means that the system is initially at rest.

If $x(t)$ is an input random process $X(t)$, then the output $y(t)$ is $Y(t)$ a random process. We would like to find the mean and autocorrelation function for the output process $Y(t)$ in terms of the mean and autocorrelation of the input process $X(t)$ whose characerization is specified in terms of its mean and autocorrelation function. In an equivalent problem for the discrete time AR system model, it has been shown that the autocorrelation and mean go through an initial transient period before a steady state is reached. This will also be shown to be true for the continuous time case as well. The steady state results for a continuous time-invariant linear system were determined in Section 5.3, and in the following the transient results are derived.

5.10.1 Mean of the Output Process

Substituting $X(t)$ as the input and $Y(t)$ as the output of Eq. (5.115), and taking the expected value, gives the following differential equation for the mean of the output process:

$$E[Y(t)] = -\sum_{k=1}^{n} a_k E\left[\frac{d^k Y(t)}{d^k t}\right] + E[X(t)], \qquad t > 0 \tag{5.117}$$

The expected value of the derivatives of the process $X(t)$ have been shown to be the derivatives of the expected value of the process as follows:

$$E\left[\frac{d^k Y(t)}{d^k t}\right] = \frac{d^k E[Y(t)]}{d^k t} \tag{5.118}$$

Substituting (5.118) into (5.117), we arrive at a differential equation in terms of the mean of the output process:

$$E[Y(t)] + \sum_{k=1}^{n} a_k \left[\frac{d^k E[Y(t)]}{d^k t}\right] = \eta_X(t), \qquad t > 0 \tag{5.119}$$

Now, taking the expected value of the initial conditions given in Eq. (5.116) gives zero conditions on the mean:

$$\left.\frac{d^k E[Y(t)]}{dt^k}\right|_{t=0} = 0 \qquad \text{for } k = 0 \text{ to } n-1 \tag{5.120}$$

Equations (5.119) and (5.120) give a differential equation with initial conditions that must be solved to obtain the mean $E[Y(t)]$ of the output process.

5.10.2 Autocorrelation of the Output Process

Similar to the procedure for obtaining the autocorrelation function for the AR process in Chapter 4, the procedure to find the output autocorrelation function $R_{XX}(t_1, t_2)$ consists of two steps: first solving for the cross-correlation function and then from that result finding the autocorrelation function. Evaluating both sides of (5.115) at t_2, multiplying by $X(t_1)$, and taking the expected value gives the cross-correlation function at t_1 and t_2 as

$$\begin{aligned} R_{XY}(t_1, t_2) &= E[X(t_1)\, Y(t_2)] \\ &= E\left[X(t_1)\left(-\sum_{k=1}^{n} a_k \frac{d^k Y(t)}{dt^k}\right)_{t=t_2}\right] + E[X(t_1)X(t_2)] \qquad t_1, t_2 > 0 \end{aligned} \tag{5.121}$$

Moving $X(t_1)$ into the summation, and taking the expected value, the first term of (5.121) becomes

$$E\left[x(t_1)\left(\frac{d^k y(t)}{d^k t}\right)_{t=t_2}\right] = \frac{d^k R_{XY}(t_1, t_2)}{d^k t_2} \tag{5.122}$$

Using (5.122) in Eq. (5.121) and rearranging yields the following differential equation for $R_{XY}(t_1, t_2)$:

$$\sum_{k=1}^{n} \frac{d^k R_{XY}(t_1, t_2)}{d^k t_2} + R_{XY}(t_1, t_2) = R_{XX}(t_1, t_2), \quad t_2 > 0 \tag{5.123}$$

The initial conditions are obtained by multiplying Eq. (5.120) by $X(t_1)$ and taking the expected value. This gives

$$\left.\frac{d^k R_{XY}(t_1, t_2)}{d^k t_2}\right|_{t_2=0} = 0 \qquad \text{for } k = 0 \text{ to } n-1 \tag{5.124}$$

Thus the solution of the differential equation (5.123) with initial conditions (5.124) gives the crosscorrelation function $R_{XY}(t_1, t_2)$ for $t_1, t_2 > 0$.

Having obtained $R_{XY}(t_1, t_2)$, we are now in position to solve for $R_{YY}(t_1, t_2)$. Multiplying (5.115) evaluated at t_1 by $Y(t_2)$ and taking the expected value gives

$$\begin{aligned} R_{YY}(t_1, t_2) &= E[Y(t_1)Y(t_2)] \\ &= E\left[Y(t_1)\left(-\sum_{k=1}^{n} a_k \frac{d^k Y(t)}{d^k t}\right)_{t=t_2}\right] + E[X(t_2)Y(t_1)], \quad t_1, t_2 > 0 \end{aligned} \tag{5.125}$$

As in the development of cross correlation, (5.132) simplifies to

$$\sum_{k=1}^{n} \frac{d^k R_{YY}(t_1, t_2)}{d^k t_1} + R_{YY}(t_1, t_2) = R_{XY}(t_1, t_2), \quad t_1 > 0 \tag{5.126}$$

with zero initial conditions:

$$\left.\frac{d^k R_{YY}(t_1, t_2)}{d^k t_1}\right|_{t_1=0} = 0 \qquad \text{for } k = 0 \text{ to } n-1 \tag{5.127}$$

Eqs. (5.126) and (5.127) give a differential equation and initial conditions to solve for $R_{YY}(t_1, t_2)$ in terms of the $R_{XY}(t_1, t_2)$ as a driving function. The following example illustrates the procedure for a simple first-order model:

EXAMPLE 5.9

Given that $y(t)$ is the solution to the following differential equation representing a transfer function $H(p) = 1/(p+1)$

$$\frac{dy(t)}{dt} = -y(t) + x(t), \quad t > 0$$

Assume that $y(t) = 0$ for all $t < 0$ and that we have the initial condition $y(0) = 0$. Let the input $x(t)$ be a random process $X(t)$ with mean and autocorrelation function given by

$$\eta_X(t) = E[X(t)] = 1, \quad R_{XX}(\tau) = 1/2\, e^{-2|\tau|} + 1$$

Find the mean η_Y of the output process $Y(t)$.

SOLUTION

From Eqs. (5.119) and (5.120), the mean, $E[Y(t)] = \eta_Y(t)$, of the output process satisfies the following differential equation and initial condition:

$$\frac{d\eta_Y(t)}{dt} + \eta_Y(t) = 1, \quad t > 0$$

$$n_Y(0) = 0$$

Upon solving the above for $\eta_Y(t)$, we have

$$\eta_Y(t) = [1 - e^{-t}]\mu(t)$$

■

EXAMPLE 5.10

Given that $y(t)$ is the solution to the following differential equation representing a transfer function $H(p) = 1/(p+1)$ and input $x(t)$:

$$\frac{dy(t)}{dt} = -y(t) + x(t), \quad t > 0$$

Assume that $y(t) = 0$ for all $t < 0$ and that we have the initial condition $y(0) = 0$. Let $x(t)$ be a random process $X(t)$ with mean and autocorrelation function given by

$$\eta_X(t) = E[X(t)] = 0, \quad R_{XX}(\tau) = \delta(\tau)$$

Find the autocorrelation function $R_{YY}(t_1, t_2)$ of the output process $Y(t)$.

SOLUTION

To obtain the $R_{YY}(t_1, t_2)$, we first find the cross correlation $R_{XY}(t_1, t_2)$ by solving a differential equation. Then from the result we find the $R_{YY}(t_1, t_2)$ by solving another differential equation. We can use the Laplace transform to solve both differential equations. From Eq. (5.123), the $R_{XY}(t_1, t_2)$ is the solution of

$$\frac{dR_{XY}(t_1, t_2)}{dt_2} + R_{XY}(t_1, t_2) = \delta(t_2 - t_1), \quad t_2 \geq 0$$

with initial conditions from (5.124) as $R_{XY}(t_1, t_2)|_{t_2=0} = 0$. Taking the Laplace transform with respect to the t_2 variable and using the initial condition results in

$$p\Phi_{XY}(t_1, p) + \Phi_{XY}(t_1, p) = e^{-pt_2}$$

Solving for $\Phi_{XY}(t_1, p)$ gives

$$\Phi_{XY}(t_1, p) = \frac{e^{-pt_1}}{p+1}$$

Taking the inverse Laplace transform gives us the cross correlation $R_{XY}(t_1, t_2)$ as

$$R_{XY}(t_1, t_2) = e^{-(t_2-t_1)}\mu(t_2 - t_1)$$

This result is the driving function for the differential equation for $R_{YY}(t_1, t_2)$ given in Eq. (5.126). So we have the following:

$$\frac{dR_{YY}(t_1, t_2)}{dt_1} + R_{YY}(t_1, t_2) = e^{-(t_2-t_1)}\mu(t_2 - t_1), \qquad t_1 \geq 0$$

Taking the Laplace transform of the equation above with respect to the t_1 variable and using the zero initial condition gives

$$\begin{aligned} p\Phi_{YY}(p, t_2) + \Phi_{YY}(p, t_2) &= e^{-t_2}\int_0^{t_2} e^{+t_1}e^{-pt_1}\,dt_1 \\ &= e^{-t_2}\left[\frac{e^{-(p-1)t_1}}{-(p-1)}\right]_0^{t_2} \\ &= e^{-t_2}\left[-\frac{e^{-(p-1)t_2}}{p-1} + \frac{1}{p-1}\right] \end{aligned}$$

Then, solving for $\Phi_{YY}(p, t_2)$, multiplying out and performing a partial fraction expansion of both terms yields

$$\begin{aligned} \Phi_{YY}(p, t_2) &= e^{-t_2}\left[-\frac{e^{-(p-1)t_2}}{p-1} + \frac{1}{p-1}\right]\frac{1}{p+1} \\ &= -\frac{e^{-pt_2}}{(p-1)(p+1)} + \frac{e}{(p-1)(p+1)} \\ &= \frac{\frac{1}{2}e^{-pt_2}}{p-1} - \frac{\frac{1}{2}e^{-pt_2}}{p+1} + \frac{\frac{1}{2}e^{-t_2}}{p-1} - \frac{\frac{1}{2}e^{-t_2}}{p+1} \end{aligned}$$

Finally, taking the inverse bilateral Laplace transform using the t_1 variable gives us the autocorrelation function $R_{XY}(t_1, t_2)$ as

$$R_{YY}(t_1, t_2) = \tfrac{1}{2}e^{(t_1-t_2)}\mu(-(t_1 - t_2)) + \tfrac{1}{2}e^{-(t_1-t_2)}\mu(t_1 - t_2) + \tfrac{1}{2}e^{(t_1-t_2)}\mu(-t_1) - \tfrac{1}{2}e^{-(t_1+t_2)}\mu(t_1)$$

Upon evaluating for $t_1 > 0$ and $t_2 > 0$ and regions $t_1 > t_2$ and $t_2 > t_1$, we find, the $R_{YY}(t_1, t_2)$ from the equation above to be

$$R_{YY}(t_1, t_2) = \begin{cases} \tfrac{1}{2}e^{-(t_1-t_2)} - \tfrac{1}{2}e^{-(t_1+t_2)}, & t_1 > t_2 \\ \tfrac{1}{2}e^{-(t_2-t_1)} - \tfrac{1}{2}e^{-(t_1+t_2)}, & t_2 > t_1 \end{cases}$$

To obtain the steady state results for the autocorrelation function, we let both t_1 and t_2 approach infinity but keep $t_1 - t_2 = \tau$. The second term goes to zero and the first term remains. Thus, for $t_1 - t_2 = \tau$, the steady state autocorrelation function $R_{YY}(\tau)$ of the output $Y(t)$ is

$$R_{YY}(\tau) = \tfrac{1}{2}e^{-|\tau|}$$

This result checks with the result obtained by using the steady state formula given in Eq. (5.18)

$$R_{YY}(\tau) = \mathscr{L}_\beta^{-1}[H(p) * \Phi_{XX}(p) * H(-p)] = \mathscr{L}_\beta^{-1}\left[\frac{1}{p+1}\ 1\ \frac{1}{-p+1}\right] = \tfrac{1}{2}e^{-|\tau|}$$

■

5.11 SUMMARY

The main topic of this chapter is the interaction of random processes with linear systems. In particular, the characterization of the output process in terms of the input process characterization. Formulas were derived for the output mean and autocorrelation function of linear time-invariant systems to a wide sense stationary input. The mean of the output was determined as a convolution of the input mean with the impulse response of the system, and the output autocorrelation function was shown to be a convolution of the input autocorrelation function, the impulse response, and the time-reversed impulse response of the system. Determination of the convolutions in many cases was best done in the Laplace domain for continuous time systems and the Z domain for discrete time systems. The output power spectral density was shown to be determined by multiplying the input power spectral density by magnitude squared of the system transfer function $H(j\omega)$ for the continuous time system and $H(e^{j\omega})$ for the discrete time system.

Formulas for the mean and autocorrelation function of the output of a linear time varying system in terms of the mean and autocorrelation function of the input process for both continuous and discrete time systems were presented.

The AR, MA, and ARMA processes generated by passing white noise through special discrete-time linear systems were analyzed using the steady state techniques described above. The results obtained for mean and autocorrelation by using the steady state results in the Z-domain were shown to be equal to those steady state results in the time domain.

In using differential equations to model continuous time invariant systems with random processes as inputs it is necessary to obtain the statistical properties of the derivatives of the input processes. Ignoring some issues with existence the auto- and cross-correlation functions for derivatives of a random process were developed using the Laplace transform domain in terms of the autocorrelation of the input process. For mathematically more rigorous presentations on this topic the interested reader could explore Papoulis [1] or Van Trees [7].

Also considered in this chapter was the transient response of a continuous linear time invariant system to an input process applied at t equal zero. For the special case where the differential equation does not contain derivatives of the input process the output mean was easily obtained as a convolution; however, the output autocorrelation function required the solution of two differential equations one to get the cross correlation function between input and output and the second with the crosscorrelation function as the input to obtain the output autocorrelation function. The solution was facilitated by using the Laplace transform. The examples illustrated that as t becomes large, the output autocorrelation and mean approach those determined using the steady state methods involving transfer functions.

The chapter concluded with a brief discussion of multiple input, multiple output systems and their output means, autocorrelation functions, and cross-correlation functions to random inputs and results were shown for wide sense stationary input processes.

PROBLEMS

5.1 Given a stationary random process $X(t)$ with mean and autocorrelation function as

$$E[X(t)] = \tfrac{1}{2}, \qquad R_X(\tau) = \tfrac{1}{4} + e^{-|\tau|}.$$

Find the mean, $E[Y(t)]$, and autocorrelation function, $R_{YY}(\tau)$, for the output random process $Y(t)$ of the system shown in Figure P5.1 with $X(t)$ as input.

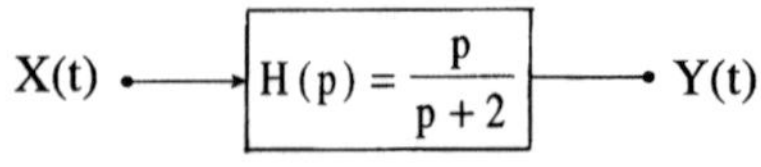

P5.1

5.2 Given the block diagram in Figure P5.2, with $X(t)$ a white random process characterized by its autocorrelation function $R_{XX}(\tau) = \delta(\tau)$, and $H_1(j\omega)$ representing a causal linear time-invariant system.

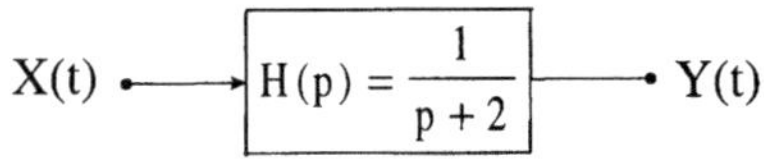

P5.2

(a) Find $E[Y(t)]$.

(b) Find $R_{XY}(t_1, t_2)$.

(c) What forms of stationarity will $Y(t)$ have if any?

5.3 Given a stationary random process $X(t)$ with $E[X(t)] = \frac{1}{2}$.

(a) Find $E[Y(t)]$ for the system given in Figure P5.3.

(b) Is the output process stationary in mean? Comment on the results.

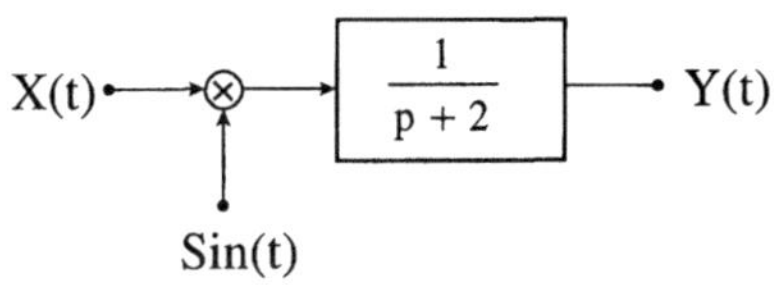

P5.3

5.4 If $R_{XX}(\tau) = \beta^2 e^{-2\lambda|\tau|}$ and $E[x(t)] = 0$. Find $E[y(t)]$ and autocorrelation function $R_{XX}(t + \tau, t)$ for the system shown in Figure P5.4.

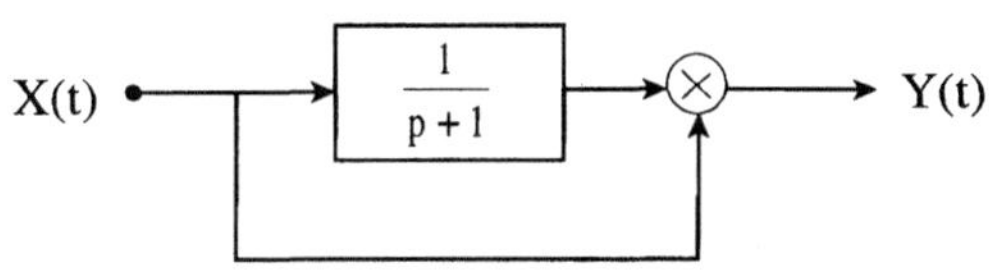

P5.4

5.5 Given the circuit diagram in Figure P5.5, assume steady state conditions, and that $X(t)$ is a random telegraph signal process with uniform density $\lambda = \frac{1}{2}$. Find:

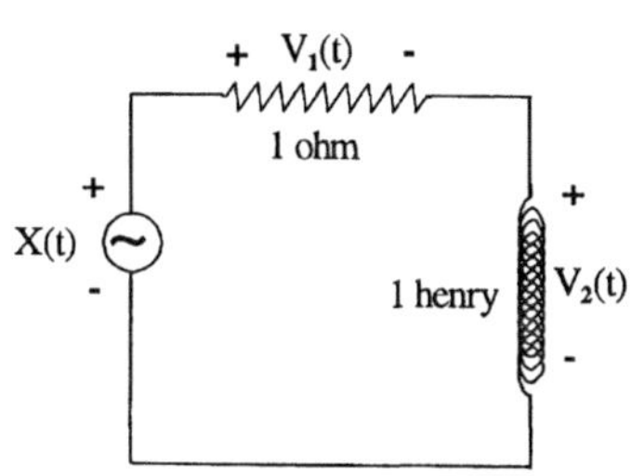

P5.5

(a) The autocorrelation function, $R_{22}(\tau)$, for the output process $V_2(t)$.
(b) The cross-correlation function, $R_{12}(\tau)$, between processes $V_1(t)$ and $V_2(t)$.
(c) The power spectral density $S_{22}(\omega)$.

5.6 In the circuit of Figure P5.6 the excitation voltage $X(t)$ is a random binary transmission process is with the autocorrelation function given in Eq (4.53).
(a) Is the process $X(t)$ wide sense stationary? Why or why not?
(b) Find the power spectral density $S_{V_2V_2}(\omega)$ of the output process $V_2(t)$.
(c) Find $S_{IV_2}(\omega)$.
(d) How would you get $R_{V_2V_2}(\tau)$?

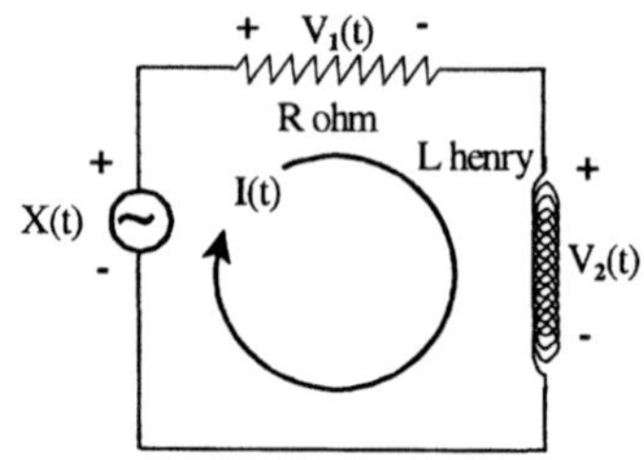

P5.6

5.7 Given a system with input process $X(t)$ and output process $Y(t)$. The system has an impulse response $h(t) = e^{-3t}\mu(t)$, and the input process has an autocorrelation function $R_{XX}(\tau) = e^{-2|\tau|}$. Define an error process $\varepsilon(t)$ by $\varepsilon(t) = X(t) - Y(t)$, and find:
(a) The autocorrelation function $R_{YY}(\tau)$.
(b) The power spectral density function $S_{YY}(\omega)$.
(c) $E[\varepsilon^2(t)]$.

5.8 If, for the system given in Figure P5.8, $X(t)$ is a wide sense stationary process with autocorrelation $R_X(\tau) = N_0/2\delta(\tau)$, find:
(a) $\phi_{XX}(\omega)$
(b) $\phi_{YY}(\omega)$

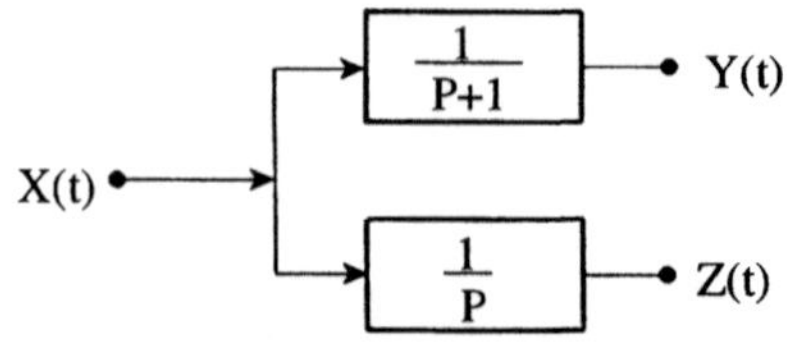

P5.8

(c) $R_{YY}(\tau)$
(d) $R_{XY}(\tau)$
(e) $R_{YZ}(\tau)$

5.9 A wide sense stationary Gaussian random process $X(t)$ with mean $E[X(t)] = m_X$ and autocorrelation function $R_X(\tau)$ is passed through two linear filters with transfer functions:

$$H_1(j\omega) = \frac{1}{j\omega + 1} \quad \text{and} \quad H_2(j\omega) = \frac{1}{j\omega + 2}$$

as shown in Figure P5.9 to give output random processes $Y(t)$ and $Z(t)$. Let $R_X(\tau) = (N_0/2)\delta(\tau)$ and $m_X = 0$.

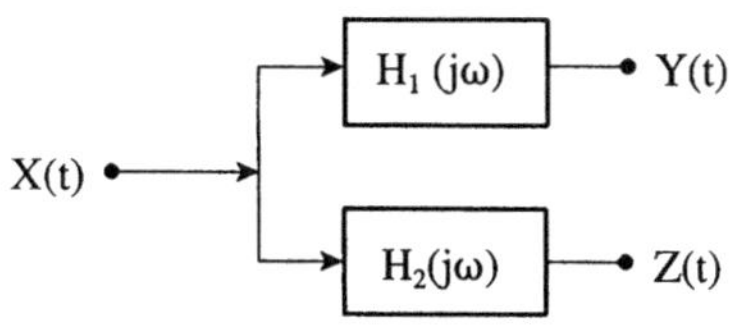

P5.9

(a) Find $R_{YY}(\tau)$ and $E[Y(t)]$.
(b) Find the power spectral density $S_{YY}(\omega)$.
(c) Are $Y(t)$ and $Z(t)$ Gaussian random processes?
(d) Find $R_{XY}(\tau)$ and $R_{YZ}(\tau)$.
(e) Find the joint density function for $Y(t_1) = Y_1$ and $Z(t_2) = Z_2$.

5.10 A stationary Gaussian random process $X(t)$ with mean $E[X(t)] = m_X$ and autocorrelation function $R_X(\tau)$ is passed through two linear filters with impulsive responses $h(t)$ and $g(t)$ yielding processes $Y(t)$ and $Z(t)$ as shown in Figure P5.10.

(a) What is the joint density function of the random variables $Y_1 \triangleq Y(t_1)$ and $Z_2 \triangleq Z(t_2)$?
(b) Evaluate $E[Y(t + \tau)X(t)]$. To what does this expression reduce when $X(t)$ is white noise?

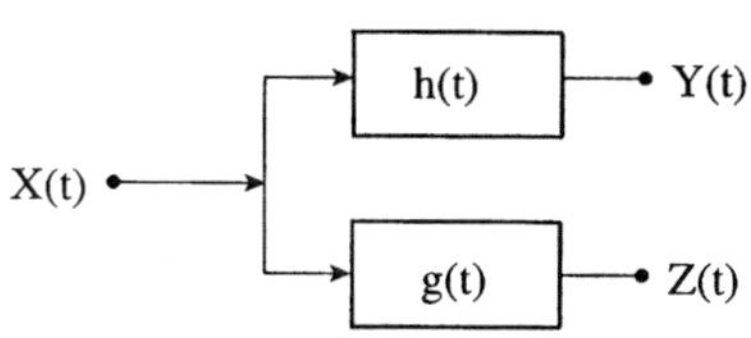

P5.10

(c) If $E[X(t)] = 0$ and $R_X(\tau) = [\sin(\pi\tau)/(\pi\tau)]^2$, find the power spectral density for $Y(t)$ when $h(t)$ is an ideal filter with transfer function $H(j\omega)$.

$$H(j\omega) = \begin{cases} 1, & \pi < |\omega| < 2\pi \\ 0, & \text{elsewhere} \end{cases}$$

5.11 Let $X(t)$ be a wide sense stationary random process with known η_X and known $R_{XX}(\tau)$. Assume that $X(t)$ is the input to the system shown in Figure P5.11. Find:

(a) $R_{ZZ}(\tau)$

(b) $R_{YZ}(\tau)$

(c) $\phi_{ZZ}(\omega)$

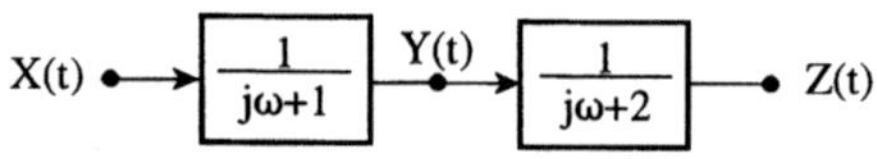

P5.11

5.12 Determine the power spectral density at points A, B, and C for the system of Figure P5.12. Let

$$\phi_{NN}(\omega) = 1, \quad R_{SS}(\tau) = e^{-|\tau|}, \quad H_2(j\omega) = \frac{1}{1+\omega^2}$$

$$h_1(t) = \begin{cases} 1, & 0 \le t \le 1 \\ 0, & \text{elsewhere} \end{cases}$$

List any assumptions that are necessary in finding your answer.

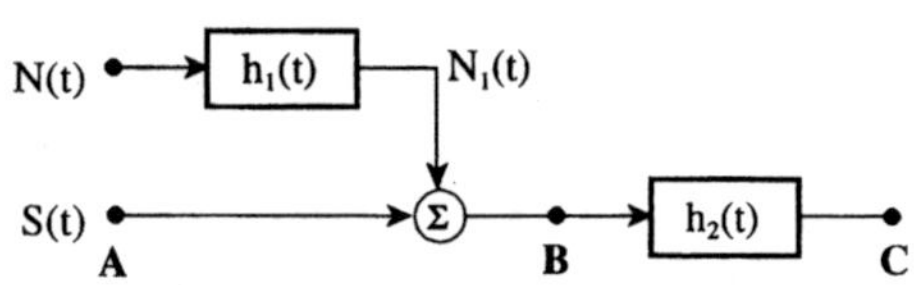

P5.12

5.13 Given random processes $X(t)$ and $N(t)$ with autocorrelation functions

$$R_{XX}(\tau) = e^{-2|\tau|} \quad \text{and} \quad R_{NN}(\tau) = \left(\frac{N_0}{2}\right)\delta(\tau)$$

Assume that $X(t)$ and $N(t)$ are independent random processes used in the system shown in Figure P5.13. Processes $X(t)$ and $N(t)$ are jointly wide sense stationary.

(a) If $X_0(t)$ and $N_0(t)$ represent the outputs of $1/(p+1)$ to $X(t)$ and $N(t)$, respectively, find the power spectral density function for the process $Z(t)$ and the signal-to-noise ratio $E[X_0^2(t)]/E[N_0^2](t)$.

(b) Find $R_{ZW}(\tau)$.

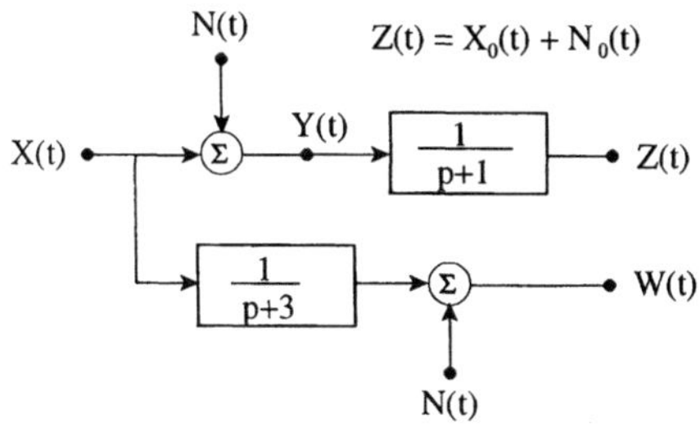

P5.13

5.14 Processes $X(t)$ and $Y(t)$ are jointly wide sense stationary with $\eta_X(t) = \eta_X$, $\eta_Y(t) = \eta_Y$, $R_{XX}(\tau) = \delta(\tau)$, $R_{YY}(\tau)$, and $R_{XY}(\tau)$ all known. They serve as inputs to the filters shown in Figure P5.14. Determine:

(a) $\eta_Q(t)$

(b) $R_{XW}(\tau)$

(c) $R_{ZZ}(\tau)$, $\phi_{ZZ}(\omega)$

(d) $R_{QQ}(\tau)$, $\phi_{QQ}(\omega)$

(e) $R_{WQ}(\tau)$

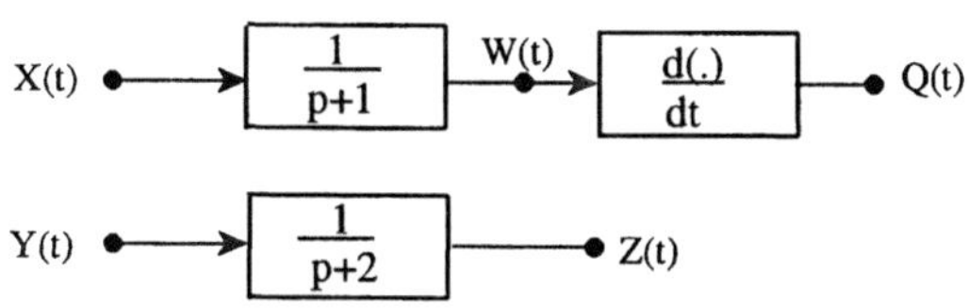

P5.14

5.15 We are given that $X(t)$ and $W(t)$ are uncorrelated jointly wide sense stationary Gaussian random processes with autocorrelation functions $R_X(\tau)$ and $R_W(\tau)$. They are the inputs to the system shown in Figure P5.15 resulting in the output $Q(t)$.

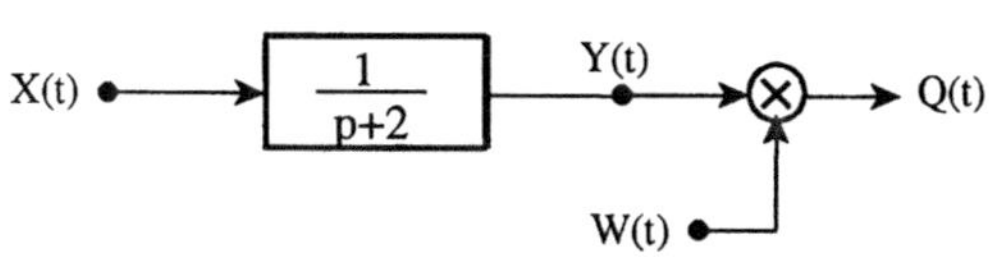

P5.15

(a) Are $X(t)$ and $W(t)$ independent random processes? Why or why not?
(b) Is $X(t)$ strict sense stationary? Why or why not?
(c) Find $R_{XW}(\tau)$.
(d) Is $Q(t)$ a Gaussian random process? Is $Y(t)$ a Gaussian random process?
(e) Find $\phi_{YY}(\omega)$, $R_{YW}(\tau)$, and $\phi_{WW}(\omega)$.
(f) Find $R_{QQ}(\tau)$ and $\phi_{QQ}(\omega)$.
(g) Find $f(x; t)$.

5.16 $X(t)$ and $Y(t)$ are jointly wide sense stationary orthogonal random processes with zero means and autocorrelation functions $R_{XX}(\tau)$ and $R_{YY}(\tau)$ given by

$$R_{XX}(\tau) = \exp(-\alpha|\tau|), \quad R_{YY}(\tau) = 1/2\delta(\tau)$$

The processes $W(t)$ and $Z(t)$ are the outputs of the filters shown in the block diagram of Figure P5.16 with $X(t)$ and $Y(t)$ as inputs.
(a) Find $\phi_{XX}(\omega)$.
(b) Find $\phi_{YY}(\omega)$.
(c) Find $\phi_{ZZ}(\omega)$.
(d) Find $R_{ZZ}(\tau)$.
(e) Find $\phi_{ZW}(\omega)$.
(f) Find $R_{WZ}(\tau)$.

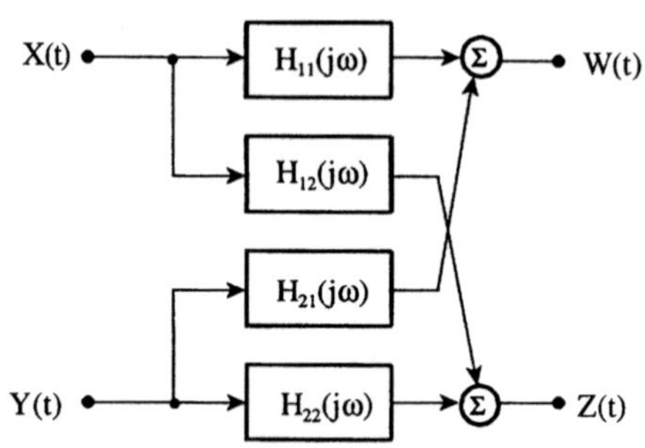

P5.16

5.17 A random "system" is defined by its impulse response $h(t) = e^{-At}u(t)$, where A is a uniform random variable on [0, 1]. Suppose that the input to such a system, $X(t)$, is a random telegraph wave process independent of A with

$$E[X(t)] = 0 \quad \text{and} \quad R_{XX}(t, u) = e^{-2|t-u|}$$

If $Y(t)$ is the resulting output process.
(a) Determine $E[Y(t)]$; also find $E[h(t)]$.
(b) Find $R_{YY}(t, u)$.

(c) Is $Y(t)$ a wide sense stationary process? If not, tell why not. If yes, determine its power spectral density.

5.18 Gaussian random process $X(t)$ is the input to the system shown and has autocorrelation function $R_{XX}(\tau)$. The FTI shown in Figure P5.18 is a finite time integrator with impulse response as follows: $h(t) = (1/T)[u(t) - u(t-T)]$.

(a) Find $R_Y(\tau)$.

(b) Find $\phi_Y(\omega)$.

(c) Find $Z(T/2)$.

(d) Find $\phi_Z(\omega)$.

(e) Find $E[Z(t)]$ and $\text{var}[Z(t)]$, the latter in integral form.

(f) What could this system be used for?

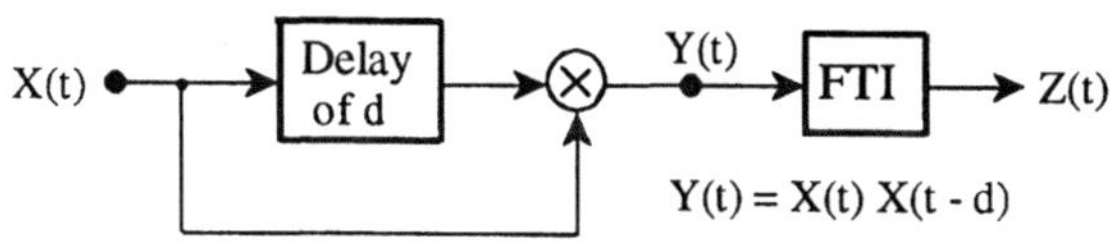

P5.18

5.19 Given a linear time-varying filter with its response to an impulse at time τ as follows:

$$h(t, \tau) = e^{-\tau t}\mu(t-\tau)\mu(\tau)$$

We know that the output $y(t)$ of the system to a given input signal is specified by

$$y(t) = \int_{-\infty}^{\infty} h(t, \tau)x(\tau)\, d\tau$$

The input $X(t)$ is a wide sense stationary random process with known mean and auto-correlation function:

$$E[X(t)] = 0, \quad R_{XX}(\tau) = e^{-\alpha|\tau|}$$

(a) Find the mean $E[Y(t)]$ of the output process.

(b) Find the autocorrelation function $R_{YY}(t_1, t_2)$ for the output process $Y(t)$.

(c) Is the output process wide sense stationary? Why or why not?

5.20 Find the cross-correlation function between input and output for the MA(2) process by using the Z-transform, and compare your result to those determined in Chapter 4.

5.21 Find and plot the autocorrelation function and power spectral density functions for the output process generated by passing a white sequence with autocorrelation function $R_{XX}[j, k] = 2\delta_{jk}$ through the following linear systems:

(a) $y[n] = -0.9y[n-1] + 2x[n-2]$for all n

(b) $y[n] = 0.9y[n-1] + 2x[n-2]$for all n

(c) $y[n] = 2x[n] - 0.9x[n-1]$for all n

(d) $y[n] = 2x[n] + 0.9x[n-1]$for all n

(e) $y[n] = 1x[n] + 1x[n-1] + 1/4x[n-2]$for all n

(f) $y[n] = -0.8y[n-1] - 0.4y[n-2] + 4x[n-2]$for all n

(g) $y[n] = -0.8y[n-1] - 0.4y[n-2] - 0.2y[n-3] + 4x[n] + 2x[n-1]$ for all n

5.22 Given a stationary random process $X(t)$ with $E[X(t)] = 1$ and autocorrelation function $R_{XX}(\tau) = 1 + \delta(\tau)$. Find:

(a) $E[Y(t)]$ for the system shown in Figure P5.22. *Note*: The process is applied at $t = 0$; therefore the mean of $Y(t)$ will be a function of time.

(b) Find the output autocorrrelation function $R_{YY}(t_1, t_2)$ for $t_1, t_2 \geq 0$.

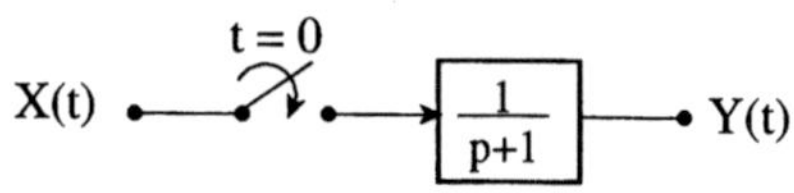

P5.22

5.23 Rework Problem 5.22 where $H(p) = p + 1$ replaces the $1/(p+1)$, and comment on your results.

5.24 Given a stationary random process $X(t)$ with $E[X(t)] = 0$ and autocorrelation function $R_{XX}(\tau) = e^{-2|\tau|}$. The $X(t)$ is applied at $t = 0$, similar to Figure P5.22, to a linear system with impulse response $h(t) = e^{-t}\mu(t)$ resulting in the output process $Y(t)$.

(a) Find $E[Y(t)]$.

(b) Find the output autocorrrelation function $R_{YY}(t_1, t_2)$ for $t_1, t_2 \geq 0$. (Not as much fun as you may think!)

(c) Show that your result of (b) in the steady state reduces to

$$R_{YY}(\tau) = \tfrac{1}{3}e^{-|\tau|} - \tfrac{1}{6}e^{-|2\tau|}$$

(d) Verify the autocorrelation of part (c) by using the formula of Eq. (5.18), which involves the bilateral Laplace transform.

5.25 Given a zero mean wide sense stationary random process $X(t)$ with autocorrelation function $R_{XX}(\tau)$. Find the autocorrelation function for the process $Y(t) = X(t) \cdot \mu(t)$, where $\mu(t)$ is the unit-step function.

REFERENCES

1. Papoulis, Athanasios, *Probability, Random Variables, and Stochastic Processes*, McGraw-Hill, 1965.
2. Davenport, Jr., William B., *Probability and Random Processes*, McGraw-Hill, 1970.
3. Shanmugan, K. Sam, and Arthur M. Breipohl, *Random Signals: Detection, Estimation, and Data Analysis*, Wiley, 1988.
4. Lathi, B. P., *An Introduction to Random Signals and Communication Theory*, International Textbook Company, 1968.
5. Thomas, John B., *An Introduction to Statistical Communication Theory*, Wiley, 1969.
6. Srinath, M. D., and P. K. Rajasekaran, *An Introduction to Statistical Signal Processing with Applications*, Wiley, 1979.
7. Van Trees, H. L., *Detection, Estimation and Modulation Theory*, Vol. 1, Wiley, 1968.

6 Nonlinear Systems: Random Processes

6.1 INTRODUCTION

In Chapter 5 the input–output characteristics were explored for a linear system excited by a random process. It was seen that the mean of the input process was sufficient to determine the mean of the output process, and correspondingly the autocorrelation function of the input was sufficient to determine the autocorrrelation function of the output process. However, there are many elements and systems, especially in communication theory, that are not linear, for example, devices like hard and soft limiters, rectifiers, modulators, and demodulators.

For these type of nonlinearities the autocorrelation functions are, in general, no longer sufficient to characterize the output autocorrelation functions, so it is necessary to introduce the concepts of higher-order correlation functions, cumulants, and higher-order spectrums. This chapter will also identify various classes of nonlinear systems and analyze them with respect to establishing input–output statistical relationships. The presentation will be guided more by what is mathematically tractable than what would be a complete and thorough investigation.

6.2 CLASSIFICATION OF NONLINEAR SYSTEMS

Currently no general theory exists that can handle all types of nonlinear systems. Therefore our presentation will contain special methods for analyzing certain classes of nonlinear systems. A hierarchical classification of nonlinear systems has been presented in Zadeh [12], and although we do not specifically use his nomenclature (type 0, type 1 etc.), the classes we present follow his lower-order classification. Our discussion will include instantaneous nonlinearities (type 0), and various cascades of linear systems and instantaneous nonlinear systems (type 1), including bilinear and trilinear cases. The chapter will conclude with a brief introduction to Volterra functionals for representing general nonlinear systems. The discussion begins with memoryless or zero-memory nonlinear systems.

6.2.1 Zero-Memory Nonlinear Systems

The zero-memory nonlinear system sometimes referred to as an **instantaneous nonlinear system**, has an input–output relationship

$$Y(t) = g(X(t)) \tag{6.1}$$

where $g(\cdot)$ is a real-valued function of one variable. The system is instantaneous in that the output at time t is determined solely by the input at time t. Several common instantaneous nonlinearities that play important roles in communication theory are the half- and full-wave rectifier, half- and full-wave square law devices, and hard and soft limiters as shown in Figure 6.1.

6.2.2 Bilinear Systems

A **bilinear system** as described by Bendat [1] is a special case of a general Volterra system (to be discussed later), and its input and output are governed by the following integral equation:

$$\begin{aligned} y(t) &\triangleq L_2[x(t)] \\ &= \int_{-\infty}^{\infty}\int_{-\infty}^{\infty} h_2(\tau_1, \tau_2)x(t-\tau_1)x(t-\tau_2)\,d\tau_1\,d\tau_2 \end{aligned} \tag{6.2}$$

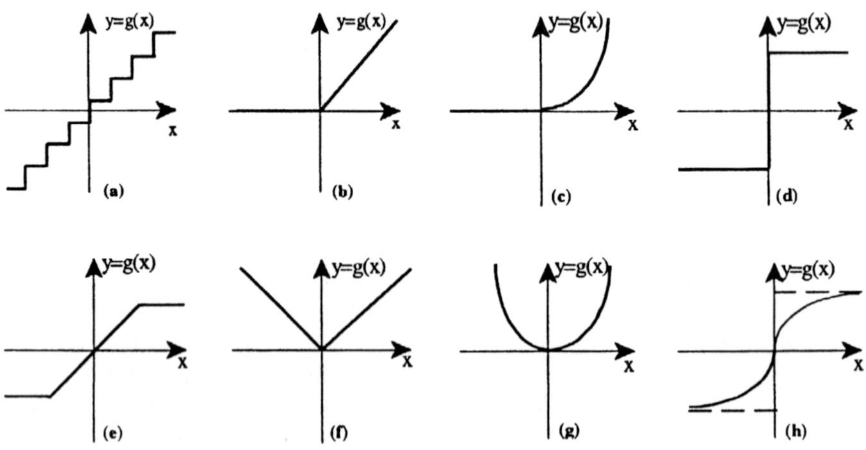

Figure 6.1 Common instantaneous nonlinearities (a) quantizer, (b) half wave rectifier, (c) square law half wave device, (d) hard limiter, (e) saturation, (f) full wave rectifier, (g) square law full wave device (h) soft limiter.

The system is specified, with respect to input and output, by $h_2(\tau_1, \tau_2)$, which is called the **time domain kernel**. Notice that if τ_1 and τ_2 are interchanged in the integral, the output can be rearranged as follows:

$$y(t) = \int_{-\infty}^{\infty} \int_{-\infty}^{\infty} h_2(\tau_2, \tau_1) x(t - \tau_1) x(t - \tau_2)\, d\tau_1\, d\tau_2 \tag{6.3}$$

Since the $y(t)$ is still the same, it is seen that the time domain kernel must be symmetrical to have a unique output, that is,

$$h_2(\tau_2, \tau_1) = h_2(\tau_2, \tau_1) \tag{6.4}$$

If the input is a sum of two inputs, $x_1(t) + x_2(t)$, the output $y(t)$ from Eq. (6.2) is

$$y(t) = \int_{-\infty}^{\infty} \int_{-\infty}^{\infty} h_2(\tau_1, \tau_2)[x_1(t - \tau_1) + x_2(t - \tau_1)][x_1(t - \tau_2) + x(t - \tau_2)]\, d\tau_1\, d\tau_2 \tag{6.5}$$

By expanding out the product and using the symmetric property of $h_2(\tau_1, \tau_2)$, we obtain an output that is the sum of responses of the system to $x_1(t)$ and $x_2(t)$ each separately and another term involving the integration of the time domain kernel and the cross product of the two signals as

$$y(t) = L_2[x_1(t)] + L_2[x_2(t)] + 2 \int_{-\infty}^{\infty} \int_{-\infty}^{\infty} h_2(\tau_1, \tau_2) x_1(t - \tau_1) x_2(t - \tau_2)\, d\tau_1\, d\tau_2 \tag{6.6}$$

Therefore, unless the integral term is zero for all time, superposition does not hold for a bilinear system, exemplifying the fact that the system is not a linear system.

Similarly it is easy to show that the response of the system to an input equal to $ax(t)$, where a is a constant is

$$L_2[ax(t)] = a^2 L_2[x(t)] \tag{6.7}$$

The two properties given in Eqs. (6.6) and (6.7) are sometimes used as an alternative definition for a bilinear system.

A bilinear system is **causal** if $h_2(\tau_1, \tau_2) = 0$ for all τ_1 and $\tau_2 < 0$. In taking the absolute values of both sides of (6.2), it is possible to see that the bilinear system will give a bounded output if the input is bounded. Thus the bilinear system will be **BIBO stable** provided that

$$\int_{-\infty}^{\infty} \int_{-\infty}^{\infty} |h_2(\tau_1, \tau_2)|\, d\tau_1\, d\tau_2 = B \qquad \text{where } B \text{ is finite} \tag{6.8}$$

If the input to the bilinear system specified in Eq. (6.2) is a delta function, the output will be

$$\begin{aligned} y(t) &= \int_{-\infty}^{\infty} \int_{-\infty}^{\infty} h_2(\tau_1, \tau_2)\delta(t-\tau_1)\delta(t-\tau_2)\, d\tau_1\, d\tau_2 \\ &= \int_{-\infty}^{\infty} h_2(t, \tau_2)\delta(t-\tau_2)\, d\tau_2 = h_2(t, t) \end{aligned} \tag{6.9}$$

Thus the response to a delta function is not the time domain kernel but is the time domain kernel evaluated on the line $\tau_1 = \tau_2 = t$. Knowing the kernel on just that line is not sufficient information for characterizing the system with respect to input and output, and thus a bilinear system cannot be identified by knowing only its impulse response as was true for a linear system.

The **frequency domain kernel** $H_2(j\omega_1, j\omega_2)$ for a bilinear system is defined as the two-dimensional Fourier transform of the time domain kernel as follows:

$$H_2(j\omega_1, j\omega_2) = \int_{-\infty}^{\infty} \int_{-\infty}^{\infty} h_2(\tau_1, \tau_2)e^{-j(\omega_1\tau_1+\omega_2\tau_2)}d\tau_1\, d\tau_2 \tag{6.10}$$

Characterizing a bilinear system means that we specify either the time domain or frequency domain kernels. Several examples of bilinear systems will be presented. They include a square law device, a square law device followed by a linear system, a linear system followed by a square law device and the cascade of linear system, a square law device and linear system. Each of these systems will be examined and the time domain kernels developed for their characterizations by using (6.10).

If the $x(t-\tau_1)$ and $x(t-\tau_2)$ are replaced in (6.10) by their inverse Fourier transforms, the output $y(t)$ is seen to be

$$\begin{aligned} y(t) &= \int_{-\infty}^{\infty} \int_{-\infty}^{\infty} h_2(\tau_1, \tau_2)\left[\frac{1}{2\pi}\int_{-\infty}^{\infty} X(j\omega_1)e^{j\omega_1(t-\tau_1)}d\omega_1\right. \\ &\quad \times \left.\frac{1}{2\pi}\int_{-\infty}^{\infty} X(j\omega_2)e^{j\omega_2(t-\tau_2)}d\omega_2\right] d\tau_1\, d\tau_2 \\ &= \frac{1}{4\pi^2}\int_{-\infty}^{\infty} \int_{-\infty}^{\infty} X(j\omega_1)X(j\omega_2) \\ &\quad \times \left[\int_{-\infty}^{\infty} \int_{-\infty}^{\infty} h_2(\tau_1, \tau_2)e^{-j\omega_1\tau_1}e^{-j\omega_2\tau_2}d\tau_1 d\tau_2\right] e^{j\omega_1 t}e^{j\omega_2 t}\, d\omega_1\, d\omega_2 \end{aligned} \tag{6.11}$$

The term in square brackets is the frequency domain kernel $H_2(j\omega_1, j\omega_2)$, and the first double integral gives the two-dimensional inverse Fourier transform of the product $X(j\omega_1)\ X(j\omega_2)$, and $H_2(j\omega_1, j\omega_2)$ but evaluated at times t_1 and t_2, both equaling t. Thus $y(t)$ can be finally written as

$$y(t) = \mathscr{F}^{-1}(X(j\omega_1)X(j\omega_2)H_2(j\omega_1, j\omega_2))]_{\substack{t_1=t\\ t_2=t}} \tag{6.12}$$

In this way it is seen not to be a product of the two-dimensional transforms, so the frequency domain kernel cannot be thought of as the frequency response of the bilinear system.

If we take the one-dimensional Fourier transform directly of the output given in the basic definition (6.2), we obtain

$$\begin{aligned}\mathscr{F}[y(t)] &= \int_{-\infty}^{\infty}\left(\int_{-\infty}^{\infty}\int_{-\infty}^{\infty} h_2(\tau_1,\tau_2)x(t-\tau_1)x(t-\tau_2)\,d\tau_1\,d\tau_2\right)e^{-j\omega t}\,dt \\ &= \int_{-\infty}^{\infty}\int_{-\infty}^{\infty} h_2(\tau_1,\tau_2)\left(\int_{-\infty}^{\infty} x(t-\tau_1)x(t-\tau_2)e^{-j\omega t}\,dt\right)d\tau_1\,d\tau_2\end{aligned} \tag{6.13}$$

The term in parentheses is the one-dimensional Fourier transform of the product of the two time-translated versions of the input signal $x(t)$ and thus is a function of both τ_1 and τ_2. Since this term in parentheses cannot be taken outside the integral signs, it is seen that the result is not the product of the transforms.

Square Law System. Let $x(t)$ and $y(t)$ represent the input and output, respectively, of a square law system as shown in Figure 6.2 and governed by the following input–output relationship:

$$y(t) = x^2(t) \tag{6.14}$$

A logical question at this time is: Is this system a bilinear system and if so what time domain kernel characterizes it? To answer this question, $x(t)$ is written in terms of a delta function as

$$x(t) = \int_{-\infty}^{\infty} x(t-\tau_1)\delta(\tau_1)\,d\tau_1 \tag{6.15}$$

Substituting this expression for $x(t)$ into (6.14) and rearranging allows the output to be written as

$$\begin{aligned}y(t) = x^2(t) &= \left(\int_{-\infty}^{\infty} x(t-\tau_1)\delta(\tau_1)\,d\tau_1\right)\left(\int_{-\infty}^{\infty} x(t-\tau_2)\delta(\tau_1)\,d\tau_2\right) \\ &= \int_{-\infty}^{\infty}\int_{-\infty}^{\infty} \delta(\tau_1)\delta(\tau_2)x(t-\tau_1)(t-\tau_2)\,d\tau_1\,d\tau_2\end{aligned} \tag{6.16}$$

It is seen that the nonlinear square law system is in the proper integral form, (6.2), and the product of the delta functions can be identified as the time domain kernel

$$h_2(\tau_1,\tau_2) = \delta(\tau_1)\delta(\tau_2) \tag{6.17}$$

The two-dimensional frequency domain kernel can be obtained by taking the Fourier transform of the time domain kernel, which for the $h_2(\tau_1,\tau_2)$ above gives

$$H(j\omega_1, j\omega_2) = 1 \qquad \text{for all } \omega_1 \text{ and } \omega_2 \tag{6.18}$$

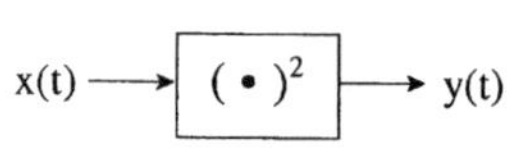

Figure 6.2 A square law device.

Care should be used in interpreting this result as the frequency response of the system for it doesn't represent the same information as the frequency response for the linear system. It is wrong to assume that it means that all signals are passed without alteration, as is the case for an all pass linear system. This is hardly the case for, if the Fourier transform of $y(t)$ is taken, it becomes

$$\mathscr{F}[y(t)] = \mathscr{F}[x^2(t)] = X(j\omega) * X(j\omega) \tag{6.19}$$

Thus the output transform is seen as the convolution of the Fourier transform of the input with itself. This will certainly give a different output transform, and as convolution has a tendency to broaden, the square law device actually creates new output frequency content outside the frequency range of the input. This result is typical of nonlinear systems in general.

Linear System Followed by a Square Law Device. Many nonlinear systems can be modeled as a nomemory nonlinear filter followed by a linear system, and the resulting system is no longer memoryless. A special case of this type of system is where the nonlinearity is a square law device as shown in Figure 6.3.

The output $y(t)$ for this case can be written in terms of the impulse response of the linear system as

$$\begin{aligned} y(t) &= \left[\int_{-\infty}^{\infty} h(\tau)x(t-\tau)\,d\tau\right]^2 \\ &= \int_{-\infty}^{\infty}\int_{-\infty}^{\infty} h(\tau_1)h(\tau_2)x(t-\tau_1)x(t-\tau_2)\,d\tau_1 d\tau_2 \end{aligned} \tag{6.20}$$

Thus the linear system followed by a square law device is a bilinear system and the time domain kernel is recognized from Eq. (6.2) as

$$h(\tau_1, \tau_2) = h(\tau_1)h(\tau_2) \tag{6.21}$$

The corresponding frequency domain kernel obtained by taking the two-dimensional transform of the time domain kernel given in Eq. (6.21), is easily seen to be the product of the frequency responses of the linear portions of the bilinear system in each of the variables as

$$H_2(j\omega_1, j\omega_2) = H(j\omega_1)H(j\omega_2) \tag{6.22}$$

This doesn't mean that the frequency response of this particular bilinear system is a product of the frequency responses of each linear part as $H_2(j\omega_1, j\omega_2)$ does not represent the frequency response of the bilinear system; it simply gives the frequency domain kernel.

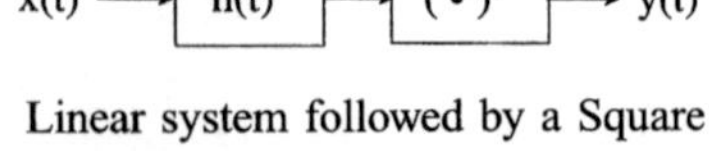

Figure 6.3 Linear system followed by a Square law device.

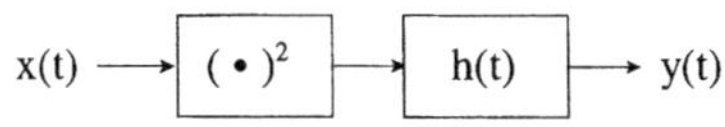

Figure 6.4 Square law device followed by a linear time invariant system.

Square Law Device Followed by a Linear System. Another combination that comes up frequently is a nonlinear system that is a no memory device followed by a linear system. The special case where the instantaneous system is a square law device is shown in Figure 6.4.

To obtain the time domain kernel for this type of nonlinear system the output of the system above is first written as

$$y(t) = \int_{-\infty}^{\infty} h(\tau_1)x^2(t-\tau_1)\,d\tau_1 \tag{6.23}$$

This can be put in the form of a bilinear system by rewriting $x^2(t)$ in terms of a delta function as

$$x^2(t-\tau_1) = \int_{-\infty}^{\infty} x(t-\tau_1)x(t-\tau_2)\delta(\tau_2-\tau_1)\,d\tau_2 \tag{6.24}$$

Substituting (6.24) into (6.23) and rearranging gives

$$y(t) = \int_{-\infty}^{\infty}\int_{-\infty}^{\infty} h(\tau_1)\delta(\tau_2-\tau_1)x(t-\tau_1)x(t-\tau_2)\,d\tau_1\,d\tau_2 \tag{6.25}$$

Thus the system specified above is a bilinear system and can be characterized by its time domain kernel as

$$h_2(\tau_1,\tau_2) = h(\tau_1)\delta(\tau_2-\tau_1) \tag{6.26}$$

The time domain kernel is seen to be nonzero only on the line $\tau_1 = \tau_2$, and thus is symmetric by force. By taking the two-dimensional transform of the time domain kernel, the corresponding frequency domain kernel is given by

$$\begin{aligned} H_2(j\omega_1, j\omega_2) &= \int_{-\infty}^{\infty}\int_{-\infty}^{\infty} h(\tau_1)\delta(\tau_2-\tau_1)e^{-j(\omega_1\tau_1+\omega_2\tau_2)}\,d\tau_1\,d\tau_2 \\ &= \int_{-\infty}^{\infty} h(\tau_1)e^{-j(\omega_1\tau_1+\omega_2\tau_2)}d\tau_1 \\ &= H(j(\omega_1+\omega_2)) \end{aligned} \tag{6.27}$$

Thus the frequency domain kernel for this type of nonlinear system is obtained by taking the Fourier transform of the linear system represented by $H(j\omega)$ and replacing the ω by the sum $\omega_1+\omega_2$.

Cascade Linear System–Square Law Device-Linear System. The cascade of a linear system, square law device, and linear system is shown in Figure 6.5. If

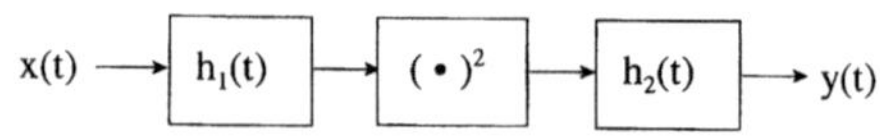

Figure 6.5 Bilinear system composed of a cascade of linear system-squarer-linear system.

$h_1(t)$ and $h_2(t)$ represent the impulse responses of the pre- and postfilter, respectively, it is possible to show with a development similar to the two preceding sections that

$$h_2(\tau_1, \tau_2) = \int_{-\infty}^{\infty} h_1(\tau_1 - \alpha)h_1(\tau_2 - \alpha)h_2(\alpha)\, d\alpha \tag{6.28}$$

Thus this cascade is a bilinear system as well.

By taking the two-dimensional Fourier transform of (6.28), we can see that the frequency domain kernel of this cascade system is given by

$$H_2(j\omega_1, j\omega_2) = H_1(j\omega_1)H_1(j\omega_2)H_2(j(\omega_1 + \omega_2)) \tag{6.29}$$

6.2.3 Trilinear Systems

The input–output relationship for a **trilinear system**, described clearly in Bendat [1], is defined as

$$\begin{aligned} y(t) &\triangleq L_3[x(t)] \\ &= \int_{-\infty}^{\infty}\int_{-\infty}^{\infty}\int_{-\infty}^{\infty} h_3(\tau_1, \tau_2, \tau_3)x(t - \tau_1)x(t - \tau_2)x(t - \tau_3)\, d\tau_1\, d\tau_2\, d\tau_3 \end{aligned} \tag{6.30}$$

where $h_3(\tau_1, \tau_2, \tau_3)$ is the **third-order time domain kernel**, and its specification characterizes the system with respect to input–output relationship.

It can be shown, with developments similar to those of the previous sections, that a cuber, and a cuber followed by a linear time-invariant system, and a linear time-invariant system followed by a cuber, and a cascade of linear system, cuber, and linear system, are all special cases of trilinear systems. These special cases are shown in Figure 6.6, and the resulting time domain kernels for these special cases are summarized below:

$$\begin{aligned} \text{Cuber:}\quad & h_3(\tau_1, \tau_2, \tau_3) = \delta(\tau_1)\delta(\tau_2)\delta(\tau_3) \\ \text{Cuber–Linear:}\quad & h_3(\tau_1, \tau_2, \tau_3) = h_2(\tau_1)\delta(\tau_1 - \tau_2)\delta(\tau_1 - \tau_3) \\ \text{Linear–cuber:}\quad & h_3(\tau_1, \tau_2, \tau_3) = h_1(\tau_1)h_1(\tau_2)h_1(\tau_3) \\ \text{Linear–cuber–linear:}\quad & h_3(\tau_1, \tau_2, \tau_3) = \int_{-\infty}^{\infty} h_1(\tau_1 - \alpha)h_1(\tau_2 - \alpha)h_1(\tau_3 - \alpha)h_2(\alpha)\, d\alpha \end{aligned} \tag{6.31}$$

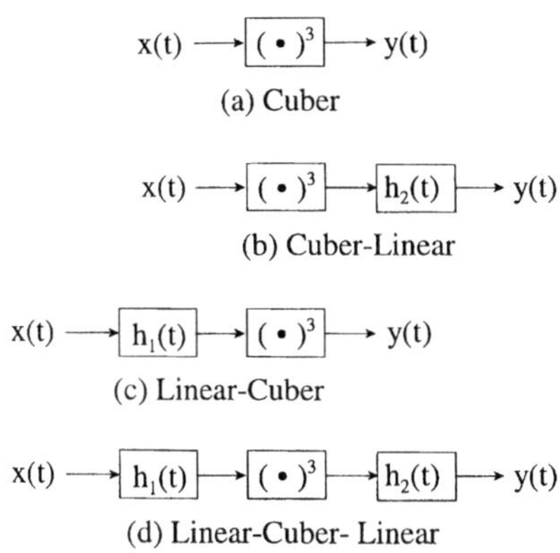

Figure 6.6 Special Trilinear systems (a) Cuber, (b) Cuber-linear system, (c) Linear system-cuber, and (d) Cascade: linear system-cuber-linear system.

Their corresponding **frequency domain kernels**, obtained by taking the three-dimensional Fourier transform of the time domain kernels can be determined as follows:

$$\begin{aligned} \text{Cuber:} \quad & H_3(j\omega_1, j\omega_2, j\omega_3) = 1 \\ \text{Cuber–linear:} \quad & H_3(j\omega_1, j\omega_2, j\omega_3) = H_2(j(\omega_1 + \omega_2 + \omega_3)) \\ \text{Linear–cuber:} \quad & H_3(j\omega_1, j\omega_2, j\omega_3) = H_1(j\omega_1)H_1(j\omega_2)H_1(j\omega_3) \\ \text{Linear–cuber–linear:} \quad & H_3(j\omega_1, j\omega_2, j\omega_3) = H_1(j(\omega_1)H_1(j\omega_2)H_1(j\omega_3) \\ & \qquad \times H_2(j(\omega_1 + \omega_2 + \omega_3)) \end{aligned} \tag{6.32}$$

6.2.4 Volterra Representation for General Nonlinear Systems

Volterra showed that the relationship between input $x(t)$ and output $y(t)$ for any nonlinear, causal, time-invariant, finite memory, analytic system, can be written as

$$\begin{aligned} y(t) = k_0 &+ \int_0^\infty k_1(\tau_1)x(t-\tau_1)\,d\tau_1 + \int_0^\infty \int_0^\infty k_2(\tau_1, \tau_2)x(t-\tau_1)x(t-\tau_2)\,d\tau_1 d\tau_2 + \cdots \\ &+ \int_0^\infty \int_0^\infty \cdots \int_0^\infty k_n(\tau_1, \tau_2, \ldots, \tau_n)x(t-\tau_1) \\ &\times x(t-\tau_2)\ldots x(t-\tau_n)\,d\tau_1 d\tau_2 \ldots d\tau_n + \cdots \end{aligned} \tag{6.33}$$

The output $y(t)$ is an infinite sum of a constant and other terms containing one, two, and k-dimensional integrals. The output is thus written as

$$y(t) = y_0 + y_1(t) + y_2(t) + \cdots + y_n(t) + \cdots \tag{6.34}$$

and can be viewed as the sum of the responses from each kernel as shown in Figure 6.7. The first term is a constant and can usually be subtracted off without loss of generality, the second term is a convolution integral representing the linear portion,

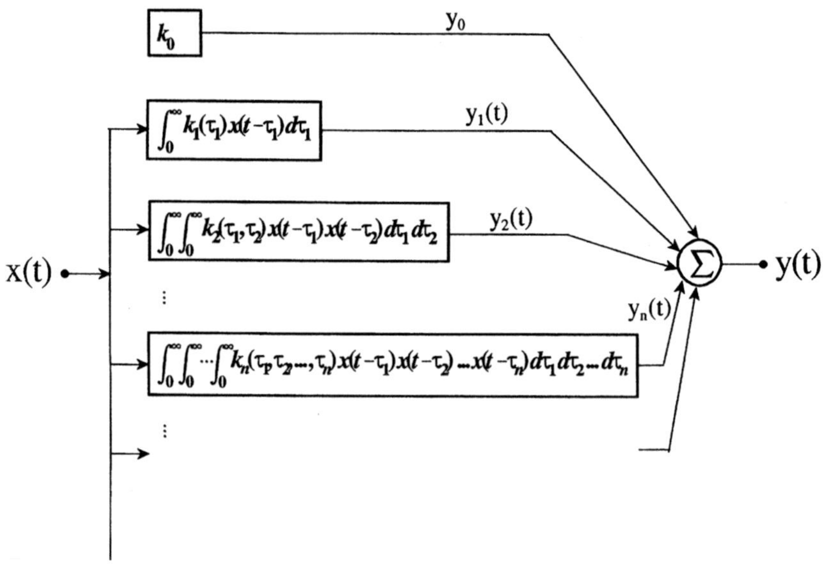

Figure 6.7 The Volterra representation of a general linear system.

and the other terms are deviations from the linear at various levels. The k_0, $k_1(\tau_1), \dots, k_n(\tau_1, \tau_2, \dots, \tau_n)$ are called the **first, second, and *n*th-order time domain kernels** of the system. From Eq. (6.33) it can be seen that knowing the kernels is sufficient information for the determination of the output for any given input.

The $y_1(t)$ part of the output appears as a convolution of the input with the first-order Volterra kernel and thus this part can be thought of as the linear part. The $y_2(t)$, taken alone can be shown to be a bilinear system. The word bilinear is used and has a precise mathematical definition as given in Section 6.2.2.

The Fourier transforms of the time domain kernels $k_1(\tau_1), k_2(\tau_1, \tau_2)$, $k_3(\tau_1, \tau_2, \tau_3), \dots$ are called **frequency domain kernels**, and the first three of them are sometimes called linear, bilinear, and trilinear frequency response functions.

The kernels are symmetric in all the tau variables, for example,

$$\begin{gathered} k_2(\tau_1, \tau_2) = k_2(\tau_2, \tau_1) \\ k_3(\tau_1, \tau_2, \tau_3) = k_3(\tau_2, \tau_1, \tau_3) = k_3(\tau_1, \tau_3, \tau_2) = k_3(\tau_3, \tau_2, \tau_1) \end{gathered} \tag{6.35}$$

This property of symmetry is a direct result of the fact that in the integral operation given in Eq. (6.33), the $x(t - \tau_k)$ can be easily reordered (associativity). Thus $k_n(\tau_1, \tau_2, \dots, \tau_n)$ must be symmetrical in all its variables. A time domain kernel is called **separable** if

$$k_n(\tau_1, \tau_2, \dots, \tau_n) = g_1(\tau_1)g_2(\tau_2)\dots g_n(\tau_n) \tag{6.36}$$

Symmetry does not necessarily imply separability.

It is important to note that the impulse response of bilinear and higher-order nonlinear systems is no longer sufficient information to determine the response to any input as was the case for a linear system. Now, if $x(t) = \delta(t)$ in (6.33), the output $y(t)$ is

$$y(t) = k_0 + k_1(t) + k_2(t, t) + \cdots + k_n(t, t, \ldots, t) + \cdots \tag{6.37}$$

Therefore each of the kernels has a contribution to the impulse response so that the total impulse response does not allow the determination of the kernels but only the sum of the kernels, and only then for the equal time values or $\tau_1 = \tau_2 = \ldots = \tau_n = t$.

A nonlinear system represented by its kernels can be symbolized by the simple block diagram shown in Figure 6.8 which conveys the same structure as that shown in Figure 6.7.

6.3 RANDOM OUTPUTS FOR INSTANTANEOUS NONLINEAR SYSTEMS

If the input to an instantaneous nonlinear system is a random process $X(t)$, the output $Y(t)$ is most often a random process as well. If we know the statistical properties of $X(t)$ of the input, a logical question is: What are the statistical properties of the output process $Y(t)$? We will partially answer this question by finding out what statistical information about the input process is necessary to determine the first-order density, mean, and autocorrelation function for the output process and give the output expressions that can be analytically obtained.

6.3.1 First-Order Density for Instantaneous Nonlinear Systems

Let the $X(t)$ and $Y(t)$ be the input and output processes, respectively, of an instantaneous nonlinear system governed by the equation

$$Y(t) = g(X(t)) \tag{6.38}$$

Assume that the first-order density, $f(x, t)$, of the input process $X(t)$ is known, and we desire the first-order density, $f(y, t)$, of the output process $Y(t)$. This problem is

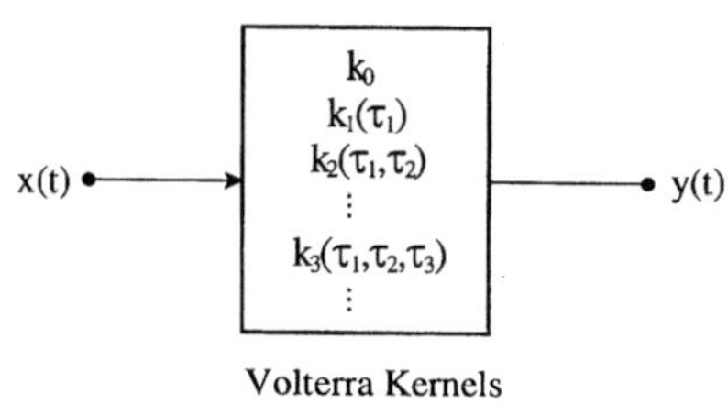

Figure 6.8 Symbolic representations of a Volterra system by block diagram.

equivalent to the single function of a single random variable defined at a given time t that was presented in Chapter 2. Thus the solution can be obtained by using any of the techniques described in that chapter. These techniques include the fundamental theorem, the distribution function approach, the auxiliary random variable method, and the Monte Carlo method. In the following example the first-order density functions for the output processes of a full-wave and half-wave square devices are presented.

EXAMPLE 6.1

The full-wave and half-wave square law device are characterized by the input–output relationships $Y(t) = X^2(t)$ and $Z(t) = X^2(t)u(t)$, respectively, and shown in Figure 6.1c and g. If the first-order density of the input process is a Gaussian random process with first-order density function $f(x_t)$ as

$$f(x_t) = \frac{1}{\sqrt{2\pi}\sigma_t} \exp\left(-\frac{x_t^2}{2\sigma_t^2}\right)$$

find the first order density for the output processes $Y(t)$ and $Z(t)$.

SOLUTION

For the **full-wave square law device** the first-order density for the random output process $Y(t)$ can be determined using the fundamental theorem for each t. If Y_t represents the random variable at time t of the output process and X_t represents the random variable of the input process evaluated at a particular time t we have

$$\begin{aligned} f(y_t) &= \left[\frac{f(x_t)|_{x_t=\sqrt{y_t}}}{|2\sqrt{y_t}|} + \frac{f(x_t)|_{x_t=-\sqrt{y_t}}}{|-2\sqrt{y_t}|}\right] u(y_t) \\ &= \frac{1}{\sigma_t\sqrt{2\pi y_t}} \exp\left(-\frac{y_t}{2\sigma_t^2}\right) u(y_t) \end{aligned}$$

For the **half-wave square law device** the flat spot for negative x gives a delta function at the origin, and the first-order density of the output process becomes

$$\begin{aligned} f(z_t) &= \left[\frac{f(x_t)|_{x_t=\sqrt{z_t}}}{|2\sqrt{z_t}|}\right] u(z_t) + \left[\int_{\infty}^{0^+} f(x_t)\, dx_t\right] \delta(z_t) \\ &= \frac{1}{2\sigma_t\sqrt{2\pi z_t}} \exp\left(-\frac{z_t}{2\sigma^2}\right) u(z_t) + \frac{1}{2}\delta(z_t) \end{aligned}$$

■

The results above, although specific for the nonlinearities given, indicate that if a Gaussian process is the input to an instantaneous nonlinearity, the output process will not be a Gaussian random process. This is in contrast to the result for a linear system where a Gaussian random process as an input produces a Gaussian random process on the output.

6.3.2 Mean of the Output Process

Let $X(t)$ represent the input and $Y(t)$ the output of an instantaneous nonlinear system characterized by $Y(t) = g(X(t))$. The mean of the output process can be calculated by a number of different methods, but in many problems it can be found conveniently by using one of the following two methods:

Method 1.

$$\eta_Y(t) = E[Y(t)] = E[g(X(t))] = \int_{-\infty}^{\infty} g(x_t) f_{X_t}(x_t)\, dx_t \tag{6.39a}$$

Method 2.

$$\eta_Y(t) = E[Y(t)] = \int_{-\infty}^{\infty} y_t f_{Y_t}(y_t)\, dy_t \tag{6.39b}$$

From both methods it is seen that the determination of the mean of the output process requires knowing the first-order density of either the input process or the output process. This means that a higher-order characterization of the input process than just the mean is required. So this result is in sharp contrast to that previously determined for linear systems where the mean of the input process was sufficient to determine the mean of the output process.

EXAMPLE 6.2

Assume that the input to a square law device described by $y = x^2$ is a random process characterized by its first-order density $f_{X_t}(x_t) = e^{-x_t}\mu(x_t)$. Determine the mean of the output process $Y(t)$ defined by $Y(t) = X^2(t)$ by both methods described above in Eqs. (6.39).

SOLUTION

Let Y_t be defined as the random variable $Y_t \triangleq Y(t)$ at time t and X_t be defined as the input random variable $X_t \triangleq X(t)$ at time t, thus giving $Y_t = X_t^2$. In using method 1, we have

$$\begin{aligned} E[Y_t] = E[X_t^2] &= \int_{-\infty}^{\infty} x_t^2 f_{X_t}(x_t)\, dx_t \\ &= \int_{-\infty}^{\infty} x_t^2 e^{-x_t}\, dx_t = 1 \end{aligned}$$

In using method 2, we must obtain the first-order density for the output process $Y(t)$. This density can be found by applying the transformation theorem as follows:

$$f_{Y_t}(y_t) = \frac{f_{X_t}(x_t)}{|2x_t|}\Bigg|_{x_t=\sqrt{y}} = \frac{1}{2\sqrt{y_t}} e^{-\sqrt{y_t}} \mu(y_t)$$

We can use this density to obtain the mean of the output process as

$$E[Y_t] = \int_{-\infty}^{\infty} y_t f_{Y_t}(y_t) dy_t$$

$$= \int_0^{\infty} y_t \frac{1}{2\sqrt{y_t}} e^{-\sqrt{y_t}}\, dy_t = 1$$

If the density of the output process is not needed, the second method does involve an unnecessary step of first finding the density, since the integral is not made any easier to evaluate. ■

6.3.3 Second-Order Densities for Instantaneous Nonlinearities

Let the random process $X(t)$ be the input to an instantaneous nonlinear system with output process $Y(t) = g(X(t))$. For the two times, t_1 and t_2, the corresponding random variables for the output are given as

$$Y(t_1) = g(X(t_1)), \quad Y(t_2) = g(X(t_2)) \tag{6.40}$$

If the random variables $X(t_1)$ and $X(t_2)$ are characterized by their joint probability density function $f(x_1, x_2, t_1, t_2)$, the basic problem of finding the second-order density of the output process can be solved by applying one of many available techniques for two functions of two random variables seen in Chapter 2. The problem is actually a simplified application, since there is no coupling for the solution of x in terms of y.

EXAMPLE 6.3

Let the random process $X(t)$, characterized by its second-order densities $f(x_1, x_2, t_1, t_2)$, be the input to a square law nonlinearity given by $y = g(x) = x^2$. Find the second-order densities for the output process $Y(t)$.

SOLUTION

For convenience, the following notation will be used: $X(t_1) = X_1, X(t_2) = X_2$, $Y(t_1) = Y_1$, and $Y(t_2) = Y_2$. The output random variables can then be written as

$$Y_1 = X_1^2, \quad Y_2 = X_2^2$$

If X_1 and X_2 are continuous random variables and $y = x^2$ is a continuous function, the two-dimensional form of the transformation theorem can be used. For y_1 and or $y_2 < 0$, there are no real roots, so the joint density function is 0. For $y_1 > 0$ and $y_2 > 0$, there are four possible pairs of solutions:

$$\begin{aligned}&(x_1 = \sqrt{y_1}, x_2 = \sqrt{y_2}), \quad (x_1 = \sqrt{y_1}, x_2 = -\sqrt{y_2})\\&(x_1 = -\sqrt{y_1}, x_2 = \sqrt{y_2}), \quad (x_1 = -\sqrt{y_1}, x_2 = -\sqrt{y_2})\end{aligned}$$

The joint density from the transformational theorem is

$$f_{Y_1Y_2}(y_1, y_2) = \sum_{\text{real roots}} \frac{f_{X_1X_2}(x_1, x_2)}{\begin{vmatrix} 2x_1 & 0 \\ 0 & 2x_2 \end{vmatrix}}$$

Substituting in real roots gives the joint density as

$$\begin{aligned}f_{Y_1Y_2}(y_1, y_2) = \frac{1}{4\sqrt{y_1y_2}}&[f_{X_1X_2}(\sqrt{y_1},\ \sqrt{y_2}) + f_{X_1X_2}(\sqrt{y_1},\ -\sqrt{y_2})\\&+ f_{X_1X_2}(-\sqrt{y_1},\ \sqrt{y_2}) + f_{X_1X_2}(-\sqrt{y_1},\ -\sqrt{y_2})]\end{aligned}$$ ■

The second-order density of the output process of an instantaneous nonlinear system is a function of only the second-order density of the input process. Thus a second-order density characterization of the input process is all that is required to get the second-order characterization of the output process.

6.3.4 Autocorrelation Function for Instantaneous Nonlinear Systems

The output autocorrelation function $R_{YY}(t_1, t_2)$ can be calculated by using the second-order density $f(x_1, x_2; t_1, t_2)$ of the input process as follows:

Method 1.

$$\begin{aligned}R_{YY}(t_1, t_2) &= E[Y(t_1)Y(t_2)] = E[g(X(t_1))g(X(t_2))]\\&= \int_{-\infty}^{\infty}\int_{-\infty}^{\infty} g(x_1)g(x_2)f(x_1, x_2; t_1, t_2)\, dx_1\, dx_2\end{aligned} \tag{6.41}$$

However, if the second-order density, $f(y_1, y_2; t_1, t_2)$, of the output happens to be known, the output autocorrrelation function can be determined by

Method 2.

$$\begin{aligned}R_{YY}(t_1, t_2) &= E[Y(t_1)Y(t_2)]\\&= \int_{-\infty}^{\infty}\int_{-\infty}^{\infty} y_1y_2f(y_1, y_2; t_1, t_2)\, dy_1\, dy_2\end{aligned} \tag{6.42}$$

As we noted in determining the mean, in general, the second-order densities of the input process are required in order for us to get the autocorrelation function of the output process for an instantaneous nonlinearity. Thus the autocorrelation function of the input process is often not sufficient for determining the autocorrelation of the output. If the input process is a wide sense stationary Gaussian process, then the mean and autocorrelation function of the input determines the second-order densities of the input and thus is sufficient information for determining the autocorrelation function of the output random process.

In the next few examples autocorrelation functions will be found for a number of common communication theory nonlinearities.

EXAMPLE 6.4

Suppose that the input $X(t)$ to a full-wave square law device is a zero mean wide sense stationary Gaussian random process characterized by its autocorrelation function $R_{XX}(\tau)$. Find the autocorrelation function $R_{YY}(\tau)$ for the output process $Y(t)$.

SOLUTION

By method 1 the autocorrelation function of the output is calculated directly in terms of the input second-order density using

$$R_{YY}(t_1, t_2) = E[X_1^2 X_2^2] = \int_{-\infty}^{\infty} \int_{-\infty}^{\infty} x_1^2 x_2^2 f_{X_1X_2}(x_1, x_2)\, dx_1\, dx_2$$

Since $X(t)$ is a Gaussian random process, the random variables $X(t_1) = X_1$ and $X(t_2) = X_2$ are jointly Gaussian. Thus the formula above for the autocorrelation is the fourth-order moment determined in Example 2.28. Recall that it is given by

$$E[X_1^2 X_2^2] = \sigma_1^2 \sigma_2^2 (1 + 2\rho_{12}^2)$$

where

$$\sigma_1^2 = E[X_1^2] - E^2[X_1] = R_{XX}(t_1, t_1) - 0 = R_{XX}(0)$$
$$\sigma_2^2 = E[X_2^2] - E^2[X_2] = R_{XX}(t_2, t_2) - 0 = R_{XX}(0)$$
$$\rho_{12} = \frac{E[X_1X_2] - E[X_1]E[X_2]}{\sigma_1\sigma_2} = \frac{R_{XX}(t_1, t_2) - 0}{R_{XX}^{1/2}(0)R_{XX}^{1/2}(0)} = \frac{R_{XX}(t_1 - t_2)}{R_{XX}(0)}$$

Substituting the expressions above for σ_1^2, σ_2^2, and ρ_{12} into the fourth-order moment equation gives the output correlation function as

$$\begin{aligned} R_{XX}^2(t_1, t_2) &= R_{XX}^2(0)(1 + R_{XX}(t_1 - t_2)/R_{XX}(0)) \\ &= R_{XX}^2(0) + R_{XX}(0)R_{XX}(t_1 - t_2) \end{aligned}$$

The second-order densities for the output of a square law device were determined in Example 6.3. The autocorrelation function of the output process can now be determined directly from that result using method 2 as follows:

$$R_{YY}(t_1, t_2) = E[Y(t_1)Y(t_2)] = \int_{-\infty}^{\infty}\int_{-\infty}^{\infty} y_1 y_2 f(y_1, y_2; t_1, t_2)\, dy_1 dy_2$$

Substituting the density function gives

$$R_{YY}(t_1, t_2) = \int_0^{\infty}\int_0^{\infty} \frac{\sqrt{y_1 y_2}}{4}[f_{X_1X_2}(\sqrt{y_1}, \sqrt{y_2}) + f_{X_1X_2}(\sqrt{y_1}, -\sqrt{y_2}) + f_{X_1X_2}(-\sqrt{y_1}, \sqrt{y_2}) + f_{X_1X_2}(-\sqrt{y_1}, -\sqrt{y_2})]\, dy_1\, dy_2$$

The joint probability density function for the random variables $X_1 = X(t_1)$ and $X_2 = X(t_2)$, defined across the zero mean wide sense stationary process $X(t)$, is Gaussian with second-order density determined from (4.133) as

$$f(x_1, x_2; t_1, t_2) = \frac{1}{2\pi|\mathbf{K}|^{1/2}} \exp\{-\tfrac{1}{2}(\mathbf{x}-\mathbf{m})^T\mathbf{K}^{-1}(\mathbf{x}-\mathbf{m})\}$$

$$\text{where} \quad \mathbf{x} = \begin{bmatrix} x_1 \\ x_2 \end{bmatrix}, \quad \mathbf{m} = \begin{bmatrix} 0 \\ 0 \end{bmatrix}, \quad \mathbf{K} = \begin{bmatrix} R_{XX}(0) & R_{XX}(t_1 - t_2) \\ R_{XX}(t_1 - t_2) & R_{XX}(0) \end{bmatrix}$$

Substituting this $f(x_1, x_2)$ into the equation above for $R_{YY}(t_1, t_2)$ gives a very messy integral that needs to be evaluated. In this example the first method takes an easier path to finding the solution for the autocorrelation function of the output process. ■

EXAMPLE 6.5

It is desired to find the output autocorrelation for a hard limiter whose nonlinearity is shown in Figure 6.1d if the input process is a zero mean wide sense stationary Gaussian random process. The process is characterized by its autocorrelation $R_{XX}(\tau)$. For the random variables $X_1 = X(t_1)$ and $X_2 = X(t_2)$, where $\tau = t_1 - t_2$ its second-order density function, $f(x_1, x_2)$ in terms of the normalized autocorrelation function is known to be

$$f(x_1, x_2) = \frac{1}{2\pi\sqrt{1 - \rho_{XX}(\tau)}} \exp\left[\frac{-(x_1^2 - 2\rho_{XX}(\tau)x_1x_2 + x_2^2)}{1 - \rho_{XX}(\tau)}\right]$$

where

$$\rho_{XX}(\tau) = \frac{R_{XX}(\tau)}{R_{XX}(0)}$$

SOLUTION

The autocorrelation function for the output of a hard limiter will be found by using Eq. (6.41):

$$
\begin{aligned}
R_{YY}(t_1, t_2) &= E[Y(t_1)Y(t_2)] = E[g(X(t_1))g(X(t_2))] \\
&= \int_{-\infty}^{\infty}\int_{-\infty}^{\infty} g(x_1)g(x_2)\,f(x_1, x_2)\,dx_1\,dx_2 \\
&= \int_{0}^{\infty}\int_{0}^{\infty} (1)(1)\,f(x_1, x_2)\,dx_1\,dx_2 + \int_{-\infty}^{0}\int_{-\infty}^{0} (-1)(-1)\,f(x_1, x_2)\,dx_1\,dx_2 \\
&\quad + \int_{-\infty}^{0}\int_{0}^{-\infty} (-1)(1)\,f(x_1, x_2)\,dx_1\,dx_2 \\
&\quad + \int_{0}^{\infty}\int_{-\infty}^{0} (1)(-1)\,f(x_1, x_2)\,dx_1\,dx_2
\end{aligned}
$$

Since $f(x_1, x_2)$ is a probability density function, it is known that it integrates to one as follows:

$$
\begin{aligned}
1 = &\int_{0}^{\infty}\int_{0}^{\infty} f(x_1, x_2)\,dx_1\,dx_2 + \int_{-\infty}^{0}\int_{-\infty}^{0} f(x_1, x_2)\,dx_1\,dx_2 \\
&+ \int_{-\infty}^{0}\int_{0}^{\infty} f(x_1, x_2)\,dx_1\,dx_2 + \int_{0}^{\infty}\int_{-\infty}^{0} f(x_1, x_2)\,dx_1\,dx_2
\end{aligned}
$$

After solving this equation for the sum of the integrals in the second and fourth quadrant and substituting the result into the equation for the autocorrelation function, we have

$$
R_{YY}(t_1, t_2) = 2\int_{0}^{\infty}\int_{0}^{\infty} f(x_1, x_2)\,dx_1\,dx_2 + 2\int_{-\infty}^{0}\int_{-\infty}^{0} f(x_1, x_2)\,dx_1\,dx_2 - 1
$$

Then, replacing x_1 by $-x_1$ and x_2 by $-x_2$ in the second integral, we find it to be equal to the first integral from the symmetry of the $f(x_1, x_2)$. So the autocorrelation becomes

$$
R_{YY}(t_1, t_2) = 4\int_{0}^{\infty}\int_{0}^{\infty} f(x_1, x_2)\,dx_1\,dx_2 - 1
$$

with $f(x_1, x_2)$ given in the problem statement. The procedure from here is a little messy involving a change of rectangular to polar coordinates, integration, and a change of variables and a final integral. Details for obtaining this integral are given in Thomas [5], where he showed the final result to be

$$
R_{YY}(\tau) = \frac{2}{\pi}\arcsin\left(\frac{R_{XX}(\tau)}{R_{XX}(0)}\right)
$$

where $\tau = t_1 - t_2$. ■

There are several other basic approaches that can be used to solve for the autocorrelation function of the output of instantaneous nonlinear systems. These include the characteristic function method, Price's theorem for Gaussian input processes, and series expansions. Thomas [5] has excellently and thoroughly developed these techniques as well as derived formulas for the autocorrelation functions of the outputs of instantaneous nonlinear systems to a Gaussian random process input when the nonlinearity is a full-wave odd, full-wave even, and half-wave νth law device given by $y = x^{\nu}$. The basic techniques he explored, however, can be applied to any type of instantaneous nonlinearity.

Although the methods are well defined, the mathematics for finding the output autocorrelation function of even simple instantaneous nonlinearities to Gaussian input processes can be tedious. The mathematics for finding output processes for non-Gaussian input processes is more complex. Indeed, since the autocorrelation function does not usually give enough information to obtain the second-order density, an entire second-order characterization of the input process is required.

6.3.5 Higher-Order Moments

Let $X(t)$ and $Y(t)$ are the input and output processes for an instantaneous nonlinear system given by $Y(t) = g(X(t))$. For a square law device we see that to get the nth-order moment of the output, we must know the input density or know the moments of the input process of order $2n$. Thus a higher-order characterization of the input process is necessary. In general, if we consider a polynomial approximation to $g(.)$, we need to know all order moments of the input process to be able to determine the nth order moments of the output process.

6.3.6 Stationarity of Output Process

Let $X(t)$ and $Y(t)$ be the input and output processes for an instantaneous nonlinear system given by $Y(t) = g(X(t))$. We are able to make a few general statements concerning the stationarity of the output process with respect to the stationarity of the input process:

(1) If $X(t)$ is stationary in the mean, the output will not necessarily be stationary in the mean.

(2) If $X(t)$ is stationary of any order n, the output process is stationary of order n.

(3) If $X(t)$ is wide sense stationary, then $Y(t)$ is not necessarily wide sense stationary.

6.4 CHARACTERIZATIONS FOR BILINEAR SYSTEMS

If the input to a bilinear system is a random process $X(t)$, the resulting output $Y(t)$ is also a random process. We now explore the statistical relationships that exist between input and output processes. Specifically, we will derive the output mean and autocorrelation function, and the cross correlation between input and output, for a causal bilinear system represented in (6.2) by the input–output relationship.

$$Y(t) = \int_0^\infty \int_0^\infty h_2(\tau_1, \tau_2) X(t - \tau_1) X(t - \tau_2) \, d\tau_1 \, d\tau_2 \tag{6.43}$$

Assume that $h_2(\tau_1, \tau_2)$ is known and that $X(t)$ is a random process characterized by its mean and autocorrelation function $E[X(t)]$ and $R_{XX}(t_1, t_2)$, respectively. The output process $Y(t)$ can be partially characterized by its mean $E[Y(t)]$ and autocorrelation function $R_{YY}(t_1, t_2)$, Also the cross-correlation function $R_{XY}(t_1, t_2)$ between the input and output is important. These partial characterizations are determined for the output of a bilinear system in the following sections.

6.4.1 Mean of the Ouput of a Bilinear System

The mean of the output process is obtained by taking the expected value of both sides of (6.43) and interchanging the order of expected value and integration to give

$$E[Y(t)] = \int_0^\infty \int_0^\infty h_2(\tau_1, \tau_2) E[X(t - \tau_1) X(t - \tau_2)] \, d\tau_1 \, d\tau_2 \tag{6.44}$$

The output mean can be determined only if we know the autocorrelation function of the input. Thus a higher-order characterization of the input, other than just the mean, is required for determining the mean of the output of a bilinear system. This is in contrast to the result for linear systems, which requires only the mean of the input process to be known to determine the mean of the output process.

If the input process is wide sense stationary with a zero mean and autocorrelation function $R_{XX}(\tau)$, then the $E[Y(t)]$ from (6.44) reduces to

$$E[Y(t)] = \int_0^\infty \int_0^\infty h_2(\tau_1, \tau_2) R_{XX}(\tau_1 - \tau_2) \, d\tau_1 \, d\tau_2 \tag{6.45}$$

6.4.2 Cross-correlation between Input and Output of a Bilinear System

The cross correlation between the input and output process can be written as

$$\begin{aligned} R_{XY}(t, t + \tau) &= E[X(t) Y(t + \tau)] \\ &= E[X(t) \int_0^\infty \int_0^\infty h_2(\tau_1, \tau_2) X(t + \tau - \tau_1) X(t + \tau - \tau_2) \, d\tau_1 \, d\tau_2] \\ &= \int_0^\infty \int_0^\infty h_2(\tau_1, \tau_2) E[X(t) X(t + \tau - \tau_1) X(t + \tau - \tau_2)] \, d\tau_1 \, d\tau_2] \end{aligned} \tag{6.46}$$

The cross-correlation function for the input and output of a bilinear system is thus a function of the third-order moments of the input process. Thus, if only the second-order moments were given, we would not have been able to determine the $R_{XY}(t, t+\tau)$.

6.4.3 Autocorrelation Function for the Output of a Bilinear System

The autocorrelation function is obtained by taking the expected value of the product of the output at time t_1 and the output at time t_2 as

$$\begin{aligned}
R_{YY}(t_1, t_2) &= E[Y(t_1)Y(t_2)] \\
&= E\left[\left(\int_0^{\infty}\int_0^{\infty} h_2(\tau_1, \tau_2)X(t_1-\tau_1)X(t_1-\tau_2)\,d\tau_1\,d\tau_2\right)\right. \\
&\quad \times \left.\left(\int_0^{\infty}\int_0^{\infty} h_2(\beta_1, \beta_2)X(t_2-\beta_1)X(t_2-\beta_2)\,d\beta_1\,d\beta_2\right)\right] \\
&= \int_0^{\infty}\int_0^{\infty}\int_0^{\infty}\int_0^{\infty} h_2(\tau_1, \tau_2)h_2(\beta_1, \beta_2)E[X(t_1-\tau_1)X(t_1-\tau_2)X(t_2-\beta_1) \\
&\quad \times X(t_2-\beta_2)]\,d\tau_1\,d\tau_2\,d\beta_1\,d\beta_2 \qquad (6.47)
\end{aligned}$$

The autocorrelation function for the output of a bilinear system is thus seen to be a function of the fourth-order moments of the input process. Clearly, a higher-order characterization of the input process—the fourth-order moments—is needed to determine just the output autocorrelation function $R_{YY}(t_1, t_2)$.

6.5 CHARACTERIZATIONS FOR TRILINEAR SYSTEMS

We now briefly look at the characteristics of the output $Y(t)$ of a trilinear system to a random input $X(t)$. In particular, the mean, cross-correlation function, and autocorrelation function are presented, and we learn that much higher-order characterizations are required to determine them.

6.5.1 Mean of the Output of a Trilinear System

The mean of the output process of a trilinear system is obtained by taking the expected value of both sides of (6.30) and interchanging the order of expected value and integration to give

$$E[Y(t)] = \int_{-\infty}^{\infty}\int_{-\infty}^{\infty}\int_{-\infty}^{\infty} h_3(\tau_1, \tau_2, \tau_3)E[x(t-\tau_1)x(t-\tau_2)x(t-\tau_3)]\,d\tau_1\,d\tau_2\,d\tau_3 \qquad (6.48)$$

The output mean can be determined only if we know the third-order moments of the input process. Thus a higher-order characterization of the input, other than just the mean and autocorrelation function, is required for determining the mean of the output of a trilinear system.

6.5.2 Cross-correlation between Input and Output of a Trilinear System

The cross correlation between the input and output process can be written by multiplying (6.30) by $Y(u)$ and taking the expected value operator through the integral signs:

$$E[X(t)Y(u)] = \int_{-\infty}^{\infty}\int_{-\infty}^{\infty}\int_{-\infty}^{\infty} h_3(\tau_1, \tau_2, \tau_3)E[X(t)X(u-\tau_1)X(u-\tau_2) \times X(u-\tau_3)]\, d\tau_1\, d\tau_2\, d\tau_3 \tag{6.49}$$

The cross-correlation function for the input and output of a bilinear system is a function of the fourth-order moments of the input process.

6.5.3 Autocorrelation Function for the Output of a Trilinear System

The autocorrelation function is obtained by taking the expected value of the product of the output at time t and the output at time u as follows:

$$\begin{aligned} E[Y(t)Y(u)] &= E\bigg[\int_{-\infty}^{\infty}\int_{-\infty}^{\infty}\int_{-\infty}^{\infty} h_3(\tau_1, \tau_2, \tau_3)X(t-\tau_1) \\ &\quad \times X(t-\tau_2)X(t-\tau_3)\, d\tau_1\, d\tau_2\, d\tau_3 \\ &\quad \times \int_{-\infty}^{\infty}\int_{-\infty}^{\infty}\int_{-\infty}^{\infty} h_3(\alpha_1, \alpha_2, \alpha_3)X(u-\alpha_1) \\ &\quad \times X(u-\alpha_2)X(u-\alpha_3)\, d\alpha_1\, d\alpha_2\, d\alpha_3\bigg] \\ &= \int_{-\infty}^{\infty}\int_{-\infty}^{\infty}\int_{-\infty}^{\infty}\int_{-\infty}^{\infty}\int_{-\infty}^{\infty}\int_{-\infty}^{\infty} h_3(\tau_1, \tau_2, \tau_3)h_3(\alpha_1, \alpha_2, \alpha_3) \\ &\quad \times E[X(t-\tau_1)\, X(t-\tau_2)\, X(t-\tau_3)\, X(u-\alpha_1)\, X(u-\alpha_2) \\ &\quad \times X(u-\alpha_3)]d\alpha_1\, d\alpha_2\, d\alpha_3\, d\tau_1\, d\tau_2\, d\tau_3 \end{aligned} \tag{6.50}$$

The autocorrelation function for the output of a trilinear system is seen to be a function of the sixth-order moments of the input process. Thus, unless these sixth-order moments of the input process are given, we would not be able to determine the autocorrelation function for the output process of a trilinear system to a random process input.

6.6 CHARACTERIZATIONS FOR VOLTERRA NONLINEAR SYSTEMS

If the input to the nonlinear system is a random process, the resulting output is a random process. We explore here the statistical relationships that exist between input and output processes. Specifically in the following example the output mean and autocorrelation function, and the cross correlation between input and output, will be derived for a system represented by a truncated Volterra expansion of the second order.

EXAMPLE 6.7

A nonlinear system is known to be characterized by a second-order Volterra expansion as

$$Y(t) = k_0 + \int_0^\infty k_1(\tau)x(t-\tau)\,d\tau + \int_0^\infty \int_0^\infty k_2(\tau_1, \tau_2)x(t-\tau_1)x(t-\tau_2)\,d\tau_1\,d\tau_2 \tag{6.51}$$

Assume that k_0, $k_1(\tau_1)$, and $k_2(\tau_1, \tau_2)$ are known and that $X(t)$ is a random process characterized by its mean and autocorrelation functions given by $E[X(t)]$ and $R_{XX}(t_1, t_2)$. Find (a) the mean $E[Y(t)]$ of the output process, (b) the cross correlation between the input and output, and (c) the autocorrelation function of the output process.

SOLUTION

(a) The mean of the output process is obtained by taking the expected value of both sides of (6.51) and interchanging the order of expected value and integration to give

$$\begin{aligned} E[Y(t)] = k_0 &+ \int_0^\infty k_1(\tau_1)E[X(t-\tau_1)]\,d\tau_1 \\ &+ \int_0^\infty \int_0^\infty k_2(\tau_1, \tau_2)E[X(t-\tau_1)X(t-\tau_2)]\,d\tau_1\,d\tau_2 \end{aligned} \tag{6.52}$$

The output mean can be determined only if we know both the mean of the input process and the autocorrelation function of the input. Thus we require a higher-order characterization of the input. This is in contrast to the result for linear systems which requires only the mean of the input process to determine the mean of the output process.

For the special case where the input process is wide sense stationary with a zero mean and autocorrelation function $R_{XX}(\tau)$, the first integral of (6.52) is zero. The expected value in the second integral can be written in terms of time difference only:

$$E[Y(t)] = k_0 + \int_0^\infty \int_0^\infty k_2(\tau_1, \tau_2)R_{XX}(\tau_1 - \tau_2)\,d\tau_1\,d\tau_2 \tag{6.53}$$

(b) The cross correlation between the input and output process can be written as follows using (6.51) and the definition of the cross-correlation function as

$$\begin{aligned} R_{XY}(t_1, t_2) &= E[X(t_1)Y(t_2)] \\ &= E[X(t_1)k_0 + X(t_1)\int_0^\infty k_1(\tau_1)X(t_2-\tau_1)\,d\tau_1 \\ &\quad + X(t_1)\int_0^\infty \int_0^\infty k_2(\tau_1, \tau_2)X(t_2-\tau_1)X(t_2-\tau_2)\,d\tau_1\,d\tau_2] \end{aligned} \tag{6.54}$$

After multiplying out and taking the expected value operator through the integral sign, we reduce (6.54) to

$$R_{XY}(t_1, t_1) = k_0 E[X(t_1)] + \int_0^\infty k_1(\tau_1)\, E[X(t_1)X(t_2 - \tau_1)]\, d\tau_1 + \int_0^\infty \int_0^\infty k_2(\tau_1, \tau_2) E[X(t_1)X(t_2 - \tau_1)X(t_2 - \tau_2)]\, d\tau_1\, d\tau_2 \tag{6.55}$$

The cross-correlation function for the input and output of a second-order Volterra system is a function of the first, second, and third-order moments of the input process.

(c) The autocorrelation function is obtained by taking the expected value of the product of the output at time t_1 and the output at time t_2 as follows:

$$\begin{aligned} R_{YY}(t_1, t_2) &= E[Y(t_1)Y(t_2)] \\ &= E\Big[\Big(k_0 + \int_0^\infty k_1(\tau_1)\, X(t_1 - \tau_1)\, d\tau_1 \\ &\quad + \int_0^\infty \int_0^\infty k_2(\tau_1, \tau_2) X(t_1 - \tau_1) X(t_1 - \tau_2)\, d\tau_1\, d\tau_2\Big) \\ &\quad \times \Big(k_0 + \int_0^\infty k_1(\beta_1)\, X(t_2 - \beta_1)\, d\beta_1 \\ &\quad + \int_0^\infty \int_0^\infty k_2(\beta_1, \beta_2) X(t_2 - \beta_1) X(t_2 - \beta_2)\, d\beta_1\, d\beta_2\Big)\Big] \end{aligned} \tag{6.56}$$

Multiplying out the terms in parentheses it is seen that the $R_{YY}(t_1, t_2)$ contains nine terms and depends on the first- through the fourth-order moments of the input process. Thus just knowing the mean and autocorrelation function is not sufficient for determining the autocorrelation function of the output process. ■

6.7 HIGHER-ORDER CHARACTERIZATIONS

We have seen that the mean, autocorrelation function, and power spectrum are sufficient for determining the mean, autocorrelation, and power spectrum for the output of linear systems to random processes. For nonlinear systems we have already noted that higher-order properties than the second-order statistics must be defined in order to determine even the output statistical properties of order two. The useful definitions of moment function, cumulant function and higher-order spectra are now presented.

6.7.1 Moment Function for Random Processes

The moment function for a random process can be considered to be an extension of the autocorrelation function. The autocorrelation function is the expected value of the product of the random variables defined at two different times, whereas the moment function is the expected value of the product of the random variables defined at more than two times. Let $t, t + \tau_1, t + \tau_2, \ldots, t + \tau_{k-1}$ be a comb of

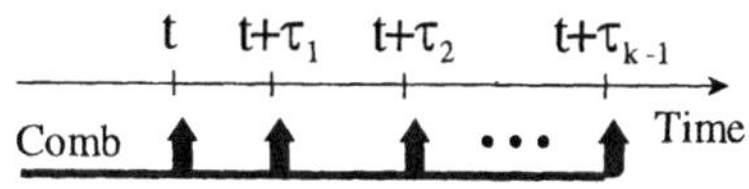

Figure 6.9 Comb of times for cummulant function definition.

times as shown in Figure 6.9, and let $X(t), X(t+\tau_1), \ldots, X(\tau_{k-1})$ represent random variables from a given random process $X(t)$. The ***k*th order moment function** $M_{XX\ldots X}(t, t+\tau_1, t+\tau_2, \ldots, t+\tau_{k-1})$ can be defined in terms of the random variables at these times as

$$M_{XX\ldots X}(t, t+\tau_1, t+\tau_2, \ldots, t+\tau_{k-1}) = E[X(t)X(t+\tau_1)X(t+\tau_2)\ldots\ X(t+\tau_{k-1})] \tag{6.57}$$

In general, the moment function is a function of the k time variables $t, \tau_1, \tau_2, \ldots, \tau_{k-1}$. We say that a random process $X(t)$ is ***k*th order moment stationary** if the expected value given in (6.57) does not depend on the t variable and is only a function of $\tau_1, \tau_2, \ldots, \tau_{k-1}$. This property is stronger than stationary in mean and autocorrelation yet weaker than stationarity of order k, which would require the equality of all density functions for the random variables at the comb of times and be independent of t.

6.7.2 Cumulant Function for Random Processes

The ***k*th order cumulant function** $C_{XX\ldots X}(t, t+\tau_1, t+\tau_2, \ldots, t+\tau_{k-1})$, for a random process $X(t)$, is defined in terms of the cumulants of the random variables $X(t), X(t+\tau_1), X(t+\tau_2), \ldots, X(t+\tau_{k-1})$ for the comb of times specified by

$$C_{XX\ldots X}(t, t+\tau_1, t+\tau_2, \ldots, t+\tau_{n-1}) \triangleq (-j)^n \left. \frac{\partial^n \ln \Phi(\omega_1, \omega_2, \ldots, \omega_n)}{\partial\omega_1 \partial\omega_2\ \ldots\ \partial\,\omega_n} \right|_{\substack{\omega_1=0\\ \omega_2=0\\ \cdots\\ \omega_n=0}} \tag{6.58}$$

where $\Phi(\omega_1, \omega_2, \ldots, \omega_k)$ is the joint characterization function for the random variables defined at the comb of times $t, t+\tau_1, t+\tau_2, \ldots, t+\tau_{k-1}$. The first-, second-, and third-order cumulant functions can be determined using results from Chapter 2 to be

$$\begin{aligned} C_{XX}(t, +\tau_1) &= E[X(t)X(t+\tau_1)] \\ C_{XXX}(t, t+\tau_1, t+\tau_2) &= E[X(t)X(t+\tau_1)X(t+\tau_2)] \\ C_{XXXX}(t, t+\tau_1, t+\tau_2, t+\tau_3) &= E[X(t)X(t+\tau_1)X(t+\tau_2)X(t+\tau_3)] \\ &\quad - C_{XX}(t, t+\tau_1)\ C_{XX}(t+\tau_2, t+\tau_3) \\ &\quad - C_{XX}(t, t+\tau_2)\ C_{XX}(t+\tau_1, t+\tau_3) \\ &\quad - C_{XX}(t, t+\tau_3)\ C_{XX}(t+\tau_1, t+\tau_2) \end{aligned} \tag{6.59}$$

With $X_1, X_2, \ldots, X_k$ the random variables defined above for the comb of times and assuming moment stationarity of the proper order, we see that the first four cumulant functions for the process $X(t)$ do not depend on t and can be written as functions of the distances between the teeth of the comb as follows:

$$\begin{aligned} C_{XX}(\tau_1) &= E[X(t)X(t+\tau_1)] \\ C_{XXX}(\tau_1, \tau_2) &= E[X(t)X(t+\tau_1)X(t+\tau_2)] \\ C_{XXXX}(\tau_1, \tau_2, \tau_3) &= E[X(t)X(t+\tau_1)X(t+\tau_2)X(t+\tau_3)] - C_{XX}(\tau_1)\, C_{XX}(\tau_2-\tau_3) \\ &\quad - C_{XX}(\tau_2)\, C_{XX}(\tau_3-\tau_1) - C_{XX}(\tau_3)\, C_{XX}(\tau_1-\tau_2) \end{aligned} \tag{6.60}$$

6.7.3 Polyspectrum for Random Processes

The power spectral density, or power spectrum, is a partial characterization of a random process and is related to the autocorrelation of a wide sense stationary process through the Fourier transform. In working with processes generated as outputs from nonlinear systems, it is necessary to include higher-order characterizations. For these problems it is useful to define the polyspectrum for a random process which is a higher-order characterization.

The kth-order polyspectrum $S_{XX\ldots X}(\omega_1, \omega_2, \ldots, \omega_{k-1})$ of a random process $X(t)$, with moment stationarity of the kth order is defined as the $(k-1)$-dimensional Fourier transform of its **kth order cumulant** function defined by Eq. (6.58).

Thus for $\boldsymbol{\omega} = [\omega_1, \omega_2, \ldots, \omega_{k-1}]^T$ and $\boldsymbol{\tau} = [\tau_1, \tau_2, \ldots, \tau_{k-1}]^T$, the $S_{XX\ldots X}(\omega_1, \omega_2, \ldots, \omega_{k-1})$ is defined by

$$S_{XX,\ldots X}(\boldsymbol{\omega}) \triangleq \int_{-\infty}^{\infty}\int_{-\infty}^{\infty}\cdots\int_{-\infty}^{\infty} C_{XX\ldots X}(\boldsymbol{\tau}) \exp\left(-j\boldsymbol{\omega}^T\boldsymbol{\tau}\right) d\boldsymbol{\tau} \tag{6.61}$$

Notice that this definition uses the Fourier transform of the cumulant function rather than the moment function. The most commonly used polyspectra are the **spectrum**, **bispectrum**, and **trispectrum**. These are given as follows:

Spectrum

$$S_{XX}(\omega_1) = \int_{-\infty}^{\infty} C_{XX}(\tau_1) \exp(-j\omega_1\tau_1)\, d\tau_1 \tag{6.62}$$

Bispectrum

$$S_{XXX}(\omega_1, \omega_2) = \int_{-\infty}^{\infty}\int_{-\infty}^{\infty} C_{XXX}(\tau_1, \tau_2) \exp(-j(\omega_1\tau_1 + \omega_2\tau_2))\, d\tau_1\, d\tau_2 \tag{6.63}$$

Trispectrum

$$\begin{aligned} S_{XXXX}(\omega_1, \omega_2, \omega_3) = \int_{-\infty}^{\infty}\int_{-\infty}^{\infty}\int_{-\infty}^{\infty} & C_{XXXX}(\tau_1, \tau_2, \tau_3) \exp(-j(\omega_1\tau_1 \\ & + \omega_2\tau_2 + \omega_3\tau_3))\, d\tau_1\, d\tau_2\, d\tau_3 \end{aligned} \tag{6.64}$$

Use of these polyspectra has led to new methods in system identification and detection algorithms. A thorough discussion of polyspectra and the use of the cumulant spectrum is presented in Nikias and Petropuou [10], Mendel [8], and Nikias and Mendel [6].

6.8 SUMMARY

It is virtually impossible for our modern-day problems to not involve nonlinear systems. Therefore the main purpose of this chapter was to present a framework for determining various characterizations of output processes of nonlinear systems with random processes as inputs. In Chapter 5 it was shown that output mean and autocorrelation function of a linear time-invariant system could be obtained knowing only the input mean and autocorrelation, respectively, and that the output spectral density could be obtained in the frequency domain by using the transfer function of the system.

It was shown in this chapter that even the simplest nonlinear system, the instantaneous nonlinear system, requires additional statistical information other than just the mean and autocorrelation function, namely the first- and second-order densities of the input process to determine the mean and autocorrelation function of the output process.

A hierarchy of nonlinear systems was presented including the instantaneous nonlinear, bilinear, trilinear, and the general Volterra system. These systems input and output relationships were described in terms of integral equations involving various time domain kernels. This was similar to the description of a linear system in terms of the convolution integral, but for these nonlinear systems the defining integral is represented by a higher-dimensional integral. Analogously, frequency domain kernels were defined that could be used to obtain higher-order spectra for the output.

The transformation theorem for random variables was the basic tool used for finding output characterizations of instantaneous nonlinear systems to random inputs. It was used to obtain the first- and second-order density functions and the mean and autocorrelation function of the output process. It was shown that a first-order density of the input is required to determine just the mean of the output process and that a second-order density of the input process is required to determine the output autocorrelation function. It was also shown that for instantaneous nonlinear systems that an nth order stationary input process produced an nth-order stationary output process. Solving for the output autocorrelation function was, in general, a very complex process even for Gausssian process inputs. Solutions for specific instantaneous nonlinearities for Gaussian random process inputs are given in Thomas [5].

The next level of nonlinearity discussed was the bilinear system. Simple bilinear systems were given as examples and involved a linear system or systems in cascade with a square law device. It was again shown that higher-order characterizations of the input process are required to determine just the output mean and autocorrelation function.

Trilinear and general Volterra nonlinear systems were then presented where inputs and outputs were modeled in terms of first-, second-, and higher-order kernels. Moment and cumulant spectra were defined and their relationships explored. For more detailed discussion of such systems, consult the excellent book by Bendat [1].

Missing from our development is the presentation of Weiner functionals which are useful for nonlinear system identification. Basically functionals are selected such that their outputs are orthogonal for Gaussian input processes. They can be shown to be related to the Volterra functionals, and this topic is nicely presented in Bendat [1] and Marmarelis [11].

PROBLEMS

6.1 Let $S(t)$ and $N(t)$ be independent Gaussian random processes with zero means and autocorrelation functions $R_{SS}(\tau)$ and $R_{NN}(\tau)$. Define a new random process $Y(t)$ by

$$Y(t) = (S(t) + N(t))^2$$

(a) Find the mean and autocorrelation function for $Y(t)$.

(b) If the power spectral densities of $S(t)$ and $N(t)$ are $S_{SS}(\omega)$ and $S_{NN}(\omega)$, respectively, find the power spectral density for the process $Y(t)$.

(c) Find the cross-correlation $R_{SY}(t_1, t_2)$.

(d) Is the output process wide sense stationary? Explain.

(e) Are the input and output processes jointly wide sense stationary? Explain.

6.2 Suppose that $X(t)$ and $Y(t)$ are real jointly wide sense stationary correlated random processes with $\eta_X, \eta_Y, R_{XX}(\tau), R_{YY}(\tau)$, and $R_{XY}(\tau)$. Define two new processes $W(t)$ and $Z(t)$ as

$$W(t) = X(t) + Y(t), \qquad Z(t) = X(t)Y(t)$$

(a) Find $E[W(t)], E[Z(t)], R_{WW}(t+\tau, t), R_{ZZ}(t+\tau, t)$, and $R_{WZ}(t+\tau)$ in terms of the given parameters.

(b) Also find the power spectral densities $S_{WW}(\omega)$ and $S_{ZZ}(\omega)$ if they can be found in terms of the parameters given.

(c) If $X(t)$ and $Y(t)$ are independent random processes, simplify your results of parts (a) and (b).

6.3 For the random processes given in Problem 6.2, suppose further that $X(t)$ and $Y(t)$ are independent random processes that are each stationary of order 2 and have first-order densities uniform on [0, 1].

(a) Find the first-order density for $Z(t)$.

(b) Is $Z(t)$ stationary of order two? Defend your answer.

6.4 Let the random process $Y(t)$ be defined as

$$Y(t) = X(t)\cos(\omega_0 t + \Theta)$$

where $X(t)$ is a wide sense stationary random process and Θ is a uniform random variable on $[0, 2\pi]$ and ω_0 is a nonrandom constant. Assume that $X(t)$ is independent of Θ for all t and that $X(t)$ is wide sense stationary with zero mean and known autocorrelation function $R_{XX}(\tau)$.

(a) Find $R_{YY}(t_1, t_2)$.

(b) Is $Y(t)$ a wide sense stationary process? Explain.

6.5 A random process $Y(t)$ is defined as

$$Y(t) = A \ \sin(X(t))$$

where $X(t)$ is the random binary transmission process defined in Chapter 4 and A is a Gaussian random variable, independent of the process $X(t)$, with zero mean and a known variance σ^2.

(a) Illustrate a couple realizations of the process $Y(t)$.

(b) Find the first-order density, $f(y; t)$, for $Y(t)$.

(c) Determine the autocorrelation function for $Y(t)$.

(d) What forms of stationarity does the process $Y(t)$ have?

(e) Repeat (a), (b), and (c) for $Y(t) = A \ \sin(\omega_0 t + X(t))$ where ω_0 is a fixed known value.

6.6 Let $X(t)$ be a Gaussian random process with first-order density function having a zero mean and known variance t. Define a new random process $Y(t)$ by $Y(t) = g(X(t))$ where $g(.)$ is as shown in Figure P6.6.

(a) Find the first-order density for $Y(t)$.

(b) Determine $E[Y(t)]$.

(c) Evaluate $E[X(t)Y(t)]$.

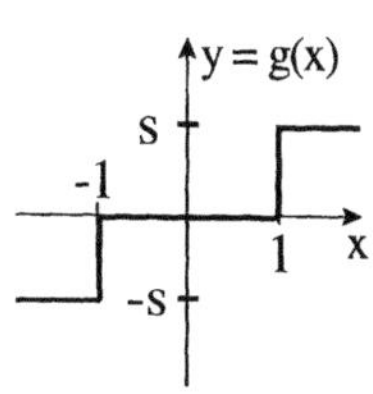

P6.6

6.7 Let $X(t)$ be the input to a nonlinear device with transfer function shown in Figure P6.7, representing a hard limiter with a dead zone. Suppose that $X(t)$

is a Gaussian random process with second-order density function $f(x_1, x_2)$ given by

$$f(x_1, x_2) = \frac{1}{2\pi\sigma^2\sqrt{1-\rho^2}} \exp \frac{-(x_1^2 - 2\rho x_1 x_2 + x_2^2)}{2\sigma^2(1-\rho^2)}$$

(a) Determine the first order density of the output process $Y(t)$.

(b) Find the output autocorrelation function $R_{YY}(t_1, t_2)$. Your answer may be left in integral form; however, describe the procedure for evaluating the integral.

(c) Is $Y(t)$ a wide sense stationary process? Why or why not?

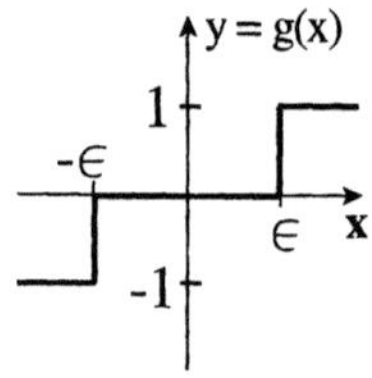

P6.7

6.8 Suppose that $X(t)$ is the binary transmission wave described in Chapter 4. Define a new random process $Y(t)$ as

$$Y(t) = e^{-\alpha X(t)}$$

(a) Find the first-order density $f(y\,;\,t)$ for $Y(t)$ the output process.

(b) Determine $E[Y(t)]$.

(c) Find the autocorrelation function $R_{YY}(t_1, t_2)$ of the output process.

(d) Is $Y(t)$ stationary in mean? wide sense stationary? strict sense stationary? Defend your answers.

6.9 Let $X(t)$, the random telegraph process, be the input to the instantaneous nonlinearity shown in Figure P6.9. $Y(t)$ is the output process.

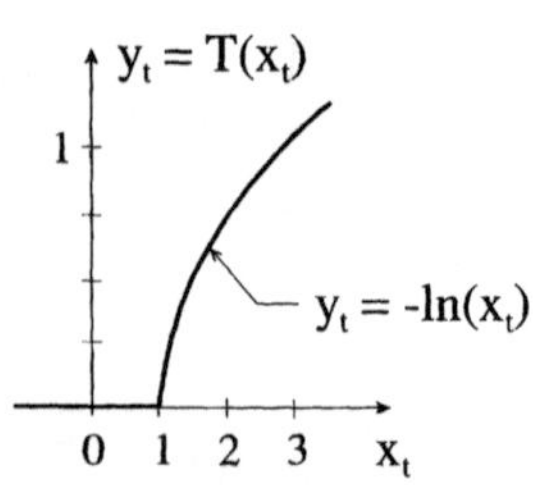

P6.9

(a) Find the first-order density function $f(y; t)$.
(b) Determine $E[Y(t)]$.
(c) Give formula for determining $R_{YY}(t_1, t_2)$.

6.10 Let $X(t)$ and $Y(t)$ be the input and output, respectively, of a nonlinear system defined by $Y(t) = g[X(t)]$, where $g[\cdot]$ is shown in Figure P6.10. Let $X(t)$ be a random process characterized by its first-order density $f(x; t)$ given by

$$f(x; t) = \frac{2}{|\cos(t)|} e^{-|x_t \cos(t)|}$$

(a) Find the mean of the input process.
(b) Describe two methods for finding the mean of the output process, and determine the mean of the output process using one of them.
(c) Determine $f(y; t)$ the first-order density of the output process.

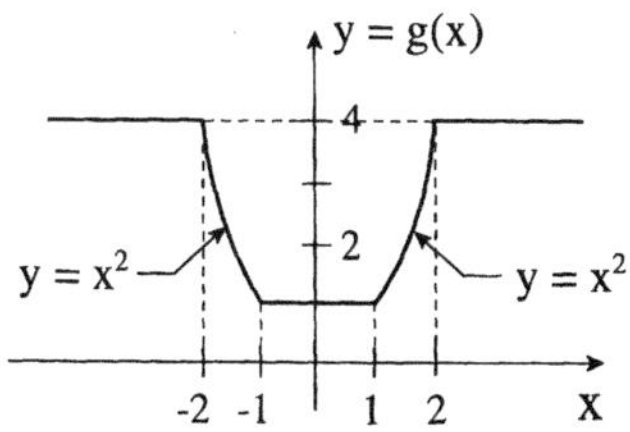

P6.10

6.11 Consider the nonlinearity defined by $y = g(x)$ which has a dead zone and saturation as shown in Figure P6.11. Let the input to this device be a Gaussian random process $X(t)$ with known mean $\eta_X(t) = 0$ and autocorrelation function $R_{XX}(\tau) = e^{-2|\tau|}$.
(a) Find the first-order density of $Y(t)$, the output process.
(b) Is the output process a Gaussian random process? Why or why not?

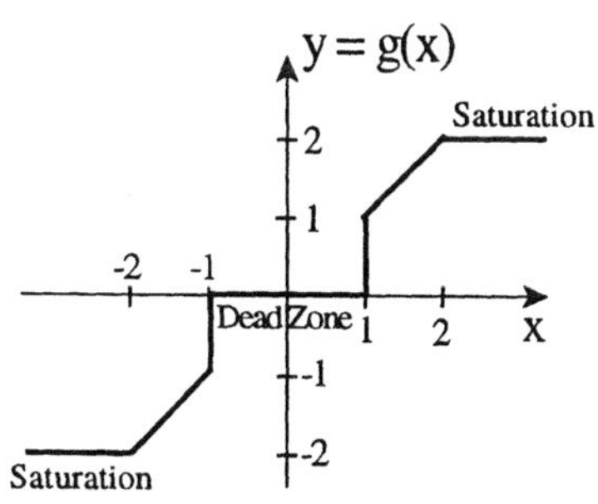

P6.11

6.12 A process $X(t)$ is a stationary real Gaussian random process with zero mean and autocorrelation function $R_{XX}(\tau) = e^{-|\tau|}$.

(a) Determine the first-order density for $X(t)$.

(b) Determine the second order density for the process.

(c) If $X(t)$ is the input to a symmetrical limiter shown in Figure P6.12, find the first order density of the output process $Y(t)$ and the mean $\eta_Y(t)$.

(d) Also find the autocorrelation function of the output process described in (c).

(e) Repeat (c) and (d) for the ideal limiter ($x_0 = 0$).

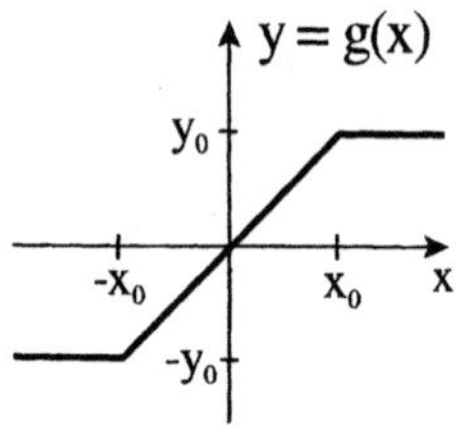

P6.12

6.13 Suppose that $X(t)$ is a Gaussian random process with $E[X(t)] = 0$ and $R_{XX}(\tau) = \cos(2\pi\tau)$. An instantaneous nonlinear system is defined by $Y(t) = g[X(t)]$ with $g(\cdot)$ as shown in Figure P6.13.

(a) Determine the first-order density of the input process $X(t)$.

(b) Find the first-order density for the output process.

(c) Is $Y(t)$ a Gaussian random process? Explain your answer.

(d) Determine $E[Y(t)]$.

(e) Find the second-order density $f(x_1, x_2; t_1, t_2)$, where $t_1 = 0$ and $t_2 = 0.5$.

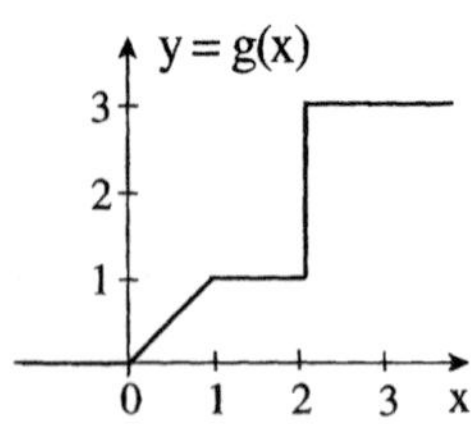

P6.13

6.14 Given that a random process $X(t)$ is a zero mean Gaussian wide sense stationary process with autocorrelation function $R_{XX}(\tau) = e^{-2|\tau|}$ for all τ. This process is passed through an ideal limiter with input–output relation shown in Figure P6.14. Let $Y(t)$ be the output process.

(a) Find the power spectral density for $X(t)$.

(b) Find the first-order density for $X(t)$.

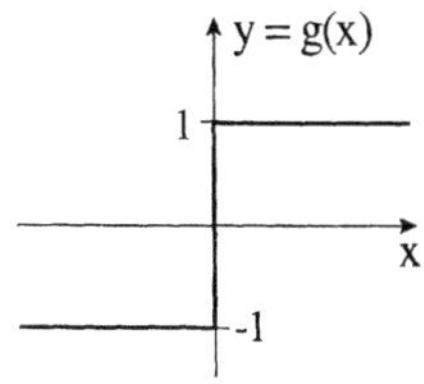

P6.14

(c) Calculate $E[Y(t)]$.

(d) Find the first-order density of the output process $Y(t)$.

(e) Find the output autocorrelation function $R_{YY}(t, u) = E[Y(t)Y(u)]$.

(f) Is the output process $Y(t)$ wide sense stationary? Why or why not?

(g) If $X(t)$ is wide sense stationary, calculate its power spectral density.

6.15 Let $X(t)$ and $Y(t)$ represent the input and output random process for the instantaneous nonlinearity $g(\cdot)$ shown in Figure P6.15. Let $X(t)$ be a Gaussian random process characterized by its mean $E[X(t)] = 0$ and autocorrelation function $R_{XX}(\tau)$ given by $R_{XX}(\tau) = \exp(-0.5|\tau|)$. Find:

(a) The power spectral density $S_{XX}(\omega)$ of the input process.

(b) The first-order density of the input process.

(c) The second-order density of the input process.

(d) The first-order density for the output process.

(e) The mean of the output process.

(f) The autocorrelation function for the output process.

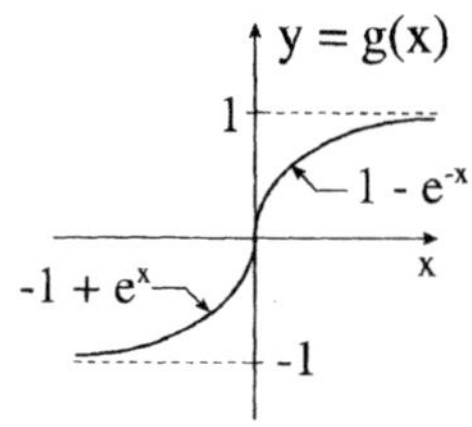

P6.15

6.16 A quantizer is described by the function $y = g(x)$ shown in Figure P6.16. Let $X(t)$, the input to the system, be a Gaussian random process with known mean and autocorrelation function as follows: $\eta_X(t) = 0$ and $R_{XX}(\tau) = d^2 e^{-\alpha|\tau|}$. Find:

(a) The first-order density $f(y_t)$ of the output process $Y(t)$.

(b) $E[Y(t)]$.

(c) The second-order density for the output process.

(d) The autocorrelation function for the output process.

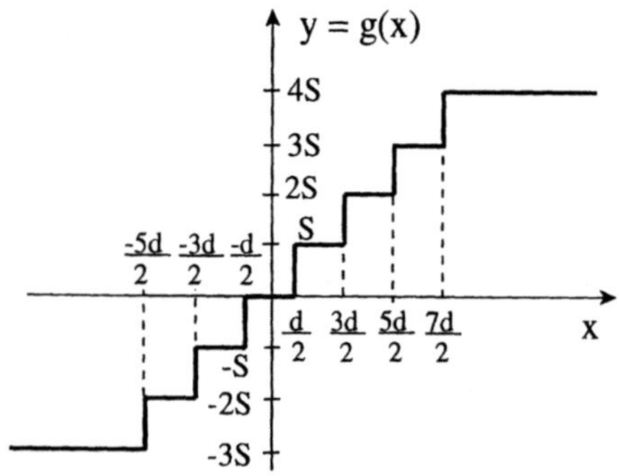

P6.16

(e) Is $Y(t)$ a wide sense stationary process? Explain.

(f) Find the cross-correlation function between input and output processes $R_{XY}(t_1, t_2)$.

(g) Define an error random process $Z(t) = X(t) - Y(t)$ and a signal-to-noise ratio $\text{SNR} = E[X^2(t)]/E[Z^2(t)]$. Determine SNR.

(h) Select d to maximize SNR for $\alpha = 2/d$.

6.17 $X(t)$ is the Wiener-Levy process with

$$f(x; t) = \frac{1}{\sqrt{2\pi\alpha t}} e^{-x^2/2\alpha t} \quad \text{and} \quad R_{XX}(t_1, t_2) = \begin{cases} \alpha t_2 & t_1 \geq t_2 \\ \alpha t_1 & t_1 \leq t_2 \end{cases}$$

Define an output process $Y(t)$ by $Y(t) = g[X(t)]$, where $g[\cdot]$ is the half-wave rectifier as shown in Figure P6.17.

(a) Find the first order density $f(y_t)$.

(b) Determine $E[Y(t)]$.

(c) Determine $R_{YY}(t_1, t_2)$ for all t_1 and t_2.

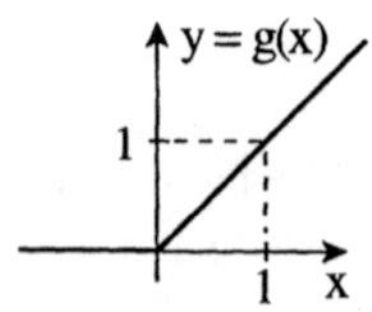

P6.17

6.18 Repeat Problem 6.17 for the full wave rectifier with nonlinearity shown in Figure P6.18.

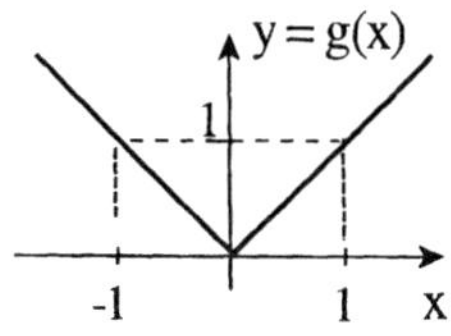

P6.18

6.19 A clutter return process $C(t)$ is defined as $C(t) = AS(t - \Lambda)$, where $S(t)$ is a wide sense stationary process with known mean η_X and autocorrelation function $R_{SS}(\tau)$. A and Λ are independent random variables, each independent of the process $S(t)$. Also given is that A is a uniform random variable on $[-0.5, 0.5]$ and Λ is a Gaussian random variable with zero mean and variance of σ^2.

(a) Find $E[C(t)]$.

(b) Determine $R_{CC}(t + \tau, t)$.

(c) Is $C(t)$ a wide sense stationary random process? Stationary of order one? Strict sense stationary? Explain your answers.

6.20 Define a process $X(t) = As(t - B)$, where $s(t)$ is a deterministic function defined by $s(t) = \mu(t) - \mu(t - 1)$ and A and B are independent random variables. A is a Gaussian random variable with zero mean and variance of 4, while B is a uniform random variable on $[0, 10]$.

(a) Illustrate a few realizations of the process.

(b) Find the mean of the process.

(c) Find the autocorrelation function for the process.

(d) Is the process Gaussian? Why or why not?

(e) Is the process wide sense stationary? Explain.

(f) Find the first-order density for the process.

6.21 Given $s(t)$ is a deterministic function in time, and B_1 and B_2 are independent random variables, each uniformly distributed on $[-1, 1]$. Suppose that A_1 and A_2 are independent Gaussian random variables with zero means and variances of one. Also assume that A_1 and A_2 are group independent of B_1 and B_2. Define a random process $X(t)$ as

$$X(t) = A_1 s(t - B_1) + A_2 s(t - B_2).$$

(a) Find the autocorrelation function for $X(t)$.

(b) Discuss the stationarity of $X(t)$.

6.22 A common problem in communications is multipath interference. If $X(t)$ is the transmitted signal (a random process), the received signal $Z(t)$ can be defined by

$$Z(t) = X(t) + A_1 X(t - t_1) + A_2 X(t - t_2)$$

The t_1 and t_2 are assumed deterministic and known (certainly an abstraction). The A_1 and A_2 are assumed to be independent uniform random variables on $[-0.5, 0.5]$ and $[-0.25, 0.25]$, respectively, and both independent of the process $X(t)$. Further assume that $X(t)$ is a wide sense stationary process with mean of zero and autocorrelation function $R_{XX}(\tau) = e^{|\tau|}$.

(a) Compute $E[Z(t)]$ and $R_{ZZ}(t, u)$.

(b) Is $Z(t)$ a wide sense stationary process? Explain.

(c) Determine the power spectral densities for $X(t)$ and $Z(t)$.

6.23 Let $G(t)$ be the square wave random process illustrated in Figure P6.23 where the phase Θ is a uniformly distributed random variable on $[0, 2\pi]$. Another wide sense stationary process $X(t)$ is characterized by its power spectral density $S_{XX}(\omega)$ as

$$S_{XX}(\omega) = \begin{cases} a^2, & |\omega| < \omega_0/4 \\ 0, & \text{otherwise} \end{cases}$$

where a is a known constant. A new process $Y(t)$ is defined as $Y(t) = G(t) \cdot X(t)$. Assume that Θ is independent of $X(t)$.

(a) Find the power spectral density $S_{YY}(\omega)$.

(b) What type of filter does one need to put $Y(t)$ through to obtain the AM modulated process $X(t)\cos(\omega_0 t)$?

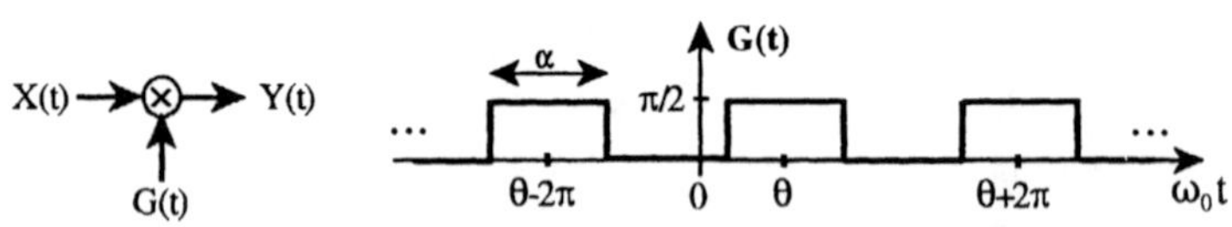

P6.23

6.24 Let $X(t)$ be the input random process to the system shown in Figure P6.24 with $Y(t)$ the output. The output of the half-wave linear rectifier device is given by $Z(t) = bX(t)\mu(X(t))$, and the ideal low pass filter has transfer function $H(j\omega)$ as

$$H(j\omega) = \begin{cases} 1, & |\omega| \leq \pi B \\ 0, & \text{otherwise} \end{cases}$$

Suppose that $X(t)$ is a Gaussian random process that has the following $S_{XX}(\omega)$:

$$S_{XX}(\omega) = \begin{cases} A, & w_c - \pi B < |\omega| < w_c + \pi B \\ 0, & \text{otherwise} \end{cases}$$

Find the power spectral density $S_{YY}(\omega)$ of the output process.

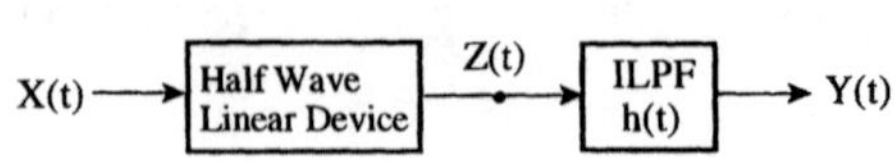

P6.24

6.25 Repeat Problem 6.24 if the half-wave device is replaced with an ideal limiter. Also evaluate the mean and variance of the limiter output process.

6.26 The input $X(t)$ to a square law device shown in Figure P6.26 is the sum of a signal $S(t)$ and $N(t)$, where the signal $S(t)$ is given by

$$S(t) = A(t)\cos(Bt + \Theta)$$

B is a known constant, Θ is a random variable uniformly distributed on $[0, 2\pi]$, and $A(t)$ is a wide sense stationary Gaussian random process, independent of Θ, with power spectral density $S_{AA}(\omega)$ as shown in the figure. The Gaussian noise process, $N(t)$, has a power spectral density $S_{NN}(\omega)$ as shown.

(a) Find the $S_{YY}(\omega)$.

(b) Find $S_{ZZ}(\omega)$ where $Z(t)$ is the output after $Y(t)$ passes through the low pass filter (LPF) which has the $H(j\omega)$ shown in the figure.

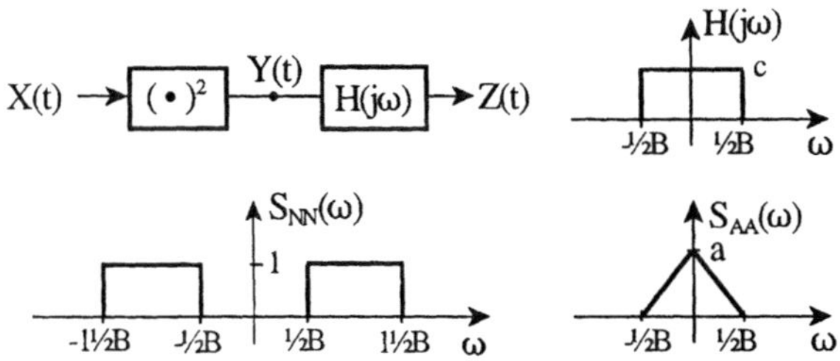

P6.26

6.27 One way of estimating the power spectral density for a process $X(t)$ at radian frequency ω_0 is shown in Figure P6.27*a*.

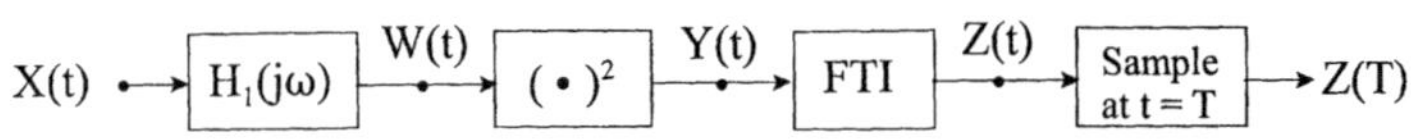

P6.27a

The first filter $H_1(j\omega)$ is a highly tuned narrow band filter about ω_0 defined by the frequency response shown in Figure P6.27*b* and the FTI called a finite time integrator is characterized by its impulse response $h(t) = \mu(t) - \mu(t - T)$. If $X(t)$ is a zero mean wide sense stationary Gaussian random process with autocorrelation function $R_{XX}(\tau) = e^{-2|\tau|}$.

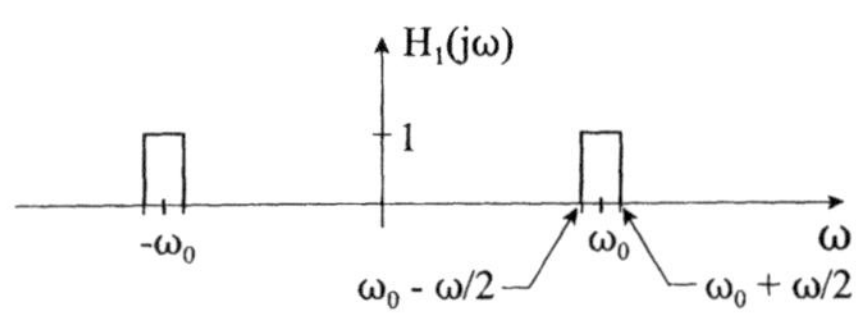

P6.27b

(a) Find $f(x_1, x_2; t_1, t_2)$, where $t_1 = 0$ and $t_2 = 2$.

(b) Determine $S_{XX}(\omega)$.

(c) Is $W(t)$ a wide sense stationary process? Why or why not?

(d) Is $W(t)$ a Gaussian process? Why or why not?

(e) Find $S_{WW}(\omega)$. Tell how you would find $R_{WW}(\tau)$.

(f) If $R_{WW}(\tau)$ is known, find $R_{YY}(\tau)$.

(g) Find $E[Z(T)]$ and $\sigma^2_{Z(\tau)}$.

(h) Determine $R_{WY}(\tau)$.

(i) Find $R_{XW}(\tau)$.

(j) How could we get the estimate of the entire spectrum using the procedure described above?

6.28 Prove the following theorem by using the characteristic function from Chapter 2. If X_1 and X_2 are zero mean jointly Gaussian random variables specified by their second-order moments $E[X_1^2]$, $E[X_2^2]$, and $E[X_1X_2]$, then the fourth-order moment $E[X_1^2X_2^2]$ can be written in terms of these second order moments as follows:

$$E[X_1^2X_2^2] = E[X_1^2]E[X_2^2] + 2E^2[X_1X_2]$$

6.29 For the following nonlinear systems prove that the time domain kernel is as shown in Eq. (6.31): **(a)** Cuber. **(b)** Cuber–linear filter. **(c)** Linear filter–cuber. **(d)** Linear filter–cuber–linear filter.

6.30 For the following nonlinear systems prove that the frequency domain kernel is as shown in Eq. (6.32): **(a)** Cuber. **(b)** Cuber–linear filter. **(c)** Linear filter–cuber. **(d)** Linear filter–cuber–linear filter.

6.31 Is there a nonlinear system that can produce a random process whose first-order density function is Gaussian but whose input process has a non-Gaussian first-order density? If so, describe, if not tell why it is not possible.

6.32 Is there a nonlinear system other than a degenerate linear whose output process has a Gaussian first-order density function when the input process has a Gaussian first-order density? Explain.

REFERENCES

1. Bendat, Julius S., *Nonlinear System Analysis and Identification from Random Data*, Wiley, 1990.
2. Papoulis, Athanasios, *Probability, Random Variables, and Stochastic Processes*, McGraw-Hill, 1965.
3. Lathi, B. P., *An introduction to Random Signals and Communication Theory*, International Textbook Company, 1968.

4. Davenport, Jr., William B., *Probability and Random Processes*, 1970.
5. Thomas, John B., *An Introduction to Statistical Communication Theory*, Wiley, 1969.
6. Nikias, Chrysostomas L., and Jerry M. Mendel, 'Signal processing with higher-order spectra', *IEEE Signal Processing Magazine*, July 1993, pp. 10–35.
7. Nikias, Chrysostomos L., and Mysor K. Raghuveer, 'Bispectrum estimation: A digital signal processing framework', *Proceedings of the IEEE*, Vol. 75, No. 7, July 1987, pp 870–890.
8. Mendel, Jerry M., 'Tutorial on higher-order statistics (spectra) in signal processing and system theory: Theoretical results and some applications', *Proceedings of the IEEE*, Vol. 79, No. 3, March 1991, pp. 278–305.
9. Bendat, Julius S., and Allan G. Piersol, *Engineering Applications of Correlation and Spectral Analysis*, Wiley, 1980.
10. Nikias, Chrysostomos L., and Petropuou, Athina P., *Higher-Order Spectrum Analysis: A Nonlinear Processing Framework*, Prentice-Hall, 1993.
11. Marmarelis, Panos Z., and Vasilis Z. Marmarelis, *Analysis of Physiological Systems—The White Noise Approach*, Plenum, 1978.
12. Zadeh, L. A., Optimum nonlinear filters, *Journal of Applied Physics*, Vol. 24, 1953, pp. 396–404.

7 Optimum Linear Filters: The Wiener Approach

7.1 OPTIMUM FILTER FORMULATION

In a general sense we think of a filter as a black box that takes information in via a continuous signal $x(t)$ or discrete time signal $x[n]$, processes it and produces a continuous output signal $y(t)$ or discretete time output $y[n]$. Conceptually this is illustrated in Figure 7.1. For example, the input signal might be the sum of a desired signal plus noise, and we would like to obtain as an output a cleaner version of the desired signal. In discussing optimum filters to perform a particular processing, we must be concerned with the following questions: What do we wish to have for the output? How are we going to process the input to obtain the desired output? What are we going to use as a performance measure for determining optimality? And what are the known statistical properties of the input and output processes? Each of these questions are discussed in turn.

7.1.1 What Is to Be Estimated?

Although the desired output can be almost anything this chapter will concentrate primarily on the following main problems: prediction, filtering, smoothing, and estimation of properties.

The process of **prediction** determines an estimate of a process $X(t)$ at time λ units in the future by processing $X(t)$ at only the current time and times in the past.

The process of **filtering** gives as an output an estimate of a given signal process from the input of another process that is somehow related to the signal process.

By **smoothing** or **interpolation** the output gives estimates of the signal process at time t, whereas the input is the signal at times other than t with some times greater than t.

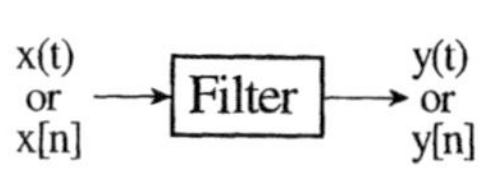

Figure 7.1 Filter input–output relationship.

Other problems fall in the category of estimating various **properties** of the input process, for example, the mean, variance, autocorrelation function, power spectral density, integral, or derivative of the input process.

7.1.2 Type of Processing

The general form of processing indicated in Figure 7.1 is a mapping from an input set containing possible signals to an output set of signals or properties. Specifically, the processing could be any combination of continuous or discrete, linear or nonlinear, time-varying or time-invariant, and causal or noncausal operations. For the most part the nonlinear processing problems are well beyond the scope of this text, and our work will concentrate on continuous and discrete linear systems that can be characterized in terms of impulse responses $h(t)$ and $h[n]$ for the time-invariant case, or an $h(t, \tau)$ and $h[m, n]$ for the time-varying case.

For the **linear continuous time-invariant case** the processing can be expressed in terms of the convolution integral as follows where $x(t)$ is the input and $y(t)$ the output:

$$y(t) = \int_{-\infty}^{\infty} h(t-\tau)x(\tau)\,d\tau = \int_{-\infty}^{\infty} h(\tau)x(t-\tau)\,d\tau \tag{7.1}$$

Similarly for the **discrete linear time-invariant case** the input–output relationship is expressed as

$$y[n] = \sum_{k=-\infty}^{\infty} h[k]x[n-k] = \sum_{k=-\infty}^{\infty} h[n-k]x[k] \tag{7.2}$$

Restricting the processing to be **causal**, equivalently realizable, $h(t)$ is zero for $t < 0$ for the continuous time case and $h[n]$ is zero for $n < 0$ for the discrete time case. Thus the output relationships above for continuous and discrete time can be modified to give

$$\begin{aligned} y(t) &= \int_{-\infty}^{t} h(t-\tau)x(\tau)\,d\tau = \int_{0}^{\infty} h(\tau)x(t-\tau)\,d\tau \\ y[n] &= \sum_{k=-\infty}^{n} h[n-k]x[k] = \sum_{k=0}^{\infty} h[k]x[n-k] \end{aligned} \tag{7.3}$$

The input–output relationship for the **continuous linear time-varying** case is described by $h(t, \tau)$, interpreted as the response of the system to an impulse at time τ, and is expressed as

$$y(t) = \int_{-\infty}^{\infty} h(t, \tau)x(\tau)\,d\tau \tag{7.4}$$

For the $h(t, \tau)$ to represent a causal system, $h(t, \tau)$ must be zero for all $t < \tau$.

Similarly the input–output relationship for the **linear discrete time-varying** case can be described in terms of its impulse response $h[m, n]$ in the following manner:

$$y[n] = \sum_{n=-\infty}^{\infty} h[m, n]x[n] \tag{7.5}$$

7.1.3 Performance Measure

The performance measure, or "yardstick", by which we judge our optimality can take many forms. Some of the most useful are forms described by the expected value of a function $g(.)$ of the error defined as the difference between what we want and what we get as a result of the processing. This can be expressed as

$$J = E[g(\text{error})] \quad \text{where}$$
$$\text{error} = \begin{pmatrix} \text{true} \\ \text{value} \end{pmatrix} - \begin{pmatrix} \text{estimated} \\ \text{value} \end{pmatrix} \tag{7.6}$$

If $g(x)$ is selected as x^2, then the performance measure J is called the **mean squared error**. As shown in Chapter 3, the mean squared error performance measure is mathematically tractable for estimating one random variable in terms of other random variables. For optimal linear estimates the solutions were obtained in terms of various second-order moments, while the nonlinear solutions required knowing or determining the conditional probability density functions. Other common choices for $g(x)$ include the absolute error and the uniform error which are shown along with the squared error in Figure 7.2. The mean squared error usually requires less statistical information and is more mathematically tractable than the others, and thus our study in this chapter will concentrate on using the mean squared error performance criterion. The other two measures of absolute and uniform error usually require much more statistical information than just the second-order characteristics as has been shown in Chapter 3.

7.1.4 Information Specified

Sufficient statistical information must be given to solve the optimum filter problem. It will be shown that for linear filtering that second-order characterizations of the input and output processes, including means, autocorrelation functions, and cross-

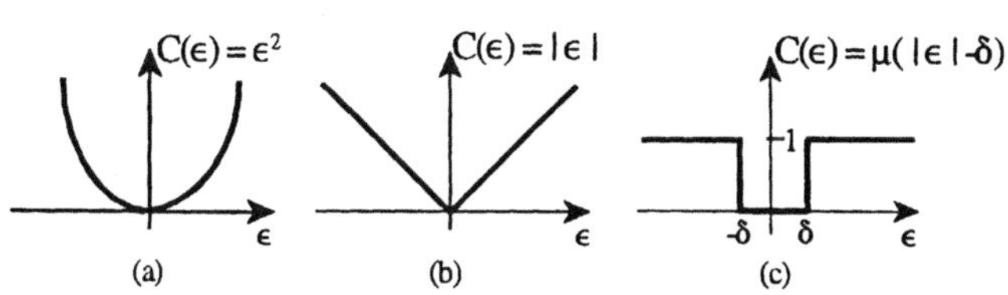

Figure 7.2 Common performance measures: (a) squared error, (b) absolute error, and (c) uniform error.

correlation functions, are sufficient to solve for the minimum mean squared error estimates. Other characterizations that might be given include various mth-order joint density functions. For obtaining optimum nonlinear filters, more information is usually required, and in many situations a total characterization is needed.

7.2 BASIC PROBLEMS

Examples are now presented for the basic types of problems, including prediction, filtering and interpolation. In these problems care will be used to emphasize answering the following four main questions:

(1) What are we going to estimate?
(2) How are we going to estimate?
(3) What performance criterion is used?
(4) What statistical information is given or required?

The basic problems presented here will be solved using an extension of the orthogonality principle described in Chapter 2. The extension is required since the orthogonality principle given earlier involves only a finite number of data points. We will be working with continuous time measurements, signals, and systems that have values for an infinite number of times or even a continuum of time.

7.2.1 Prediction of a Random Process

A wide sense stationary random process $X(t)$ is partially characterized with mean zero and autocorrelation function given by $R_{XX}(\tau)$. The problem is to predict what the random process will be at time $t_N + \lambda$, given the random variables $X(t_1), X(t_2), \ldots, X(t_N)$, where $t_1 < t_2 < \cdots < t_N$. The form of the processing selected is a linear function of all above random variables and thus $\hat{X}(t_N + \lambda)$, the estimator of $X(t_N + \lambda)$, is given by

$$\hat{X}(t_N + \lambda) = \sum_{i=1}^{N} a_i X(t_i) \tag{7.7}$$

The performance measure J is selected as the mean squared error, which is given by

$$J = E[(X(t_N + \lambda) - \hat{X}(t_N + \lambda))^2] \tag{7.8}$$

The problem then is to select the a_i in Eq. (7.7) such that J given in (7.8) is minimized.

SOLUTION

The conditions for using the orthogonality principle are satisfied, so the error is orthogonal to the data. The optimum a_i must satisfy

$$E\left[\left(X(t_N+\lambda)-\sum_{i=1}^{N} a_i X(t_i)\right)X(t_k)\right]=0, \qquad k=1,2,\ldots,N \tag{7.9}$$

Multiplying through, taking the expected value using the wide sense stationary property of $X(t)$, and rearranging, the a_i are found to be the solution of the following set of linear equations:

$$R_{XX}(t_N+\lambda-t_k)=\sum_{i=1}^{N} a_i R_{XX}(t_i-t_k), \qquad k=1,2,\ldots,N \tag{7.10}$$

By the other part of the orthogonality principle, the minimum J or minimum mean squared error can be written as follows

$$\begin{aligned} J_{\min} \triangleq e_{\text{mmse}} &= E\left[\left(X(t_N+\lambda)-\sum_{i=1}^{N} a_i X(t_i)\right)X(t_N+\lambda)\right] \\ &= R_{XX}(0)-\sum_{i=1}^{N} a_i R_{XX}(t_N+\lambda-t_i) \end{aligned} \tag{7.11}$$

From this equation for the minimum mean squared error, we see that if $R_{XX}(t_N+\lambda-t_i)$ is zero for any t_i, then the reduction of the error for using that term is zero. Thus if $X(t_i)$ is orthogonal to $X(t_N+\lambda)$, the error will not be reduced.

7.2.2 Filtering out Noise

A random process $X(t)$ is composed of the sum of a zero mean signal process $S(t)$ and a zero mean noise process $N(t)$. The characterizations for the signal and noise processes are specified by known autocorrelation functions $R_{SS}(t, u)$ and $R_{NN}(t, u)$ and cross-corrrelation function $R_{SN}(t, u)$. We desire an instantaneous linear time-varying amplifier to estimate $S(t)$ from $X(t)$ at each time t, that is, we do not use past or future values of $X(t)$, only the current value. Thus the form of the estimator is given by

$$\hat{S}(t)=a(t)X(t) \tag{7.12}$$

The performance measure selected for this problem is the mean squared error, which is given by

$$J(t)=E[(S(t)-a(t)X(t))^2] \tag{7.13}$$

It is desired to find $a(t)$ such that $J(t)$ is minimized for each t.

SOLUTION

Again the solution can be obtained by using the orthogonality principle:

$$E[(S(t) - a(t)X(t))X(t)] = 0 \qquad \text{for all } t \tag{7.14}$$

After taking the expected value and solving for $a(t)$, the optimum solution is given by

$$a(t) = \frac{R_{SX}(t, t)}{R_{XX}(t, t)} \tag{7.15}$$

where $R_{SX}(t, t)$ and $R_{XX}(t, t)$ are cross-correlation and correlation functions described by $R_{SX}(t, t) = E[S(t)X(t)]$ and $R_{XX}(t, t) = E[X(t)X(t)]$. The minimum J for this problem is the minimum mean squared error given by

$$\begin{aligned} J_{\min}(t) &= E[(S(t) - a(t)X(t))\, S(t)] \\ &= R_{SS}(t, t) - a(t)R_{XS}(t, t) \end{aligned} \tag{7.16}$$

Special Case 1. If the $S(t)$ and $N(t)$ are **uncorrelated** with **zero means**, the cross-correlation function $R_{SN}(t, u) = 0$ for all t and u. If also $S(t)$ and $N(t)$ are **jointly wide sense stationary** processes, then the optimum gain $a(t)$ is independent of time as seen from (7.15), and the minimum mean squared error will also not be a function of time t. For this case the gain and error can be determined from (7.15) and (7.16), with a little work, to be

$$\begin{aligned} a(t) &= \frac{R_{SS}(t, t)}{R_{SS}(t, t) + R_{NN}(t, t)} = \frac{R_{SS}(0)}{R_{SS}(0) + R_{NN}(0)} \\ J_{\min}(t) &= (1 - a(t))R_{SS}(t, t) = \frac{R_{NN}(0)R_{SS}(0)}{R_{SS}(0) + R_{NN}(0)} \end{aligned} \tag{7.17}$$

Special Case 2. If $S(t)$ and $N(t)$ have **equal power** ($R_{SS}(0) = R_{NN}(0)$), than (7.17) gives $a(t) = \frac{1}{2}$ for all t, and the minimum mean squared error is $J_{\min}(t) = \frac{1}{2}R_{SS}(0)$. This result is reasonable from an intuitive standpoint as it says that the estimate just uses half of the measurement, considering that the other half comes from the noise. If $R_{SS}(0) = 2R_{NN}(0)$, then we find that $a(t) = \frac{2}{3}$.

In many problems the estimator of the signal process is desired not just in terms of the instantaneous measurement but in terms of a continuous set of measurements. Such problems for estimating $S(t)$ using all $X(\alpha)$ for $\alpha \leq t$ will be presented in Section 7.3.

7.2.3 Interpolation for Random Processes

A zero mean, wide sense stationary random process $X(t)$ is characterized by its autocorrelation function $R_{XX}(\tau)$. If $X(t)$ is sampled at $t = 0$ and 1, the random variables $X(0)$ and $X(1)$ are obtained. It is desired to estimate $X(t)$ for t between

0 and 1, which corresponds to an **interpolation**. The form of the estimator will be linear and could vary with the time. The interpolation is performed as follows:

$$\hat{X}(t) = a_0(t)X(0) + a_1(t)X(1) \tag{7.18}$$

The mean squared error performance criterion J is specified, and it is desired to find $a_0(t)$ and $a_1(t)$ such that J is minimized.

SOLUTION

As before, since the problem is one of minimizing the mean squared error for a linear estimator, the solution is obtained by using the orthogonality principle. Thus $a_0(t)$ and $a_1(t)$ must satisfy two equations:

$$\begin{aligned} R_{XX}(t) &= a_0(t)R_{XX}(0) + a_1(t)R_{XX}(1) \\ R_{XX}(t-1) &= a_0(t)R_{XX}(1) + a_1(t)R_{XX}(0) \end{aligned} \tag{7.19}$$

After carrying out the algebra, we have for $a_0(t)$ and $a_1(t)$,

$$\begin{aligned} a_0(t) &= \frac{R_{XX}(0)R_{XX}(t) - R_{XX}(1)R_{XX}(t-1)}{R_{XX}^2(0) - R_{XX}^2(1)} \\ a_1(t) &= \frac{R_{XX}(0)R_{XX}(t-1) - R_{XX}(1)R_{XX}(t)}{R_{XX}^2(0) - R_{XX}^2(1)} \end{aligned} \tag{7.20}$$

Special Case. Consider the special case where $R_{XX}(\tau) = \exp(-\mid \tau \mid)$ and $t = \frac{1}{2}$, which is halfway between 0 and 1. It is easily shown that $a_0(\frac{1}{2})$ and $a_1(\frac{1}{2})$ are

$$a_0(\tfrac{1}{2}) = a_1(\tfrac{1}{2}) = \frac{e^{-\frac{1}{2}}}{1 + e^{-1}} = 0.4434094$$

The coefficients $a_0(\frac{1}{2})$ and $a_1(\frac{1}{2})$ are not equal to $\frac{1}{2}$, which is the value normally used for ordinary linear interpolation for deterministic functions. Thus we see that autocorrelation function of the process $X(t)$ plays a key role in giving different weights for interpolation of random processes.

7.2.4 Estimation of Properties of Random Processes

In the previous examples we have been concerned with estimates of the signal process at various times using given observation times. In certain problems we wish to estimate properties of the signal process rather than the signal process itself. For example, we might wish to estimate the mean, autocorrelation, or various first- and higher-order density functions. For these problems different assumptions are made concerning the statistical properties of the process, and the framework is slightly different. Such problems will not be considered in our presentation, however some problems of this type are relegated to homework.

7.3 THE WIENER FILTER

Given two zero mean random processes $G(t)$ and $X(t)$ that are somehow statistically related. The basic problem is that of trying to find the best, in the minimum mean squared error sense, estimator of $G(t)$ by performing a linear operation on an available $X(t)$ throughout the interval $a < t < b$. The $X(t)$ is referred to as the **data process** or observation process. For example, $X(t)$ might be the sum of a signal process $S(t)$, and a noise process $N(t)$ and $G(t)$ could equal $S(t)$ or some linear function of $S(t)$. The statistical relationship between $G(t)$ and $X(t)$ will in general be specified with the cross-correlation function $R_{XY}(\alpha, \beta)$ of the processes and the autocorrelation functions $R_{XX}(\alpha, \beta)$, and $R_{YY}(\alpha, \beta)$ for the processes, or by giving statistical information that would allow the determination of these autocorrelation and cross-correlation functions. The best, in the mean squared error sense, estimator of $G(t)$ knowing $X(t)$ throughout the time interval (a, b) is desired. Furthermore we will restrict the form of the estimator $\hat{G}(t)$ to be a linear function of the data and expressed in the following form:

$$\hat{G}(t) = \int_a^b h(t, \beta) X(\beta)\, d\beta \tag{7.21}$$

In the following sections the solutions for several special cases of $h(t, \tau)$, a, and b are investigated and the resulting optimum solutions are collectively referred to as **Weiner filters** in honor of Norbert Wiener [9]. His pioneering research along with the independent development by Kolmogorov [10] are considered major contributions to the field of random process estimation.

7.3.1 Wiener-Kolmogorov Filter (Finite Interval)

Given two zero mean random processes $G(t)$ and $X(t)$, where $G(t)$ and $X(t)$ are characterized by their cross-correlation and autocorrelation functions. Consider that $X(t)$ is the received process or **data** that is somehow statistically related to $G(t)$ the **desired** process. We want the best estimator of $G(t)$, knowing $X(t)$ throughout the time interval (a, b) where a and b are finite numbers and where "best" is in the sense of minimizing the mean squared error $e = E[(G(t) - \hat{G}(t))^2]$. The form of the estimator is as follows:

$$\hat{G}(t) = \int_a^b h(t, \beta) X(\beta)\, d\beta \tag{7.22}$$

That is, our estimator is a time-varying, perhaps unrealizable, linear filter with fixed finite memory. By extending the orthogonality principle to the noncountable case, we may state that $h(t, \eta)$ must be such that the error is orthogonal to random variables $X(\alpha)$ in the data as follows:

$$E[(G(t) - \hat{G}(t))X(\alpha)] = 0 \qquad \text{for all } a < \alpha < b \tag{7.23}$$

The corresponding minimum mean squared error $e_{\text{mmse}}(t)$ can also be determined from the orthogonal principle as

$$e_{\text{mmse}}(t) = E[(G(t) - \hat{G}(t))G(t)] \tag{7.24}$$

After substituting (7.22) into (7.23), multiplying out and evaluating the expected value operator, we have

$$\begin{aligned} &E[G(t)X(\alpha) - \int_a^b h(t,\beta)X(\beta)X(\alpha)\,d\beta] = 0 \qquad \text{for } a < \alpha < b \quad \text{or} \\ &R_{GX}(t,\alpha) - \int_a^b h(t,\beta)R_{XX}(\beta,\alpha)\,d\beta = 0 \qquad \text{for } a < \alpha < b \end{aligned} \tag{7.25}$$

As $R_{GX}(t,\alpha)$, and $R_{XX}(\beta,\alpha)$ are known, the integral equation above can be solved to obtain $h(t,\alpha)$. The existence of such a solution depends, in general, on the properties of $R_{GX}(t,\alpha)$ and $R_{XX}(\beta,\alpha)$. For most of our problems the solution exists, and the minimum mean squared error can be written by simplifying (7.24) using the optimal filter $h(t,\tau)$ to give

$$e_{\text{mmse}}(t) = R_{GG}(t,t) - \int_a^b h(t,\beta)R_{GX}(t,\beta)\,d\beta \tag{7.26}$$

7.3.2 Linear Time-Invariant Noncausal Filter

For the case where a **linear time-invariant noncausal filter** is required and the data set is assumed to be available for all time, the estimate can be written as the following convolution integral:

$$\hat{G}(t) = \int_{-\infty}^{\infty} h(t-\beta)X(\beta)\,d\beta = \int_{-\infty}^{\infty} h(\beta)X(t-\beta)\,d\beta \tag{7.27}$$

where $h(t)$ is the impulse response of the noncausal filter. Using Eq. (7.23), an application of the orthogonality principle, and the $\hat{G}(t)$ above, we can show, by repeating the steps leading to (7.25), that the optimal filter must satisfy the integral equation

$$\begin{aligned} &E[G(t)X(\alpha) - \int_{-\infty}^{\infty} h(t-\beta)X(\beta)X(\alpha)\,d\beta] = 0 \qquad \text{for } -\infty < \alpha < \infty \quad \text{or} \\ &R_{GX}(t,\alpha) - \int_{-\infty}^{\infty} h(t-\beta)R_{XX}(\alpha,\beta)\,d\beta = 0 \qquad \text{for } -\infty < \alpha < \infty \end{aligned} \tag{7.28}$$

Now the minimum mean squared error from Eq. (7.24) can be simplified to

$$e_{\text{mmse}}(t) = R_{GG}(t,t) - \int_{-\infty}^{\infty} h(t-\beta)R_{GX}(t,\beta)\,d\beta \tag{7.29}$$

The **optimal time-invariant noncausal filter**, denoted by its impulse response $h(t)$, is the solution of Eq. (7.28), and its performance is given by the mean squared error shown in Eq. (7.29).

Special Case (Jointly Wide Sense Stationary Processes). If $X(t)$ and $G(t)$ are jointly wide sense stationary, their autocorrelation and cross-correlation can be written as a function of time difference as

$$R_{XX}(\alpha, \beta) = R_{XX}(\alpha - \beta), \quad R_{GX}(\alpha, \beta) = R_{GX}(\alpha - \beta) \tag{7.30}$$

Thus (7.28) can be written as

$$R_{GX}(t - \alpha) - \int_{-\infty}^{\infty} h(t - \beta) R_{XX}(\alpha - \beta)\, d\beta = 0 \qquad \text{for } -\infty < \alpha < \infty \tag{7.31}$$

This equation can be put into a simpler form by making a couple of simple changes of variables. First let $t - \alpha = \tau$ in Eq. (7.31) to give

$$R_{GX}(\tau) - \int_{-\infty}^{\infty} h(\tau + \alpha - \beta) R_{XX}(\alpha - \beta)\, d\beta = 0 \qquad \text{for } -\infty < \tau < \infty \tag{7.32}$$

Then let $\theta = -(\alpha - \beta)$ and the integral equation, whose solution is the **optimal time-invariant noncausal filter for jointly wide sense stationary processes,** $h(t)$, is easily seen to be

$$R_{GX}(\tau) - \int_{-\infty}^{\infty} h(\tau - \theta) R_{XX}(\theta)\, d\theta = 0 \qquad \text{for } -\infty < \tau < \infty \tag{7.33}$$

The corresponding minimum mean squared error for the **optimal time-invariant noncausal filter for jointly wide sense stationary processes** is determined from (Eq. 7.29) as

$$\begin{aligned} e_{\text{mmse}}(t) &= R_{GG}(0) - \int_{-\infty}^{\infty} h(t - \beta) R_{GX}(t - \beta)\, d\beta \\ &= R_{GG}(0) - \int_{\infty}^{-\infty} h(\theta) R_{GX}(\theta)\, d\theta \end{aligned}$$

Thus the error for the joint wide sense stationary case is seen to be independent of the time of the estimates.

Actually the solution of (7.33) for the optimal filter transfer function $H(p)$ can be obtained in a straightforward manner by using the bilateral Laplace transform. First let

$$\Phi_{GX}(p) = \mathscr{L}_{\beta}[R_{GX}(\tau)] \quad \text{and} \quad \Phi_{XX}(p) = \mathscr{L}_{\beta}[R_{XX}(\tau)] \tag{7.35}$$

and then take the bilateral Laplace transform of both sides of Eq. (7.33) to find

$$\Phi_{GX}(p) - H(p)\Phi_{XX}(p) = 0 \tag{7.36}$$

The transfer function of the **optimal time-invariant noncausal filter, $H(p)$, for jointly wide sense stationary processes** is now obtained by solving Eq. (7.36) for $H(p)$ as

$$H(p) = \frac{\Phi_{GX}(p)}{\Phi_{XX}(p)} \tag{7.37}$$

It is seen that the transfer function of the optimal filter given in (7.37) is the ratio of the cross-spectral density for the desired process $G(t)$ and the measurement process

$X(t)$ and the spectral density of the measurement process in the bilateral Laplace domain. The impulse response $h(t)$ can be found by taking the inverse bilateral Laplace transform of $H(p)$.

EXAMPLE 7.1

Given a random process $X(t)$ that is the sum of a zero mean signal process $S(t)$ with autocorrelation function $R_{SS}(\tau) = (15/2)e^{-|\tau|}$ and a zero mean noise process $N(t)$ with autocorrelation function $R_{NN}(\tau) = (21/4)e^{-2|\tau|}$. Assume that $S(t)$ and $N(t)$ are orthogonal. Our objective is to obtain the best estimate of $S(t)$ for all time t by operating on $X(t)$ with a linear time-invariant noncausal filter. "Best" is in the sense of minimizing the mean squared error. Find $H(p)$, $h(t)$, the form of the estimator, and the corresponding minimum mean squared error.

SOLUTION

From the previous section, $H(p)$ was found to be

$$H(p) = \frac{\Phi_{GX}(p)}{\Phi_{XX}(p)}$$

where $\Phi_{GX}(p) = \mathscr{L}_\beta(R_{GX}(\tau))$ and $\Phi_{XX}(p) = \mathscr{L}_\beta(R_{XX}(\tau))$. Since we wish to estimate $S(t)$, we have $G(t) = S(t)$ and the observed data is $X(t) = S(t) + N(t)$.

Using the fact that the $S(t)$ and $N(t)$ are orthogonal the cross-correlation function $R_{GX}(\tau)$ is determined as

$$\begin{aligned} R_{GX}(\tau) &= E[G(t+\tau)X(t)] = E[S(t+\tau)(S(t)+N(t))] \\ &= R_{SS}(\tau) \end{aligned}$$

Similarly $R_{XX}(\tau)$ is found to be

$$\begin{aligned} R_{XX}(\tau) &= E[X(t+\tau)X(t)] = E[(S(t+\tau)+N(t+\tau))(S(t)+N(t))] \\ &= R_{SS}(\tau) + R_{NN}(\tau) \end{aligned}$$

The $\Phi_{GX}(p)$ and $\Phi_{XX}(p)$ are now easily found from the given autocorrelation and cross-correlation functions as

$$\Phi_{GX}(p) = \mathscr{L}[R_{GX}(\tau)] = \mathscr{L}[\tfrac{15}{2}e^{-|\tau|}] = \frac{15}{-p^2+1}$$

$$\begin{aligned} \Phi_{XX}(p) &= \mathscr{L}[R_{XX}(\tau)] = \mathscr{L}[\tfrac{15}{2}e^{-|\tau|} + \tfrac{21}{4}e^{-2|\tau|}] \\ &= \frac{36(-p^2+9/4)}{(-p^2+1)(-p^2+4)} \end{aligned}$$

Finally the transfer function, $H(p)$, for the optimal noncausal filter is determined from (7.37) as

$$H(p) = \frac{\Phi_{GX}(p)}{\Phi_{XX}(p)} = \frac{5(-p^2+4)}{12(-p^2+9/4)}$$

The impulsive response $h(t)$ of the optimal filter characterized by $H(p)$ can be found by taking the inverse bilateral Laplace transform of the equation above for $H(p)$ to give $h(t)$ as

$$h(t) = \mathscr{L}^{-1}[H(p)] = \tfrac{5}{12}\delta(t) + \tfrac{35}{144}e^{-3|\tau|/2}$$

Using the impulse response above and the linear form of processing, a convolution of $h(t)$ and $X(t)$, which is given in Eq. (7.27), we can write the $\hat{S}(t)$ as

$$\hat{S}(t) = \tfrac{5}{12}X(t) + \tfrac{35}{144}\int_{-\infty}^{\infty} e^{-3|\tau|/2}X(t-\tau)\,d\tau$$

The corresponding minimum mean squared error can be found from (7.34) as

$$\begin{aligned} e_m(t) &= R_{GG}(0) - \int_{-\infty}^{\infty} h(\theta)R_{GX}(\theta)\,d\theta = R_{SS}(0) - \int_{-\infty}^{\infty} h(\theta)R_{SS}(\theta)\,d\theta \\ &= \tfrac{15}{2} - \int_{-\infty}^{\infty} (\tfrac{5}{12}\delta(\theta) + \tfrac{35}{144}e^{-|\theta|3/2})\tfrac{15}{2}e^{-|\theta|}d\theta \end{aligned}$$ ■

7.3.3 Linear Time-Invariant Causal Filter

In the previous section an optimum nonrealizable (noncausal) linear filter was obtained. If processing is to take place in real time, it would be more realistic to obtain an **optimum causal filter** (output at anytime t depends on only the past and present values of the input). The linear time-invariant causal filter input–out relationship is given by the familiar convolution integral:

$$\begin{aligned} \hat{G}(t) &= \int_{-\infty}^{t} h(t-\beta)X(\beta)\,d\beta \quad \text{or} \\ &= \int_{0}^{\infty} h(\beta)X(t-\beta)\,d\beta \end{aligned} \tag{7.38}$$

Such an optimum causal filter is the main topic of this section.

Given a received signal process $X(t)$ and a desired process $G(t)$, where $X(t)$ and $G(t)$ are jointly wide sense stationary random processes and both have zero means. We desire the best linear estimate of $G(t)$, knowing $X(t)$ for all time in the range $(-\infty, t)$. This can be accomplished by passing $X(t)$ through a linear time invariant causal system.

The optimal linear causal filter with impulse response $h(t)$ can be found similarly to the procedure used in the previous section for the noncausal filter by using the orthogonality principle. We have that the error is orthogonal to the data $X(\alpha)$ on interval $(-\infty, t)$ as follows:

$$E[(G(t) - \int_0^\infty h(\beta)X(t-\beta)\,d\beta)X(\alpha)] = 0 \qquad \text{for } -\infty < \alpha \leq t \tag{7.39}$$

Multiplying out and taking the expected value using the joint wide sense stationary assumption gives

$$R_{GX}(t-\alpha) - \int_0^\infty h(\beta)R_{XX}(t-\beta-\alpha)\,d\beta = 0 \qquad \text{for } -\infty < \alpha \leq t \tag{7.40}$$

If we make a change of variables $t - \alpha = \tau$, the equation above can be written as

$$R_{GX}(\tau) - \int_0^\infty h(\beta)R_{XX}(\tau-\beta)\,d\beta = 0 \qquad \text{for } -\infty < \alpha \leq \tau + \alpha \tag{7.41}$$

Rewriting the interval of validity of (7.41) by subtracting α from the inequalities and changing the dummy variable in the integral, the impulse response, $h(t)$, of the **optimal causal time-invariant filter for estimation of $G(t)$ from the data $X(t)$ for jointly wide sense stationary zero mean processes** is seen to satisfy the integral equation

$$R_{GX}(\tau) - \int_0^\infty R_{XX}(\tau-\alpha)h(\alpha)\,d\alpha = 0 \qquad \text{for } \tau \geq 0 \tag{7.42}$$

The integral equation (7.42) for $h(t)$, the impulse response of the optimal filter, is called the **Wiener-Hopf integral equation**. The information required to solve this equation is $R_{GX}(\tau)$ and $R_{XX}(\tau)$. Because $R_{XX}(\tau)$ is a positive definite function, the equation is assured of having a solution.

In solving Eq. (7.42), we see that the bilateral Laplace transform (discussed in Appendix A) is both convenient and expedient. Because the integral equation is only defined for $\tau \geq 0$, the solution is not direct but is obtained in a roundabout way by defining a new function $q(\tau)$ for all τ as follows:

$$q(\tau) = R_{GX}(\tau) - \int_0^\infty R_{XX}(\tau-\alpha)h(\alpha)\,d\alpha \qquad \text{for all } \tau \tag{7.43}$$

From (7.42) we see that $q(\tau)$ is zero for $\tau > 0$, while it does have nonzero values for $\tau \leq 0$. Taking the bilateral Laplace transform of (7.43) yields

$$Q(p) = \Phi_{GX}(p) - \Phi_{XX}(p)H(p) \tag{7.44}$$

Since $R_{XX}(\tau)$ is an even function in τ, $\Phi_{XX}(p)$ can be broken up in the following way to facilitate the solution for $H(p)$:

$$\Phi_{XX}(p) = \Phi_{XX}^{+}(p)\Phi_{XX}^{-}(p) \tag{7.45}$$

where $\Phi_{XX}^{+}(p)$ contains only **poles and zeros in the left half-plane** (LHP) and $\Phi_{XX}^{-}(p)$ contains only **poles and zeros in the right half-plane** (RHP).

This process of breaking $\Phi_{XX}(p)$ into $\Phi^+_{XX}(p)$ and $\Phi^-_{XX}(p)$ is called **spectral factorization**, and it plays a primary role in obtaining the solution to the Wiener-Hopf integral equation. It allows us to rewrite (7.44) as

$$Q(p) = \Phi_{GX}(p) - \Phi^+_{XX}(p)\Phi^-_{XX}(p)H(p) \tag{7.46}$$

Dividing through by $\Phi^-_{XX}(p)$ yields

$$\frac{Q(p)}{\Phi^-_{XX}(p)} = \frac{\Phi_{GX}(p)}{\Phi^-_{XX}(p)} - \Phi^+_{XX}(p)H(p) \tag{7.47}$$

Upon rearranging, we have

$$H(p)\Phi^+_{XX}(p) = \frac{-Q(p)}{\Phi^-_{XX}(p)} + \frac{\Phi_{GX}(p)}{\Phi^-_{XX}(p)} \tag{7.48}$$

Taking the inverse bilateral Laplace transform of both sides yields

$$\mathscr{L}_\beta^{-1}[H(p)\Phi^+_{XX}(p)] = \mathscr{L}_\beta^{-1}\left[\frac{-Q(p)}{\Phi^-_{XX}(p)} + \frac{\Phi_{GX}(p)}{\Phi^-_{XX}(p)}\right] \tag{7.49}$$

Since $H(p)$ and $\Phi^+_{XX}(p)$ both have only LHP poles, the inverse of the left-hand side is zero for negative τ. As $q(\tau)$ is zero for positive τ, it has poles only in the right half-plane, and because $\Phi^-_{XX}(p)$ has all its zeros in the right half-plane, the term $Q(p)/\Phi^-_{XX}(p)$ has all its poles in the right half-plane and is thus zero for positive τ. Therefore taking the unilateral Laplace of both sides yields

$$H(p)\Phi^+_{XX}(p) = \mathscr{L}_u[\mathscr{L}_\beta^{-1}(\Phi_{GX}(p)/\Phi^-_{XX}(p))] \tag{7.50}$$

Finally Eq. (7.50) can be solved for $H(p)$, the transfer function for the **optimal causal time-invariant linear filter for jointly wide sense stationary processes**, to obtain

$$H(p) = \frac{\mathscr{L}_u[\mathscr{L}_\beta^{-1}(\Phi_{GX}(p)/\Phi^-_{XX}(p))]}{\Phi^+_{XX}(p)} \tag{7.51}$$

Special Case **($\Phi_{XX}(p)$ and $\Phi_{XX}(p)$ Rational Expressions in p).** The $H(p)$ in (7.51) is useful in many cases. However, a simpler form can be found for certain special cases. If $\Phi_{GX}(p)$ and $\Phi^-_{XX}(p)$ are **rational in p**, then $\Phi_{GX}(p)/\Phi^-_{XX}(p)$ can be expanded in a partial fraction expansion as follows:

$$\frac{\Phi_{GX}(p)}{\Phi^-_{XX}(p)} = \left\{\frac{\Phi_{GX}(p)}{\Phi^-_{XX}(p)}\right\}^+ + \left\{\frac{\Phi_{GX}(p)}{\Phi^-_{XX}(p)}\right\}^- \tag{7.52}$$

where

$$\left\{\frac{\Phi_{GX}(p)}{\Phi_{XX}^-(p)}\right\}^+ \quad \text{is the part of the partial fraction expansion contributed by the poles in LHP (the positive time part)} \tag{7.53}$$

$$\left\{\frac{\Phi_{GX}(p)}{\Phi_{XX}^-(p)}\right\}^- \quad \text{is the part of the partial fraction expansion contributed by the poles in RHP (the negative time part)} \tag{7.54}$$

In Eqs. (7.53) and (7.54) RHP and LHP are right and left half-planes, respectively. Using (7.52) in (7.51), where the negative time part goes away, we observe the transfer function $H(p)$ of the optimal filter for the case of **linear time-invariant causal with jointly wide sense stationary processes and rational $\Phi_{GX}(p)$ and $\Phi_{XX}^-(p)$ to be**

$$H(p) = \frac{\left\{\dfrac{\Phi_{GX}(p)}{\Phi_{XX}^-(p)}\right\}^+}{\Phi_{XX}^+(p)} \tag{7.55}$$

Thus the **causal Wiener filter for rational spectra** is seen to be composed of a **spectral factorization** and a **partial fraction expansion**, while for nonrational spectra a unilateral Laplace of a bilateral inverse Laplace transform is required.

The **minimum mean squared error**, or performance, for the **optimal time-invariant causal Wiener filter for not jointly wide sense stationary processes** can be determined from the other part of the orthogonality principle as follows:

$$\begin{aligned} e_{\text{mmse}} &= E[(G(t) - \int_{-\infty}^{t} X(\tau)h(t-\tau)\,d\tau)G(t)] \\ &= R_{GG}(t,t) - \int_{-\infty}^{t} R_{GX}(t,\tau)h(t-\tau)\,d\tau \end{aligned} \tag{7.56}$$

Special Case (Zero Mean, Jointly Wide Sense Stationary Processes). Assuming jointly wide sense stationary processes $X(t)$ and $G(t)$ and letting $t - \tau = \alpha$, we can simplify Eq. (7.56) to

$$e_{\text{mmse}} = R_{GG}(0) - \int_0^{\infty} R_{GX}(\alpha)h(\alpha)\,d\alpha \tag{7.57}$$

An example is now presented to illustrate the use of (7.55) and (7.57).

EXAMPLE 7.2 (CAUSAL WIENER FILTER—Rational Spectra)

We are given a random process $X(t)$ that is the sum of a zero mean signal process $S(t)$ with an autocorrelation function $R_{SS}(\tau) = (15/2)e^{-|\tau|}$ and a zero mean process $N(t)$ with an autocorrelation function $R_{NN}(\tau) = (21/4)e^{-2|\tau|}$. Assume that $S(t)$ and $N(t)$ are orthogonal. Our objective is to obtain the best estimator of $S(t)$ at time t by operating on $X(t)$ with a linear time-invariant causal filter with transfer function $H(p)$. "Best" will be in the sense of minimizing the mean squared error. Find

$H(p)$, $h(t)$, the form of the estimator, and the associated minimum mean squared error or performance of the estimator.

SOLUTION

From the previous section $H(p)$ can be found as

$$H(p) = \frac{\{\Phi_{GX}(p)/\Phi_{XX}^{-}(p)\}^{+}}{\Phi_{XX}^{+}(p)}$$

Associating what we wish to estimate $G(t)$ with $S(t)$ and what we observe $X(t)$ with $S(t) + N(t)$, we easily find the necessary cross- and autocorrelation functions to be

$$\begin{aligned} R_{GX}(\tau) &= E[G(t+\tau)X(t)] = E[S(t+\tau)(S(t)+N(t))] = R_{SS}(\tau) \\ R_{XX}(\tau) &= E[X(t+\tau)X(t)] = E[(S(t+\tau)+N(t+\tau))(S(t)+N(t))] \\ &= R_{SS}(\tau) + R_{NN}(\tau) \end{aligned}$$

Using the given $R_{SS}(\tau)$ and $R_{NN}(\tau)$ and taking the bilateral Laplace transforms, we have $\Phi_{GX}(p)$ and $\Phi_{XX}(p)$ as

$$\begin{aligned} \Phi_{GX}(p) &= \mathscr{L}[R_{GX}(\tau)] = \mathscr{L}[(\tfrac{15}{2})e^{-|\tau|}] = \frac{15}{-p^2+1} \\ \Phi_{XX}(p) &= \mathscr{L}[R_{XX}(\tau)] = \mathscr{L}[(\tfrac{15}{2})e^{-|\tau|} + (\tfrac{21}{4})e^{-2|\tau|}] \\ &= \frac{36(-p^2+\frac{9}{4})}{(-p^2+1)(-p^2+4)} \end{aligned}$$

First, the $\Phi_{XX}(p)$ from above is spectrally factored as

$$\Phi_{XX}(p) = \frac{6(-p+\frac{3}{2})}{(-p+1)(-p+2)} \cdot \frac{6(p+\frac{3}{2})}{(-p+1)(-p+2)}$$

The corresponding $\Phi_{XX}^{+}(p)$ and $\Phi_{XX}^{-}(p)$ are then given by

$$\Phi_{XX}^{+}(p) = \frac{6(p+3/2)}{(p+1)(p+2)} \quad \text{and} \quad \Phi_{XX}^{-}(p) = \frac{6(-p+3/2)}{(-p+1)(-p+2)}$$

Next, the numerator for the transfer function $H(p)$ of the optimal filter is determined as

$$\begin{aligned} \left\{\frac{\Phi_{GX}(p)}{\Phi_{XX}^{-}(p)}\right\}^{+} &= \left\{\frac{15/(-p^2+1)}{6(-p+\frac{3}{2})/(-p+2)(-p+1)}\right\}^{+} = \frac{15}{6}\left\{\frac{(-p+2)}{(-p+\frac{3}{2})(p+1)}\right\}^{+} \\ &= \frac{15}{6}\left\{\frac{A}{-p+\frac{3}{2}} + \frac{B}{p+1}\right\}^{+} = \frac{15}{6}\frac{B}{p+1} = \frac{15}{6}\frac{3}{\frac{5}{2}(p+1)} = \frac{3}{p+1} \end{aligned}$$

Finally, the optimal filter $H(p)$ can be written from Eq. (7.55) as

$$H(p) = \frac{3(p+1)(p+2)}{(p+1)6(p+\frac{3}{2})} = \frac{1}{2}\frac{p+2}{p+\frac{3}{2}}$$

From the transfer function, $H(p)$, the optimal filter impulsive response of the optimal causal filter is easily seen to be

$$h(t) = \mathscr{L}^{-1}[H(p)] = \tfrac{1}{2}\,\delta(t) + \tfrac{1}{4}\,e^{-3/2t}\mu(t)$$

This result, compared with the impulsive response of the optimum noncausal filter determined previously in Example 7.1, illustrates the fact that the optimum causal filter is not just the positive time truncated version of the optimum noncausal filter.

From (7.38) and the impulse response above, the form of the processing can be put in the time domain as

$$\hat{S}(t) = \tfrac{1}{2}X(t) + \tfrac{1}{4}\int_{-\infty}^{t} e^{-3(\tau-t)/2}X(\tau)\,d\tau$$

From Eq. (7.57) the minimum mean squared error for the optimal linear time-invariant causal filter is given by

$$e_{\text{mmse}} = R_{GG}(0) - \int_0^\infty R_{GX}(\alpha)h(\alpha)\,d\alpha$$

The first term is determined from the problem statement by setting $G(t) = S(t)$:

$$R_{GG}(0) = R_{SS}(\tau)\,|_{\tau=0} = \tfrac{15}{2}e^{-|\tau|}\,|_{\tau=0} = \tfrac{15}{2}$$

The second term is evaluated by using the $R_{GX}(\tau)$ and the optimal $h(t)$ previously determined as

$$\begin{aligned}\int_0^\infty R_{GX}(\alpha)h(\alpha)\,d\alpha &= \int_0^\infty R_{SS}(\alpha)h(\alpha)\,d\alpha \\ &= \int_0^\infty \tfrac{15}{2}e^{-|\tau|}(\tfrac{1}{2}\,\delta(\alpha) + \tfrac{1}{4}\,e^{-3/2})^{\alpha}\mu(\alpha)\,d\alpha \\ &= \tfrac{15}{4} + \int_0^\infty \tfrac{15}{8}e^{-5/2\alpha}\mu(\alpha)\,d\alpha = \tfrac{9}{2}\end{aligned}$$

Combining the two results above in the equation for e_{mms}, we determine the minimum means squared error as

$$e_{\text{mmse}} = \tfrac{15}{2} - \tfrac{9}{2} = 3$$

Thus the mean squared error has been reduced from 7.5 to 3 by use of the causal linear time-invariant filtering operation. ■

7.3.4 Pure Prediction

The problem of pure prediction is find the best estimator of $S(t+\lambda)$, $\lambda > 0$, having available $S(t)$ for the present and past times. Specifically, we wish to find the optimal realizable linear filter to give the minimum mean squared error estimate of $S(t+\lambda)$ from $S(t)$ up to time t. The $S(t)$ is assumed to be a **wide sense stationary process with zero mean** for our discussion. Therefore the $G(t)$ and $X(t)$ of Eq. (7.51) are recognized as

$$G(t) = S(t+\lambda), \quad X(t) = S(t) \tag{7.58}$$

From Eq. (7.51) we have, for a nonrational power spectrum of $X(t)$, that the optimal filter transfer function $H(p)$ is

$$H(p) = \mathscr{L}_u(\mathscr{L}_\beta^{-1}[\Phi_{GX}(p)/\Phi_{XX}^-(p)])/\Phi_{XX}^+(p) \tag{7.59}$$

The cross- and autocorrelation functions $R_{GX}(\tau)$ and $R_{XX}(\tau)$ necessary to obtain the $\Phi_{GX}(p)$ and $\Phi_{XX}(p)$ are determined as

$$\begin{aligned} R_{GX}(\tau) &= E[G(t+\tau)X(t)] = E[S(t+\lambda+\tau)S(t)] = R_{SS}(\tau+\lambda) \\ R_{XX}(\tau) &= E[X(t+\tau)X(t)] = R_{SS}(\tau) \end{aligned} \tag{7.60}$$

Therefore, for the cross- and autocorrelation functions above the corresponding transforms become

$$\begin{aligned} \Phi_{GX}(p) &= \mathscr{L}_\beta[R_{GX}(\tau)] = \mathscr{L}_\beta[R_{SS}(\tau+\lambda)] = e^{\lambda p}\Phi_{SS}(p) \\ \Phi_{XX}(p) &= \mathscr{L}_\beta[R_{SS}(\tau)] = \Phi_{SS}(p) = \Phi_{SS}^+(p)\Phi_{SS}^-(p) \end{aligned} \tag{7.61}$$

where the $\Phi_{SS}(p)$ has been spectrally factored, with $\Phi_{SS}^+(p)$ having all its poles and zeros in the left half-plane. The term in parentheses of (7.59) is determined as

$$\mathscr{L}_\beta^{-1}\left[\frac{\Phi_{GX}(p)}{\Phi_{XX}^-(p)}\right] = \mathscr{L}_\beta^{-1}\left[\frac{e^{\lambda p}\Phi_{SS}(p)}{\Phi_{SS}^-(p)}\right] = \mathscr{L}_\beta^{-1}\left[e^{\lambda p}\Phi_{SS}^+(p)\right] \tag{7.62}$$

From (7.59) the optimal filter transfer function $H(p)$ can then be written as

$$H(p) = \frac{\mathscr{L}_u\left(\mathscr{L}_\beta^{-1}\left[e^{\lambda p}\Phi_{SS}^+(p)\right]\right)}{\Phi_{XX}^+(p)} \tag{7.63}$$

Special Case 1 ($\Phi_{SS}(p)$ Rational Expression in p). Assuming that $\Phi_{SS}(p)$ is rational with n unique poles p_i, we can write $\Phi_{SS}^+(p)$ as

$$\Phi_{SS}^+(p) = \sum_{i=1}^{n} \frac{a_i}{p-p_i} \tag{7.64}$$

If the poles are not unique, then the partial fraction expansion will necessarily involve coefficients for powers of $1/(p-p_i)$. Isolating the case of unique poles and substituting (7.64) into the term in parentheses of (7.63) gives

$$\mathscr{L}_\beta^{-1}\left[e^{\lambda p}\sum_{i=1}^{n}\frac{a_i}{p-p_i}\right] = \sum_{i=1}^{n} a_i e^{+p_i(t+\lambda)}\mu(t+\lambda) \tag{7.65}$$

The unilateral Laplace of (7.65) is obtained as follows:

$$\begin{aligned}\mathscr{L}_u\left(\mathscr{L}_\beta^{-1}[e^{\lambda p}\Phi_{SS}(p)]\right) &= \int_0^{\infty}\left[\sum_{i=1}^{n} a_i e^{+p_i(t+\lambda)}\right] e^{-pt}\,dt \\ &= \sum_{i=1}^{n} a_i e^{+p_i\lambda}\frac{1}{p-p_i}\end{aligned} \tag{7.66}$$

After substituting (7.66) into (7.63), we determine the transfer function $H(p)$ for the **optimal linear time-invariant causal filter for pure prediction of a wide sense zero mean stationary process $S(t)$** with rational $\Phi_{SS}(p)$ to be

$$H(p) = \frac{\left[\sum_{i=1}^{n} a_i e^{p_i\lambda}/(p-p_i)\right]}{\Phi_{SS}^{+}(p)} \tag{7.67}$$

where the a_i are the coefficients in the partial fraction expansion of $\Phi_{XX}^{+}(p)$. These coefficients are determined as follows:

$$a_i = [(p-p_i)\Phi_{SS}^{+}(p)]_{p=p_i} \tag{7.68}$$

Special Case 2 **($\Phi_{SS}(p) = K/\prod_{i=1}^{n}(-p^2+c_i^2)$).** The formula (7.67) is good for pure prediction when the power spectral density is rational in p and there are no repeated poles. The result can be further simplified if the power spectral density is of the following form:

$$\Phi_{SS}(p) = \frac{K}{\prod_{i=1}^{n}(-p^2+c_i^2)} = \frac{\sqrt{K}}{\prod_{i=1}^{n}(p+c_i)}\,\frac{\sqrt{K}}{\prod_{i=1}^{n}(-p+c_i)} \tag{7.69}$$

With this $\Phi_{SS}(p)$, Eq. (7.67) can be written as

$$\begin{aligned}H(p) &= \left[\sum_{i=1}^{n}\frac{a_i e^{p_i\lambda}}{p+c_i}\right]\frac{\prod_{i=1}^{n}(p+c_i)}{\sqrt{K}} \\ &= \sum_{i=1}^{n} a_i d_i(p) e^{p_i\lambda}\end{aligned} \tag{7.70}$$

$$\text{where}\quad d_i(p) = \frac{\prod_{j=1, j\neq i}^{n}(p+c_j)}{\sqrt{K}} = \sum_{j=1}^{n-1} r_j p^j \tag{7.71}$$

Result (7.70) can be simplified to give the optimal transfer function $H(p)$ for the **pure prediction of a zero mean wide sense stationary process with the special rational power spectral density of Eq. (7.69)** as

$$H(p) = \sum_{i=0}^{n-1} K_i(\lambda)p^i \tag{7.72}$$

where the $K_i(\lambda)$ are functions of λ for a fixed λ and are a result of the partial fraction expansion and related simplification.

It is interesting to look at the impulse response of such a filter. Taking the inverse Laplace transform gives the impulse response of the optimal predictor as

$$h(t) = \sum_{i=0}^{n-1} K_i(\lambda)\delta^i(t)$$

$$\text{where} \quad \delta^i(t) = \frac{d^i\delta(t)}{d^i t} \tag{7.73}$$

The optimal filter implementation is obtained by convolving $h(t)$ with $S(t)$. Thus the **optimal pure predictor for a wide sense stationary zero mean process with $\Phi_{SS}(,p)$ given in (7.69)** can be written in terms of $S(t)$ and its derivatives as

$$\hat{S}(t+\lambda) = \sum_{i=0}^{n-1} K_i(\lambda)\frac{d^i S(t)}{d^i t} \tag{7.74}$$

■

Other types of prediction problems that can be formulated include (a) impure prediction with rational $\Phi_{SS}(p)$, (b) impure prediction for non rational $\Phi_{SS}(p)$, (c) pure prediction with nonrational $\Phi_{SS}(p)$ and (d) pure prediction with rational $\Phi_{SS}(p)$ but with multiple poles. An example for the first type is now presented while the other types appear in the homework.

EXAMPLE 7.3 (Pure Prediction Rational Spectra)

Suppose that we need to find the estimate of $S(t+\lambda)$ knowing $S(t)$ for the semi-infinite past to and including t. Assume that $S(t)$ is a zero mean wide sense stationary process with known power spectral density $S_{SS}(\omega)$ given as

$$\mathscr{F}[R_{SS}(\tau)] = \frac{1}{(\omega^2+1)(\omega^2+9)}$$

Find $H(p)$, the transfer function of a linear time-invariant filter, to provide the best (in the mean squared error sense) estimate of $S(t+\lambda)$. Also find the corresponding impulse response, and the form of the optimum estimator.

SOLUTION

We obtain $\Phi_{SS}(p)$ by replacing ω by p/j in the power spectral density above.

$$\Phi_{SS}(p) = \frac{1}{(-p^2+1)(-p^2+9)} = \frac{1}{(-p+1)(-p+3)} \cdot \frac{1}{(p+1)(p+3)}$$

The $\Phi_{SS}^{+}(p)$ can be expanded in a partial fraction expansion as

$$\Phi_{SS}^{+}(p) = \frac{a_1}{p+1} + \frac{a_2}{p+3}$$

where the coefficients of the partial fraction expansion are determined as

$$a_1 = \left.\frac{1}{p+3}\right|_{p=-1} = 1/2$$

$$a_2 = \left.\frac{1}{p+1}\right|_{p=-3} = -1/2$$

Now from (7.67) and (7.68) the transfer function $H(p)$ for the optimal filter is

$$H(p) = \left[\frac{\frac{1}{2}e^{-\lambda}}{p+1} + \frac{-\frac{1}{2}\,e^{-3\lambda}}{p+3}\right] \cdot \frac{(p+1)(p+3)}{1}$$

Multiplying through and simplifying gives

$$H(p) = K_1(\lambda)p + K_2(\lambda)$$

$$\text{where} \quad K_1(\lambda) = \tfrac{1}{2}e^{-\lambda} - \tfrac{1}{2}e^{-3\lambda} \quad \text{and} \quad K_2(\lambda) = \tfrac{3}{2}e^{-\lambda} - \tfrac{1}{2}e^{-3\lambda}$$

The impulsive response for the optimal system is obtained by taking the inverse of $H(p)$, and $\hat{S}(t+\lambda)$ is obtained by convolving $h(t)$ with the input process $S(t)$. That is,

$$h(t) = \mathscr{L}^{-1}(H(p)) = K_1(\lambda)\frac{d\delta(t)}{dt} + K_2(\lambda)\delta(t)$$

and

$$\hat{S}(t+\lambda) = h(t) * S(t) = K_1(\lambda)\frac{dS(t)}{dt} + K_2(\lambda)S(t)$$

From the result above we see that the estimator $\hat{S}(t+\lambda)$ of $S(t+\lambda)$ is a linear combination of $S(t)$ and the derivative of $S(t)$ as shown in Figure 7.3. This means that to get the best estimate of $S(t+\lambda)$, we need to know only the values of the process and its derivative at time t, and do not need to know the values of $S(t)$ for times less than the present time t if we had the derivative. For example, $\lambda = 1/2$, then $K_1(1/2) = 0.1917$ and $K_2(1/2) = 0.79823$, and the predicted value becomes $S(t+1/2) = 0.1917dS(t)/dt + 0.79823S(t)$. ■

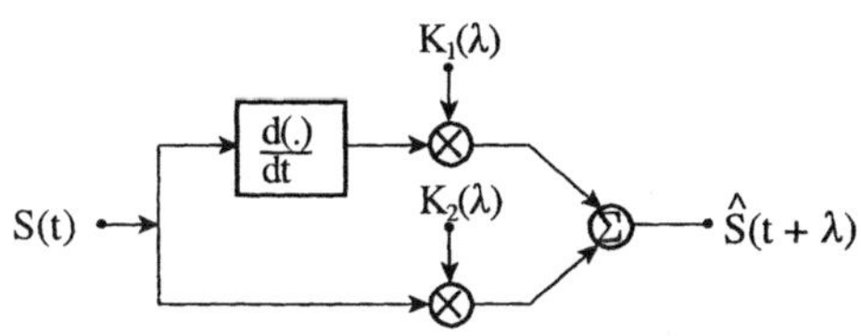

Figure 7.3 Pure predictor for Example 7.3.

7.4 THE DISCRETE WIENER FILTER

We have available a random process $X[n]$ and desire the best estimate of a related process $G[n]$. Let $\hat{G}[n]$ be the estimator of $G[n]$ and the result of passing $X[n]$ through a discrete time filter. The word "best" is meant in the sense of minimizing the mean squared error where the error at each time index is

$$e[n] = G[n] - \hat{G}[n] \tag{7.75}$$

Therefore we wish to find the discrete time filter that minimizes the mean squared error for every n given by

$$E[e^2[n]] = E[(G[n] - \hat{G}[n])^2] \tag{7.76}$$

If we let $h[n]$ represent the impulse response of a linear shift invariant noncausal discrete time filter, then we have the estimate of $G[n]$ given by

$$\hat{G}[n] = \sum_{k=-\infty}^{\infty} h[k]X[n-k] \tag{7.77}$$

While, if we let $h[n]$ represent the impulse response of a causal linear shift-invariant discrete time filter, the lower limit of the summation is zero and $\hat{G}[n]$ is given by

$$\hat{G}[n] = \sum_{k=0}^{\infty} h[k]X[n-k] \tag{7.78}$$

The solutions for the optimal causal and noncausal filters are now presented. They parallel the development given earlier for continuous time systems.

7.4.1 Linear Shift-Invariant Noncausal Filter

In assuming that the measurement or data process $X[n]$ and the desired process $G[n]$ both have zero means, the orthogonality principle states that the error is orthogonal to the data. The error can be written as

$$E[(G[n] - \hat{G}[n])X[j]] = 0, \quad \text{for } -\infty < j < \infty \tag{7.79}$$

The corresponding minimum mean squared error for the optimal estimator from the orthogonality principle is the expected value of the product of the error and the desired process to estimate. This can be written as

$$e_{\text{mmse}}[n] = E[(G[n] - \hat{G}[n])G[n]] \tag{7.80}$$

Substituting (7.77) into (7.79) and taking the expected value yields

$$R_{GX}[n,j] - \sum_{k=-\infty}^{\infty} h[k]R_{XX}[n-k,j] = 0, \quad -\infty < j < \infty \tag{7.81}$$

Thus the impulse response $h[k]$ for the **optimal linear time-invariant filter for estimating a zero mean process $G(t)$ from a zero mean process $X(t)$** is the solution of (7.81) for a given $R_{GX}[m,n]$ and $R_{XX}[m,n]$

Special Case (Jointly Wide Sense Stationary Processes). If we assume that $G[n]$ and $X[n]$ are **jointly wide sense stationary** random processes with known autocorrelation and cross-correlation functions, the solution of (7.81) can be obtained using the Z-transform. This is analogous to using the Laplace transform for the continuous time case. Using the wide sense stationary properties (7.81) can be simplified to

$$R_{GX}[n-j] - \sum_{k=-\infty}^{\infty} h[k]R_{XX}[n-k-j] = 0, \quad -\infty < j < \infty \tag{7.82}$$

For convenience, let $n - j = m$, to give

$$R_{GX}[m] - \sum_{k=-\infty}^{\infty} h[k]R_{XX}[m-k] = 0, \quad -\infty < m < \infty \tag{7.83}$$

The system function, $H(z)$, for the impulse response $h[n]$ can be solved for by taking the Z-transform of (7.83) and rearranging:

$$H(z) = \frac{\psi_{GX}(z)}{\psi_{XX}(z)} \tag{7.84}$$

where $\psi_{GX}(z) = Z(R_{GX}[m])$ and $\psi_{XX}(z) = Z(R_{XX}[m])$. This $H(z)$ is is the optimal system function for a **linear time-invariant noncausal filter** for estimating one process $G[n]$ in terms of another $X[n]$ when $G[n]$ and $X[n]$ are jointly wide sense stationary.

The result (7.84) for the optimum discrete time shift-invariant linear noncausal filter is the counterpart to the results shown in Eq. (7.37) for the continuous time solution. By simplifying (7.80), we can write the corresponding minimum mean squared error for this optimal filter $H(z)$ using the jointly wide sense stationary properties of $G[n]$ and $X[n]$:

$$e_{\text{mmse}}[n] = R_{GG}[0] - \sum_{k=-\infty}^{\infty} h[k]R_{GX}[k] \tag{7.85}$$

The minimum mean squared error is seen to be independent of the value of n. If the processes are **not jointly wide sense stationary** $e_{\text{mmse}}[n]$ can be found from (7.80) to be

$$e_{\text{mmse}}[n] = R_{GG}[n, n] - \sum_{k=-\infty}^{\infty} h[k]R_{GX}[n, n-k] \tag{7.86}$$

7.4.2 Linear Shift-Invariant Causal Filter

Again, assuming $X[n]$ and $G[n]$ both have zero means, and that searching for a **linear shift-invariant causal filter** characterized by (7.78), we are able to write, from the orthogonality principle, that the error is orthogonal to the data, yielding

$$E[(G[n] - \hat{G}[n])X[j]] = 0, \quad \text{for } 0 \leq j \leq \infty \tag{7.87}$$

Substituting (7.78) into (7.87), taking expected values, and simplifying yields the following equation of which $h[n]$ is the solution:

$$R_{GX}[n,j] - \sum_{k=0}^{\infty} h[k]R_{XX}[n-k,j] = 0, \quad 0 \le j < \infty \tag{7.88}$$

The minimum mean squared error corresponding to the optimal $h[n]$ is the expected value of the product of the error and the process we are estimating:

$$e_{\text{mmse}}[n] = E[(G[n] - \hat{G}[n])G[n]] \tag{7.89}$$

This equation for the minimum mean squared error can be simplified further by substituting (7.77) for $\hat{G}[n]$ and taking the resulting expected values to yield

$$e_{\text{mmse}}[n] = R_{GG}[n,n] - \sum_{k=-\infty}^{\infty} h[k]R_{GX}[n,n-k] \tag{7.90}$$

Special Case 1 (Jointly Wide Sense Stationary Process). Assuming that $G[n]$ and $X[n]$ are jointly wide sense stationary processes, we have from Eq. (7.88) the following:

$$R_{GX}[n-j] - \sum_{k=0}^{\infty} h[k]R_{XX}[n-k-j] = 0, \quad 0 \le j < \infty \tag{7.91}$$

By letting $n - j = m$, the equation above can be rewritten as

$$R_{GX}[m] - \sum_{k=0}^{\infty} h[k]R_{XX}[m-k] = 0, \quad m = 0, 1, \ldots \tag{7.92}$$

Equation (7.92) is the discrete equivalent of the Weiner-Hopf integral equation given previously for the continuous time case. Its solution gives the **optimal discrete-time causal linear time-invariant filter** for estimating $G[n]$ from $X[n]$. The development for the solution $h[k]$ follows the procedure given previously for the continuous time case but uses the bilateral Z-transform instead of the bilateral Laplace transform.

Define an artificial sequence $f[j]$ for all j by the following:

$$f[j] \triangleq R_{GX}[j] - \sum_{k=0}^{\infty} h[k]R_{XX}[j-k] \tag{7.93}$$

Taking the Z-transform of both sides of (7.93) yields

$$F(z) = \psi_{GX}(z) - H(z)\psi_{XX}(z) \tag{7.94}$$

where $\psi_{GX}(z) = Z_{\beta}(R_{GX}[j])$ and $\Psi_{XX}(z) = Z_{\beta}(R_{XX}[j])$. Since $h[n]$ is causal, the bilateral Z-transform of $h[n]$ is the same as the unilateral transform. Therefore

$$H(z) = Z_{\beta}(h[k]) = Z_{\mu}(h[k]). \tag{7.95}$$

From (7.93) we know that $f[j]$ must equal 0 for $j \ge 0$, though it may have nonzero values for $j \le 0$. To facilitate the solution, $\psi_{XX}(z)$ is factored into two terms:

$$\psi_{XX}(z) = \psi_{XX}^{+}(z)\psi_{XX}^{-}(z) \tag{7.96}$$

Here $\psi_{XX}^{+}(z)$ contains **all the poles and zeros** of $\psi_{XX}(z)$ **inside** the unit circle and $\psi_{XX}^{-}(z)$ contains **all the poles and zeros of** $\psi_{XX}(z)$ **outside** that unit circle.

This process is again called **spectral factorization** and it corresponds to the procedure for the continuous time case. As an intermediate step (7.94) is divided by $\psi_{XX}^{-}(z)$ to give

$$\frac{F(z)}{\psi_{XX}^{-}(z)} = -H(z)\psi_{XX}^{+}(z) + \frac{\psi_{GX}(z)}{\psi_{XX}^{-}(z)} \tag{7.97}$$

Now taking the inverse bilateral Z-transform of both sides of Eq. (7.97) gives

$$Z_{\beta}^{-1}[F(z)/\psi_{XX}^{-}(z)] = -Z_{\beta}^{-1}[H(z)\psi_{XX}^{+}(z)] + Z_{\beta}^{-1}[\psi_{GX}(z)/\psi_{XX}^{-}(z)] \tag{7.98}$$

Since $f[n]$ is zero for positive time indexes, the poles of $F(z)$ are outside the unit circle, and since $\psi_{XX}^{-}(z)$ has all its zeros outside the unit circle, their quotient has only poles outside the unit circle, the inverse of the first term of (7.98) is zero for positive n. Also, since $H(z)$ is stable and causal, it has poles only inside the unit circle, while $\psi_{XX}^{+}(z)$, by definition, has only poles inside the unit circle, therefore the second term in (7.98) represents a positive time signal. In light of these results the unilateral Z-transform of the Eq. (7.98) gives

$$H(z)\psi_{XX}^{+}(z) = Z_u[Z_{\beta}^{-1}(\psi_{GX}(z)/\psi_{XX}^{-}(z))] \tag{7.99}$$

This equation is easily solved for $H(z)$ to yield the **optimum linear causal shift-invariant filter *H*(*z*)** for jointly wide sense zero mean processes as

$$H(z) = \frac{Z_u[Z_{\beta}^{-1}(\psi_{GX}(z)/\psi_{XX}^{-}(z))]}{\psi_{XX}^{+}(z)} \tag{7.100}$$

Equation (7.100) is similar to the corresponding one for the continuous time case, Eq. (7.51), except that the Z-transform is used instead of the Laplace transform.

***Special Case 2* (*Rational Spectra*).** If $\psi_{GX}(z)$ and $\psi_{XX}(z)$ are rational functions in z, then the optimal causal filter with transfer function $H(z)$ can be simplified by paralleling the development given for the continuous time case to

$$H(z) = \left\{\frac{\psi_{GX}(z)}{\psi_{XX}^{-}(z)}\right\}^{+} \Big/ \psi_{XX}^{+}(z) \tag{7.101}$$

where $\{\alpha(z)\}^{+}$ is the part of the partial fraction expansion of $\alpha\{z\}$ that corresponds to positive time, in other words, the Z-transform of the **positive time** part.

An example of the use of Eq. (7.101) is now presented.

EXAMPLE 7.4 (Causal Discrete Wiener Filter)

Suppose that the received process $X[n]$ is obtained by sampling the continuous signal $X(t)$ at times $t = nT$, where $X(t)$ is composed of the sum of a signal process $S(t)$ and a noise process $N(t)$. It is desired to obtain an estimate of $S[n]$ at times n

where $S[n]$ equals $S(t)$ evaluated at times $t = nT$. The autocorrelation functions for $S(t)$ and $N(t)$ are given as $R_{SS}(\tau) = \frac{1}{2}e^{-|\tau|}$ and $R_{NN}(\tau) = \frac{1}{4}e^{-2|\tau|}$, and the processes $S(t)$ and $N(t)$ are assumed to be orthogonal. Let T, the time between samples, equal 0.69314718. Find the optimum realizable digital filter characterized by $H(z)$ that gives us the best estimate of $S[n]$, where "best" is in minimum mean squared error sense.

SOLUTION

Since the spectra are rational the conditions are satisfied to use Eq. (7.101) where $G[n] = S[n]$ is the signal to be estimated and $X[n] = S[n] + N[n]$ is the data or measurement process. The first step is to find the respective autocorrelation and cross-correlation functions and then to take the Z-transform. We have

$$\begin{aligned} R_{GX}[k] &= E[G[j+k]X[j]] = E[S[j+k](S[j] + N[j])] \\ &= R_{SS}[k] \end{aligned}$$

The last step in the equation above is possible because of the assumption that $S[n]$ and the noise process $N[n]$ are orthogonal, that is $E[S[k]N[j]]$ is zero for all j and k. This fact is also used to obtain the autocorrelation function of the measurement or data process as follows:

$$\begin{aligned} R_{XX}[k] &= E(S[j+k] + N[j+k])(S[j] + N[j]) \\ &= R_{SS}[k] + R_{NN}[k] \end{aligned}$$

Then the Z-transform of the cross-correlation and autocorrelation functions can be found as

$$\begin{aligned} \psi_{GX}(z) &= Z(R_{GX}[k]) = Z(R_{SS}(kT)) = Z(\tfrac{1}{2}e^{-|kT|}) \\ &= \frac{1}{2}\left(\frac{z}{z - e^{-T}} - \frac{z}{z - e^{T}}\right) = \frac{-0.75z}{(z-0.5)(z-2)} \end{aligned}$$

and

$$\begin{aligned} \psi_{XX}(z) &= \tfrac{1}{2}\left(\frac{z}{z - e^{-T}} - \frac{z}{z - e^{T}}\right) + \tfrac{1}{4}\left(\frac{z}{z - e^{-2T}} - \frac{z}{z - e^{2T}}\right) \\ &= \frac{-0.75z}{(z-\frac{1}{2})(z-2)} - \frac{0.9375z}{(z-\frac{1}{4})(z-4)} = \frac{-1.6875z(z-0.34044)(z-2.93733)}{(z-\frac{1}{2})(z-2)(z-\frac{1}{4})(z-4)} \\ &= \frac{0.787143(z-0.34044)z}{(z-\frac{1}{4})(z-\frac{1}{2})} \cdot \frac{0.787143(z^{-1}-0.34044)z^{-1}}{(z^{-1}-\frac{1}{4})(z^{-1}-\frac{1}{2})} \end{aligned}$$

From $\psi_{XX}(z)$ the spectral factorization procedure leads to

$$\psi_{xx}^{+}(z) = \frac{0.787143(z-0.34044)z}{(z-\frac{1}{4})(z-\frac{1}{2})}, \quad \psi_{xx}^{-}(z) = \frac{0.787143(z^{-1}-0.34044)z^{-1}}{(z^{-1}-\frac{1}{4})(z^{-1}-\frac{1}{2})}$$

We have the term in brackets of (7.100) as

$$Z_{\beta}^{-1}(\psi_{GX}(z)/\psi_{XX}^{-}(z)) = Z_{\beta}^{1}\left(\frac{-0.75z}{(z-\frac{1}{2})(z-2)}\,\frac{(z^{-1}-\frac{1}{4})(z^{-1}-\frac{1}{2})}{0.787143(z^{-1}-0.340445)z^{-1}}\right)$$

$$= Z_{\beta}^{-1}\left(\frac{0.34984z(z-4)}{(z-\frac{1}{2})(z-2.93733)}\right)$$

The quantity in the big parentheses above is then broken up into positive time and negative time portions as follows:

$$\frac{0.34984(z-4)z}{(z-\frac{1}{2})(z-2.93733)} = \frac{Az}{z-\frac{1}{2}} + \frac{Bz}{z-2.93733}$$

Therefore

$$Z_u\left[Z_{\beta}^{-1}\left(\frac{0.34984(z-4)z}{(z-\frac{1}{2})(z-2.93733)}\right)\right] = Z_u\left[Z_{\beta}^{-1}\left(\frac{Az}{z-\frac{1}{2}} + \frac{Bz}{z-2.93733}\right)\right]$$
$$= \frac{Az}{z-\frac{1}{2}}$$

where

$$A = \left.\frac{0.34984(z-4)}{(z-2.93733)}\right|_{z=\frac{1}{2}} = 0.50237$$

Finally, the transfer function $H(z)$ for the optimal discrete time causal shift-invariant filter can be written from Eq. (7.100) as

$$H(z) = \frac{0.50237z}{(z-\frac{1}{2})\psi_{XX}^{+}(z)} = \frac{0.63781(z-\frac{1}{4})}{z-0.340445}$$

The unit-sample response $h[n]$ for the optimal filter can be obtained by taking the inverse Z-transform of $H(z)$, which gives

$$h[n] = 0.468367\delta[n] + 0.169445(0.340445)^n u[n]$$

The solution for the continuous time problem can be shown to be

$$h_0(t) = 0.581139\delta(t) + 0.243416e^{-1.581139t}\mu(t)$$

If we sample this $h_0(t)$ at $t = nT = 0.69314718n$ and use the weight of the impulse at the origin, we obtain

$$h_0(nT) = 0.581139\delta[n] + 0.243416e^{-1.58113883n(0.69314718)}\mu[n]$$
$$= 0.581139\delta[n] + 0.243416(0.334218)^n\mu[n]$$

It should be noticed that $h[n]$ is not just the sampled version of the optimal continuous time filter. ■

7.4.3 Discrete Time Pure Prediction

Suppose that $X[n]$, a discrete time zero mean wide sense stationary random process with known autocorrelation function $R_{XX}[n]$, is observed from $-\infty$ to the present time n. It is desired to find the optimal (in the minimum mean squared error sense) causal linear time-invariant discrete time filter for estimating $X[n+j]$, where j is a positive number. This process of predicting j units into the future of the process $X[n]$ without any interfering noise is called **pure prediction**.

Thus the desired signal $G[n]$ is $X[n+j]$, and the data are $X[k]$, for $k \leq n$, giving the optimal filter transfer function $H(z)$ as previously determined by (7.100) as follows:

$$H(z) = \frac{Z_u[Z_\beta^{-1}(\psi_{GX}(z)/\psi_{XX}^-(z))]}{\psi_{XX}^+(z)} \tag{7.102}$$

To obtain the optimal filter, we must determine $\psi_{XX}(z)$ and $\psi_{GX}(z)$, which are the Z-transforms of the autocorrelation of $X[n]$ given by $R_{XX}[n]$ and the cross correlation of $G[n]$ and $X[n]$. The cross correlation function $R_{GX}[n]$ is determined as

$$R_{GX}[n] = E[G[m+n]X[m]] = E[X[m+n+j]X[m]] = R_{XX}[n+j] \tag{7.103}$$

The needed $\psi_{GX}(z)$ is obtained by taking the Z-transform of $R_{GX}[n]$ as follows:

$$\psi_{GX}(z) = Z(R_{GX}[n]) = Z(R_{XX}(n+j)) = z^j\psi_{XX}(z) \tag{7.104}$$

The term in square brackets of (7.102) is found using the spectral factorization of $\psi_{XX}(z)$, $\psi_{XX}(z) = \psi_{XX}^+(z)\psi_{XX}^-(z)$, to be

$$Z_\beta^{-1}\left(\frac{\psi_{GX}(z)}{\psi_{XX}^-(z)}\right) = Z_\beta^{-1}\left(\frac{z^j\psi_{XX}(z)}{\psi_{XX}^-(z)}\right) = Z_\beta^{-1}(z^j\psi_{XX}^+(z)) = Z_\beta^{-1}(\psi_{XX}^+(z))\,|_{n\to n+j} \tag{7.105}$$

If no assumptions are placed on the $\psi_{XX}^+(z)$, then the transform $H(z)$ for the **optimal linear time-invariant causal prediction system** is obtained from (7.102) and (7.105) as

$$H(z) = \frac{Z_u\left[Z_\beta^{-1}(\psi_{XX}^+(z))\,|_{n\to n+j}\right]}{\psi_{XX}^+(z)} \tag{7.106}$$

This result can be further simplified for the special case where the $\psi_{XX}(z)$ is a rational expression in z.

Special Case (Pure Prediction with Rational Spectra). For pure prediction with rational $\psi_{XX}(z)$, Eq. (7.106) can be simplified further for the special case where $\psi_{XX}^+(z)$ has unique poles z_i and thus expressible in the following form:

$$\psi_{XX}^+(z) = \sum_{i=1}^{K} \frac{a_i z}{z - z_i} \tag{7.107}$$

Using (7.107), we find the bracketed term in the numerator of (7.106) to be

$$\left[Z_\beta^{-1}(\psi_{XX}^+(z))\,|_{n\to n+j}\right] = Z_\beta^{-1}\left(\sum_{i=1}^{K}\frac{a_i z}{z-z_i}\right)|_{n\to n+j} = \sum_{i=1}^{K} a_i z_i^n u[n]\,|_{n\to n+j}$$
$$= \sum_{i=1}^{K} a_i z_i^j z_i^n u[n+j] \tag{7.108}$$

Taking the unilateral Z-transform of (7.108) and dividing by $\psi_{XX}^+(z)$, which is given in (7.107), the transfer function $H(p)$ from (7.106) of the **optimal causal linear shift-invariant pure predictor for rational spectra $H(z)$ is**

$$H(z) = \frac{\sum_{i=1}^{K} a_i z_i^j [z/(z-z_i)]}{\sum_{i=1}^{K} [a_i z/(z-z_i)]} \tag{7.109}$$

Example 7.5 illustrates the use of this formula for finding the optimal $H(z)$ for the case of unique poles.

If there are repeated roots in $\psi_{XX}^+(z)$, then the above formula would need to be modified by using properties of the inverse transform for repeated roots.

EXAMPLE 7.5

Given a random process characterized by its autocorrelation function $R_{XX}[k] = (1/2)^{|k|}$. Find the optimal, in the minimum mean squared error sense, causal linear filter for predicting $X[n+j]$ from the data $X[m]$ for $m = -\infty$ to $m = n$.

SOLUTION

Since $\psi_{XX}(z)$ is a rational expression in z, Eq. (7.109) can be directly applied where

$$\psi_{XX}(z) = Z_\beta(R_{XX}[n]) = Z_\beta[(1/2)^{|k|}]$$
$$= \frac{z}{z-\frac{1}{2}} - \frac{z}{z-2} = \frac{-\frac{3}{2}z}{(z-\frac{1}{2})(z-2)}$$

The $\psi_{XX}(z)$ can now be spectrally factorized as follows:

$$\psi_{XX}^+(z) = \frac{-\frac{3}{2}z}{(z-\frac{1}{2})(z-2)} = \frac{z\sqrt{\frac{3}{2}}}{(z-\frac{1}{2})}\cdot\frac{-\sqrt{\frac{3}{2}}}{(z-2)}$$

The $\psi_{XX}^{+}(z)$ is recognized as the first term in the product and $\psi_{XX}^{-}(z)$ is the second and from (7.109) the $H(z)$ for the optimum predictor becomes

$$H(z) = \frac{\sqrt{\frac{3}{2}}(\frac{1}{2})^j[z/(z-\frac{1}{2})]}{\sqrt{(\frac{3}{2})z/(z-\frac{1}{2})}} = (\tfrac{1}{2})^j$$

Thus the predictor form from the output of the filter given by $H(z)$ is just

$$\hat{X}[n+j] = (\tfrac{1}{2})^j X[n]$$

We see that the optimum predictor depends only on the data at time equals n and not on any of the other $X[m]$ for $m < n$. This will not be the case for impure prediction or prediction in noise. Note that as the distance into the future, j, of the prediction increases, the scale factor becomes smaller as it is the jth power of $\frac{1}{2}$ approaching zero as j approaches ∞. ■

7.5 OPTIMAL LINEAR SYSTEM OF PARAMETRIC FORM

In the previous sections on optimal linear time-invariant filters for minimizing the mean squared error, formulas were developed for the filters, and their form depended on the process characteristics. In some cases a filter of a particular form might be used because of filter complexity or hardware or software limitations, thus leading to a suboptimum filter. For these type of problems, transfer functions of a parametric form might be required. Typical simple parametric forms involving rational expressions in p, the Laplace variable, that could be used are

$$\begin{aligned}
H(p) &= \frac{1}{p+a} \triangleq H(p; a) \\
H(p) &= \frac{a}{p+b} \triangleq H(p; a, b) \\
H(p) &= \frac{ap+b}{p^2+cp+d} \triangleq H(p; a, b, c, d) \\
H(p) &= \frac{c(p+a)}{(p+b)(p+d)} \triangleq H(p; a, b, c, d) = H(p; \mathbf{a})
\end{aligned} \tag{7.110}$$

If the parametric form includes the form of the Weiner filter, then the filter obtained by optimization with respect to the parameters will give the Wiener filter. Thus the resulting filter is no longer suboptimum.

Suppose that we are given two jointly wide sense stationary zero mean random processes $X(t)$ and $G(t)$ that are somehow related. This relationship is expressed in terms of knowing, or being able to determine, their autocorrelation functions $R_{XX}(\tau)$ and $R_{GG}(\tau)$, and their cross-correlation function $R_{GX}(\tau)$. We would like to find a linear causal time-invariant filter with given form of transfer function $H(p; a_1,$

$a_2, \ldots, a_M) \triangleq H(p; \mathbf{a})$ that, when operating on $X(t)$, will give us the best filter, where "best" is in terms of minimizing the mean squared error given by

$$J = E[(G(t) - \hat{G}(t))^2] \tag{7.111}$$

The estimator of $G(t)$, called $\hat{G}(t)$, can be expressed in terms of the convolution of $X(t)$ with the impulse response of the linear filter as follows:

$$\begin{aligned} \hat{G}(t) &= X(t) * h(t; a_1, a_2 \ldots, a_M) \triangleq X(t) * h(t; \mathbf{a}) \\ \text{where} \quad h(t; \mathbf{a}) &= \mathscr{L}^{-1}[H(p; \mathbf{a})] \end{aligned} \tag{7.112}$$

There are a couple of ways that we can find this optimum filter: (a) We could find the optimal Wiener filter as if there were no constraints, and if that filter can be expressed in terms of the $H(p; \mathbf{a})$ assumed, the optimum filter of that form has been found. (b) We could use a direct optimization technique on the performance J with respect to the parameters $\mathbf{a}$ of the filter.

The approach used in this section will be the second, the direct minimization of the mean squared error given in (7.111). However, the minimization will take place in the Laplace domain rather than in the time domain. Expanding J into products and carrying out the expected value operator of (7.111), we can write

$$\begin{aligned} J &= E[(G(t) - \hat{G}(t))^2] \\ &= E[G(t)G(t) - G(t)\hat{G}(t) - \hat{G}(t)G(t) + \hat{G}(t)\hat{G}(t)] \\ &= R_{GG}(0) - R_{G\hat{G}}(0) - R_{\hat{G}G}(0) + R_{\hat{G}\hat{G}}(0) \end{aligned} \tag{7.113}$$

But these autocorrelations and cross correlations can be written in terms of their Laplace transforms as

$$J = \mathscr{L}^{-1}\big(\Phi_{GG}(p)) - \Phi_{G\hat{G}}(p) - \Phi_{\hat{G}G}(p) + \Phi_{\hat{G}\hat{G}}(p)\big) \,|_{\tau=0} \tag{7.114}$$

where the $\Phi_{GG}(p)$, $\Phi_{G\hat{G}}(p)$, $\Phi_{\hat{G}G}(p)$, and $\Phi_{\hat{G}\hat{G}}(p)$ are bilateral Laplace transforms of the respective autocorrelation and cross-correlation functions. They are determined as follows:

$$\begin{aligned} \Phi_{GG}(p) &= \Phi_{GG}(p) \\ \Phi_{\hat{G}G}(p) &= H(p; \mathbf{a})\Phi_{XG}(p) \\ \Phi_{G\hat{G}}(p) &= \Phi_{GX}(p)H(-p; \mathbf{a}) \\ \Phi_{\hat{G}\hat{G}}(p) &= H(p; \mathbf{a})\Phi_{XX}(p)H(-p; \mathbf{a}) \end{aligned} \tag{7.115}$$

Substituting (7.115) into (7.114) and rearranging, we can express the mean squared error for the **suboptimum** filter $H(p; \mathbf{a})$ as

$$\begin{aligned} J(\mathbf{a}) = {} & \mathscr{L}^{-1}\big(\Phi_{GG}(p) + H(p; \mathbf{a})\Phi_{XX}(p)H(-p; \mathbf{a})\big)|_{\tau=0} \\ & - \mathscr{L}^{-1}\big(H(p; \mathbf{a})\Phi_{XG}(p) + \Phi_{GX}(p)H(-p; \mathbf{a})\big)|_{\tau=0} \end{aligned} \tag{7.116}$$

The optimum filter with respect to the parameter vector $\mathbf{a}$ can then be obtained by minimizing (7.116) using any parameter minimization procedure. As we will demonstrate next, the following special case of estimating a signal in additive noise, which comes up quite frequently, will result in a simpler expression for $J(\mathbf{a})$.

Special Case. If the problem is one of estimating $G(t) = S(t)$ from a process $X(t) = S(t) + N(t)$ where the signal $S(t)$ and $N(t)$ are zero mean uncorrelated random processes, then the $J(\mathbf{a})$ can be simplified using

$$\begin{aligned} \Phi_{GG}(p) &= \Phi_{SS}(p) \\ \Phi_{XG}(p) &= \Phi_{GX}(p) = \Phi_{SS}(p) \\ \Phi_{XX}(p) &= \Phi_{SS}(p) + \Phi_{NN}(p) \end{aligned} \tag{7.117}$$

to give

$$\begin{aligned} J(a) = {} & \mathscr{L}^{-1}\big(\Phi_{SS}(p) + H(p;\mathbf{a})[\Phi_{SS}(p) + \Phi_{NN}(p)]H(-p;\mathbf{a})\big)\big|_{\tau=0} \\ & - \mathscr{L}^{-1}\big(H(p;\mathbf{a})\Phi_{SS}(p) - \Phi_{SS}(p)H(-p;\mathbf{a})\big)\big|_{\tau=0} \end{aligned} \tag{7.118}$$

After multiplying out, factoring, and rearranging, we easily see the $J(\mathbf{a})$ to be the sum of two terms:

$$\begin{aligned} J(\mathbf{a}) = {} & \mathscr{L}^{-1}\big([1 - H(p;\mathbf{a})]\Phi_{SS}(p)[1 - H(-p;\mathbf{a}])\big)\big|_{\tau=0} \\ & + \mathscr{L}^{-1}\big(H(p;\mathbf{a})\Phi_{NN}(p)H(-p;\mathbf{a})\big)\big|_{\tau=0} \end{aligned}$$

Finally, by the definition of the inverse bilateral transform evaluated at $\tau = 0$, the formula above can be written in terms of a complex integration:

$$\begin{aligned} J(\mathbf{a}) = {} & \frac{1}{2\pi j}\int_{\sigma-j\infty}^{\sigma+j\infty} [1 - H(p;\mathbf{a})]\Phi_{SS}(p)[1 - H(-p;\mathbf{a})]dp \\ & + \frac{1}{2\pi j}\int_{\sigma-j\infty}^{\sigma+j\infty} H(p;\mathbf{a})\Phi_{NN}(p)H(-p;\mathbf{a})dp \end{aligned}$$

Equation (7.120) is an alternative formula for the additive uncorrelated noise case. Now, if $H(p;\mathbf{a})$, $\Phi_{SS}(p)$, and $\Phi_{SS}(p)$ are special rational functions of p, the $J(a)$ can be determined by using a table given by Nichols [15] or a procedure given in Lanning and Battin [5]. The table can be essentially bypassed by using partial fraction expansions and then evaluating the inverse at $\tau = 0$ as indicated in (7.119).

Once the performance $J(\mathbf{a})$ is determined as a function of $\mathbf{a}$, any minimization procedure can be used to perform the minimization. **Necessary conditions** for a minimum are that $\nabla_{\mathbf{a}}(J(\mathbf{a})) = \mathbf{0}$. An example is now presented to illustrate the procedure for the special case of additive white noise.

EXAMPLE 7.7

Given the process $X(t)$ which is the sum of a signal process $S(t)$ and a noise process $N(t)$ as follows: $X(t) = S(t) + N(t)$ with $E[S(t)] = E[N(t)] = 0$ and $E[S(t)N(u)] = 0$ for all t and u. Assume that the autocorrelation functions for the signal and noise processes are given by $R_{SS}(\tau) = e^{-|\tau|}$ and $R_{NN}(\tau) = \delta(\tau)$. We would like to estimate $S(t)$ by passing $X(t)$ through a linear time-invariant causal filter specified by its transfer function $H(p)$ given in the parametric form by $H(p) = 1/(p+a)$. Find the a that minimizes the mean squared error $J = E[(S(t) - \hat{S}(t))^2]$.

SOLUTION

The mean squared error of Eq. (7.119) provides the starting place since the signal and noise are additive and uncorrelated. After substituting the given information into (7.119), we have the mean squared error as

$$\begin{aligned} J(a) &= \mathscr{L}^{-1}\left(\left[1 - \frac{1}{p+a}\right]\frac{2}{-p^2+1}\left[1 - \frac{1}{-p+a}\right]\right)\bigg|_{\tau=0} \\ &+ \mathscr{L}^{-1}\left(\frac{1}{p+a}1\frac{1}{-p+a}\right)\bigg|_{\tau=0} \\ &\triangleq I_1 + I_2 \end{aligned}$$

The inverse transform of the first term can be obtained by first performing a partial fraction expansion as follows:

$$\frac{(p+a-1)2(-p+a-1)}{(p+a)(-p^2+1)(-p+a)} = \frac{A}{p+a} + \frac{B}{p+1} + \frac{C}{-p+1} + \frac{D}{-p+a}$$

where the A, B, C, and D are determined as

$$\begin{aligned} A &= \frac{(p+a-1)2(-p+a-1)}{(-p^2+1)(-p+a)}\bigg|_{p=-a} = \frac{1-2a}{a(-a^2+1)} \\ B &= \frac{(p+a-1)2(-p+a-1)}{(p+a)(-p+1)(-p+a)}\bigg|_{p=-1} = \frac{-a^2+2a}{-a^2+1} \\ C &= B \\ D &= A \end{aligned}$$

Then the inverse Laplace transform is obtained and evaluated for $\tau = 0$ using the positive time answer only, to give

$$\begin{aligned} I_1 &= \left(\frac{1-2a}{a(-a^2+1)}e^{-\tau}\mu(\tau) + \frac{-a^2+2a}{-a^2+1}e^{-a\tau}\mu(\tau)\right)\bigg|_{\tau=0} \\ &= \frac{1-2a+2a^2-a^3}{a(-a^2+1)} \end{aligned}$$

The second inverse Laplace transform, I_2, is similarly obtained in the following equations:

$$I_2 = \mathscr{L}^{-1}\left(\frac{A}{p+a} + \frac{B}{-p+a}\right)\Big|_{\tau=0}$$
$$= \frac{1}{2a}e^{-a\tau}\mu(\tau)|_{\tau=0} = \frac{1}{2a}$$

The performance $J(a)$ is finally determined to be the following function of a:

$$J(a) = I_1 + I_2 = \frac{1 - 2a + 2a^2 - a^3}{a(-a^2+1)} + \frac{1}{2a}$$
$$= \frac{1.5 - 2a + 1.5a^2 - a^3}{-a^3 + a}$$

To minimize $J(a)$ we could take the partial of $J(a)$ with respect to a and set it equal to zero to find the optimal value. By computer program the optimum a is $a_{\text{opt}} = 2.419$ with a mean squared error for that value of a as $J(2.419) = 0.74264$. Thus the best filter of the form $H(p) = 1/(p+a)$ is

$$H(p) = \frac{1}{p + 2.419} \qquad \text{with } J = 0.74264$$

Solving the problem above for the Wiener filter gives $H_{\text{opt}}(p)$ as

$$H_{\text{opt}}(p) = \frac{0.3205}{p + 1.73205} \qquad \text{with } e_{\text{mmse}} = 0.73205$$

Since the Wiener filter has an amplitude that is not 1, the filter form of $H(p) = 1/(p+a)$ does not contain the Weiner filter. Thus $H(p)$ represents a suboptimum filter in the minimum mean squared error sense. For this problem it is noted that the use of the suboptimum filter loses very little in performance over the optimal filter. ■

7.6 SUMMARY

In this chapter we have presented the framework for optimal linear estimation of one random process in terms of another random process, where optimality was with respect to the minimum mean squared error performance measure. The basic problems of prediction, filtering, interpolation, and estimation of properties were presented.

Both continuous time and discrete time causal and noncausal Wiener filters were presented, where the solution was obtained by using the orthogonality principle. It was shown for jointly wide sense stationary desired random process $G(t)$ and observed random process $X(t)$ that the optimal linear system could be described knowing only the means, autocorrelation function, and cross-correlation function of the two processes. The optimal filter was expressed in terms of spectral and cross-spectral densities, and thus a steady state and frequency domain determination. The form of the optimum filters for rational and nonrational spectra were given, and the case of rational spectrum led to the expression of the optimal filter in terms of a spectral factorization and a partial fraction expansion.

If the processes are not jointly wide sense stationary, then the optimal filter was no longer time invariant, and its solution involved solving an integral equation for the case of continuous time and a summation equation for discrete time. The optimal discrete time filter for time-varying statistics was very complex. A solution for this type of problems is postponed to Chapter 8, where the Kalman filter is presented, and the analysis is done primarily in the time domain by using state space techniques.

Formulas for obtaining the minimum mean squared error for the optimal continuous and discrete time Wiener filters were derived. The error could be calculated in the time domain or alternatively determined in the Laplace or Z-transform domains.

The final topic of this chapter involved a procedure for obtaining suboptimal filters of a given specific parametric form. The first possible solution was to simply find the optimal Wiener filter, and if it could be expressed in the fixed form given, the resulting suboptimum filter would be optimal. If, however, the optimal Weiner filter could not be represented in the given form, then the mean squared error must first be found in terms of the parameters of the fixed form. Then any parameter optimization technique could be used to find the optimum parameters of the fixed form. A simple formula was presented for evaluating the mean squared error provided that the spectral and cross-spectral information was rational and the observation was signal plus noise.

PROBLEMS

7.1 A random signal $X(t)$ has an autocorrelation function $R_{XX}(\tau) = \cos 2\pi\tau$ and $E[X(t)] = 0$. We wish to estimate $X(t)$ at $t = \frac{1}{2}$ as a linear combination of samples of $X(t)$ at $t = 0$ and $t = \frac{1}{4}$.

$$\hat{X}(\tfrac{1}{2}) = a_1 X(0) + a_2 X(\tfrac{1}{4})$$

Find a_1 and a_2 such that the mean squared error is minimized and evaluate the minimum mean squared error. What type of an estimation problem have we described? Is their anything peculiar about your solution? Give a plausible meaning for your results.

7.2 Say we observe a zero mean random process $S(t)$ at $t = 0$ and $t = 1$. $R_{SS}(\tau)$ is known and given by $R_{SS}(\tau) = (\frac{1}{2})^{|\tau|}$.

(a) Find the optimum, in the minimum mean squared error sense, linear estimate of $S(\frac{1}{2})$ and the corresponding minimum mean squared error.

(b) Repeat (a) for estimating $S(\frac{3}{4})$.

(c) Repeat (a) for estimating $S(2)$.

7.3 A zero mean random process $X(t)$ is characterized by its autocorrelation function $R_{XX}(\tau) = e^{-|\tau|}$. We sample the process at $t = 0, 1, 2$. To get the random variables $X(0), X(1), X(2)$, we want to use these samples to find the best linear estimate of the random variable $X(1\ \frac{1}{2})$, where "best" is in the sense of minimizing the mean squared error. Use the orthogonality principle to find this estimate, and determine the resulting mean squared error.

7.4 Given two processes $S(t)$ and $N(t)$ that are uncorrelated zero mean random processes with known autocorrelation functions $R_{SS}(\tau)$ and $R_{NN}(\tau)$, respectively, given as

$$R_{SS}(\tau) = e^{-|\tau|} \quad \text{and} \quad R_{NN}(\tau) = \tfrac{1}{2}\, e^{-5|\tau|}$$

Suppose that we receive $X(t) = S(t) + N(t)$.

(a) If we observe $X(0)$ and $X(1)$, find the best linear estimate of

$$I = \int_0^1 S(t)\,dt$$

where "best" is in the sense of minimizing the mean squared error and calculate the minimum mean squared error.

(b) If we observe

$$Z = \int_0^1 X(t)\,dt$$

find the best linear estimate of I given in part (a) from Z, where "best" is in the minimum mean squared error sense, and calculate the minimum mean squared error for your estimator.

7.5 We desire the estimate of a signal $S(t)$ for $a < t < b$ by

$$\hat{S}(t) = k(t) \int_a^t X(t)\,dt$$

where $X(t) = S(t) + N(t)$ and $S(t)$ and $N(t)$ are orthogonal processes with autocorrelation functions $R_S(\tau)$ and $R_N(\tau)$. Find $k(t)$ such that $E[(S(t) - \hat{S}(t))^2]$ is a minimum.

7.6 Given a random process $X(t)$ with an autocorrelation function and mean as $R_{XX}(\tau) = e^{-|\tau|}$, $E[X(t)] = 0$. Estimate I defined as follows:

$$I = \int_0^1 X(t)\,dt$$

from $X(0), X(\alpha), X(1)$ by a linear combination, that is $\hat{I} = aX(0) + bX(\alpha) + cX(1)$. Select α such that the mean squared error is minimized. Compare your results to those of Gauss quadrature for integration of functions that are not processes, and compare a, b, and c for $\alpha = \frac{1}{2}$ to the coefficients for the Simpsons rule for the evaluation of integrals.

7.7 We are given a zero mean random process $S(t)$ with autocorrelation function $R_{SS}(t, u) = (\frac{1}{2})^{|t|}(\frac{1}{2})^{|u|}$. We want to estimate the integral of the process from 0 to 1 by a linear combination of the endpoints of $S(t)$ as follows:

$$\int_0^1 S(t)\,dt = I \quad \text{where} \quad \hat{I} = a_0 S(0) + a_1 S(1)$$

We want a_0 and a_1 such that $E[(I - \hat{I})^2]$ is minimized.

(a) Find a_0 and a_1.

(b) Determine the mean squared error for the estimator in part (a).

7.8 The random process $S(t)$ is defined by $S(t) = A\ \cos(2\pi t + \Theta)$, where Θ and A are independent random variables with Θ uniform on 0 to 2π, and A has a zero mean and variance of $\frac{1}{2}$. Assume that $X(t) = S(t) + N(t)$, where $N(t)$ is a colored noise process independent of $S(t)$ with autocorrelation function

$$R_{NN}(\tau) = \tfrac{1}{2} e^{-|\tau|}$$

We wish to find the best linear estimate of $S(t)$ at the points $t = \frac{1}{2}$ and $t = 1\frac{1}{2}$ in terms of $X(0)$ and $X(1)$.

(a) Find $\hat{S}(\frac{1}{2})$ and associated minimum mean squared error.

(b) Find $\hat{S}(1\frac{1}{2})$ and associated minimum mean squared error.

(c) Discuss your results.

7.9 Say we are given $X(t) = S(t) + N(t)$, $R_{SS}(\tau) = e^{-|\tau|}$, $R_{NN}(\tau) = \frac{1}{2}\delta(\tau)$ and that $S(t)$ and $N(t)$ are orthogonal random processes.

(a) Estimate $S(1)$ and $S(2)$ by linear combinations of $X(0)$ and $X(3)$ such that I given below is minimized:

$$I = E[(S(1) - \hat{S}(1))^2] + E[(S(2) - \hat{S}(2))^2]$$

(b) Calculate I for results obtained in (a).

7.10 Given a random process $X(t)$ with autocorrelation function $R_{XX}(\tau) = 1/(1 + \tau^2)$ and mean $E[X(t)] = 0$. We wish to estimate

$$I = \int_{-1}^{1} X(t)\,dt$$

from two distinct data points taken somewhere in the interval $[-1, 1]$ by a linear combination of data points

$$\hat{I} = a_1 X(t_1) + a_2 X(t_2)$$

(a) Select a_1, t_1, a_2 and t_2 such that we minimize the mean squared error.

(b) Compare your weights and t's with those for the Gauss quadrature calculation of the integral. Discuss your results.

(c) Suppose that $X(t) = A + Bt$, where A and B are zero mean independent Gaussian random variables with equal variances $\sigma^2 = 1$. Repeat (a) and (b).

7.11 Say we are given a discrete time random process $X(n)$ that has an autocorrelation function $R_{XX}(k) = (\frac{1}{2})^{|k|}$ and a mean of zero. We would like to predict $X(n)$ as a linear function of the previous L (finite) values (i.e., $X(n-1), X(n-2), \ldots, X(n-L)$) in such a way that we minimize the mean squared error.

(a) Solve for the optimum predictor coefficients for the special cases $L = 1, L = 2$ and $L = 3$, and determine the mean squared error for each case. Do you notice anything unusual about your answers.

(b) Generalize your result if you can.

7.12 Given the zero mean process $X(t)$ with autocorrelation function $R_{XX}(\tau)$:

$$R_{XX}(\tau) = e^{-|\tau|} + e^{-2|\tau|}$$

We have $X(t)$ from $-\infty$ to t, and we wish to predict the process $X(t)$ at $t+1$. Find the optimum Wiener prediction filter.

7.13 Given the zero mean random processes $S(t)$ and $N(t)$ with the following characterizations:

$$R_{SS}(\tau) = e^{-|\tau|} \quad \text{and} \quad R_{NN}(\tau) = \frac{1}{2}\,\delta(\tau)$$

Assume that $S(t)$ and $N(t)$ are uncorrelated random processes, and that we observe $X(t)$ given by $X(t) = S(t) + N(t)$. We wish to process $X(t)$ using a linear time-invariant filter to obtain the best estimate of $S(t)$, where "best" is in the sense of minimum mean squared error.

(a) Find the optimal causal filter for estimating $S(t)$, determine its transfer function $H(p)$, its impulse response $h(t)$, and its frequency response $|H(j\omega)|$. Also express the processing described in integral form.

(b) Calculate the mean squared error for your filter of (a).

(c) Find the optimal noncausal filter for estimating $S(t)$, giving $H(p)$, $h(t)$, and $|H(j\omega)|$. Describe the optimal processing in an integral form.

(d) Determine the minimum mean squared error for filter in (c).

(e) Compare your results for the causal and noncausal case.

7.14 Given $X(t) = S(t) + N(t)$ where $S(t)$ and $N(t)$ are zero mean wide sense stationary orthogonal random processes with $R_{SS}(\tau)$ and $R_{NN}(\tau)$ given by

$$R_{SS}(\tau) = e^{-|\tau|} \quad \text{and} \quad R_{NN}(\tau) = e^{-\frac{1}{4}|\tau|}$$

We desire the best estimate of $S(t)$, in the minimum mean squared error sense, by filtering $X(t)$.

(a) Find the best causal linear time invariant filter for estimating $S(t)$ from $X(t)$. Determine the mmse for your filter.

(b) Plot the power spectral densities for $S(t)$ and $N(t)$ and $X(t)$. Also plot the frequency response for the filter determined in (a).

(c) Determine the optimal linear time invariant non causal filter for estimating $S(t)$, and calculate the minimum mean squared error for your filter.

7.15 Given a received signal $X(t) = F(S(t), N(t)) = S(t) + N(t)$ where $S(t)$ and $N(t)$ are independent zero mean random processes with $\Phi_{NN}(p)$ and $\Phi_{SS}(p)$, given by

$$\Phi_{NN}(p) = \frac{1}{-p^2+1} + 2 \quad \text{and} \quad \Phi_{SS} = \frac{5}{-p^2+1}$$

Find the causal linear filter $H(p)$ that minimizes the mean squared error for the estimation of $S(t)$.

7.16 A random process $S(t)$ is defined by $S(t) = A \cos(2\pi t + \Theta)$. Assume the Θ to be uniform on $[0, 2\pi]$ and A to be Gaussian with mean of zero and variance σ^2. Say we observe $X(t)$ where $X(t)$ is $X(t) = S(t) + N(t)$ and $N(t)$ is a white noise process independent of $S(t)$. Find the optimal causal and noncausal filters for estimating $S(t)$ from $X(t)$.

7.17 Given a zero mean random process $X(t)$ with known autocorrelation function as

$$R_{XX}(\tau) = e^{-\alpha|\tau|} \cos(2\pi\tau)$$

(a) Find the optimal causal linear time-invariant filter for estimating $X(t+d)$ from $\{X(\beta) : -\infty < \beta \leq t\}$ for $d = \frac{1}{4}, \frac{1}{2}$, and 1.

(b) Determine the minimum mean squared error for your results at $d = \frac{1}{4}, \frac{1}{2}$, and 1.

(c) Find the limit of your answer in (b) as $\alpha \to 0$. If the limit exists, what would it mean about predicting sinusoidal random processes.

7.18 Given $X(t) = S(t) + N(t)$ where $S(t)$ and $N(t)$ are zero mean orthogonal random processes characterized by

$$\Phi_{ss}(p) = \frac{2}{1-p^2} \quad \text{and} \quad \Phi_{nn}(p) = 2$$

Knowing the values of $X(t)$ from $-\infty$ to t, we wish the best mean squared estimate of $S(t+3)$ to accomplish this estimation.

(a) Find the optimal causal filter $H_c(p)$.

(b) Find the optimal noncausal linear filter $H_{nc}(p)$.

7.19 Given $X(t) = S(t) + N(t)$, and $E[S(t)] = E\{N(t)] = 0$, $R_{SN}(\tau) = 0$, $R_{SS}(\tau)$, and $R_{NN}(\tau)$ given by

$$R_{SS}(\tau) = \tfrac{15}{2} e^{-|\tau|} \quad \text{and} \quad R_{NN}(\tau) = \tfrac{21}{4} e^{-2|\tau|}$$

(a) Find the optimal causal filter for estimating $S(t + \lambda)$ for $\lambda = 1$.

(b) Find the mean squared error for your estimate.

(c) Find the optimal noncausal filter for estimating $N(t)$. Is this the same as operating on $X(t)$ using the optimal noncausal filter for estimating $S(t)$ and then subtracting the results from $X(t)$?

7.20 Given the random process $X(t)$ which is the sum of signal and noise processes $S(t)$ and $N(t)$ with the following properties:

(1) $E[S(t)] = E[N(t)] = 0$, $R_{SS}(\tau) = e^{-|\tau|}$, and $R_{NN}(\tau) = \frac{1}{2}\,\delta(\tau)$.

(2) $S(t)$ and $N(t)$ are uncorrelated random processes.

(3) $X(t)$ is received on the interval $(-\infty, t)$.

We want to process $X(t)$ using a linear time invariant filter to obtain the best (in the mmse sense) estimate of $S(t + \lambda)$ where $\lambda = 0.69314718$. This is an example of impure prediction.

(a) Find the optimal causal filter for estimating $S(t + 0.69314718)$.

(b) Using your optimal filter from (a) give $\hat{S}(t + \lambda)$ in terms of an integral of $X(t)$ and the impulse response.

(c) Calculate the minimum mean squared error for your optimal filter.

7.21 Given $X(t) = S(t) + N(t)$, and $E[S(t)] = E\{N(t)] = 0$, $R_{SN}(\tau) = 0$, $R_{SS}(\tau)$, and $R_{NN}(\tau)$ given by

$$R_{SS}(\tau) = \tfrac{15}{2} e^{-|\tau|} \quad \text{and} \quad R_{NN}(\tau) = \tfrac{21}{4}\,\delta(\tau)$$

(a) Find the optimal causal filter for estimating $S(t + \lambda)$ for $\lambda = 1$.

(b) Find the mean squared error for your estimate of part (a).

7.22 The process $X(t)$ is a zero mean random process with a "Butterworth" power spectral density $\Phi_{XX}(\omega)$ given by

$$\phi_{xx}(\omega) = \frac{1}{1 + \omega^4}$$

Say we have available data from $-\infty$ up to t and desire the best estimate of $X(t + 1)$, where the "best" is in the minimum mean squared error sense by using a linear time-invariant causal filter. Find the filter, give the formula for the calculation of the mean squared error, and evaluate.

7.23 Assume $X(t) = S(t) + N(t)$ where $X(t)$ and $N(t)$ are orthogonal zero mean random processes with autocorrelation functions $R_{SS}(\tau)$ and $R_{NN}(\tau)$ written as

$$R_{SS}(\tau) = e^{-2|\tau|}, \qquad R_{NN}(\tau) = \tfrac{1}{2}\,\delta(\tau)$$

(a) Find $H(p)$ and the impulse response $h(t)$ of the optimal causal linear filter for estimating $S(t+1)$.

(b) Calculate the minimum mean squared error for the filter determined in (a).

7.24 Assume $X(t) = S(t) + N(t)$ with $S(t)$ and $N(t)$ independent random processes characterized by their power spectral densities $\phi_{SS}(\omega)$ and $\phi_{NN}(\omega)$ as follows:

$$\phi_{SS}(\omega) = \frac{2b}{\omega^2 + b^2}, \qquad \phi_{nn}(\omega) = \frac{1}{\omega^2 + 1}$$

We desire the best linear estimate of $S(t+\lambda)$ in the minimum mean squared error sense.

(a) Find causal $H(p)$ and the corresponding minimum mean squared error in terms of b (optimum filtering and prediction).

(b) Comment on the selection of b to give the minimum mean squared error.

7.25 Let $X(t)$ be the sum of a signal process $S(t)$ and noise process $N(t)$ where $S(t)$ and $N(t)$ are independent zero mean random processes with autocorrelation functions $R_{SS}(\tau)$ and $R_{NN}(\tau)$ given by

$$R_{SS}(\tau) = \tfrac{15}{2}\,e^{-|\tau|} \quad \text{and} \quad R_{NN}(\tau) = \tfrac{21}{4}\,e^{-2|\tau|}$$

(a) Find the optimal causal filter for estimating $G(t)$, where

$$G(t) = S(t) * h_k(t) \quad \text{and} \quad H_k(p) = \frac{1}{p+3}$$

(b) Repeat (a) for an optimal noncausal filter.

(c) For assumptions above find the best linear causal and noncausal estimate of $S(t)$ instead of $G(t)$, and determine mean squared errors.

(d) Discuss commutativity of estimating linear operations on signals for the causal and noncausal cases.

7.26 Rework parts (a), (b), (c), and (d) of Problem 7.25 for $R_{NN}(\tau) = (21/4)\,\delta(\tau)$.

7.27 Let $X(t)$ be a random process composed of the sum of a zero mean signal process $S(t)$ and a zero mean noise process $N(t)$ as $X(t) = S(t) + N(t)$. Assume we have the autocorrelation functions for the processes given as

$$R_{SS}(\tau) = e^{-|\tau|} \quad \text{and} \quad R_{NN}(\tau) = \tfrac{1}{4}\,\delta(\tau)$$

Find the best linear causal and noncausal filters for estimating the $dS(t)/dt$ from $X(t)$, where "best" is in the minimum mean squared error sense.

7.28 Given a zero mean Gaussian random process $X(t)$ characterized by its power spectral density

$$\phi_{SS}(\omega) = e^{-\omega^2}$$

(a) Estimate $Z(t)$ where

$$Z(t) = \int_{t-1}^{t} S^2(t)\,dt$$

by a causal linear filter operating on $X(t) = S(t) + N(t)$ to minimize the mean squared error where $N(t)$ is a Gaussian white noise process with $R_{NN}(\tau) = \frac{1}{2}\delta(\tau)$.

(b) Calculate minimum mean squared error, and compare with mmse obtained in (a) for the following nonlinear estimator of $S(t)$:

$$\hat{S}(t) = \int_{t-1}^{t} X^2(t)\,dt$$

7.29 Given two zero mean orthogonal random processes $S(t)$ and $N(t)$ with power spectral densities

$$\phi_{SS}(\omega) = \frac{3}{1+\omega^2} \quad \text{and} \quad \phi_{NN}(\omega) = 1$$

Let the observed random process be $X(t) = S(t) + N(t)$. Find:

(a) The causal Wiener filter for estimating $S(t)$ from $X(t)$.

(b) The noncausal Wiener filter for estimating $S(t)$ from $X(t)$.

(c) The causal Wiener filter for estimating $dS(t)/dt$.

(d) The noncausal Wiener filter for estimating $dS(t)/dt$.

(e) Comment on commutativity of estimators and linear functions of estimators.

(f) The minimum mean squared error for (b).

7.30 Given the process $X(t) = S(t) + N(t)$ where $S(t)$ and $N(t)$ are zero mean independent random processes with known autocorrelation functions

$$R_{SS}(\tau) = e^{-|\tau|} \quad \text{and} \quad R_{NN}(\tau) = \tfrac{1}{2}\,\delta(\tau)$$

(a) Find the optimal causal and noncausal Wiener filters that operate on $X(t)$ for estimating a process $Z(t)$ defined by

$$Z(t) = \int_{-\infty}^{t} S(t)\,dt$$

(b) Are your filters in (a) the same as finding the Wiener filter for estimating $S(t)$ and then passing the resulting output to an integrator? Answer for both causal and noncausal filters.

7.31 Given the received signal $X(t) = S(t) + N(t)$. Assume that $S(t)$ and $N(t)$ are zero mean random processes with autocorrelation functions given by

$$R_{SS}(\tau) = e^{-|\tau|} \quad \text{and} \quad R_{NN}(\tau) = \delta(\tau).$$

Assume that $S(t)$ and $N(t)$ are orthogonal, and find the noncausal linear filter $H(p)$ that gives the best, in the minimum mean squared error sense, estimate of $Y(t)$, the smoothed version of $S(t)$ given by

$$Y(t) = \frac{1}{2\varepsilon}\int_{t-\varepsilon}^{t-\varepsilon} S(\alpha)\,d\alpha$$

7.32 Given a zero mean random process $X(t)$ with autocorrelation function $R_{XX}(\tau)$ as

$$R_{XX}(\tau) = e^{-|\tau|}$$

(a) Find the best, in the minimum mean squared error sense, linear causal filter for estimating $Z(t)$, where $Z(t)$ is given by

$$Z(t) = \int_{t+1}^{t+2} X(\beta)\,d\beta$$

(b) Calculate the mean squared error for the filter obtained in (a).

7.33 Define a process $G(t)$ by

$$G(t) = \int_{t}^{t+1} X(\beta)\,d\beta$$

where $X(t)$ is a zero mean wide sense stationary random process with autocorrelation function $R_{XX}(\tau)$ given by

$$R_{XX}(\tau) = \begin{cases} 1- \mid \tau \mid, & \mid \tau \mid \le 1 \\ 0, & \mid \tau \mid > 1 \end{cases}$$

(a) Find the optimal, in the minimum mean squared error sense, causal linear filter for estimating $G(t)$ given the data $X(t)$ from $(-\infty, t]$.

(b) Calculate the mean squared error for the optimal filter.

7.34 Given a zero mean wide sense stationary Gaussian random process with autocorrelation function $R_{XX}(\tau) = \frac{1}{2}e^{-|\tau|}$. Say we receive $Y(t) = X(t) + N(t)$ where $N(t)$ is a zero mean Gaussian white noise process with autocorrelation function $R_{NN}(\tau) + \frac{1}{4}\delta(\tau)$. Assume that $N(t)$ is independent of $X(t)$. We desire the best estimate, in the minimum mean squared error sense, of $X^2(t)$ by operating on $Y(t)$.

(a) Find the optimal causal linear time invariant filter for estimating $X^2(t)$ from $Y(t)$. Calculate the minimum mean squared error.

(b) Find the optimal noncausal linear time-invariant filter for estimating $X^2(t)$.

(c) Is the optimal causal linear time-invariant filter for estimating $X^2(t)$ the same as the optimal linear time invariant filter for estimating $X(t)$ followed by a squaring operation? If not, which one gives the smallest minimum mean squared error?

7.35 Given a received signal $X(t) = F[S(t), N(t)] = S(t)N(t)$ where $S(t)$ and $N(t)$ are independent random processes with power spectral densities given by

$$\phi_{nn}(\omega) = \frac{1}{\omega^2 + 4}, \quad \phi_{ss}(\omega) = \frac{2}{\omega^2 + 1}$$

Find the optimal **(a)** causal and **(b)** noncausal linear filter $H(p)$ that minimizes the mean squared error for the estimation of $S(t)$ from $X(t)$.

7.36 Given $X(t) = S(t) + N(t)$ where $S(t)$ is a nonstationary process with auto-correlation function $R_{SS}(t, u)$ and $N(t)$, independent of $S(t)$, is a stationary process with auto-correlation function $R_{NN}(t - u)$.

(a) Derive the integral equation for the best linear estimator of $S(t)$ at time t given all values of $X(t)$ for t in interval $(-\infty, t]$ by using a linear time-varying filter with system function $h(t, u)$.

(b) Determine the mean squared error for the optimum filter.

(c) Repeat (a) and (b) using the interval $(-\infty, \infty)$.

7.37 Find the system function $H(p)$ for a noncausal time-invariant Wiener filter to estimate $S(t)$ under from $X(t)$ under the following conditions. $S(t)$ is independent of $N_1(t)$ and $N_2(t)$, and $X(t)$ is given by $X(t) = S(t)N_1(t) + N_2(t)$. Assume further that

$$E[S(t)] = 0, \quad E[N_1(t)] = \eta_1, \quad \text{and} \quad E[N_2(t)] = \eta_2$$
$$E[N_1(t+\tau)N_1(t)] = R_1(\tau) \quad \text{and} \quad E[N_2(t+\tau)N_2(t)] = R_2(\tau)$$

Assume that $N_1(t)$ and $N_2(t)$ have cross correlation $E[N_1(t+\tau)N_2(t)] = R_{12}(\tau)$.

(a) Express $H(p)$ in terms of $n_1, n_2, \Phi_{SS}(p), \Phi_{11}(p), \Phi_{22}(p)$, and $\Phi_{12}(p)$.

(b) Find the minimum mean squared error e_{mse} for the filter determined in part (a).

7.38 Given a zero mean wide sense stationary process $S(t)$ that has autocorrelation function $R_{ss}(\tau) = e^{-|\tau|}$. We are not able to measure $S(t)$ directly but measure

$$Z(t) = S(t) + \tfrac{1}{2} S(t-1) \qquad \text{(this is an example of multipath)}$$

(a) Find the optimal (minimum mean squared error sense) linear noncausal filter for estimating $S(t)$ from $Z(t)$.

(b) Begin the process of finding the optimal causal filter (mmse sense), and perform the spectral factorization part. Continue as far as you can in finding the optimal filter. Discuss any problems encountered.

(c) For the optimal noncausal filter found in part (a), determine the minimum mean square error.

7.39 Assume $S(t)$ and $N(t)$ are zero mean orthogonal random processes with given autocorrelation functions $R_{SS}(\tau)$ and $R_{NN}(\tau)$ as

$$R_{SS}(\tau) = \tfrac{15}{2} e^{|\tau|} \quad \text{and} \quad R_{NN}(\tau) = \tfrac{21}{4}\, \delta(\tau)$$

We wish to find the best estimate of $S(t)$ at time t by operating on $X(t) = S(t) + N(t)$ with a linear filter of the form $H(p) = 1/(p + T)$. This is optimization with respect to a parameter T. Find T to give the minimum mean squared error, and calculate the minimum mean squared error for that T. Compare your result with that of the optimum causal filter and comment on differences.

7.40 Rework Problem 7.40 with $R_{NN}(\tau) = (21/4)e^{-2|\tau|}$.

7.41 Given $X_1(t) = S(t) + N_1(t)$ and $X_2(t) = S(t) + N_2(t)$ where $S(t), N_1(t)$, and $N_2(t)$ are zero mean independent random processes with specified $R_{SS}(\tau), R_{11}(\tau)$, and $R_{22}(\tau)$. We want to operate on $X_1(t)$ and $X_2(t)$ with linear time-invariant filters to obtain the best estimator of the signal $S(t)$, where best is in the sense of minimizing the mean squared error.

(a) Find the optimal noncausal filters, specified by their impulse responses $h_1(t)$ and $h_2(t)$, for the estimator $\hat{S}(t)$ given by

$$\hat{S}(t) = \int_{-\infty}^{\infty} X_1(\tau)h_1(t-\tau)\, d\tau + \int_{-\infty}^{\infty} X_2(\tau)h_2(t-\tau)\, d\tau$$

(b) Find the optimal causal filters, specified by their impulse responses $h_1(t)$, and $h_2(t)$, for the estimator of $S(t)$ such that the mean squared error is minimized for $\hat{S}(t)$ given by

$$\hat{S}(t) = \int_{-\infty}^{t} X_1(\tau)h_1(t-\tau)\, d\tau + \int_{-\infty}^{t} X_2(\tau)h_2(t-\tau)\, d\tau$$

7.42 Consider the system shown in Figure P7.43 with input $X(t) = S(t) + N(t)$, system transfer function $H(p) = G/(1 + pT)$, and output $Y(t) = \hat{S}(t)$. The

$S(t)$ and $N(t)$ are zero mean independent random processes with autocorrelation functions given by

$$R_{SS}(\tau) = e^{-2|\tau|} \quad \text{and} \quad R_{NN}(\tau) = \tfrac{1}{4}\delta(t)$$

(a) Find G and T such that $E[(S(t) - \hat{S}(t))^2]$ is a minimized for $\hat{S}(t) = X(t) * h(t)$ where $h(t) = \mathscr{L}^{-1}[H(p)]$.

(b) Compare this result to the optimal causal Wiener filter for estimating $S(t)$.

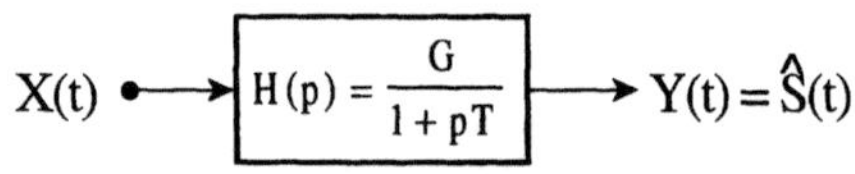

P7.42

7.43 Rework Problem 7.42 using $H(p) = 1/(p + T)$.

7.44 Optimum linear filtering with constraints. Define the constraint on $h(t)$ as

$$\int_{-\infty}^{\infty} h^2(t)\, dt = E_h$$

Find the optimal noncausal, linear filter with impulse response $h(t)$ for estimating $S(t)$ from $X(t) = S(t) + N(t)$, where "best" is in the minimum mean squared error sense. Assume that $S(t)$ and $N(t)$ have known $R_{SS}(\tau)$ and $R_{NN}(\tau)$ and that $S(t)$ and $N(t)$ are independent zero mean random processes.

7.45 Given a random process $X(t)$ that is the sum of a zero mean signal process $S(t)$ with autocorrelation function $R_{SS}(\tau) = (15/2)e^{-|\tau|}$ and a zero mean noise process $N(t)$ with autocorrelation function $R_{NN}(\tau) = (21/4)e^{-2|\tau|}$. Assume that $S(t)$ and $N(t)$ are orthogonal random processes and that $X(t)$ is available only at the sample instants kT, where $T = 0.693147$.

(a) Find the optimum realizable digital filter characterized by $H(z)$ that gives us the best estimate of $S(nT)$, where "best" is in the mean squared error sense, and determine the minimum mean squared error.

(b) Compare this with the optimum continuous causal filter for estimating $S(t)$. Is the discrete time filter the sampled version of the continuous filter? Could you have anticipated this result?

(c) Repeat (a) and (b) for $T = 0.10536$, and discuss the effect of decreasing the time between samples.

7.46 Given a random process $X[n]$ that is the sum of a zero mean signal process $S[n]$ with autocorrelation function $R_{SS}[k] = (\frac{1}{2})^{|k|}$ and a zero mean noise process $N[n]$ with autocorrelation function $R_{NN}[k] = (\frac{1}{4})^{|k|}$. Assume that $S[n]$ and $N[n]$ are orthogonal random processes.

(a) Find the optimum realizable digital filter characterized by $H(z)$ that gives us the best estimate of $S[n]$, where "best" is in the minimum mean squared error sense, and determine the corresponding minimum mean squared error.

(b) Find the optimum noncausal digital filter characterized by $H(z)$ that gives us the best estimate of $S[n]$, where "best" is in the minimum mean squared error sense and determine the corresponding minimum mean squared error.

7.47 A random process $S(t)$ is characterized by its mean and autocorrelation functions as follows:

$$\eta_S(t) = 0, \quad R_{SS}(\tau) = e^{-|\tau|}$$

We wish to estimate $S(t+1)$ from $S(t)$ on the interval $(t-1, t)$. This is a sliding memory predictor. Assume that the form of the estimator is as follows:

$$\hat{S}(t+1) = \int_{t-1}^{t} h(t-\tau)S(\tau)\,d\tau$$

We want to find $h(t)$ such that the mean squared error as follows is minimized:

$$e_{\text{ms}} = E[(S(t+1) - \hat{S}(t+1))^2]$$

(a) Use the orthogonality principle to derive an integral equation whose solution will be the optimal filter.

(b) Also from the orthogonality principle obtain a formula for the minimum mean squared error for the optimum filter.

7.48 Given

$$I = \frac{1}{2\pi}\int_{-\infty}^{\infty} \frac{\omega^2 + a^2}{(\omega^2 + 1/a^2)(\omega^2 + 4)}\,d\omega$$

Evaluate I in terms of a. What value of a gives the maximum value of I?

REFERENCES

1. Papoulis, Athanasios, *Probability, Random Variables, and Stochastic Processes*, McGraw-Hill, 1965; 2nd ed., 1984.
2. Davenport, Jr., Wilbur B, and William L. Root, *An Introduction to the Theory of Random Signals in Noise*, McGraw-Hill, 1958.
3. Lathi, B. P., *An introduction to Random Signals and Communication Theory*, International Textbook Company, 1968.
4. Thomas, John B., *An Introduction to Statistical Communication Theory*, Wiley, 1969.

5. Laning, Jr., J. Halcombe, and Richard H. Battin, *Random Processes in Automatic Control*, McGraw-Hill, 1956.
6. Mendel, Jerry M., *Lessons in Digital Estimation Theory*, Prentice-Hall, 1987.
7. Wainstein, L. A., and V. D. Zubakov, *Extraction of Signals from Noise*, trans. by Richard A. Silverman. Prentice-Hall, 1962.
8. Yaglom, A. M., *An Introduction to the Theory of Stationary Random Functions*, Dover, 1962.
9. Wiener, N., *Extrapolation, Interpolation, and Smoothing of Stationary Time Series*, Wiley, 1949.
10. Kolmogorov, A. N., Interpolation and extrapolation of stationary time series, *Bulletin Academy of Sciences, USSR Math Series*, Vol 5, 1941.
11. Mendel, Jerry M., *Discrete Techniques of Parameter Estimation*, Dekker, 1973.
12. Kailath, T., *Lectures on Linear Least-Squares Estimation*, Springer, 1976.
13. Brown, Robert Grover, *Introduction to Random Signal Analysis and Kalman Filtering*, Wiley, 1983.
14. Sage, Andrew P., and James L. Melsa, *Estimation Theory with Applications to Communications and Control*, McGraw-Hilll, 1971.
15. James, H. M., N. B. Nichols, and R. S. Phillips, *Theory of Servo Mechanisms*, MIT Radiation Laboratory Series, Vol. 25, McGraw-Hill, 1947.

8 Optimum Linear Systems: The Kalman Approach

8.1 INTRODUCTION

In the previous chapter optimal linear systems were obtained from the classical Wiener framework to perform operations of filtering prediction and smoothing. The optimal filter was determined in terms of various cross-correlation and autocorrelation functions, or equivalently cross and auto spectral densities, and thus can be considered primarily a frequency domain approach. The Weiner results presented were primarily for wide sense stationary processes; however, for nonstationary processes integral equations were obtained for the solutions. For the case of vector process estimation, the spectral factorization problem for the estimation of wide sense stationary processes proved to be mathematically intractable, and for estimation of nonstationary processes, the resulting integral equations could not be solved by the standard Laplace transform techniques and their solutions were virtually impossible to obtain.

The Kalman filter addresses both the problem of nonstationarity and vector estimation, and it gives the solution in the time domain by virtue of formulating the problem in the state space. Thus rather than the autocorrelation functions or power spectral densities being given, the required information for the Kalman filter solution is given in the state variable format. This information is provided in terms of the state transition, excitation, and measurement matrices of the given state space formulation. This chapter begins with a brief review of discrete time systems and progresses through the description of the various algorithms that are collectively called the Kalman filter. It will also be shown that under steady state conditions, the Kalman and Wiener filters are equivalent for the wide sense stationary case.

In this chapter lowercase letters in the latter part of the alphabet, although not bolded, will indicate vectors. This is because almost all signals indicated will be time-varying vectors or matrices, so bolding the letter would be very distractive. Capital Greek letters will indicate matrices. No distinction will be made between random vectors and deterministic ones because of cumbersomeness of using capital letters; however, for the most part, vectors will be random. Also the traditional square brackets [ ] used in digital signal processing to indicate discrete time signals will be replaced by ( ) where the arguments used are lowercase letters in the middle of the alphabet like j, k, l, m, or n.

8.2 DISCRETE TIME SYSTEMS

A discrete time linear system with vector input or excitation vector $w(k)$ and output $y(k)$ can be globally illustrated as in Figure 8.1. The system can be further described internally in terms of two equations, called the **state equation** and **output equation**:

State Equation:

$$x(k+1) = \Phi(k+1, k)x(k) + \Gamma(k+1, k)w(k), \quad k = 0, 1, \ldots \tag{8.1a}$$

Output Equation:

$$y(k+1) = H(k+1)x(k+1) \tag{8.1b}$$

In these two equations, the following definitions are used:

$x(k)$ is an $n \times 1$ vector called the **state vector** at time k.

$\Phi(k+1, k)$ is an $n \times n$ matrix called the **state transition matrix** from time k to time $k+1$.

$\Gamma(k+1, k)$ is an $n \times p$ matrix called the **excitation transition matrix** from time k to time $k+1$.

$w(k)$ is a $p \times 1$ vector called the **excitation vector** at time k.

$x(0)$ is an $n \times 1$ vector called the initial condition or **initial state vector** at time $k = 0$.

$H(k)$ is an $m \times n$ matrix called the **output transition matrix** at time k.

$y(k+1)$ is an $m \times 1$ vector called the **output vector** at time $k+1$.

Equation (8.1a) gives a recursive formula for calculating the state of the system at time index $k+1$ as a function of the state of the system at time index k and the excitation vector at time index k. A vector block diagram is shown in Figure 8.2, which illustrates the linear operations on the state and excitation vectors that produce the next state vector.

The solid line indicates a vector operation, and the blocks represent premultiplication by a matrix, the circles a summation of vector components, and the delay represents a one sample delay in time of each component.

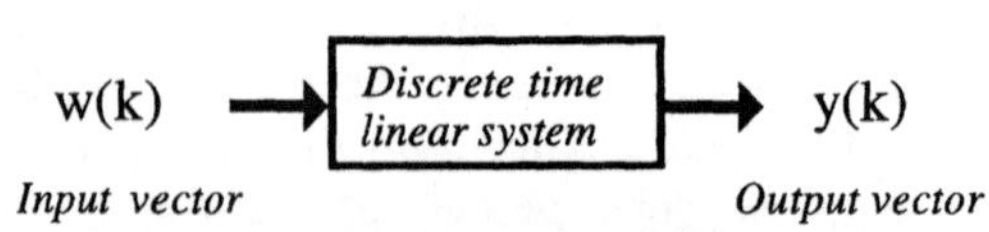

Figure 8.1 Input–output representation of a discrete-time linear system.

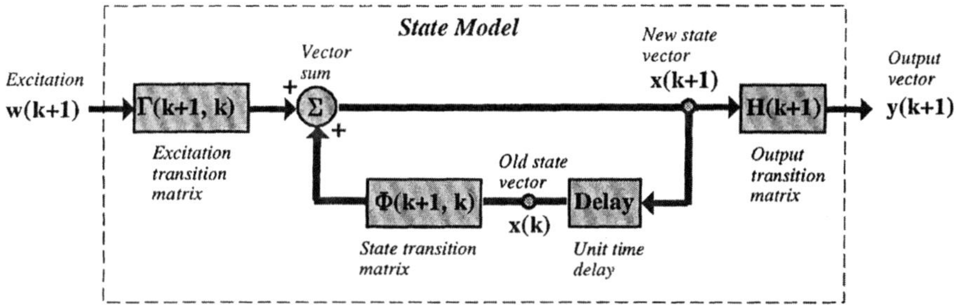

Figure 8.2 Vector block diagram showing the next state and output equations of a discrete-time linear system.

The state $x(k)$ of the system at time k can be obtained in a nonrecursive fashion by solving Eq. (8.1a) for $x(k)$. This solution will be obtained by sequential evaluating $x(k)$. For example, evaluating (8.1a) at $k = 0$ allows us to solve for $x(1)$ as

$$x(1) = \Phi(1, 0)x(0) + \Gamma(1, 0)w(0)$$

Similarly at $k = 1$ we have

$$x(2) = \Phi(2, 1)x(1) + \Gamma(2, 1)w(1)$$

Substituting $x(1)$ given before in the equation above, we see that

$$x(2) = \Phi(2, 1)\Phi(1, 0)x(0) + \Phi(2, 1)\Gamma(1, 0)w(0) + \Gamma(2, 1)w(1)$$

Now substituting the $x(2)$ above into Eqs. (8.1a) evaluated at $k = 3$ gives

$$\begin{aligned} x(3) = \Phi(3, 2)\{&\Phi(2, 1)\Phi(1, 0)x(0) + \Phi(2, 1)\Gamma(1, 0)w(0) \\ &+ \Gamma(2, 1)w(1)\} + \Gamma(3, 2)w(2) \end{aligned}$$

It is useful to define $\Phi(k, j)$, the transition matrix from time k to time j, as

$$\Phi(k, j) = \Phi(k, k-1) \cdot \Phi(k-1, k-2) \ldots \Phi(j+1, j) \qquad (8.2)$$

By this definition of $\Phi[k, j]$, $x(3)$ can be written as

$$x(3) = \Phi(3, 0)x(0) + \Phi(3, 1)\Gamma(1, 0)w(0) + \Phi(3, 2)\Gamma(2, 1)w(1) + \Phi(3, 3)\Gamma(3, 2)w(2)$$

Continuing this process, it can be shown that $x(k)$, at any time k, can be written in terms of the initial state vector, state transition matrix, excitation vector, and excitation transition matrix as follows:

$$x(k) = \Phi(k, 0)x(0) + \sum_{i=1}^{k} \Phi(k, i)\Gamma(i, i-1)w(i-1) \qquad k = 1, 2, \ldots \qquad (8.3)$$

This equation (8.3) describes $x(k)$ in terms of the product of the state transition from k to 0 and the initial condition plus a weighted sum of the excitation vector from index 0 to index $k-1$ where the weighting is a function of $\Phi(k, i)$ and $\Gamma(i, i-1)$.

8.2.1 State Dynamics with Random Excitations

In many problems the excitation vector sequence, $w(k)$, is a random process and the initial condition a random vector. Thus the state vector is a vector random process. We will now find the first- and second-order statistical properties of the state vector, namely the mean and covariance matrix.

Taking the expected value of both sides of Eqs. (8.1a), assuming that $\Phi(k+1, k)$ and $\Gamma(k+1, k)$ are deterministic, gives the following vector difference equation for the mean vector:

$$E[x(k+1)] = \Phi(k+1, k)E[x(k)] + \Gamma(k+1, k)E[w(k)], \quad k = 0, 1, \ldots \tag{8.4}$$

The initial condition for the difference equation is the vector $E[x(0)]$.

Following the steps described earlier for solving for $x(k)$, we can solve the difference equation for the expected value or mean of the random state vector. This solution can be obtained also by taking the expected value of both sides of (8.3).

$$E[x(k)] = \Phi(k, 0)E[x(0)] + \sum_{i=1}^{k} \Phi(k, i)\Gamma(i, i-1)E[w(i-1)], \quad k = 1, 2, \ldots \tag{8.5}$$

The covariance matrix for $x(k)$ provides variances and covariances of the components of the state vector $x(k)$ as a function of time k, and thus it provides information concerning the spread of each component in the state vector. Formulas for the covariance matrices as a function of k will now be presented. For convenience, we use the notation $E[x(k)] = \bar{x}(k)$ and $E[w(k)] = \bar{w}(k)$ in the derivation of the covariance matrix $P(k)$ of the state vector. If $x(k)$ is an n vector, then $\bar{x}[0]$ is an n vector and the covariance matrix $P(0)$ for the state at time zero, $x(0)$, will be an $n \times n$ matrix. To simplify the results, we will assume that the initial condition vector $x(0)$, with mean $\bar{x}(0)$ and covariance matrix $P(0)$, is uncorrelated with the excitation vector process $w(k)$. This assumption can be written as

$$E[(x(0) - \bar{x}(0))(w(k) - \bar{x}(k))^T] = 0, \quad k = 0, 1, 2, \ldots \tag{8.6}$$

Furthermore, in the development to follow, the excitation vector process, a p vector, is assumed a **white vector process** that is defined by

$$E[(w(k) - \bar{w}(k))(w(j) - \bar{w}(j))^T] = \begin{cases} Q(k), & j = k \\ 0, & j \neq k \end{cases} \tag{8.7}$$

where $Q(k)$ is an $n \times n$ covariance matrix for the vector excitation process. This definition of a white vector process is consistent with the fact that random vectors at different time indexes are uncorrelated but allow the components of the vector process at any given k to possibly be correlated. Thus $Q(k)$ need not be a diagonal matrix.

The **state covariance matrix** at index $k+1$ is defined as

$$P(k+1) \triangleq E[(x(k+1) - \bar{x}(k+1))(x(k+1) - \bar{x}(k+1))^T] \tag{8.8}$$

With the assumptions above it will be shown that the state covariance matrix $P(k)$ satisfies the recursive matrix equation.

$$P(k+1) = \Phi(k+1,k)P(k)\Phi^T(k+1,k) + \Gamma(k+1,k)Q(k)\Gamma^T(k+1,k), \quad k = 0, 1, \ldots \tag{8.9}$$

with initial condition $P(0)$.

The proof of (8.9) can be obtained in a straightforward but tedious fashion. As the proof is somewhat constructive, giving some interesting auto- and cross-correlation results along the way, it is now presented.

The proof begins with the definition of the state covariance matrix Eq. (8.8) and uses the expected value or mean of the state as given in (8.6). Substituting Eqs. (8.1a) and (8.4) into (8.8) yields.

$$\begin{aligned} P(k+1) = E[&\{\Phi(k+1,k)(x(k) - \bar{x}(k)) + \Gamma(k+1,k)(w(k) - \bar{w}(k))\} \\ &\times \{\Phi(k+1,k)(x(k) - \bar{x}(k)) + \Gamma(k+1,k)(w(k) - \bar{w}(k))\}^T] \end{aligned} \tag{8.10}$$

Expanding the matrix product indicated by the $\{\ldots\}$ and taking the deterministic $\Phi(k+1,k)$ and $\Gamma(k+1,k)$ outside the expected value, (8.10) can be written as the sum of four terms as follows:

$$\begin{aligned} P(k+1) = {} & \Phi(k+1,k)E[(x(k) - \bar{x}(k))(x(k) - \bar{x}(k))^T]\Phi^T(k+1,k) \\ & + \Phi(k+1,k)E[(x(k) - \bar{x}(k))(w(k) - \bar{w}(k))^T]\Gamma^T(k+1,k) \\ & + \Gamma(k+1,k)E[(w(k) - \bar{w}(k))(x(k) - \bar{x}(k))^T]\Phi^T(k+1,k) \\ & + \Gamma(k+1,k)E[(w(k) - \bar{w}(k))(w(k) - \bar{w}(k))^T]\Gamma^T(k+1,k) \end{aligned} \tag{8.11}$$

To finish the proof, we must show that the middle two terms of (8.11) are zero.

First let us look at the correlation between $x(k) - \bar{x}(k)$ and $w(k) - \bar{w}(k)$. Substituting (8.3) and (8.5) for $x(k)$ and $\bar{x}(k)$ into $E[(x(k) - \bar{x}(k))(w(k) - \bar{w}(k))^T]$ gives

$$\begin{aligned} & E[(x(k) - \bar{x}(k))(w(k) - \bar{w}(k))^T] \\ & \quad = \Phi(k,0)E[(x(0) - \bar{x}(0))(w(k) - \bar{w}(k))^T] \\ & \qquad + \sum_{i=1}^{k} \Phi(k,i)\Gamma(i,i-1)E[(w(i-1) - \bar{w}(i-1))(w(k) - \bar{w}(k))^T] \end{aligned} \tag{8.12}$$

By the assumption (8.6), the first term on the right side of (8.12) is zero. Also since the summation of the second term goes only to k, the expected value of each term of the summation is zero by the assumption (8.7) and thus

$$E[(x(k) - \bar{x}(k))(w(k) - \bar{w}(k))^T] = 0 \tag{8.13}$$

This states that the $x(k) - \bar{x}(k)$ and $w(k) - \bar{w}(k)$ are orthogonal. Taking the transpose of (8.13), we also see that the third term of (8.11) is also zero and thus (8.9) is verified.

8.2.2 Markov Sequence Model

In this section we define the Markov sequence model that is used in the development of the Kalman filter. The vector **Markov sequence** $x(k)$ is defined as the output of a dynamic system to a vector white noise process $w(k)$ as

$$x(k+1) = \Phi(k+1, k)x(k) + \Gamma(k+1, k)w(k), \quad k = 0, 1, \ldots \tag{8.14}$$

In Eq. (8.14) the $\Phi[k+1, k]$ is an n by n matrix, $\Gamma[k+1, k]$ is an n by p matrix, and $w(k)$ is a p by 1 white vector sequence characterized by its covariance matrix $Q(k)$ as follows:

$$E[(w(j) - \bar{w}(j))(w(k) - \bar{w}(k))^T] = \begin{cases} Q(k), & k = j \\ 0, & k \neq j \end{cases} \tag{8.15}$$

Since $w(k)$ is a p by 1 vector, the covariance matrix, $Q(k)$, is a p by p positive definite matrix but is not necessarily diagonal at each k as we will allow correlation between components of the $w(k)$ vector with this model. The fact that $Q(k)$ is a function of k allows certain nonstationary processes to be included in the model.

The initial condition vector $x(0)$ is assumed to be a random n vector characterized by its n by one mean vector, $E[x(0)]$, and n by n covariance matrix, $P(0)$, which are defined by

$$\begin{aligned} E[x(0)] &= \bar{x}(0) \\ E[(x(0) - \bar{x}(0))(x(0) - \bar{x}(0))^T] &= P(0) \end{aligned} \tag{8.16}$$

Furthermore we will assume that the white sequence $w(k)$ is uncorrelated for all k with the initial condition vector $x(0)$, that is,

$$E[(x(0) - \bar{x}(0))(w(k) - \bar{w}(k))^T] = 0, \qquad k = 0, 1, \ldots \tag{8.17}$$

This assumption is reasonable for most situations and will simplify considerably the work and the results that follow.

8.2.3 Observation Model

The state of the system (usually a random vector) is not directly available to us but is usually observed through a linear operation and contaminated with additive noise. So a reasonable model for the observation or measurement is

$$z(k+1) = H(k+1)x(k+1) + v(k+1), \qquad k = 0, 1, \ldots \tag{8.18}$$

where

$z(k+1)$ is an m vector specifying the **observation (measurement) vector** at time $k+1$.

$x(k+1)$ is an n vector representing the **state vector** of the system at time $k+1$.

$H(k+1)$ is an $m \times n$ matrix called the **measurement matrix** at time $k+1$.

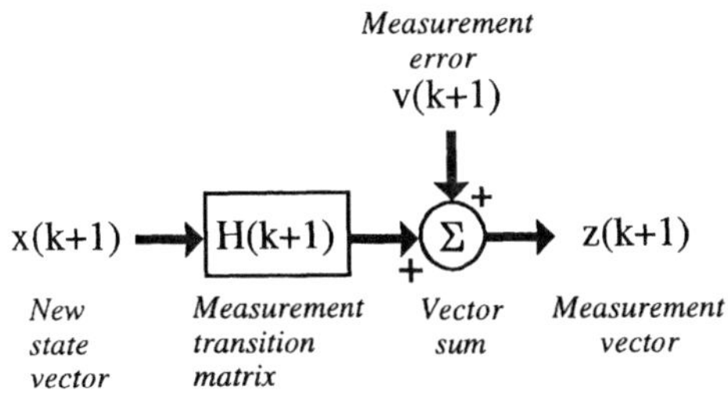

Figure 8.3 Vector block diagram for the observation model.

$v(k+1)$ is an m vector representing the **measurement error** at time $k+1$.
$x(0)$ is the **initial condition** with mean $\bar{x}(0)$ and covariance matrix $P(0)$.

A vector block diagram for the observation process is shown in Figure 8.3. The **observation error vector** $v(k+1)$ is assumed to be a white random process with mean and covariance matrix defined by

$$E[v(k+1)]] = \bar{v}(k+1)$$
$$E[(v(j+1) - \bar{v}(j+1))(v(k+1) - \bar{v}(k+1))^T] = \begin{cases} R(k+1), & k = j \\ 0, & k \neq j \end{cases} \tag{8.19}$$

The $R(k)$ is an m by m covariance matrix that is not necessarily diagonal and could be a function of the time index k. Thus $v(k)$, the measurement error vector, is not necessarily required to be a stationary process.

Assume further, and quite reasonably so, that $v(k+1)$ is uncorrelated with the state initial condition, $x(0)$, which can be expressed by

$$E[(x(0) - \bar{x}(0))(v(k+1) - \bar{v}(k+1))^T] = 0, \quad k = 0, 1, \ldots \tag{8.20}$$

and that $v(k+1)$ is uncorrelated with $w(j+1)$ for all j and k as follows:

$$E[(w(j+1) - \bar{w}(j+1))(v(k+1) - \bar{v}(k+1))^T] = 0 \tag{8.21}$$

8.3 BASIC ESTIMATION PROBLEM

It is assumed that the state of the system that we desire to estimate is modeled as a Gaussian Markov sequence given by the following matrix difference equation from Eq. (8.14) with properties given in Section 8.2.2:

$$x(k+1) = \Phi(k+1, k)x(k) + \Gamma(k+1, k)w(k), \quad k = 0, 1, \ldots \tag{8.22}$$

In most cases the states of the system themselves may not be directly measurable without error. We will assume that we have available $z(k+1)$, where $z(k+1)$ is given by

$$z(k+1) = H(k+1)x(k+1) + v(k+1), \quad k = 0, 1, \ldots \tag{8.23}$$

whose properties of $v(k+1)$ are given in Section 8.2.3. The $z(k+1)$ is called the **observation or measurement vector**, which is assumed to be an m vector, $H(k+1)$ is an m by n matrix relating the observation vector to the state vector, and $v(k+1)$ is an m vector representing the **observation or measurement error**.

8.3.1 Problem Formulation

The basic problem of estimation can now be formulated. Given the observation vectors at times $1, 2, \ldots, j$ as $z(1), z(2), \ldots, z(j)$, determine an estimate of the state vector $x(k)$ as a function of those observations. Let $\hat{x}(k \mid j)$ represent the estimate of $x(k)$ given $z(1), z(2), \ldots, z(j)$, that is,

$$\hat{x}(k \mid j) = g[z(1), z(2), \ldots, z(j)] \tag{8.24}$$

Depending on the relative positions of the time indexes k and j, we can identify three major types of estimation problems:

If $k > j$, the problem is **prediction**.
If $k = j$, the problem is **filtering**.
If $k < j$, the problem is **smoothing** or interpolation.

In making the estimate of $x(k)$ given j observation vectors, an error is usually made. Define this **estimation error** vector $\varepsilon(k \mid j)$ as follows:

$$\varepsilon(k \mid j) \triangleq \tilde{x}(k \mid j) = x(k) - \hat{x}(k \mid j) \tag{8.25}$$

Because $x(k)$ and $\hat{x}(k \mid j)$, which is a function of $z(j)$, are random vectors, this error being the difference is also a random vector. A convenient measure of performance for the estimates is the expected value of some function C of the error:

$$J = E[C(\varepsilon(k \mid j))] \tag{8.26}$$

We desire then to find a function $g(.)$, shown in Eq. (8.24), of the measurements $z(0), z(1), \ldots, z(j)$ such that J is minimized. Such an estimate is called the best or **optimal estimate** with respect to J. Different selections of $C(.)$ may result in different estimates. For the purpose of this chapter, $C(\varepsilon(k \mid j))$ will be assumed to be $C(\varepsilon(k \mid j)) = \varepsilon^T(k \mid j)\varepsilon(k \mid j)$, which gives rise to the minimum mean squared error estimates.

The choice of the $g(.)$ can also result in different types of estimators including linear, nonlinear, time-invariant, and time-varying solutions. We will concentrate on linear time-varying estimates as discussed in the following section.

8.3.2 Linear Estimation with Minimum Mean Squared Error Criterion

In many cases a linear estimate is desired. It will be shown, first, that it is easy to implement, second, that theoretical results can usually be obtained for the performance before the procedure is implemented, and third, that only the first- and second-order properties of the measurement errors and state vector are needed to determine the estimator.

Assume that the $g(.)$ is a linear function of the measurement vectors given in Eq. (8.23) represented by a matrix multiplication of the observations $z(1), z(2), \ldots, z(j)$. For convenience let $z_*(j)$ be the $j \cdot m$ column vector of measurements composed of the stacked observation vectors $z(k)$ at times $k = 1, 2, \ldots, j$ given by

$$z_*(j) = \begin{bmatrix} z(1) \\ z(2) \\ \vdots \\ z(j) \end{bmatrix} \tag{8.27}$$

Since each $z(k)$ is an m vector, $z_*(j)$ will be a $j \cdot m$ column vector, and its size will increase as j is increased. We can therefore write our estimate $\hat{x}(k \mid j)$ as

$$\hat{x}(k \mid j) = A \cdot z_*(j) + \mathbf{b} \tag{8.28}$$

Where A is an n by $j \cdot m$ matrix called the **coefficient matrix** and **b** is an n by 1 **constant** vector. Assume for convenience that $x(k)$ and $v(k)$ are zero mean. If this is not the case, we may subtract off the mean vectors from the observation vectors and work with the new measurements without loss in generality. For the mean squared error performance criterion, assume that $C(x)$ in (8.26) is expressed by $C(x) = x^T x$. Thus J becomes

$$J = E[\tilde{x}^T(k \mid j)\tilde{x}(k \mid j)] \tag{8.29}$$

By this assumption, **b** can be shown to be a zero vector, and the problem reduces to finding a matrix A such that J is minimized. Since $x(k)$ and $z(k)$ are zero-mean stochastic sequences, the orthogonality principle can be applied to yield

$$E[(x(k) - \hat{x}(k \mid j))z_*^T(j)] = 0 \tag{8.30}$$

where the zero matrix on the right side is an n by $m \cdot k$ zero matrix with elements all zero. Also from the orthogonality principle the minimum mean squared error is given by

$$e_{\text{mmse}} = E[(x(k) - \hat{x}(k \mid j))x^T(k)] \tag{8.31}$$

Substituting (8.28) into (8.30) with $\mathbf{b} = \mathbf{0}$, we have

$$E[(x(k) - Az_*(j))z_*^T(j)] = 0 \tag{8.32}$$

Taking the expected value of the bracketed term and rearranging gives

$$E[x(k)z_*^T(j)] = A \cdot E[z_*(j)z_*^T(j)] \tag{8.33}$$

Finally, solving Eq. (8.33) for the coefficient matrix A yields

$$A = P_{x(k)z_*(j)} \cdot P^{-1}_{z_*(j)z_*(j)} \tag{8.34}$$

$$\text{where} \quad P_{x(k)z_*(j)} = E[x(k)z_*^T(j)] \quad \text{and} \tag{8.35}$$
$$P_{z_*(j)z_*(j)} = E[z_*(j)z_*^T(j)]$$

The **optimal linear estimate $\hat{x}(k \mid j)$ in the minimum mean squared error sense** can thus be written as

$$\hat{x}(k \mid j) = P_{x(k)z_*(j)} \cdot P^{-1}_{z_*(j)z_*(j)} \cdot z_*(j) \tag{8.36}$$

To get the estimate, we must invert $P_{z_*(j)z_*(j)}$ which is a jm by jm matrix. As j the number of observation vectors increases, this matrix size increases and for most large j becomes impractical to invert. What is desired from an engineering point of view are efficient and practical algorithms for evaluating optimum estimates $\hat{x}(k \mid j)$. We will see later in this chapter that if certain assumptions are placed on properties of the measurement error vector $v(k+1)$ and the message model source $u(k)$ that sequential algorithms, the Kalman filter [14], exist for practical implementation of the estimates. It has been shown by Meditch [1] that the estimate forms are optimal for other forms of L and that the linear estimates are also optimum nonlinear filters if the processes are all Gaussian.

8.4 OPTIMAL FILTERED ESTIMATE

One of the most basic problems in communication theory is the filtering of a noisy signal to obtain a cleaner version of the signal. In general, the signal could represent a voltage, current, position, velocity, acceleration, or any combination. In Chapter 7 the optimal linear filter under the Wiener framework was found. Procedures were described for the estimation of a single scalar signal, assuming that the signal and noise are modeled as wide sense stationary processes characterized by their autocorrelation functions and power spectral densities. The extension to the nonstationary case and vector processes was not accomplished because of a complexity that borders on impossibility. The Kalman filter can easily handle nonstationary and vector processes because of its basic state space formulation. The signals and nonstationarities must be, however, within the signal model and linear state space formulation, and the noise process must be of the additive form and in particular assumed to be white.

8.4.1 The Kalman Filter

In the Kalman formulation we desire the best linear estimate of the state vector $x(k)$ as a function of the observation vectors $z(1), z(2), \ldots, z(k)$, where best is assumed to be in the minimum mean squared error sense. It can be shown that the Kalman filter is optimum under a wider class of performance functions than the mmse function.

Let $\hat{x}(k \mid k)$ represent the estimate of the state at time index k given observations $z(1), z(2), \ldots, z(k)$. The Kalman filter thus gives us a running estimate of the state at time index k as a function of the observations as shown conceptually in Figure 8.4.

The Kalman filter and associated sequential algorithms for prediction and smoothing will be presented, theorem like, specifying clearly the given assumptions, the form of the optimal filter, and the performance associated with the optimum filter. Proofs will not be given. The overall presentation of results summarizes those given in Meditch [1]. Students interested in seeing the proofs can find them in many references, in particular, Meditch [1] and Kalman and Bucy [13].

KALMAN FILTER SUMMARY

Assumptions

The state of the system is governed by the model in Section 8.2.2. It is assumed that $\Phi(k+1, k), \Gamma(k+1, k), P(0)$, and $Q(k)$ are known and deterministic. It is assumed that the observation model is given in Section 8.2.3 with $H(k+1)$ and $R(k)$ known and deterministic. These are summarized as follows:

$$\begin{aligned} x(k+1) &= \Phi(k+1, k)x(k) + \Gamma(k+1, k)w(k), \qquad k = 0, 1, \ldots \\ z(k+1) &= H(k+1)x(k+1) + v(k+1) \end{aligned} \tag{8.37}$$

Kalman Filter Algorithm: Filtered Estimate

The optimal linear filtered estimate $\hat{x}(k+1 \mid k+1)$ as shown by Kalman [13] to be governed by the recursive matrix formula for

$$\begin{aligned} \hat{x}(k+1 \mid k+1) = {} & \Phi(k+1, k)\hat{x}(k \mid k) + K(k+1)[z(k+1) - H(k+1) \\ & \times \Phi(k+1, k)\hat{x}(k \mid k)] \\ & \text{with initial condition } \hat{x}(0 \mid 0) = 0, \quad k \geq 0 \end{aligned} \tag{8.38}$$

where $K(k+1)$ is called the **Kalman gain**.

Kalman Gain

The Kalman gain is given by

$$\begin{aligned} K(k+1) = {} & P(k+1 \mid k)H^T(k+1) \\ & \times [H(k+1)P(k+1 \mid k)H^T(k+1) + R(k+1)]^{-1} \quad k = 0, 1, \ldots \end{aligned} \tag{8.39}$$

Notice that the Kalman gain is written in terms of a $P(k+1\,|\,k)$, which is the single-step prediction error covariance matrix, which is now defined.

Single-Step Prediction Error Covariance Matrix (Single-Step Performance)

$$P(k+1\,|\,k) = \Phi(k+1,k)P(k\,|\,k)\Phi^T(k+1,k) + \Gamma(k+1,k)Q(k)\Gamma^T(k+1,k)$$
$$\text{with initial condition } P(0\,|\,0) = P(0), \qquad k = 0, 1, \ldots \tag{8.40}$$

Error Covariance Matrix for Filtered Error (Covariance Recursion)

$$P(k+1\,|\,k+1) = [I - K(k+1)H(k+1)]P(k+1\,|\,k), \qquad k = 0, 1, \ldots \tag{8.41}$$

Performance (Covariance Matrix for Kalman Filtered Estimate of the State)

$$P(k+1\,|\,k+1) \triangleq E[(x(k+1) - \hat{x}(k+1\,|\,k+1))(\hat{x}(k+1) - \hat{x}(k+1\,|\,k+1))^T], \qquad k = 0, 1, \ldots \tag{8.42}$$

The diagonal components of $P(k+1\,|\,k+1)$, the error covariance matrix, are the variances of the estimates of each individual component of the state vector $x(k+1)$, thus providing a measure of performance of the component estimates determined by the Kalman filter. This measure of effectiveness is available from Eq. (8.41) and can be calculated before any measurements are taken or processed. The off-diagonal

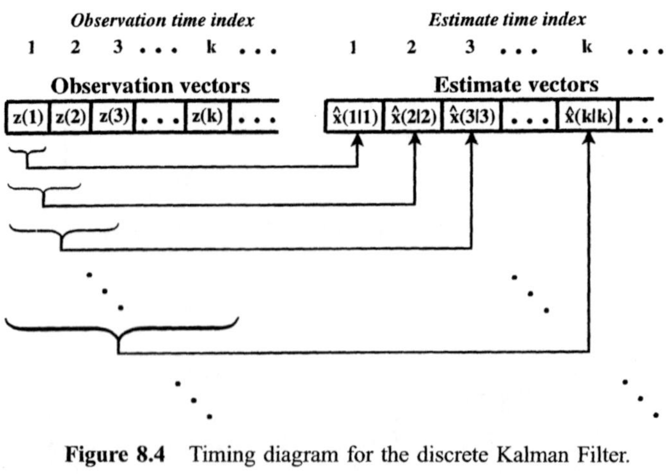

Figure 8.4 Timing diagram for the discrete Kalman Filter.

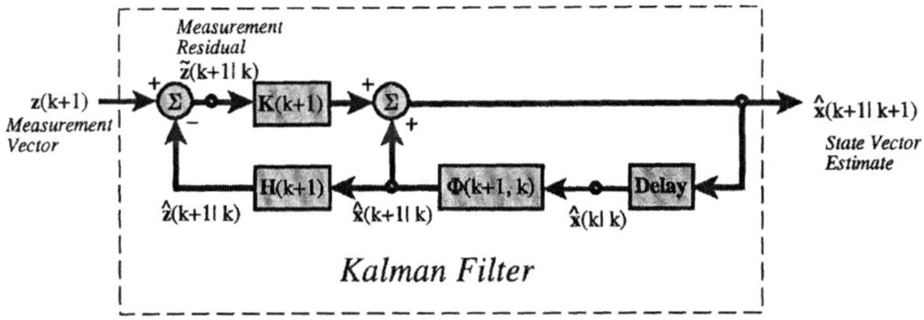

Figure 8.5 Vector block diagram for the Kalman Filter in terms of the measurement residual.

components give the covariances of the error vector components. The state vector block diagram for the Kalman filtered estimate given in Eq. (8.38) is shown in Figure 8.5.

An alternative form of the Kalman filter can be derived from (8.38) by grouping together the $\hat{x}(k\,|\,k)$ terms to get

$$\begin{aligned}\hat{x}(k+1\,|\,k+1) = {} & [\Phi(k+1, k) - K(k+1)H(k+1)\Phi(k+1, k)]\hat{x}(k\,|\,k) \\ & + K(k+1)z(k+1) \qquad\qquad k \geq 0 \\ & \text{with initial condition } \hat{x}(0\,|\,0) = 0,\end{aligned} \tag{8.43}$$

This implementation of the Kalman filter is shown in Figure 8.6 as a first-order vector difference equation with the measurement vector as the driving function.

In the following sections the structure and the internal workings of the Kalman filter are explored.

8.4.2 The Anatomy of the Kalman Filter

From Eq. (8.38) we saw that the filtered estimate at time $k+1$ in terms of $k+1$ measurements is a function of the optimum estimate at time k using k measurements

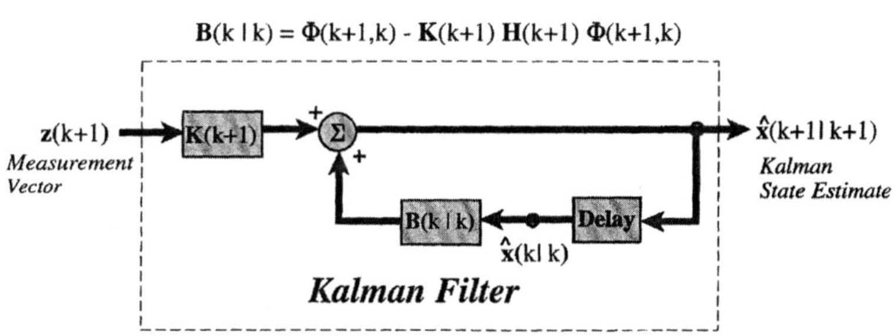

Figure 8.6 Alternative vector form of the Kalman Filter.

and the new measurement at time $k+1$ as shown in Figure 8.7. The **measurement residual** $\tilde{z}(k+1\,|\,k)$ is defined by the error in the predicted measurement as follows:

$$\tilde{z}(k+1\,|\,k) = z(k+1) - \hat{z}(k+1\,|\,k) \tag{8.44}$$

In Figure 8.7 sections of the filter equation are identified that illustrate how the Kalman filter works as a predictor-corrector. The filter first predicts the new state (single-step prediction) and then corrects it by adding the product of the Kalman gain and the measurement residual. The measurement residual is seen as the difference between the new measurement and the single-step prediction of the measurement. If they are the same then the residual will be zero and no correction is applied.

8.4.3 Physiology of the Kalman Filter

From Eq. (8.38) it is seen that the new estimate is a function of the state transition matrix, $\Phi(k+1\,|\,k)$, the measurement matrix, $H(k+1)$, the old estimate $\hat{x}(k\,|\,k)$, the new measurement $z(k+1)$, and the Kalman gain $K(k+1)$. Of these only the Kalman gain needs to be calculated. The Kalman gain matrix (possibly different at each time index) is determined by the coupled Eqs. (8.39) to (8.41), which specify a recursive evaluation of $P(k+1\,|\,k)$, $K(k+1)$, and $P(k+1\,|\,k+1)$. This recursion displayed in Figure 8.8 shows that the $P(k+1\,|\,k), K(k+1)$, and $P(k+1\,|\,k+1)$ must be calculated **in that order**.

In Eq. (8.39) to (8.41) notice that $P(k+1\,|\,k), K(k+1)$, and $P(k+1\,|\,k+1)$ do not depend on the measurements or estimates. Thus they can be calculated "off line" before they are needed, or in "real time" as the measurements come in. If they are calculated ahead of time there is no overhead for their calculation, but storage must be provided for all the $K(k+1)$ determined for all k that the filter is to be used. If we calculate them between the presentation of observations, the acceptable sample rate may have to be reduced because of the extra computational load.

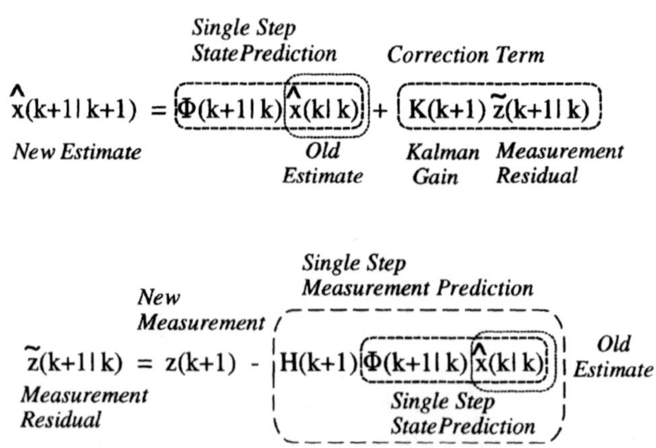

Figure 8.7 Predictor-corrector structure of the Kalman filtering equation.

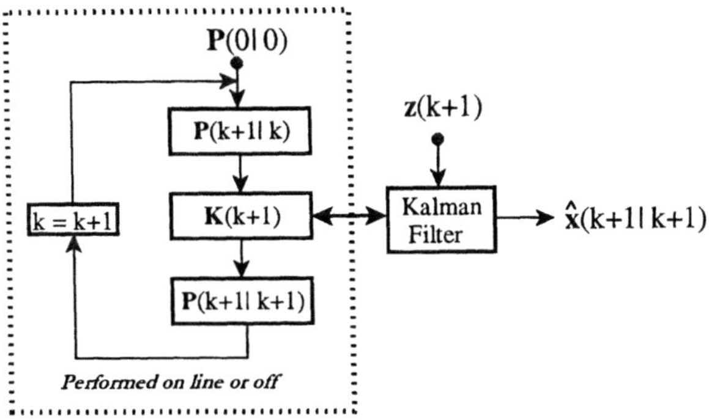

Figure 8.8 Recursive calculation procedure for the prediction error covariance, the Kalman gain, and the filtered error covariance matrices.

It is also interesting to note that the performance characteristics given by $P(k+1\,|\,k+1)$, the error covariance matrix for estimating the state, can be determined without a single measurement. Thus the overall performance of the Kalman filter can be easily computed before any implementation.

A simple one-dimensional or scalar filtering problem is now presented.

EXAMPLE 8.1

The message and observation models are given as follows

$$\begin{aligned} x(k+1) &= -0.7x(k) + w(k), \quad k \geq 0 \\ z(k+1) &= x(k+1) + v(k+1) \end{aligned}$$

Assume that R and Q are given by

$$\begin{aligned} E[w(j)w(k)] &= Q\delta(j-k) = \delta(j-k) \\ E[v(j)v(k)] &= R\delta(j-k) = 1/2\delta(j-k) \end{aligned}$$

Find the Kalman filter for estimating the signal $x(k+1)$ from the given $k+1$ measurements.

SOLUTION

Since the state vector is just one dimensional, the Kalman filter in the scalar form of Eq. (8.38) can be written as follows, with initial condition $\hat{x}(0\,|\,0) = 0$:

$$\hat{x}(k+1\,|\,k+1) = -0.7\hat{x}(k\,|\,k) + K(k+1)[z(k+1) + 0.7\hat{x}(k\,|\,k)], \quad k = 0, 1, \ldots$$

where the $K(k+1)$, $P(k+1\,|\,k)$, and $P(k+1\,|\,k+1)$ are determined from (8.39) to (8.41) for $k = 0, 1, 2, \ldots$ as

$$\begin{aligned} P(k+1\,|\,k) &= -0.7P(k\,|\,k)(-0.7) + 0.5 \qquad \text{with } P(0\,|\,0) = P(0) \\ K(k+1) &= P(k+1\,|\,k)[P(k+1\,|\,k) + 0.5]^{-1} \\ P(k+1\,|\,k+1) &= [1 - K(k+1)]P(k+1\,|\,k) \end{aligned}$$

For the order shown above and $P(0\,|\,0) = 10$, the values for the $P(k+1\,|\,k)$, $K(k+1)$, and $P(k+1\,|\,k+1)$ are calculated. The graph of each is plotted in Figure 8.9.

Notice that the Kalman gain is dynamic for the first few terms but quickly reaches a steady state result of 0.5603575. The $P(k+1\,|\,k)$ approaches a steady state value of 0.6372876. This gives the single-step prediction error. The $P(k+1\,|\,k+1)$ starts out high and approaches quickly a steady state value of 0.2801788 which represents the mean squared error for the filtered estimate in the steady state.

Also note that the form of the Kalman filter is determined without the knowledge of the measurements and thus can be computed offline if desired before any measurements are taken. ■

EXAMPLE 8.2

In this example we begin with an AR(2) model for a signal $x(k)$, which is given by

$$x(k) = -\tfrac{3}{4}x(k-1) - \tfrac{1}{2}x(k-2) + 2w(k), \qquad k \geq 0$$

with $E[w(k)w(j)] = \delta(j-k)$. We observe $x(k)$ in additive white noise $v(k)$, with an autocorrelation function $R_{VV}(k-j) = \delta(j-k)$. We want to estimate the scalar signal $x(k)$ using a Kalman filter. Find the message and observation models for this estimation problem, and specify the Kalman filter that performs the desired estimation of $x(k)$.

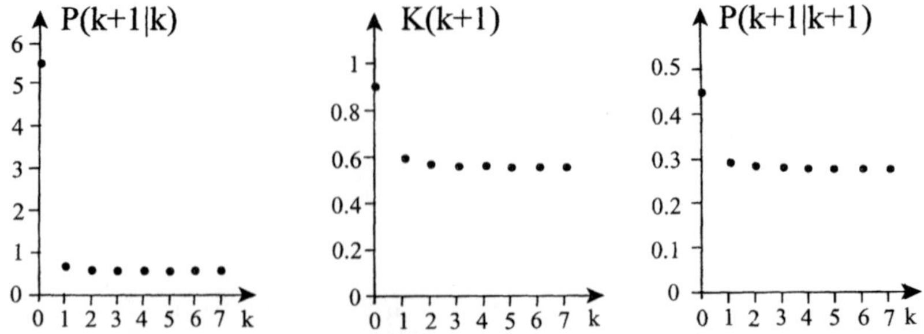

Figure 8.9 Plots of $P(k+1\,|\,k)$, $K(k+1)$, and $P(k+1\,|\,k+1)$ for Example 8.1.

SOLUTION

Although the signal we wish to estimate is a scalar, the state variable formulation results in a vector formulation because of the delay of two samples in the model. To obtain the state variable representation, let the first state variable be $x_1(k) = x(k)$, what we wish to estimate, and the second be $x_2(k) = x(k-1)$. This choice of state variables facilitates the development of the state model. Any other assignment can be made giving a different filter form, yet the resulting filter will give equivalent performance. Since the Kalman filter estimates the state of the system, we will simultaneously estimate both $x(k)$ and $x(k-1)$. Evaluating the AR(2) model given above at $k+1$ gives

$$x(k+1) = -\tfrac{3}{4}x(k) - \tfrac{1}{2}x(k-1) + 2w(k+1), \qquad k \geq 0$$

By the definitions for $x_1(k)$ and $x_2(k)$, this equation can be rewritten as

$$x_1(k+1) = -\tfrac{3}{4}x_1(k) - \tfrac{1}{2}x_2(k) + 2w(k+1)$$

The second equation for the state model is obtained from the definitions of $x_1(k)$ and $x_2(k)$ as

$$x_2(k+1) = x_1(k)$$

These two equations can be put in the standard state variable form, resulting in the following message model:

$$\begin{bmatrix} x_1(k+1) \\ x_2(k+1) \end{bmatrix} = \begin{bmatrix} -\frac{3}{4} & -\frac{1}{2} \\ 1 & 0 \end{bmatrix} \begin{bmatrix} x_1(k) \\ x_2(k) \end{bmatrix} + \begin{bmatrix} 2 \\ 0 \end{bmatrix} w(k+1)$$

Since $v(k+1)$ is the additive noise and we observe only $z(k+1) = x(k+1) + v(k+1)$, the measurement model can be written as

$$z(k+1) = \begin{bmatrix} 1 & 0 \end{bmatrix} \begin{bmatrix} x_1(k+1) \\ x_2(k+1) \end{bmatrix} + v(k+1)$$

The model is now in the standard form for application of the Kalman filter. The model parameters are

$$\Phi = \begin{bmatrix} -\frac{3}{4} & -\frac{1}{2} \\ 1 & 0 \end{bmatrix}, \quad \Gamma = \begin{bmatrix} 2 \\ 0 \end{bmatrix}, \quad H = \begin{bmatrix} 1 & 0 \end{bmatrix}$$

The Q associated with the message model $w(k)$ and the R associated with the additive measurement noise model $v(k)$ are, respectively, $Q = 1$ and $R = 1$ from the given autocorrelation functions of $w(k)$ and $v(n)$. To complete the model, initial conditions on the state at the origin and the state covariance at the origin must be assigned. These initial values are assigned as follows:

$$\begin{bmatrix} x_1(0) \\ x_2(0) \end{bmatrix} = \begin{bmatrix} 0 \\ 0 \end{bmatrix}, \quad P(0|0) = \begin{bmatrix} 10 & 0 \\ 0 & 10 \end{bmatrix}$$

$P(0\,|\,0)$ is arbitrarily selected, usually with large values on the main diagonal, whereas the zero vector for $x(0)$ comes from the fact that the signal has a zero mean, since it is generated from a white source which implies a zero mean.

From the assumptions above and the definitions in Eqs. (8.39), (8.40), and (8.41), the $P(k+1\,|\,k)$, $K(k+1)$, and $P(k+1\,|\,k+1)$ for the Kalman filter are determined as follows:

$$\begin{bmatrix} p_{11}(k+1\,|\,k) & p_{12}(k+1\,|\,k) \\ p_{21}(k+1\,|\,k) & p_{22}(k+1\,|\,k) \end{bmatrix} = \begin{bmatrix} -\frac{3}{4} & -\frac{1}{2} \\ 1 & 0 \end{bmatrix} \begin{bmatrix} p_{11}(k\,|\,k) & p_{12}(k\,|\,k) \\ p_{21}(k\,|\,k) & p_{22}(k\,|\,k) \end{bmatrix}$$
$$\times \begin{bmatrix} -\frac{3}{4} & -\frac{1}{2} \\ 1 & 0 \end{bmatrix} + \begin{bmatrix} 2 \\ 0 \end{bmatrix} 1 [2 \quad 0]$$

$$\begin{bmatrix} k_1(k+1) \\ k_2(k+1) \end{bmatrix} = \begin{bmatrix} p_{11}(k+1\,|\,k) & p_{12}(k+1\,|\,k) \\ p_{21}(k+1\,|\,k) & p_{22}(k+1\,|\,k) \end{bmatrix} \begin{bmatrix} 1 \\ 0 \end{bmatrix}$$
$$\times \left([1 \quad 0] \begin{bmatrix} p_{11}(k+1\,|\,k) & p_{12}(k+1\,|\,k) \\ p_{21}(k+1\,|\,k) & p_{22}(k+1\,|\,k) \end{bmatrix} \begin{bmatrix} 1 \\ 0 \end{bmatrix} + 1 \right)^{-1}$$
$$= \begin{bmatrix} p_{11}(k+1\,|\,k)/(1+p_{11}(k+1\,|\,k)) \\ p_{21}(k+1\,|\,k)/(1+p_{11}(k+1\,|\,k)) \end{bmatrix}$$

$$\begin{bmatrix} p_{11}(k+1\,|\,k+1) & p_{12}(k+1\,|\,k+1) \\ p_{21}(k+1\,|\,k+1) & p_{22}(k+1\,|\,k+1) \end{bmatrix} = \left(\begin{bmatrix} 1 & 0 \\ 0 & 1 \end{bmatrix} - \begin{bmatrix} k_1(k+1) \\ k_2(k+1) \end{bmatrix} [1 \quad 0] \right)$$
$$\times \begin{bmatrix} p_{11}(k\,|\,k) & p_{12}(k\,|\,k) \\ p_{21}(k\,|\,k) & p_{22}(k\,|\,k) \end{bmatrix}$$
$$= \begin{bmatrix} (1-k_1(k+1)) & 0 \\ -k_2(k+1) & 1 \end{bmatrix}$$
$$\times \begin{bmatrix} p_{11}(k\,|\,k) & p_{12}(k\,|\,k) \\ p_{21}(k\,|\,k) & p_{22}(k\,|\,k) \end{bmatrix}$$

The Kalman filter in terms of $P(k+1\,|\,k)$, $K(k+1)$, and $P(k+1\,|\,k+1)$ and the model parameters from (8.38) becomes

$$\begin{bmatrix} \hat{x}_1(k+1) \\ \hat{x}_2(k+1) \end{bmatrix} = \begin{bmatrix} -\frac{3}{4} & -\frac{1}{2} \\ 1 & 0 \end{bmatrix} \begin{bmatrix} \hat{x}_1(k) \\ \hat{x}_2(k) \end{bmatrix} + \begin{bmatrix} k_1(k+1) \\ k_2(k+1) \end{bmatrix}$$
$$\times \left(z(k+1) - [1 \quad 0] \begin{bmatrix} -\frac{3}{4} & -\frac{1}{2} \\ 1 & 0 \end{bmatrix} \begin{bmatrix} \hat{x}_1(k) \\ \hat{x}_2(k) \end{bmatrix} \right)$$

By collecting the terms for $\hat{x}_1(k)$ and $\hat{x}_2(k)$ and simplifying the equation above we can show the Kalman filter to be

$$\begin{bmatrix} \hat{x}_1(k+1) \\ \hat{x}_2(k+1) \end{bmatrix} = \begin{bmatrix} -\frac{3}{4}+\frac{3}{4}k_1(k+1) & -\frac{1}{2}+\frac{1}{2}k_1(k+1) \\ 1+\frac{3}{4}k_2(k+1) & \frac{1}{2}k_2(k+1) \end{bmatrix} \begin{bmatrix} \hat{x}_1(k) \\ \hat{x}_2(k) \end{bmatrix} + \begin{bmatrix} k_1(k+1) \\ k_2(k+1) \end{bmatrix} z(k+1)$$

In this form the new vector estimate is written as a linear function of the previous estimates plus a linear function of the new measurement. Thus the Kalman filter recursively calculates both the estimate of $x(k+1)$, which is $\hat{x}_1(k+1)$ and $x(k)$ which is $\hat{x}_2(k+1)$ given the $k+1$ measurements. ■

The procedure described in establishing the state model that of using delayed versions of the message signal as states can be used for an AR model of any order. An AR(p) model leads to a p by p state transition matrix. The state assignments for an ARMA(p, q) model can be done in many different ways, all of which lead to alternative, but equivalent, forms for estimating the ARMA signal in additive noise.

8.5 OPTIMAL PREDICTION

Prediction is the process of estimating a process in future time from current and past measurements. There are many practical problems that require prediction. For example, we may have data representing measurements of position, velocity, and acceleration of an object in space and desire to "predict" where the object will be at a certain time into the future, and also what its velocity and acceleration are at that time. Another example is that we may have measurements of a random processes and wish to estimate the value of the process at times advanced from the present like a stock market projection. There are two main basic types of predictors based on where in the future the predictions occur; they are (a) fixed lead prediction and (b) fixed point prediction.

8.5.1 Fixed Lead Prediction

Suppose that we have available measurements $z(1), z(2), \ldots, z(k)$ that are somehow statistically related to the state vector, $x(k)$, for all k, and that we desire the **fixed lead prediction** or estimation of the state vector $x(k+N)$ given by $\hat{x}(k+N\,|\,k)$. The N, which is a fixed integer, represents the **lead time** for the prediction. In fixed lead prediction, each time we accept a new measurement, we are always predicting a fixed amount ahead in time using all the measurements. This is illustrated in Figure 8.10. From the figure we see that at time index k, we observe $z(1), z(2), \ldots, z(k)$ to estimate $x(k+N)$. Similarly, at time index $k+1$, we use all the previous measurements $z(1), z(2), \ldots, z(k)$ and the new measurement $z(k+1)$ to estimate $x(k+1+N)$. This process is continued for all time indexes.

If we restrict the form of the predictor to be linear, as in the linear state model, and assume that our observations are the sum of linearly transformed states and a white noise process, then the optimal fixed lead predictor can be shown to be expressed in the following summary. The presentation is composed of three main parts: basic assumptions, the optimal algorithm, and the performance of the optimal predictor measured by the mean squared error.

OPTIMAL FIXED LEAD PREDICTOR SUMMARY

Assumptions

Let the state of the system be governed by the Markov sequence model given in section 8.2.2 with known deterministic $\Phi(k+1,k)$, $\Gamma(k+1,k)$, $P(0)$, and $Q(k)$, and let the observation model be as given in Section 8.2.3 with known deterministic $H(k+1)$ and $R(k)$. These assumptions are summarized as follows:

$$\begin{aligned} x(k+1) &= \Phi(k+1,k)x(k) + \Gamma(k+1,k)w(k), \qquad k = 0, 1, \ldots \\ z(k+1) &= H(k+1)x(k+1) + v(k+1) \end{aligned} \tag{8.45}$$

Fixed Lead Prediction Algorithm

The **optimal fixed lead predictor** $\hat{x}(k+N\,|\,k)$ with **lead time** N can be shown to be

$$\begin{aligned} &\hat{x}(k+N\,|\,k) = \Phi(k+N,k)\hat{x}(k\,|\,k), \qquad k = 0, 1, \ldots \\ &\text{where} \quad \Phi(k+N,k) = \Phi(k+N, k+N-1) \\ &\qquad\qquad \Phi(k+N-1, k+N-2)\ldots\Phi(k+1,k) \end{aligned} \tag{8.46}$$

and $\hat{x}(k\,|\,k)$ is the optimal linear filtered estimate of the state at time index k obtained from the Kalman filter (Eq. 8.38) using initial conditions $\hat{x}(0\,|\,0)$ and $P(0\,|\,0) = P(0)$.

Performance

The performance specified by the error covariance matrix can be shown from Meditch [1] to be

$$\begin{aligned} P(k+N\,|\,k) &= \Phi(k+N,k)P(k\,|\,k)\Phi^T(k+N,k) \\ &+ \sum_{i=k+1}^{k+N} \Phi(k+N,i)\Gamma(i,i-1)Q(i-1)\Gamma^T(i,i-1)\Phi^T(k+N,i) \end{aligned} \tag{8.47}$$

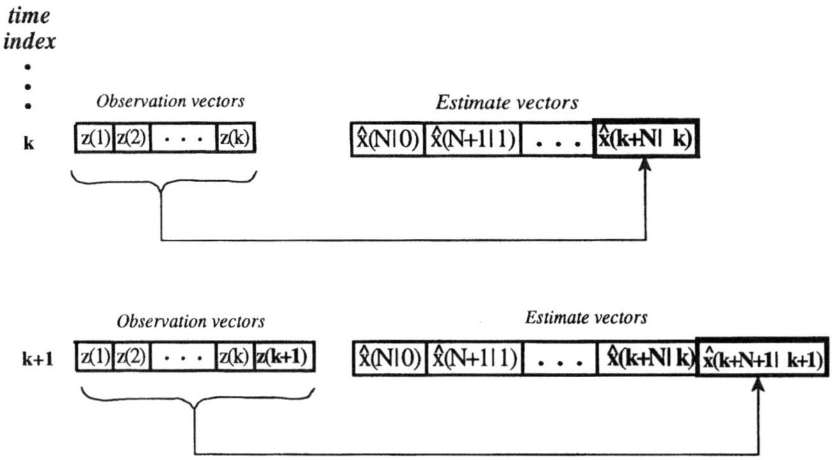

Figure 8.10 Fixed lead prediction using all data in the past.

Recursive formulas can be derived for the determination of the sum term as well as for the generation of the $\Phi(k+N, k)$ term from previous terms if desired, but as this can be determined off line, it is not necessary unless implementing directly on line.

8.5.2 Fixed Lead Prediction (Sliding Window)

Because of memory limitations there are cases where only a certain portion of the past measurements are used to perform the estimation. The most common of this type is the **sliding window fixed lead prediction** which uses only the previous W indexed measurements, where W is the length of the sliding window. Such algorithms will not be presented in our discussion, and Van Trees [16] can be consulted for a thorough discussion of sliding window estimators.

8.5.3 Fixed Point Prediction

In **fixed point prediction** a time index is selected or fixed, say at N, and estimates of the state at time N are made using successive observations up to that time. This process generates a running estimate of the state at time $N, \hat{x}(N \mid k)$ for $k = 0, 1, 2, \ldots, N-1$, as illustrated in Figure 8.11.

At time index 1 we observe $z(1)$ and predict $x(N)$ with $\hat{x}(N \mid 1)$, and next for time index 2 we use both observations $z(1)$ and $z(2)$ to predict $x(N)$ with $\hat{x}(N \mid 2)$ This process is continued until we observe $z(1), z(2), \ldots, Z(N-1)$ to predict $x(N)$ with $\hat{x}(N \mid N-1)$. If further observations are included, the problem is no longer that of prediction but that of filtering for $k = N$ and smoothing for $k > N$.

The summation term can be obtained recursively by subtracting the relevant term in the summation each time from the previous sum if desired.

OPTIMUM FIXED POINT PREDICTOR SUMMARY

Assumptions: Same as for Fixed Lead Predictor

The state of the system is governed by the model given in Section 8.2.2 with known deterministic $\Phi(k+1,k), \Gamma(k+1,k), P(0)$, and $Q(k)$, and the observation model is as given in Section 8.2.3 with known deterministic $H(k+1)$ and $R(k)$. These are summarized as follows:

$$\begin{aligned} x(k+1) &= \Phi(k+1,k)x(k) + \Gamma(k+1,k)w(k), \qquad k = 0,1,\ldots \\ z(k+1) &= H(k+1)x(k+1) + v(k+1) \end{aligned} \tag{8.48}$$

Fixed Point Predictor Algorithm

The **optimal fixed point predictor** for estimating $x(N)$, given measurements up to time k is defined by $\hat{x}(N\,|\,k)$, where N is a fixed constant. The algorithm for obtaining $\hat{x}(N\,|\,k)$ follows:

$$\begin{aligned} &\hat{x}(N\,|\,k) = \Phi(N\,|\,k)\hat{x}(k\,|\,k), \qquad k = 0,1,\ldots,N-1 \\ &\text{where} \quad \Phi(N\,|\,k) = \Phi(N,N-1)\Phi(N-1,N-2)\ldots\Phi(k+1,k) \end{aligned} \tag{8.49}$$

The $\hat{x}(k\,|\,k)$ in the predictor algorithm is the optimal linear filtered estimate of the state at time index k obtained from the Kalman filter (Eq. 8.38) with $\hat{x}(0\,|\,0) = 0$ and $P(0\,|\,0) = P(0)$.

Performance: Error Covariance Matrix

$$\begin{aligned} P(N\,|\,k) &= \Phi(N,k)P(k\,|\,k)\Phi^T(N,k) \\ &\quad + \sum_{i=k+1}^{N} \Phi(N,i)\Gamma(i,i-1)Q(i-1)\Gamma^T(i,i-1)\Phi^T(N,i) \end{aligned} \tag{8.50}$$

8.6 OPTIMAL SMOOTHING

Smoothing is the term used to describe estimation of the state of the system at time index k using measurements from the beginning at 1 up to the time index j where j is greater than k. The estimator uses measurements both before and after k and thus is inherently a noncausal form of processing. Because of this extra flexibility, there are several forms of smoothing algorithms including: fixed interval, fixed point, and fixed lag smoothing.

In **fixed interval smoothing** the purpose is to determine the estimate of the state at all time indexes $k = 0,1,2,\ldots,N-1$ for a fixed interval length of N. This is a nonreal time form of processing.

For **fixed point smoothing** we desire the estimate of the state at some time index N using all measurements up to time index $N+1$, then up to $N+2$ and so on. In this way we obtain running estimates of the state at time N as we accept more and

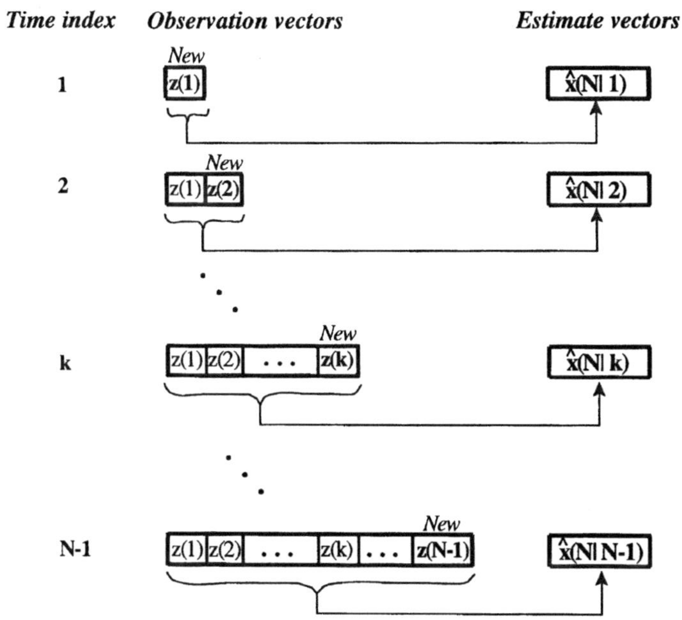

Figure 8.11 Fixed point prediction timing diagram.

more measurements. Although noncausal in the sense of providing estimates at times in terms of future measurements, the algorithm can be developed to operate in pseudo real time in a recursive fashion as the measurements come in.

Fixed lag smoothing gives the smoothed estimate at successive time indexes k in terms of all measurements up to $k + L$, where L represents the **time lag** behind the measurements. Fixed lag smoothing can also be obtained in a recursive fashion and in real time if we accept a time lag.

Each of these forms of smoothing is explored in the following sections as the details of the optimal smoothing algorithms are presented.

8.6.1 Fixed Interval Smoothing

In fixed interval smoothing we desire the estimates of the state of the system for all values of time index, $k = 0, 1, \ldots, N - 1$, where N is the length of the fixed time interval over which data are processed. Conceptually this is shown in Figure 8.12.

The optimal linear fixed interval smoothing estimators, where optimality is in the sense of minimizing the mean squared error, can be obtained by a development similar to that used for obtaining the filtered estimate in Section 8.4.1. However this procedure requires the calculation of a different coefficient matrix for each k in evaluating $\hat{x}(k\,|\,N)$ and correspondingly the evaluation of inverses of large matrices for large N. It does not take a very large N to make this particular approach not only computationally intensive but virtually impossible.

In the literature, recursive algorithms for calculating the smoothed estimates have been developed because of the unique framework induced by the state and measurement models. Most of the algorithms are seen to include the determination of the Kalman filtered estimates in a forward recursive formulation and the determination of the smoothed estimates in a backward recursive formulation. The following presentation is given without proof and summarizes the algorithm given by Meditch [1].

OPTIMAL FIXED INTERVAL SMOOTHING SUMMARY

Assumptions

The state of the system is governed by the model given in Section 8.2.2 with known deterministic $\Phi(k+1,k)$, $\Gamma(k+1,k)$, $P(0)$, and $Q(k)$, and the observation model is as given in Section 8.2.3 with known deterministic $H(k+1)$ and $R(k)$. These are summarized as follows:

$$\begin{aligned} x(k+1) &= \Phi(k+1,k)x(k) + \Gamma(k+1,k)w(k), \qquad k = 0, 1, \ldots \\ z(k+1) &= H(k+1)x(k+1) + v(k+1) \end{aligned} \tag{8.51}$$

Algorithm (Fixed Interval Smoothing)

$$\begin{aligned} \hat{x}(k\,|\,N) &= \hat{x}(k\,|\,k) + A(k)[\hat{x}(k+1\,|\,N) - \Phi(k+1,k)\hat{x}(k\,|\,k)], \\ &\quad k = N-1, N-2, \ldots, 0 \\ &\quad \text{with initial condition } \hat{x}(N\,|\,N) \end{aligned} \tag{8.52}$$

The **smoothing gain**, $A(k)$, is determined as follows:

$$A(k) = P(k\,|\,k)\Phi^T(k+1,k)P^{-1}(k+1\,|\,k) \tag{8.53}$$

where $P(k\,|\,k)$ and $P(k+1\,|\,k)$ are the same as those determined from the Kalman filter. Thus $A(k)$ can be calculated by using the recursive calculation of $P(k\,|\,k)$ and $P(k+1\,|\,k)$ as shown in Figure 8.8 and described by Eqs. (8.39) to (8.41).

Performance

The error covariance matrix $P(k\,|\,N)$ defined by

$$P(k\,|\,N) = E[(x(k) - \hat{x}(k\,|\,N))(x(k) - \hat{x}(k\,|\,N))^T] \tag{8.54}$$

is determined by a backward recursion as follows:

$$\begin{aligned} P(k\,|\,N) &= P(k\,|\,k) + A(k)[P(k+1\,|\,N) - P(k+1\,|\,k)]A^T(k), \\ &\quad k = N-1, N-2, \ldots, 0 \\ &\quad \text{with } P(N\,|\,N) \text{ as the initial condition} \end{aligned} \tag{8.55}$$

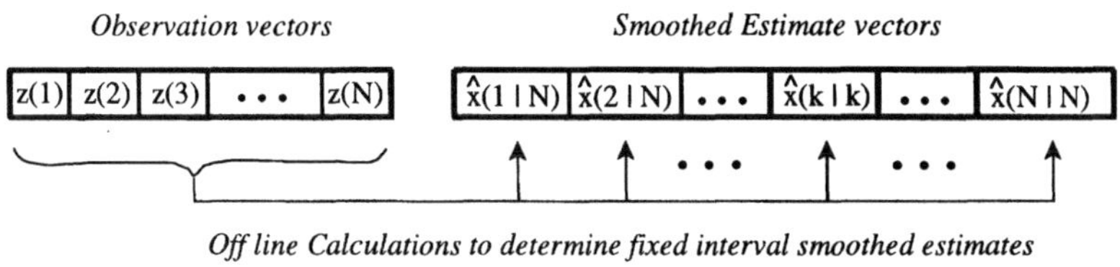

Figure 8.12 Conceptual diagram for fixed interval smoothing.

The $P(N\,|\,N)$ is obtained in the forward calculation from the Kalman filter. From examination of the algorithm's equations (8.52 and 8.55), we see that it uses measurements in a forward recursion to obtain the filtered estimate for all k indexes in the entire fixed interval N, and then uses these Kalman filtered estimates in a backward recursion to generate the smoothed estimates at time indices $N-1, N-2$, and so on. This procedure is shown conceptually in Figure 8.13. In the figure the **KF** represents the Kalman filtering operation described in Eqs. (8.38) to (8.41), and the **SM** represents the smoothing algorithm just described in Eq. (8.52). Note that the initial condition for the forward direction is $\hat{x}(0\,|\,0)$, which is a zero vector, and that the initial condition in the backward direction is $\hat{x}(N\,|\,N)$, the filtered estimate at time index N, which uses all the data up to and including time index N.

The performance of the optimum smoothed estimates is given in Eq. (8.55) as it gives the error covariance matrix for the estimates $\hat{x}(k\,|\,N)$. This performance can be calculated "off line" as it is not a function of the measurements as was the case for

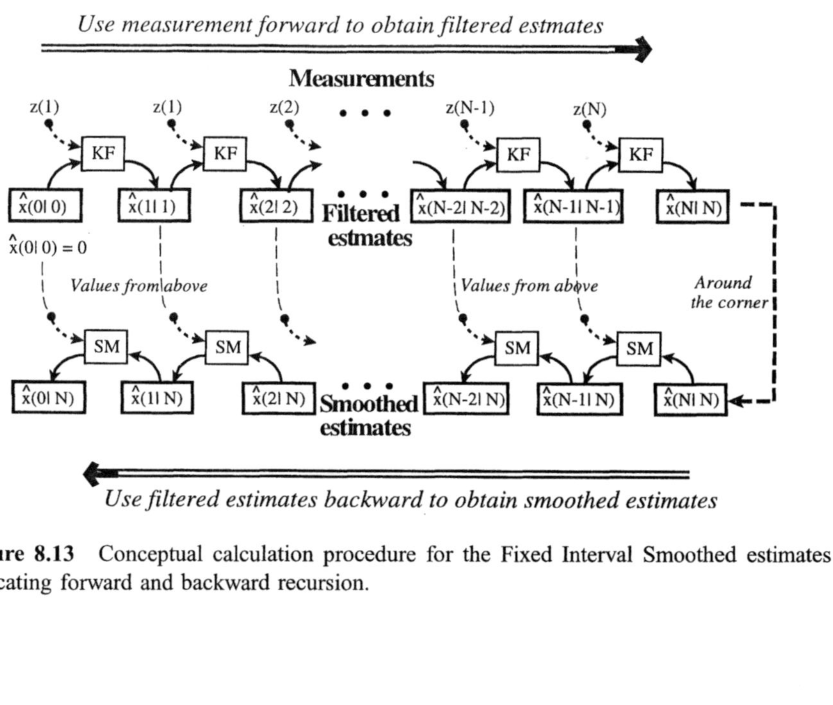

Figure 8.13 Conceptual calculation procedure for the Fixed Interval Smoothed estimates indicating forward and backward recursion.

the Kalman filter performance. By virtue of having additional measurements in the calculation of the estimates, the entries on the diagonals of $P(k|N)$ are less than the corresponding entries in $P(k|k)$. The difference of the corresponding elements on the diagonals, $P(k|N) - P(k|k)$, represents the gain in performance for estimating the components of the state vector $x(k)$ by using the smoothing algorithm rather than the filtered algorithm.

8.6.2 Fixed Point Smoothing

Fixed point smoothing is the process of estimating the state of the system at some fixed time index K as the observation index j increases from $K+1$. This overall operation is indicated in Figure 8.14.

In this figure the first line indicates that $z(1), z(2), \ldots, z(K+1)$ are used to estimate $x(K)$ by $\hat{x}(K|K+1)$, the second line indicates that $z(1), z(2), \ldots, z(K+1)$ are used to estimate $x(K)$ by $\hat{x}(K|K+2)$, and so on. This process is continued as long as there are new inputs.

One way of implementing fixed point smoothing is by performing successively the Kalman filter to obtain $\hat{x}(K+1|K+1)$, and then the fixed interval smoothing algorithm of size $K+1$ to obtain $\hat{x}(K|K+1)$ by iterating one step back from the filtered estimate. Similarly, to get $\hat{x}(K|K+2)$, we perform successively the Kalman filter to obtain $\hat{x}(K+2|K+2)$, and then the fixed interval smoothing algorithm of size $K+2$ to obtain $\hat{x}(K|K+2)$ by iterating two steps back from the filtered estimate. As further measurements are taken in, it is necessary to iterate further and further back. A more efficient algorithm has been developed (see Meditch

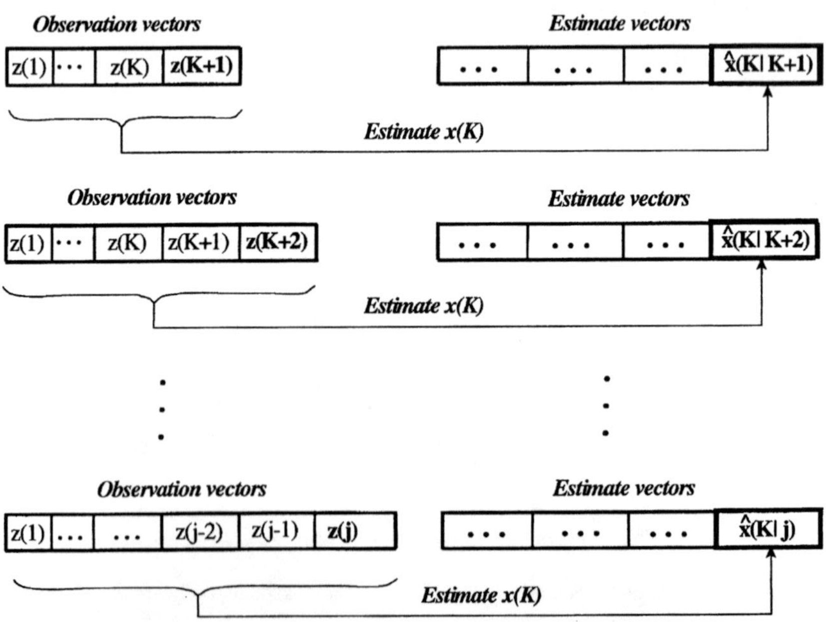

Figure 8.14 Timing diagram for the fixed point smoothing algorithm.

[1]), and it is now described without proof in the following fixed point presentation. This algorithm determines the required estimates recursively using the previous estimate at the fixed point and the two filtered estimates at the running indexes j and $j-1$ as shown in Eq. (8.56).

FIXED POINT SMOOTHING SUMMARY

Assumptions

The state of the system is governed by the model given in Section 8.2.2 with known deterministic $\Phi(k+1, k)$, $\Gamma(k+1, k)$, $P(0)$, and $Q(k)$, and the observation model is as given in Section 8.2.3 with known deterministic $H(k+1)$ and $R(k)$. These are summarized as follows:

Fixed Point Smoothing Algorithm

$$\begin{aligned} \hat{x}(K \mid j) &= \hat{x}(K \mid j-1) + W(j)H^T(j)R^{-1}(j) \\ &\quad \cdot [z(j) - H(j)\Phi(j, j-1)\hat{x}(j-1 \mid j-1)], \\ j &= K+1, K+2, \ldots \\ &\text{where } W(j) = W(j-1)\Phi^T(j, j-1)[I - S(j)P(j \mid j)] \end{aligned} \tag{8.56}$$

The initial condition is $W(K) = P(K \mid K)$.

Performance

The error covariance matrix $P(K \mid j)$ is defined by

$$P(K \mid j) \triangleq E[(x(K) - \hat{x}(K \mid j))(x(K) - \hat{x}(K \mid j)^T)] \tag{8.57}$$

It is determined from

$$\begin{aligned} P(K \mid j) &= P(K \mid j-1) - W(j)[S(j)P(j \mid j-1)S^T(j)]W^T(j) \\ &\text{where} \quad S(j) = H^T(j)R^{-1}(j)H(j) \end{aligned} \tag{8.58}$$

8.6.3 Fixed Lag Smoothing

Fixed lag smoothing gives the estimate of the state of the system at time index k as a result of measurements taken up to the time index $k+L$, where L is a fixed lag and k runs from one to infinite. This process is illustrated in Figure 8.15.

It is possible to obtain the fixed smoothed lag filter by using the fixed interval smoothing algorithm for each new iteration and a step back to obtain the smoothed

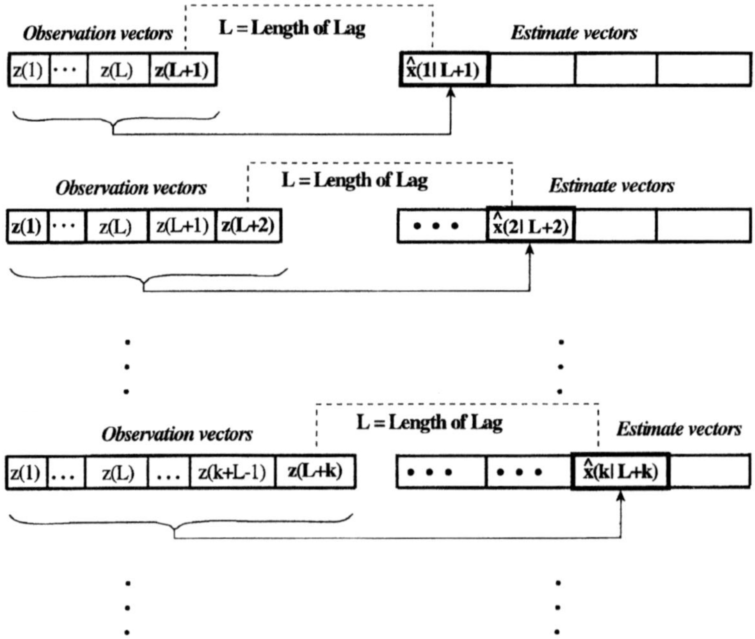

Figure 8.15 Timing diagram for the fixed lag smoothed estimates.

estimate at the desired lag. Recursive algorithms have been developed for the fixed lag smoothed estimate, and in general, they are quite complex. If desired consult Meditch [1] for details.

8.7 STEADY STATE EQUIVALENCE OF THE KALMAN AND WIENER FILTERS

In general, the Kalman filter is different from the Wiener filter, although both are optimum linear filters for estimating signals. A big difference is in the main formulation where the Kalman filter is described for time index n greater than or equal to zero, whereas the Weiner filter assumes that the processes are defined for all indexes of time. The resulting Kalman filter thus contains an initial transient as part of its solution, while the Weiner filter is assumed to be steady state. The standard Weiner filter works for estimating random processes that are wide sense stationary, while the Kalman filter inherently estimates nonstationary processes. The Kalman filter is described in a state variable form. Thus it is able to handle optimum estimation of vector processes, while the Wiener filter is very cumbersome mathematically, bordering on impossibility, for handling vector processes.

The following example will illustrate a form of equivalence of the Kalman steady state filter and the Weiner filters for estimating a signal process in noise with comparatively equal first-order scalar models, each described in a different framework.

8.7.1 Kalman Filter Formulation

Suppose that a scalar random process $x(k)$ is observed in additive white noise and that the signal and observation are mathematically modeled as follows:

$$\begin{aligned} x(k+1) &= \tfrac{1}{2}x(k) + w(k), \qquad k \geq 0 \\ z(k+1) &= x(k+1) + v(k+1), \qquad k \geq 0 \end{aligned} \tag{8.59}$$

The following assumptions are made with respect to the characterization of the $w(k)$, the signal generation white noise process, and $v(k)$, the additive white noise process:

$$\begin{aligned} &(1)\ E[w(k)] = E[v(k)] = 0 \\ &(2)\ E[[w(j)w(k)] = Q = \tfrac{3}{4}\,\delta_{jk} \\ &(3)\ E[v(j)v(k)] = R = \tfrac{1}{2}\,\delta_{jk} \\ &(4)\ E[w(j)v(k)] = 0 \qquad \text{for all } j \text{ and } k \end{aligned} \tag{8.60}$$

These assumptions are consistent with those for the Kalman filter. They are described in terms of expected values of various time domain products. We wish to find the optimal estimate $\hat{x}(k+1)$ of the signal process by a linear filter where optimality is in the minimum mean squared error sense.

From (8.59) and (8.60) the given values of the message and observation models are seen to be not a function of k and determined as $\Phi = \frac{1}{2}, \Gamma = 1, H = 1, R = \frac{1}{2}$, and $Q = \frac{3}{4}$. Substituting these parameters into Eq. (8.38) gives the Kalman filter as

$$\begin{aligned} \hat{x}(k+1\,|\,k+1) = \tfrac{1}{2} + K(k+1)\big[z(k+1) - \tfrac{1}{2}\hat{x}(k\,|\,k)\big], \qquad k \geq 0 \\ \text{with initial condition } \hat{x}(0\,|\,0) = 0 \end{aligned} \tag{8.61}$$

where $K(k+1), P(k+1\,|\,k)$, and $P(k+1\,|\,k+1)$ are given from (8.39 to 8.41) by

$$\begin{aligned} K(k+1) &= P(k+1\,|\,k)\big[P(k+1\,|\,k) + \tfrac{1}{2}\big]^{-1}, \qquad k = 0, 1, \ldots \\ P(k+1\,|\,k) &= \tfrac{1}{2}P(k\,|\,k)\tfrac{1}{2} + \tfrac{3}{4} \qquad \text{with } P(0\,|\,0) = 0 \\ P(k+1\,|\,k+1) &= [1 - K(k+1)]P(k+1\,|\,k) \end{aligned} \tag{8.62}$$

Remember that the order of computation is $P(k\,|\,k) \to P(k+1\,|\,k) \to K(k+1) \to P(k+1\,|\,k+1)$ starting with $k = 0$ and continuing with $k = 1, 2, \ldots$. If it is assumed that a steady state has been reached, then the $K(k+1), P(k+1\,|\,k)$, and $P(k\,|\,k)$ all will be constants and defined as follows:

$$\lim_{k\to\infty} K(k+1) \triangleq K, \quad \lim_{k\to\infty} P(k+1\,|\,k) \triangleq P_1, \quad \lim_{k\to\infty} P(k+1\,|\,k+1) \triangleq P_0 \tag{8.63}$$

Substituting the steady state filter parameters from (8.63) into (8.62) gives the following:

$$\begin{aligned} K &= P_1/(P_1 + \tfrac{1}{2}) \\ P_1 &= \tfrac{1}{4} P_0 + \tfrac{3}{4} \\ P_0 &= [1 - K]P_1 \end{aligned} \tag{8.64}$$

Equation (8.64) gives us three nonlinear equations in terms of K, P_1, and P_0. They can be solved to give

$$P_0 = 0.3117, \quad P_1 = 0.8279, \quad K = 0.6235 \tag{8.65}$$

Using the steady state filter values of (8.65) in (8.61), gives the steady state Kalman filter as follows:

$$\hat{x}(k+1\,|\,k+1) = \tfrac{1}{2}\hat{x}(k\,|\,k) + 0.6224[z(k+1) - \tfrac{1}{2}\hat{x}(k\,|\,k)] \tag{8.66}$$

The transfer function $H(z)$ for the Kalman filter can be determined by taking the Z-transform of (8.66) to yield

$$z\hat{X}(z) = \tfrac{1}{2}\hat{X}(z) + 0.6235\left[zZ(z) - \tfrac{1}{2}\hat{X}(z)\right] \tag{8.67}$$

Solving for $\hat{X}(z)$ over $Z(z)$, which is the transfer function $H(z)$ for the steady state Kalman filter for this problem, we obtain

$$H(z) \triangleq \frac{\hat{X}(z)}{Z(z)} = \frac{0.6235z}{z - 0.188} \tag{8.68}$$

We now pose the problem so that a Wiener filter can be obtained and find the optimal filter using the Wiener formulation.

8.7.2 Wiener Filter Formulation

The Wiener filter is an optimum linear time-invariant filter for estimating one signal process $x(n)$ from an observed process $z(n)$, where optimality is in the minimum mean squared error sense. The information required to obtain the optimal filter is in terms of the autocorrelations and cross correlations of the processes involved, or equivalently in the frequency domain in terms of the auto- and cross-spectral densities of the processes. It is assumed that the observed process $z(k)$ is the sum of the signal process $x(k)$ and a white noise process $v(k)$ as

$$z(k) = x(k) + v(k) \tag{8.69}$$

The difference between this part of the Wiener formulation and the Kalman filter formulation (8.59) is that (8.69) is assumed to hold for all time rather than just at times greater than or equal zero. To make the models otherwise equal, we need to select the properties of $v(k)$ and $x(k)$ to be equivalent to the Kalman formulation, that is,

$$\begin{aligned}&(1)\ E[x(k)] = E[v(k)] = 0\\ &(2)\ E[v(j)v(k) = R = \tfrac{1}{2}\,\delta_{jk}\\ &(3)\ E[x(j)v(k)] = 0 \text{ for all } j \text{ and } k\end{aligned} \tag{8.70}$$

The third assumption is that the signal process $x(j)$ is uncorrelated with the noise process $v(k)$, which is similar to the uncorrelated property of the $w(k)$ in the Kalman formulation. It is also required that the signal process $x(k)$ be somehow equivalent to the state variable representation given in the Kalman formulation. In general, the Kalman filter has a time-varying autocorrelation function as a result of the governing state model starting at zero. So, for equivalence of this example, it is necessary to assign the steady state autocorrelation function to the signal model for the Wiener formulation. The steady state autocorrelation function for the state model can be shown to be $R_{XX}(k) = (\frac{1}{2})^{|k|}$, so the last assumption for the Wiener formulation is

$$(4)\ E[x(j+k)x(j)] = R_{XX}(k) = (\tfrac{1}{2})^{|k|} \qquad \text{for all } k \tag{8.71}$$

Thus we specify the autocorrelation function or equivalently the power spectral density of the random process $x(k)$ rather than the state model for its generation from white noise.

We would like to find the discrete time Wiener filter for estimating the $x(k)$. From Chapter 7 (Eq. 7.101), the optimal discrete-time filter for estimating the desired $G[n] = x(n)$ from the observed $Z(n)$, each with rational spectra, has a transfer function $H(z)$ given by

$$H(z) = \frac{\{\Psi_{XZ}(z)/\Psi_{XX}(z)\}^{+}}{\Psi_{XX}^{+}} \tag{8.72}$$

where $\Psi_{XZ}(z) = Z_\beta(R_{XZ}(k))$ and $\Psi_{XX}(z) = Z_\beta(R_{XX}(k))$, which are determined as

$$\begin{aligned}\Psi_{ZZ}(z) &= Z_\beta(R_{ZZ}(k)) = Z_\beta(R_{XX}(k) + R_{VV}(k))\\ &= \frac{-1.5z}{(z-2)(z-\frac{1}{2})} + \frac{1}{2}\end{aligned} \tag{8.73}$$

$$\begin{aligned}\Psi_{XZ}(z) &= Z_\beta(R_{XZ}(k)) = Z_\beta((\tfrac{1}{2})^{|k|})\\ &= \frac{-1.5z}{(z-2)(z-\frac{1}{2})}\end{aligned} \tag{8.74}$$

The $\Psi_{XZ}(z)$ is then spectrally factorized as follows:

$$\begin{aligned}\Psi_{ZZ}(z) &= \Psi_{ZZ}^{+}(z)\Psi_{ZZ}^{-}(z)\\ &= \frac{1.15(z-0.188)}{(z-\frac{1}{2})}\cdot\frac{1.15(z^{-1}-0.188)}{(z^{-1}-\frac{1}{2})}\end{aligned} \tag{8.75}$$

The next step is to obtain the + time part of the partial fraction expansion:

$$\left\{\frac{\Psi_{XZ}(z)}{\Psi_{XX}(z)}\right\}^{+} = \left\{\frac{-1.5z}{(z-2)(z-\frac{1}{2})} \cdot \frac{(z^{-1}-\frac{1}{2})}{1.15(z^{-1}-0.188)}\right\}^{+}$$
$$= -3.45z\left\{\frac{A}{z-\frac{1}{2}} + \frac{B}{z-5.28}\right\}^{+} \tag{8.76}$$
$$= -3.45z\frac{A}{z-\frac{1}{2}} = 0.717\frac{z}{z-\frac{1}{2}}$$

Finally, substituting (8.76) into (8.72), we find the transfer function $H(z)$ for the optimum linear causal time-invariant filter to be

$$H(z) = 0.717\frac{z}{z-\frac{1}{2}} \cdot \frac{(z-\frac{1}{2})}{1.15(z-0.188)}$$
$$= \frac{0.6235z}{z-0.188} \tag{8.77}$$

Thus the $H(z)$ determined using the Kalman filter steady state formulation (8.68) and the $H(z)$ determined using the Wiener filter formulation (8.77) are the same, which establishes the equivalence for this simple example. This type of steady state equivalence for wide sense stationary processes can also be established in general, but the proof is beyond the scope of our development for this chapter.

8.8 SUMMARY

In this chapter optimal discrete time linear systems were presented where optimality was with respect to the minimum mean squared error criterion. The optimal filters were written in terms of the parameters of the state variable representation of the message generation model and the parameters of an additive white noise observation model. The main result, the Kalman filter, was originally presented by R. E. Kalman [13,14].

Our presentation begins with the establishment of the basic structure of the message and observation models. The message model was described in a state variable representation as the output of a linear, possibly time-varying, system to a white noise source input. This was the Markov sequence model. The observation model was assumed as a sum of a weighted white noise process and a linear combination of the states.

For these message and observation models the important problems of prediction, fitering, and smoothing were explored, and the procedures and formulas were presented for the optimum filters and their mean squared error performances.

The Kalman filter was presented without proof, and several examples were given to describe the process of obtaining the Kalman gain which, along with the filter form, determined the optimal filter. The filter form was expressed is in terms of the original message state model properties and was shown to be in a prediction-correction form. The first example presented was a one-dimensional filtering problem, while the second was the estimation of an AR(2) random process in additive noise. The state vector formulations for these examples were obtained by a reasonable assignment of the state variables, and this indicates the generality of the vector Markov message model.

Although our presentation did not include proofs, various different proofs for the Kalman filter exist in the literature. The interested reader should consult Kalman [14], Meditch [1], and Sage and Melsa [5] for their detailed versions.

The basic optimal filter structures, procedures, and performances for fixed interval fixed point, and fixed lag smoothing were presented without proof in a format similar to that given by Meditch [1]. Also presented, without proof, were optimal filters for obtaining the fixed point and fixed lead prediction estimates.

The chapter concludes with an example that illustrates a form of equivalence of the steady state Kalman filter and the Wiener filter. Since the Kalman filter is a time domain presentation and the Wiener filter is frequency based, care was needed in formulating the equivalent problems.

The presentation in this chapter has been limited to the discrete time Kalman filter formulation. A parallel theory is available for the continuous time Kalman filter. For a discussion of the continuous time Kalman filter, the reader can benefit from presentations by Kalman [13], Brown [10], Sage and Melsa [5], and Van Trees [16]. In many cases it is convenient to implement the continuous Kalman filter in discrete time allowing computer implementation. This usually requires obtaining a discrete time model for the continuous message model. Various techniques for handling this type of modeling have been established by the modern-day control theory community, and a good starting point for this area can be found in Cadzow and Martens [12].

The Kalman filter has been shown to be a little more general in terms of problems it can solve. Extensions include those for different performance measures than the mean squared error and different statistical representations than Gaussian. These properties are nicely presented in the text by Meditch [1].

Also the Kalman filter has been extended to nonlinear state models, and the procedures and formulas for implementation are given by Sage and Melsa [5], Anderson and Moore [4], and Candy [9]. The resulting filter structure, called the extended Kalman filter, is no longer linear and also not necessarily the optimum nonlinear filter either. Optimum nonlinear filters are outside the scope of this book.

PROBLEMS

8.1 A system model is given as

$$x_1(k+1) = x_1(k) + x_2(k) + w(k), \qquad k = 0, 1, \ldots$$
$$x_2(k+1) = x_2(k) + w(k)$$

with the following assumptions:

(1) Initial condition is known and given by $x(0) = [0, 0]^T$.

(2) $w(k)$ is a scalar zero mean white Gaussian sequence with a known autocorrelation function given by $E[w(j)w(k)] = \delta_{jk}$.

We are given the measurement model as

$$z_1(k+1) = x_1(k+1) + v_1(k+1), \qquad k = 0, 1, \ldots$$
$$z_2(k+1) = x_2(k+1) + v_2(k+1)$$

with the following assumptions:

(3) $v_1(k+1)$ and $v_2(k+1)$ are independent zero mean white Gaussian sequences with the properties

$$E[v_1(j)v_1(k)] = \delta_{jk} \quad \text{and} \quad E[v_2(j)v_2(k)] = 2\delta_{jk} \qquad \text{for} \quad k = 0, 1, 2 \ldots$$

(4) $w(k)$ is uncorrelated with both measurement noise processes $v_1(k+1)$ and $v_2(k+1)$ for all j and k.

$$E[w(j)v_1(k)] = 0 \quad \text{and} \quad E[w(j)v_2(k)] = 0$$
$$\text{for} \quad j = 0, 1, 2 \ldots, k = 1, 2, \ldots$$

We wish to find the formulas governing the Kalman filter for the case of estimating the state vector $x(k+1) = [x_1(k+1), x_2(k+1)]^T$ from the measurement vector $z(k+1)$ given by $z(k+1) = [z_1(k+1), z_2(k+1)]^T$. Specifically work the following parts:

(a) Determine the Kalman gain matrix and thus the recursive relation for the optimal estimate of $x(k+1)$ in terms of the measurement vectors $z(1), z(2), \ldots, z(k+1)$.

(b) Illustrate by a scalar block diagram (not a vector block diagram) the formulas for calculating the estimates of the components $x_1(k+1)$ and $x_2(k+1)$ of the state vector $x(k+1)$.

(c) Use the simulated data provided to determine the estimate, $\hat{x}(k \mid k)$ of the state vector $x(k)$ for $k = 0, 1, \ldots, 10$, and plot $\hat{x}_1(k \mid k)$ and $\hat{x}_2(k \mid k)$ for $k = 1, 2, \ldots, 10$.

(d) In general, the true values of the state vector are not available. However, for the purpose of illustration the state vector components are given in Tables P8.1 and P8.2. Plot the true values and the estimates for $x_1(k)$ determined in (c) on the same graph for $k = 1, 2, \ldots, 10$. Repeat for $x_2(k)$ and its estimate. Compute and plot on separate graphs the error, $x_i(k) - \hat{x}_i(k \mid k)$ for each component $i = 1, 2$ as k goes from 1 to 10.

(e) Using the error covariance matrix for the state determined by the Kalman filter, as k goes from 1 to 10, plot $E[\varepsilon_1^2(k \mid k)]$ and $E[\varepsilon_2^2(k \mid k)]$, where $\varepsilon_1(k \mid k)] = x_1(k) - \hat{x}_1(k \mid k)$ and $\varepsilon_2(k \mid k) = x_2(k) - \hat{x}_2(k \mid k)$.

(f) Comment on the relationship between your errors of part (d) and the variances of part (e).

8.2 In this problem the fixed interval smoothed estimates of the state vector are required using the data and results from Problem 8.1.

TABLE P8.1 Measurement values for Problems 8.1 through 8.3

k Time Index	$z_1(k)$ Measurement Component	$z_2(k)$ Measurement Component
1	3.29691969	2.10134294
2	3.38736515	0.47540797
3	7.02830641	3.17688898
4	9.71212521	2.49811140
5	11.42018315	2.91992424
6	15.97870583	6.17307616
7	22.06934285	5.42519274
8	28.30212781	3.05365741
9	30.44683831	5.98051141
10	38.75875595	4.51016361

TABLE P8.2 Actual states from simulation for problems 8.1 through 8.3

k Time Index	$x_1(k)$ True State	$x_2(k)$ True State
0	0.0000000000	0.000000000
1	1.65428714	1.65428714
2	3.50300702	1.84871988
3	5.997852924	2.47552222
4	9.15040740	3.17187816
5	12.50873910	3.35833170
6	16.92192594	4.41318684
7	21.34483352	4.42290758
8	25.89335144	4.54851792
9	31.54135330	5.64800186
10	36.93605670	5.394470340

(a) Compute the fixed interval smoothed estimates $\hat{x}(k\,|\,N)$ for $k = 1$ to 10, where $N = 10$, and plot.

(b) Compute and plot $\tilde{x}(k\,|\,10)$ for $k = 1$ to 10. Compare your results with those of 8.1(d). Comment on the results.

(c) Determine the error covariance matrices for the fixed interval smoothed estimates of the state for $k = 1$ to 10, and compare the results with those determined in part (b). Comment on the statistical regularity.

(d) What do the off-diagonal terms of the covariance matrix tell us about our estimates?

8.3 We desire $\hat{x}(5\,|\,k)$, the running predicted estimate of $x(5)$, as k goes from 0 to 4, and then the filtered estimate, $\hat{x}(5\,|\,5)$, followed by the fixed point smoothed estimate $\hat{x}(5\,|\,k)$ as k varies from 6 to 10. Use the data and results from Problems 8.1 and 8.2.

(a) Compute $\hat{x}(5\,|\,k)$ for $k = 1$ to 10, and plot each component as a function of k. Indicate on your plots with a dotted line the actual state value for all k (use actual data given).

(b) Compute and plot each component of $\tilde{x}(5\,|\,k)$ for $k = 1$ to 10. Discuss your results with reference to the respective error covariance matrix components for the estimate.

(c) Determine the steady state error covariance matrix for the Kalman filtered estimate of the state vector $x(k)$ defined as

$$\lim_{j\to\infty} E[\tilde{x}(5\,|\,j)\tilde{x}^T(5\,|\,j)]$$

and discuss the relationship between this matrix and the limited accuracy of the estimation of the state components.

8.4 A system and measurement model for a moving object in a two-dimensional space are given below. The $x_1(k)$ and $x_2(k)$ could represent the position in an x and y direction at time index k, while $x_3(k)$ and $x_4(k)$ could represent the velocities in the x and y direction, respectively, at time index k. In the equation the $\alpha = 0.97979589$.

$$\begin{bmatrix} x_1(k+1) \\ x_2(k+1) \\ x_3(k+1) \\ x_4(k+1) \end{bmatrix} = \begin{bmatrix} 1 & 0 & 0.5 & 0 \\ 0 & 1 & 0 & 0.5 \\ 0 & 0 & \alpha & 0 \\ 0 & 0 & 0 & \alpha \end{bmatrix} \begin{bmatrix} x_1(k) \\ x_2(k) \\ x_3(k) \\ x_4(k) \end{bmatrix} + \begin{bmatrix} 0.25 & 0 \\ 0 & 0.25 \\ 1 & 1 \\ 0 & 1 \end{bmatrix} \begin{bmatrix} w_1(k) \\ w_2(k) \end{bmatrix}, \quad k = 0, 1, \ldots$$

$$\begin{bmatrix} z_1(k+1) \\ z_2(k+1) \end{bmatrix} = \begin{bmatrix} 1 & 0 & 0 & 0 \\ 0 & 1 & 0 & 0 \end{bmatrix} \begin{bmatrix} x_1(k+1) \\ x_2(k+1) \\ x_3(k+1) \\ x_4(k+1) \end{bmatrix} + \begin{bmatrix} v_1(k) \\ v_2(k) \end{bmatrix}, \quad k = 1, 2$$

Accordingly, we have the following assumptions:

(1) Initial condition $x(0)$ is a zero mean Gaussian random vector with covariance matrix given as $P(0) = 25\, I_4$. Let $\hat{x}(0\,|\,0) = [0, 0, 0, 0]^T$

(2) $w_1(k)$ and $w_2(k)$ are zero mean uncorrelated white Gaussian random sequences with known autocorrelation functions given by

$$E[w_1(j)w_1(k)] = 4\delta_{jk} \quad \text{and} \quad E[w_2(j)w_2(k)] = 1\delta_{jk} \quad \text{for } j, k = 0, 1, 2, \ldots$$

(3) $v_1(k+1)$ and $v_2(k+1)$ are independent zero mean white Gaussian sequences with the properties

$$E[v_1(j)v_1(k)] = 16\delta_{jk} \quad \text{and} \quad E[v_2(j)v_2(k)] = 16\delta_{jk} \qquad \text{for } j, k = 1, 2, \ldots$$

(4) $w_1(j)$ and $w_2(j)$ are zero mean and orthogonal to both measurement noise processes $v_1(k)$ and $v_2(k)$ for all j and k

$$E[w_1(j)v_1(k)] = 0, \quad E[w_1(j)v_2(k)] = 0, \quad E[w_2(j)v_1(k)] = 0,$$
$$E[w_2(j)v_2(k)] = 0$$

Using the state and measurement models above along with the assumptions specified and Tables P8.4 and P8.5 perform the following tasks:

(a) Use the measurements shown in the tables along with your generated $K(k)$ of the Kalman filter to estimate the current states of the system for $k = 1, 2, \ldots, 10$. This will give you estimates of the position $(\hat{x}_1(k|k)$ and $(\hat{x}_2(k|k)$ and the velocity $(\hat{x}_3(k|k))$ and $\hat{x}_4(k|k))$ in each direction from the measurement data.

(b) Plot the actual position in a reduced state space (two dimensions) from the information on the data and the estimated position on the same plot. (Ordinarily you do not have access to the true location.)

(c) Plot each of the following position errors for $k = 1, 2, \ldots, 10$.

$$\tilde{x}_i(k|k) \triangleq x_i(k) - \hat{x}_i(k|k), \qquad i = 1, 2.$$

These plots give the position estimate errors for your data.

(d) Compare the magnitude of the error at each point for $k = 1, 2, \ldots, 10$ with the corresponding entry in the error covariance matrix and comment on statistical regularity.

(e) Plot the estimates of the velocities in each direction. (Note that even though the velocities were not measured directly, you are able to estimate them.)

TABLE P8.4 Measurements for Problem 8.4 and 8.5

k Time Index	$z_1(k)$ Measurement	$z_2(k)$ Measurement
1	12.79067	12.96666
2	7.464684	22.80052
3	20.39026	31.67403
4	22.77083	25.7334
5	33.19124	49.1115
6	32.40265	44.7559
7	33.2771	58.48168
8	33.66809	64.372
9	40.74091	67.70678
10	37.3577	78.19464

8.5 Using the models, assumptions, data, and results from Problem 8.4

(a) Obtain $\hat{x}(6\,|\,k)$, $k = 0, 1, \ldots, 8$. Plot $\hat{x}_1(6\,|\,k)$ and $\hat{x}_2(6\,|\,k)$ for $k = 0, 1, \ldots, 8$ showing the actual states $x_1(6)$ and $x_2(6)$ as dotted line for all k.

(b) Would you expect your estimate $\hat{x}_1(6\,|\,3)$ or your estimate $\hat{x}_1(6\,|\,8)$ to be a more reliable estimate of $x_1(6)$? Back up your answer by referring to properties of the error covariance matrices.

TABLE P8.5 Actual state components $x(k)$ for Problems 8.4 and 8.5

$x(1)$	$x(2)$	$x(3)$	$x(4)$	$x(5)$
6.694718	12.6676	17.92915	22.21501	25.57497
9.848555	18.04262	26.11322	33.72115	40.93572
12.57683	11.06058	9.762154	7.18408	6.110596
16.09116	16.35988	15.5919	14.52476	14.04006

$x(6)$	$x(7)$	$x(8)$	$x(9)$	$x(10)$
28.96234	32.47833	34.80732	36.00437	37.45878
48.3139	55.55404	62.22753	68.55708	74.76478
7.315652	6.600291	2.582289	2.582289	3.620353
15.18898	13.46473	12.95716	12.09928	12.48703

REFERENCES

1. Meditch, J. S., *Stochastic Optimal Linear Estimation and Control*, McGraw-Hill, 1969.
2. Mendel, Jerry M., *Discrete Techniques of Parameter Estimation*, Dekker, 1973.
3. Kailath, T., *Lectures on Linear Least-Squares Estimation*, Springer 1976.
4. Anderson, Brian D. O. and John B. Moore, *Optimal Filtering*, Prentice-Hall, 1979.
5. Sage, Andrew P., and James L. Melsa, *Estimation Theory with Applications to Communications and Control*, McGraw-Hill, 1971.
6. Srinath, M. D, P. K. Rajasekaran, and R. Viswanathan, *Statistical Signal Processing with Applications*, Prentice-Hall, 1996.
7. Brown, Robert Grover, and Patrick Y. C. Hwang, *Introduction to Random Signals and Applied Kalman Filtering*, 2nd Ed., Wiley, 1992.
8. Mendel, Jerry M., *Lessons in Digital Estimation Theory*, Prentice-Hall, 1987.
9. Candy, James V., *Signal Processing the Model-Based Approach*, McGraw-Hill, 1986.
10. Brown, Robert Grover, *Introduction to Random Signal Analysis and Kalman Filtering*, Wiley, 1983.
11. Gupta, S. C., *Transform and State Variable Methods in Linear Systems*, Wiley, 1966.
12. Cadzow, James A., and H. R. Martens, *Discrete-Time Computer Control Systems*, Prentice-Hall, 1970.
13. Kalman, R. E. and R. S. Bucy, 'New results in linear filtering and prediction theory', *Transactions of the ASME Journal of Basic Engineering.*, Vol. 83, 1961, p. 95.

14. Kalman, R. E., A new approach to linear filtering and prediction problems, *ASME Journal of Basic Engineering*, March 1960, pp. 35–45.
15. Gelb, Arthur, Ed., *Applied Optimal Estimation*, MIT Press, 1975.
16. Van Trees, H. L., *Detection, Estimation and Modulation Theory*, Vol. 1, Wiley, 1968.

9 Detection Theory: Discrete Observation

9.1 BASIC DETECTION PROBLEM

In this chapter decision rules for a specified performance measure will be formulated for detection problems that have a statistical framework. By a statistical framework it is meant that the underlying probability density functions for the observed vectors and the class or hypotheses a priori probabilities are known. Possible performance measures include a posteriori probability, probability of error, and cost per decision. We will see that likelihood ratios carry the information necessary to make optimum classifications and that other so-called sufficient statistics can be found for many problems.

Many problems involve just two signal classes, but most problems require classification of multiple signal classes such as recognition of handwritten numerals and recognition of vowel sounds. The observed vectors are rarely one dimensional, and in most cases they are finite dimensional. Problems that contain observations over a continuous interval are of infinite dimension.

Decision rules will be developed in this chapter for the two-class case with extensions to the K-class case for a single observation and a vector observation, while the continuous or infinite observation case will be presented in Chapter 10.

As will be shown, for the two-class case, the optimum decision rule, under rather wide assumptions and performance measures, is a likelihood ratio test. For the M-class problem, the optimum decision rule can be put in a form of linear equations involving $M - 1$ likelihood ratios.

The basic problems presented in this chapter are considered **hypotheses testing problems**. For the two-class case we have two possible hypotheses, call them C_1 and C_2. In pattern recognition we might use the word **classes**, in radar problems the words **target and no target**, and for binary communication the words **binary 0 or 1** to represent the hypotheses. Throughout this chapter the word "class" will mainly be used, favoring the pattern recognition notation but with the assumption that it is interchangeable with hypothesis. A measurement process results in a measurement, or **observation** X, or in most cases results in an **observation vector** $\mathbf{X} = [X_1, X_2, \ldots, X_n]^T$. We are to decide which of the two hypotheses the observation $\mathbf{x}$ comes from.

Information concerning the two hypotheses will take the form of known statistical properties, the most common assignment being the hypotheses or **class conditional probability density functions** $p(\mathbf{x} \mid C_1)$ and $p(\mathbf{x} \mid C_2)$ and the probabilities of each of the hypotheses before we take the observation called the **a priori probabilities** $P(C_1)$ and $P(C_2)$. This assignment will sometimes be summarized for the two-hypothesis case as

$$\begin{aligned} C_1\colon \quad & \mathbf{X} \sim p(\mathbf{x} \mid C_1), \quad P(C_1) \\ C_2\colon \quad & \mathbf{X} \sim p(\mathbf{x} \mid C_2), \quad P(C_2) \end{aligned} \tag{9.1}$$

For the **M-hypotheses** case we have M possible **hypotheses**, call them C_1, $C_2, \ldots, C_M$ and the information can be summarized in the following way:

$$\begin{aligned} C_1\colon \quad & \mathbf{X} \sim p(\mathbf{x} \mid C_1), \quad P(C_1) \\ C_2\colon \quad & \mathbf{X} \sim p(\mathbf{x} \mid C_2), \quad P(C_2) \\ \vdots \quad & \quad \vdots \qquad\qquad\quad \vdots \\ C_M\colon \quad & \mathbf{X} \sim p(\mathbf{x} \mid C_M), \quad P(C_M) \end{aligned} \tag{9.2}$$

The **observation space** $\mathscr{X}$ is defined as the set of all possible values of the vector $\mathbf{x}$. A method for making the assignment of a hypothesis for every possible $\mathbf{x} \in \mathscr{X}$ is called a **decision rule**. The decision rule is thus a mapping from $\mathscr{X}$, the observation space, to a **decision space** C, which consists of the set of all possible hypotheses C_i. The symbol d_j will mean that C_j was selected by the decision rule.

9.2 MAXIMUM A POSTERIORI DECISION RULE

The maximum a posteriori (MAP) decision rule is based on the concept that we should assign an observed value $\mathbf{x}$ to the class it most likely came from. We calculate the a posteriori probability that the observation $\mathbf{x}$ comes from each class and then select the class that has the highest probability. The decision or classifier rule presented is based on the following underlying assumptions:

(1) **Conditional probability density functions**, $p(\mathbf{x} \mid C_1)$ and $p(\mathbf{x} \mid C_2)$, for the observation vector $\mathbf{X}$ under each of the classes are known.
(2) **A priori probability of each class** $P(C_i)$ is known.

The **MAP decision rule** can be stated as follows: If $\mathbf{x}$ is the observed vector, it is classified as coming from class C_j if

$$P(C_j \mid \mathbf{x}) > P(C_i \mid \mathbf{x}) \qquad \text{for all } i \neq j \tag{9.3}$$

If equalities occur, the class will be assigned randomly from those classes.

In the following sections the two-class and M-class cases are presented in detail.

9.2.1 Two-Class Problem (MAP)

If there are only two classes, the **maximum a posteriori decision rule** can be written as follows:

$$\begin{aligned}
\text{If} \quad P(C_1 \mid \mathbf{x}) > P(C_2 \mid \mathbf{x}) \quad & \text{then decide } \mathbf{x} \text{ comes from } C_1 \\
< \cdots \quad & \text{then decide } \mathbf{x} \text{ comes from } C_2 \\
= \cdots \quad & \text{then randomly decide}
\end{aligned} \tag{9.4}$$

The **a posteriori probabilities** can be determined in terms of the given information by using the following form of Bayes' rule:

$$P(C_1 \mid \mathbf{x}) = \frac{p_X(\mathbf{x} \mid C_1)P(C_1)}{p(\mathbf{x})} \tag{9.5}$$

where the probability density function $p(\mathbf{x})$ of the observed pattern vector $\mathbf{X}$ can be determined by the total probability theorem as

$$p(\mathbf{x}) = p(\mathbf{x} \mid C_1)P(C_1) + p(\mathbf{x} \mid C_2)P(C_2) \tag{9.6}$$

After applying Bayes's rule to both sides of the decision rule given in Eq. (9.4), we have the following **equivalent rule**:

$$\begin{aligned}
\text{If} \quad \frac{p(\mathbf{x} \mid C_1)P(C_1)}{p(\mathbf{x})} > \frac{p(\mathbf{x} \mid C_2)P(C_2)}{p(\mathbf{x})} \quad & \text{then decide } \mathbf{x} \text{ comes from } C_1 \\
< \quad \ldots \quad & \text{then decide } \mathbf{x} \text{ comes from } C_2 \\
= \quad \ldots \quad & \text{then randomly decide } C_1 \text{ or } C_2
\end{aligned} \tag{9.7}$$

When the two sides are equal, we flip an honest coin. If it is a head, we decide C_1; if it is a tail, we decide C_2. Assuming that $p(\mathbf{x})$ is nonzero, we can rearrange (9.7) to obtain the following **equivalent decision rule**:

$$\begin{aligned}
\text{If} \quad \frac{p(\mathbf{x} \mid C_1)}{p(\mathbf{x} \mid C_2))} > \frac{P(C_2)}{P(C_1)} \quad & \text{then decide } C_1 \\
< \ \ldots \quad & \text{then decide } C_2 \\
= \ \ldots \quad & \text{then randomly decide } C_1 \text{ or } C_2
\end{aligned} \tag{9.8}$$

Define the left side of the decision rule as the **likelihood ratio** $l(\mathbf{x})$ and the right side as the **threshold** η as follows:

$$l(\mathbf{x}) = \frac{p(\mathbf{x} \mid C_1)}{p(\mathbf{x} \mid C_2)}, \quad \eta_{\mathrm{MAP}} = \frac{P(C_2)}{P(C_1)} \tag{9.9}$$

By these definitions, the maximum a posteriori classification rule becomes a **likelihood ratio test** (LRT) given by

$$\begin{aligned}
\text{If} \quad l(\mathbf{x}) > \eta_{\mathrm{MAP}} \quad & \text{then decide } C_1 \\
< \eta_{\mathrm{MAP}} \quad & \text{then decide } C_2 \\
= \eta_{\mathrm{MAP}} \quad & \text{then decide randomly}
\end{aligned} \tag{9.10}$$

The likelihood ratio is a scalar function of the observation vector **x** for the two-class case. A typical likelihood function for a scalar observation x is shown in Figure 9.1.

An example of a classification rule determined by using the maximum a posteriori rule will now be presented.

EXAMPLE 9.1

Say we are to make the classification that a person is a male or female using the single measurement of height of the individual by the maximum a posteriori decision rule. We will assume that the a priori probabilities of males and females are known as well as the conditional probability densities of their heights. For the purpose of this example, we will assume that the conditional probability densities for height x are Gaussian with means of 5 and 6 for females and males, respectively, and common variances of 1. We will also assume that the a priori probabilities of males and females are both equal to $\frac{1}{2}$. The problem is to formulate the maximum a posteriori classification rule that uses the observed or measured height of an individual as the only feature.

SOLUTION

The MAP decision rule is a likelihood ratio test, Eq. (9.10), where the likelihood ratio is given by

$$l(x) = \frac{p(x\,|\,\text{male})}{p(x\,|\,\text{female})} = \frac{(1/\sqrt{2\pi}1)\exp(-\frac{1}{2}(x-6)^2)}{(1/\sqrt{2\pi}1)\exp(-\frac{1}{2}(x-5)^2)}$$

The threshold eta is as follows:

$$\eta_{\text{MAP}} = \frac{P(\text{female})}{P(\text{male})} = \frac{\frac{1}{2}}{\frac{1}{2}} = 1$$

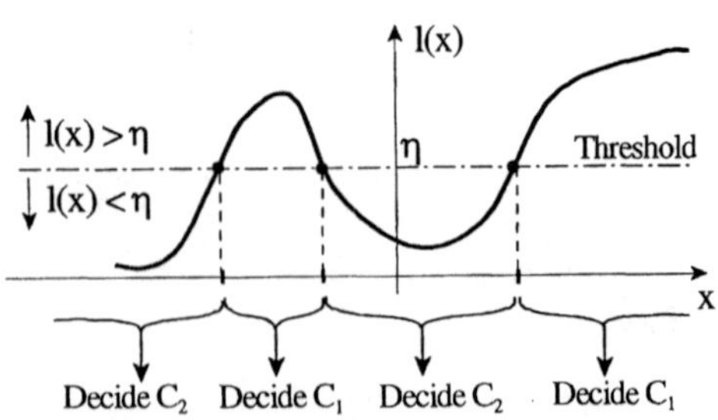

Figure 9.1 Typical likelihood ratio function and the corresponding induced decision regions in the observation space.

Using this threshold and likelihood ratio, we can write the maximum a posteriori decision rule from Eq. (9.10) as

$$\begin{aligned} \text{If} \quad l(x) = \exp(-\tfrac{1}{2}(x-6)^2 + \tfrac{1}{2}(x-5)^2) &> 1 \quad \text{then decide } x \text{ is male} \\ &< 1 \quad \text{then decide } x \text{ is female} \\ &= 1 \quad \text{then decide randomly} \end{aligned}$$

The decision regions and threshold in the likelihood ratio space are shown in Figure 9.2(a).

For this problem equivalent decision rules can be determined in intermediate as well as the observation space. Taking the ln of both sides of the decision rule and simplifying the exponents results in the following equivalent decision rule in the observation space:

$$\begin{aligned} \text{If} \quad x &> 5.5 \quad \text{then decide } x \text{ is male} \\ &< 5.5 \quad \text{then decide } x \text{ is female} \\ &= 5.5 \quad \text{then decide randomly} \end{aligned}$$

Thus the MAP decision rule is: If the height of the person tested is greater than 5.5, the person is classified as male, whereas if the height is less than 5.5, the person is classified as female. From an intuitive point of view this is a pleasing result. However, the rule is not perfect as it will lead to errors. ■

9.2.2 *M*-Class Problem (MAP)

The ***M*-class maximum a posteriori decision rule** for classifying the observed vector $\mathbf{x}$ can be stated as follows:

$$\begin{aligned} &\text{If} \quad P(C_j \mid \mathbf{x}) > P(C_i \mid \mathbf{x}) \qquad \text{for all } i \neq j \\ &\text{then decide } \mathbf{x} \text{ comes from } C_j \end{aligned} \tag{9.11}$$

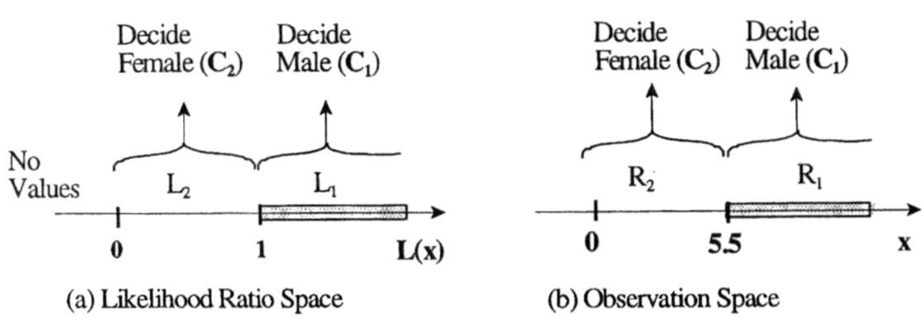

Figure 9.2 Decision regions in the likelihood ratio space and the observation space.

If equalities occur, the class will be assigned randomly from those classes with equal maximum values. The a posteriori probabilities can be determined in terms of the given information by using the following form of the Bayes rule:

$$P(C_i \mid \mathbf{x}) = \frac{p(\mathbf{x} \mid C_i)P(C_i)}{p(\mathbf{x})} \tag{9.12}$$

where the probability density function $p(\mathbf{x})$ for the observed vector $\mathbf{x}$ can be determined by the total probability theorem

$$p(\mathbf{x}) = \sum_{k=1}^{K} p(\mathbf{x} \mid C_k)P(C_k) \tag{9.13}$$

After applying Bayes's rule for all conditional probabilities the decision rule given in Eq. (9.11) yields the following **equivalent decision rule**:

$$\begin{aligned}&\text{If} \quad \frac{p(\mathbf{x} \mid C_j)P(C_j)}{p(\mathbf{x})} > \frac{p(\mathbf{x} \mid C_i)P(C_i)}{p(\mathbf{x})} \qquad \text{for all } i \neq j \\ &\text{then decide } \mathbf{x} \text{ comes from } C_j\end{aligned} \tag{9.14}$$

If a tie occurs for the largest value, randomly decide between the classes tied.

Assuming that $p(\mathbf{x})$ is nonzero, we can multiply Eq. (9.14) by $p(\mathbf{x})$ to obtain another **equivalent decision rule** as

$$\begin{aligned}&\text{If} \quad p(\mathbf{x} \mid C_j)P(C_j) > p(\mathbf{x} \mid C_i)P(C_i) \qquad \text{for all } i \neq j \\ &\text{then decide } \mathbf{x} \text{ comes from } C_j\end{aligned} \tag{9.15}$$

If a tie occurs for the largest value, randomly decide between the classes tied.

The form of the decision rule given above is usually the easiest to work from. However, another form of the multiple class maximum a posteriori decision can be written in terms of likelihood ratios similar to the one defined in Eq. (9.9). Using class M, an arbitrary choice, we divide each side by $p(\mathbf{x} \mid C_M)$ to obtain M likelihood ratios as follows:

$$L_i(x) = \frac{p(\mathbf{x} \mid C_i)}{p(\mathbf{x} \mid C_M)} \qquad \text{for } i = 1, 2, \ldots, M \tag{9.16}$$

Note that $L_M(\mathbf{x}) = 1$. Thus an **equivalent decision rule** in a likelihood ratio space can be described using these likelihood ratios and Eq. (9.15) as follows:

$$\begin{aligned}&\text{If} \quad L_j(\mathbf{x})P(C_j) > L_i(\mathbf{x})P(C_i) \qquad \text{for all } j \neq i, \quad i = 1, 2, \ldots, M \\ &\text{then decide } \mathbf{x} \text{ comes from class } C_j\end{aligned} \tag{9.17}$$

If a tie occurs for the largest value, randomly decide between the classes tied. The decision rule is thus seen to be described by a set of linear equations in the likelihood ratio space defined by Eq. (9.16). This space is of dimension $M - 1$, one less than the number of classes.

EXAMPLE 9.2

An observed pattern X is one dimensional and has the following class conditional densities:

$$p(x \mid C_1) = 2\exp(-2x)\mu(x)$$
$$p(x \mid C_2) = \exp(-x)\mu(x)$$
$$p(x \mid C_3) = \tfrac{1}{2}\exp(-\tfrac{1}{2}x)\mu(x)$$

Assume that the a priori probabilities of each class are

$$P(C_1) = 0.2, \quad P(C_2) = 0.3, \quad P(C_3) = 0.5$$

(a) Calculate the a posteriori probabilities for each of the three classes for the observed value $x = \frac{1}{2}$. For this value of x, what class would be assigned using the maximum a posteriori decision rule?

(b) Find the maximum a posteriori decision rule for all x in the observation space.

SOLUTION

To obtain the a posteriori probabilities, it is necessary to compute the total probability density function for the pattern vector x using (9.13) and evaluate it at $x = \frac{1}{2}$ as follows:

$$\begin{aligned} p_X(x)|_{x=1/2} &= [2\exp(-2x)P(C_1) + \exp(-x)P(C_2) + \tfrac{1}{2}\exp(-\tfrac{1}{2}x)P(C_3)]|_{x=1/2} \\ &= 0.1471518 + 0.1819592 + 0.1947002 \\ &= 0.5238112 \end{aligned}$$

(a) The a posteriori probabilities are then easily calculated using (9.12) as

$$P(C_1 \mid x = \tfrac{1}{2}) = 2\exp(-2(\tfrac{1}{2}))0.2/0.5238112 = 0.280925$$
$$P(C_2 \mid x = \tfrac{1}{2}) = \exp(-\tfrac{1}{2})0.3/0.5238112 = 0.347376$$
$$P(C_3 \mid x = \tfrac{1}{2}) = \tfrac{1}{2}\exp(-\tfrac{1}{2}(\tfrac{1}{2}))0.5/0.5238112 = 0.371699$$

Since $P(C_3 \mid x = \frac{1}{2})$ is the largest, the class assigned using the maximum a posteriori decision rule is class C_3.

(b) To obtain the decision rule for all x, Eq. (9.15) can be used. If x is greater than or equal to zero this gives

$$p(x \mid C_1)P(C_1) = 2\exp(-2x)0.2 = 0.4\exp(-2x)$$
$$p(x \mid C_2)P(C_2) = \exp(-x)0.3 = 0.3\exp(-x)$$
$$p(x \mid C_3)P(C_3) = \tfrac{1}{2}\exp(-\tfrac{1}{2}x)0.5 = \tfrac{1}{4}\exp(-\tfrac{1}{2}x)$$

Graphs of these weighted conditional densities are shown on the same axis in Figure 9.3. The decision regions correspond to the values of x for which the functions are the largest and are identified in the figure. The boundaries of the decision regions shown below are obtained by setting the weighted conditional densities equal to each other as follows:

$$0.4\exp(-2x) = 0.3\exp(-x) \qquad -x = 0.28768$$
$$0.4\exp(-2x) = \tfrac{1}{4}\exp(-\tfrac{1}{2}x) \qquad -x = 0.31334$$
$$0.3\exp(-x) = \tfrac{1}{4}\exp(-\tfrac{1}{2}x) \qquad -x = 0.36464$$

The decision rule is if $x \in R_i$ decide C_i where R_i are shown in Figure 9.3. ■

9.3 MINIMUM PROBABILITY OF ERROR CLASSIFIER

When the probability of error is the selected performance measure, the resulting decision or classification rule is the **minimum probability of error decision rule**. The underlying assumptions are that the conditional probability density functions for the pattern vector for each class and the a priori probabilities of each class are known.

9.3.1 Two-Class Problem (MPE)

The minimum probability of error decision rule divides the pattern space into two regions R_1 and R_2 such that the total probability of error is minimized. The regions R_1 and R_2 are mutually exclusive and exhaustive as shown in Figure 9.4. The regions are not necessarily simply connected.

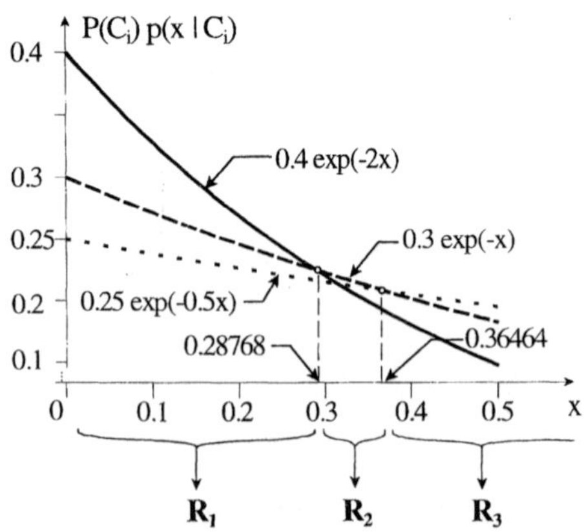

Figure 9.3 Determination of decision regions for Example 9.2.

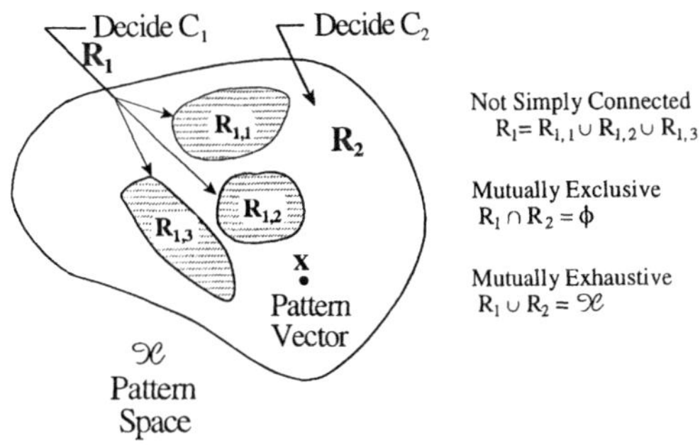

Figure 9.4 Properties of decision regions.

The total probability of error is given by

$$P(\text{error}) = P(\text{error} \mid C_1)P(C_1) + P(\text{error} \mid C_2)P(C_2) \tag{9.18}$$

Each of these conditional errors can then be written in terms of the decision regions R_1 and R_2 as follows:

$$\begin{aligned} P(\text{error} \mid C_1) &= P(\text{decide } C_2 \mid C_1) \\ &= \int\int \cdots \int_{R_2} p(\mathbf{x} \mid C_1)\, d\mathbf{x} \\ P(\text{error} \mid C_2) &= P(\text{decide } C_1 \mid C_2) \\ &= \int\int \cdots \int_{R_1} p(\mathbf{x} \mid C_2)\, d\mathbf{x} \end{aligned} \tag{9.19}$$

The first integral above can be rewritten in terms of R_1 as follows:

$$\begin{aligned} P(\text{error} \mid C_1) &= \int\int \cdots \int_{S-R_1} p(\mathbf{x} \mid C_1)\, d\mathbf{x} \\ &= 1 - \int\int \cdots \int_{R_1} p(\mathbf{x} \mid C_1)\, d\mathbf{x} \end{aligned} \tag{9.20}$$

Substituting the conditional probabilities of error, $P(\text{error} \mid C_1)$ from (9.20), and $P(\text{error} \mid C_2)$ from (9.19), into the total probability of error in (9.18) gives

$$P(\text{error}) = P(C_1) + \int\int \cdots \int_{R_1} [-p(\mathbf{x} \mid C_1)P(C_1) + p(\mathbf{x} \mid C_2)P(C_2)]\, d\mathbf{x} \tag{9.21}$$

Since $P(C_1)$ is fixed, it can be discarded, and the $P(\text{error})$ is minimized by selecting R_1 such that the integral is minimized. This can be done by putting all values of $\mathbf{x}$ into R_1 that give negative values of the integrand, thus reducing the error. The values of **x** that give positive values of the integrand would not be put in R_1 as this would increase the error. This is expressed as follows:

$$x \in R_1 \quad \text{if} \quad [-p(\mathbf{x} \mid C_1)P(C_1) + p(\mathbf{x} \mid C_2)P(C_2)] > 0 \tag{9.22}$$

Rearranging the above by putting the first part on the right side of the equality and dividing through gives R_1 to be selected as follows:

$$x \in R_1 \quad \text{if} \quad \frac{p(\mathbf{x} \mid C_1)}{p(\mathbf{x} \mid C_2)} > \frac{P(C_2)}{P(C_1)} \triangleq \eta_{\text{MPE}} \tag{9.23}$$

Thus the **minimum probability of error decision rule** is a likelihood ratio test with threshold equal to the ratio of the a priori probabilities of each class:

$$\begin{aligned} \text{If} \quad \frac{p(\mathbf{x} \mid C_1)}{p(\mathbf{x} \mid C_2)} &> \frac{P(C_2)}{P(C_1)} \triangleq \eta_{\text{MPE}} && \text{then decide } C_1 \\ &< && \text{then decide } C_2 \\ &= && \text{then decide randomly} \end{aligned} \tag{9.24}$$

Notice that this is exactly the same decision rule obtained using the maximum a posteriori criterion, Eq. (9.10). Thus, for the two-class case, it has been shown that **minimizing** the probability of error leads to the same decision rule as **maximizing** the a posteriori probability.

EXAMPLE 9.3

Find the minimum probability of error classifier for the assumptions given in Example 9.1, and evaluate the total probability of error for that decision rule.

SOLUTION

The minimum probability of error classifier is the same likelihood ratio test given in Example 9.1. The performance in terms of the probability of error will now be determined. The types of errors are (a) classifying a female as a male and (b) classifying a male as a female. These errors are the integrals of the conditional densities over the regions as shown in Figure 9.5. Using the decision rule given in Eq. (9.24), we see that the probability of classifying a female as a male is the integral of $p(x \mid \text{female})$ over the region classified as male (from 5.5 to infinite), while the probability of classifying a male a female is the integral of $p(x \mid \text{male})$ over the

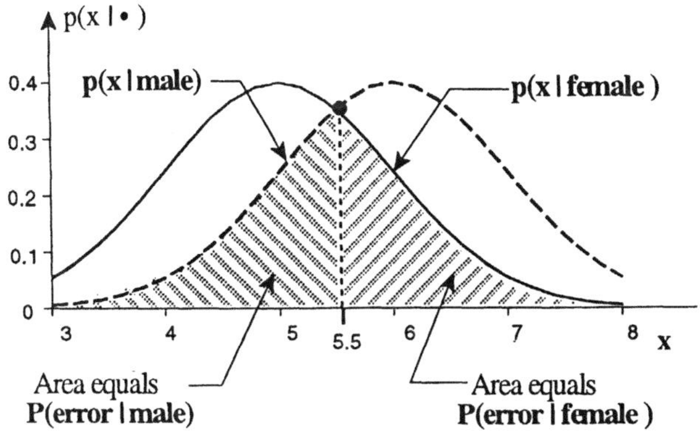

Figure 9.5 Probabilities of error for Example 9.3.

region classified as female (from $-\infty$ to the threshold 5.5). These error probabilities can be calculated in the following way:

$$
\begin{aligned}
P(\text{error}\,|\,\text{female}) &= P(\text{deciding male}\,|\,\text{person is female}) \\
&= \int_{5.5}^{\infty} p(x\,|\,\text{female})\,dx = \int_{5.5}^{\infty} \frac{1}{\sqrt{2\pi}1} \exp\left[\frac{-(x-5)^2}{2}\right] dx \\
&= \int_{(5.5-5)/1}^{\infty} \frac{1}{\sqrt{2\pi}} \exp\left[-\frac{y^2}{2}\right] dy = 1 - \Phi(\tfrac{1}{2}) \\
&= 0.3085
\end{aligned}
$$

$$
\begin{aligned}
P(\text{error}\,|\,\text{male}) &= P(\text{deciding female}\,|\,\text{person is male}) \\
&= \int_{-\infty}^{5.5} p(x\,|\,\text{male})\,dx = \int_{-\infty}^{5.5} \frac{1}{\sqrt{2\pi}1} \exp\left[\frac{-(x-6)^2}{2(1)}\right] dx \\
&= \int_{-\infty}^{(5.5-6)/1} \frac{1}{\sqrt{2\pi}} \exp\left[-\frac{y^2}{2}\right] dy = \Phi(-\tfrac{1}{2}) \\
&= 0.3085
\end{aligned}
$$

The total probability of error, from Eq. (9.18), is then given by

$$
\begin{aligned}
P(\text{error}) &= P(\text{error}\,|\,\text{female})P(\text{female}) + P(\text{error}\,|\,\text{male})P(\text{male}) \\
&= (0.3085)\tfrac{1}{2} + (0.3085)\tfrac{1}{2} = 0.3085
\end{aligned}
$$

For this problem the probability of error is quite large. Therefore, using height alone to make the decision does not yield a low probability of error classifier. It does, however, give the lowest probability of error of any decision rule just using height. Performance can be improved by adding different measurements like weight, ankle size, or waist size. Although these measurements may not provide totally independent information from height, they do provide some correlated information that should result in improved performance. ■

9.3.2 *M*-Class Problem (MPE)

For the M-class case the minimum probability of error decision rule divides the pattern space into M regions $R_1, R_2, \ldots, R_M$ such that the total probability of error is minimized. The regions $R_1, R_2, \ldots, R_M$ selected must be mutually exclusive and exhaustive and representationally shown in Figure 9.6. The regions are not necessarily simply connected.

The total probability of error is given by

$$P(\text{error}) = \sum_{k=1}^{M} P(\text{error} \mid C_k) P(C_k) \tag{9.25}$$

Each of the conditional errors can be written in terms of their corresponding decision regions R_i as follows:

$$\begin{aligned} P(\text{error} \mid C_i) &= 1 - P(\text{correct} \mid C_i) \\ &= 1 - \int\int \cdots \int_{R_i} p(\mathbf{x} \mid C_i)\, dx \end{aligned} \tag{9.26}$$

Substituting the conditional error probabilities from (9.26) into the total probability of error given into (9.25) and rearranging gives

$$\begin{aligned} P(\text{error}) &= \sum_{k=1}^{M} P(C_k) - \sum_{k=1}^{M} \int\int \cdots \int_{R_k} p(\mathbf{x} \mid C_k) P(C_k)\, d\mathbf{x} \\ &= 1 - \sum_{k=1}^{M} \int\int \cdots \int_{R_k} p(\mathbf{x} \mid C_k) P(C_k)\, d\mathbf{x} \end{aligned} \tag{9.27}$$

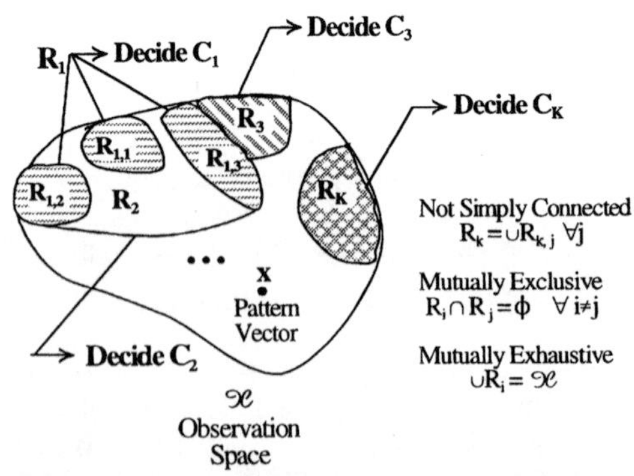

Figure 9.6 Properties of decision regions.

Since the first sum equals one, a fixed number, the $P(\text{error})$ is minimized by selecting $\mathbf{x}$ to be a member of R_k such that the integrand term is maximized. That is $\mathbf{x} \in R_k$ if its $p(\mathbf{x} \mid C_k)P(C_k)$ is the largest. This leads to the following **minimum probability of error decision rule**:

$$\begin{aligned} &\text{If} \quad p(\mathbf{x} \mid C_k)P(C_k) > p(\mathbf{x} \mid C_j)P(C_j) \qquad \text{for } j = 1, 2, \ldots, M \text{ and } j \neq k \\ &\text{then decide that } \mathbf{x} \text{ comes from } C_k \end{aligned} \tag{9.28}$$

As before, for the other decision rules, ties can be resolved randomly.

Thus minimizing the probability of error for the M-class case gives a decision rule that maximizes the product of the a posteriori probability density and the a priori class probability and thus is equivalent to the MAP decision rule. The following example illustrates the use of (9.28) to establish the decision regions for a three-class problem.

EXAMPLE 9.4

Consider a three-class problem with the following class conditional density functions:

$$\begin{aligned} p(x \mid C_1) &= \frac{1}{\sqrt{2\pi}} \exp\left\{ -\frac{(x-1)^2}{2} \right\} \\ p(x \mid C_2) &= \exp(-x)\mu(x) \\ p(x \mid C_3) &= \exp(x)\mu(-x) \end{aligned}$$

Assume equally likely classes. Find the regions in the observation space which the minimum probability of error decision rule would assign to each of the three classes.

SOLUTION

Since the $P(C_i)$ are all equal, it is seen from (9.28) that the minimum probability of error decision rule selects the decision regions R_1, R_2, and R_3 in the following way:

$$\begin{aligned} &x \in R_k \quad \text{if} \quad p(x \mid C_k) > p(x \mid C_j) \\ &\text{for all } j = 1, 2, 3, \ j \neq k \end{aligned}$$

The decision regions can be determined by plotting the conditional probability density functions $p(x \mid C_k)$ for $k = 1, 2$, and 3 on the same x-axis, and finding the regions for which the highest values occur as shown in Figure 9.7.

Regions R_1, R_2, and R_3 are determined as the values of x where $p(x \mid C_1)$, $p(x \mid C_2)$, and $p(x \mid C_3)$ are the largest, respectively. The boundary values between R_1 and R_2 are determined by setting $p(x \mid C_1) = p(x \mid C_2)$ and solving for the intersections. For positive x this gives

$$\frac{1}{\sqrt{2\pi}} e^{-(x-1)^2/2} = e^{-x}$$

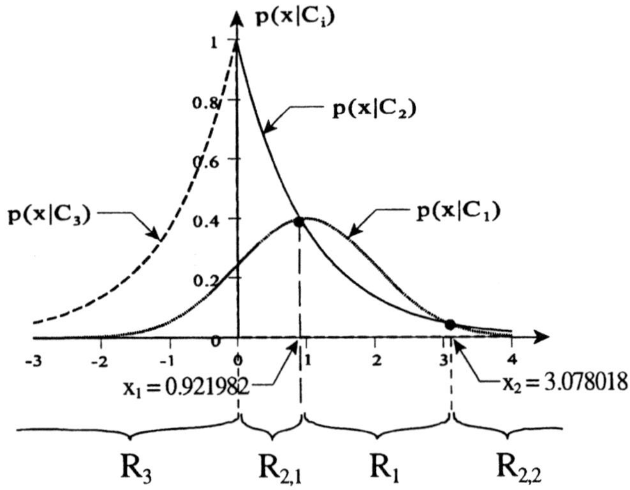

Figure 9.7 Determination of decision regions for Example 9.4.

Taking the ln of both sides and simplifying gives a quadratic equation with solutions as

$$-x^2 + 4x - 2.837877 = 0$$
$$x_1 = 0.921982 \quad \text{and} \quad x_2 = 3.078018$$

Therefore $R_1 = \{x\colon 0.921982 < x < 3.078018\}$.

For all negative x, $p(x \mid C_3)$ is greater than both $p(x \mid C_2)$ and $p(x \mid C_3)$, so

$$R_3 = \{x\colon x < 0\}$$

R_2 is the remainder of the values of x not included in regions R_1 or R_3 and is a multiply connected region given by the union of two sets:

$$\begin{aligned} R_2 &= R_{2,1} \cup R_{2,2} \\ &= \{x\colon 0 < x < 0.921982\} \cup \{x\colon x > 3.078018\} \end{aligned}$$

The values of x on the boundaries $x = 0, 0.921982$, and 3.078018 are decided randomly as follows:

If $x = 0$ decide randomly between C_2 and C_3
$x = 0.921982$ decide randomly between C_1 and C_2
$x = 3.078018$ decide randomly between C_1 andC_2 ■

9.4 BAYES DECISION RULE

In certain problems it is not just the probability of error that is important but the type of error. For example, it may be more costly to call a rotten egg okay when it is rotten than to call a good egg rotten. In one case the cost of the error is the cost of the egg, whereas the other type of error results in consumer dissatisfaction and eventual loss of customers. The assignment of realistic costs is not necessarily analytical but usually based on subjective information. Once these costs are defined, we are in the position of being able to make our classifications in such a way that we minimize the average cost per decision. The Bayes decision rule does just that.

Assumptions need to be made concerning the conditional probability density functions of the pattern vector, the a priori probabilities of the pattern classes, and the cost for each type of decision. The following sections develop the Bayes decision rules for the two and M-class cases.

9.4.1. Two-Class Problem (Bayes)

Assumptions. Let C_1 and C_2 represent the two classes. Assume that class dependent conditional density functions, $p(\mathbf{x} \mid C_1)$ and $p(\mathbf{x} \mid C_2)$, for the pattern vector $\mathbf{X}$ and that the a priori probabilities of each class, $P(C_1)$ and $P(C_2)$, are known.

Let C_{ij} be the **cost** associated with the decision of classifying a pattern $\mathbf{x}$ as coming from class i given that it is from class C_j. These costs are assumed known and are all greater than or equal zero. The **average cost for a decision**, sometimes called **risk**, is defined as follows:

$$\begin{aligned}\text{Risk} = {} & P(\text{deciding } C_1 \mid C_1)P(C_1)C_{11} + P(\text{deciding } C_2 \mid C_2)P(C_2)C_{22} \\ & + P(\text{deciding } C_1 \mid C_2)P(C_2)C_{12} + P(\text{deciding } C_2 \mid C_1)P(C_1)C_{21}\end{aligned} \tag{9.29}$$

Decision Rule. The Bayes decision rule minimizes the average cost defined above. The probability of deciding class C_i given class C_j can be obtained by integrating the conditional density over the decision region R_i for $i, j = 1, 2$ as follows:

$$P(\text{deciding } C_i \mid C_j) = \int\!\!\int \cdots \int_{R_i} p(\mathbf{x} \mid C_j)\, d\mathbf{x} \tag{9.30}$$

After substituting the integrals given in (9.30) into (9.29) and collecting the terms that have a common region of integration, we obtain, the risk as

$$\begin{aligned}\text{Risk} = {} & \int\!\!\int \cdots \int_{R_1} [P(C_1)C_{11}p(\mathbf{x} \mid C_1) + P(C_2)C_{12}p(\mathbf{x} \mid C_2)]\, d\mathbf{x} \\ & + \int\!\!\int \cdots \int_{R_2} [P(C_2)C_{22}p(\mathbf{x} \mid C_2) + P(C_1)C_{21}p(x \mid C_1)]\, d\mathbf{x}\end{aligned} \tag{9.31}$$

The problem of minimizing the risk is that of selecting regions of integration R_1 and R_2 such that the risk is minimized. This requires that $\mathbf{x}$ has to be assigned to the region for which the integrand is the smallest. Thus $\mathbf{x}$ is put in region R_1 if the first integrand is smaller than the second, that is,

$$P(\omega_1)C_{11}p(\mathbf{x}\,|\,C_1) + P(C_2)C_{12}p(\mathbf{x}\,|\,C_2) < P(C_2)C_{22}p(\mathbf{x}\,|\,C_2) + P(C_1)C_{21}p(\mathbf{x}\,|\,C_1) \tag{9.32}$$

If the inequality is reversed, $\mathbf{x}$ is put in region R_2, if equal, it will not matter for continuous conditional probability density functions which region it is put in, so the assignment is arbitrary. Equation (9.32) can be rearranged by collecting common factors times the conditional densities to give

$$P(C_1)(C_{11} - C_{21})p(\mathbf{x}\,|\,C_1) < P(C_2)(C_{22} - C_{12})p(\mathbf{x}\,|\,C_2) \tag{9.33}$$

Since C_{11}, the cost for a correct classification of class C_1, is usually less than C_{21}, the cost for an incorrect classification of C_2, the division of both sides of (9.33) by the first two terms of the left side reverses the inequality. Then dividing both sides by $p(\mathbf{x}\,|\,C_2)$ establishes the **Bayes decision rule** for the two-class case as follows:

$$\begin{array}{llll} \text{If} & \dfrac{p_X(\mathbf{x}\,|\,C_1)}{p_X(\mathbf{x}\,|\,C_2)} > \dfrac{(C_{22} - C_{12})P(C_2)}{(C_{11} - C_{21})P(C_1)} \triangleq \eta_{\text{BAYES}} & \text{then decide } C_1 \\ & \qquad < \qquad \cdots & \text{then decide } C_2 \\ & \qquad = \qquad \cdots & \text{then decide randomly} \end{array} \tag{9.34}$$

Thus the Bayes decision rule is seen to be a likelihood ratio test, with the threshold determined as a function of the a priori probabilities and the specified costs. If $C_{11} = C_{22} = 0$, no cost is associated with a correct decision, and $C_{12} = C_{21} = 1$, the costs associated with the errors are equal and equal to 1; the Bayes risk given in (9.29) is the same as the minimum probability of error. Such a cost assignment is called a **zero-one cost assignment**. Thus the minimum probability of error classifier is a special case of the Bayes classifier.

The general expression for risk given in (9.29) can be rewritten in a form involving only correct decisions by using the following relationships between incorrect and correct decisions:

$$\begin{aligned} P(\text{deciding } C_2\,|\,C_1) &= 1 - P(\text{deciding } C_1\,|\,C_1) \\ P(\text{deciding } C_1\,|\,C_2) &= 1 - P(\text{deciding } C_2\,|\,C_2) \end{aligned} \tag{9.35}$$

Substituting the formulas of (9.35) into (9.29), the risk can be written in terms of the correct decisions as follows:

$$\begin{aligned} \text{Risk} = {} & P(C_2)C_{12} + P(C_1)C_{21} \\ & + (C_{11} - C_{21})P(\text{decide } C_1\,|\,C_1)P(C_1) \\ & + (C_{22} - C_{12})P(\text{decide } C_2\,|\,C_2)P(C_2) \end{aligned} \tag{9.36}$$

Similarly another form of the risk can be obtained in terms of only the probabilities of errors by substituting for the correct decisions, to give

$$\begin{aligned}\text{Risk} = {} & P(C_1)C_{11} + P(C_2)C_{22} \\ & + (C_{21} - C_{11})P(\text{decide } C_2 \,|\, C_1)P(C_1) \\ & + (C_{12} - C_{22})P(\text{decide } C_1 \,|\, C_2)P(C_2)\end{aligned} \tag{9.37}$$

These alternative forms for the Bayes risk can be used to evaluate the risk more efficiently than the form given in (9.29) which involves four integrals. An example follows that finds the Bayes decision rule and calculates its associated risk.

EXAMPLE 9.5

Given the class conditional densities $p(x\,|\,C_1)$ and $p(x\,|\,C_2)$ as

$$p(\mathbf{x}\,|\,C_1) = \exp(-x)\mu(x)$$
$$p(\mathbf{x}\,|\,C_2) = 2\exp(-2x)\mu(x)$$

Assume that the a priori probabilities and costs are known and given by

$$P(C_1) = \tfrac{1}{3}, \quad P(C_2) = \tfrac{2}{3}$$
$$C_{21} = 2, \quad C_{12} = 3, \quad C_{11} = C_{22} = 0$$

(a) Find the test that minimizes the Bayes risk.
(b) Simplify your test to the observation space.
(c) Calculate the Bayes risk for the Bayes decision rule determined in part (a).

SOLUTION

(a) The optimum test in the Bayes sense is a likelihood ratio test with likelihood ratio and threshold given in (9.34) as

$$\eta_{\text{BAYES}} = \frac{(C_{22} - C_{12})\,P(C_2)}{(C_{11} - C_{21})\,P(C_1)} = \frac{(0-3)\,\frac{2}{3}}{(0-2)\,\frac{1}{3}} = 3$$

$$l(x) = \frac{p(x\,|\,C_1)}{p(x\,|\,C_2)} = \frac{\exp(-x)}{2\exp(-2x)}\,\mu(x) = \tfrac{1}{2}\exp(x)\mu(x)$$

The Bayes decision rule from (9.34) can thus be written for $x \geq 0$, as follows:

$$\text{If} \quad \tfrac{1}{2}\exp(x) \begin{cases} > 3 & \text{decide class } C_1 \\ < 3 & \text{decide class } C_2 \\ = 3 & \text{decide randomly } C_1 \text{ or } C_2 \end{cases}$$

Notice that for $x < 0$ no decision rule is required, or we could just simply choose randomly between the two classes for any of those values to define a decision rule for all x.

(b) The rule above can be rewritten in the observation space by multiplying both sides by two and taking the ln of both sides to give

$$\begin{aligned} \text{If} \quad x &> \ln 6 \quad \text{decide class } C_1 \\ &< \ln 6 \quad \text{decide class } C_2 \\ &= \ln 6 \quad \text{decide randomly } C_1 \text{ or } C_2 \end{aligned}$$

(c) In general, to calculate the Bayes risk, we would evaluate the four conditional probabilities given in (9.29), the two probabilities of correct decisions in (9.36), or the two probabilities of incorrect decisions given in (9.37). For this problem the costs associated with making a correct decision are zero, so the conditional probabilities of error are all that need be determined. They are as follows:

$$P(\text{deciding } C_1 \mid C_2) = \int_{R_1} p(\mathbf{x} \mid C_2)\, dx = \int_{\ln 6}^{\infty} 2\exp(-2x)\mu(x)\, dx = \frac{1}{36}$$

$$P(\text{deciding } C_2 \mid C_1) = \int_{-\infty}^{\ln 6} p(\mathbf{x} \mid C_1)\, dx = \int_{0}^{\ln 6} \exp(-x)\, dx = \frac{5}{6}$$

The risk can now be determined from Eq. (9.29) as

$$\begin{aligned} \text{Risk} &= P(\text{decide } C_2 \mid C_1)P(C_1)C_{21} + P(\text{decide } C_1 \mid C_2)P(C_2)C_{12} \\ &= \left(\frac{1}{36}\right)\left(\frac{2}{3}\right)3 + \left(\frac{5}{6}\right)\left(\frac{1}{3}\right)2 = \frac{11}{18} \end{aligned}$$

Thus the average cost is 11/18 units/decision. ■

9.4.2 *M*-Class Bayes Decision Rule

The Bayes decision rule can be extended to the multiple-class case, M hypotheses, by making a number of assumptions and defining a performance measure as follows:

Assumptions

(a) The a priori probabilities of the classes $P(C_1), P(C_2), \ldots, P(C_M)$ are known.
(b) The conditional probability densities for the pattern vector $\mathbf{x}$ given the classes are known and given by $p(\mathbf{x} \mid C_1), p(\mathbf{x} \mid C_2), \ldots, p(\mathbf{x} \mid C_M)$.
(c) The costs, C_{ij}, of choosing C_i when correct class is C_j are assigned and known for all i and $j \in \{1, 2, \ldots, M\}$.

Performance Measure. The Bayes approach to the multiple-class case uses an extension of the average cost or risk defined in (9.29) for the two-class case. This performance measure, risk, for the M-class case can be defined as

$$\text{Risk} \triangleq \sum_{i=1}^{M} \sum_{j=1}^{M} C_{ij} P(\text{deciding } C_i \mid C_j) P(C_j) \tag{9.38}$$

Decision Rule. If R is the total sample space of observed pattern vectors, then the Bayes decision rule selects the mutually exhaustive and exclusive decision regions $R_1, R_2, \ldots, R_M$ such that the average cost or risk is minimized. These decision regions will now be shown to be polyhedral regions in the space of M likelihood functions.

The probability of deciding C_i given that C_j is the true class can be written as

$$P(\text{decide } C_i \mid C_j) = \int\int \cdots \int_{R_i} p(\mathbf{x} \mid C_j)\, d\mathbf{x} \tag{9.39}$$

After substituting Eq. (9.39) into (9.38) and separating out the costs with identical indexes, we can write the risk in terms of M integrals as

$$\begin{aligned} \text{Risk} &= \sum_{i=1}^{M} \sum_{j=1}^{M} C_{ij} P(C_j) \int\int \cdots \int_{R_i} p(\mathbf{x} \mid C_j)\, d\mathbf{x} \\ &= \int\int \cdots \int_{R_1} \sum_{j=1}^{M} C_{1j} p(\mathbf{x} \mid C_j) P(C_j) \\ &\quad + \int\int \cdots \int_{R_2} \sum_{j=1}^{M} C_{2j} p(\mathbf{x} \mid C_j) P(C_j) \\ &\quad + \cdots \\ &\quad + \int\int \cdots \int_{R_M} \sum_{j=1}^{M} C_{Mj} p(\mathbf{x} \mid C_j) P(C_j) \end{aligned} \tag{9.40}$$

The assignment of each $\mathbf{x}$ to a decision region is to be made such that the risk is minimized. This can be done by defining new functions $y_i(\mathbf{x})$, the integrands of Eq. (9.40):

$$y_i(\mathbf{x}) \triangleq \sum_{j=1}^{M} C_{ij} p(\mathbf{x} \mid C_j) P(C_j), \qquad i = 1, 2, \ldots, M \tag{9.41}$$

Since $\mathbf{x}$ can be assigned to only one of the regions, the R_i are mutually exclusive. From the risk equation, (9.40), we see that the risk will be minimized if the regions R_j are selected in the following way: $\mathbf{x} \in R_j$ if $y_j(x) < y_i(x)$ for $i = 1, 2, \ldots, M$; $i \neq j$, where ties are resolved randomly among those tied.

We can now write the **M-class Bayes decision rule** as

$$\begin{aligned} &\text{If} \quad y_j(\mathbf{x}) < y_i(\mathbf{x}) \qquad \text{for } i = 1, 2, \ldots, M;\ i \neq j \\ &\text{then decide } \mathbf{x} \text{ comes from } C_j \end{aligned} \tag{9.42}$$

The ties can be resolved randomly between tying classes.

Each decision region can thus be seen to be a set of $M - 1$ inequalities in the space of functions $y_i(\mathbf{x})$ or alternatively can be viewed as a set of linear inequalities, boundaries specified by hyperplanes in the space of functions $p(\mathbf{x} \mid C_i)P(C_i)$. Thus each of these spaces can be considered possible feature spaces for the purpose of classification. The following example illustrates the procedure in finding the Bayes decision regions for a three-class problem.

EXAMPLE 9.6

The observation $\mathbf{x}$ under each of three classes, C_1, C_2, and C_3, is assumed to be Gaussian with known mean and variance. The class a priori probabilities and conditional probability density functions are given as

$$\begin{aligned} C_1&: \quad X \sim N(1, 1), \quad P(C_1) = 0.2 \\ C_2&: \quad X \sim N(2, 1), \quad P(C_2) = 0.3 \\ C_3&: \quad X \sim N(3, 1), \quad P(C_3) = 0.5 \end{aligned}$$

Assume that the Bayes costs are known and have the following values:

$$\begin{array}{lll} C_{11} = 0.1 & C_{12} = 1.2 & C_{13} = 1.3 \\ C_{21} = 2.1 & C_{23} = 0.2 & C_{23} = 2.3 \\ C_{31} = 3.1 & C_{32} = 3.2 & C_{33} = 0.3 \end{array}$$

Find the optimum decision rule where optimality is in the Bayes sense.

SOLUTION

The class conditional probability densities under each of the three classes are given as

$$p(\mathbf{x} \mid C_i) = \frac{1}{\sqrt{2\pi}} \exp\left\{ -\frac{(x - m_i)^2}{2} \right\}, \qquad i = 1, 2, 3$$

where the m_i are 1, 2, and 3, respectively. From Eqs. (9.41) and (9.42) of the previous section the optimum decision rule selects the class with the smallest value of $y_i(x)$, $i = 1, 2, 3$, where the $y_i(x)$ are given by

$$y_i(x) = \sum_{j=1}^{3} C_{ij} p(\mathbf{x} \mid C_j) P(C_j), \qquad i = 1, 2, 3$$

With the costs given in the problem formulation, the $y_i(x)$ can be shown to be

$$y_1(x) = 0.02p(\mathbf{x} \mid C_1) + 0.36p(\mathbf{x} \mid C_2) + 0.65p(\mathbf{x} \mid C_3)$$
$$y_2(x) = 0.42p(\mathbf{x} \mid C_1) + 0.06p(\mathbf{x} \mid C_2) + 1.15p(\mathbf{x} \mid C_3)$$
$$y_3(x) = 0.62p(\mathbf{x} \mid C_1) + 0.96p(\mathbf{x} \mid C_2) + 0.15p(\mathbf{x} \mid C_3)$$

If S_1 represents the region in the conditional probability density space for which we decide class C_1, then this region is obtained by setting $y_1(x) < y_2(x)$ and $y_1(x) < y_3(x)$. Therefore region S_1 is the intersection of two regions determined by the inequalities as functions of the conditional probability density functions as follows:

$$0.02p(\mathbf{x} \mid C_1) + 0.36p(\mathbf{x} \mid C_2) + 0.65p(\mathbf{x} \mid C_3) < 0.42p(\mathbf{x} \mid C_1) + 0.06p(\mathbf{x} \mid C_2) + 1.15p(\mathbf{x} \mid C_3)$$

AND

$$0.02p(\mathbf{x} \mid C_1) + 0.36p(\mathbf{x} \mid C_2) + 0.65p(\mathbf{x} \mid C_3) < 0.62p(\mathbf{x} \mid C_1) + 0.96p(\mathbf{x} \mid C_2) + 0.15p(\mathbf{x} \mid C_3)$$

By subtracting the right side from the left side of both of the inequalities above, we can obtain the region S_1 for deciding class C_1 in the space of conditional densities with respect to zero as

$$S_1\colon \begin{cases} -0.4p(\mathbf{x} \mid C_1) + 0.3p(\mathbf{x} \mid C_2) - 0.5p(\mathbf{x} \mid C_3) < 0 & \text{AND} \\ -0.6p(\mathbf{x} \mid C_1) - 0.6p(\mathbf{x} \mid C_2) + 0.5p(\mathbf{x} \mid C_3) < 0 & \end{cases}$$

The corresponding inequalities representing decision regions S_2 and S_3 for classes C_2 and C_3 can be similarly obtained as

$$S_2\colon \begin{cases} -0.2p(\mathbf{x} \mid C_1) - 0.9p(\mathbf{x} \mid C_2) + p(\mathbf{x} \mid C_3) < 0 & \text{AND} \\ -0.4p(\mathbf{x} \mid C_1) + 0.3p(\mathbf{x} \mid C_2) - 0.5p(\mathbf{x} \mid C_3) < 0 & \end{cases}$$

$$S_3\colon \begin{cases} 0.6p(\mathbf{x} \mid C_1) + 0.6p(\mathbf{x} \mid C_2) - 0.5p(\mathbf{x} \mid C_3) < 0 & \text{AND} \\ 0.2p(\mathbf{x} \mid C_1) + 0.9p(\mathbf{x} \mid C_2) - p(\mathbf{x} \mid C_3) < 0 & \end{cases}$$

The decision region S_1 specified above is shown in the space of conditional densities in Figure 9.8, and the other two decision regions S_2 and S_3 could similarly be determined in terms of intersections of regions in the space of conditional densities. The decision can be made in this space, which is a transformation of the original observation space, through the conditional densities or in the original one-dimensional observation space. Note that the space determined by the conditional densities is of dimension 3, whereas the original pattern space for this problem is of dimension 1.

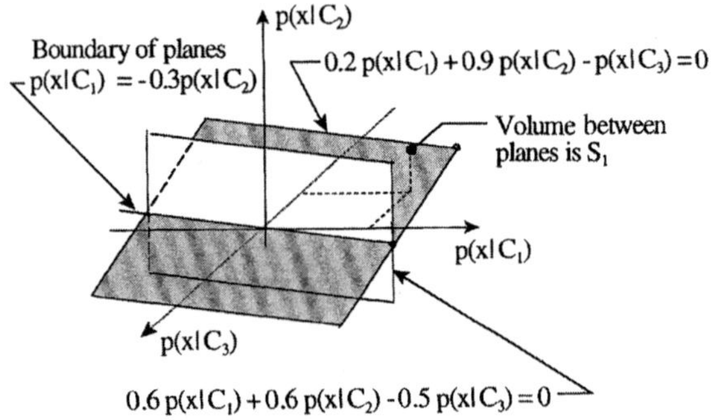

Figure 9.8 Decision region S_1 in the space of $p(x \mid C_i)$, $i = 1, 2, 3$ for Example 9.7.

The decision rule can be further simplified to obtain a decision rule in terms of the original observation space if desired. For this problem it can be simplified to a feature space by substituting the given conditional probability density functions into the inequalites in the space of conditional densities. For example, the region R_1, in the observation space, where we decide class C_1 can be written as the set of all x that satisfy the following two inequalities:

$$-0.4 \frac{1}{\sqrt{2\pi}} \exp\left\{-\frac{(x-1)^2}{2}\right\} + 0.3 \frac{1}{\sqrt{2\pi}} \exp\left\{-\frac{(x-2)^2}{2}\right\}$$
$$- 0.5 \frac{1}{\sqrt{2\pi}} \exp\left\{-\frac{(x-3)^2}{2}\right\} < 0$$

$$-0.6 \frac{1}{\sqrt{2\pi}} \exp\left\{-\frac{(x-1)^2}{2}\right\} - 0.6 \frac{1}{\sqrt{2\pi}} \exp\left\{-\frac{(x-2)^2}{2}\right\}$$
$$+ 0.5 \frac{1}{\sqrt{2\pi}} \exp\left\{-\frac{(x-3)^2}{2}\right\} < 0$$

These two inequalities are nonlinear in the original observation space and taking the log will not simplify the results because of the additive terms. Defining the previous inequalities by $g_1(x) < 0$ and $g_2(x) < 0$, we can plot them with respect to x and the corresponding decision region, R_1 in the observation space, for deciding C_1 can be identified as the intersection of the regions in x for each inequality in the observation space as shown in Figure 9.9. For the particular cost functions and a priori probabilities specified in the problem the other decision regions can be obtained similarly resulting in regions R_2 and R_3 for deciding classes C_2 and C_3, respectively, as shown in Figure 9.10.

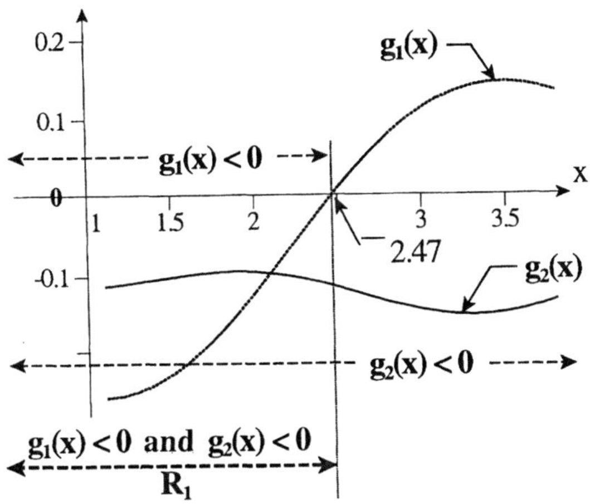

Figure 9.9 Determination of decision region R_1 for Example 9.6.

It has thus been shown that the optimum decision rule is a multiple-threshold test in the observation space which can be written as follows:

If	$x < 2.47$	decide class C_1
	$2.47 < x < 2.91$	decide class C_2
	$x > 2.91$	decide class C_3
	$x = 2.47$	decide randomly between the classes C_1 and C_2
	$x = 2.91$	decide randomly between the classes C_2 and C_3 ■

The Bayes decision rule for the multiple-class case can also be written in terms of likelihood ratios similar to the development for the minimum probability of error classifier. Using class M, an arbitrary choice, we can define a set of likelihood ratios $L_i(\mathbf{x})$ as follows:

$$L_i(\mathbf{x}) \triangleq \frac{p(\mathbf{x} \mid C_i)}{p(\mathbf{x} \mid C_M)} \qquad \text{for } i = 1, 2, \ldots, M \tag{9.43}$$

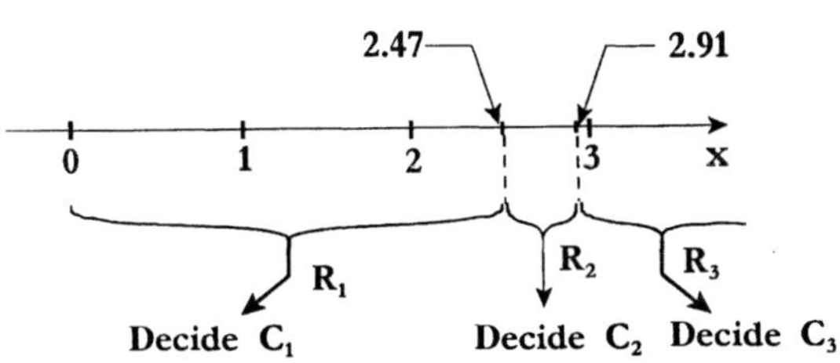

Figure 9.10 Decision regions for Example 9.6.

Note that $L_M(\mathbf{x}) = 1$ with this definition. An optimum test equivalent to Eqs. (9.41) and (9.42) in this space of likelihood ratios can now be described. Define a new set of statistics $v_i(x)$ by dividing the $y_i(\mathbf{x})$ of Eq. (9.41) by $p(\mathbf{x} \mid C_M)$ as

$$v_i(x) \triangleq \frac{y_i(\mathbf{x})}{p(\mathbf{x} \mid C_M)} = \sum_{j=1}^{M} C_{ij} L_j(\mathbf{x}) P(C_j), \quad i = 1, 2, \ldots, M \tag{9.44}$$

From (9.44) an equivalent decision rule is thus one of selecting the class with the **smallest** $v_i(\mathbf{x})$:

$$\text{If} \quad \sum_{j=1}^{M} C_{ij} L_j(x) P(C_j) < \sum_{j=1}^{M} C_{kj} L_j(x) P(C_j) \qquad \text{for all } k \neq i,\ k = 1, 2, \ldots, M$$
$$\text{then decide class } C_i \tag{9.45}$$

The boundaries of the decision rule are thus seen to be described by inequalities that are linear in the likelihood ratio space defined by (9.43). This likelihood ratio space is of dimension $K - 1$. An example of finding decision regions for the multiple class case in a likelihood ratio space and a feature space is now presented.

EXAMPLE 9.7

The random variable X under each of three classes is assumed Gaussian with known mean, variance, and a priori probabilities as follows:

$$\begin{aligned} C_1&: X \sim N(m_1, \sigma^2), \quad P(C_1) = \tfrac{1}{3} \\ C_2&: X \sim N(m_2, \sigma^2), \quad P(C_2) = \tfrac{1}{3} \\ C_3&: X \sim N(m_3, \sigma^3), \quad P(C_3) = \tfrac{1}{3} \end{aligned}$$

Assume $m_1 < m_2 < m_3$ and that the Bayes costs are known with all costs for errors being 1 and all costs for correct classifications being zero as follows:

$$\begin{aligned} C_{ij} &= 1 \quad \text{for } i, j = 1, 2, 3;\ i \neq j \\ C_{ii} &= 0 \quad \text{for } i = 1, 2, 3 \end{aligned}$$

Assume that n independent observations of $\mathbf{X}$ are taken resulting in an $n \times 1$ observation vector $\mathbf{x} = [x_1, x_2, \ldots, x_n]^T$. Find the optimum classification rule where optimality is in the Bayes sense in (a) conditional density space, (b) a sufficient statistic space, and (c) in a likelihood ratio space.

SOLUTION

The probability densities of the observation vector under each of the three classes can be written as a product of the original class conditional densities, since the observations are independent, as follows:

$$p(\mathbf{x} \mid C_i) = \prod_{j=1}^{n} p(x_j \mid C_i) = \frac{1}{(2\pi)^{n/2}\sigma^n} \exp\left\{ -\sum_{j=1}^{n} \frac{(x_j - m_i)^2}{2\sigma^2} \right\}, \quad i = 1, 2, 3$$

(a) From Eq. (9.41) the optimum decision rule calculates $y_i(\mathbf{x})$ given by

$$y_i(\mathbf{x}) = \sum_{j=1}^{M} C_{ij} p(\mathbf{x} \mid C_j) P(C_j), \quad i = 1, 2, \ldots, M$$

and selects the class with the smallest value. Using the assigned costs, we show the $y_i(\mathbf{x})$ to be as follows

$$y_1(\mathbf{x}) = \tfrac{1}{3} p(\mathbf{x} \mid C_2) + \tfrac{1}{3} p(\mathbf{x} \mid C_3)$$
$$y_2(\mathbf{x}) = \tfrac{1}{3} p(\mathbf{x} \mid C_1) + \tfrac{1}{3} p(\mathbf{x} \mid C_3)$$
$$y_3(\mathbf{x}) = \tfrac{1}{3} p(\mathbf{x} \mid C_1) + \tfrac{1}{3} p(\mathbf{x} \mid C_2)$$

If R_1 represents the region for which we decide class C_1, then this region is obtained by setting $y_1(\mathbf{x}) < y_2(\mathbf{x})$ and $y_1(\mathbf{x}) < y_3(\mathbf{x})$. The region R_1, which is the intersection of the two regions specified by inequalities, can be written in terms of the conditional probability density functions as follows:

$$y_1(\mathbf{x}) < y_2(\mathbf{x}) \quad \rightarrow \quad p(\mathbf{x} \mid C_2) + p(\mathbf{x} \mid C_3) < p(\mathbf{x} \mid C_1) + p(\mathbf{x} \mid C_3)$$
$$y_1(\mathbf{x}) < y_3(\mathbf{x}) \quad \rightarrow \quad p(\mathbf{x} \mid C_2) + p(\mathbf{x} \mid C_3) < p(\mathbf{x} \mid C_1) + p(\mathbf{x} \mid C_2)$$

Simplifying the above two inequalities gives the region R_1 for the optimum decision rule as the set of x that satisfy the following two inequalities:

$$R_1\colon \quad p(\mathbf{x} \mid C_2) < p(\mathbf{x} \mid C_1) \quad \text{AND} \quad p(\mathbf{x} \mid C_3) < p(\mathbf{x} \mid C_1)$$

Similarly the regions for class C_2 and class C_3 can be obtained as R_2 and R_3 as

$$R_2\colon \quad p(\mathbf{x} \mid C_1) < p(\mathbf{x} \mid C_2) \quad \text{AND} \quad p(\mathbf{x} \mid C_3) < p(\mathbf{x} \mid C_1)$$
$$R_3\colon \quad p(\mathbf{x} \mid C_1) < p(\mathbf{x} \mid C_3) \quad \text{AND} \quad p(\mathbf{x} \mid C_2) < p(\mathbf{x} \mid C_3)$$

The three regions are shown in the space of conditional densities in Figure 9.11. The decision can be made in this space (a feature space), which is described by coordinates that are functions of the original observation space. Note that the decision space is of dimension 3, whereas the original pattern space is of dimension n.

(b) For this problem the decision rule can be simplified to another feature space by substituting the given conditional probability density functions. For

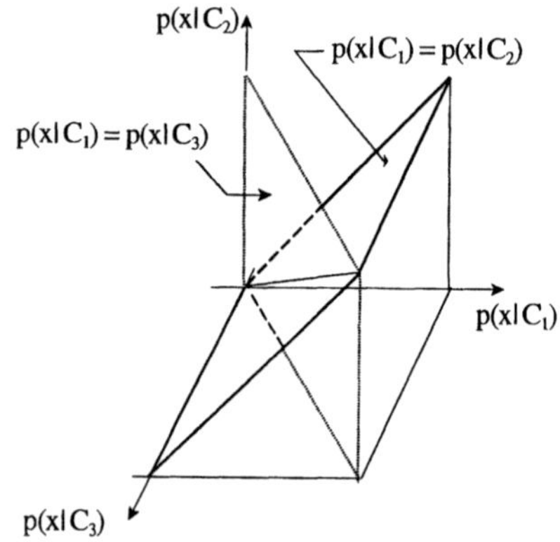

Figure 9.11 Decision regions in the space of $p(x \mid C_i)$, $i = 1, 2, 3$ for Example 9.7.

example, the inequalities that lead to the decision region for class C_2 can be shown from the region R_2 to be

$$\frac{1}{(2\pi)^{n/2}\sigma^n} \exp\left\{-\sum_{i=1}^{n} \frac{(x_i - m_1)^2}{2\sigma^2}\right\} < \frac{1}{(2\pi)^{n/2}\sigma^n} \exp\left\{-\sum_{i=1}^{n} \frac{(x_i - m_2)^2}{2\sigma^2}\right\}$$

$$\frac{1}{(2\pi)^{n/2}\sigma^n} \exp\left\{-\sum_{i=1}^{n} \frac{(x_i - m_3)^2}{2\sigma^2}\right\} < \frac{1}{(2\pi)^{n/2}\sigma^n} \exp\left\{-\sum_{i=1}^{n} \frac{(x_1 - m_2)^2}{2\sigma^2}\right\}$$

Canceling constant terms, taking the ln of both sides, we can simplify the inequalities above further in terms of a statistic s defined by

$$s = \frac{1}{n} \sum_{i=1}^{n} x_i$$

The resulting region S_2 in the s-space, such that C_2 is decided, can be shown to be

$$S_2: \quad \frac{m_1 + m_2}{2} < s \quad \text{AND} \quad s < \frac{m_2 + m_3}{2}$$

In a similar fashion the other decision regions become

$$S_1: \quad s < \frac{m_1 + m_2}{2}$$

$$S_3: \quad s > \frac{m_2 + m_3}{2}$$

These decision regions are shown in the s-space in Figure 9.12. Note that the decision rule for this problem reduces to a particular linear function of the observation vector components, namely the sample mean. Intuitively this answer seems reasonable as the decision thresholds are at the midpoints between the means of the respective classes.

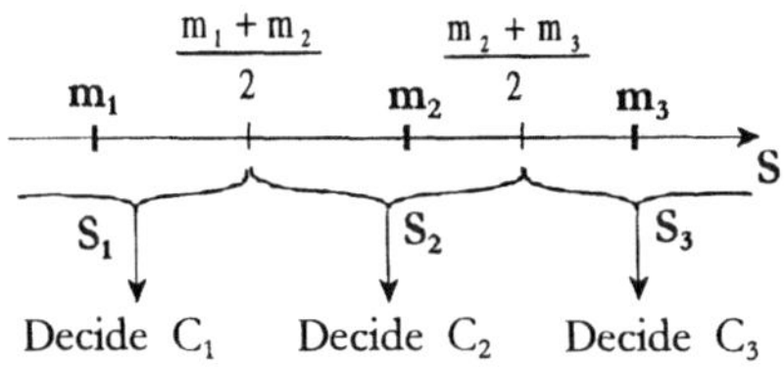

Figure 9.12 Decision regions in the sufficient statistic space s for Example 9.7.

(c) Alternatively the decision regions can be found in a two-dimensional space consisting of likelihood ratios. Define $L_1(\mathbf{x})$ and $L_2(\mathbf{x})$ as

$$L_1(\mathbf{x}) = \frac{p(\mathbf{x}\,|\,C_1)}{p(\mathbf{x}\,|\,C_3)}, \quad L_2(\mathbf{x}) = \frac{p(\mathbf{x}\,|\,C_2)}{p(\mathbf{x}\,|\,C_3)}$$

Dividing each of the inequalities determined earlier labeled as R_1, R_2, and R_3 by $p(x\,|\,C_3)$ and recognizing the respective likelihood ratios, we can determine the decision regions $\mathscr{L}_1$, $\mathscr{L}_2$, and $\mathscr{L}_3$ in the space of likelihood ratios as follows:

$$\mathscr{L}_1\text{:}\quad L_2(\mathbf{x}) < L_1(\mathbf{x}) \quad \text{AND} \quad 1 < L_1(\mathbf{x})$$
$$\mathscr{L}_2\text{:}\quad L_1(\mathbf{x}) < L_2(\mathbf{x}) \quad \text{AND} \quad 1 < L_2(\mathbf{x})$$
$$\mathscr{L}_3\text{:}\quad L_1(\mathbf{x}) < 1 \quad \text{AND} \quad L_2(\mathbf{x}) < 1$$

The decision regions in the likelihood ratio space are shown in Figure 9.13. Notice that for this space the decision boundaries are straight lines and that the decision space is of dimension 2. Solving the multiple class with this type of formulation yields a decision space of dimension one less than the total number of classes. ■

Changing this problem by assigning different costs will not yield such a simple result but will lead to a set of simultaneous linear equations. Changing the measure-

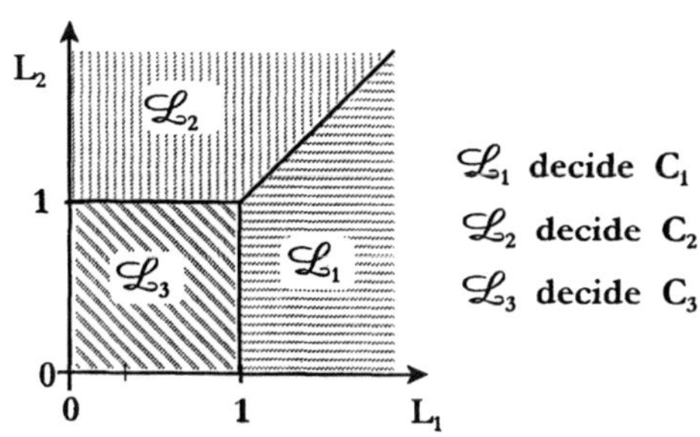

Figure 9.13 Decision regions in the space of likelihood ratios for Example 9.7.

ment variances, however, would cause the decision region boundaries to be quadratics in the original pattern space.

9.5 SPECIAL CASES FOR THE MULTIPLE-CLASS PROBLEM (BAYES)

For multiple-class problems the Bayes optimal decision rule, which includes the minimum probability of error and the maximum a posteriori as special cases, calculated $y_i(\mathbf{x})$ given in Eq. (9.41) for $i = 1, 2, \ldots, M$ for each class C_i and selected the class C_k for which $y_k(\mathbf{x}) < y_i(\mathbf{x})$ for all $i = 1, 2, \ldots, M$ except $i = k$. The $y_i(\mathbf{x})$ are as follows:

$$y_i(\mathbf{x}) = \sum_{j=1}^{M} C_{ij} p(\mathbf{x} \mid C_j) P(C_j), \qquad i = 1, 2, \ldots, M \tag{9.46}$$

The form of the optimum classifier is shown in Figure 9.14. The form of the classifier for special cases including equal costs, zero-one costs, equiprobable classes and Gaussian statistics are now presented.

9.5.1 Special Case 1 (Minimum Probability of Error)

For the special case where $C_{ij} = 1$ and $C_{jj} = 0$ for all i and j, that is, each type of error has equal importance, and zero cost is associated with a correct decision, the optimum classification rule in the Bayes sense is equivalent to the minimum probability of error decision rule. The idea is to select the class for which $y_i(\mathbf{x})$, with the appropriate costs above, is a minimum. Using $y_i(\mathbf{x})$ from Eq. (9.46) with the costs specified gives

$$y_i(\mathbf{x}) = \sum_{j=1, j\neq i}^{M} p(\mathbf{x} \mid C_j) P(C_j), \qquad i = 1, 2, \ldots, M \tag{9.47}$$

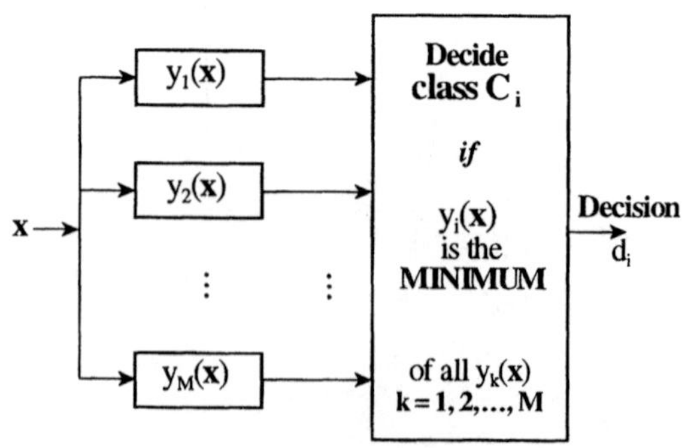

Figure 9.14 Form of optimum classifier in the Bayes sense for the M-class case.

Adding and subtracting $p(\mathbf{x}\,|\,C_i)P(C_i)$ on the right side of the equation above results in

$$y_i(\mathbf{x}) = \sum_{j=1}^{M} p(\mathbf{x}\,|\,C_j)P(C_j) - p(\mathbf{x}\,|\,C_i)P(C_i), \qquad i = 1, 2, \ldots, M \tag{9.48}$$

Since the summation terms of Eq. (9.48) for each i are the same, an equivalent test can be formulated just in terms of second term defined by $s_i(\mathbf{x})$ as follows:

$$s_i(\mathbf{x}) \triangleq p(\mathbf{x}\,|\,C_i)P(C_i), \qquad i = 1, 2, \ldots, M \tag{9.49}$$

But minimizing $y_i(\mathbf{x})$ is the same as maximizing $s_i(\mathbf{x})$, because of the negative sign, so the optimal test is to select the class C_k for which $s_k(\mathbf{x})$ is the largest. This can be written in terms of the $s_i(\mathbf{x})$ as follows: Choose class C_k if for all $k \neq i$ that

$$s_k(\mathbf{x}) > s_i(\mathbf{x}), \qquad i = 1, 2, \ldots, M; i \neq k \tag{9.50}$$

The optimal test in terms of the $s_i(\mathbf{x})$ is shown in Figure 9.15 for the special case $C_{ij} = 1$ and $C_{jj} = 0$.

9.5.2 Special Case 2 (Minimum Probability of Error–Equal A Priori Probabilities)

Assume that the costs are the same as in case 1, $C_{ij} = 1$ and $C_{jj} = 0$. Further assume that the a priori probabilities of the classes are all equal:

$$P(C_1) = P(C_2) = \cdots = P(C_M) = \frac{1}{M} \tag{9.51}$$

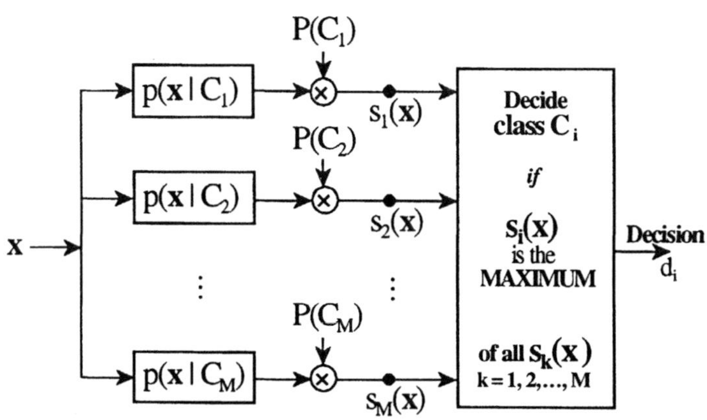

Figure 9.15 Special Case 1: Optimum Bayes classifier with $C_{ii} = 0$ and $C_{ij} = 1$ for $i \neq j$, (minimum probability of error) in terms of the $s_i(\mathbf{x})$.

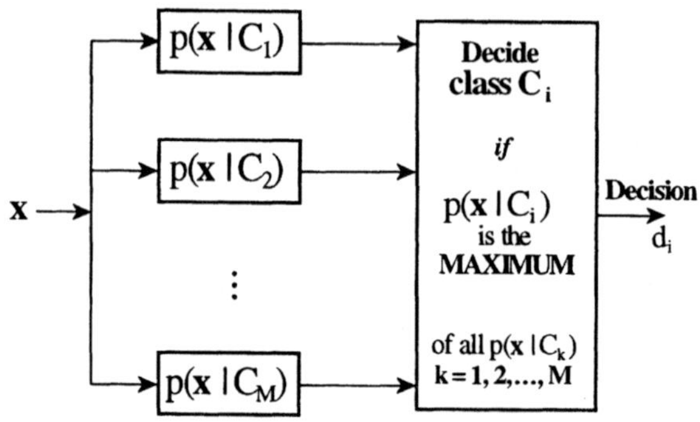

Figure 9.16 Optimum classifier for Special Case 2: $C_{ij} = 1$ for $i \neq j$, $C_{jj} = 0$, and $P(C_j) = 1/M$ for $i, j = 1, 2, \ldots, M$. (Minimum Probability of error and equal apriori probabilities).

Since the multiplicative factors of (9.49) are now all the same, they can be left off without affecting the decision rule, thus leading to the optimum classifier shown in Figure 9.16 and decision rule described by

$$\text{Choose } C_k \quad \text{if} \quad p(\mathbf{x}\,|\,C_k) > p(\mathbf{x}\,|\,C_k) \qquad \text{for all } i = 1, 2, \ldots, M;\ i \neq k \tag{9.52}$$

It is seen that the optimal classifier for special case 2 evaluates the conditional probability density function $p(\mathbf{x}\,|\,C_i)$ for the pattern vector $\mathbf{x}$ and each i and selects the class that gives the maximum value. This can be visualized by graphing all the conditional densities on the same $\mathbf{x}$ axis and selecting the regions corresponding to the highest value as shown in Figure 9.17 for a scalar x.

For a two-dimensional pattern vector $\mathbf{x}$, regions in the pattern space are similarly defined and conceptually can be found for any dimensional pattern space, although they cannot be realistically visualized. An example in a one-dimensional pattern space is presented to illustrate the special case 2 optimum classifier.

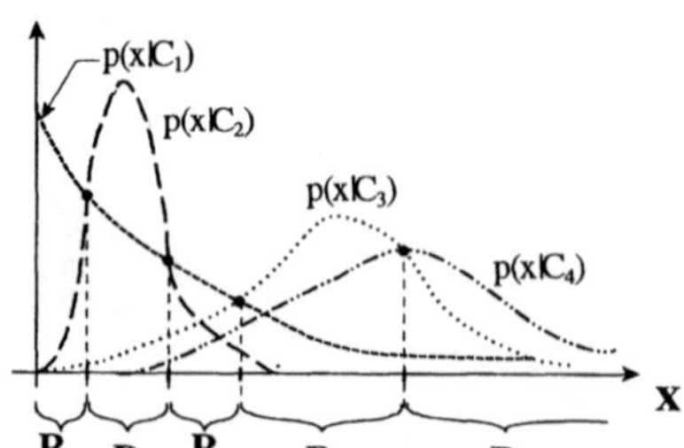

Figure 9.17 An example of selection of regions for the Minimum Probability of error classifier with equal a priori class probabilities (Special Case 2) for a one dimensional observation x.

EXAMPLE 9.8

Consider three classes represented by the following conditional probability density functions where $\mu(x)$ is the unit-step function:

$$p(\mathbf{x}\,|\,C_1) = 3e^{-3x}\mu(x)$$
$$p(\mathbf{x}\,|\,C_2) = 2e^{-2x}\mu(x)$$
$$p(\mathbf{x}\,|\,C_3) = e^{-x}\mu(x)$$

Assume that the a priori probabilities of the classes are equal and that the zero-one cost assignment is made.

(a) Find the optimum classifier in the Bayes sense using a single observed value.
(b) Determine the decision regions R_1, R_2, and R_3 in the observation space for the classifier determined in (a). If x is in R_i, then decide x is from class C_i, $i = 1, 2, 3$.
(c) Calculate $P(\text{correct}\,|\,C_i)$, the probability that your classifier is correct given that the pattern is from class C_i.
(d) Calculate $P(\text{error}\,|\,C_1)$, the probability that your classifier is in error given that the pattern is from class C_1.
(e) Give the formula for calculating the total probability of error for your classifier.
(f) Repeat (a) and (b) if a total of N independent measurements are made.

SOLUTION

(a) The optimum classifier for this problem belongs to the special case 1, and it computes $s_i(\mathbf{x})$ for $i = 1, 2, 3$ (Eq. 9.49) and selects the class corresponding to the maximum value. Since the a priori probabilities $P(C_i)$ are all equal, the optimal classifier need only evaluate the conditional probability density functions for the observed pattern. Thus the optimal classifier appears in Figure 9.18.

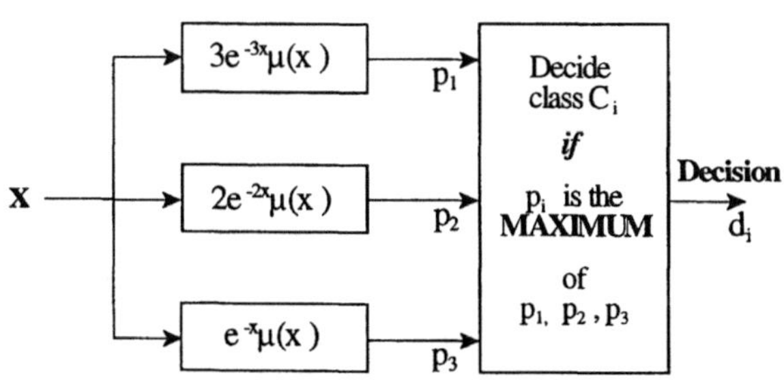

Figure 9.18 Optimum classifier for Example 9.8.

(b) For $x \geq 0$ the region R_1 will be composed of all x such that $p(x|C_1) > p(x|C_2)$ and $p(x|C_1) > p(x|C_3)$. Region R_1 is described as follows:

$$\begin{aligned} R_1 &= \{x\colon 3e^{-3x} > 2e^{-2x} \quad \text{AND} \quad 3e^{-3x} > e^{-x}\} \\ &= \{x\colon x < -\ln\tfrac{2}{3} \quad \text{AND} \quad x < -\tfrac{1}{2}\ln\tfrac{1}{3}\} \\ &= \{x\colon 0 \leq x < 0.40546\} \end{aligned}$$

In a similar fashion R_2 and R_3 can be found to be

$$\begin{aligned} R_2 &= \{x\colon 0.40546 < x < 0.693147\} \\ R_3 &= \{x\colon x > 0.693147\} \end{aligned}$$

The decision regions are shown in Figure 9.19 for the observation space with the understanding that values of x on the boundaries, 0.40546 and 0.69314, are classified randomly between the classes on either side.

(c) The probability that the decision is correct given that the observation is from C_1 can be obtained by integrating over the conditional probability density:

$$\begin{aligned} P(\text{correct}\,|\,C_1) = P(\text{deciding } C_1\,|\,C_1) &= \int_{R_1} p(\mathbf{x}\,|\,C_1)\,dx \\ &= \int_0^{0.40546} 3e^{-3x}\,dx = 0.\overline{703} \end{aligned}$$

Similarly we can find the other conditional probabilities of being correct:

$$\begin{aligned} P(\text{correct}\,|\,C_2) = P(\text{deciding } C_2\,|\,C_2) &= \int_{R_2} p(\mathbf{x}\,|\,C_2)\,dx \\ &= \int_{0.40546}^{0.693147} 2e^{-2x}\,dx = 0.194449 \end{aligned}$$

$$\begin{aligned} P(\text{correct}\,|\,C_3) = P(\text{deciding } C_3\,|\,C_3) &= \int_{R_3} p(\mathbf{x}\,|\,C_3)\,dx \\ &= \int_{0.693147}^{\infty} e^{-x}\,dx = 0.5 \end{aligned}$$

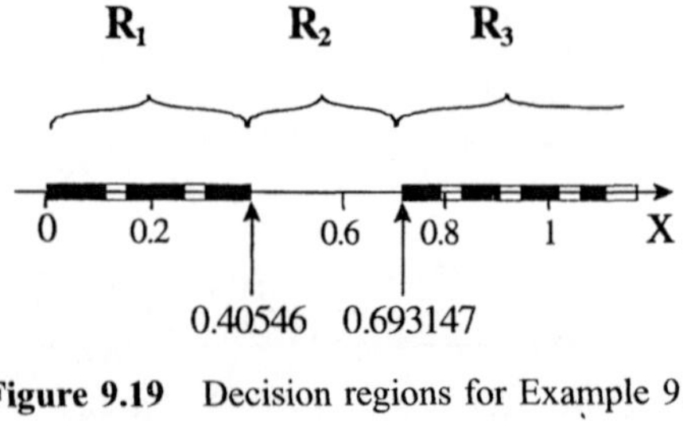

Figure 9.19 Decision regions for Example 9.8.

(d) The probability of error given that the observation is from C_1 can be determined in terms of the probability of being correct is calculated as

$$P(\text{error} \mid C_1) = 1 - P(\text{correct} \mid C_1) = 0.\overline{296}$$

Notice that it is easier to get this probability as above rather than calculating the sum of all possible errors: deciding C_2 given C_1 and deciding C_3 given C_1.

(e) The total probability of error is obtained from the total probability theorem:

$$\begin{aligned} P(\text{error}) &= P(\text{error} \mid C_1)P(C_1) + P(\text{error} \mid C_2)P(C_2) + P(\text{error} \mid C_3)P(C_3) \\ &= [1 - P(\text{correct} \mid C_1)]P(C_1) + [1 - P(\text{correct} \mid C_2)]P(C_2) \\ &\quad + [1 - P(\text{correct} \mid C_3)]P(C_3) \end{aligned}$$

Using the conditional probabilities determined in part (c), we can determine the total probability of error as $P(\text{error}) = 0.533935$. This is a rather high probability of error. However, it is the best in the Bayes sense that can be obtained for the defined problem. ■

9.6 NEYMAN-PEARSON CLASSIFIER

In many problems and in the radar problem, in particular, relative costs for making decisions cannot be realistically assigned. Also the a priori probabilities of the pattern classes are usually very lopsided and meaningful values are difficult, if not impossible, to define. For these problems the Neyman-Pearson classifier rule can provide a meaningful result without the a priori probabilities and costs.

9.6.1 Two-Class Case

For the two-class problem an observed pattern is to be classified as coming from class C_1, called the target, or from class C_2, no target. There are four possible results of making a decision:

Detection. Deciding there is a target, C_1, when there is a target, C_1.

False Alarm. Deciding there is a target, C_1, when there is not a target, C_2.

Miss. Deciding there is not a target, C_2, when there is a target, C_1.

Correct dismissal. Deciding there is not a target, C_2, when there is not a target, C_2.

Definitions for the associated probabilities of error for each of the actions above are given below:

$$\begin{aligned} P_{\text{D}} &= P(\text{deciding } C_1 \mid C_1) = \text{Probability of detection} \\ P_{\text{FA}} &= P(\text{deciding } C_1 \mid C_2) = \text{Probability of false alarm} \\ P_{\text{M}} &= P(\text{deciding } C_2 \mid C_1) = \text{Probability of a miss} \\ P_{\text{CD}} &= P(\text{deciding } C_2 \mid C_2) = \text{Probability of correct dismissal} \end{aligned} \tag{9.53}$$

The probability of a miss, P_M, is sometimes referred to as the probability an **error of the first kind** and called e_1, and the probability of false alarm, P_{FA}, is referred to as the probability of an **error of the second kind**, e_2. The probabilities of P_M, P_D, P_{FA}, and P_{CD} are related by the following formulas:

$$P_M = 1 - P_D, \quad P_{FA} = 1 - P_{CD} \tag{9.54}$$

What is desired is a decision rule to minimize P_M, which is equivalent to maximizing P_D, and at the same time to minimize P_{FA}. However, these are conflicting objectives. The P_{FA} can be minimized to zero by never saying there is a target, which causes the P_M to go to one since we will always miss a target. Conversely, P_M can be minimized to 0 by always saying there is a target that causes the P_{FA} to go to 1 since every time there is no target we say target. The Neyman-Pearson decision rule strikes a compromise and tries to minimize the probability of a miss, P_M, or equivalently maximize the probability of detection, subject to the constraint that the $P_{FA} < e_0$, where e_0 is some acceptable false alarm weight set by the user.

Assumptions. It will be assumed that the conditional probability densities $p(\mathbf{x} \mid C_1)$ and $p(\mathbf{x} \mid C_2)$ are known and that an acceptable false alarm rate, e_0, is assigned. **No** a priori probabilities or costs are specified.

Neyman-Pearson Decision Rule. The Neyman-Pearson rule minimizes the P_M (same as maximizing the P_D) subject to the constraint $P_{FA} \le \epsilon_0$, the acceptable false alarm rate. The solution can be obtained by using a Lagrangian multiplier and minimizing a performance J defined by

$$\begin{aligned} J &= P_m + \lambda(P_{FA} - \epsilon_0) \\ &= \iint \cdots \int_{R_2} p(\mathbf{x} \mid C_1)\, dx + \lambda\left(\iint \cdots \int_{R_1} p(\mathbf{x} \mid C_2)\, d\mathbf{x} - \epsilon_0\right) \end{aligned} \tag{9.55}$$

The first integral P_M can be written as 1 minus the integral over the compliment of R_2 as follows:

$$\iint \cdots \int_{R_2} p(\mathbf{x} \mid C_1)\, d\mathbf{x} = 1 - \iint \cdots \int_{R_1} p(\mathbf{x} \mid C_2)\, d\mathbf{x} \tag{9.56}$$

Substituting (9.56) into (9.55) and rearranging gives

$$\begin{aligned} J &= 1 - \iint \cdots \int_{R_1} p(\mathbf{x} \mid C_1)\, d\mathbf{x} + \lambda\left(\iint \cdots \int_{R_1} p(\mathbf{x} \mid C_2)\, d\mathbf{x} - \epsilon_0\right) \\ &= 1 - \epsilon_0 \lambda + \iint \cdots \int_{R_1} [\lambda p(\mathbf{x} \mid C_2) - p(\mathbf{x} \mid C_1)]\, d\mathbf{x} \end{aligned} \tag{9.57}$$

As the first part of J is fixed, it can be seen that to minimize J, x should be assigned to R_1 if the integrand of the integral is negative, that is,

$$\mathbf{x} \in R_1 \quad \text{if} \quad \lambda p(\mathbf{x} \mid C_2) - p(\mathbf{x} \mid C_1) < 0 \tag{9.58}$$

Rearranging the preceding equation gives the **Neyman-Pearson decision rule** as

$$\begin{aligned}
\text{If} \quad l(\mathbf{x}) = \frac{p(\mathbf{x} \mid C_1)}{p(\mathbf{x} \mid C_2)} &> \lambda \triangleq \eta_{\mathrm{NP}} \quad \text{then decide } C_1 \\
&< \quad \ldots \quad \text{then decide } C_2 \\
&= \quad \ldots \quad \text{then decide randomly}
\end{aligned} \tag{9.59}$$

Thus it is seen that the Neyman-Pearson decision rule is a likelihood ratio test with the threshold equal to the Lagrangian multiplier. The region R_1 as a function of λ must be adjusted so that the constraining equality holds:

$$\int\int \cdots \int_{R_1(\lambda)} p(\mathbf{x} \mid C_2)\, d\mathbf{x} = \epsilon_0 \tag{9.60}$$

Equivalently, the constraining equality can be written in terms of the conditional density of the likelihood ratio as follows:

$$\int_{\eta_{\mathrm{NP}}}^{\infty} p(l \mid C_2)\, dl = \epsilon_0 \tag{9.61}$$

9.6.2 Receiver Operating Characteristic

For most problems this η_{NP} of Eq. (9.161) would not be an easy value to find and what is done in practice is to vary η_{NP} (the threshold) from zero to infinity and determine the operating point $(P_{\mathrm{FA}}, P_{\mathrm{D}})$ as an implicit function of this threshold. The plot of these resulting operating points is called the **receiver operating characteristic** (ROC) and a typical ROC is shown in Figure 9.20.

The receiver operating characteristic of a Neyman-Pearson decision rule can be shown to exhibit a number of special properties, some of which are listed below.

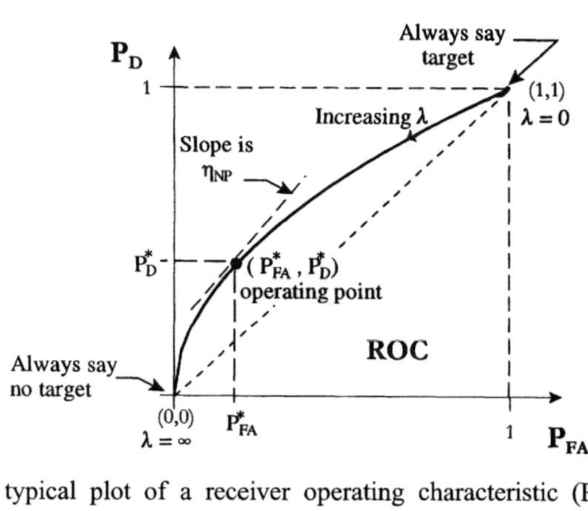

Figure 9.20 A typical plot of a receiver operating characteristic (ROC) as an implicit function of the threshold lambda for the Neyman-Pearson decision rule.

(1) The point $P_{\text{FA}} = 1, P_{\text{D}} = 1$ indicates the case of always deciding there is a target.
(2) The point $P_{\text{FA}} = 0, P_{\text{D}} = 0$ indicates the case of always deciding there is no target.
(3) The ROC curve as λ varies from 0 to ∞ is bounded from below as it always lies above the line $P_{\text{D}} = P_{\text{FA}}$.
(4) The derivative of the curve at an operating point gives the threshold λ for the likelihood ratio test:

$$\left[\frac{dP_{\text{D}}}{dP_{\text{FA}}}\right]_{\text{operating point}} = \lambda \qquad \text{for that operating point} \tag{9.62}$$

(5) The ROC curve is a nondecreasing function of P_{FA}.
(6) The slope of the curve is always ≥ 0.

The receiver operating characteristic provides a visualization of the improvement in probability of detection as the acceptable false alarm rate is increased. The steeper the curve, the higher is the probability of detection that can be obtained by a small increase in P_{FA}. However, if the slope is small, a large increase in P_{FA} is required to obtain a significant improvement in performance P_{D}.

For certain problems, especially radar and communications-related problems, a family of receiver operating characteristics is presented as a function of some system parameters. The most common is the signal to noise ratio.

An example showing the details of determining the receiver operating characteristic is now given for a scalar observation.

EXAMPLE 9.9

Under C_1 and C_2 the observation X is exponentially distributed as

$$p(x \mid C_1) = \exp(-x)\mu(x)$$
$$p(x \mid C_2) = 2\exp(-2x)\mu(x)$$

Find the Neyman-Pearson decision rule including threshold for the test for an acceptable false alarm rate of 0.01. Find the probability of detection for this operating point. Plot the receiver operating characteristic as the threshold is varied.

SOLUTION

The Neyman-Pearson decision rule is a likelihood ratio test with threshold η_{NP} given in Eq. (9.59). Using the given conditional probability density functions, we determine the likelihood ratio as

$$l(x) = \frac{p(x \mid C_1)}{p(x \mid C_2)} = \frac{\exp(-x)\mu(x)}{2\ \exp(-2x)\mu(x)}$$
$$= \tfrac{1}{2}\ \exp(x)\mu(x)$$

Therefore the likelihood ratio test for $x \geq 0$ becomes

$$\text{If} \quad \tfrac{1}{2}\exp(x) > \eta_{\text{NP}} \quad \text{decide } C_1 \text{(target)}$$
$$< \quad \text{decide } C_2 \text{ (no target)}$$
$$= \quad \text{decide randomly}$$

An equivalent test in the observation space can be obtained by multiplying both sides of the inequality by 2 and taking the natural log of both sides for x greater than or equal zero to give the following:

$$\text{If} \quad x > \ln(2\eta_{\text{NP}}) \quad \text{decide } C_1 \text{ (target)}$$
$$< \quad \text{decide } C_2 \text{ (no target)}$$
$$= \quad \text{decide randomly}$$

To obtain the Neyman-Pearson threshold, η_{NP}, we must equate the false alarm rate to the specified false alarm value. In the observation space the resulting equation for η_{NP} of the test above becomes

$$P_{\text{FA}} = P(\text{decide } C_1 \text{ (target)} \mid C_2 \text{ (no target)})$$
$$= \int_{\ln(2\eta_{\text{NP}})}^{\infty} 2\exp(-2x)\mu(x)\, dx$$
$$= \exp(-2\ln(2\eta_{\text{NP}})) = \left(\frac{1}{2\eta_{\text{NP}}}\right)^2$$

Setting $P_{\text{FA}} = 0.01$ gives $(\frac{1}{2}\eta_{\text{NP}})^2 = 0.01$, which is easily solved to give $\eta_{\text{NP}} = 5$.

The probability of detection, P_{D}, is easily seen as

$$P_{\text{D}} = P(\text{decide } C_1 \text{ (target)} \mid C_1 \text{ (target)})$$
$$= \int_{\ln(2\eta_{\text{NP}})}^{\infty} \exp(-x)\mu(x)\, dx$$
$$= \exp(-\ln(2\eta_{\text{NP}})) = \frac{1}{2\eta_{\text{NP}}}$$

Thus the operating point for $\eta_{\text{NP}} = 5$ is $(P_{\text{FA}} = 0.01, P_{\text{D}} = 0.1)$. This point is shown on the ROC of Figure 9.21.

If P_{FA}, or equivalently, η_{NP}, is varied, the receiver operating characteristic for this problem can be determined as shown in Figure 9.21. In most cases the relationship between P_{D} and P_{FA} cannot be found analytically. However, for this particular

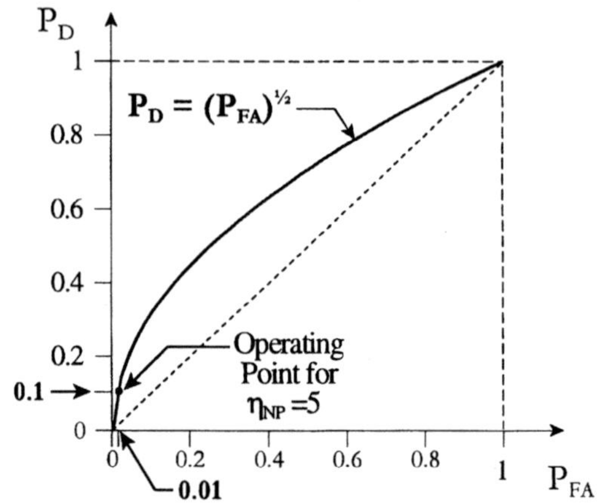

Figure 9.21 Receiver operating characteristic (ROC) for Example 9.9.

problem it is possible to write the P_D in terms of the P_{FA}, which is easily seen from the equations above for P_D and P_{FA} that $P_D = (P_{FA})^{1/2}$. ■

9.7 GENERAL CALCULATION OF PROBABILITY OF ERROR

A general decision rule can be specified directly in the pattern space, in the likelihood ratio space, or in some feature space. The decision regions for each of these spaces are represented in Figure 9.22. In the pattern spaces R_1 and R_2 are the decision regions for class C_1 and C_2, respectively, in the feature space, Y_1 and Y_2

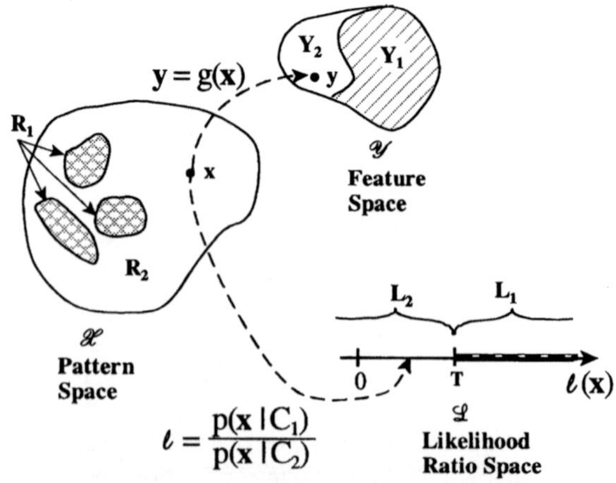

Figure 9.22 Decision rule in different spaces for the two class problem.

are the decision regions for class C_1 and C_2, respectively; and in the likelihood ratio space, L_1 and L_2 are the decision regions for class C_1 and C_2, respectively. The probability of error for a given decision rule can be calculated in any of these spaces in terms of the decision regions.

9.7.1 In Likelihood Ratio Space

For the two-pattern class problem, the likelihood ratio is one dimensional and the decision rule is that values of $l(x)$ greater than the threshold T are assigned to the L_1 decision region for class C_1 and values of $l(x)$ less than T are assigned to the L_2 decision region for class C_2. The probabilities of each kind of error can then be calculated as follows:

$$\begin{aligned} P(\text{error} \mid C_2) &= P(\text{decide } C_1 \mid C_2) = \int_T^{\infty} p(l \mid C_2)\, dl \\ P(\text{error} \mid C_1) &= P(\text{decide } C_2 \mid C_1) = \int_0^{T} p(l \mid C_1)\, dl \end{aligned} \tag{9.63}$$

The evaluation is a simple integral type. However, the conditional densities of the likelihood ratio $l(\mathbf{x})$ are not usually known, and they must be obtained by any one of the density transformation techniques of random variables—the fundamental theorem, the distribution method, Monte Carlo method, and the like.

9.7.2 In Pattern Space

In the pattern space the evaluation of the probabilities of error can be fairly complicated if the pattern space is of dimension n higher than two as the integrals become n-dimensional and the regions of integration are not necessarily simply connected. The error probabilities in the pattern space are given as follows:

$$\begin{aligned} P(\text{error} \mid C_2) &= P(\text{decide } C_1 \mid C_2) \\ &= \int\int \cdots \int_{R_1} p(\mathbf{x} \mid C_2)\, d\mathbf{x} \\ P(\text{error} \mid C_1) &= P(\text{decide } C_2 \mid C_1) \\ &= \int\int \cdots \int_{R_2} p(\mathbf{x} \mid C_1)\, d\mathbf{x} \end{aligned} \tag{9.64}$$

Although the conditional densities are known, the actual decision regions in the pattern space reflected back from the likelihood space may be difficult to obtain.

9.7.3 In Feature Space

A feature space is obtained as a transformation, $y = g(\mathbf{x})$, defined on the observation space. It is usually of lower dimension than the observation space but not necessarily the size of the likelihood ratio space. In most cases the feature space will be determined by a transformation $g(\cdot)$, which is a sufficient statistic for the problem. The regions are usually simpler than those given in the observation space but perhaps more complicated dimensionwise than the likelihood space. The error probabilities can be determined with formulas similar to those in the pattern space as follows:

$$\begin{aligned} P(\text{error}\,|\,C_2) &= P(\text{decide } C_1\,|\,C_2) \\ &= \iint \cdots \int_{Y_1} p(\mathbf{y}\,|\,C_2)\,d\mathbf{y} \\ P(\text{error}\,|\,C_1) &= P(\text{decide } C_2\,|\,C_1) \\ &= \iint \cdots \int_{Y_2} p(\mathbf{y}\,|\,C_1)\,d\mathbf{y} \end{aligned} \tag{9.65}$$

The conditional densities for the feature vector usually must be determined by using the transformational methods. An example illustrating the calculation of the error probabilities in the three spaces is presented next.

EXAMPLE 9.10

The pattern vectors in each of two classes C_1 and C_2 are assumed to be Gaussian with known mean and covariance matrices as follows:

$$p(\mathbf{x}\,|\,C_1) = p(x_1, x_2\,|\,C_1) = \frac{1}{2\pi}\exp\left[\frac{-x_1^2 - x_2^2}{2}\right]$$

$$p(\mathbf{x}\,|\,C_2) = p(x_1, x_2\,|\,C_2) = \frac{1}{2\pi}\exp\left[\frac{-(x_1 - 2)^2 - (x_2 - 2)^2}{2}\right]$$

Assume that the a priori probabilities of the two classes are given by

$$P(C_1) = P(C_2) = \tfrac{1}{2}$$

Find the minimum probability of error decision rule for classifying a given pattern vector $\mathbf{x} = (x_1, x_2)^T$ into one of the two classes C_1 or C_2. Express the decision rule in the likelihood ratio space, a sufficient statistic feature space, and the observation space, and calculate the probabilities of errors in each of the spaces.

SOLUTION

The minimum probability of the error decision rule is a likelihood ratio test with the threshold and likelihood ratio given in Eq. (9.24). The likelihood ratio using the conditional densities given is

$$l(\mathbf{x}) = \frac{p(x_1, x_2 \mid C_1)}{p(x_1, x_2 \mid C_2)} = \frac{\left(\frac{1}{2\pi}\right)\exp[-(x_1^2 + x_2^2)/2]}{\left(\frac{1}{2\pi}\right)\exp\{-[(x_1 - 2)^2 + (x_2 - 2)^2]/2\}}$$

$$= \exp[-2(x_1 + x_2) + 4]$$

The threshold for the test becomes

$$\eta_{\text{MPE}} = \frac{P(C_2)}{P(C_1)} = \frac{\frac{1}{2}}{\frac{1}{2}} = 1$$

By this η_{MPE}, the optimum test in the likelihood ratio space is

$$\begin{aligned} \text{If} \quad l(\mathbf{x}) &> 1 \quad \text{decide } C_1 \\ &< 1 \quad \text{decide } C_2 \\ &= 1 \quad \text{randomly decide} \end{aligned}$$

An equivalent test in the observation space can be obtained by substituting the $l(\mathbf{x}) = \exp[-2(x_1 + x_2) + 4]$ into the decision rule above, taking the ln of both sides of the equality and inequalities, multiplying by $-\frac{1}{2}$, thus changing the direction of the inequality, and adding 2 to both sides to give

$$\begin{aligned} \text{If} \quad x_1 + x_2 &< 2 \quad \text{decide } C_1 \\ &> \quad \text{decide } C_2 \\ &= \quad \text{randomly decide} \end{aligned}$$

An equivalent test in a feature space, and in this problem a sufficient statistic space, can be found by using the sum of the observations, $y = x_1 + x_2$, resulting in

$$\begin{aligned} \text{if} \quad y &< 2 \quad \text{decide } C_1 \\ &> \quad \text{decide } C_2 \\ &= \quad \text{randomly decide} \end{aligned}$$

Since the decision depends only on the sum of x_1 and x_2, it means that we do not need to know the actual values of the components x_1 and x_2, only their sum—a sufficient statistic—to make the decision. The decision regions for each of these equivalent tests are shown in Figure 9.23. Notice that in the feature and likelihood space the decision regions are portions of infinite lines, whereas in the pattern space the decision regions are infinite half-planes.

The probabilities of each type of error and the total probability of error will now be determined in each of the spaces considered above.

Feature Space. To evaluate the probabilities of error in the feature space requires knowing the conditional densities of the feature $Y = X_1 + X_2$ under each pattern class C_1 and C_2. Since X_1 and X_2 are independent Gaussian random variables, their sum is a Gaussian random variable conditioned on each class. The conditional

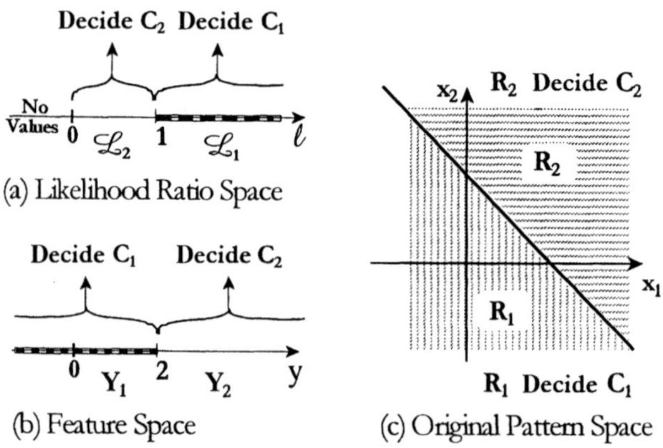

Figure 9.23 Decision regions for (a) the likelihood ratio space, (b) a feature space, and (c) the pattern space for Example 9.10.

probability density function for Y under C_1 is $N(0, 2)$ and under C_2 is $N(4, 2)$. Using the regions for the decision rule shown in Figure 9.23b as Y_1 and Y_2, the error probabilities are determined as follows:

$$P(\text{error} \mid C_2) = \int_{Y_1} p(y \mid C_2)\, dy = \int_{-\infty}^{2} \frac{1}{\sqrt{2\pi}\sqrt{2}} \exp\left[-\frac{(y-4)^2}{4}\right] dy$$

$$= \Phi\left(\frac{-2}{\sqrt{2}}\right) = \Phi(-\sqrt{2})$$

$$P(\text{error} \mid C_1) = \int_{Y_2} p(y \mid C_1)\, dy = \int_{2}^{\infty} \frac{1}{\sqrt{2\pi}\sqrt{2}} \exp\left(-\frac{y^2}{4}\right) dy$$

$$= 1 - \Phi\left(\frac{-2}{\sqrt{2}}\right) = \Phi(\sqrt{2})$$

Observation Space. Calculation of the error probabilities in the observation space involves conditional probability density functions and integration over the regions given in Figure 9.23c. The probabilities can be written as follows:

$$P(\text{error} \mid C_2) = \iint_{R_1} p(\mathbf{x} \mid C_2)\, d\mathbf{x}$$

$$= \int_{-\infty}^{+\infty} \int_{-\infty}^{2-x_2} \frac{1}{2\pi} \exp\left(-\frac{(x_1-2)^2 + (x_2-2)^2}{2}\right) dx_1\, dx_2$$

$$P(\text{error} \mid C_1) = \iint_{R_2} p(\mathbf{x} \mid C_1)\, dx$$

$$= \int_{-\infty}^{\infty} \int_{2-x_2}^{\infty} \frac{1}{2\pi} \exp\left(-\frac{x_1^2 + x_2^2}{2}\right) dx_1\, dx_2$$

The completion of the evaluation of these integrals is not an easy task. The purpose here is to show the evaluation formula in each of the spaces rather than all the mathematical mechanics. The integral can be solved by changing the two-dimensional integral to a one-dimensional integral using a suitable transformation. For this problem the transformation $y = x_1 + x_2$ will actually expedite the solution changing the integrals to those in the feature space.

Likelihood Ratio Space. In order to evaluate the probabilities of error in the likelihood ratio space the conditional densities of L under C_1 and C_2 must be determined. This can be accomplished by using an exponential function of a sum of random variables and the transformation theorem.

The random variable $Z = -2(X_1 + X_2) + 4$ under C_1 is normally distributed with mean of 4 and a variance of $4(\text{var } X_1 + \text{var } X_2) = 8$, while under C_2 it is normally distributed with mean of -4 and a variance of 8. The density of $L = \exp(-Z)$ under each of the classes can then be obtained using the transformation theorem. For class C_1 the conditional density on L becomes

$$p(l \mid C_1) = \left[p(z \mid C_1)/\,|(d/dz)\exp(z)|\right]_{z=\ln l}\,\mu(l)$$
$$= \frac{(1/\sqrt{2\pi 8}\exp\left[-(\ln l - 4)^2/2(8)\right]}{l}\,\mu(l)$$

Using this density, we find that the probability of error given C_1 becomes

$$P(\text{error} \mid C_1) = \int_0^1 p(l \mid C_1)\,dl = \int_0^1 \frac{1}{\sqrt{2\pi 8}\,l}\exp\left[-\frac{(\ln l - 4)^2}{16}\right]dl$$
$$= \int_{-\infty}^{4} \frac{1}{\sqrt{2\pi 8}}\exp\left(-\frac{w^2}{16}\right)dw = \Phi\left(\frac{4}{\sqrt{8}}\right) = \Phi(\sqrt{2})$$

The last integral is obtained by substituting $w = \ln l - 4$ in the previous integral to obtain the Φ function.

Similarly the $p(l \mid C_2)$ and $P(\text{error} \mid C_2)$ can be evaluated as

$$p(l \mid C_2) = \left[\frac{p(z \mid C_2)}{|(d/dz)\exp(z)|}\right]_{z=\ln l}\mu(l)$$
$$= \frac{(1/\sqrt{2\pi 8})\exp\left[-(\ln l + 4)^2/2(8)\right]}{|\exp(\ln l)|}\,\mu(l)$$

$$P(\text{error} \mid C_2) = \int_1^{\infty} p(l \mid C_2)dl = \int_1^{\infty} \frac{1}{\sqrt{2\pi 8 l}}\exp\left[-\frac{(\ln l + 4)^2}{16}\right]dl$$
$$= \int_{-\infty}^{4} \frac{1}{\sqrt{2\pi 8}}\exp\left(-\frac{w^2}{16}\right)dw = \Phi\left(\frac{4}{\sqrt{8}}\right) = \Phi(\sqrt{2}) \quad ■$$

9.8 GENERAL GAUSSIAN PROBLEM

Gaussian or normal random variables pervade the literature. There are many reasons for this fact. Perhaps the most important is that the Gaussian assumption makes many problems mathematically tractable. Another important reason is that many physical situations involve the sum of independent random variables, which results in approximately Gaussian random variables as stated through various central limit theorems. Also biological evolution creates situations that tend to be symmetrically distributed around some mean value that are approximately Gaussian.

Some of the examples shown in the preceding sections were Gaussian, and the optimal classifiers were obtained using the general form of the decision rule. Since a simplification usually occurs when the statistics are Gaussian, the purpose of this section is to develop the solutions for the general Gaussian problem for a number of special cases. This way the results can be used without returning each time to the general formulas to determine the decision rule. Thus a solution is immediately at hand if the underlying conditions are satisfied.

Before developing the optimum classifiers, we provide a brief introduction to Gaussian pattern vectors for the 1-, 2-, and N-dimensional case.

9.8.1 Gaussian Pattern Vectors

A random variable X is said to be **Gaussian** or normal if its probability density function is given by

$$p(x) = \frac{1}{\sqrt{2\pi}\sigma} \exp\left[\frac{-(x-m)^2}{2\sigma^2}\right] \tag{9.66}$$

The parameters m and σ^2 are the **mean** and **variance** of the random variable.

Two random variables X_1 and X_2 are **jointly Gaussian** if their probability density function can be written as

$$p(x_1, x_2) = \frac{1}{k} \exp\left\{\frac{-1}{2(1-\rho^2)}\left[\frac{(x_1-m_1)^2}{\sigma_1^2} - \frac{2\rho(x_1-m_1)(x_2-m_2)}{\sigma_1\sigma_2} + \frac{(x_2-m_2)^2}{\sigma_2^2}\right]\right\} \tag{9.67}$$

where k, **a normalizing constant**, is given by

$$k = 2\pi\sigma_1\sigma_2\sqrt{1-\rho^2} \tag{9.68}$$

and ρ, the **correlation coefficient**, is defined by

$$\rho \triangleq \frac{E[(X_1-m_1)(X_2-m_2)]}{\sigma_1\sigma_2} \tag{9.69}$$

When more than two jointly normal random variables are considered, it becomes convenient to use a vector formulation. Random variables $X_1, X_2, \ldots, X_N$, or in vector notation $\mathbf{X} = [X_1, X_2, \ldots, X_N]^T$, are **jointly Gaussian** if their joint probability density function can be written in the form

$$p(\mathbf{x}) \triangleq p(x_1, x_2, \ldots, x_N) = \frac{1}{(2\pi)^{N/2} |\mathbf{K}|^{1/2}} \exp\left[\frac{-(\mathbf{x}-\mathbf{m})^T \mathbf{K}^{-1}(\mathbf{x}-\mathbf{m})}{2}\right] \quad (9.70)$$

where

$$\mathbf{m} = E[\mathbf{X}] = [m_1 \quad m_2 \quad \ldots \quad m_N]^T$$
$$\mathbf{K} = E[(\mathbf{X}-\mathbf{m})(\mathbf{X}-\mathbf{m})^T]$$

In the equation above $E[\cdot]$ is the expected value operator. For the random vector **X**, the **m** is the **mean vector** and the matrix **K** is called the **covariance matrix**. The covariance matrix has a number of properties. Some of these are as follows:

(1) If **X** is an $n \times 1$ vector, then **K** is an $N \times N$ matrix.
(2) **K** is symmetric.
(3) **K** is invertable; thus $\mathbf{K}^{-1}$ exists.
(4) **K** is positive definite and thus has positive eigen values.
(5) **K** is diagonally dominant.
(6) The diagonal components of **K** are the variances of the components of the **X** vector.
(7) The off-diagonal terms of **K** are the covariances of the component random variables in all combinations.

Instead of writing the long form of the probability, (9.70), it is convenient to use a short-hand notation and write that $\mathbf{X} \sim N(\mathbf{m}, \mathbf{K})$. This form means that **X** is a Gaussian random vector with mean vector **m**, covariance matrix **K**, and the joint density given in Eq. (9.70).

9.8.2 Two-Class General Gaussian Problem (Bayes)

Say that for class C_1, $\mathbf{X} \sim N(\mathbf{m}_1, \mathbf{K}_1)$, and for class C_2, $\mathbf{X} \sim N(\mathbf{m}_2, \mathbf{K}_2)$. We wish to find the **Bayes optimum classifier** assuming the class a priori probabilities $P(C_1)$ and $P(C_2)$, and costs C_{11}, C_{12}, C_{21}, and C_{22}, are specified. We have previously seen in the general case that the Bayes' decision rule is a likelihood ratio test with likelihood ratio and threshold given in Eq. (9.34). Using the vector form of the densities, we can write the likelihood ratio as

$$l(\mathbf{x}) = \frac{p(\mathbf{x}|C_1)}{p(\mathbf{x}|C_2)} = \frac{[1/(2\pi)^{N/2}|\mathbf{K}_1|^{1/2}]\exp[-\frac{1}{2}(\mathbf{x}-\mathbf{m}_1)^T\mathbf{K}_1^{-1}(\mathbf{x}-\mathbf{m}_1)]}{[1/(2\pi)^{N/2}|\mathbf{K}_2|^{1/2}]\exp[-\frac{1}{2}(\mathbf{x}-\mathbf{m}_2)^T\mathbf{K}_2^{-1}(\mathbf{x}-\mathbf{m}_2)]}$$
$$= \frac{|\mathbf{K}_2|^{1/2}}{|\mathbf{K}_1|^{1/2}} \exp[-\tfrac{1}{2}(\mathbf{x}-\mathbf{m}_1)^T\mathbf{K}_1^{-1}(\mathbf{x}-\mathbf{m}_1) + \tfrac{1}{2}(\mathbf{x}-\mathbf{m}_2)^T\mathbf{K}_2^{-1}(\mathbf{x}-\mathbf{m}_2)] \quad (9.71)$$

We take the ln of the likelihood ratio and threshold and rearrange to obtain an equivalent test as

$$\text{If} \quad -(\mathbf{x}-\mathbf{m}_1)^T\mathbf{K}_1^{-1}(\mathbf{x}-\mathbf{m}_1)+(\mathbf{x}-\mathbf{m}_2)^T\mathbf{K}_2^{-1}(\mathbf{x}-\mathbf{m}_2) \begin{array}{ll} > T & \text{decide } C_1 \\ < & \text{decide } C_2 \\ = & \text{decide randomly} \end{array}$$

$$\text{where} \quad T = 2\ln\eta_{\text{BAYES}} + \ln\frac{|\mathbf{K}_1|}{|\mathbf{K}_2|} \quad \text{and} \quad \eta_{\text{BAYES}} = \frac{(C_{22}-C_{12})}{(C_{11}-C_{21})}\frac{P(C_2)}{P(C_1)} \tag{9.72}$$

The expression on the left-hand side of the inequality is a quadratic function of the observation vector and the quantity on the right side is a threshold that is a function of the determinants of the covariance matrices and the Bayes threshold defined in terms of the costs and the a priori probabilities. An alternative interpretation of the decision rule can be expressed in terms of a generalized distance measure called the Mahalanobis distance between two vectors.

The **Mahalanobis distance** between two $n \times 1$ vectors $\mathbf{x}$ and $\mathbf{y}$ with respect to an $n \times n$ positive definite symmetric matrix $\mathbf{A}$ is defined as

$$d_{\text{MAH}}(\mathbf{x},\mathbf{y}) = (\mathbf{x}-\mathbf{y})^T\mathbf{A}(\mathbf{x}-\mathbf{y}) \tag{9.73}$$

When $\mathbf{A} = \mathbf{I}$, an $n \times n$ identity matrix, the $d_{\text{MAH}}(\mathbf{x},\mathbf{y}) = d_{\text{Euc}}^2(\mathbf{x},\mathbf{y})$, the square of the standard **Euclidean distance** between the vectors x and y which is defined by

$$d_{\text{Euc}}(\mathbf{x},\mathbf{y}) = [(\mathbf{x}-\mathbf{y})^T(\mathbf{x}-\mathbf{y})]^{1/2} \tag{9.74}$$

The decision rule given in (9.72) can thus be interpreted in terms of comparing to a threshold the difference of the Mahalanobis distance between $\mathbf{x}$ and $\mathbf{m}_2$ with matrix $\mathbf{K}_2^{-1}$ and the Mahalanobis distance between $\mathbf{x}$ and $\mathbf{m}_1$ with a matrix $\mathbf{K}_1^{-1}$.

Special Case 1: Equal Covariance Matrices (Optimum Decision Rule). For the special case of equal covariance matrices, that is, if $\mathbf{K}_1 = \mathbf{K}_2 = \mathbf{K}$, the decision rule given in Eq. (9.72) can be easily simplified to

$$\text{If} \quad 2\mathbf{m}_1^T\mathbf{K}^{-1}\mathbf{x} - \mathbf{m}_1^T\mathbf{K}^{-1}\mathbf{m}_1 - 2\mathbf{m}_2^T\mathbf{K}^{-1}\mathbf{x} + \mathbf{m}_2^T\mathbf{K}^{-1}\mathbf{m}_2 \begin{array}{ll} > 2\ln\eta_{\text{BAYES}} & \text{decide } C_1 \\ < & \text{decide } C_2 \\ = & \text{decide randomly} \end{array} \tag{9.75}$$

The decision rule above can be further simplified by factoring the left side of the equation, moving various constant terms to the right, and finally dividing by two, to give the following equivalent decision rule:

$$\begin{aligned}\text{If}\quad (\mathbf{m}_1 - \mathbf{m}_2)^T\mathbf{K}^{-1}\mathbf{x} &> T_1 \quad \text{decide } C_1\\ &< \quad \text{decide } C_2\\ &= \quad \text{decide randomly}\\ \text{where}\quad T_1 = \ln \eta_{\text{BAYES}} &+ \tfrac{1}{2}(\mathbf{m}_1^T\mathbf{K}^{-1}\mathbf{m}_1 - \mathbf{m}_2^T\mathbf{K}^{-1}\mathbf{m}_2)\end{aligned} \tag{9.76}$$

Thus for the case of equal covariances it has been shown that the decision rule compares a linear function of the observation vector components with a threshold and bases the decision on whether the linear function is greater than, less than, or equal to the threshold.

Special Case 2: Equal Scaled Identity Covariance Matrices (Optimum Decision Rule). Let $\mathbf{K}_1 = \mathbf{K}_2 = \sigma^2\mathbf{I}$, where $\mathbf{I}$ is an $N \times N$ identity matrix. This corresponds to the case where the feature vectors components are uncorrelated and each has the same variance σ^2. Substituting this covariance matrix into the decision rule expressed in Eq. (9.76) and simplifying gives the decision rule as

$$\begin{aligned}\text{If}\quad (\mathbf{m}_1 - \mathbf{m}_2)^T\mathbf{x} &> T_2 \quad \text{decide } C_1\\ &< \quad \text{decide } C_2\\ &= \quad \text{decide randomly}\\ \text{where}\quad T_2 = \sigma^2 \ln \eta_{\text{BAYES}} &+ \tfrac{1}{2}(\mathbf{m}_1^T\mathbf{m}_1 - \mathbf{m}_2^T\mathbf{m}_2)\end{aligned} \tag{9.77}$$

The optimum classifier for this special case is also seen to be linear in the observation vector components with weighting coefficients equal to the difference of the means. The threshold, equal to T_2, is a function of η_{BAYES} and the dot product of the means with themselves.

Classifier Implementations for Case 2. Before the implementations are presented the useful definitions of discriminant function and correlation will be given. A **discriminant function** is defined as any function of the measurement vector $\mathbf{x}$ that can be used to facilitate the decision process. Usually it is a sufficient statistic, but it may be used in collaboration with other discriminant functions to be equivalent to a sufficient statistic. In the general minimum probability of error optimum detector given in Eq. (9.28) we can view the products $p(x \mid C_i)P(C_i)$ as discriminant functions. The **correlation of vectors x and y** is defined as the dot product of the two vectors indicated by the vector product as $\mathbf{x}^T\mathbf{y}$.

The decision rule given in Eq. (9.77) can be implemented in terms of discriminant functions in many different ways. The most common implementations are (a) a single discriminant function, (b) comparative distance operation, (c) maximum correlation structure, and (d) single correlation structure. Each of the implementations will now be presented.

Single-Discriminant Function. The **single-discriminant function** form of the optimum Bayes decision rule shown in Eq. (9.77) is as follows:

$$\begin{aligned} \text{If} \quad g(\mathbf{x}) &> 0 \quad \text{decide } C_1 \\ &< \quad\;\; \text{decide } C_2 \\ &= \quad\;\; \text{decide randomly} \end{aligned} \tag{9.78}$$

$$\text{where} \quad g(\mathbf{x}) = (\mathbf{m}_1 - \mathbf{m}_2)^T\mathbf{x} - \sigma^2 \ln \eta_{\text{BAYES}} - \tfrac{1}{2}(\mathbf{m}_1^T\mathbf{m}_1 - \mathbf{m}_2^T\mathbf{m}_2)$$

The $g(\mathbf{x})$ is seen as a correlation of the difference of the mean vectors $(\mathbf{m}_1 - \mathbf{m}_2)$, with $\mathbf{x}$ plus a constant.

Two-Discriminant Functions. The classifier given in Eq. (9.77) can be rearranged in terms of two-discriminant functions in many different ways, one of which is

$$\begin{aligned} \text{If} \quad g_1(\mathbf{x}) &> g_2(\mathbf{x}) \quad \text{decide } C_1 \\ &< \qquad\quad\; \text{decide } C_2 \\ &= \qquad\quad\; \text{decide randomly} \end{aligned} \tag{9.79}$$

$$\begin{aligned} \text{where} \quad g_1(\mathbf{x}) &= \mathbf{m}_1^T\mathbf{x} - \tfrac{1}{2}\sigma^2 \ln \eta_{\text{BAYES}} - \tfrac{1}{2}\mathbf{m}_1^T\mathbf{m}_1 \\ g_2(\mathbf{x}) &= \mathbf{m}_2^T\mathbf{x} + \tfrac{1}{2}\sigma^2 \ln \eta_{\text{BAYES}} - \tfrac{1}{2}\mathbf{m}_2^T\mathbf{m}_2 \end{aligned}$$

In this form the $g_1(\mathbf{x})$ and $g_2(\mathbf{x})$ contain terms $\mathbf{m}_1^T\mathbf{x}$ and $\mathbf{m}_2^T\mathbf{x}$, which can be interpreted as correlations or dot products of the observation vector with the respective mean vectors.

Correlation Structure. The decision rule of Eq. (9.77) can be manipulated to give the decision rule in terms of the correlations of $\mathbf{x}$ with the means $\mathbf{m}_1$ and $\mathbf{m}_2$ as

$$\begin{aligned} \text{If} \quad \mathbf{m}_1^T\mathbf{x} - \mathbf{m}_2^T\mathbf{x} &> T \quad \text{decide } C_1 \\ &< \quad\;\; \text{decide } C_2 \\ &= \quad\;\; \text{decide randomly} \end{aligned} \tag{9.80}$$

$$\text{where} \quad T = \sigma^2 \ln \eta_{\text{BAYES}} + \tfrac{1}{2}(\mathbf{m}_1^T\mathbf{m}_1 - \mathbf{m}_2^T\mathbf{m}_2)$$

Comparative Distance Classifier. The starting point for determining the comparative distance classifier is Eq. (9.72), the general form of the decision rule before algebraic simplifications were performed. Letting $K = \sigma^2 I$, where I is the identity matrix, the optimum decision rule for special case 2 given in (9.72) can be written as follows:

$$\begin{aligned} \text{If} \quad -(\mathbf{x} - \mathbf{m}_1)^T(\mathbf{x} - \mathbf{m}_1) + (\mathbf{x} - \mathbf{m}_2)^T(\mathbf{x} - \mathbf{m}_2) &> 2\sigma^2 \ln \eta_{\text{BAYES}} \quad \text{decide } C_1 \\ &< \qquad\qquad\qquad\; \text{decide } C_2 \\ &= \qquad\qquad\qquad\; \text{decide randomly} \end{aligned} \tag{9.81}$$

The left side of the equation above is the difference of the squared Euclidean distances from $\mathbf{x}$ to $\mathbf{m}_1$ and $\mathbf{m}_2$, respectively, while the right side represents the threshold, which is a function of σ^2 and η_{BAYES}. Each of these forms of the optimum classifier are shown in Figure 9.24.

9.8.3 *M*-Class General Gaussian Problem (Bayes)

For the M-class Gaussian case it is assumed that $C_i: X \sim N(\mathbf{m}_i, \mathbf{K}_i)$ and that $\mathbf{m}_i, \mathbf{K}_i$, and $P(C_i)$ are known for $i = 1, 2, \ldots, M$. From Eq. (9.42) the optimum Bayes decision rule selects the class corresponding to the **minimum** of the $y_i(\mathbf{x})$ defined in (9.41) by

$$y_i(\mathbf{x}) = \sum_{j=1}^{M} C_{ij} p(\mathbf{x} \,|\, C_j) P(C_j), \qquad i = 1, 2, \ldots, M \tag{9.82}$$

These $y_i(\mathbf{x})$ are functions of the costs and conditional densities and are thus weighted sums of the Gaussian exponential forms. In general, this form with the exponentials does not lead to any further simplification.

Special cases of the optimum classifier for the M-class general Gaussian occur frequently. Cases 3, 4, and 5 to follow are Bayes tests for **zero-one cost functions**. These are equivalent to a minimum probability of error detector, for various types of covariance matrices: different, identical, and scaled identity, respectively.

Special Case 3: Gaussian with Zero-One Cost Assignment (Optimum Decision Rule). The optimum decision rule for the minimum probability of the error performance measure, a Bayes test with a zero-one cost assignment ($C_{ij} = 1$ for $i \neq j$ and $C_{ii} = 0$ for all $i, j = 1, 2, \ldots, M$), will now be described. Conditional probability density functions for the pattern vector under each class are assumed Gaussian with known mean vectors $\mathbf{m}_i$ and covariance matrices $\mathbf{K}_i$ for $i = 1, \ldots, M$, and we can denote that under class C_i: $\mathbf{x} \sim N(\mathbf{m}_i, \mathbf{K}_i)$. The optimal Bayes classifier for this case is specified by the optimum decision rule, Eq. (9.50). It selects the class for which $s_i(\mathbf{x})$, defined below, is a **maximum**:

$$s_i(\mathbf{x}) = p(\mathbf{x} \,|\, C_i) P(C_i), \qquad i = 1, 2, \ldots, M \tag{9.83}$$

Substituting the Gaussian densities in the above expression and taking the ln of the $s_i(\mathbf{x})$ yields

$$\begin{aligned} \ln s_i(\mathbf{x}) &= \ln P(C_i) + \ln p(\mathbf{x} \,|\, C_i), \qquad i = 1, 2, \ldots, M \\ &= \ln P(C_i) - \ln[(2\pi)^{M/2}] - \ln[|\mathbf{K}_i|^{1/2}] + \ln[\exp(-\tfrac{1}{2}(\mathbf{x} - \mathbf{m}_i)^T \mathbf{K}_i^{-1} (\mathbf{x} - \mathbf{m}_i))] \end{aligned} \tag{9.84}$$

As the $\ln(\cdot)$ operator preserves the maximum selection, the $\ln(s_i(\mathbf{x}))$ can used for the comparison, and since $\ln[(2\pi)^{M/2}]$ is common to all of the $\ln[s_i(\mathbf{x})]$, it can be ignored

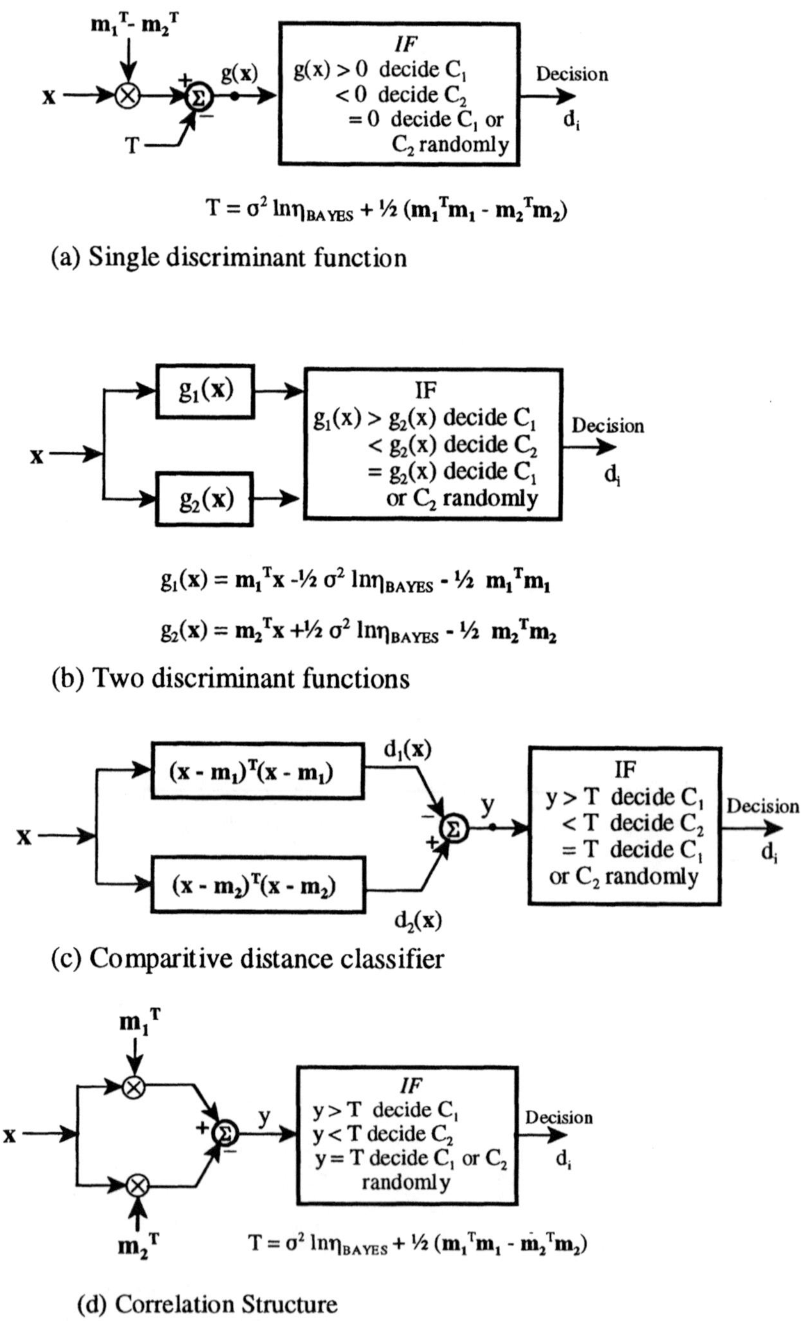

Figure 9.24 Block diagrams for Special Case 2. (a) Single discriminant function, (b) Two discriminant functions, (c) Comparative distance classifier, and (d) Correlation structure.

for purposes of determining the maximum. Thus the optimum classifier would use the terms

$$\ln P(C_i) - \tfrac{1}{2}\ln|\mathbf{K}_i| - \tfrac{1}{2}(\mathbf{x}-\mathbf{m}_i)^T\mathbf{K}_i^{-1}(\mathbf{x}-\mathbf{m}_i), \qquad i = 1, 2, \ldots, M \tag{9.85}$$

If the terms above are multiplied by a -2, the maximum selector turns into a **minimum selector** of a set of functions $Q_i(\mathbf{x})$ given by

$$Q_i(\mathbf{x}) = (\mathbf{x}-\mathbf{m}_i)^T\mathbf{K}_i^{-1}(\mathbf{x}-\mathbf{m}_i) - 2\ln P(C_i) + \ln|\mathbf{K}_i|, \qquad i = 1, 2, \ldots, M \tag{9.86}$$

The optimum decision rule for the zero-one Gaussian case is shown in Figure 9.25. The figure shows that the optimum classifier for the Gaussian problem involves the calculation of the Mahalanobis distances between the pattern vector $\mathbf{x}$ to be classified and the mean vectors $\mathbf{M}_i$ of the respective classes using the inverse covariance matrices $\mathbf{K}_i^{-1}$ as factors. Since the decision rule involves only quadratic functions, it is called a **quadratic classifier**.

If it is assumed further that the covariance matrices are equal, the optimal decision rule can be put in a simpler form as seen in the following special case 4.

Special Case 4: Gaussian with Zero-One Cost and Equal Covariance Matrices (Optimum Decision Rule). Assume that $\mathbf{X} \sim N(\mathbf{m}_i, \mathbf{K}_i)$ under each class C_i, $C_{ij} = 1$, and $C_{jj} = 0$ for all not equal i and j, and further assume that the covariance matrices are identical: $\mathbf{K}_1 = \mathbf{K}_2 = \cdots = \mathbf{K}_M = \mathbf{K}$. Each quadratic function $Q_i(\mathbf{x})$ of Eq. (9.86) can be expanded as

$$\begin{aligned} Q_i(\mathbf{x}) &= (\mathbf{x}-\mathbf{m}_i)^T\mathbf{K}^{-1}(\mathbf{x}-\mathbf{m}_i) - 2\ln P(C_i) + \ln|\mathbf{K}|, \qquad \text{for } i = 1, \ldots, M \\ &= \mathbf{x}^T\mathbf{K}^{-1}\mathbf{x} - 2\mathbf{m}_i^T\mathbf{K}^{\mathrm{rm}-1}\mathbf{x} + \mathbf{m}_i^T\mathbf{K}^{-1}\mathbf{m}_i - 2\ln P(C_i) + \ln|\mathbf{K}| \end{aligned} \tag{9.87}$$

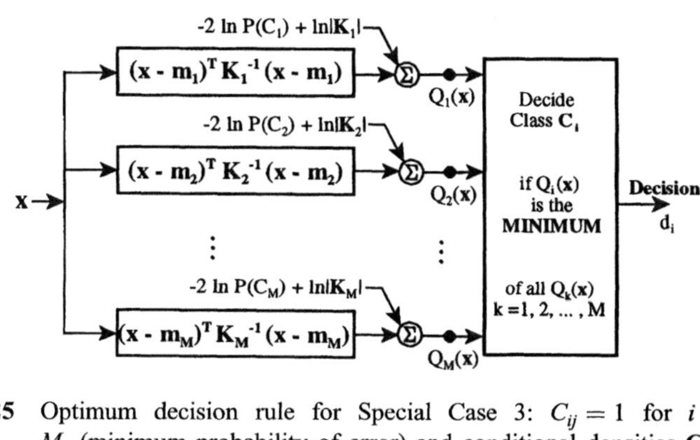

Figure 9.25 Optimum decision rule for Special Case 3: $C_{ij} = 1$ for $i \neq j$, $C_{jj} = 0$, $i, j = 1, 2, \ldots, M$, (minimum probability of error) and conditional densities **Gaussian** with known means $\mathbf{m}_i$ and covariance matrices $\mathbf{K}_i$.

As the first term of (9.87) would be the same for each branch, it can be ignored in the calculation. Also the part of the bias, terms ln $|\mathbf{K}_i|$, are all the same and can be dropped. Finally, if each branch is divided by -2, the minimum selector must be changed to a maximum selector, and thus the optimum decision rule computes the $L_i(\mathbf{x})$ defined by

$$L_i(\mathbf{x}) = \mathbf{m}_i^T\mathbf{K}^{-1}\mathbf{x} - \tfrac{1}{2}\mathbf{m}_i^T\mathbf{K}^{-1}\mathbf{m}_i + \ln P(C_i), \qquad i = 1, \ldots, M \tag{9.88}$$

and selects the class that corresponds to the **maximum** of the $L_i(\mathbf{x})$.

The resulting structure of the optimum decision rule, shown in Figure 9.26, is called a **maximum correlation decision rule**. The first blocks compute the correlation or dot product of $\mathbf{x}$, the observed pattern vector, with $\mathbf{m}_i^T\mathbf{K}^{-1}$, a transformed version of the conditional mean, and then a bias is added before the maximization takes place.

Special Case 5: Gaussian, Zero-One Cost, $\mathbf{K}_i = \sigma^2\mathbf{I}$ ***for all*** $i = 1, 2, \ldots, M$ ***(Optimum Decision Rule).*** The same conditions are assumed as for case 4, with the additional assumption that the components of the pattern vector are uncorrelated. In other words, $\mathbf{K} = \sigma^2\mathbf{I}$, where $\mathbf{I}$ is the identity matrix. Replacing $\mathbf{K}^{-1}$ by $\mathbf{I}/\sigma^2$ in Eq. (9.88), and multiplying by σ^2, we have a new $L_i'(\mathbf{x})$ defined by

$$L_i'(\mathbf{x}) = \mathbf{m}_i^T\mathbf{x} - \tfrac{1}{2}\,\mathbf{m}_i^T\mathbf{m}_i + \sigma^2 \ln P(C_i), \qquad i = 1, 2, \ldots, M \tag{9.89}$$

The optimum decision rule for this case computes a correlation with each of the mean vectors, adds a bias and then selects the class corresponding to the **maximum** value. This decision rule is shown in Figure 9.27.

Special Case 6: Gaussian, Zero-One Cost, Equal Diagonal $\mathbf{K}_i$, $i = 1, 2, \ldots, M$ ***(Optimum Decision Rule).*** The same basic conditions are assumed as for case 4, with the additional assumption that the components of the pattern vector are uncor-

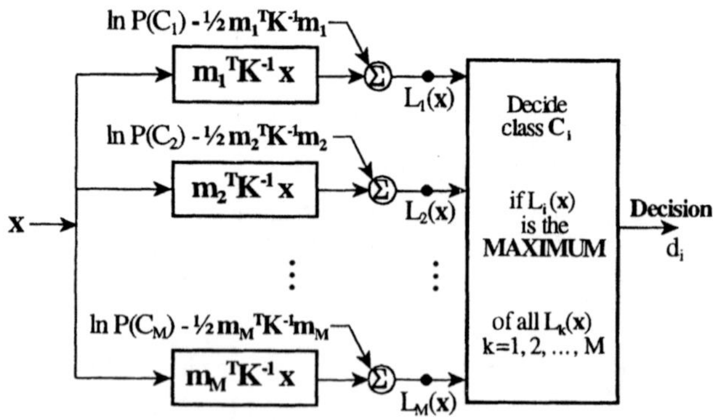

Figure 9.26 Optimum classifier for Case 4: $C_{ij} = 1$, $C_{jj} = 0, i, j = 1, 2, \ldots, M (i \neq j)$, $x \sim N(m_i, K)$, (minimum probability of error equal covariance matrices). A maximum correlation structure with warped versions of the means and different biases.

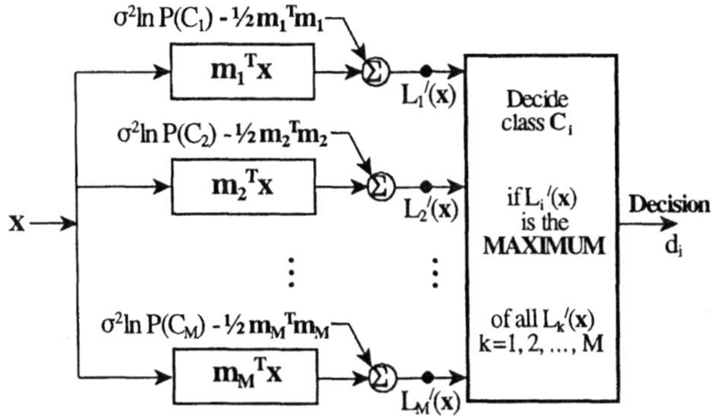

Figure 9.27 Optimum classifier for Case 5: $C_{ij} = 1$ for $i \neq j$, $C_{jj} = 0, \mathbf{x} \sim N(\mathbf{m}_i, \sigma^2\mathbf{I})$, $i, j = 1, \ldots, M$. (Minimum Probability of error Gaussian with equal scaled identity covariance matrices).

related but the variances are different for each component. In other words, $\mathbf{K} = \text{diag}[\sigma_1^2, \sigma_2^2, \ldots, \sigma_n^2]$. The optimum classifier for this case computes a weighted correlation with the mean vector, adds to it a bias, and then selects the class corresponding to the maximum value.

Example 9.11 shows the details of determining the optimum decision rule and calculating the probability of error for case 5.

EXAMPLE 9.11

Given the following three pattern classes, multiple observation problem with $P(C_1) = P(C_2) = P(C_3) = \frac{1}{3}$. The pattern vector $\mathbf{x}$ has two components as defined below for each of the classes.

$$\begin{aligned}
C_1&: \quad x_k = \sin(\tfrac{1}{2}\pi k) + n_k, && k = 0, 1 \\
C_2&: \quad x_k = \cos(\tfrac{1}{2}\pi k) + n_k, && k = 0, 1 \\
C_3&: \quad x_k = \sin(\tfrac{1}{2}\pi k) + \cos(\tfrac{1}{2}\pi k) + n_k, && k = 0, 1
\end{aligned}$$

Assume the n_k are independent identically distributed Gaussian random variables with zero means and unity variances. This problem is an abstraction of a tri-phase modulation problem.

(a) Find the minimum probability of error decision rule.

(b) Illustrate your decision regions in the observation space.

(c) Suppose that you are presented with the pattern vector $\mathbf{x} = [0.4, 0.7]^T$. Classify this vector using the decision rule found in (b).

(d) Find the probability of error for the decision rule determined in (b).

SOLUTION

(a) Substituting in $k = 0$ and $k = 1$ results in the sines and cosines being either ones or zeros, and the resulting pattern vector $\mathbf{X} = [X_1, X_2]^T$ under each of the classes is Gaussian with mean vectors and covariance matrices as follows:

$$C_1:\quad \mathbf{X} \sim N(\mathbf{m}_1, \mathbf{K}_1), \quad \mathbf{m}_1 = \begin{bmatrix} 0 \\ 1 \end{bmatrix}, \quad \mathbf{K}_1 = \begin{bmatrix} 1 & 0 \\ 0 & 1 \end{bmatrix}$$

$$C_2:\quad \mathbf{X} \sim N(\mathbf{m}_2, \mathbf{K}_2), \quad \mathbf{m}_2 = \begin{bmatrix} 1 \\ 0 \end{bmatrix}, \quad \mathbf{K}_2 = \begin{bmatrix} 1 & 0 \\ 0 & 1 \end{bmatrix}$$

$$C_3:\quad \mathbf{X} \sim N(\mathbf{m}_3, \mathbf{K}_3), \quad \mathbf{m}_3 = \begin{bmatrix} 1 \\ 1 \end{bmatrix}, \quad \mathbf{K}_3 = \begin{bmatrix} 1 & 0 \\ 0 & 1 \end{bmatrix}$$

The minimum probability of error classifier is the same as the Bayes decision rule for the zero-one cost assignment. Since the conditional probability density functions are Gaussian, and the covariance matrices are identical under all patterns, the optimum decision rule is that given under special case 5. For this case the optimum decision rule computes $L_i'(\mathbf{x})$, for $i = 1, 2$, and 3, and selects the pattern class corresponding to the maximum value where $L_i'(\mathbf{x})$ from Eq. (9.89) are

$$L_i'(\mathbf{x}) = \mathbf{m}_i^T \mathbf{x} - \tfrac{1}{2}\, \mathbf{m}_i^T \mathbf{m}_i + \sigma^2 \ln P(C_i), \qquad i = 1, 2, 3$$

Using the mean vectors and covariance matrices above, with $\sigma^2 = 1$, we determine the $L_i'(\mathbf{x})$ as

$$L_1'(\mathbf{x}) = \begin{bmatrix} 0 & 1 \end{bmatrix} \begin{bmatrix} x_1 \\ x_2 \end{bmatrix} - \tfrac{1}{2} \begin{bmatrix} 0 & 1 \end{bmatrix} \begin{bmatrix} 0 \\ 1 \end{bmatrix} + \ln\tfrac{1}{3} = x_2 - \tfrac{1}{2} + \ln\tfrac{1}{3}$$

$$L_2'(\mathbf{x}) = \begin{bmatrix} 1 & 0 \end{bmatrix} \begin{bmatrix} x_1 \\ x_2 \end{bmatrix} - \tfrac{1}{2} \begin{bmatrix} 1 & 0 \end{bmatrix} \begin{bmatrix} 1 \\ 0 \end{bmatrix} + \ln\tfrac{1}{3} = x_1 - \tfrac{1}{2} + \ln\tfrac{1}{3}$$

$$L_3'(\mathbf{x}) = \begin{bmatrix} 1 & 1 \end{bmatrix} \begin{bmatrix} x_1 \\ x_2 \end{bmatrix} - \tfrac{1}{2} \begin{bmatrix} 1 & 1 \end{bmatrix} \begin{bmatrix} 1 \\ 1 \end{bmatrix} + \ln\tfrac{1}{3} = x_1 + x_2 + \ln\tfrac{1}{3} - 1$$

As $-\frac{1}{2} + \ln(\frac{1}{3})$ is common to all $L_k'(\mathbf{x})$, it can be dropped from each $L_k'(\mathbf{x})$, and the decision rule can be written in terms of selecting the class corresponding to the maximum of $s_i(\mathbf{x})$ defined by

$$s_1(\mathbf{x}) = x_2$$
$$s_2(\mathbf{x}) = x_1$$
$$s_3(\mathbf{x}) = x_1 + x_2 - \tfrac{1}{2}$$

The decision rule then becomes

If $s_1(\mathbf{x}) > s_2(\mathbf{x})$ AND $s_1(\mathbf{x}) > s_3(\mathbf{x})$ then decide C_1
If $s_2(\mathbf{x}) > s_1(\mathbf{x})$ AND $s_2(\mathbf{x}) > s_3(\mathbf{x})$ then decide C_2
If $s_3(\mathbf{x}) > s_1(\mathbf{x})$ AND $s_3(\mathbf{x}) > s_2(\mathbf{x})$ then decide C_3

with random decisions on the boundaries.

(b) Substituting the $s_i(\mathbf{x})$ thus determined into the decision rule above, we can determine the optimum decision rule in the observation space. If R_1 represents the region for which we decide C_1, the first line of the decision rule above gives

$$\begin{aligned} R_1 &= \{\mathbf{x}: x_2 > x_1 \quad \text{AND} \quad x_2 > x_1 + x_2 - \tfrac{1}{2}\} \\ &= \{\mathbf{x}: x_2 > x_1 \quad \text{AND} \quad x_1 < \tfrac{1}{2}\} \end{aligned}$$

The two straight lines of R_1 above specify the boundary for region R_1. Similarly R_2 and R_3 can be determined as

$$\begin{aligned} R_2 &= \{\mathbf{x}: x_2 < x_1 \quad \text{AND} \quad x_2 < \tfrac{1}{2}\} \\ R_3 &= \{\mathbf{x}: x_1 > \tfrac{1}{2} \quad \text{AND} \quad x_2 > \tfrac{1}{2}\} \end{aligned}$$

The decision regions for optimum decision rule are shown in Figure 9.28.

(c) The pattern vector $\mathbf{x} = [0.4 \;\; 0.7]^T$ is in R_1. Therefore $\mathbf{x}$ would be classified as coming from class C_1.

(d) The probability of error for the three- and higher-class cases is usually more efficiently calculated in terms of the probability of being correct. Using this approach the probability of error for the decision rule given in part (b) can be written

$$\begin{aligned} P(\text{error}) &= 1 - P(\text{correct}) \\ &= 1 - P(\text{correct} \mid C_1)P(C_1) - P(\text{correct} \mid C_2)P(C_2) \\ &\quad - P(\text{correct} \mid C_3)P(C_3) \end{aligned}$$

The second part of the equation above was obtained by using the total probability theorem, and it is composed of conditional probabilities of being

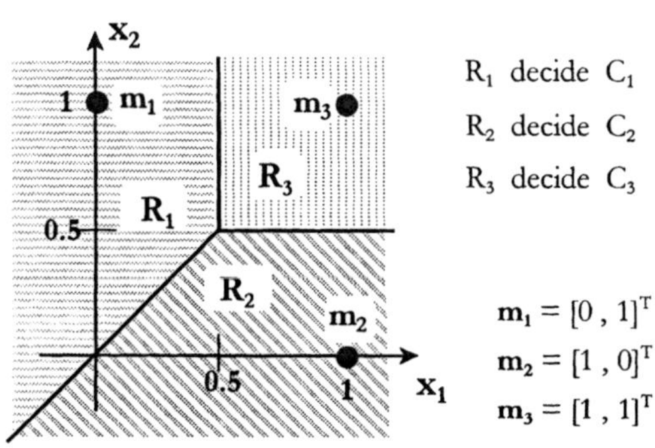

Figure 9.28 Decision regions in the observation space for Example 9.11.

correct. These conditional probabilities are given as

$$P(\text{correct}\,|\,C_i) = P(\text{deciding } C_i\,|\,C_i)$$
$$= \iint_{R_i} p(\mathbf{x}\,|\,C_i)\,d\mathbf{x}, \qquad i = 1, 2, 3$$

where $p(\mathbf{x}\,|\,C_i)$ are Gaussian densities with the given mean vectors and covariance matrices and the decision regions R_i are those determined earlier and shown in Figure 9.28.

Each of the three conditional probabilities of being correct are now determined by performing the corresponding two dimensional integrals.

$$P(\text{correct}\,|\,C_1) = \iint_{R_1} p(\mathbf{x}\,|\,C_1)\,d\mathbf{x}$$
$$= \int_{-\infty}^{1/2}\int_{x_1}^{\infty} \frac{1}{2\pi}\exp\left\{-\frac{x_1^2}{2} - \frac{(x_2-1)^2}{2}\right\} dx_2\,dx_1$$
$$= \int_{-\infty}^{1/2} \frac{1}{\sqrt{2\pi}}\exp\left\{-\frac{x_1^2}{2}\right\}\int_{x_1}^{\infty}\frac{1}{\sqrt{2\pi}}\exp\left\{-\frac{(x_2-1)^2}{2}\right\} dx_2\,dx_1$$

The second integral can be written in terms of the $\Phi(.)$ function to give

$$P(\text{correct}\,|\,C_1) = \int_{-\infty}^{1/2}\frac{1}{\sqrt{2\pi}}\exp\left\{-\frac{x_1^2}{2}\right\}(1-\Phi(x_1-1))dx_1$$

The integral above does not have an analytical solution since the $\Phi(.)$ does not have an analytical form. Therefore we need to evaluate it using the Gaussian distribution table and any standard method of numerical integration. Carrying out this integration, we get

$$P(\text{correct}\,|\,C_1) = 0.62$$

Similarly the $P(\text{correct}\,|\,C_2)$ must be obtained numerically. It can be determined by

$$P(\text{correct}\,|\,C_2) = \iint_{R_2} p(\mathbf{x}\,|\,C_2)\,d\mathbf{x}$$
$$= \int_{-\infty}^{1/2}\int_{x_2}^{\infty}\frac{1}{2\pi}\exp\left\{-\frac{(x_1-1)^2}{2} - \frac{(x_2)^2}{2}\right\} dx_1\,dx_2$$
$$= \int_{-\infty}^{1/2}\exp\left\{-\frac{(x_2)^2}{2}\right\}\int_{x_2}^{\infty}\frac{1}{\sqrt{2\pi}}\exp\left\{-\frac{(x_1-1)^2}{2}\right\} dx_1\,dx_2$$

Carrying out the integration as before, the $P(\text{correct}\,|\,C_2)$ reduces to

$$P(\text{correct}\,|\,C_1) = 0.62.$$

This result is equal to the $P(\text{correct} \mid C_1)$ and could have been obtained by symmetry in the x_1, x_2 space for the means m_1 and m_2 and the regions R_1 and R_2.

The $P(\text{correct} \mid C_3)$ can be obtained analytically, since the region of integration is not on the diagonal and is given by

$$\begin{aligned}
P(\text{correct} \mid C_3) &= \iint_{R_3} p(\mathbf{x} \mid C_3)\, d\mathbf{x} \\
&= \int_{1/2}^{\infty} \int_{1/2}^{\infty} \frac{1}{2\pi} \exp\left\{ -\frac{(x_1-1)^2}{2} - \frac{(x_2-1)^2}{2} \right\} dx_2\, dx_1 \\
&= \int_{1/2}^{\infty} \frac{1}{\sqrt{2\pi}} \exp\left\{ -\frac{(x_1-1)^2}{2} \right\} dx_1 \\
&\quad \times \int_{1/2}^{\infty} \frac{1}{\sqrt{2\pi}} \exp\left\{ -\frac{(x_2-1)^2}{2} \right\} dx_2 \\
&= (1 - \Phi(-\tfrac{1}{2}))^2 = 0.47817
\end{aligned}$$

Using these values for the conditional correct probabilities, we can finally write the total error as

$$\begin{aligned}
P(\text{error}) &= 1 - [P(\text{correct} \mid C_1)P(C_1) + P(\text{correct} \mid C_2)P(C_2) \\
&\quad + P(\text{correct} \mid C_3)P(C_3)] \\
&= 1 - [(0.62)\tfrac{1}{3} + (0.62)\tfrac{1}{3} + (0.47817)\tfrac{1}{3}] = 0.42728 \qquad \blacksquare
\end{aligned}$$

9.8.4 Two-Class General Gaussian Performance

Performance for a classifier can be specified by the probability of error or the Bayes risk. As the risk is a function of the error probabilities, (Eq. 9.37), this section will concentrate on the evaluation of error probabilities for the two-class classifier for the general Gaussian case.

Assume for class C_1 that $\mathbf{x} \sim N(\mathbf{m}_1, \mathbf{K}_1)$, for class C_2, $\mathbf{x} \sim N(\mathbf{m}_2, \mathbf{K}_2)$, and that the class a priori probabilities, $P(C_1)$ and $P(C_2)$, and costs C_{11}, C_{12}, C_{21}, and C_{22}, are specified. The Bayes decision rule was shown to be a likelihood ratio test with likelihood ratio and threshold given in Eq. (9.72) as

$$\text{If} \quad -(\mathbf{x}-\mathbf{m}_1)^T \mathbf{K}_1^{-1}(\mathbf{x}-\mathbf{m}_1) + (\mathbf{x}-\mathbf{m}_2)^T \mathbf{K}_2^{-1}(\mathbf{x}-\mathbf{m}_2) \begin{array}{ll} > T & \text{decide } C_1 \\ < & \text{decide } C_2 \\ = & \text{decide randomly} \end{array}$$

$$\text{where} \quad T = 2 \ln \eta_{\text{BAYES}} + \ln \frac{|\mathbf{K}_1|}{|\mathbf{K}_2|} \tag{9.90}$$

The expression on the left-hand side of the inequality is a quadratic function of the random vector **X**, and the quantity on the right side is a threshold value determined by the statistical properties of the pattern vector under each class, the costs, and the a priori probabilities. The computation of the error probabilities for the general case is extremely complicated and is not presented. However, two special cases deserve analytical development because of their realistic assumptions.

Special Case 1: Equal Covariance Matrices (Performance). For the special case of equal covariance matrices, that is, $\mathbf{K}_1 = \mathbf{K}_2 = \mathbf{K}$, the decision rule was seen in Eq. (9.76) to be written in terms of a test statistic z as

$$\begin{aligned} \text{If} \quad z \triangleq (\mathbf{m}_1 - \mathbf{m}_2)^T \mathbf{K}^{-1} \mathbf{x} &> T_1 \quad \text{decide } C_1 \\ &< \quad \text{decide } C_2 \\ &= \quad \text{decide randomly} \\ \text{where} \quad T_1 = \ln \eta_{\text{BAYES}} &+ \tfrac{1}{2}(\mathbf{m}_1^T \mathbf{K}^{-1} \mathbf{m}_1 - \mathbf{m}_2^T \mathbf{K}^{-1} \mathbf{m}_2) \end{aligned} \tag{9.91}$$

For the case of equal covariances it is shown that the optimum decision rule is a linear one comparing z, a weighted sum of the observation vector components, to a threshold. The test statistic z is a scalar since $(\mathbf{m}_1 - \mathbf{m}_2)^T$ is $1 \times n$, $\mathbf{K}^{-1}$ is $n \times n$, and $\mathbf{x}$ is $n \times 1$. To find the probability of error, it is necessary to determine the conditional densities of z for each of the two classes. Since $\mathbf{x}$ is a Gaussian random vector under both classes, z is scalar Gaussian because it is a linear combination of the vector components namely $(\mathbf{m}_1 - \mathbf{m}_2)^T \mathbf{K}^{-1}$ times the $\mathbf{x}$ vector. The Gaussian densities for Z under both classes can then be specified once the conditional means and conditional variances for Z are determined.

Under class C_1 the conditional expected value is determined as

$$\begin{aligned} E[Z \mid C_1] &= E[(\mathbf{m}_1 - \mathbf{m}_2)^T \mathbf{K}^{-1} \mathbf{x} \mid C_1] \\ &= (\mathbf{m}_1 - \mathbf{m}_2)^T \mathbf{K}^{-1} E[\mathbf{x} \mid C_1] \\ &= (\mathbf{m}_1 - \mathbf{m}_2)^T \mathbf{K}^{-1} \mathbf{m}_1 \end{aligned} \tag{9.92}$$

In the calculations above the expected value operator can be taken through the linear operation on $\mathbf{x}$ because $\mathbf{m}_1$, $\mathbf{m}_2$, and $\mathbf{K}^{-1}$ are nonrandom. The variance for Z conditioned on class C_1 can be determined using Eq. (9.91) and (9.92) as

$$\begin{aligned} \sigma^2_{Z \mid C_1} &= E[\{Z - E[Z \mid C_1]\}^2] \\ &= E[\{(\mathbf{m}_1 - \mathbf{m}_2)^T \mathbf{K}^{-1} \mathbf{x} - (\mathbf{m}_1 - \mathbf{m}_2)^T \mathbf{K}^{-1} \mathbf{m}_1\}^2 \mid C_1] \end{aligned} \tag{9.93}$$

Factoring out the $(\mathbf{m}_1 - \mathbf{m}_2)^T \mathbf{K}^{-1}$, implementing the squared operation by taking the part in parenthesis times its transpose (alright since it is a scalar), and recognizing that $\mathbf{K}^{-1}$ is symmetric, $\mathbf{K}^{-1} = (\mathbf{K}^{-1})^T$, we determine the conditional variance as

$$\sigma^2_{Z \mid C_1} = E[(\mathbf{m}_1 - \mathbf{m}_2)^T \mathbf{K}^{-1} (\mathbf{x} - \mathbf{m}_1)(\mathbf{x} - \mathbf{m}_1)^T \mathbf{K}^{-1} (\mathbf{m}_1 - \mathbf{m}_2) \mid C_1] \tag{9.94}$$

Taking the expected value operator through the linear operations and recognizing the covariance matrix in the middle which cancels with one of the $\mathbf{K}^{-1}$ leads to

$$\begin{aligned}\sigma_{Z|C_1}^2 &= (\mathbf{m}_1 - \mathbf{m}_2)^T\mathbf{K}^{-1}\mathbf{E}[(\mathbf{x} - \mathbf{m}_1)(\mathbf{x} - \mathbf{m}_1)^{\mathbf{T}}]\mathbf{K}^{-1}(\mathbf{m}_1 - \mathbf{m}_2) \,|\, C_1] \\ &= (\mathbf{m}_1 - \mathbf{m}_2)^T\mathbf{K}^{-1}(\mathbf{m}_1 - \mathbf{m}_2)\end{aligned} \tag{9.95}$$

We have just shown that Z conditioned on C_1 is Gaussian with the conditional mean and conditional variance shown below

$$C_1: \quad Z \sim N((\mathbf{m}_1 - \mathbf{m}_2)^T\mathbf{K}^{-1}\mathbf{m}_1, (\mathbf{m}_1 - \mathbf{m}_2)^T\mathbf{K}^{-1}(\mathbf{m}_1 - \mathbf{m}_2)) \tag{9.96}$$

Following the derivations just presented for the determination of the conditional mean and variance under class C_1, we determine the conditional mean and variance for class C_2 as

$$E[Z \,|\, C_2] = (\mathbf{m}_1 - \mathbf{m}_2)^T\mathbf{K}^{-1}\mathbf{m}_2 \tag{9.97}$$

$$\begin{aligned}\sigma_{Z|C_2}^2 &= E[(z - E[Z \,|\, C_2])^2 \,|\, C_2] \\ &= (\mathbf{m}_1 - \mathbf{m}_2)^T\mathbf{K}^{-1}(\mathbf{m}_1 - \mathbf{m}_2)\end{aligned} \tag{9.98}$$

Thus under C_2 we have that Z is a Gaussian random variable with mean given by (9.97) and variance by (9.98), which can be summarized as follows:

$$C_2: \quad Z \sim N((\mathbf{m}_1 - \mathbf{m}_2)^T\mathbf{K}^{-1}\mathbf{m}_2, (\mathbf{m}_1 - \mathbf{m}_2)^T\mathbf{K}^{-1}(\mathbf{m}_1 - \mathbf{m}_2)) \tag{9.99}$$

The total probability of error can now be written in terms of the conditional probabilities of error as

$$P(\text{error}) = P(\text{error} \,|\, C_1)P(C_1) + P(\text{error} \,|\, C_2)P(C_2) \tag{9.100}$$

where

$$\begin{aligned}P(\text{error} \,|\, C_1) &= \int_{-\infty}^{T} p(z \,|\, C_1)\, dz \\ P(\text{error} \,|\, C_2) &= \int_{T}^{\infty} p(z \,|\, C_2)\, dz\end{aligned} \tag{9.101}$$

The $P(\text{error} \,|\, C_1)$ can be found by using the class conditional density of Z as Gaussian with mean $(\mathbf{m}_1 - \mathbf{m}_2)^T\mathbf{K}^{-1}\mathbf{m}_1$ and variance $(\mathbf{m}_1 - \mathbf{m}_2)^T\mathbf{K}^{-1}(\mathbf{m}_1 - \mathbf{m}_2)$. Making a transformation of variables and using the $\Phi(.)$ function, we determine the conditional probability of error as

$$\begin{aligned}P(\text{error} \,|\, C_1) &= \Phi\left(\frac{T - E[Z \,|\, C_1]}{\sigma_{Z|C_1}}\right) \\ &= \Phi\left(\frac{\frac{1}{2}(\boldsymbol{m}_1^T\mathbf{K}^{-1}\mathbf{m}_1 - \mathbf{m}_2^T\mathbf{K}^{-1}\mathbf{m}_2) + \ln \eta_{\text{BAYES}} - (\mathbf{m}_1 - \mathbf{m}_2)^T\mathbf{K}^{-1}\mathbf{m}_1}{\sqrt{(\mathbf{m}_1 - \mathbf{m}_2)\mathbf{K}^{-1}(\mathbf{m}_1 - \mathbf{m}_2)}}\right) \\ &= \Phi\left(-\frac{1}{2}\, d(\mathbf{m}_1, \mathbf{m}_2; \mathbf{K}) + \frac{\ln \eta_{\text{BAYES}}}{d(\mathbf{m}_1, \mathbf{m}_2; \mathbf{K})}\right)\end{aligned} \tag{9.102}$$

where

$$d(\mathbf{m}_1, \mathbf{m}_2; \mathbf{K}) = \sqrt{(\mathbf{m}_1 - \mathbf{m}_2)^T \mathbf{K}^{-1} (\mathbf{m}_1 - \mathbf{m}_2)}$$

Using the class conditional density for C_2, we similarly determine the $P(\text{error} \mid C_2)$ as

$$\begin{aligned} &P(\text{error} \mid C_2) \\ &= 1 - \Phi\left(\frac{T - E[Z \mid C_2]}{\sigma_{Z \mid C_2}}\right) \\ &= 1 - \Phi\left(\frac{\frac{1}{2}(\mathbf{m}_1^T \mathbf{K}^{-1} \mathbf{m}_1 - \mathbf{m}_2^T \mathbf{K}^{-1} \mathbf{m}_2) + \ln \eta_{\text{BAYES}} - (\mathbf{m}_1 - \mathbf{m}_2)^T \mathbf{K}^{-1} \mathbf{m}_2}{\sqrt{(\mathbf{m}_1 - \mathbf{m}_2) \mathbf{K}^{-1} (\mathbf{m}_1 - \mathbf{m}_2)}}\right) \\ &= 1 - \Phi\left(\frac{1}{2} d(\mathbf{m}_1, \mathbf{m}_2; \mathbf{K}) + \frac{\ln \eta_{\text{BAYES}}}{d(\mathbf{m}_1, \mathbf{m}_2; \mathbf{K})}\right) \end{aligned} \tag{9.103}$$

Finally, we can write the total probability of error as

$$\begin{aligned} P(\text{error}) = {} & \Phi\left(-\frac{1}{2} d(\mathbf{m}_1, \mathbf{m}_2; \mathbf{K}) + \frac{\ln \eta_{\text{BAYES}}}{d(\mathbf{m}_1, \mathbf{m}_2; \mathbf{K})}\right) P(C_1) \\ & + \left[1 - \Phi\left(\frac{1}{2} d(\mathbf{m}_1, \mathbf{m}_2; \mathbf{K}) + \frac{\ln \eta_{\text{BAYES}}}{d(\mathbf{m}_1, \mathbf{m}_2; \mathbf{K})}\right)\right] P(C_2) \end{aligned} \tag{9.104}$$

where

$$d(\mathbf{m}_1, \mathbf{m}_2; \mathbf{K}) = \sqrt{(\mathbf{m}_1 - \mathbf{m}_2)^T \mathbf{K}^{-1} (\mathbf{m}_1 - \mathbf{m}_2)}$$

EXAMPLE 9.12

Find the probability of error for the special case where $\eta_{\text{BAYES}} = 1$. One way in which this can occur is a zero-one cost function and equal probabilities.

SOLUTION

Substituting $\eta_{\text{BAYES}} = 1$ into (9.102) and using the property that $\Phi(-x) = 1 - \Phi(x)$, we simplify the conditional error given C_1 to

$$P(\text{error} \mid C_1) = 1 - \Phi\left(\tfrac{1}{2}\sqrt{(\mathbf{m}_1 - \mathbf{m}_2)^T \mathbf{K}^{-1} (\mathbf{m}_1 - \mathbf{m}_2)}\right) \tag{9.105}$$

Similarly substituting $\eta_{\text{BAYES}} = 1$ into (9.103), we simplify the conditional error under C_2 to

$$P(\text{error} \mid C_2) = 1 - \Phi\left(\tfrac{1}{2}\sqrt{(\mathbf{m}_1 - \mathbf{m}_2)^T \mathbf{K}^{-1} (\mathbf{m}_1 - \mathbf{m}_2)}\right) \tag{9.106}$$

With these conditional errors, the total error the total probability of error given in (9.104) for the case of **zero-one cost and equal probable classes** case becomes

$$P(\text{error}) = 1 - \Phi\left(\tfrac{1}{2}\sqrt{(\mathbf{m}_1 - \mathbf{m}_2)^T \mathbf{K}^{-1}(\mathbf{m}_1 - \mathbf{m}_2)}\right) \tag{9.107}$$

Thus it is seen that the probability of error is a function of the Mahalanobis distance between the means of the two classes. Since $\Phi(x)$ is a monotonic function of its argument, the larger the argument, the larger the function, and thus the probability of error is smaller. This result seems intuitively pleasing based on the subjective notion that the wider the separation of the means, the better we can do in terms of classification. The distance between the means for this case, however, is the warped distance with respect to the covariance matrix. ■

Special Case 2: Equal Scaled Identity Covariance Matrices (Performance). For the case of equiprobable classes, Gaussian conditional densities and equal scaled identity covariance matrices ($\mathbf{K}_1 = \mathbf{K}_2 = \sigma^2\mathbf{I}$, where $\mathbf{I}$ is an $n \times n$ identity matrix), the probability of error can be simplified for either a minimum probability of error decision rule or equivalently a Bayes rule with a zero-one cost assignment. This corresponds to the case where the observed vector components are uncorrelated, and each has the same variance σ^2. Since the components are uncorrelated and Gaussian, they are also statistically independent. Substituting the given covariance matrices into Eq. (9.104), the probability of error can be determined as

$$\begin{aligned} P(\text{error}) = {} & \Phi\left(-\frac{1}{2\sigma}\, d(\mathbf{m}_1, \mathbf{m}_2; \mathbf{I}) + \frac{\sigma \ln \eta_{\text{BAYES}}}{d(\mathbf{m}_1, \mathbf{m}_2; \mathbf{I})}\right) P(C_1) \\ & + \left[1 - \Phi\left(\frac{1}{2\sigma}\, d(\mathbf{m}_1, \mathbf{m}_2; \mathbf{I}) + \frac{\sigma \ln \eta_{\text{BAYES}}}{d(\mathbf{m}_1, \mathbf{m}_2; \mathbf{I})}\right)\right] P(C_2) \end{aligned} \tag{9.108}$$

where

$$d(\mathbf{m}_1, \mathbf{m}_2; \mathbf{I}) = \sqrt{(\mathbf{m}_1 - \mathbf{m}_2)^T (\mathbf{m}_1 - \mathbf{m}_2)}$$

Example 9.13 below shows a further simplification when $\eta_{\text{BAYES}} = 1$.

EXAMPLE 9.13

Determine the probability of error for the Gaussian case 2 problem (equal scaled identity covariance matrices) when $P(C_1) = \frac{1}{2}$ and $P(C_2) = \frac{1}{2}$, which for a zero-one cost assignment gives $\eta_{\text{BAYES}} = 1$.

SOLUTION

Substituting $\eta_{\text{BAYES}} = 1$ into Eq. (9.108) and using the properties of $\Phi(-x)$, we easily simplify the probability of error to

$$
\begin{aligned}
P(\text{error}) &= 1 - \Phi\left(\frac{1}{2\sigma}\sqrt{(\mathbf{m}_1 - \mathbf{m}_2)^T(\mathbf{m}_1 - \mathbf{m}_2)}\right) \\
&= 1 - \Phi\left(\frac{1}{2\sigma}\sqrt{\sum_{i=1}^{n}(m_{1i} - m_{2i})^2}\right)
\end{aligned}
\tag{9.109}
$$

Note that the larger the Euclidean distance between the class mean vectors, the smaller is the probability of error, as expected. ■

Special Case 3: Equal Diagonal Covariance Matrices (Performance). Another common case is where the observations are uncorrelated but have different variances. For this case $\mathbf{K}_1 = \mathbf{K}_2 = \text{diag}[\sigma_1^2, \sigma_2^2, \ldots, \sigma_n^2]$, and the probability of error can also be determined from Eq. (9.104), when $\eta_{\text{BAYES}} = 1$ to be

$$
\begin{aligned}
P(\text{error}) &= 1 - \Phi\left(\frac{1}{2}\sqrt{(\mathbf{m}_1 - \mathbf{m}_2)^T\mathbf{K}^{-1}(\mathbf{m}_1 - \mathbf{m}_2)}\right) \\
&= 1 - \Phi\left(\frac{1}{2}\sqrt{\sum_{i=1}^{n}\frac{(m_{1i} - m_{2i})^2}{\sigma_i^2}}\right)
\end{aligned}
\tag{9.110}
$$

Notice here that the probability of error is a function of the sum of the squared differences of the class mean components, which are individually scaled by their respective variances σ_i^2.

EXAMPLE 9.14

Suppose that we are given that $X \sim N(\mathbf{m}_1, \mathbf{K}_1)$ under class C_1 and $X \sim N(\mathbf{m}_2, \mathbf{K}_2)$ under class C_2, where $\mathbf{m}_1 = [1, -2, 1]^T$ and $\mathbf{m}_2 = [-1, -1, -2]^T$. Calculate the probability of error for the optimum detector in the Bayes sense, using a zero-one cost function, for the following cases:

(a) $\mathbf{K}_1 = \mathbf{K}_2 = 2\mathbf{I}$ and $P(C_1) = P(C_2) = \frac{1}{2}$.
(b) $\mathbf{K}_1 = \mathbf{K}_2 = 4\mathbf{I}$ and $P(C_1) = \frac{1}{3}$ and $P(C_2) = \frac{2}{3}$.
(c) $P(C_1) = P(C_2) = \frac{1}{2}$ and $\mathbf{K}_1 = \mathbf{K}_2 = \text{diag}[2, 4, 16]$.
(d) $P(C_1) = \frac{1}{2}, P(C_2) = \frac{1}{2}$, and

$$
\mathbf{K}_1 = \mathbf{K}_2 = \begin{bmatrix} 1 & -\frac{1}{2} & 0 \\ -\frac{1}{2} & 2 & -\frac{1}{4} \\ 0 & -\frac{1}{4} & 1 \end{bmatrix}
$$

SOLUTION

(a) Since $\mathbf{K}_1$ and $\mathbf{K}_2$ are equal to a scaled identity, case 2 conditions are satisfied, and thus the P(error) is determined from (9.108) as

$$P(\text{error}) = 1 - \Phi\left(\frac{1}{2\sigma}\sqrt{(\mathbf{m}_1 - \mathbf{m}_2)^T(\mathbf{m}_1 - \mathbf{m}_2)}\right)$$

$$= 1 - \Phi\left(\frac{1}{2\sqrt{2}}\left([2 \quad -1 \quad 3]\begin{bmatrix} 2 \\ -1 \\ 3 \end{bmatrix}\right)^{1/2}\right)$$

$$= 1 - \Phi(1.322876) = 0.09292$$

(b) Since $\mathbf{K}_1$ and $\mathbf{K}_2$ are equal to a scaled identity, $4\mathbf{I}$, case 2 conditions are satisfied, and the P(error) is determined from (9.108) with $\ln \eta_{\text{BAYES}} = \ln(P(C_2)/P(C_1)) = \ln 2$, the $\mathbf{m}_1$ and $\mathbf{m}_2$ given above, and $d(\mathbf{m}_1, \mathbf{m}_2; \mathbf{I}) = \sqrt{14}$ as

$$P(\text{error}) = \Phi\left(-\frac{\sqrt{14}}{4} + \frac{2\ln 2}{\sqrt{14}}\right)\frac{1}{3} + \left(1 - \Phi\left(\frac{\sqrt{14}}{4} + \frac{2\ln 2}{\sqrt{14}}\right)\right)\frac{2}{3}$$

$$= \Phi(-0.565)\frac{1}{3} + (1 - \Phi(1.306))\frac{2}{3} = 0.1592$$

(c) Since the a priori probabilities are equal, and the covariance matrices are equal and diagonal, we have special case 3 where the probability of error can be obtained from Eq. (9.110) as

$$P(\text{error}) = 1 - \Phi\left(\frac{1}{2}\sqrt{\sum_{i=1}^{n}\frac{(m_{1i} - m_{2i})^2}{\sigma_i^2}}\right)$$

$$= 1 - \Phi\left(\frac{1}{2}\sqrt{\frac{(2)^2}{2} + \frac{(-1)^2}{4} + \frac{(3)^2}{16}}\right)$$

$$= 1 - \Phi(0.838525) = 0.20087$$

(d) Since $\mathbf{K}_1$ and $\mathbf{K}_2$ are equal but not scaled identity or diagonal, we have case 1. With equal probabilities and zero-one cost assignment, the P(error) can be written from (9.104) as

$$P(\text{error}) = 1 - \Phi\left(\tfrac{1}{2}\sqrt{(\mathbf{m}_1 - \mathbf{m}_2)^T\mathbf{K}^{-1}(\mathbf{m}_1 - \mathbf{m}_2)}\right)$$

$$= 1 - \Phi\left(\tfrac{1}{2}\left([2 \quad -1 \quad 3]\begin{bmatrix} 1 & -\frac{1}{2} & 0 \\ -\frac{1}{2} & 2 & -\frac{1}{4} \\ 0 & -\frac{1}{4} & 1 \end{bmatrix}^{-1}\begin{bmatrix} 2 \\ -1 \\ 3 \end{bmatrix}\right)^{1/2}\right)$$

$$= 1 - \Phi(1.82574) = 0.03002$$ ■

9.8.5 *M*-Class General Gaussian Performance

The probability of error for the general Gaussian case is difficult to find analytically as it involves solving n-dimensional integrals for which no analytical result is available. In this section the approach to finding the probability of error will be examined and results given for a couple of special cases.

The conditional probability densities for the pattern vector under each of M classes are assumed Gaussian with known mean vectors $\mathbf{m}_i$ and covariance matrices $\mathbf{K}_i$ for $i = 1, \ldots, M$. That is, under class C_i, $\mathbf{x} \sim N(\mathbf{m}_i, \mathbf{K}_i)$. As was shown earlier, in Eq. (9.86), the optimal classifier for this general Gaussian case is a **minimum** selector of the $Q_i(\mathbf{x})$ given by

$$Q_i(\mathbf{x}) = (\mathbf{x} - \mathbf{m}_i)^T \mathbf{K}_i^{-1}(\mathbf{x} - \mathbf{m}_i) - 2\ln P(C_i) + \ln|\mathbf{K}_i|, \quad i = 1, 2, \ldots, M \tag{9.111}$$

The probability of error conditioned on class C_1 can be written in terms of the conditional density for the $Q_i(\mathbf{x})$ and the conditional probability of being correct as follows:

$$\begin{aligned} P(\text{error}\,|\,C_1) &= 1 - P(\text{correct}\,|\,C_1) \\ &= 1 - \iint \cdots \int_{\substack{q_1 < q_2 \\ q_1 < q_3 \\ \cdots \\ q_1 < q_n}} p(q_1, q_2, q_3, \ldots, q_n\,|\,C_1)\, dq_2\, dq_3 \ldots dq_n\, dq_1 \end{aligned} \tag{9.112}$$

where $p(q_1, q_2, q_3, \ldots, q_n\,|\,C_1)$ is the conditional density of $Q_1(\mathbf{x}), Q_2(\mathbf{x}), \ldots, Q_n(\mathbf{x})$ on class C_1. The Q_ks are quadratic functions of the $\mathbf{x}$ vector components, which are Gaussian. To find their joint density, in general, we require methods for evaluating M functions of n random variables by the transformation theorem. The individual densities for the $Q_i(\mathbf{x})$ are also not Gaussian, since they are quadratic functions of the components of $\mathbf{x}$. Without simplifying assumptions on the covariance matrices, we cannot analytically determined the integral (9.112). The conditional probabilities of error for the other classes are analytically represented similarly to that given in (9.112) where the C_1 is replaced by C_i.

If it is assumed further that the covariance matrices are equal (special case 4), the optimal classifier can be put in a linear form as seen in the following special case which will allow a simplification in the calculation of the error probability.

Special Case 4 (Performance). Assume that $X \sim N(\mathbf{m}_i, \mathbf{K}_i))$ under each class $C_i, i = 1, 2, \ldots, M$, with zero-one cost assignment ($C_{ij} = 1, C_{ii} = 0$) and equal covariance matrices $\mathbf{K}_1 = \mathbf{K}_2 = \cdots = \mathbf{K}_M = \mathbf{K}$. The optimum classifier in the Bayes and minimum probability error sense (from Eq. 9.88) selects the class that corresponds to the **maximum** of $Z_i(\mathbf{x})$ defined by

$$\mathbf{Z}_i(x) = \mathbf{m}_i^T \mathbf{K}^{-1}\mathbf{x} - \tfrac{1}{2}\mathbf{m}_i^T \mathbf{K}^{-1}\mathbf{m}_i + \ln P(C_i), \quad i = 1, \ldots, M \tag{9.113}$$

The conditional probability of error given each class is more conveniently calculated in terms of the conditional probabilities of being correct. For example, if the correct class is C_1, then the conditional probability of error can be written similar to

(9.112), but with a region of integration containing greater than inequalities because of the decision rule selects the maximum:

$$P(\text{error} \mid C_1) = 1 - P(\text{correct} \mid C_1)$$
$$= 1 - \int\int \cdots \int_{\substack{z_1 > z_2 \\ z_1 > z_3 \\ \cdots \\ z_1 > z_n}} p(z_1, z_2, z_3, \ldots, z_n \mid C_1)\, dz_1\, dz_2\, dz_3 \ldots dz_n \tag{9.114}$$

The $Z_i(\mathbf{x})$ are jointly Gaussian, since they are a linear functions of pattern vector components that are jointly Gaussian. The mean vector and covariance matrix for the $\mathbf{Z} = [Z_1 Z_2 \ldots Z_M]^T$ vector can be obtained by using the transformation of jointly Gaussian random variables.

Under class C_1 the $E[Z_i(\mathbf{x}) \mid C_1]$ can be obtained by taking the conditional expected value of $Z_i(\mathbf{x})$ in Eq. (9.113) to give

$$E[Z_i(\mathbf{x}) \mid C_1] = \mathbf{m}_i^T \mathbf{K}^{-1} \mathbf{m}_1 - \tfrac{1}{2} \mathbf{m}_i^T \mathbf{K}^{-1} \mathbf{m}_i + \ln P(C_i), \quad i = 1, \ldots, M \tag{9.115}$$

The i, jth component of the conditional covariance matrix, $v_{ij}^{(1)}$, under class C_1 is defined as

$$v_{ij}^{(1)} = E[(Z_i - E[Z_i \mid C_1])(Z_j - E[Z_j \mid C_1]) \mid C_1] \tag{9.116}$$

Substituting the Z_i and Z_j, from Eq. (9.113), and the $E[Z_i(\mathbf{x}) \mid C_1]$ and $E[Z_j(\mathbf{x}) \mid C_1]$ from Eq. (9.115) into (9.116), we see that the ijth component of the covariance matrix reduces to

$$\begin{aligned} v_{ij}^{(1)} &= E[\mathbf{m}_i^T \mathbf{K}^{-1}(\mathbf{x} - \mathbf{m}_1) \cdot \mathbf{m}_j^T \mathbf{K}^{-1}(\mathbf{x} - \mathbf{m}_1) \mid C_1] \\ &= E[\mathbf{m}_i^T \mathbf{K}^{-1}(\mathbf{x} - \mathbf{m}_1) \cdot (\mathbf{x} - \mathbf{m}_1)^T (\mathbf{K}^{-1})^T \mathbf{m}_j \mid C_1] \\ &= \mathbf{m}_i^T \mathbf{K}^{-1} E[(\mathbf{x} - \mathbf{m}_1) \cdot (\mathbf{x} - \mathbf{m}_1)^T \mid C_1] \mathbf{K}^{-1} \mathbf{m}_j \\ &= \mathbf{m}_i^T \mathbf{K}^{-1} \mathbf{m}_j \end{aligned} \tag{9.117}$$

Similarly the $v_{ij}^{(k)}$, the covariance entry for Z conditioned on class C_k, can be derived to give exactly the same results given in (9.117). Thus it is seen that the covariance matrices for Z conditioned on each class are the same regardless of the class. Notice that in general, these covariance matrices, although identical, are not diagonal or scaled identity matrices as they have off-diagonal terms. Therefore the Z_i are correlated and thus are not independent. Unless further restrictions are placed on the mean vectors and covariance matrices under the classes, the integral given in Eq. (9.114) cannot be simplified further and must usually be evaluated numerically.

Special Case 5 (Performance). Special case 5 has the same conditions as for case 4 but with the additional assumption that the components of the pattern vector are uncorrelated, and thus the covariance matrix is a scaled identity matrix. In other words, $\mathbf{K} = \sigma^2\mathbf{I}$, where $\mathbf{I}$ is an $n \times n$ identity matrix. Replacing $\mathbf{K}^{-1}$ by $\mathbf{I}/\sigma^2$ in Eq. (9.113), multiplying by σ^2, and simplifying leads to the test statistics $L_i'(\mathbf{x})$ defined by

$$L_i'(\mathbf{x}) = \mathbf{m}_i^T\mathbf{x} - \frac{\mathbf{m}_i^T\mathbf{m}_i}{2} + \sigma^2 \ln P(C_i), \qquad i = 1, \ldots, M \tag{9.118}$$

The optimum classifier for this case computes a correlation or vector dot product of $\mathbf{x}$ with each mean vector $\mathbf{m}_i$ adds a bias and then selects the class corresponding to the **maximum** value. The mean of the $L_i(\mathbf{x})$ conditioned on class C_k is as follows:

$$E[L_i'(\mathbf{x}) \mid C_k] = \mathbf{m}_i^T\mathbf{m}_k - \frac{\mathbf{m}_i^T\mathbf{m}_i}{2} + \sigma^2 \ln P(C_i), \qquad i = 1, \ldots, M \tag{9.119}$$

The components $q_{ij}^{(k)}$ of the conditional covariance matrix for $L'(\mathbf{x})$ are determined from Eqs. (9.118) and (9.119) similar to the development of (9.117) or by substitution of the $\mathbf{K} = \sigma^2\mathbf{I}$ directly into (9.117) to give

$$\begin{aligned} q_{ij}^{(k)} &= E[\mathbf{m}_i^T(\mathbf{x} - \mathbf{m}_1)\mathbf{m}_j^T(\mathbf{x} - \mathbf{m}_1) \mid C_k] \\ &= E[\mathbf{m}_i^T(\mathbf{x} - \mathbf{m}_1)(\mathbf{x} - \mathbf{m}_1)^T\mathbf{m}_j \mid C_k] \\ &= \mathbf{m}_i^T E[(\mathbf{x} - \mathbf{m}_1)(\mathbf{x} - \mathbf{m}_1)^T \mid C_k]\mathbf{m}_j \\ &= \sigma^2\mathbf{m}_i^T\mathbf{m}_j, \qquad i,j = 1, 2, \ldots, M \\ & \qquad\qquad\qquad i \neq j \end{aligned} \tag{9.120}$$

A simplification occurs if the mean vectors are pairwise orthonormal or orthogonal and the a priori probabilities are equal as will be seen in the following example.

EXAMPLE 9.15

Find the probability of error for the special case 5, where $\mathbf{K} = \sigma^2\mathbf{I}$, $P(C_i)$ are equal, and the conditional mean vectors are orthonormal which can be expressed as

$$\begin{aligned} &\mathbf{m}_i^T\mathbf{m}_j = 0, \qquad i \neq j \\ &\mathbf{m}_i^T\mathbf{m}_i = 1, \qquad i = j \\ &\text{for all } i, j = 1, 2, \ldots, M \end{aligned} \tag{9.121}$$

SOLUTION

Using the a priori probabilities and orthogonality property specified, we see that the test statistics $L_i'(\mathbf{x})$ from Eq. (9.118) have the same last two terms, and thus the optimum test can be reduced to a set of test statistics, by subtracting them off. This gives a new set of statistics $R_i(\mathbf{x})$ defined by

$$R_i(\mathbf{x}) = \mathbf{m}_i^T\mathbf{x}, \qquad i = 1, 2, \ldots, M \tag{9.122}$$

Under class C_1 the conditional mean of $R_i(\mathbf{x})$ becomes

$$E[R_i(\mathbf{x})\,|\,C_1] = E[\mathbf{m}_i^T\mathbf{x}\,|\,C_1] = \mathbf{m}_i^T\mathbf{m}_1 = \delta_{i1} = \begin{cases} 1, & i=1 \\ 0, & i \neq 1 \end{cases} \tag{9.123}$$

and the covariance terms conditioned on C_1 are seen, from (9.120), to be

$$\begin{aligned} q_{ij}^{(1)} &\triangleq E[(R_i(\mathbf{x}) - E[R_i(\mathbf{x})\,|\,C_1])(R_j(\mathbf{x}) - E[R_j(\mathbf{x})\,|\,C_1])\,|\,C_1] \\ &= \sigma^2\mathbf{m}_i^T\mathbf{m}_j = \delta_{ij} \end{aligned} \tag{9.124}$$

Thus for all $i \neq j$, $q_{ij}^{(1)} = 0$, which means that the $R_i(\mathbf{x})$ are uncorrelated, and since they are Gaussian, they are independent. Similar to the development above it is easy to show that under the other classes $q_{ij}^{(k)} = 0$ as well, and thus that the $R_i(\mathbf{x})$ are conditionally independent.

One part of the total probability of error is the probability of error conditioned on class C_1, which is given by

$$\begin{aligned} P(\text{error}\,|\,C_1) &= 1 - P(\text{correct}\,|\,C_1) \\ &= 1 - \iint \cdots \int_{\substack{r_1>r_2 \\ r_1>r_3 \\ \cdots \\ r_1>r_n}} p(r_1, r_2, r_3, \ldots, r_n)\,dr_1\,dr_2\,dr_3 \ldots dr_n \end{aligned} \tag{9.125}$$

Similar formulas exist for each $P(\text{error}\,|\,C_k)$ with the region and class conditional density changed. Since the random variables are conditionally independent, the joint density conditioned on C_1 is just the product of the conditional marginal densities, and since the densities are identical, the conditional error can be simplified as follows:

$$\begin{aligned} P(\text{error}\,|\,C_1) &= 1 - \iint \cdots \int_{\substack{r_1>r_2 \\ r_1>r_3 \\ \cdots \\ r_1>r_n}} p(r_1\,|\,C_1)p(r_2\,|\,C_1)\ldots, p(r_M\,|\,C_1)\,dr_1\,dr_2 \ldots dr_M \\ &= 1 - \int_{-\infty}^{\infty} p(r_1\,|\,C_1) \prod_{i=2}^{M} \left(\int_{-\infty}^{r_1} p(r_i\,|\,C_1)\,dr_i \right) dr_1 \end{aligned} \tag{9.126}$$

From (9.123) and (9.124) we have conditioned on C_1 that $R_1 \sim N(1, \sigma^2)$ and for all other $i \neq 1$ that $R_i \sim N(0, \sigma^2)$. Each of the integrals in the product are identical and can be written in terms of the $\Phi(.)$ function. However, no analytical form exists, and this allows the conditional error to be more conveniently written as

$$\begin{aligned} P(\text{error}\,|\,C_1) &= 1 - \int_{-\infty}^{\infty} \frac{1}{\sqrt{2\pi}\sigma} \exp\left\{ -\frac{(r_1-1)^2}{2\sigma^2} \right\} \Phi^{M-1}\left(\frac{r_1}{\sigma}\right) dr_1 \\ &\text{where} \quad \Phi(x) = \int_{-\infty}^{x} \frac{1}{\sqrt{2\pi}} \exp\left(-\frac{x^2}{2}\right) dx \end{aligned} \tag{9.127}$$

The conditional error probabilities, $P(\text{error} \mid C_k)$, for the other $C_k, k = 2, 3, \ldots, M$ by a similar development used for C_1, give the same result as (9.127). Therefore the total probability of error using the total probability theorem and the fact that $P(C_i) = 1/M$ is the same as the individual conditional errors, that is,

$$P(\text{error}) = 1 - \int_{-\infty}^{\infty} \frac{1}{\sqrt{2\pi}\sigma} \exp\left\{ -\frac{(r_1 - 1)^2}{2\sigma^2} \right\} \Phi^{M-1}\left(\frac{r_1}{\sigma}\right) dr_1 \tag{9.128}$$

This error can be determined numerically once σ and M are given. ■

9.9 COMPOSITE HYPOTHESES

Suppose that we must decide which of two hypotheses that an observation vector **x** comes from. In the previous sections we explored this problem where the known information about the two classes or hypotheses consists of the class conditional probability density function, the a priori probabilities of each class, and a specified performance measure. In certain problems this information may not be entirely specified but given as a function of a parameter or parameter vector that is unknown to us. A hypothesis is called a **composite hypothesis** if it is a function of some unknown parameter or parameter vector. For example, under a hypothesis the observed vector might be sinusoidal with unknown frequency and phase.

We will distinguish two main types of composite hypotheses problems: (1) the parameter, or parameter vector, is unknown but random with known probability density function and (2) the parameter or parameter vector is unknown and deterministic. Other types could include those for which the unknown parameter or parameter vectors may be random but with unknown densities or known density forms with unknown deterministic parameters, or the unknown parameter or parameter vector is deterministic but confined to a given range of values. The following discussion concentrates on the first two main types of problems as the others may not have clearly defined avenues to their solutions.

9.9.1 Composite Hypotheses with Random Parameters

For the two-class problems we can formulate the composite hypotheses problem as follows for random vectors **a** and **b**:

$$\begin{aligned} C_1&: \mathbf{x} \sim p(x \mid \mathbf{a}, C_1) \quad \text{where } p_A(\mathbf{a}) \text{ is known} \\ C_2&: \mathbf{x} \sim p(x \mid \mathbf{b}, C_2) \quad \text{where } p_B(\mathbf{b}) \text{ is known} \end{aligned} \tag{9.129}$$

It will be assumed that $P(C_1)$ and $P(C_2)$ are known. Our task is to find the optimum decision rule where optimality is in the minimum probability of error sense or in the Bayes minimum average cost sense if the various costs are additionally provided.

Notice that for this composite hypotheses case that the class conditional probability densities, not a function of the parameters, can be obtained as follows:

$$\begin{aligned} p(\mathbf{x}\,|\,C_1) &= \int_{-\infty}^{\infty}\int_{-\infty}^{\infty}\cdots\int_{-\infty}^{\infty} p(\mathbf{x}\,|\,\mathbf{a},\,C_1)p_A(\mathbf{a})\,d\mathbf{a} \\ p(\mathbf{x}\,|\,C_2) &= \int_{-\infty}^{\infty}\int_{-\infty}^{\infty}\cdots\int_{-\infty}^{\infty} p(\mathbf{x}\,|\,\mathbf{b},\,C_2)p_B(\mathbf{b})\,d\mathbf{b} \end{aligned} \tag{9.130}$$

Thus the problem can be treated as a simple hypothesis-testing problem for which the solution is known as a likelihood ratio test given by

$$\text{If} \quad l(\mathbf{x}) = \frac{p(\mathbf{x}\,|\,C_1)}{p(\mathbf{x}\,|\,C_2)} = \frac{\int_{-\infty}^{\infty}\int_{-\infty}^{\infty}\cdots\int_{-\infty}^{\infty} p(\mathbf{x}\,|\,\mathbf{a},\,C_1)p_A(\mathbf{a})\,d\mathbf{a}}{\int_{-\infty}^{\infty}\int_{-\infty}^{\infty}\cdots\int_{-\infty}^{\infty} p(\mathbf{x}\,|\,\mathbf{b},\,C_2)p_B(\mathbf{b})\,d\mathbf{b}} \begin{array}{ll} > T & \text{decide } C_1 \\ < & \text{decide } C_2 \\ = & \text{decide randomly} \end{array} \tag{9.131}$$

and the T is determined by whichever performance measure is used.

9.9.2 Composite Hypotheses with Deterministic Parameters

The composite hypotheses problem with deterministic unknown parameter or pattern vector can be formulated as follows

$$\begin{aligned} C_1&: \mathbf{x} \sim p(\mathbf{x};\,\mathbf{a}\,|\,C_1) \\ C_2&: \mathbf{x} \sim p(\mathbf{x};\,\mathbf{b}\,|\,C_2) \end{aligned} \tag{9.132}$$

where $p(\mathbf{x};\,\mathbf{a}\,|\,C_1)$ and $p(\mathbf{x};\,\mathbf{b}\,|\,C_2)$ are the known conditional probability density functions in terms of the unknown parameter vectors **a** and **b**, respectively. It is also assumed that the a priori probabilities $p(C_1)$ and $p(C_2)$ are known and that we would like to find the optimum decision rule where optimality is in the minimum probability of error sense or in the Bayes sense if costs are given.

For this composite hypotheses case we are not able to obtain the conditional densities by integrating as we did before for the case of unknown random parameter vectors, since the **a** and **b** are not random. The general approach for solving this type of problem is not clear-cut. We could just solve the problem using the parameters as deterministic in terms of a likelihood ratio test and hope that the test including threshold could be determined without the parameters. Impossible as this may sound, there are cases where this happens. Consult Van Trees [1] for discussions of such uniformly most powerful (UMP) tests. In most problems this will not be true, and we must look for alternative methods for solving the composite hypotheses problem with unknown deterministic parameter vectors. The most popular approach to this type of problem is the generalized likelihood test, which will now be presented.

Basically the generalized likelihood ratio approach for solving the composite hypotheses problem is to estimate the parameter vector by using the maximum likelihood estimation procedure for estimating nonrandom parameters and then to substitute these estimates, as if known, into the likelihood ratio test. Thus the decision rule called the **generalized likelihood ratio test** is given by

$$\text{If} \quad l_g(\mathbf{x}) = \frac{p(\mathbf{x};\, \hat{\mathbf{a}}_{ml}(\mathbf{x})\,|\,C_1)}{p(\mathbf{x};\, \hat{\mathbf{b}}_{ml}(\mathbf{x})\,|\,C_2)} \begin{array}{ll} > T & \text{decide } C_1 \\ < & \text{decide } C_2 \\ = & \text{decide randomly} \end{array} \tag{9.133}$$

where $\hat{\mathbf{a}}_{ml}(\mathbf{x})$ is the maximum likelihood estimate of $\mathbf{a}$ under C_1 and $\hat{\mathbf{b}}_{ml}(\mathbf{x})$ is the maximum likelihood estimate of $\mathbf{b}$ under C_2. The threshold T is a determined from the a priori probabilities if using the minimum probability of error performance or from the a priori costs and a priori probabilities if using a Bayes performance measure. The maximum likelihood estimators under each class are given by

$$\begin{aligned} \hat{\mathbf{a}}_{ml}(\mathbf{x}) &= \max_{\mathbf{a}} p(\mathbf{x};\ \mathbf{a}\,|\,C_1) \\ \hat{\mathbf{b}}_{ml}(\mathbf{x}) &= \max_{\mathbf{b}} p(\mathbf{x};\ \mathbf{b}\,|\,C_2) \end{aligned} \tag{9.134}$$

EXAMPLE 9.16

We are given the following composite hypotheses testing problem with scalar observations and scalar parameters:

$$\begin{aligned} C_1&: \quad x \sim p(x\,|\,a, C_1) \quad \text{where } p_A(a) \text{ is known} \\ C_2&: \quad x \sim p(x\,|\,b, C_2) \quad \text{where } p_B(b) \text{ is known} \end{aligned} \tag{9.135}$$

The conditional densities under C_1 and C_2 in (9.135) are known to be Gaussian, each with known variances of σ^2 and unknown random means of A and B, respectively. The means A and B are Gaussian with means of 1 and -1 and equal variances of 1. Thus the given probability density functions $p(x\,|\,a, C_2)$, $p(x\,|\,a, C_2)$, $p_A(a)$, and $p_B(b)$ are as follows:

$$p(x\,|\,a, C_1) = \frac{1}{\sqrt{2\pi}\sigma} \exp\left\{-\frac{(x-a)^2}{2\sigma^2}\right\}, \quad p_A(a) = \frac{1}{\sqrt{2\pi}} \exp\left\{-\frac{(a-1)^2}{2}\right\}$$

$$p(x\,|\,b, C_2) = \frac{1}{\sqrt{2\pi}\sigma} \exp\left\{-\frac{(x-b)^2}{2\sigma^2}\right\}, \quad p_B(b) = \frac{1}{\sqrt{2\pi}} \exp\left\{-\frac{(b+1)^2}{2}\right\}$$

The hypotheses are composite since the conditional densities are functions of the unknown random parameters. If the a priori probabilities of each hypothesis are equal, find the optimum test in the minimum probability of error sense for distinguishing which class, C_1 or C_2, a single observation x is from.

SOLUTION

Since the unknown means are random, the optimum test is a likelihood ratio test given by (9.131) where $T = P(C_2)/P(C_1) = 1$. From (9.130) we have the $p(x \mid C_i)$ given as follows:

$$p(\mathbf{x} \mid C_1) = \int_{-\infty}^{\infty} \frac{1}{\sqrt{2\pi}\sigma} \exp\left\{-\frac{(x-a)^2}{2\sigma^2}\right\} \frac{1}{\sqrt{2\pi}} \exp\left\{-\frac{(a-1)^2}{2}\right\} da$$

$$p(\mathbf{x} \mid C_2) = \int_{-\infty}^{\infty} \frac{1}{\sqrt{2\pi}\sigma} \exp\left\{-\frac{(x-b)^2}{2\sigma^2}\right\} \frac{1}{\sqrt{2\pi}} \exp\left\{-\frac{(b+1)^2}{2}\right\} da$$

The first integral can be evaluated by rewriting the integrand in terms of a Gaussian density, thus giving an integral of 1 with the other factors determined by the rewriting. First adding the exponents and then rearranging, we have the following:

$$p(\mathbf{x} \mid C_1) = \frac{1}{\sqrt{2\pi}\sigma} \int_{-\infty}^{\infty} \frac{1}{\sqrt{2\pi}} \exp\left\{-\frac{(x-a)^2 - \sigma^2(a-1)^2}{2\sigma^2}\right\} da$$

The term in { } can be rewritten to be a function of a as

$$p(\mathbf{x} \mid C_1) = \frac{1}{\sqrt{2\pi}\sigma} \exp\left\{-\frac{(x^2+\sigma^2)}{2\sigma^2}\right\}$$
$$\times \int_{-\infty}^{\infty} \frac{1}{\sqrt{2\pi}} \exp\left\{\frac{-(a^2 - 2a(x+\sigma^2)/(1+\sigma^2))}{2\sigma^2/(1+\sigma^2)}\right\} da$$

By completing the square in the { } the integral becomes 1 times $\sigma/\sqrt{(1+\sigma^2)}$, and we are left with

$$p(\mathbf{x} \mid C_1) = \frac{1}{\sqrt{2\pi}\sigma} \frac{\sigma}{\sqrt{1+\sigma^2}} \exp\left\{-\frac{(x^2+\sigma^2)^2}{2\sigma^2}\right\} \exp\left\{\frac{-(x+\sigma^2)}{2\sigma^2(1+\sigma^2)}\right\}$$

Adding the terms in the exponent and simplifying finally yields

$$p(\mathbf{x} \mid C_1) = \frac{1}{\sqrt{2\pi(1+\sigma^2)}} \exp\left\{-\frac{(x-1)^2}{2(1+\sigma^2)}\right\}$$

Thus it is seen that the observation x under C_1 is normally distributed with mean of 1 and a variance of $(1+\sigma^2)$. Similarly the integral for the conditional density under C_2 can be shown to be

$$p(\mathbf{x} \mid C_2) = \frac{1}{\sqrt{2\pi(1+\sigma^2)}} \exp\left\{-\frac{(x+1)^2}{2(1+\sigma^2)}\right\}$$

Finally, from (9.131) the likelihood ratio test can be written as

$$\text{If} \quad l(x) = \frac{p(\mathbf{x}\,|\,C_1)}{p(\mathbf{x}\,|\,C_2)} = \frac{[1/\sqrt{2\pi(1+\sigma^2)}]\exp\{-(x-1)^2/2(1+\sigma^2)\}}{[1/\sqrt{2\pi(1+\sigma^2)}]\exp\{-(x+1)^2/2(1+\sigma^2)\}}$$

$$> \frac{P(C_2)}{P(C_1)} \quad \text{decide } C_1$$

Simplifying and taking the ln of both sides the test very nicely reduces to

$$\begin{aligned} \text{If} \quad x &> \tfrac{1}{2}(1+\sigma^2)\ln\left(\frac{P(C_2)}{P(C_1)}\right) && \text{decide } C_1 \\ &< && \text{decide } C_2 \\ &= && \text{decide randomly} \end{aligned}$$

■

EXAMPLE 9.17

Suppose that an observation under C_1 is Gaussian with an unknown deterministic mean $a > 0$ and known variance σ^2, while under C_2 it is Gaussian with unknown mean $b < 0$ and known variance σ^2. Assume that the a priori probabilities are known as $P(C_1) = \frac{1}{3}$ and $P(C_2) = \frac{2}{3}$ and $\sigma^2 = 1$. Find the optimum decision rule where optimality is in the minimum probability of error sense.

SOLUTION

This is a composite hypothesis testing problem, since the densities are not totally known but are functions of the parameters a and b. Since the unknown parameters are deterministic, one way to solve the problem is to use the generalized likelihood ratio test given in (9.133) where maximum likelihood estimates of the parameters are found (9.134) and substituted into the likelihood ratio test as if known.

The maximum likelihood estimator for the parameter a under C_1 is the value of a that maximizes $p(x; a\,|\,C_1)$ which is thought as a function of a as follows:

$$p(x; a\,|\,C_1) = \frac{1}{\sqrt{2\pi}\sigma}\exp\left\{-\frac{(a-x)^2}{2\sigma^2}\right\}$$

For $x > 0$ the maximum will occur at x, whereas if $x < 0$, the maximum will occur at 0 since $a \geq 0$. Therefore we can write

$$\hat{a}_{\text{ml}}(x) = \begin{cases} 0, & x < 0 \\ x, & x \geq 0 \end{cases}$$

Similarly the maximum likelihood estimator for the parameter b under C_2 is the value of b that maximizes $p(x; b \mid C_2)$ which is thought as a function of b as follows:

$$p(x; b \mid C_2) = \frac{1}{\sqrt{2\pi}\sigma} \exp\left\{ -\frac{(b-x)^2}{2\sigma^2} \right\}$$

For $x \geq 0$ the maximum will occur at 0, since $b \leq 0$, whereas if $x < 0$, the maximum will occur at x. Therefore we can write

$$\hat{b}_{\text{ml}}(x) = \begin{cases} x, & x < 0 \\ 0, & x \geq 0 \end{cases}$$

The generalized likelihood ratio test given in (9.133), using the maximum likelihood estimates gives us

$$\text{If} \quad l_g(x) = \exp\left\{ \frac{(\hat{a}_{\text{ml}} - \hat{b}_{\text{ml}})x}{\sigma^2} + \frac{(\hat{b}_{\text{ml}}^2 - \hat{a}_{\text{ml}}^2)}{2\sigma^2} \right\} \begin{matrix} > \\ < \\ = \end{matrix} \frac{P(C_2)}{P(C_1)} \quad \begin{matrix} \text{decide } C_1 \\ \text{decide } C_2 \\ \text{decide randomly} \end{matrix}$$

Substituting $\hat{a}_{\text{ml}}$ and $\hat{b}_{\text{ml}}$ determined above $\sigma^2 = 1$, $P(C_1) = \frac{1}{3}$, and $P(C_2) = \frac{2}{3}$ into $l_g(x)$ and simplifying, the decision reduces the rule to the following:

$$\text{If} \quad x \begin{matrix} > \\ < \\ = \end{matrix} 1.17741 \quad \begin{matrix} \text{decide } C_1 \\ \text{decide } C_2 \\ \text{decide randomly} \end{matrix}$$

■

9.10 SUMMARY

In this chapter optimum decision rules were determined using maximum a posteriori, minimum probability of error, and Bayes risk performance measures for the two-class and multi-class cases. Also presented was the Neyman-Pearson decision rule used in classical radar problems and statistics. These rules could be interpreted in a number of different spaces, in particular, the original sample space, the likelihood ratio space, and other sufficient statistic spaces. Methods for obtaining the performance for the decision rules in terms of the error probabilities were presented in all of these spaces, and the choice of space depended on the particular problem being solved. The original observation or sample space was shown to be usually N dimensional, whereas the sufficient statistic was usually of a lower dimensional space, and the likelihood ratio space could be as low as one dimensional for the two-class case and $M-1$ dimension for the M-class case.

It is important to emphasize that for the two class case the **likelihood ratio test** was optimum for MAP, MPE, Bayes, and Neyman-Pearson criteria. That this is true is quite remarkable! It has been shown by other researchers that this is true for even a wider class of performance measures than those discussed here, in particular, the maximin criterion and others. For the M-class case it was shown that the optimum decision rules under MAP, MPE, and Bayes criteria can be expressed in terms of $M - 1$ likelihood ratios.

Decision rules were also presented for several different special cases that allowed simplification of the decision rule. In particular, the two-class and M-class cases for observations that had Gaussian statistics were presented. These special case solutions provide convenient starting places for solutions of problems falling in those categories, relieving us of the details of the general solution. The correlator was shown to play a significant role in the optimum solutions, and various forms were discussed. Also the determination of performance in terms of probability of error for the special Gaussian cases was presented.

The choice of the performance measure to use for a given problem is dependent on a number of important assumptions. It is not that any of the performance measures is better than the other, but that they involve different rules based on different conditions and assumptions. In general, if errors were subjectively determined to be equal and the a priori probabilities are known, the MPE decision rule is convenient and meaningful. Whereas, if the errors are of different costs, the Bayes rule seems appropriate. These costs must again be based on subjective evaluation and cannot be analytically determined.

For the two-class case when reasonable costs cannot be assigned and a priori probabilities are unable to be assigned, as in radar and sonar problems, the Neyman-Pearson rule was shown to be appropriate.

We saw that optimum tests for cases where the class conditional densities are unknown or of a known parametric form, with the parameters either unknown or random with known statistical properties or unknown and deterministic, lead to different forms of decision rules and strategies. These optimum tests were not included in this chapter because of the complex level of background material needed for a useful presentation. Complete presentations are available in a number of references: Van Trees [1], McDonough and Whalen [4], and Helstrom [7].

Not covered in this chapter, because of space limitations are tests that are independent of the distributions and densities (distributional free); these tests have been developed by sound mathematical statistical methods as well as ad hoc methods. The interested reader should consult Srinath, Rajasekaran, and Viswanathan [5] for an introduction to this important area. Also not covered were optimal sequential tests and data fusion techniques. For sequential tests, consult Wald [9] or Wetherill and Glazebrook [10], and for data fusion or distributed decisions, look at Teeney and Sandell [11] or Therrien [12].

Classical decision theory and hypotheses testing provide useful methods for solving many classes of problems. Their use and application depend inevitably on the assumptions made, many of which turn out to be subjective assignments. The success of the application of these methods hinges on these assignments, and there is no substitute for good engineering and statistical assumptions.

PROBLEMS

9.1 Let the observed X from class C_1 be Gaussian with a mean of -1 and a variance of 1 and X from class C_2 be Gaussian with a mean of $+1$ and a variance of 4.

(a) If the classes are equally probable, compute the a posteriori probabilities of each class for the observed value of $x = 1$. Using the maximum a posteriori decision rule, which class would $x = 1$ be assigned to?

(b) Find the optimum decision rule, in the maximum a posteriori probability sense, for classifying an observed value of x, when the a priori probabilities of class 1 and 2 are equal. (Show decision regions in the observation space.)

(c) Find the optimum decision rule, in the maximum a posteriori sense, for classifying an observed value x when the a priori probabilities of class 1 and 2 are 0.7 and 0.3, respectively, and show the decision region in the observation space.

9.2 Let the observed pattern vector $\mathbf{X} = [X_1, X_2]^T$ have the following class conditional densities: Under class C_1, $\mathbf{X}$ is normally distributed with mean vector $[0\ \ 0]^T$ and covariance matrix $\mathbf{K}_1 = \mathbf{I}$, where $\mathbf{I}$ is a 2×2 identity matrix and under class C_2, $\mathbf{X}$ is normally distributed with mean vector $[1\ \ 1]^T$ and covariance matrix $\mathbf{K}_2 = 4\mathbf{I}$.

(a) For the observed value $\mathbf{x} = [1\ \ 0]^T$, compute the a posteriori probabilities of each class.

(b) Find the optimum decision rule, in the maximum a posteriori sense, in the two-dimensional pattern space when the a priori probabilities of the two classes are equal.

(c) Repeat (a) and (b) for the case where the a priori probabilities of the two classes are 0.4 and 0.6 for class C_1 and class C_2, respectively.

9.3 Let the observed X under each of two classes C_1 and C_2 be from the finite mixture densities given below:

$$p(x\,|\,C_1) = \frac{0.5\exp(-x^2/2) + 0.25\exp(-(x-2)^2/8)}{\sqrt{2\pi}}$$

$$p(x\,|\,C_2) = \frac{0.25\exp(-x^2/8) + 0.5\exp(-(x-2)^2/2)}{\sqrt{2\pi}}$$

(a) Find the likelihood ratio for the maximum a posteriori decision rule when the a priori probabilities of the two classes are equal, and plot it for all x.

(b) Determine the threshold for your test, and from the plot in (a) show the final decision rule in the observation space for all x.

(c) Repeat (a) and (b) for the case that $P(C_1) = 0.3$ and $P(C_2) = 0.7$.

(d) Compare your results determined in (a) and (b) with those determined in (c), and tell the affect that the a priori probabilities have on the decision rule.

9.4 For a particular two-pattern case the likelihood ratio test is

$$\text{If} \quad l(x) \begin{cases} > T & \text{decide } C_1 \\ < T & \text{decide } C_2 \\ = T & \text{decide } C_1 \text{ or } C_2 \text{ randomly} \end{cases}$$

The ln of the likelihood ratio is shown in Figure P9.4 where x is the measurement. Show the decision regions in the observation space x for the likelihood ratio test with thresholds of $\ln T$ as follows: **(a)** $\ln T = 5$, **(b)** $\ln T = 3$, **(c)** $\ln T = 1$, **(d)** $\ln T = 0$, **(e)** $\ln T = -1$, **(f)** $\ln T = -4$.

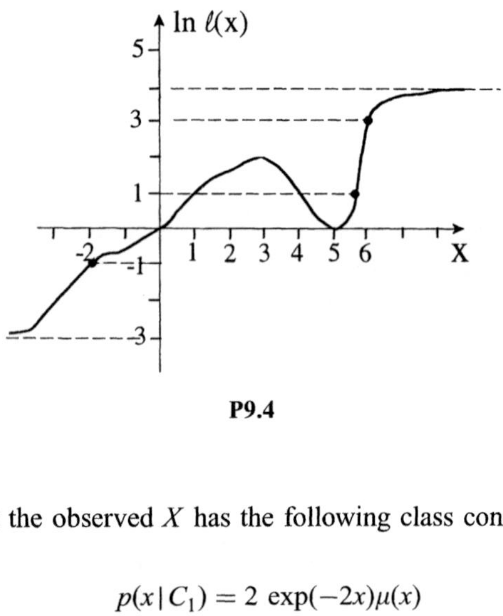

P9.4

9.5 Assume that the observed X has the following class conditional densities:

$$p(x \mid C_1) = 2\exp(-2x)\mu(x)$$
$$p(x \mid C_2) = \exp(-x)\mu(x)$$
$$p(x \mid C_3) = 0.5\exp(-0.5x)\mu(x)$$

Assume further that the probabilities of each class are equal.

(a) Calculate the a posteriori probabilities for each of the three classes for the observed value $x = 0.5$, and determine the class from the observation using the MAP decision rule.

(b) Find the maximum a posteriori decision rule in the observation space.

9.6. The observed pattern value X could belong to one of two classes C_1 or C_2 with class conditional probability densities $p(x \mid C_1)$ and $p(x \mid C_2)$ as follows:

$$p(x \mid C_1) = x \exp\left(-\frac{x^2}{2}\right)\mu(x)$$

$$p(x \mid C_2) = \frac{x}{4} \exp\left(-\frac{x^2}{8}\right)\mu(x)$$

Assume that the two classes are equally probable.

(a) Specify the optimal decision rule in the likelihood ratio space, where optimality is in terms of minimizing the probability of error.

(b) Simplify your decision rule to give the decision regions in the observation space.

(c) Calculate the probability of deciding class C_1 given that the observation is from class C_2 for the decision rule determined in (b).

(d) Calculate the probability of deciding class C_2 given that the observation is from class C_1 for the decision rule determined in (b).

(e) Calculate the total probability of error for the decision rule determined in (b).

(f) Calculate the probability of deciding class C_1 given that the observation is from class C_1 for the decision rule determined in (b).

(g) Calculate the probability of deciding class C_2 given that the observation is from class C_2 for the decision rule determined in (b).

9.7 Use the a priori class probabilities and the class conditional densities from Problem 9.6 for each of two independent observations X_1 and X_2.

(a) Find the optimum classifier in the likelihood ratio space where optimality is in the sense of minimizing the total probability of error.

(b) Express the decision rule determined in (a) in the observation space (x_1, x_2).

(c) Give the integrals necessary and calculate the conditional error probabilities.

(d) Evaluate the total probability of error for this decision rule.

9.8 Under class C_1 the observed pattern value X is normally distributed with mean of zero and variance of 1, while under class C_2, X is normally distributed with a mean of zero and a variance of 4. Assume that $P(C_1) = \frac{3}{8}$ and $P(C_2) = \frac{5}{8}$.

(a) Find the optimum decision rule in the sense of minimizing the total probability of error in terms of the likelihood ratio $l(x)$.

(b) Find the probability density function for L defined by $L = l(X)$ conditioned on the fact the observed X is from class C_1.

(c) For the decision rule obtained in (a) and using the conditional density in (b), find the probability of deciding class C_2 given that the observation is from class C_1.

(d) Find the probability density function for L defined by $L = l(X)$ conditioned on the fact the observed X is from class C_2.

(e) For the decision rule obtained in (a) and using the conditional density in (d) find the probability of deciding class C_1 given that the observation is from class C_2.

9.9 The observation value for X for a two class problem has the following class conditional probability density functions:

$$p(x \mid C_1) = \exp(-x)\mu(x)$$
$$p(x \mid C_2) = 0.5\exp(-\mid x - 1 \mid)$$

Assume that the a priori probabilities of the classes are given by $P(C_1) = \frac{3}{8}$ and $P(C_2) = \frac{5}{8}$.

(a) Find the optimum decision rule using the minimum probability of error criteria. Reduce the decision rule to the observation space.

(b) Calculate the probability of error for your decision rule.

(c) Suppose that three independent observations were obtained of x with the densities above for each class given. Repeat (a) for this new three-dimensional pattern space.

9.10. The observation value for X for a two-class problem has the following class conditional probability density functions:

$$p(x \mid C_1) = 0.5[\mu(x + 1) - \mu(x - 1)]$$
$$p(x \mid C_2) = 2\exp(-2\mid x \mid)$$

where $\mu(x)$ is the unit-step function

Assume that the a priori probabilities of the classes are given by $P(C_1) = 0.4$ and $P(C_2) = 0.6$.

(a) Find the optimum decision rule using the minimum probability of error criterion. Reduce the decision rule to the observation space.

(b) Calculate the probability of deciding class C_2 given that the true class is C_1 for your decision rule.

(c) Calculate the a posteriori probability of class C_1 given the observed value $x = 0$.

(d) Find the probability density function for the likelihood ratio conditioned on class C_1.

9.11 The observed pattern value X could belong to one of two classes C_1 and C_2 with class conditional probability densities $p(x \mid C_1)$ and $p(x \mid C_2)$ as follows:

$$p(x \mid C_1) = \frac{1}{\sqrt{2\pi}} \exp\left(-\frac{x^2}{2}\right)$$

$$p(x \mid C_2) = \frac{1}{2} \frac{1}{\sqrt{2\pi}} \exp\left(-\frac{(x+1)^2}{2}\right) + \frac{1}{2} \frac{1}{\sqrt{2\pi}} \exp\left(-\frac{(x-1)^2}{2}\right)$$

Assume that the two classes are equally probable.

(a) Specify the optimal decision rule in the likelihood ratio space, where optimality is in terms of minimizing the probability of error.

(b) Simplify your decision rule to give the decision regions in the observation space.

(c) Calculate the probability of deciding class C_1 given that the observation is from class C_2 for the decision rule determined in (b).

(d) Calculate the probability of deciding class C_2 given that the observation is from class C_1 for the decision rule determined in (b).

(e) Calculate the total probability of error for the decision rule determined in (b).

9.12 Repeat Problem 9.11 if the class a priori probabilities are $P(C_1) = \frac{3}{4}$, $P(C_2) = \frac{1}{4}$, and the conditional densities are the following finite mixtures:

$$p(x \mid C_1) = \left(\tfrac{3}{5}\right) 2 \exp(-2x)\mu(x) + \left(\tfrac{2}{5}\right) 3 \exp(-3x)\mu(x)$$
$$p(x \mid C_2) = \left(\tfrac{2}{3}\right) \exp(-x)\mu(x) + \left(\tfrac{1}{3}\right) 2 \exp(-2x)\mu(x)$$

9.13 Repeat Problem 9.11 if the class a priori probabilities are $P(C_1) = \frac{2}{3}$, $P(C_2) = \frac{1}{3}$, and the conditional densities are the following finite mixtures:

$$p(x \mid C_1) = \left(\tfrac{3}{5}\right) 2 \exp(-2x)\mu(x) + \left(\tfrac{2}{5}\right) 3 \exp(-3x)\mu(x)$$
$$p(x \mid C_2) = \exp(-x)\mu(x)$$

9.14 The observation X for a two-class problem has the following class conditional class probability densities:

$$p(x \mid C_1) = \exp(-x)\mu(x)$$
$$p(x \mid C_2) = \tfrac{1}{2}\exp(-\lvert x - 1 \rvert)$$

Assume that the a priori probabilities of the classes are given by $P(C_1) = \frac{3}{8}$ and $P(C_2) = \frac{5}{8}$, similar to Problem 9.9, but with the additional cost assignments as follows:

$$c_{11} = 0.1, \quad c_{12} = 2.1, \quad c_{21} = 1.1, \quad c_{22} = 0.1$$

(a) Find the Bayes decision rule for deciding between the two classes.

(b) Calculate the Bayes risk for the decision rule of (a).

(c) Repeat part (a) for the case of two independent observations.

9.15 In a two-class problem the N observed samples x_i constituting the pattern vector are from independent and identically distributed random variables with the following class conditional densities:

$$p(x_i \mid C_j) = \frac{1}{\sqrt{2\pi}\sigma_j} \exp\left\{ -\frac{(x_i - m_j)^2}{2\sigma_j^2} \right\} \quad j = 1, 2$$

$$i = 1, 2, \ldots, N$$

$$\text{where} \quad \sigma_1^2 = 1, \sigma_2^2 = 2, m_1 = -1, \text{ and } m_2 = 1$$

Assume that $P(C_1) = P(C_2) = \frac{1}{2}$.

(a) Find the optimum test in the minimum probability of error sense.

(b) Express your results in terms of s_1 and s_2 defined below:

$$s_1 = \frac{1}{N} \sum_{i=1}^{N} x_i \quad \text{and} \quad s_2 = \frac{1}{N} \sum_{i=1}^{N} x_i^2$$

and determine the decision regions in the s_1-s_2 space. For this problem s_1 and s_2 are sufficient statistics to make the decision as they are all that is necessary to make the optimum decision and knowing the observation vector explicitly is not necessary.

9.16 The observed pattern value X could belong to one of two classes C_1 (target) and C_2 (no target) with class conditional probability densities $p(x \mid C_1)$ and $p(x \mid C_2)$ as follows:

$$p(x \mid C_1) = x \exp(-\tfrac{1}{2} x^2)\mu(x)$$
$$p(x \mid C_2) = \tfrac{1}{4} x \exp(-\tfrac{1}{8} x^2)\mu(x)$$

(a) Specify the optimal decision rule in the Neyman-Pearson sense.

(b) Simplify your decision rule to give the decision regions in the observation space for a false alarm rate of 0.1.

(c) Calculate the probability of a detection for the decision rule in (b) and specify the operating point for your decision rule of (b).

(d) For the decision rule found in (a), determine the receiver operating characteristic P_D vs P_{FA}.

9.17 We have two pattern classes described as follows:

$$\text{Class } C_1: \quad X = A + N$$
$$\text{Class } C_2: \quad X = B + N$$

Assume that A and B are each independent of N. The probability density functions for each of the random variables are

$$p(a) = \exp(-a)\mu(a)$$
$$p(n) = \exp(-n)\mu(n)$$
$$p(b) = 2\exp(-2b)\mu(b)$$

Furthermore assume a zero-one cost assignment and equally probable classes, and find the optimum decision rule where optimality is in the Bayes sense. Reduce your decision rule to the form: If $g(x) > 0$, decide class C_1; if $g(x) < 0$, decide class C_2; and if $g(x) = 0$, decide randomly.

9.18 The observation value for X for a two-class problem has the following class conditional probability density functions:

$$p(x\,|\,C_1) = \frac{1}{\sqrt{2\pi}}\exp\left(-\frac{(x-1)^2}{2}\right)$$
$$p(x\,|\,C_2) = 0.5[\mu(x-1) - \mu(x-3)]$$

where $\mu(x)$ is the unit-step function

The following decision rule, **not necessarily optimal**, is used

If $x \le 2$ decide class C_1; otherwise, decide class C_2

Assume that the a priori probabilities of the classes are $P(C_1) = \frac{1}{3}$ and $P(C_2) = \frac{2}{3}$.

(a) Determine the probability of deciding class C_1 given that the true class is C_2 for the given decision rule.

(b) Determine the probability of deciding class C_2 given that the true class is C_2 for the given decision rule.

(c) Find the total probability of error.

9.19 For a two-class problem the observation value for X has the following class conditional probability density functions:

$$p(x\,|\,C_1) = \frac{1}{\pi(1+x^2)}$$
$$p(x\,|\,C_2) = \frac{1}{\pi(1+(x-1)^2)}$$

Assume that the a priori probabilities of the classes are given by $P(C_1) = \frac{1}{3}$ and $P(C_2) = \frac{2}{3}$ and that the Bayes costs are known as $C_{11} = C_{22} = 0$ and $C_{12} = C_{21} = 1$.

(a) Find the optimum decision rule in the Bayes sense.

(b) Reduce the decision rule to the observation space.

(c) Give in precise integral form the probability of deciding class C_2 given that the true class is C_1 for your decision rule.

(d) Calculate the a posteriori probability of class C_1 as a function of x and evaluate for observed value $x = 0$.

9.20 The observation value X for a two-class problem has the following class conditional probability density functions:

$$p(x \mid C_1) = \frac{1}{\pi(1 + x^2)}$$
$$p(x \mid C_2) = \frac{2}{\pi(4 + x^2)}$$

Assume that the a priori probabilities of the classes are given by $P(C_1) = 0.4$ and $P(C_2) = 0.6$.

(a) Find the optimum decision rule using the minimum probability of error criteria. Reduce the decision rule to the observation space.

(b) Calculate the probability of deciding class C_2 given that the true class is C_1 for your decision rule.

(c) Calculate the a posteriori probability of class C_1 given the observed value x and evaluate at $x = 0$.

(d) Find the probability density function for the likelihood ratio conditioned on class C_1.

(e) Give the form of the optimum decision rule in the minimum probability of error sense if N independent observations of X were taken.

9.21 Rework Problem 9.20, except find the optimum decision rule in the Bayes sense where the costs are given by $C_{11} = C_{22} = 0$, $C_{12} = 1.5$, and $C_{21} = 1$.

9.22 Rework Problem 9.20, except find the optimum decision rule in the Bayes sense where the costs are given by $C_{11} = C_{22} = 0$, $C_{12} = 1$, and $C_{21} = 1.5$.

9.23 Consider a three-class problem with the following class conditional density functions:

$$p(x \mid C_1) = \frac{2}{\pi(4 + (x - 1)^2)}$$
$$p(x \mid C_2) = \frac{2}{\pi(4 + x^2)}$$
$$p(x \mid C_3) = \frac{2}{\pi(4 + (x + 1)^2)}$$

With equally likely classes, find the region in the observation space where the minimum probability of error decision rule would decide each of the classes.

9.24 Two pattern classes are described as

$$\begin{aligned} \text{Class } C_1: \quad X_1 &= A + N_1 \\ X_2 &= 2A + N_2 \\ \text{Class } C_2: \quad X_1 &= B + N_1 \\ X_2 &= 2B + N_2 \end{aligned}$$

with $A \sim N(1, 1)$ and $B \sim N(-1, 1)$. Assume that A and B are each independent of N_1 and N_2 and that N_1 and N_2 are statistically characterized as follows:

$$C_1: \quad \begin{bmatrix} N_1 \\ N_2 \end{bmatrix} \sim N\left(\begin{bmatrix} 0 \\ 0 \end{bmatrix}, \begin{bmatrix} 1 & 0 \\ 0 & 2 \end{bmatrix} \right)$$

$$C_2: \quad \begin{bmatrix} N_1 \\ N_2 \end{bmatrix} \sim N\left(\begin{bmatrix} 1 \\ 0 \end{bmatrix}, \begin{bmatrix} 2 & 0 \\ 0 & 4 \end{bmatrix} \right)$$

Assume a zero-one cost assignment and equally probable classes.

(a) Find the optimum detector where optimality is in the Bayes sense. Reduce your decision rule to the observation space.

(b) Find a sufficient statistic for this problem if one exists. If one doesn't exist, say why it doesn't exist.

9.25 Consider a three-class problem with the following class conditional density functions and a priori probabilities:

$$p(x \mid C_1) = \mu(x) - \mu(x - 1)$$
$$p(x \mid C_2) = [\mu(x + 2) - \mu(x - 2)]/4$$
$$p(x \mid C_3) = 2 \exp(-2x)\mu(x)$$

where $\mu(x)$ is the unit-step function and $P(C_1) = \frac{1}{4}$,

$P(C_2) = \dfrac{1}{4}$, and $P(C_3) = \frac{1}{2}$

(a) Find the optimum decision rule in the minimum probability of error sense, and reduce it to the observation space.

(b) Calculate the probability of being correct using your decision rule.

9.26 Two pattern classes are described as follows:

$$C_1: \quad \begin{bmatrix} X_1 \\ X_2 \end{bmatrix} \sim N\left(\begin{bmatrix} 0 \\ 0 \end{bmatrix}, \begin{bmatrix} 1 & 0 \\ 0 & 1 \end{bmatrix} \right)$$

$$C_2: \quad \begin{bmatrix} X_1 \\ X_2 \end{bmatrix} \sim N\left(\begin{bmatrix} 0 \\ 0 \end{bmatrix}, \begin{bmatrix} 1 & 0 \\ 0 & 4 \end{bmatrix} \right)$$

Assume a zero-one cost assignment and equally probable classes.

(a) Find the optimum detector where optimality is in the Bayes sense, and reduce your decision rule to the observation space.

(b) Give a sufficient statistic for this problem. If one doesn't exist, say why it doesn't.

(c) Determine the probability of error for your classifier.

9.27 Assume that the observed pattern vector x has the following class conditional densities:

$$p(x\,|\,C_1) = \tfrac{1}{2}(\mu(x) - \mu(x-2))$$
$$p(x\,|\,C_2) = \exp(-x)\mu(x)$$
$$p(x\,|\,C_3) = \exp(x-1)\mu(-(x-1))$$

Assume further that the probabilities of each class are equal. Find the minimum probability of error decision rule in the observation space.

9.28 We have two pattern classes described as follows (which is an abstraction of a channel diversity problem):

$$\text{Class } C_1\!: \quad \mathbf{X} = [X_1, X_2]^T, \quad X_1 = A + N_1, \quad X_2 = A + N_2$$
$$\text{Class } C_2\!: \quad \mathbf{X} = [X_1, X_2]^T, \quad X_1 = B + N_1, \quad X_2 = B + N_2$$

Assume under C_1 that A, N_1, and N_2 are independent random variables and under C_2 that B, N_1, and N_2 are independent random variables. The probability density functions for each of the random variables are

$$p_A(a) = \exp(-a)\mu(a), \quad p_{N_1}(n_1) = \exp(-n_1)\mu(n_1)$$
$$p_B(b) = 2\exp(-2b)\mu(b), \quad p_{N_1}(n_2) = \exp(-n_2)\mu(n_2)$$

Assume a zero-one cost assignment and equally probable classes.

(a) Find the optimum detector where optimality is in the Bayes sense. Reduce your decision rule to the observation space.

(b) Find a sufficient statistic for this problem if one exists. If not, say why it doesn't exist.

9.29 Assume that the random variable X has the following class conditional densities:

$$p(x\,|\,C_1) = 2\exp(-2x)\mu(x)$$
$$p(x\,|\,C_2) = \exp(-x)\mu(x)$$
$$p(x\,|\,C_3) = \tfrac{1}{2}\exp(-\tfrac{1}{2}x)\mu(x)$$

Assume that two independent observations, X_1 and X_2, are taken to form an observed pattern vector $\mathbf{X}$. Assume further that the probabilities of each class are $P(C_1) = 0.3$, $P(C_2) = 0.3$, and $P(C_3) = 0.4$ and that the relative cost matrix is given by

$$C = \begin{bmatrix} 0 & 1 & 1 \\ 1.1 & 0 & 1.1 \\ 1.2 & 1.2 & 0 \end{bmatrix}$$

(a) Find the optimum test in the Bayes sense, and specify the decision rule in the (x_1, x_2) observation space.

(b) Are there a set of sufficient statistics for this problem? If your answer is yes, describe the statistics.

(c) Calculate the Bayes risk for your decision rule.

9.30 Assume the following three-class, two-observation problem

$$C_1: \quad \mathbf{X} \sim N\left(\begin{bmatrix} 0 \\ 0 \end{bmatrix}, \begin{bmatrix} 1 & 0 \\ 0 & 1 \end{bmatrix}\right)$$

$$C_2: \quad \mathbf{X} \sim N\left(\begin{bmatrix} 1 \\ 0 \end{bmatrix}, \begin{bmatrix} 2 & 0 \\ 0 & 2 \end{bmatrix}\right)$$

$$C_3: \quad \mathbf{X} \sim N\left(\begin{bmatrix} 0 \\ 1 \end{bmatrix}, \begin{bmatrix} 3 & 0 \\ 0 & 3 \end{bmatrix}\right)$$

With equally likely classes, find the region in the observation space where the minimum probability of error decision rule would decide class C_1.

9.31 We have two pattern classes described as follows:

$$\text{Class } C_1: \quad \mathbf{X} = [X_1, X_2, X_3]^T, \quad X_1 = A + N_1, \quad X_2 = 2A + N_2, \quad X_3 = 3A + N_3$$

$$\text{Class } C_2: \quad \mathbf{X} = [X_1, X_2, X_3]^T, \quad X_1 = B + N_1, \quad X_2 = 2B + N_2, \quad X_3 = 3B + N_3$$

Where under C_1, the A, N_1, N_2, and N_3 are independent Gaussian random variables with $A \sim N(1, 1)$ and $N_k \sim N(0, 1), k = 1, 2, 3$.

Where under C_2, the B, N_1, N_2, and N_3 are independent Gaussian random variables with $B \sim N(-1, 1)$ and $N_k \sim N(0, 1), k = 1, 2, 3$.

Assume that A and B are independent random variables, that the a priori probabilities of each class are equal, and that the Bayes cost assignment is given by $C_{11} = C_{22} = 0$, $C_{12} = C_{21} = 1$.

Find the optimum detector where optimality is in the Bayes sense. Reduce your decision to the observation space. (*Hint*: This is a multiple-observation problem where the observations are not independent.)

9.32 Let the observed pattern vector $\mathbf{X} = [X_1, X_2]^T$ have the following class conditional densities: Under class C_1, X is normally distributed with mean vector $[0\ 0]^T$ and covariance matrix $\mathbf{K}_1 = \mathbf{I}$, where $\mathbf{I}$ is a 2×2 identity matrix, and under class C_2, $\mathbf{X}$ is normally distributed with mean vector $[1\ \ 1]^T$ and covariance matrix $\mathbf{K}_2 = \mathbf{I}$.

(a) Find the optimum test in the Bayes sense for the zero-one cost assignment when the a priori probabilities of the classes are equal. Reduce your decision rule to the observation space.

(b) For your decision rule in (a), calculate the probability of making an error.

(c) Repeat (a) and (b) with same assumptions except that $\mathbf{K}_1$ and $\mathbf{K}_2$ are given by

$$\mathbf{K}_1 = \begin{bmatrix} 1 & 0.5 \\ 0.5 & 1 \end{bmatrix}, \quad \mathbf{K}_2 = \begin{bmatrix} 1 & -0.5 \\ -0.5 & 1 \end{bmatrix}$$

(d) Compare your results of (b) and (c), and make a statement concerning the effect that correlation can have on performance.

9.33 Let $\mathbf{X}$ be the observed pattern vector $[X_1, X_2]^T$. Under C_1, $\mathbf{X}$ is normally distributed with mean vector $[0\ 0]^T$ and covariance matrix $\mathbf{K}_1$, and under C_2, $\mathbf{X}$ is normally distributed with mean vector $[1\ \ 1]^T$ and covariance matrix $\mathbf{K}_2$, where $\mathbf{K}_1$ and $\mathbf{K}_2$ are given by

$$\mathbf{K}_1 = \begin{bmatrix} 1 & r \\ r & 1 \end{bmatrix}, \quad \mathbf{K}_2 = \begin{bmatrix} 1 & r \\ r & 1 \end{bmatrix}$$

(a) Find the optimum test in the Bayes sense for the zero-one cost assignment when the a priori probabilities of the classes are equal. Reduce your decision rule to the observation space.

(b) For your decision rule in (a), calculate the probability of making an error, and plot the probability of error as r is varied from 1 to -1. Comment on your results and on the part that correlation can have on performance.

9.34 Let the observed pattern vector $[X_1, X_2]^T$ have the following class conditional densities: Under class C_1, $\mathbf{X}$ is normally distributed with mean vector $[0\ 0]^T$ and covariance matrix $\mathbf{K}_1 = \mathbf{I}$, where $\mathbf{I}$ is a 2×2 identity matrix, and under class C_2, $\mathbf{X}$ is normally distributed with mean vector $[1\ \ 1]^T$ and covariance matrix $\mathbf{K}_2 = 4\mathbf{I}$.

Find the optimum decision rule in the two-dimensional pattern space, where optimality is in the Bayes sense for $C_{12} = C_{21} = 0$ and $C_{11} = C_{22} = 1$ and when the a priori probabilities of the two classes are equal. Also calculate the probability of error for your decision rule.

9.35 Assume the following three class, three-observation problem:

$$C_1: \quad \mathbf{X} \sim N\left(\begin{bmatrix} 0 \\ 1 \\ 1 \end{bmatrix}, \begin{bmatrix} 1 & 0 & 0 \\ 0 & 1 & 0 \\ 0 & 0 & 1 \end{bmatrix}\right),$$

$$C_2: \quad \mathbf{X} \sim N\left(\begin{bmatrix} 1 \\ 1 \\ 0 \end{bmatrix}, \begin{bmatrix} 2 & 0 & 0 \\ 0 & 2 & 0 \\ 0 & 0 & 2 \end{bmatrix}\right)$$

$$C_3: \quad \mathbf{X} \sim N\left(\begin{bmatrix} 1 \\ 0 \\ 1 \end{bmatrix}, \begin{bmatrix} 3 & 0 & 0 \\ 0 & 3 & 0 \\ 0 & 0 & 3 \end{bmatrix}\right)$$

With equally likely classes, find the region in the observation space for which the minimum probability of error decision rule would decide class C_1.

9.36 Assume we are given that $\mathbf{X} \sim N(\mathbf{m}_1, \mathbf{K}_1)$ under class C_1 and $\mathbf{X} \sim N(\mathbf{m}_2, \mathbf{K}_2)$ under class C_2, where $\mathbf{m}_1 = [1, -2, 1]^T$ and $\mathbf{m}_2 = [-1, -1, -2]^T$. Calculate the probability of error for the optimum detector in the Bayes sense with the zero-one cost function for the following cases:

(a) $\mathbf{K}_1 = \mathbf{K}_2 = 2\mathbf{I}$ and $P(C_1) = P(C_2) = \frac{1}{2}$.

(b) $\mathbf{K}_1 = \mathbf{K}_2 = 4\mathbf{I}$ and $P(C_1) = \frac{1}{3}$ and $P(C_2) = \frac{2}{3}$.

(c) $P(C_1) = P(C_2) = \frac{1}{2}$ and $\mathbf{K}_1 = \mathbf{K}_2 = \text{diag}[2, 4, 16]$.

(d) $P(C_1) = \frac{1}{2}, P(C_2) = \frac{1}{2}$, and

$$\mathbf{K}_1 = \mathbf{K}_2 = \begin{bmatrix} 1 & -\frac{1}{2} & 0 \\ -\frac{1}{2} & 2 & -\frac{1}{4} \\ 0 & -\frac{1}{4} & 1 \end{bmatrix}.$$

REFERENCES

1. Van Trees, H. L., *Detection, Estimation and Modulation Theory*, Vol. 1, Wiley, 1968.
2. Selin, Ivan, *Detection Theory*, Princeton University Press, 1965.
3. Young, Tzay Y., and Thomas W. Calvert, *Classification, Estimation, and Pattern Recognition*, Elsevier, 1974.
4. McDonough, Robert N., and Anthony D. Whalen, *Detection of Signal in Noise*, Academic Press, 1995.
5. Srinath, M. D., P. K. Rajasekaran, and R. Viswanathan, *Statistical Signal Processing with Applications*, Prentice-Hall, 1996.
6. Helstrom, Carl W., *Statistical Theory of Signal Detection*, 2nd ed., Pergamon, 1968.
7. Helstrom, Carl W., *Elements of Signal Detection and Estimation*, Prentice-Hall, 1995.
8. Poor, H. Vincent, *An Introduction to Signal Detection and Estimation*, 2nd ed., Springer, 1994.

9. Wald, Abraham, *Sequential Analysis*, Wiley, 1947.
10. Wetherill, G. Barrie, and Glazebrook, *Sequential Methods in Statistics*, 3rd ed., Chapman and Hall, 1986.
11. Tenney, R. R., and N. R. Sandell, Detection with distributed sensors, *IEEE Transactions on Aerospace and Electronic Systems*, Vol. 17, No. 4, July, 1981, p 501–509.
12. Therrien, Charles W., *Decision, Estimation, and Classification*, Wiley, 1989.

10 Detection Theory: Continuous Observation

10.1 CONTINUOUS OBSERVATIONS

The methods presented in the previous sections used a vector of observations. More often the problems we are to solve measure a continuum rather than a finite set. For example, the observed signal $x(t)$ might be recorded for all values of t in the interval from 0 to T. Approaches to the continuous problem are based on expansion of those signals into a weighted sum of CON (**complete orthonormal**) basis functions. This changes the noncountable problem to a countable infinite one that can be approached by modifying the techniques previously discussed for the finite-dimensional vector case.

For the binary case the problem can be formulated as follows. The observation for class C_1 and class C_2 is $x(t)$ over the interval $[0, T]$ as below:

$$\begin{aligned} C_1:\quad x(t) &= s_1(t) + N(t), \quad 0 \le t \le T \\ C_2:\quad x(t) &= s_2(t) + N(t), \quad 0 \le t \le T \end{aligned} \tag{10.1}$$

Problems that are mathematically tractable include the special case where $N(t)$ is a realization of a **Gaussian white noise** random process with known autocorrelation function $R_{NN}(\tau) = \frac{1}{2}N_0\delta(\tau)$ and power spectral density function $\phi_{NN}(\omega) = \frac{1}{2}N_0$ for all ω. The representation above is called an **additive white Gaussian noise** (AWGN) channel, where $x(t)$ is the received signal, $s_i(t)$ are the signals that are sent, and $N(t)$ is a realization from a white Gaussian noise process. There are varied levels of complexity depending on the assumptions placed on $s_1(t)$ and $s_2(t)$. A few of these are now listed in increasing levels of complexity.

Level 1. $s_1(t)$ and $s_2(t)$ are known deterministic signals.

Example: $s_1(t) = t, \quad s_2(t) = t^2, \qquad 0 \le t \le T$

Level 2. $s_1(t)$ and $s_2(t)$ are of known deterministic signal form but depend on an unknown deterministic parameter.

Example: $s_1(t) = e^{-at}, \quad s_2(t) = \cos(bt), \qquad 0 \le t \le T$

where a and b are deterministic and unknown.

Level 3. $s_1(t)$ and $s_2(t)$ are unknown deterministic signals.

Level 4. $s_1(t)$ and $s_2(t)$ are of known deterministic form with random parameters that have known probability density functions.

Example: $s_1(t) = A\cos 2\pi t, \quad s_2(t) = B\sin 2\pi t, \qquad 0 \le t \le T$

where $A \sim N(0, 1), B \sim N(-1, 1)$.

Level 5. $s_1(t)$ and $s_2(t)$ are of known deterministic form with random parameters that have unknown probability density functions.

Example: $s_1(t) = A\cos 2\pi t, \quad s_2(t) = B\sin 2\pi t, \quad 0 \le t \le T$

with $p(a, b)$ unknown.

Level 6. $s_1(t)$ and $s_2(t)$ are realizations from random processes with known statistical characterizations.

Level 7. $s_1(t)$ and $s_2(t)$ are realizations from random processes with unknown statistical characterizations.

Our discussion begins with level 1 problems and the two class case with AWGN and progresses through levels 2, 4, and 6 including the multiclass cases. Other levels will not be examined as they become totally separate areas of investigation outside the general framework of this text.

10.2 DETECTION OF KNOWN SIGNALS IN WHITE GAUSSIAN NOISE

We begin our discussion by looking at a special subset of the two-class continuous observation problem. In particular, we consider those problems for the case of deterministic signals in additive white noise (AWGN).

10.2.1 Two-Class Continuous Observations (AWGN)

Suppose that one of two possible **known deterministic signals** $s_1(t)$ and $s_2(t)$ is observed after being passed through an additive white Gaussian noise channel. The continuous observation signal $x(t)$ under each class is given by

$$\begin{aligned} C_1:&\quad x(t) = s_1(t) + N(t), \qquad 0 \le t \le T \\ C_2:&\quad x(t) = s_2(t) + N(t), \qquad 0 \le t \le T \end{aligned} \tag{10.2}$$

It will be assumed that $P(C_1)$ and $P(C_2)$ are known and $N(t)$ is a white process characterized by a zero mean and autocorrelation function $R_{XX}(\tau) = \frac{1}{2}N_0\delta(\tau)$. The problem can be changed to a countable observation vector by expanding $x(t)$ using a complete orthonormal (CON) set $\{\phi_i(t)\colon i = 1, 2, \ldots\}$ on the interval $[0, T]$.

A set $\{\phi_i(t)\colon i = 1, 2, \ldots\}$ on the interval $[0, T]$ is a **complete orthonormal** set if

$$\int_0^T \phi_i(t)\phi_j(t)\,dt = \delta_{ij} \qquad \text{for } i, j = 1, 2, \ldots \tag{10.3}$$

and there exists no function of time that cannot be expanded in terms of those basis functions.

Let x_i be the coefficient of the ith basis function for the expansion of $x(t)$.

$$x_i = \int_0^T x(t)\phi_i(t)\,dt \tag{10.4}$$

and thus the observation $x(t)$ can be written as

$$x(t) = \sum_{i=1}^{\infty} x_i \phi_i(t) \tag{10.5}$$

As $x(t)$ is a random process (a random variable at each t), the infinite series expansion could converge to $x(t)$ in the mean squared error sense. This form of convergence is expressed by

$$x(t) = \underset{k\to\infty}{\text{l.i.m.}} \sum_{i=1}^{k} x_i \phi_i(t) \qquad \text{for all } t \in [0, T] \tag{10.6}$$

where l.i.m. is **limit in the mean** defined by

$$\lim_{k\to\infty} E\left[\left|x(t) - \sum_{i=1}^{k} x_i \phi_i(t)\right|^2\right] = 0 \qquad \text{for all } t \in [0, T] \tag{10.7}$$

Let s_{1i} and s_{2i} be the coefficients of $s_1(t)$ and $s_2(t)$ expanded in terms of the basis functions $\phi_i(t)$ as

$$\begin{aligned} s_{1i} &= \int_0^T s_1(t)\phi_i(t)\,dt, \qquad i = 1, 2, \ldots \\ s_{2i} &= \int_0^T s_2(t)\phi_i(t)\,dt, \qquad i = 1, 2, \ldots \end{aligned} \tag{10.8}$$

Define N_i, the coefficient of $N(t)$ expanded in terms of the same basis functions, as

$$N_i = \int_0^T N(t)\phi_i(t)\,dt, \qquad i = 1, 2, \ldots \tag{10.9}$$

The N_i will be Gaussian random variables since they are integrals of a weighted $N(t)$, which is a Gaussian random process (integrals of random processes are random variables). Using these definitions, (10.8) and (10.9), for s_{1i}, s_{2i}, and N_i, we write the equivalent countably infinite problem to the continuous one given in equation Eq. (10.2) as

$$\begin{aligned} C_1&: \quad \mathbf{x} = \mathbf{s}_1 + \mathbf{N} \\ C_2&: \quad \mathbf{x} = \mathbf{s}_2 + \mathbf{N} \end{aligned} \tag{10.10}$$

where $\mathbf{s}_1$, $\mathbf{s}_2$, and $\mathbf{N}$ are infinite-dimensional vectors given by

$$\mathbf{s}_1 = \begin{bmatrix} s_{11} \\ s_{12} \\ \vdots \\ s_{1k} \\ \vdots \end{bmatrix}, \quad \mathbf{s}_2 = \begin{bmatrix} s_{21} \\ s_{22} \\ \vdots \\ s_{2k} \\ \vdots \end{bmatrix}, \quad \text{and} \quad \mathbf{N} = \begin{bmatrix} N_1 \\ N_2 \\ \vdots \\ N_k \\ \vdots \end{bmatrix} \tag{10.11}$$

The $\mathbf{s}_1$ and $\mathbf{s}_2$ are infinite-dimensional deterministic vectors since $s_1(t)$ and $s_2(t)$ are known deterministic signals, while $\mathbf{N}$ is an infinite-dimensional random Gaussian vector. The conditional densities for $\mathbf{x}$ under both classes will now be found. Since under both classes the observation vector $\mathbf{x}$ is the sum of a deterministic vector and a Gaussian random vector, it will also be a Gaussian random vector. The probability densities conditioned on the classes can be specified by finding their corresponding mean vectors and covariance matrices.

First, under C_1 we have the mean vector

$$E[\mathbf{x} \,|\, C_1] = E[(\mathbf{s}_1 + \mathbf{N}) \,|\, C_1] = E[\mathbf{s}_1 \,|\, C_1] + E[\mathbf{N} \,|\, C_1] \tag{10.12}$$

Since $\mathbf{s}_1$ is a deterministic vector, its expected value will be just $\mathbf{s}_1$. From Eq. (10.9), the ith component of $\mathbf{E}[\mathbf{N} \,|\, C_1]$ is determined by taking the expected value operator through the linear operation as

$$\begin{aligned} E[N_i \,|\, C_1] &= E\left[\int_0^T N(t)\,\phi_i(t)\,dt \,|\, C_1\right] \\ &= \int_0^T E[N(t) \,|\, C_1]\phi_i(t)\,dt \end{aligned} \tag{10.13}$$

But since $N(t)$ is a zero mean random process, zero for all t by the white noise assumption, the integral from 0 to T is also zero. Thus $E[N_i \,|\, C_1] = 0$ for all i, and we see from (10.12) that

$$E[\mathbf{x} \,|\, C_1] = \mathbf{s}_1 \tag{10.14}$$

Similarly it can be shown that $E[\mathbf{x} \,|\, C_2] = \mathbf{s}_2$.

The components of the covariance matrix under each of the two classes will now be determined. For class C_1 the ijth component of the covariance matrix for $\mathbf{x}$ is

$$k_{ij}^{(1)} = E[(x_i - E[x_i \,|\, C_1])(x_j - E[x_j \,|\, C_1])] \tag{10.15}$$

After substituting x_i and $E[x_i \,|\, C_1]$ from (10.10) and (10.14), respectively, (10.15), we have

$$\begin{aligned} k_{ij}^{(1)} &= E[(s_{1i} + N_i - s_{1i})(s_{1j} + N_j - s_{1j}) \,|\, C_1] \\ &= E[N_i N_j \,|\, C_1] \end{aligned} \tag{10.16}$$

Replacing the N_i and N_j above with (10.9) gives

$$\begin{aligned} E[N_i N_j \mid C_1] &= E\left[\left(\int_0^T N(t)\phi(t)\,dt\right)\left(\int_0^T N(u)\phi_j(u)\,du\right)\middle| C_1\right] \\ &= \int_0^T \int_0^T E[N(t)N(u)]\phi_i(t)\phi_j(u)\,du \end{aligned} \tag{10.17}$$

Now, after substituting the autocorrelation function of the white noise $N(t)$ into (10.17), rearranging, and using the sampling property of the delta function, we obtain

$$\begin{aligned} E[N_i N_j \mid C_1] &= \int_0^T \left(\int_0^T \frac{N_0}{2}\phi_i(t)\delta(t-u)\,dt\right)\phi_j(u)\,du \\ &= \int_0^T \frac{N_0}{2}\phi_i(u)\phi_j(u)\,du \end{aligned} \tag{10.18}$$

Then, using the orthonormality property of the basis functions, (10.3), we reduce the equation above to

$$E[N_i N_j \mid C_1] = \frac{N_0}{2}\delta_{ij} \tag{10.19}$$

where δ_{ij} is the **Kronecker delta** given by

$$\delta_{ij} = \begin{cases} 1, & i = j \\ 0, & i \neq j \end{cases} \tag{10.20}$$

Substituting (10.19) into (10.16) finally gives the ijth component of the covariance matrix for $\mathbf{x}$ conditioned on C_1 as

$$k_{ij}^{(1)} = \frac{N_0}{2}\delta_{ij} \tag{10.21}$$

and determines $\mathbf{K}_1$ as an infinite-dimensional diagonal matrix with diagonal elements equal to $\frac{1}{2}N_0$.

Similarly under class C_2 we find, by repeating the steps of (10.14) to (10.21), that $\mathbf{K}_2$ is a diagonal matrix with all diagonal elements equal to $\frac{1}{2}N_0$. Thus the original problem can be formulated in terms of the infinite-dimensional Gaussian problem as

$$\begin{aligned} C_1:&\quad \mathbf{x} \sim N[\mathbf{s}_1, \mathbf{K}_1] \\ C_2:&\quad \mathbf{x} \sim N[\mathbf{s}_2, \mathbf{K}_2] \end{aligned} \tag{10.22}$$

where $\mathbf{s}_1$ and $\mathbf{s}_2$ represent the coefficient vectors of the expansion of the deterministic signals $s_1(t)$ and $s_2(t)$, respectively, given in (10.11) and $\mathbf{K}_1 = \mathbf{K}_2 = \frac{1}{2}N_0\mathbf{I}_\infty$ are the observation $\mathbf{x}$ conditional covariance matrices. Using special case 5, with zero-one costs, from Eq. (9.89), since the problem is a one with equal scaled identity covariance matrices, the optimum decision rule simplifies to a maximum selector as described next.

Select class C_k if $L'_k(\mathbf{x})$ is the largest of $L'_i(\mathbf{x})$ for $i = 1, 2$, where $L'_i(\mathbf{x})$ are defined by

$$L'_i(\mathbf{x}) = \mathbf{s}_i^T \mathbf{x} - \frac{1}{2}\,\mathbf{s}_i^T \mathbf{s}_i + \frac{N_0}{2}\ln P(C_i), \qquad i = 1, 2 \tag{10.23}$$

The block diagram for the optimal detector which implements the decision rule given in (10.23) appears in matrix and vector form in Figure 10.1. The following simple example shows the procedure for solving this type of problem.

EXAMPLE 10.1

Known deterministic signals, $s_1(t)$ and $s_2(t)$, are observed over the time interval $[0, T]$ in an additive white Gaussian noise channel with power spectral density of $N_0/2$ for all frequencies. The two-class problem can be written as

$$\begin{aligned} C_1&:\quad x(t) = s_1(t) + N(t), \qquad 0 \le t \le T \\ C_2&:\quad x(t) = s_2(t) + N(t), \qquad 0 \le t \le T \end{aligned} \tag{10.24}$$

Let $s_1(t)$ and $s_2(t)$ be orthogonal, which means that

$$\int_0^T s_1(t)s_2(t)\,dt = 0 \tag{10.25}$$

with known finite energies E_1 and E_2 as follows:

$$\int_0^T s_1^2(t)\,dt = E_1 \quad \text{and} \quad \int_0^T s_2^2(t)\,dt = E_2 \tag{10.26}$$

Assume further that the a priori probabilities of the two classes are given by $P(C_1)$ and $P(C_2)$ and that a zero-one cost function is used. Find the optimal detector in the Bayes sense.

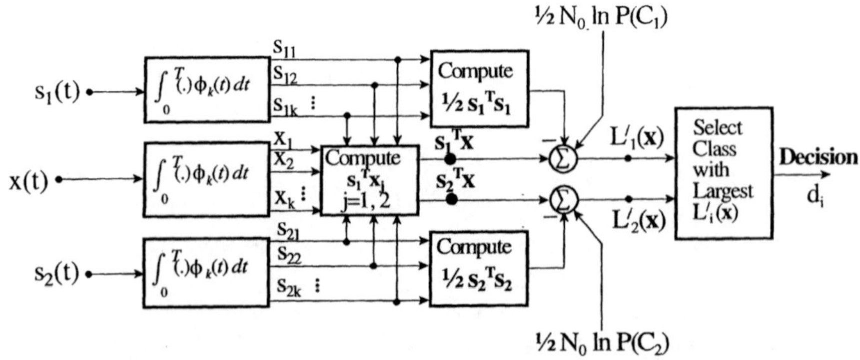

Figure 10.1 Optimum detector for the continuous observation AWGN binary case with deterministic $s_1(t)$ and $s_2(t)$.

SOLUTION

The first step is to change the continuous time observation problem to a countable infinite-dimensional vector problem by using a complete orthonormal set of basis functions. As the results previously obtained do not depend on the choice of the $\phi_i(t)$, only that they are orthonormal, we are free to select any CON basis. Certain choices lead to very simple solutions, since they reduce the infinite-dimensional vectors to effective finite-dimensional vectors. For example, the choice of a Fourier basis for expanding pulselike waveforms would need a countably infinite number of coefficients, whereas expansion of the same waveforms in a Walsh basis may need only a finite number of coefficients. With this in mind, the first two basis functions for our example may be selected as

$$\phi_1(t) = \frac{1}{\sqrt{E_1}} s_1(t) \quad \text{and} \quad \phi_2(t) = \frac{1}{\sqrt{E_2}} s_2(t) \tag{10.27}$$

and the remaining functions $\phi_k(t)$ for $k \geq 3$ are selected to form a CON set. Their actual functions will not play a part in the solution. Only the fact that they are orthonormal will be used, so they will not have to be described further. Expanding $x(t)$ with these basis functions over the interval $[0, T]$ gives the following equivalent problem where x is a countable infinite size vector:

$$\begin{aligned} C_1\colon &\quad \mathbf{x} \sim N(\mathbf{m}_1, \mathbf{K}_1) \\ C_2\colon &\quad \mathbf{x} \sim N(\mathbf{m}_2, \mathbf{K}_2) \end{aligned} \tag{10.28}$$

The density functions, from which the solution can be obtained, are Gaussian and thus specified once $\mathbf{m}_1$, $\mathbf{m}_2$, $\mathbf{K}_1$, and $\mathbf{K}_2$ are determined.

Under C_1 the $x_1, x_2, \ldots$ can be determined as

$$x_i = \int_0^T x(t)\phi_i(t)\,dt, \qquad i = 1, 2, \ldots \tag{10.29}$$

For class C_1 the class conditional expected value of x_i will give the ith components of the mean vector $\mathbf{m}_1$ as

$$\begin{aligned} m_{1i} &\triangleq E[x_i \mid C_1] = E\left[\int_0^T x(t)\phi_i(t)\,dt \mid C_1\right] \\ &= \int_0^T E[s_1(t) + N(t) \mid C_1]\phi_i(t)\,dt \\ &= \int_0^T s_1(t)\phi_i(t)\,dt + \int_0^T E[N(t) \mid C_1]\phi_i(t)\,dt \end{aligned} \tag{10.30}$$

But $N(t)$ is a realization from a white Gaussian noise, which implies that $E[N(t) \mid C_1] = 0$. So the second integral of (10.30) is zero, and $E[x_i \mid C_1]$ becomes

$$m_{1i} = E[x_i \mid C_1] = \int_0^T s_1(t)\phi_i(t)\,dt = s_{1i} \tag{10.31}$$

Using the $\phi_i(t)$ assigned in this problem, (10.27), we determine the m_{1i} for $i = 1, 2, \ldots,$ as follows:

$$m_{11} = \int_0^T s_1(t) \frac{1}{\sqrt{E_1}} s_1(t)\, dt = \sqrt{E_1} \quad \text{by energy assumptions}$$

$$m_{12} = \int_0^T s_1(t) \frac{1}{\sqrt{E_2}} s_2(t)\, dt = 0 \quad \text{for all } i \geq 2's \text{ by orthogonality of } s_1(t) \text{ and } s_2(t)$$

$$m_{1i} = \int_0^T s_1(t)\phi_i(t)\, dt = \int_0^T \sqrt{E_1}\phi_1(t)\phi_i(t)\, dt = 0 \tag{10.32}$$

by orthonormality property of the $\phi_i(t)$

Therefore the $\mathbf{m}_1$ vector can be written as the infinite-dimensional vector

$$\mathbf{m}_1 = \begin{bmatrix} \sqrt{E_1} \\ 0 \\ 0 \\ \vdots \end{bmatrix} \triangleq \mathbf{s}_1 \tag{10.33}$$

The covariance matrix for the observation vector $\mathbf{x}$ under class C_1 must now be determined. Let $k_{ij}^{(1)}$ be the i, j entry in the covariance matrix $\mathbf{K}_1$ for class C_1 defined as follows:

$$k_{ij}^{(1)} = E[(x_i - E[x_i \mid C_1])(x_j - E[x_j \mid C_1]) \mid C_1] \tag{10.34}$$

Using the white noise assumption and results previously determined (10.21) for this type of problem, we write the observation covariance matrix $\mathbf{K}_1$ under class C_1 is an infinite-dimension diagonal matrix with all diagonal entries equal to $N_0/2$ and off-diagonal terms zero as follows:

$$\mathbf{K}_1 = \frac{N_0}{2} \begin{bmatrix} 1 & 0 & 0 & \cdots \\ 0 & 1 & 0 & \cdots \\ 0 & 0 & 1 & \cdots \\ \vdots & \vdots & \vdots & \ddots \end{bmatrix} = \frac{N_0}{2} \mathbf{I}_\infty \tag{10.35}$$

Under class C_2, $\mathbf{m}_2$ and $\mathbf{K}_2$ can be found by repeating the steps used to obtain $\mathbf{m}_1$ and $\mathbf{K}_1$ as follows:

$$\begin{aligned} m_{21} &= \int_0^T s_2(t) \frac{s_1(t)}{\sqrt{E_1}} = 0 \\ m_{22} &= \int_0^T s_2(t) \frac{s_2(t)}{\sqrt{E_2}}\, dt = \sqrt{E_2} \\ &\vdots \\ m_{2j} &= \int_0^T s_2(t)\phi_j(t)\, dt = 0 \qquad \text{for } j > 2 \end{aligned} \tag{10.36}$$

Therefore the mean $\mathbf{m}_2$ under class C_2 becomes

$$\mathbf{m}_2 = \begin{bmatrix} 0 \\ \sqrt{E_2} \\ 0 \\ \vdots \end{bmatrix} \triangleq \mathbf{s}_2 \tag{10.37}$$

and the covariance matrix $\mathbf{K}_2$ can be shown, by repeating the steps used in finding $\mathbf{K}_1$, to be

$$\mathbf{K}_2 = \frac{N_0}{2} \begin{bmatrix} 1 & 0 & 0 & \cdots \\ 0 & 1 & 0 & \cdots \\ 0 & 0 & 1 & \cdots \\ \vdots & \vdots & \vdots & \ddots \end{bmatrix} = \frac{N_0}{2} \mathbf{I}_\infty \tag{10.38}$$

Since $\mathbf{x}$ is a Gaussian vector under each class, with the mean vectors and covariance matrices determined above, the optimal detector is that of special case 5. It selects the class corresponding to the largest of the $L_i'(\mathbf{x})$ given by

$$L_i'(\mathbf{x}) = \mathbf{m}_i^T \mathbf{x} - \frac{1}{2} \mathbf{m}_i^T \mathbf{m}_i + \frac{N_0}{2} \ln P(C_i) \qquad \text{for } i = 1, 2 \tag{10.39}$$

Substituting the $\mathbf{m}_i$ from (10.33) into (10.39) gives $L_1'(\mathbf{x})$ as

$$\begin{aligned} L_1'(x) &= [\sqrt{E_1} \quad 0 \quad 0 \quad \cdots] \begin{bmatrix} x_1 \\ x_2 \\ x_3 \\ \vdots \end{bmatrix} - \frac{1}{2} [\sqrt{E_1} \quad 0 \quad 0 \quad \cdots] \begin{bmatrix} \sqrt{E_1} \\ 0 \\ 0 \\ \vdots \end{bmatrix} + \frac{1}{2} N_0 \ln P(C_1) \\ &= \sqrt{E_1} x_1 - \frac{1}{2} E_1 + \frac{1}{2} N_0 \ln P(C_1) \end{aligned} \tag{10.40}$$

Notice that in the vector product the components x_i for $i > 1$ are multiplied by zero and therefore do not appear in the final results.

Similarly $L_2'(\mathbf{x})$ is obtained as

$$\begin{aligned} L_2'(\mathbf{x}) &= [0 \quad \sqrt{E_2} \quad 0 \quad \cdots] \begin{bmatrix} x_1 \\ x_2 \\ x_3 \\ \vdots \end{bmatrix} - \frac{1}{2} [0 \quad \sqrt{E_2} \quad 0 \quad \cdots] \begin{bmatrix} 0 \\ \sqrt{E_2} \\ 0 \\ \vdots \end{bmatrix} + \frac{1}{2} N_0 \ln P(C_2) \\ &= \sqrt{E_2} x_2 - \frac{1}{2} E_2 + \frac{1}{2} N_0 \ln P(C_2) \end{aligned} \tag{10.41}$$

After substituting the x_1 and x_2 from (10.29) into (10.40) and (10.41), we have the optimal decision rule to select the class with largest value of $L_i'(\mathbf{x})$, where

$$\begin{aligned} L_1'(\mathbf{x}) &= \sqrt{E_1} \int_0^T x(t)\phi_1(t)\,dt - \frac{1}{2} E_1 + \frac{1}{2} \ln P(C_1) \\ L_2'(\mathbf{x}) &= \sqrt{E_2} \int_0^T x(t)\phi_2(t)\,dt - \frac{1}{2} E_2 + \frac{1}{2} \ln P(C_2) \end{aligned} \tag{10.42}$$

The optimum decision rule implemented in terms of the $\phi_1(t)$ and $\phi_2(t)$ is shown in Figure 10.2.

For this problem an alternative form of the optimum classifier can also be conveniently expressed in terms of the original signals $s_1(t)$ and $s_2(t)$. Substituting the first two basis functions from (10.27), $\phi_1(t) = s_1(t)/\sqrt{E_1}$ and $\phi_2(t) = s_2(t)/\sqrt{E_2}$, into (10.42) reduces the $L_1'(\mathbf{x})$ and $L_2'(\mathbf{x})$ to

$$\begin{aligned} L_1'(\mathbf{x}) &= \int_0^T x(t)s_1(t)\,dt - \frac{1}{2} E_1 + \frac{1}{2} \ln P(C_1) \\ L_2'(\mathbf{x}) &= \int_0^T x(t)s_2(t)\,dt - \frac{1}{2} E_2 + \frac{1}{2} \ln P(C_2) \end{aligned} \tag{10.43}$$

This form of the optimum decision rule is shown in Figure 10.3, and each branch represents a correlation of the observed signal $x(t)$ with the known signals for that class.

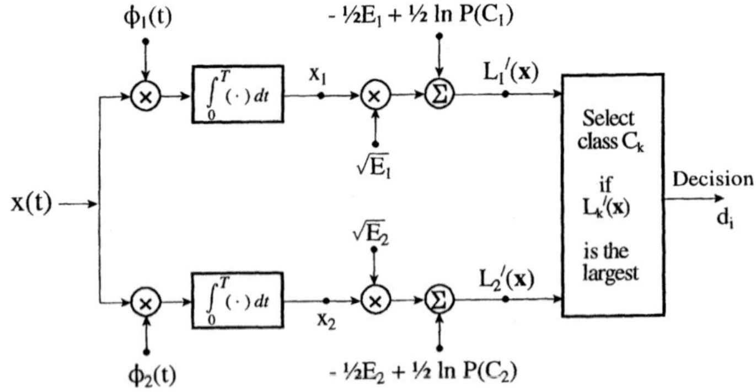

Figure 10.2 Optimum decision rule in the Minimum probability of error sense for deterministic orthogonal signals, Example 10.1, in terms of orthonormal basis functions $\varphi_1(t)$ and $\varphi_2(t)$.

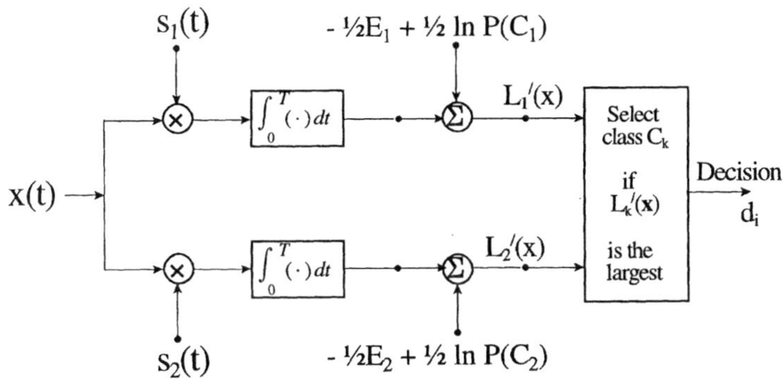

Figure 10.3 Optimum detector for the case of orthogonal signals $s_1(t)$ and $s_2(t)$ in an additive white Gaussian noise channel in terms of original signals.

The optimal detector can also be put in the form of a single correlation. The optimal decision rule computes $L_1'(\mathbf{x})$ and $L_2'(\mathbf{x})$ and selects class corresponding the larger of the two. Using (10.43), we can write

$$\text{If} \quad L_1'(\mathbf{x}) > L_2'(\mathbf{x}), \quad \text{then decide } C_1$$

$$\int_0^T x(t)s_1(t)\,dt - \frac{1}{2}E_1 + \frac{1}{2}\ln P(C_1) > \int_0^T x(t)s_2(t)\,dt - \frac{1}{2}E_2 + \frac{1}{2}\ln P(C_2) \qquad (10.44)$$

After moving the terms on the right side of the inequality to the left and factoring, we obtain an equivalent decision rule as

$$\text{If} \quad v = \int_0^T x(t)[s_1(t) - s_2(t)]\,dt + b > 0 \quad \text{then decide } C_1$$

$$< \quad \text{then decide } C_2 \qquad (10.45)$$

where

$$b = -\frac{1}{2}E_1 + \frac{1}{2}E_2 + \frac{1}{2}\ln\frac{P(C_1)}{P(C_2)}$$

Ties can be resolved randomly.

The decision rule above is thus seen to be a correlation of $x(t)$ with $s_1(t) - s_2(t)$, the difference of the two known deterministic signals as shown in Figure 10.4. ■

When $s_1(t)$ and $s_2(t)$ are **not** orthogonal on the interval of observation, the choice of the orthonormal basis functions can no longer be just $s_1(t)$ and $s_2(t)$. One way to solve the problem is to choose $\phi_1(t) = s_1(t)/E_1^{1/2}$ and $\phi_2(t)$ to be the normalized version of $s_2(t)$ minus the part correlated to $s_1(t)$, which gives $\phi_2(t) = [s_2(t) - E_{12}s_1(t)]/(E_2 - 2E_{12}^2 + E_{12}^2E_1)^{1/2}$. This approach is basically the first step in the Gram-Schmidt orthonormalization procedure which will be used to expedite the solution for the multiple-class case; it is thoroughly discussed in the

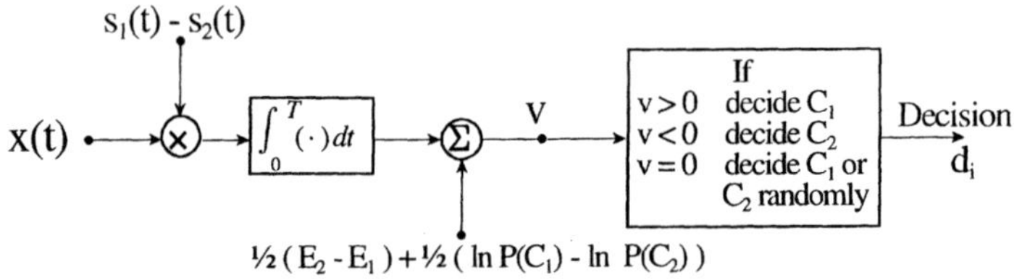

Figure 10.4 Optimum detector for case of deterministic orthogonal signals $s_1(t)$ and $s_2(t)$, in terms of a single correlation with the difference of the two signals.

next section. Extending these two basis functions to give a CON, we can then apply the same procedure described in Example 10.1 for the detection of orthogonal signal $s_1(t)$ and $s_2(t)$. The only changes made will be that the mean vectors are different, which changes slightly the structure of the optimum detector.

10.2.2 Multiple-Class Continuous Observation (AWGN)

The multiple-class continuous observation case for AWGN can be written as

$$\begin{aligned} C_1:&\quad x(t) = s_1(t) + N(t), &\quad 0 \le t \le T \\ C_2:&\quad x(t) = s_2(t) + N(t), &\quad 0 \le t \le T \\ &\quad\vdots &\quad\vdots \\ C_M:&\quad x(t) = s_M(t) + N(t), &\quad 0 \le t \le T \end{aligned} \tag{10.46}$$

It will be assumed that the $s_i(t)$ are known deterministic signals, that the a priori probability of each class is known and given by $P(C_i)$, and that $N(t)$ is a realization from a white Gaussian random process with zero mean and autocorrrelation function given by $R_{NN} = \frac{1}{2}N_0\delta(\tau)$. We wish to find the optimum decision rule where optimality is in the minimum probability of error sense or in the Bayes sense for only a zero-one cost assignment.

The general approach to solving this continuous observation case will be to expand the continuous signals in terms of a complete orthonormal set of basis functions. Selection of the basis functions is critical to the simplification of the problem one is solving. In general, we would like to select K basis functions for $K \le M$ so that all the signals $s_i(t)$, $i = 1, 2, \ldots, K$ can be written as a linear combination of a finite number of basis functions. It is convenient in many problems to use the Gram-Schmidt orthonormalization procedure for selecting the basis functions that fit the problem you are solving. We now present the Gram-Schmidt procedure followed by an example illustrating the procedure and finally present an example of a multiple-class problem that illustrates the selection of basis functions and the use of the resulting basis functions to obtain the optimum classifier.

Gram-Schmidt Orthonormalization Procedure. Given a set of signals $s_1(t)$, $s_2(t), \ldots, s_K(t)$, we wish to find an orthonormal set of functions $\phi_1(t)$, $\phi_2(t), \ldots, \phi_K(t)$ that can serve as a basis for representing all the $s_k(t)$, $k = 1, 2, \ldots, K$ on the interval 0 to T. The following procedure described by Hancock and Wintz [16] does not necessarily give a unique set of basis functions as the original set of signals can be reordered and the technique applied to the new order, which may give a different set of orthonormal functions.

Let $f_1(t) = s_1(t)$, then select the **first orthonormal basis function** $\phi_1(t)$ to be

$$\phi_1(t) = \frac{f_1(t)}{\sqrt{F_1}} \quad \text{where} \quad F_1 = \int_0^T s_1^2(t)\,dt \tag{10.47}$$

Next let

$$f_2(t) = s_2(t) - a_{21}\phi_1(t) \quad \text{where} \quad a_{21} = \int_0^T s_2(t)\phi_1(t)\,dt \tag{10.48}$$

Thus $f_2(t)$ is the signal $s_2(t)$ minus the component of $s_2(t)$ in $\phi_1(t)$. The **second orthonormal basis function** $\phi_2(t)$ is defined as the normalized version of $f_2(t)$:

$$\phi_2(t) = \frac{f_2(t)}{\sqrt{F_2}} \quad \text{where} \quad F_2 = \int_0^T f_2^2(t)\,dt \tag{10.49}$$

Selection in this fashion guarantees both the normal and orthogonal properties necessary for the $\phi_i(t)$ to be a part of a CON basis function.

Similarly let $f_3(t)$, the residual signal after its components in both the $\phi_1(t)$ and $\phi_2(t)$ directions have been subtracted off, be defined as

$$f_3(t) = s_3(t) - a_{31}\phi_1(t) - a_{32}\phi_2(t) \tag{10.50}$$

where

$$a_{31} = \int_0^T s_3(t)\phi_1(t)\,dt \quad \text{and} \quad a_{32} = \int_0^T s_3(t)\phi_2(t)\,dt \tag{10.51}$$

Then the **third orthonormal basis function** is defined as

$$\phi_3(t) = \frac{f_3(t)}{\sqrt{F_3}} \quad \text{where} \quad F_3 = \int_0^T f_3^2(t)\,dt \tag{10.52}$$

The process above is continued until the last or Kth step with resulting $f_K(t)$ as

$$f_K(t) = s_K(t) - a_{K1}\phi_1(t) - a_{K2}\phi_2(t) - \cdots - a_{K,K-1}\phi_{K-1}(t) \tag{10.53}$$

$$a_{Kj} = \int_0^T s_K(t)\phi_j(t)\,dt, \qquad j = 1, 2, \ldots, K-1 \tag{10.54}$$

The last or ***K*****th orthonormal basis function** $\phi_K(t)$ is defined as the normalized version of $f_K(t)$ as follows:

$$\phi_K(t) = \frac{f_K(t)}{\sqrt{F_K}} \quad \text{where} \quad F_K = \int_0^T f_K^2(t)\,dt \tag{10.55}$$

If at anytime during the procedure $f_K(t) = 0$, it means that $s_K(t)$ can be expressed as a linear combination of the preceding basis functions and thus can be skipped, without generating a new basis function. Thus the procedure will give K basis functions if the signals are linearly independent and fewer than K basis functions if the signals are not linearly independent. The following examples illustrate both of these cases.

EXAMPLE 10.2

Use the Gram-Schmidt procedure to find an orthonormal basis on the interval [0, 1] for the $s_1(t)$, $s_2(t)$, and $s_3(t)$ given in Figure 10.5 using the order $s_1(t)$, $s_2(t)$, and $s_3(t)$.

SOLUTION

For the order $s_1(t)$, $s_2(t)$, and $s_3(t)$. Let $f_1(t) = s_1(t)$, which results in $\phi_1(t)$ being

$$\phi_1(t) = \frac{f_1(t)}{[\int_0^T f_1^2(t)\,dt]^{1/2}} = \frac{s_1(t)}{[\int_0^T s_1^2(t)\,dt]^{1/2}} = \frac{2}{2} = 1, \qquad 0 \le t \le 1$$

The second basis function is found by using $s_2(t)$ to define the new function $f_2(t)$ as follows:

$$\text{Let} \quad f_2(t) = s_2(t) - a_{21}\phi_1(t)$$

$$\text{where} \quad a_{21} = \int_0^1 s_2(t)\phi_1(t)\,dt = \int_0^1 2t(1)\,dt = 1$$

Thus $f_2(t)$ can be simplified to

$$f_2(t) = 2t - 1(1) = 2t - 1, \qquad 0 \le t \le 1$$

Now normalize by finding

$$\int_0^1 f_2^2(t)\,dt = \int_0^1 (2t-1)^2 dt = \frac{1}{3}$$

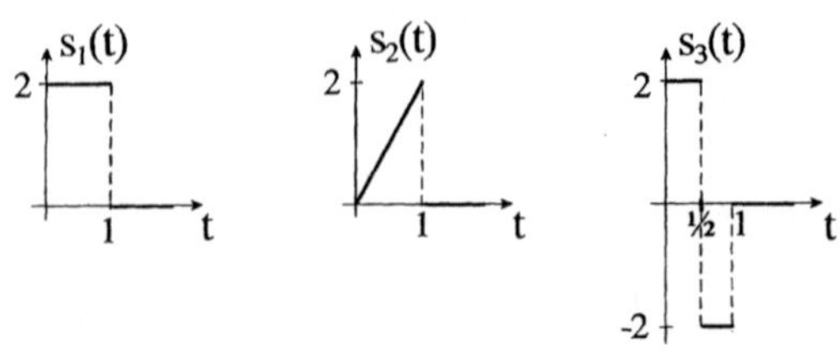

Figure 10.5 Signals $s_1(t)$, $s_2(t)$, and $s_3(t)$ for Example 10-2.

Thus the second orthonormal basis function $\phi_2(t)$ becomes

$$\phi_2(t) = \frac{f_2(t)}{\sqrt{1/3}} = \sqrt{3}(2t - 1), \qquad 0 \le t \le 1$$

The last step in the process is to let $f_3(t) = s_3(t) - a_{31}\phi_1(t) - a_{32}\phi_2(t)$, where

$$a_{31} = \int_0^1 s_3(t)\phi_1(t)\,dt = \int_0^{1/2} (2)1\,dt + \int_{1/2}^1 (-2)1\,dt = 0$$

$$a_{32} = \int_0^1 s_3(t)\phi_2(t)\,dt = \int_0^{1/2} 2\sqrt{3}(2t - 1)\,dt - \int_{1/2}^1 2\sqrt{3}(2t - 1)\,dt = -\sqrt{3}$$

Thus

$$\begin{aligned} f_3(t) &= s_3(t) + \sqrt{3}\cdot\sqrt{3}(2t - 1) \\ &= \begin{cases} 2 + 3(2t - 1) = -1 + 6t, & 0 \le t \le \frac{1}{2} \\ -2 + 3(2t - 1) = -5 + 6t, & \frac{1}{2} \le t \le 1 \end{cases} \end{aligned}$$

So the third orthonormal basis function is found by normalizing $f_3(t)$ to give

$$\phi_3(t) = \frac{f_3(t)}{\sqrt{\int_0^1 f_3^2(t)}} = \frac{f_3(t)}{1} = \begin{cases} -1 + 6t, & 0 \le t \le \frac{1}{2} \\ -5 + 6t, & \frac{1}{2} < t \le 1 \end{cases}$$

The three orthonormal functions $\phi_1(t)$, $\phi_2(t)$, and $\phi_3(t)$ have thus been determined for this example, and they are illustrated in Figure 10.6. Notice that for this problem the maximum number 3 of basis functions was obtained.

Also, if the order used in the procedure had been $s_1(t)$, $s_3(t)$, and $s_2(t)$, normalized versions of $s_1(t)$ and $s_3(t)$ would have been the first two orthonormal functions, both pulse type waveforms, thus leading to a slightly "simpler" set of orthonormal functions. Unfortunately, there is no prescribed procedure for selecting whatever we would mean by the simplest set. ■

Example 10.3 shows a problem for which the number of basis functions is fewer than the maximum number.

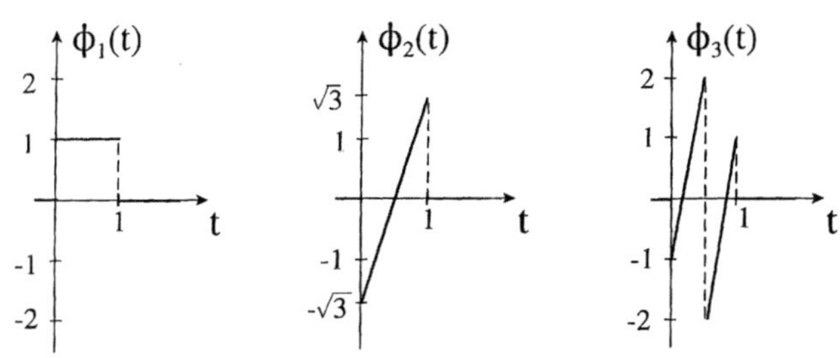

Figure 10.6 Orthonormal set for the $s_1(t)$, $s_2(t)$, and $s_3(t)$ of Example 10-2.

EXAMPLE 10.3

Use the Gram-Schmidt procedure on the functions $s_1(t), s_2(t)$, and $s_3(t)$ on the interval [0, 1] where we have the following relationships: $s_2(t) = -s_1(t)$ and $s_3(t) = 0$.

SOLUTION

By the order $s_1(t), s_2(t)$, and $s_3(t)$ we see that the first basis function $\phi_1(t)$ is

$$\phi_1(t) = \frac{1}{\sqrt{E_1}}\, s_1(t) \quad \text{where } E_1 = \int_0^1 s_1^2(t)\, dt$$

Using $s_2(t)$, we find that

$$a_{21} = \int_0^1 s_2(t)\phi_1(t)\, dt = \int_0^1 -s_1(t)\, \frac{1}{\sqrt{E_1}}\, s_1(t)\, dt = -\sqrt{E_1}$$

which leads to $f_2(t)$ as

$$f_2(t) = s_2(t) - (-\sqrt{E_1}\phi_1(t)) = -s_1(t) + \sqrt{E_1}\, \frac{1}{\sqrt{E_1}}\, s_1(t) = 0$$

This means that $s_2(t)$ is a linear function of $\phi_1(t)$. This is no real surprise, since $s_2(t) = -s_1(t) = -\sqrt{E_1}\phi_1(t)$, so no new basis function is added. Since $s_3(t) = 0 \cdot \phi_1(t)$, it is also a linear function of $\phi_1(t)$, so no new basis function is required. Thus only one basis function $\phi_1(t)$ is required. ■

The use of a CON set to solve a particular multiple-class problem where the signals are not orthogonal signals is now presented. For the next example on quadrature phase shift keying, the procedure will yield fewer basis functions than the four original signals, namely 2 basis functions for 4 original signals. The example will be worked in considerable detail to illustrate the steps necessary in the process of obtaining the optimum decision rule for multi-class AWGN systems and typical performance evaluation in terms of probability of error.

EXAMPLE 10.4

Assume we have quadrature phase shift modulation signal classes expressed as follows:

$$\begin{aligned}
C_1&: \quad x(t) = +A\ \cos w_c t + A\ \sin w_c t + N(t), && 0 \le t \le T\\
C_2&: \quad x(t) = -A\ \cos w_c t + A\ \sin w_c t + N(t), && 0 \le t \le T\\
C_3&: \quad x(t) = +A\ \cos w_c t - A\ \sin w_c t + N(t), && 0 \le t \le T\\
C_4&: \quad x(t) = -A\ \cos w_c t - A\ \sin w_c t + N(t), && 0 \le t \le T
\end{aligned} \tag{10.56}$$

Let $N(t)$ be a realization from a Gaussian white noise process with autocorrelation function $R_{NN}(\tau) = \frac{1}{2}\delta(\tau)$. Assume that w_c and A are known constants and that each of the classes is equally probable. Find the optimal detector in the Bayes sense for a zero-one cost assignment where T is an integer number of periods of $\sin w_c t$ and calculate the probability of error.

SOLUTION

The first step in the procedure is the selection of the orthonormal basis for expansion. If the Grahm-Schmidt procedure is used it would lead to the first two orthonormal basis functions as normalized versions of $\sin w_c t + \cos w_c t$ and $\sin w_c t - \cos w_c t$.

For this problem it is more convenient from a mathematical viewpoint to select normalized versions of $\sin w_c t$ and $\cos w_c t$ directly as the first two orthonormal basis functions with remaining harmonics being used to specify the remaining members of the complete orthonormal set, namely the standard Fourier basis. The energy E_1 and E_2 for $\cos w_c t$ and $\sin w_c t$ on interval $(0, T)$ are

$$\begin{aligned} E_1 &= \int_0^T \cos^2 w_c t\, dt = \int_0^T \frac{1}{2}(1 + \cos 2w_c t)\, dt \\ &= \frac{1}{2}T - \int_0^T \frac{1}{2}\cos 2w_c t\, dt \end{aligned} \tag{10.57}$$

Since the integral of an even number of periods of a cosine wave is zero, $E_1 = \frac{1}{2}T$.

Similarly it can be shown that $E_2 = \frac{1}{2}T$. Thus the first two orthonormal basis functions are

$$\begin{aligned} \phi_1(t) &= \sqrt{\frac{2}{T}}\cos w_c t, \qquad 0 \le t \le T \\ \phi_2(t) &= \sqrt{\frac{2}{T}}\sin w_c t, \qquad 0 \le t \le T \end{aligned} \tag{10.58}$$

The remaining orthonormal functions can be defined as the normalized harmonics of w_c for $k = 2, 3, \ldots$:

$$\begin{aligned} \phi_{2k-1}(t) &= \sqrt{\frac{2}{T}}\cos(2k-1)w_c t, \qquad 0 \le t \le T \\ \phi_{2k}(t) &= \sqrt{\frac{2}{T}}\sin 2kw_c t, \qquad 0 \le t \le T \end{aligned} \tag{10.59}$$

The $\phi_{2k}(t)$ and $\phi_{2k-1}(t)$ for $k > 2$ will be shown not to contribute to the final results.

Under C_1 we can find an infinite-dimensional vector equivalent to the original problem where components of the observation vector have the following expected value:

$$E[x_k \mid C_1] = E\left[\int_0^T x(t)\phi_k(t)\,dt \mid C_1\right] = \int_0^T E[s_1(t) + N(t) \mid C_1]\phi_k(t)\,dt$$
$$= \int_0^T s_1(t)\phi_k(t)\,dt + \int_0^T E[N(t) \mid C_1]\phi_k(t)\,dt \tag{10.60}$$

The second integral is zero since $E[N(t)] = 0$. Therefore the conditional expected value of the observation x_k becomes

$$E[x_k \mid C_1] = \int_0^T s_1(t)\phi_k(t)\,dt \triangleq s_{1k} \tag{10.61}$$

The s_{1k} will now be evaluated starting with s_{11} as

$$s_{11} = \int_0^T s_1(t)\phi_1(t)\,dt = \int_0^T (A\cos w_c t + A\sin w_c t)\sqrt{\frac{2}{T}}\cos w_c t\,dt$$
$$= A\sqrt{\frac{2}{T}}\int_0^T \cos^2 w_c t\,dt + A\sqrt{\frac{2}{T}}\int_0^T \sin w_c t\cos w_c t\,dt \tag{10.62}$$

In this expression the first integral is easily determined as $T/2$, while the second integral is zero, since it is an integral of a sinusoid over an integer number of full periods. Thus

$$s_{11} = A\sqrt{\frac{T}{2}} \tag{10.63}$$

In following the steps above we can easily determine the s_{12} as $s_{12} = A\sqrt{T/2}$.

For $k > 2$ the $s_{1,2k}$ are found similarly as follows:

$$s_{1,2k} = \int_0^T s_1(t)\phi_{2k}(t)\,dt$$
$$= \int_0^T (A\cos w_c t + A\sin w_c t)\sqrt{\frac{2}{T}}\sin 2kw_c t\,dt \tag{10.64}$$

Since $\sin 2kw_c t$ is orthogonal to both $\sin w_c t$ and $\cos w_c t$ on the interval $[0, T]$ by virtue of all being in the Fourier basis, we have $s_{1,2k} = 0$.

The $s_{1,2k-1} = 0$ for $k > 2$ by the same argument. Thus, in summary, the vector $\mathbf{s}_1$ becomes

$$\mathbf{s}_1 = \begin{bmatrix} A\sqrt{\frac{T}{2}} \\ A\sqrt{\frac{T}{2}} \\ 0 \\ 0 \\ \vdots \end{bmatrix} \tag{10.65}$$

By a similar development for the other classes the signal vectors can be found as

$$\mathbf{s}_2 = \begin{bmatrix} -A\sqrt{\frac{T}{2}} \\ A\sqrt{\frac{T}{2}} \\ 0 \\ 0 \\ \vdots \end{bmatrix}, \quad \mathbf{s}_3 = \begin{bmatrix} A\sqrt{\frac{T}{2}} \\ -A\sqrt{\frac{T}{2}} \\ 0 \\ 0 \\ \vdots \end{bmatrix}, \quad \mathbf{s}_4 = \begin{bmatrix} -A\sqrt{\frac{T}{2}} \\ -A\sqrt{\frac{T}{2}} \\ 0 \\ 0 \\ \vdots \end{bmatrix} \tag{10.66}$$

Since the $E[N_i \mid C_1] = 0$, the expected values of the observation vectors conditioned under each class are as follows:

$$\begin{aligned} C_1&: \quad E[\mathbf{x} \mid C_1] = \mathbf{s}_1 \\ C_2&: \quad E[\mathbf{x} \mid C_2] = \mathbf{s}_2 \\ C_3&: \quad E[\mathbf{x} \mid C_3] = \mathbf{s}_3 \\ C_4&: \quad E[\mathbf{x} \mid C_4] = \mathbf{s}_4 \end{aligned} \tag{10.67}$$

The covariance matrix under class C_1 can be determined by finding the covariances of observation vector components. Define $k_{ij}^{(1)}$ as follows:

$$k_{ij}^{(1)} \triangleq E\big[(x_i - E[x_i \mid C_1])(x_j - E[x_j \mid C_1]) \mid C_1\big] \tag{10.68}$$

The ijth components of the covariance matrix of the observation vector $\boldsymbol{x}$ under class C_1 are determined by the procedure shown in (10.15) through (10.21) as

$$\begin{aligned} k_{ij}^{(1)} &= E[(s_{1i} + N_i - s_{1i})(s_{1j} + N_i - s_{1j}) \mid C_1] = E[N_i N_j \mid C_1] \\ &= E\left[\int_0^T N(t)\phi_i(t)\,dt \int_0^T N(t)\phi_j(u)\,du \mid C_1\right] \\ &= \int_0^T \left(\int_0^T E[N(t)N(u) \mid C_1]\phi_i(t)\phi_j(u)\,dt\right) du \\ &= \int_0^T \left(\int_0^T \frac{N_0}{2}\delta(t-u)\phi_i(t)\phi_j(u)\,dt\right) du = \int_0^T \frac{N_0}{2}\phi_i(u)\phi_j(u)\,du \end{aligned} \tag{10.69}$$

Then, using the orthonormal property of the $\phi_k(t)$s, we find that the integral is δ_{ij}. Thus we have

$$k_{ij}^{(1)} = \frac{N_0}{2}\delta_{ij} \tag{10.70}$$

where δ_{ij} is the **Kronecker delta function.**

The covariance matrix, K_1, under class C_1 is clearly a scaled diagonal identity matrix given by

$$\mathbf{K}_1 = \frac{N_0}{2}\mathbf{I}_\infty \tag{10.71}$$

Similarly under the other classes it is easily shown that

$$\mathbf{K}_2 = \frac{N_0}{2}\mathbf{I}_\infty, \quad \mathbf{K}_3 = \frac{N_0}{2}\mathbf{I}_\infty, \quad \text{and} \quad \mathbf{K}_4 = \frac{N_0}{2}\mathbf{I}_\infty \tag{10.72}$$

The original problem can now be written in an infinite-dimensional form as

$$\begin{aligned} &C_1\colon\ \mathbf{x} \sim N[\mathbf{s}_1, \mathbf{K}] \\ &C_2\colon\ \mathbf{x} \sim N[\mathbf{s}_2, \mathbf{K}] \\ &C_3\colon\ \mathbf{x} \sim N[-\mathbf{s}_2, \mathbf{K}] \\ &C_4\colon\ \mathbf{x} \sim N[-\mathbf{s}_1, \mathbf{K}] \end{aligned} \tag{10.73}$$

where $\mathbf{s}_1$ and $\mathbf{s}_2$ are given by (10.65) and (10.66) and $\mathbf{K} = (N_0/2)\mathbf{I}_\infty$.

The optimal decision rule can finally be determined by using the special case of scaled equal covariance matrices and Gaussian random vectors from (9.89) with $\sigma^2 = \frac{1}{2}N_0$. The decision rule selects the class with the maximum of the $L_i'(\mathbf{x})$, which are given by

$$L_i'(\mathbf{x}) = \mathbf{m}_i^T\mathbf{x} - \frac{1}{2}\mathbf{m}_i^T\mathbf{m}_i + \frac{1}{2}N_0 \ln P(C_i), \quad i = 1, 2, 3, 4 \tag{10.74}$$

Substituting the mean from (10.65) into (10.74) for $i = 1$, we have

$$L_1'(\mathbf{x}) = \begin{bmatrix} A\sqrt{\frac{T}{2}} & A\sqrt{\frac{T}{2}} & 0 & \cdots \end{bmatrix} \begin{bmatrix} x_1 \\ x_2 \\ x_3 \\ \vdots \end{bmatrix} - \frac{1}{2} \begin{bmatrix} A\sqrt{\frac{T}{2}} & A\sqrt{\frac{T}{2}} & 0 & \cdots \end{bmatrix} \begin{bmatrix} A\sqrt{\frac{T}{2}} \\ A\sqrt{\frac{T}{2}} \\ 0 \\ \vdots \end{bmatrix} + \frac{1}{2} N_0 \ln P(C_1) \tag{10.75}$$

Since the row vectors have all zeros past the second component, the product of the infinite vectors gives

$$L_1'(\mathbf{x}) = A\sqrt{\frac{T}{2}}x_1 + A\sqrt{\frac{T}{2}}x_2 - \frac{1}{2}A^2T + \frac{1}{2}N_0 \ln P(C_1) \tag{10.76}$$

Notice that x_3, x_4, and so on, do not appear in the simplification, and thus the decision rule only needs to consider the first two components of the observation vector (the coefficients of the expansion of $x(t)$ in terms of the first two functions of the orthonormal basis).

Similarly the other $L_i'(\mathbf{x})$ for $i = 2, 3$, and 4 can be determined as

$$\begin{aligned} L_2'(\mathbf{x}) &= -A\sqrt{\frac{T}{2}}x_1 + A\sqrt{\frac{T}{2}}x_2 - \frac{1}{2}A^2T + \frac{1}{2}N_0 \ln P(C_2) \\ L_3'(\mathbf{x}) &= +A\sqrt{\frac{T}{2}}x_1 - A\sqrt{\frac{T}{2}}x_2 - \frac{1}{2}A^2T + \frac{1}{2}N_0 \ln P(C_3) \\ L_4'(\mathbf{x}) &= -A\sqrt{\frac{T}{2}}x_1 - A\sqrt{\frac{T}{2}}x_2 - \frac{1}{2}A^2T + \frac{1}{2}N_0 \ln P(C_4) \end{aligned} \tag{10.77}$$

For equal a priori probabilities the last two terms of (10.76) and (10.77) are equal, and they can be dropped from the $L_i'(\mathbf{x})$ above. This simplifies the optimum decision rule, which computes the $L_i''(\mathbf{x})$ below and selects the class corresponding to the largest value:

$$\begin{aligned} L_1''(\mathbf{x}) &= +A\sqrt{\frac{T}{2}}x_1 + A\sqrt{\frac{T}{2}}x_2 \\ L_2''(\mathbf{x}) &= -A\sqrt{\frac{T}{2}}x_1 + A\sqrt{\frac{T}{2}}x_2 \\ L_3''(\mathbf{x}) &= +A\sqrt{\frac{T}{2}}x_1 - A\sqrt{\frac{T}{2}}x_2 \\ L_4''(\mathbf{x}) &= -A\sqrt{\frac{T}{2}}x_1 - A\sqrt{\frac{T}{2}}x_2 \end{aligned} \tag{10.78}$$

Since the $A\sqrt{T/2}$ is a positive constant multiplier, it can also be dropped because multiplication does not change which one is largest. This simplifies the optimum test to that of selecting the class with the largest of the following $L_i'''(\mathbf{x})$

$$\begin{aligned} L_1'''(\mathbf{x}) &= x_1 + x_2 \\ L_2'''(\mathbf{x}) &= -x_1 + x_2 \\ L_3'''(\mathbf{x}) &= x_1 - x_2 \\ L_4'''(\mathbf{x}) &= -x_1 - x_2 \end{aligned} \tag{10.79}$$

Where x_1 and x_2 are the coefficients of the expansion of the observation $x(t)$ on the interval $[0, T]$ given by

$$\begin{aligned} x_1 &= \int_0^T x(t)\phi_1(t)\,dt = \int_0^T x(t)\sqrt{\frac{2}{T}}\cos w_c t\,dt \\ x_2 &= \int_0^T x(t)\phi_2(t)\,dt = \int_0^T x(t)\sqrt{\frac{2}{T}}\sin w_c t\,dt \end{aligned} \tag{10.80}$$

The optimal decision rule for the quadrature phase shift keying problem is illustrated in Figure 10.7.

We can view the decision regions for this optimum decision rule in the space of x_1 and x_2, our "new observation" space. For example, R_1 is the region for which $L_1'''(\mathbf{x})$ is greater than all other $L_i'''(\mathbf{x})$ defined by the following regions in the x_1, x_2 space

$$\begin{aligned} L_1'''(\mathbf{x}) > L_2'''(\mathbf{x}) &\rightarrow x_1 + x_2 > -x_1 + x_2 \rightarrow x_1 > 0 \\ L_1'''(\mathbf{x}) > L_3'''(\mathbf{x}) &\rightarrow x_1 + x_2 > x_1 - x_2 \rightarrow x_2 > 0 \\ L_1'''(\mathbf{x}) > L_4'''(\mathbf{x}) &\rightarrow x_1 + x_2 > -x_1 - x_2 \rightarrow x_1 > -x_2 \end{aligned} \tag{10.81}$$

The intersection of the three regions specified by the inequalities above determine the quadrant region R_1 as shown in Figure 10.8. The regions for deciding C_2, C_3, and

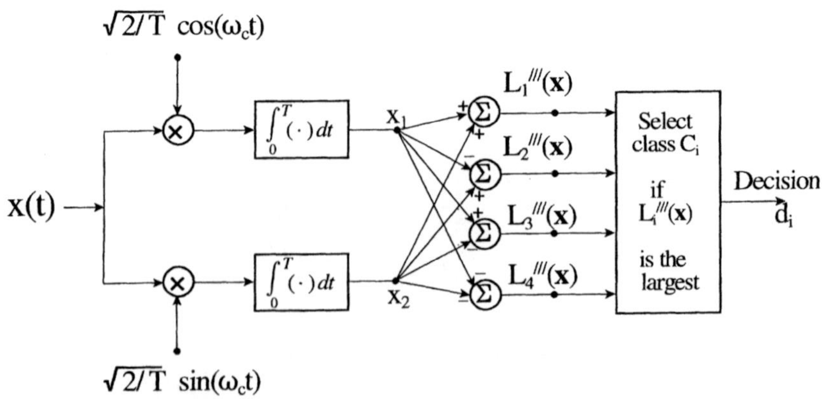

Figure 10.7 Optimal detector for QPSK, Example 10.4.

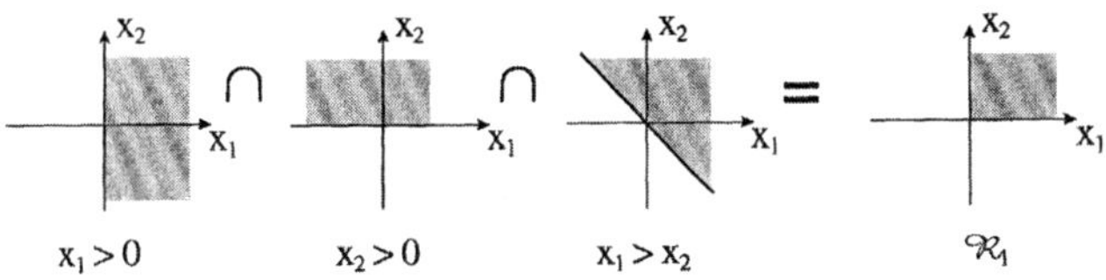

Figure 10.8 Determination of Region R_1 for the optimum decision rule for Example 10.4.

C_4 can be similarly determined by inequalities like (10.81) and the final regions $\mathcal{R}_2$, $\mathcal{R}_3$, and $\mathcal{R}_4$ are shown in Figure 10.9.

Probability of Error for Example 10.4. The probability of error for the decision rule shown in Figure 10.9 for Example 10.4 can be calculated similar to the procedure shown in Example 9.11. The probability of error is known as

$$P(\text{error}) = \sum_{i=1}^{4} P(\text{error} \mid C_i)P(C_i) \tag{10.82}$$

As before, for the multiple class case the conditional probability of error is more easily determined by calculating the conditional probability of being correct as

$$\begin{aligned} P(\text{error}) &= 1 - P(\text{correct}) \\ &= 1 - \sum_{i=1}^{4} P(\text{correct} \mid C_i)P(C_i) \end{aligned} \tag{10.83}$$

By symmetry of the problem, we see that the probability of being correct conditioned on each class are equal and also that for this example the a priori probabilities of each class were given as equal. So the $P(\text{error})$ above simplifies to

$$\begin{aligned} P(\text{error}) &= 1 - \sum_{i=1}^{4} P(\text{correct} \mid C_i)\,\frac{1}{4} \\ &= 1 - P(\text{correct} \mid C_1) \end{aligned} \tag{10.84}$$

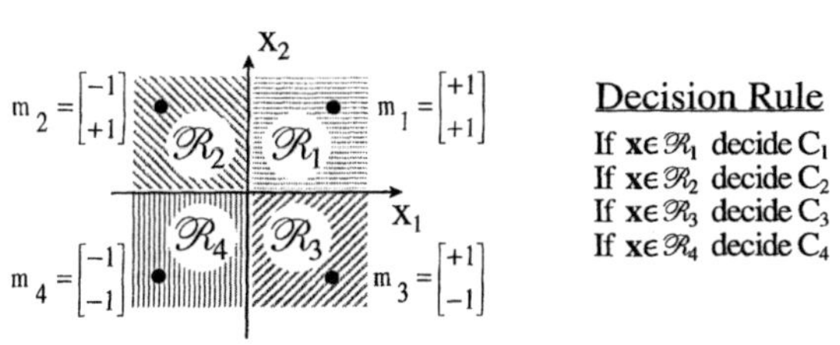

Figure 10.9 Decision regions for the optimum decision rule for Example 10.4.

The $P(\text{correct} \mid C_1)$ is the integral of the conditional density of x_1 and x_2 over the region $\mathcal{R}_1$ shown in Figure 10.9 and is given by

$$P(\text{correct} \mid C_1) = \int_{\mathcal{R}_1} p(x_1, x_2 \mid C_1)\, dx_1\, dx_2 \tag{10.85}$$

The extracted x_1 and x_2 have been previously shown to be jointly Gaussian, since they are integrals of Gaussian random processes, and have means, $E[x_1 \mid C_1] = E[x_2 \mid C_2] = A\sqrt{T/2}$, and a covariance matrix, $\frac{1}{2}N_0\mathbf{I}_2$. Thus the $P(\text{correct} \mid C_1)$ can be written as

$$\begin{aligned} P(\text{correct} \mid C_1) &= \int_0^\infty \int_0^\infty \frac{1}{2\pi \cdot N_0/2} \exp\left\{\frac{-(x_1 - A\sqrt{T/2})^2 - (x_2 - A\sqrt{T/2})^2}{2 \cdot N_0/2}\right\} dx_1\, dx_2 \\ &= \left(\int_0^\infty \frac{1}{\sqrt{2\pi N_0/2}} \exp\left\{\frac{-(x_1 - A\sqrt{T/2})^2}{2N_0/2}\right\} dx_1\right)^2 \end{aligned} \tag{10.86}$$

The term inside the parentheses can be written in terms of the $\Phi(\cdot)$, and thus the probability of being correct conditioned on C_1 simplifies as

$$\begin{aligned} P(\text{correct} \mid C_1) &= \left(1 - \Phi\left(\frac{0 - A\sqrt{T/2}}{\sqrt{N_0/2}}\right)\right)^2 \\ &= \left(1 - \Phi\left(-A\sqrt{T/N_0}\right)\right)^2 = \Phi^2\left(A\sqrt{T/N_0}\right) \end{aligned} \tag{10.87}$$

Therefore we can finally write that the probability of error from (10.84) as

$$P(\text{error}) = 1 - \Phi^2\left(A\sqrt{T/N_0}\right) \tag{10.88}$$

We see that this formula bears out our intuition that increasing T or decreasing N_0 will result in a smaller error. If we define TA^2/N_0 as a **signal to noise ratio** (SNR), the probability of error can be written as $P(\text{error}) = 1 - \Phi^2(\sqrt{SNR})$. ■

10.3 DETECTION OF KNOWN SIGNALS IN NONWHITE GAUSSIAN NOISE (ANWGN)

In the preceding sections results were shown for the two- and multi-class cases that are AWGN based, namely for the pattern classes that are deterministic signals observed in additive white Gaussian noise. If any of these conditions on the noise are relaxed, special techniques must be used to solve the problems. While the same structures and formulas could be used as before, no claim of optimality can be made. In this section we will modify the techniques previously presented to find optimal detectors for additive nonwhite noise (ANWGN) problems.

There are two basic approaches for determining the optimum decision rules for the detection of known signals in nonwhite Gaussian noise. The first approach, and the main approach used in this and following sections, is the application of the

Karhunen-Loeve (KL) expansion. The second approach is the use of whitening filters to transform the noise into white noise and then to solve the resulting "bleached" problem. This second approach is thoroughly explored by Van Trees [1] and others and exchanges the problem of finding the KL expansion with that of finding the bleaching filter.

White noise occupies the entire spectrum. With additive nonwhite Gaussian noise it is possible to obtain perfect detection with a probability of error equal to zero should the deterministic signals have frequency content in a band of frequencies not occupied by the nonwhite noise. Such cases are called singular detection and are discussed by Van Trees [1]. In this chapter we will be looking only at nonsingular detection in ANWGN.

10.3.1 Two-Class Detection in ANWGN

We begin by considering a special subset of the two-class continuous observation problem. In particular, we are interested in addressing those problems of the detection of deterministic signals in additive nonwhite Gaussian noise (ANWGN).

Suppose that one of two possible **known deterministic signals** $s_1(t)$ and $s_2(t)$ is observed over the interval $[0, T]$ after being passed through an **additive nonwhite Gaussian noise** channel. The continuous observation signal $x(t)$ under each class is given by

$$\begin{aligned} C_1:&\quad x(t) = s_1(t) + N(t), \quad 0 \le t \le T \\ C_2:&\quad x(t) = s_2(t) + N(t), \quad 0 \le t \le T \end{aligned} \tag{10.89}$$

We will assume that $P(C_1)$ and $P(C_2)$ are known and that $N(t)$ is a nonwhite Gaussian process characterized by a zero mean and autocorrelation function $R_{XX}(t, u)$. The result presented does not require that the $N(t)$ be wide sense stationary, but it will be assumed that it has a zero mean for convenience in the development. If it does not have a zero mean, the problem can be changed to one with a zero mean by subtracting the mean for the observation, giving a new observation.

We want to find the optimum decision rule in the minimum probability of error sense. In the previous sections we changed the continuous time observations into an infinite-dimensional vector of Gaussian components by using an orthonormal expansion whose choice depended on the deterministic signals to simplify the solution. If we expand (10.89) using a CON basis, the equivalent problem is

$$\begin{aligned} C_1:&\quad \mathbf{x} = \mathbf{s}_1 + \mathbf{N} \\ C_2:&\quad \mathbf{x} = \mathbf{s}_2 + \mathbf{N} \end{aligned} \tag{10.90}$$

where the $\mathbf{s}_1$ and $\mathbf{N}$ have components s_{1i} and N_i as follows:

$$s_{1i} = \int_0^T s_1(t)\phi_i(t)\,dt, \quad N_i = \int_0^T N(t)\phi_i(t)\,dt \tag{10.91}$$

The N_i and N_j are Gaussian random variables, since they are integrals of the Gaussian random process $N(t)$. However, with a nonwhite spectral density they are not necessarily independent for every choice of basis functions. For them to be uncorrelated, the basis functions must be chosen to satisfy $E[N_iN_j \mid C_k] = 0$ for $k = 1, 2$, where $E[N_iN_j \mid C_k]$ is given as

$$\begin{aligned} E[N_iN_j \mid C_k] &= E\left[\left(\int_0^T N(t)\phi_i(t)\,dt\right)\left(\int_0^T N(u)\phi_j(u)\,du\right)\bigg| C_k\right] \\ &= \int_0^T \int_0^T E[N(t)N(u) \mid C_k]\phi_i(t)\phi_j(u)\,dt\,du \end{aligned} \tag{10.92}$$

Since the $N(t)$ is the same for each class, the expected value is the autocorrelation function $R_{NN}(\tau)$. So the $E[N_iN_j \mid C_k]$ becomes

$$E[N_iN_j \mid C_k] = \int_0^T \int_0^T R_{NN}(t, u)\phi_i(t)\phi_j(u)\,dt\,du \tag{10.93}$$

To make the N_i and N_j independent, we only need to show that they are uncorrelated, since uncorrelated and Gaussian implies independence. To make them uncorrelated, we set the $E[N_iN_j \mid C_k]$ equal to zero, since the conditional means are zero. This results in the following condition for $i \neq j$:

$$\int_0^T \int_0^T R_{NN}(t, u)\phi_i(t)\phi_j(u)\,dt\,du = 0 \tag{10.94}$$

If the basis functions are chosen to satisfy this equation the solution will be simplified. Note that this expansion is based on the autocorrelation of the noise rather than the properties of the deterministic signals as used for the AWGN case. Equation (10.94) can be rearranged as

$$\int_0^T \phi_i(t)\left(\int_0^T R_{NN}(t, u)\phi_j(u)\,du\right)dt = 0 \tag{10.95}$$

We see from Eq. (10.95) that if the term in brackets equals $\lambda_j\phi_j(t)$, then the condition will be satisfied for all $i \neq j$. This is the exact condition that the Karhunen-Loeve expansion requires, and we will give further properties and explanations later so as not to detract from this development. The solution is well known in the fields of mathematics and physics to be the set of eigen functions $\{\phi_j(t)\colon j = 1, 2, \ldots\}$ associated with set of eigen values $\{\lambda_j\colon j = 1, 2, \ldots\}$ for the kernel $R_{NN}(\tau)$. These eigen values and functions are related in the following way:

$$\int_0^T R_{XX}(t, u)\phi_j(u)\,du = \lambda_j\phi_j(t) \tag{10.96}$$

This equation is similar in concept to that of eigen values and vectors associated with a matrix. However, here the function $\phi_j(t)$ gives a scaled version of the function when operated on by the integration of its product with the kernel and integration over the interval $[0, T]$. Let us assume that the basis functions have been chosen to satisfy this equation. We will now determine the mean vector and covariance matrix for the **N** vector conditioned on both classes. From (10.91) we know that for each component of the $\mathbf{N}$ vector $E[N_j \mid C_k] = 0$, since $E[N(t)] = 0$ in the integrand gives a zero integral under both classes. The expected value of the $\mathbf{N}$ vector conditioned on class C_k then can be written as

$$E[\mathbf{N} \mid C_k] = \mathbf{0}, \qquad k = 1, 2 \tag{10.97}$$

To obtain the conditional covariance matrices, we first obtain the diagonal elements by letting $i = j$ in Eq. (10.93):

$$\begin{aligned} E[N_i^2 \mid C_k] &= \int_0^T \int_0^T R_{NN}(t, u)\phi_i(t)\phi_i(u)\, dt\, du \\ &= \int_0^T \phi_i(t)\left(\int_0^T R_{NN}(t, u)\phi_i(u)\, du\right) dt \end{aligned} \tag{10.98}$$

Substituting the eigen function property for the Karhunen-Loeve CON basis, Eq. (10.96), in the term in parentheses of (10.98), we obtain

$$\begin{aligned} E[N_i^2 \mid C_k] &= \int_0^T \phi_i(t)\left(\int_0^T R_{NN}(t, u)\phi_i(u)\, du\right) dt \\ &= \int_0^T \phi_i(t)\big(\lambda_i\phi_i(t)\big)\, dt \end{aligned} \tag{10.99}$$

By the orthonormality property of the Karhunen-Loeve basis functions, the integral above gives λ_i. So the diagonal elements of the conditional covariance matrices under each class are

$$E[N_i^2 \mid C_k] = \lambda_i, \qquad k = 1, 2;\ i = 1, 2, \ldots \tag{10.100}$$

The off-diagonal components of the conditional covariance matrices under each class are zero from the assumed property given in (10.94). Thus the infinite-dimensional conditional covariance matrices are diagonal, with diagonal components equaling the eigen values. Thus the vector observation $\mathbf{x}$ in (10.90) under each class is Gaussian, and the mean vectors and covariance matrices determined above give the following infinite-dimensional vector formulation of our problem:

$$\begin{aligned} &C_1\colon \quad \mathbf{x} \sim N(\mathbf{s}_1, \mathbf{K}_d) \\ &C_2\colon \quad \mathbf{x} \sim N(\mathbf{s}_2, \mathbf{K}_d) \\ &\text{where} \quad \mathbf{K}_d = \operatorname{diag}(\lambda_1, \lambda_2, \ldots, \lambda_\infty) \end{aligned} \tag{10.101}$$

These conditions are exactly the same as the Gaussian special case 4, in Chapter 9, so the optimum decision rule from Eq. (9.88) can be used with $\mathbf{K} = \mathbf{K}_d$ and $\mathbf{m}_i = \mathbf{s}_i$. The resulting optimal decision rule computes $Z_1(x)$ and $Z_2(x)$ and selects the class corresponding to the maximum of the two where the $Z_i(x)$ are given by

$$Z_i(\mathbf{x}) = \mathbf{s}_i^T \mathbf{K}_d^{-1}\mathbf{x} - \frac{1}{2}\mathbf{s}_i^T \mathbf{K}_d^{-1}\mathbf{s}_i + \ln P(C_i) \qquad \text{for } i = 1, 2 \tag{10.102}$$

Substituting $\mathbf{K}_d^{-1} = \text{diag}(1/\lambda_1, 1/\lambda_2, \ldots)$ into (10.102), and multiplying out, gives the $Z_1(x)$ and $Z_2(x)$ as

$$Z_i(\mathbf{x}) = \sum_{j=1}^{\infty} \frac{s_{ij}\mathbf{x}_j}{\lambda_j} - \frac{1}{2}\sum_{j=1}^{\infty} \frac{s_{ij}^2}{\lambda_j} + \ln P(C_i), \qquad i = 1, 2 \tag{10.103}$$

The structure of the optimum decision rule is similar to that shown in Figure 10.1 except for the $1/\lambda_i$ scaling of the branches resulting from the diagonal elements of K_d. This optimal decision rule for the ANWGN case can be implemented as shown in Figure 10.10.

As shown in Example 10.1 the problem simplified for the AWGN problem to a finite number of branches by choosing the orthonormal basis to contain normalized versions of signals $s_1(t)$ and $s_2(t)$. Since the basis functions are now determined by the Karhunen-Loeve basis, this simplification will not occur unless $s_1(t)$ and $s_2(t)$ happen to be in the KL basis or can be written as a linear combination of a finite number of the KL basis functions. A brief discussion of the KL expansion will now be presented followed by an example of finding the eigen functions and values for a given $R_{XX}(t, u)$.

10.3.2 The Karhunen-Loeve Expansion

The following theorem from Papoulis [6] gives conditions for convergence and expansion of a random process in terms of a set of orthonormal functions.

Theorem. Given that $x(t)$ is a mean square continuous zero mean random process with finite $E[x^2(t)]$ for all t, it can be expanded in terms of an orthonormal basis

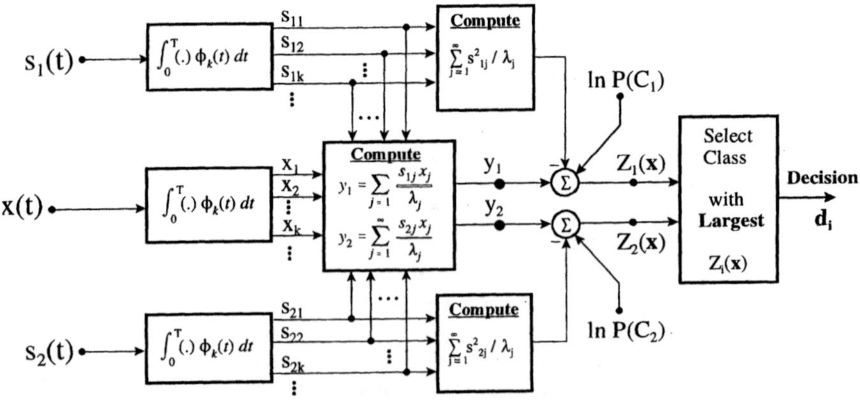

Figure 10.10 Optimum detector for the continuous observation ANWGN binary case with deterministic $s_1(t)$ and $s_2(t)$.

$\{\phi_j(t): j = 1, 2, \ldots\}$ on the interval $[0, T]$ such that we have convergence in the mean square sense as follows:

$$x(t) = \underset{J\to\infty}{\text{l.i.m.}} \sum_{j=1}^{J} a_j \phi_j(t) \tag{10.104}$$

where l.i.m. is the limit in the mean square sense as follows:

$$\lim_{J\to\infty} E\left[\left(x(t) - \sum_{j=1}^{J} a_j \phi_j(t)\right)^2\right] = 0 \qquad \text{for all } t \in [0, T] \tag{10.105}$$

The **Karhunen-Loeve basis** is the set $\{\phi_j(t): j = 1, 2, \ldots\}$ such that the $\phi_j(t)$ are solutions of the integral equation

$$\int_0^T R_{XX}(t, u)\phi_j(u)\, du = \lambda_j \phi_j(t) \tag{10.106}$$

The $R_{XX}(t, u)$ is called the **kernel** of the integral equation, and λ_j and $\phi_j(t)$ are called the **eigen values** and **eigen functions**, respectively, for the integral equation (10.106). A kernel is defined as **positive definite** if

$$\int_0^T \int_0^T g(t) R_{XX}(t, u) g(u)\, dt\, du > 0 \tag{10.107}$$

for all $g(t)$, not identically equal to zero and square integrable on $[0, T]$. If the integral equals zero for some nonidentically zero $g(t)$ that is square integrable on $[0, T]$, then the kernel is called **positive semidefinite**.

A kernel $R_{XX}(t, u)$ is called a **separable kernel** if it can be written as a finite number of products of eigen functions in u and in t as follows:

$$R_{XX}(t, u) = \sum_{i=1}^{N} b_i \phi_i(t)\phi_i(u) \tag{10.108}$$

Choosing the basis functions as the eigen functions of the integral equation it can be shown that the following properties hold for the coefficients in the Karhunen-Loeve expansion of a random process:

(1) The coefficients a_i and a_j of the KL expansion are random variables such that $E[a_i a_j] = 0$ for all $i \neq j$. Thus, if $X(t)$ is a Gaussian random process, then the coefficients are also independent random variables.
(2) If $R_{XX}(t, u)$ is positive definite, then the set of eigen functions $\phi_j(t)$ is a complete orthonormal set on the interval $[0, T]$.
(3) $E[a_i^2] = \lambda_i$ for all i.
(4) $E[X^2(t)] = \sum_{j=1}^{\infty} \lambda_j$.
(5) If $R_{XX}(t, u)$ is **positive semidefinite**, then the set of eigen functions $\phi_j(t)$ do not form a complete orthonormal set on the interval $[0, T]$, and they would need to be augmented with additional normal functions to make a CON.

(6) If $R_{XX}(t, u)$ is **positive definite**, it can always be written in terms of the eigen functions and values as

$$R_{XX}(t, u) = \sum_{i=1}^{\infty} \lambda_i \phi_i(t)\phi_i(u) \tag{10.109}$$

(7) If $R_{XX}(t, u)$ is **positive semidefinite**, then $R_{XX}(t, u)$ is a separable kernel and can be expressed as a finite sum of eigen functions as follows:

$$R_{XX}(t, u) = \sum_{i=1}^{N} \lambda_i \phi_i(t)\phi_i(u) \tag{10.110}$$

Details for these properties are given in Van Trees [1]. An example of a Karhunen-Loeve expansion for a random process with known positive semidefinite autocorrelation function is now presented.

EXAMPLE 10.5

A zero mean nonstationary random process $X(t)$ is defined by

$$X(t) = A_1 s_1(t) + A_2 s_2(t)$$

where $s_1(t)$ and $s_2(t)$ are known deterministic signals while A_1 and A_2 are independent Gaussian random variables characterized by $A_1 \sim N(0, \sigma_1^2)$ and $A_2 \sim N(0, \sigma_2^2)$. The auto-correlation function $R_{XX}(t, u)$ for the process $X(t)$ can easily be shown to be

$$R_{XX}(t, u) = \sigma_1^2 s_1(t)s_1(u) + \sigma_2^2 s_2(t)s_2(u)$$

This $X(t)$ could model a two-clutter pulse interference or signal. Assume, for convenience, that

$$\int_0^T s_1^2(t)\,dt = 1, \quad \int_0^T s_2^2(t)\,dt = 1, \quad \text{and} \quad \int_0^T s_1(t)s_2(t)\,dt = E_{12}$$

Find the Karhunen-Loeve basis functions and associated eigen values for expanding the process $X(t)$ on the interval $[0, T]$.

SOLUTION

To get the basis functions, we must solve the integral equation (10.106). Notice that the kernel is a sum of two products and will turn out to be separable. Because of this, the solution for the eigen values and functions is simplified. The overall approach will be to substitute the autocorrelation function into the integral equation, recognize the form of the solution, substitute this solution for the basis functions back into the integral equation, and match up the signals.

The approach begins by substituting the covariance function above into the integral equation (10.106) to get

$$\int_0^T [\sigma_1^2 s_1(t)s_1(u) + \sigma_2^2 s_2(t)s_2(u)]\phi_j(u)\, du = \lambda_1 \phi_j(t)$$

After multiplying out we can rearrange the above equation as

$$\sigma_1^2 s_1(t) \int_0^T s_1(u)\phi_j(u)\, du + \sigma_2^2 s_2(t) \int_0^T s_2(u)\phi_j(u)\, du = \lambda \phi_j(t)$$

Notice that the eigen values are just numbers, as are the integrals once the $\phi_j(t)$ are known. Thus the form of $\phi_j(t)$ is seen to be the weighted sum of $s_1(t)$ and $s_2(t)$ as follows:

$$\phi_j(t) = C_1 s_1(t) + C_2 s_2(t)$$

Because $\phi_j(t)$ is an eigen function, it is by definition that we have

$$\int_0^T \phi_j^2(t)\, dt = 1$$

Substituting the form of the eigen function into this equation, multiplying out, and rearranging gives the following constraining equation that C_1 and C_2 must satisfy:

$$C_1^2 + 2C_1C_2E_{12} + C_2^2 = 1$$

A second equation in C_1 and C_2 will be obtained by substituting the form for the eigen function back into the original integral equation to get

$$\int_0^T [\sigma_1^2 s_1(t)s_1(u) + \sigma_2^2 s_2(t)s_2(u)][C_1 s_1(u) + C_2 s_2(u)]\, du = \lambda_j[C_1 s_1(t) + C_2 s_2(t)]$$

Integrating out the terms on the left, bringing the terms on the right to the left, and collecting like terms the equation above can be written as

$$(\sigma_1^2 C_1 + \sigma_1^2 C_2 E_{12} - \lambda_j C_1)s_1(t) + (\sigma_2^2 E_{12} C_1 + \sigma_2^2 C_2 - \lambda_j C_2)s_2(t) = 0$$

Since $s_1(t)$ and $s_2(t)$ are not identical and not identically equal to zero, the only way for the equation to be satisfied for all t is for both terms in parentheses to equal zero or

$$(\sigma_1^2 C_1 + \sigma_1^2 C_2 E_{12} - \lambda_j C_1) = 0, \quad (\sigma_2^2 E_{12} C_1 + \sigma_2^2 C_2 - \lambda_j C_2) = 0$$

Solving the first equation above for λ_j gives the eigen value as

$$\lambda_j = \sigma_1^2 + \frac{\sigma_1^2 E_{12} C_2}{C_1}$$

Substituting this result into the second equation and rearranging gives the following equation that the C_1 and C_2 must satisfy:

$$\sigma_2^2 E_{12} C_1 + \sigma_2^2 C_2 - \frac{C_2(\sigma_1^2 C_1 + \sigma_1^2 C_2 E_{12})}{C_1} = 0$$

Then, multiplying through by C_1 and rearranging, we reduce this equation to

$$E_{12}\sigma_2^2 C_1^2 + (\sigma_2^2 - \sigma_1^2) C_1 C_2 - E_{12}\sigma_1^2 C_2^2 = 0$$

which together with the constraining equation determined earlier as

$$C_1^2 + 2C_1 C_2 E_{12} + C_2^2 = 1$$

gives the solution for the C_1 and C_2.

Since the two equations are quadratic in C_1 and C_2, it is possible to have as many as four solutions giving two eigen values and functions. Although they look simple, a closed form analytical solution was not obtained for these equations.

Consider the special case where $\sigma_1^2 = 1$, $\sigma_2^2 = 2$, and $E_{12} = -0.5$, to be less abstract. For this case a computer solution gives the C_1, C_2, and the corresponding eigen values λ_j as follows:

$$\begin{array}{lll} C_1 = 1.115355 & C_2 = 0.8164968 & \lambda_1 = 0.6339747 \\ C_1 = -1.115355 & C_2 = -0.8164968 & \lambda_1 = 0.6339747 \\ C_1 = -0.298858 & C_2 = 0.8164968 & \lambda_2 = 2.366018 \\ C_1 = 0.298858 & C_2 = -0.8164968 & \lambda_2 = 2.366018 \end{array}$$

Notice that the first two solutions are negatives of each other as are the third and fourth. Therefore selecting the first and third gives one possible solution with the following eigen functions and eigen values:

$$\begin{array}{ll} \phi_1(t) = 1.115355 s_1(t) + 0.8164968 s_2(t), & \lambda_1 = 0.6339747 \\ \phi_2(t) = -0.298858 s_1(t) + 0.8164968 s_2(t), & \lambda_2 = 2.366018 \end{array}$$

Selecting any combination of one of the functions with eigen value $\lambda_1 = 0.6339747$ and one with eigen value $\lambda_2 = 2.366018$ will give other alternative solutions.

As a check on the solution given above, it is easy to show that

$$\int_0^T \phi_1(t)\phi_2(t)\,dt = 0$$

It is also interesting to note that the covariance function is of separable kernel form and can be written as

$$K_{XX}(t, u) = \lambda_1 \phi_1(t)\phi_1(t) + \lambda_2 \phi_2(t)\phi_2(t)$$

Substituting the $\phi_1(t)$ and $\phi_2(t)$ determined above into the equation above, the $K_{XX}(t, u)$ is seen to be

$$\begin{aligned}K_{XX}(t, u) &= \lambda_1(1.115355 s_1(t) + 0.8164968 s_2(t))(1.115355 s_1(u) + 0.8164968 s_2(u)) \\ &\quad + \lambda_2(-0.298858 s_1(t) + 0.8164968 s_2(t))(-0.298858 s_1(u) \\ &\quad + 0.8164968 s_2(u))\end{aligned}$$

Finally substituting in the corrresponding eigen values determined above, multiplying out, and collecting like terms gives the covariance function as

$$K_{XX}(t, u) = 1 s_1(t) s_1(u) + 2 s_2(t) s_2(u)$$

This is the given covariance function. ■

In general, the analytical solution for the Karhunen-Loeve basis can be found by first changing the integral equation into a differential equation, and then solving the differential equation for the form of the eigen functions, substituting these functional forms back into the integral equation, and matching up functions to obtain equations for the eigen values. The procedure, although straightforward, is very messy and time-consuming to carry out, and the resulting equations usually must be solved numerically.

Examples of this procedure are given in Davenport and Root [7] and Van Trees [1] for the simplest form of autocorrelation function $R_{XX}(\tau) = Pe^{-\alpha|\tau|}$ and an expansion on the interval $[-T, T]$. Their results show that the eigen functions are nonharmonically related sines and cosines given by

$$\phi_j(t) = \begin{cases} C_j \cos b_j t, & j = 1, 3, 5, \ldots \\ S_j \sin b_j t, & j = 2, 4, 6, \ldots \end{cases} \tag{10.111}$$

where the C_j and S_j are as follows:

$$C_j = T^{-1/2}\left(1 + \frac{\sin 2b_j T}{2b_j T}\right)^{-1/2}, \quad S_j = T^{-1/2}\left(1 - \frac{\sin 2b_j T}{2b_j T}\right)^{-1/2} \tag{10.112}$$

and the b_j are the numerical solutions, respectively, of the equations

$$\tan b_j T - \frac{\alpha}{b_j} = 0, \quad \tan b_j T + \frac{b_j}{\alpha} = 0 \tag{10.113}$$

The eigen values corresponding to the $\cos(b_j t)$ and $\sin(b_j t)$ were shown to be the values of the power spectral density at the corresponding frequency or

$$\lambda_j = \frac{2P\alpha}{\alpha^2 + b_j^2} \tag{10.114}$$

To determine the KL expansion, we find that for most practical problems a numerical solution may be the only feasible method of solution. For the special case of wide sense stationary processes where the power spectral density is a rational expression in ω^2 and the expansion is for the entire interval $-\infty$ to ∞, Davenport and Root [7] give a Fourier transform method for solving for the eigen functions and values, but the application is cumbersome even for moderate order rational expressions.

10.4 DETECTION OF KNOWN SIGNALS IN COMBINATION OF WHITE AND NONWHITE GAUSSIAN NOISE (AW&NWGN)

In many practical cases the observed signal is an additive combination of white and nonwhite Gaussian noise and a deterministic signal. With this additive white noise component we can circumvent the problem of singularity and obtain meaningful solutions. Our discussion of this type of problem begins with the two class case.

10.4.1 Two-Class AW&NWGN Detection

Suppose that one of two possible known deterministic signals $s_1(t)$ and $s_2(t)$ is observed after being passed through an **additive white and nonwhite Gaussian noise** channel. The continuous observation signal $x(t)$ under each class is given by

$$\begin{aligned} C_1:&\quad x(t) = s_1(t) + N_C(t) + N_W(t), \qquad 0 \le t \le T \\ C_2:&\quad x(t) = s_2(t) + N_C(t) + N_W(t), \qquad 0 \le t \le T \end{aligned} \tag{10.115}$$

Assume that $P(C_1)$ and $P(C_2)$ are known, $N_C(t)$ is nonwhite Gaussian process characterized by an autocorrelation function $R_{CC}(t, u)$, and $N_W(t)$ is a zero mean white Gaussian process with autocorrelation function $R_{WW}(t, u) = \frac{1}{2}N_0\delta(t - u)$. Assume also that $N_C(t)$ and $N_W(t)$ are independent random processes. The result to be presented does not require that the $N_C(t)$ be wide sense stationary, but we will assume that it has a zero mean for convenience in the development. If it does not have a zero mean, the problem can be changed to one with a zero mean by subtracting the mean for the observation, giving a new observation.

Our objective is to find the optimum decision rule in the minimum probability of error sense. In the previous sections the continuous time observations were first changed into an infinite-dimensional vector of Gaussian components by using an orthonormal expansion whose choice depended on the deterministic signals used to simplify the solution. However, for this problem a different orthonormal basis will be used. If we expand the $x(t)$ given in (10.115) using an arbitrary CON basis $\{\phi_i(t)\colon i = 1, 2, \ldots\}$, then the equivalent countably infinite formulation becomes

$$\begin{aligned} C_1:&\quad \mathbf{x} = \mathbf{s}_1 + \mathbf{N}_C + \mathbf{N}_W \\ C_2:&\quad \mathbf{x} = \mathbf{s}_2 + \mathbf{N}_C + \mathbf{N}_W \end{aligned} \tag{10.116}$$

The $\mathbf{s}_1$, $\mathbf{s}_2$, $\mathbf{N}_C$, and $\mathbf{N}_W$ have components s_{1i}, s_{2i}, N_{Ci}, and N_{Wi} as coefficients in their expansion in the basis functions as

$$\begin{aligned} s_{1i} &= \int_0^T s_1(t)\phi_i(t)\,dt, \quad N_{Ci} = \int_0^T N_C(t)\phi_i(t)\,dt, \quad i = 1, 2, \ldots \\ s_{2i} &= \int_0^T s_2(t)\phi_i(t)\,dt, \quad N_{Wi} = \int_0^T N_W(t)\phi_i(t)\,dt, \quad i = 1, 2, \ldots \end{aligned} \tag{10.117}$$

The N_{Ci} and N_{Wi} are Gaussian random variables since they are integrals of the Gaussian random processes $N_C(t)$ and $N_W(t)$, respectively, and thus $\mathbf{N}_C$ and $\mathbf{N}_W$ are infinite-dimensional Gaussian random vectors. The vectors $\mathbf{N}_C$ and $\mathbf{N}_W$ are independent, since the $N_C(t)$ and $N_W(t)$ are independent and functions of independent random variables are independent. It is convenient to select the Karhunen-Loeve basis for $N_C(t)$ as the CON set. In this way the components of $\mathbf{N}_C$ are uncorrelated and the components of $\mathbf{N}_W$ are uncorrelated, since they are uncorrelated for any choice of orthonormal basis functions where $N_W(t)$ is a Gaussian white random process.

The observation vector $\mathbf{x}$ under each class is Gaussian and is characterized by its mean vector and covariance matrix. Using (10.117), we see that

$$E[N_{Ci} \mid C_1] = E[N_{Ci} \mid C_2] = 0 \tag{10.118}$$

since $E[N_C(t) \mid C_1] = E[N_C(t) \mid C_1] = 0$ for all t. Similarly

$$E[N_{Wi} \mid C_1] = E[N_{Wi} \mid C_2] = 0 \tag{10.119}$$

since $E[N_W(t) \mid C_1] = E[N_W(t) \mid C_1] = 0$ for all t. Thus the mean vectors of $\mathbf{N}_C$ and $\mathbf{N}_W$ conditioned on each class can be written as

$$\begin{aligned} E[\mathbf{N}_C \mid C_1] &= \mathbf{0}, \qquad E[\mathbf{N}_W \mid C_1] = \mathbf{0} \\ E[\mathbf{N}_C \mid C_2] &= \mathbf{0}, \qquad E[\mathbf{N}_W \mid C_2] = \mathbf{0} \end{aligned} \tag{10.120}$$

Taking the expected value of Eq. (10.116), and using (10.120), we find that the means for the observation $\mathbf{x}$ conditioned on each class become

$$E[\mathbf{x} \mid C_1] = \mathbf{s}_1, \quad E[\mathbf{x} \mid C_2] = \mathbf{s}_2 \tag{10.121}$$

Subtracting these conditional mean vectors from Eq. (10.116), and since $N_C(t)$ and $N_W(t)$ are independent, it is easy to show $\mathbf{K}_1 = \mathbf{K}_2$ and that the conditional covariance matrices are

$$\mathbf{K}_1 = \mathbf{K}_2 = \mathbf{K}_{CC} + \mathbf{K}_{WW} \tag{10.122}$$

Since the CON set used to generate $\mathbf{N}_C$ was the Karhunen-Loeve basis for $N_C(t)$, the $\mathbf{K}_{CC}$ is a diagonal matrix with components equal to the eigen values associated with the eigen functions for the kernel $R_{CC}(t, u)$

$$\mathbf{K}_{CC} = \operatorname{diag}(\lambda_1, \lambda_2, \ldots, \lambda_k, \ldots) \tag{10.123}$$

Because $N_W(t)$ is white noise the $\mathbf{K}_{WW}$ is a diagonal matrix with diagonal components equal to $\frac{1}{2}N_0$ as follows:

$$\mathbf{K}_{WW} = \operatorname{diag}\left(\frac{N_0}{2}, \frac{N_0}{2}, \ldots, \frac{N_0}{2}, \ldots\right) \tag{10.124}$$

Substituting the above $\mathbf{K}_{WW}$ and $\mathbf{K}_{CC}$ into (10.122) finally gives the $\mathbf{K}_1$ and $\mathbf{K}_2$ as follows:

$$\begin{aligned}\mathbf{K}_1 &= \mathbf{K}_2 \triangleq \mathbf{K} \\ &= \operatorname{diag}\left(\lambda_1 + \frac{N_0}{2}, \lambda_2 + \frac{N_0}{2}, \ldots, \lambda_k + \frac{N_0}{2}, \ldots\right)\end{aligned} \tag{10.125}$$

From (10.116) $\mathbf{x}$ is an infinite-dimensional Gaussian vector with conditional means $\mathbf{s}_1$ and $\mathbf{s}_2$ that are the coefficient vectors for the expansion of $s_1(t)$ and $s_2(t)$ in the Karhunen-Loeve basis for $N_C(t)$, and conditional covariance matrices $\mathbf{K}$ given by Eq. (10.125). Thus an equivalent representation of our problem can be stated as follows:

$$\begin{aligned}C_1: &\quad \mathbf{x} \sim N(\mathbf{s}_1, \mathbf{K}) \\ C_2: &\quad \mathbf{x} \sim N(\mathbf{s}_2, \mathbf{K})\end{aligned} \tag{10.126}$$

Using the results for Gaussian vectors and equal covariance matrices from (9.88), the optimum test in the minimum probability of error (MPE) sense allow us to compute the following $L_i(\mathbf{x})$ for $i = 1, 2$ and select the class corresponding to the larger value:

$$L_i(\mathbf{x}) = \mathbf{s}_i^T \mathbf{K}^{-1} \mathbf{x} - \tfrac{1}{2} \mathbf{s}_i^T \mathbf{K}^{-1} \mathbf{s}_i + \ln P(C_i) \qquad \text{for } i = 1, 2 \tag{10.127}$$

Substituting the $\mathbf{K}$ matrix of (10.125) into the $L_i(\mathbf{x})$ above gives the optimum test statistics:

$$L_i(\mathbf{x}) = \sum_{j=1}^{\infty} s_{ij} x_j \frac{1}{\lambda_j + (\frac{1}{2})N_0} - \frac{1}{2} \sum_{j=1}^{\infty} s_{ij}^2 \frac{1}{\lambda_j + (\frac{1}{2})N_0} + \ln P(C_i), \qquad i = 1, 2 \tag{10.128}$$

The optimum test is to select the class that corresponds to the greater value of the $L_i(\mathbf{x})$. The block diagram for the implementation of this optimum decision rule is shown in Figure 10.11. The optimum detector of Figure 10.11 is slightly simplified when the nonwhite Gaussian component is of separable kernel, as seen in the next section.

10.4.2 Two-Class AW&NWGN Detection (Separable Kernel)

When the nonwhite component $N_C(t)$ is of separable kernel, its autocorrelation function can be written, for a finite integer J, in the following way:

$$R_{XX}(t, u) = \sum_{j=1}^{J} \lambda_j \phi_j(t) \phi_j(u) \tag{10.129}$$

where the λ_i and $\phi_i(t)$ are the eigen values and functions of $R_{CC}(t, u)$. We would like to find the optimal detector for this case. Therefore the $\mathbf{K}_{CC}$ has zero entries along the main diagonal past the Jth component as

$$\mathbf{K}_{CC} = \operatorname{diag}\left(\lambda_1 + \frac{N_0}{2}, \lambda_2 + \frac{N_0}{2}, \ldots, \lambda_J + \frac{N_0}{2}, 0, \ldots, 0, \ldots\right) \tag{10.130}$$

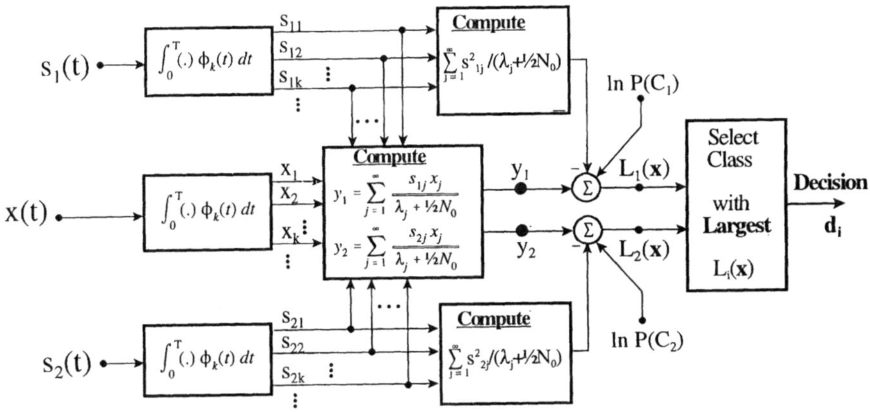

Figure 10.11 Optimum detector for the two class case with deterministic signals $s_1(t)$ and $s_2(t)$ in additive white, $R_{WW}(t, u) = \frac{1}{2}N_0\,\delta(t-u)$, and nonwhite, $R_{CC}(t, u)$

and the covariance matrix $\mathbf{K}$ of the previous section simplifies slightly to

$$\mathbf{K} = \text{diag}\left(\lambda_1 + \frac{N_0}{2}, \lambda_2 + \frac{N_0}{2}, \ldots, \lambda_J + \frac{N_0}{2}, \frac{N_0}{2}, \ldots, \frac{N_0}{2}, \ldots\right) \tag{10.131}$$

Substituting this K into Eq. (10.127) gives

$$\begin{aligned} L_i(\mathbf{x}) = {} & \sum_{j=1}^{J} s_{ij}x_j \frac{1}{\lambda_j + (1/2)N_0} + \sum_{j=J+1}^{\infty} s_{ij}x_j \frac{2}{N_0} \\ & -\frac{1}{2}\sum_{j=1}^{J} s_{ij}^2 \frac{1}{\lambda_j + (1/2)N_0} - \frac{1}{2}\sum_{j=J+1}^{\infty} s_{ij}^2 \frac{2}{N_0} + \ln P(C_i), \qquad i = 1, 2 \end{aligned} \tag{10.132}$$

and the optimal classifier selects the class corresponding to the larger of the $L_i(\mathbf{x})$. This result is similar to the general result shown in Figure 10.11 up to J and slightly different for the remainder of the sum.

10.5 OPTIMUM CLASSIFIER FOR GENERAL GAUSSIAN PROCESSES (TWO-CLASS DETECTION)

There are many problems that can be formulated where we do not have the additive structure given in the previous sections. A basic problem formulation for the two-class case is now presented for this slightly more general type of problem.

Suppose that one of two possible random processes are observed over the interval $[0, T]$. The continuous observation signal $x(t)$ under each class is given by:

$$\begin{aligned} C_1&: \quad x(t) = Y_1(t), \qquad 0 \le t \le T \\ C_2&: \quad x(t) = Y_2(t), \qquad 0 \le t \le T \end{aligned} \tag{10.133}$$

It will be assumed that $P(C_1)$ and $P(C_2)$ are known, that $Y_1(t)$ is a realization from a zero mean Gaussian random process characterized by its autocorrelation function $R_{11}(t, u)$, and that $Y_2(t)$ is a realization from a zero mean Gaussian random process characterized by its autocorrelation function $R_{22}(t, u)$. Neither of the processes is necessarily a white process.

Our objective is to find the optimum decision rule in the minimum probability of error sense. In the previous sections the continuous time observations were first changed in to an infinite-dimensional vector of Gaussian components by an orthonormal expansion whose choice depended on the deterministic signals or the nonwhite spectral density to simplify the solution. However, for this problem no deterministic signals are identified, and there is not necessarily a white noise component. Since there are two different processes, the choice of one orthonormal set may make uncorrelated components for one class but not for the other. So a logical question is, Which orthonormal set provides a convenient solution? If we expand the $x(t)$ given in (10.133) using an arbitrary CON basis $\{\phi_i(t): i = 1, 2, \ldots\}$, then the equivalent countably infinite formulation becomes

$$\begin{aligned} C_1&: \quad \mathbf{x} = \mathbf{y}_1 \\ C_2&: \quad \mathbf{x} = \mathbf{y}_2 \end{aligned} \tag{10.134}$$

where the $\mathbf{y}_1$ and $\mathbf{y}_2$ have components y_{1i} and y_{2i} as the coefficients in their expansion in the basis functions as follows:

$$\begin{aligned} y_{1i} &= \int_0^T Y_1(t)\phi_i(t)\,dt, \qquad i = 1, 2, \ldots \\ y_{2i} &= \int_0^T Y_2(t)\phi_i(t)\,dt, \qquad i = 1, 2, \ldots \end{aligned} \tag{10.135}$$

The y_{1i} and y_{2i} are Gaussian random variables, since they are integrals of the Gaussian random processes $Y_1(t)$ and $Y_2(t)$, respectively, and thus $\mathbf{y}_1$ and $\mathbf{y}_2$ are infinite-dimensional Gaussian random vectors. It is convenient to select the Karhunen-Loeve basis for $Y_1(t)$ as the CON set. In this way the components of $\mathbf{y}_1$ are uncorrelated while the components of $\mathbf{y}_2$ are not necessarily uncorrelated.

The observation vector $\mathbf{x}$ under each class is Gaussian and is thus characterised by its mean vector and covariance matrix. Using (10.135), we see that $E[y_{1i} \mid C_1] = 0$, since $E[Y_1(t) \mid C_1] = 0$ for all t. Similarly $E[y_{1i} \mid C_1] = 0$, since $E[Y_1(t) \mid C_1] = 0$ for all t. Thus the mean vectors of $\mathbf{y}_1$ and $\mathbf{y}_2$ conditioned on each class can be written as

$$E[\mathbf{y}_1 \mid C_1] = \mathbf{0}, \qquad E[\mathbf{y}_2 \mid C_2] = \mathbf{0} \tag{10.136}$$

Taking the expected value of Eq. (10.134), and using (10.135), we find that the mean for the observation $\mathbf{x}$ conditioned on each class becomes

$$E[\mathbf{x} \mid C_1] = \mathbf{0}, \qquad E[\mathbf{x} \mid C_2] = \mathbf{0} \tag{10.137}$$

Since the CON set used to generate $\mathbf{y}_1$ was the Karhunen-Loeve basis for $Y_1(t)$, the $\mathbf{K}_1$ for the observation $\mathbf{x}$ under class 1 is a diagonal matrix with components equal to the eigen values associated with the eigen functions for the kernel $R_{11}(t, u)$:

$$\mathbf{K}_1 = \text{diag}(\lambda_1, \lambda_2, \ldots, \lambda_k, \ldots) \tag{10.138}$$

However, the covariance matrix for the observation $\mathbf{x}$ under class C_2 is not necessarily diagonal but given by K_2. The covariance matrices $\mathbf{K}_1$ and $\mathbf{K}_2$ are not necessarily equal, so the optimum decision rule using the general Gaussian case from (9.72) is as follows:

$$\text{If} \quad -(\mathbf{x}-\mathbf{m}_1)^T\mathbf{K}_1^{-1}(\mathbf{x}-\mathbf{m}_1) + (\mathbf{x}-\mathbf{m}_2)^T\mathbf{K}_2^{-1}(\mathbf{x}-\mathbf{m}_2) \begin{array}{ll} > T & \text{decide } \omega_1 \\ < & \text{decide } \omega_2 \\ = & \text{decide randomly} \end{array}$$

$$\text{where} \quad T = 2\ln\eta_{\text{BAYES}} + \ln\frac{|\mathbf{K}_1|}{|\mathbf{K}_2|} \quad \text{and} \quad \eta_{\text{BAYES}} = \frac{(C_{22}-C_{12})\,P(C_2)}{(C_{11}-C_{21})\,P(C_1)} \tag{10.139}$$

Using a zero-one cost assignment and the fact that $\mathbf{m}_1$ and $\mathbf{m}_2$ are zero, we can simplify the decision rule to

$$\text{If} \quad -\mathbf{x}^T\mathbf{K}_1^{-1}\mathbf{x} + \mathbf{x}^T\mathbf{K}_2^{-1}\mathbf{x} \begin{array}{ll} > T & \text{decide } C_1 \\ < & \text{decide } C_2 \\ = & \text{decide randomly} \end{array}$$

$$\text{where} \quad T = 2\ln\eta_{\text{BAYES}} + \ln\frac{|\mathbf{K}_1|}{|\mathbf{K}_2|} \quad \text{and} \quad \eta_{\text{BAYES}} = \frac{(C_{22}-C_{12})\,P(C_2)}{(C_{11}-C_{21})\,P(C_1)} \tag{10.140}$$

Thus the optimum classifier in the Bayes sense is a messy quadratic function, and the inverse of the infinite-dimensional nondiagonal $\mathbf{K}_2$ is virtually impossible to obtain in general.

10.6 DETECTION OF KNOWN SIGNALS WITH RANDOM PARAMETERS IN ADDITIVE WHITE GAUSSIAN NOISE

Certain problems in radar and communication theory can be modeled by random processes where some signal parameters are random with known joint probability density function. For example, the received signal might have a random amplitude but otherwise be deterministic; in radar the signal returned or observed may have both a random amplitude and random phase with specified joint probability density functions. In the following sections detection problems of this particular type will be explored. It will be shown that some problems involving known Gaussian parameters can be solved similar to the procedure discussed for detection of known deterministic signals in additive white and nonwhite Gaussian channels, whereas non-Gaussian problems are, for the most part, intractable.

In general, we can write the detection of known signals for the **two-class case with random parameters in additive white Gaussian noise** as follows:

$$\begin{aligned} C_1:&\quad x(t) = s_1(t, \mathbf{a}_1) + N_W(t), \qquad 0 \le t \le T \\ C_2:&\quad x(t) = s_2(t, \mathbf{a}_2) + N_W(t), \qquad 0 \le t \le T \end{aligned} \tag{10.141}$$

Here $s_1(t, \cdot)$ and $s_2(t, \cdot)$ are known deterministic functions and $\mathbf{a}_1$ and $\mathbf{a}_2$ are random vectors given by

$$\mathbf{a}_1 = [A_1, A_2, \ldots, A_J]^T, \quad \mathbf{b}_1 = [B_1, B_2, \ldots, B_K]^T \tag{10.142}$$

with known probability density functions $p(\mathbf{a}_1 \mid C_1)$ and $p(\mathbf{a}_2 \mid C_2)$. It will be further assumed that the a priori probabilities of the classes, $P(C_1)$ and $P(C_2)$, are known and a performance measure of probability of error is specified. We would like to find the optimum decision rule for classifying $x(t)$ as coming from C_1 or C_2 where optimality is in the minimum probability of error sense.

A few examples of possible $s_1(t, \mathbf{a})$ are

(1) $s_1(t, A_1) = A_1 \sin \Omega_c t$ where A_1 is random and Ω_c is a known constant
(2) $s_1(t, A_1, A_2) = A_1 \sin(\Omega_c t + A_2)$ where A_1 and A_2 are random and Ω_c is a known constant
(3) $s_1(t, A_1, A_2) = A_1 \exp(-A_2 t)$ where A_1 and A_2 are random
(4) $s_1(t, A_1, A_2, A_3) = A_1 \cos(A_2 t + A_3)$ where A_1, A_2, and A_3 are random

Not all functions of random variables yield a tractable solution, for if $s_1(t, \mathbf{a})$ is not a Gaussian process, it is very difficult if not impossible to obtain an analytical solution. Thus rather than solve the general problem formulated, we will be content to present a number of representative problems that have tractable solutions and at the same time present useful and meaningful situations.

10.6.1 Detection of Signal with Random Amplitude in AWGN

Perhaps the simplest problem with a random parameter is where the amplitude of a deterministic signal is random. For this case $s(t, \mathbf{a}) = As_1(t, A)$, which can be formulated as

$$\begin{aligned} C_1:&\quad x(t) = As_1(t) + N_W(t), \qquad 0 \le t \le T \\ C_2:&\quad x(t) = N_W(t), \qquad 0 \le t \le T \end{aligned} \tag{10.143}$$

where $s_1(t)$ is an arbitrary but known deterministic signal, A is a Gaussian random variable with zero mean and variance σ_A^2, and $N_W(t)$ is a realization from a stationary white Gaussian process with autocorrelation function $R_{N_W N_W}(\tau) = \frac{1}{2} N_0 \, \delta(\tau)$. It will be assumed that the $P(C_1)$ and $P(C_2)$, the a priori probabilities of the classes, are known and that we desire the optimum detector in the sense of minimizing the probability of error. It will also be assumed that the random variable A is independent of the random process $N_W(t)$.

The general procedure in finding the optimum detector is the selection of a complete orthonormal set and an expansion of $x(t)$ on the interval $[0, T]$ using those basis functions followed by the use of previous Gaussian results for a countable observation space.

Again the choice of the CON is problem specific, and for this problem the obvious choice is to select $\phi_1(t)$ as the normalized version of $s_1(t)$:

$$\phi_1(t) = \frac{1}{\sqrt{E_1}} s_1(t) \quad \text{where } E_1 = \int_0^T s_1^2(t)\,dt \tag{10.144}$$

The other basis functions $\phi_i(t)$, $i = 1, 2, \ldots$, are arbitrarily selected to form a CON and will be shown not to affect the solution. Expanding the observation $x(t)$ over the interval $[0, T]$ leads to the countable vector observation

$$\begin{aligned} C_1&: \quad \mathbf{x} = \mathbf{S} + \mathbf{N} \\ C_2&: \quad \mathbf{x} = \mathbf{N} \end{aligned} \tag{10.145}$$

In the expression above the **S** and **N** are infinite-dimensional vectors. They have as elements the coefficients for the orthonormal expansion which are given as follows:

$$S_i = \int_0^T A s_1(t)\phi_i(t)\,dt, \quad N_i = \int_0^T N_W(t)\phi_i(t)\,dt, \qquad i = 1, 2, \ldots \tag{10.146}$$

Since $\phi_1(t)$ is $s_1(t)/\sqrt{E_1}$, the **S** vector has zeros for all coefficients except the first, which is as follows:

$$S_1 = \int_0^T A s_1(t)\phi_i(t)\,dt = \int_0^T A s_1(t) s_1(t) \frac{1}{\sqrt{E_1}}\,dt = A\sqrt{E_1} \tag{10.147}$$

Thus the **S** vector can be written as $S = [\sqrt{E_1}A, 0, 0, \ldots]^T$. It is seen that the first entry in **S** is a scaled version of the Gaussian random variable A, and since A was independent of $N_W(t)$, it will also be independent of its integrals with the $\phi_i(t)$ and thus independent of the components of **N**. The components of **x** under class C_1 are Gaussian, since they are sums of Gaussian random vectors. Thus the probability density of **x** can be characterized by its mean vector and covariance matrix as

$$\begin{aligned} C_1&: \quad \mathbf{x} \sim N(\mathbf{m}_1, \mathbf{K}_1) \\ C_2&: \quad \mathbf{x} \sim N(\mathbf{m}_2, \mathbf{K}_2) \end{aligned} \tag{10.148}$$

where $\mathbf{m}_1$ and $\mathbf{m}_2$ are easily determined by taking the expected value of **x** under both classes as

$$\begin{aligned} \mathbf{m}_1 &= [\sqrt{E_1}E[A] + E[N_1], E[N_2], E[N_3], \ldots]^T = \mathbf{0} \\ \mathbf{m}_2 &= [E[N_1], E[N_2], E[N_3], \ldots]^T = \mathbf{0} \end{aligned} \tag{10.149}$$

Under class C_1 the **S** and **N** are independent, so the covariance matrix of **x** becomes the sum of the covariance matrices of **S** and **N** as

$$\mathbf{K}_1 = \mathbf{K}_{SS} + \mathbf{K}_{NN} = \text{diag}\left[\sigma_A^2 + \frac{N_0}{2}, \frac{N_0}{2}, \frac{N_0}{2}, \ldots\right] \tag{10.150}$$

Under class C_2 the covariance matrix for **x** is simply the covariance matrix for **N** or

$$\mathbf{K}_2 = \mathbf{K}_{\mathbf{NN}} = \operatorname{diag}\left[\frac{N_0}{2}, \frac{N_0}{2}, \frac{N_0}{2}, \ldots\right] \tag{10.151}$$

The equivalent Gaussian formulation has diagonal covariance matrices, but they are not equal. Therefore the optimum decision rule for the general Gaussian case from (9.72) is

$$\begin{aligned} \text{If} \quad -(\mathbf{x}-\mathbf{m}_1)^T\mathbf{K}_1^{-1}(\mathbf{x}-\mathbf{m}_1) + (\mathbf{x}-\mathbf{m}_2)^T\mathbf{K}_2^{-1}(\mathbf{x}-\mathbf{m}_2) &> T \quad \text{decide } C_1 \\ &< \quad\;\; \text{decide } C_2 \\ &= \quad\;\; \text{decide randomly} \end{aligned}$$

$$\text{where} \quad T = 2\ln\frac{P(C_2)}{P(C_1)} + \ln\frac{|\mathbf{K}_1|}{|\mathbf{K}_2|} \tag{10.152}$$

Substituting the conditional mean vectors and covariance matrices from (10.149) to (10.151) into (10.152) the decision rule can be simplified as

$$\begin{aligned} \text{If} \quad -\mathbf{x}^T\mathbf{K}_1^{-1}\mathbf{x} + \mathbf{x}^T\mathbf{K}_2^{-1}\mathbf{x} &> T \quad \text{decide } C_1 \\ &< \quad\;\; \text{decide } C_2 \\ &= \quad\;\; \text{decide randomly} \\ \text{where} \quad T = 2\ln\frac{P(C_2)}{P(C_1)} &+ \ln\left(\sigma_A^2 + \frac{N_0}{2}\right) \end{aligned} \tag{10.153}$$

The left side of the decision rule can further be simplified by using the inverses of the covariance matrices:

$$\begin{aligned} -\mathbf{x}^T\mathbf{K}_1^{-1}\mathbf{x} + \mathbf{x}^T\mathbf{K}_2^{-1}\mathbf{x} &= -\mathbf{x}^T\operatorname{diag}\left[\frac{1}{\sigma_A^2 + (N_0/2)}, \frac{2}{N_0}, \frac{2}{N_0}, \ldots\right]\mathbf{x} \\ &\quad + \mathbf{x}^T\operatorname{diag}\left[\frac{2}{N_0}, \frac{2}{N_0}, \frac{2}{N_0}, \ldots\right]\mathbf{x} \\ &= \frac{2\sigma_A^2}{N_0(\sigma_A^2 + N_0/2)}x_1^2 \end{aligned} \tag{10.154}$$

After dividing through by the coefficient of the x_1^2 term in (10.154), we can rewrite the optimum decision rule from (10.153) as

$$\begin{aligned} \text{If} \quad x_1^2 > T_3 \triangleq \frac{N_0(\sigma_A^2 + N_0/2)}{2\sigma_A^2}\left(2\ln\frac{P(C_2)}{P(C_1)} + \ln\left(\sigma_A^2 + \frac{N_0}{2}\right)\right) &\quad \text{decide } C_1 \\ < &\quad \text{decide } C_2 \\ = &\quad \text{decide randomly} \end{aligned} \tag{10.155}$$

The optimum decision rule can be implemented as shown in Figure 10.12.

In the preceding section there was only one unknown random parameter in C_1 and none in class C_2. If there are random parameters in both classes, the procedure

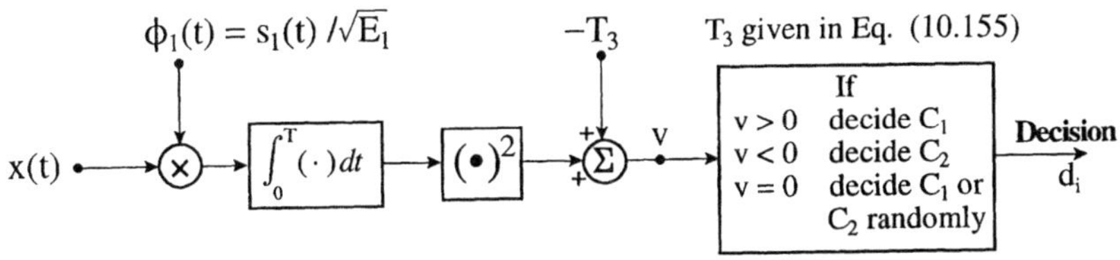

Figure 10.12 Optimum decision rule for detecting a signal $s_1(t)$ with random amplitude $A, A \sim N(0, \sigma_A^2)$, in additive white Gaussian noise of power spectral density $\frac{1}{2}N_0$.

described above does not change and there are essentially no difficulties should the parameters be linear and Gaussian. ■

10.6.2 Detection of Sinusoid with Random Amplitude and Phase

The following topic looks at a possible case involving two unknown random parameters where the densities of the random parameters are not Gaussian yet through the signal nonlinearity produces Gaussian random parameters. The problem is a basic one in communications, and it is to decide whether or not a received signal $x(t)$ contains a known sinusoidal signal that has a random phase and amplitude when observed in white Gaussian noise $N_W(t)$:

$$\begin{aligned} C_1&: \quad x(t) = A\sin(\omega_C t + \Theta) + N_W(t), \qquad 0 \le t \le T \\ C_2&: \quad x(t) = N_W(t), \qquad 0 \le t \le T \end{aligned} \tag{10.156}$$

where A and Θ are independent random variables. $N_W(t)$ is a realization of a white Gaussian wide sense stationary noise process with autocorrelation function $R_{WW}(\tau) = \frac{1}{2}N_0\,\delta(\tau)$. The random variable A is Raleigh distributed with parameter α and the Θ is uniformly distributed on $[-\pi, \pi]$ with the following probability density functions:

$$p_A(a) = \frac{1}{\alpha^2}\, a \exp\left(-\frac{1}{2\alpha^2}a^2\right)\mu(a), \quad p_\Theta(\theta) = \frac{1}{2\pi}\,[\mu(\theta+\pi) - \mu(\theta-\pi)] \tag{10.157}$$

It will be assumed that $\omega_C T$ is an integer multiple of 2π and that A and Θ are independent of the noise process $N_W(t)$ for all $t \in [0, T]$. We would like to find the optimal decision rule where optimality is in the minimum probability of error sense.

If $A\sin(\omega_C t + \Theta)$ is expanded out, we have under C_1 that $x(t)$ on $[0, T]$ is

$$\begin{aligned} x(t) &= A\sin(\omega_C t + \Theta) + N_W(t) \\ &= A\cos\Theta\ \sin(\omega_C t) + A\sin\Theta\cos(\omega_C t) + N_W(t) \\ &= B_1\sin(\omega_C t) + B_2\cos(\omega_C t) + N_W(t) \end{aligned} \tag{10.158}$$

Where B_1 and B_2 are random variables defined as follows:

$$B_1 = A\cos\Theta, \quad B_2 = A\sin\Theta \tag{10.159}$$

It can be easily shown and is well known that the transformation (two functions of two random variables) causes B_1 and B_2 to be independent Gaussian random variables distributed as

$$B_1 \sim N(0, \sigma_1^2 = \alpha^2), \quad B_2 \sim N(0, \sigma_2^2 = \alpha^2) \tag{10.160}$$

Here σ_1^2 and σ_2^2 were determined from the fact that A and Θ are independent random variables as follows:

$$\begin{aligned} \sigma_1^2 &= E[B_1^2] = E[A^2 \cos^2 \Theta] = E[A^2]E[\tfrac{1}{2}(1 - \cos 2\Theta)] = \tfrac{1}{2}E[A^2] = \alpha^2 \\ \sigma_2^2 &= E[B_2^2] = E[A^2 \ \sin^2 \Theta] = E[A^2]E[\tfrac{1}{2}(1 + \cos 2\Theta)] = \tfrac{1}{2}E[A^2] = \alpha^2 \end{aligned} \tag{10.161}$$

Substituting the results of (10.158) into (10.156) allows the observation $x(t)$ to be written for each class as

$$\begin{aligned} &C_1: \quad x(t) = B_1 \sin(\omega_C t) + B_2 \cos(\omega_C t) + N_W(t), \qquad 0 \le t \le T \\ &C_2: \quad x(t) = N_W(t), \qquad 0 \le t \le T \end{aligned} \tag{10.162}$$

In treating the sum of sinusoids as a nonwhite random process, we have the obvious choice for the CON set as the Fourier series basis

$$\phi_1(t) = \sqrt{\frac{2}{T}} \sin(\omega_C t), \quad \phi_2(t), = \sqrt{\frac{2}{T}} \cos(\omega_C t), \qquad 0 \le t \le T \tag{10.163}$$

and for $k = 2, 3, \ldots,$

$$\phi_{2k-1}(t) = \sqrt{\frac{2}{T}} \sin((2k-1)\omega_C t), \quad \phi_{2k}(t) = \sqrt{\frac{2}{T}} \cos(2k\omega_C t), \qquad 0 \le t \le T$$

Expanding the continuous observation $x(t)$ on that interval using the Fourier basis gives the countably infinite equivalent vector problem as follows:

$$\begin{aligned} &C_1: \quad \mathbf{x} = \mathbf{s} + \mathbf{c} + \mathbf{N} \\ &C_2: \quad \mathbf{x} = \mathbf{N} \end{aligned} \tag{10.164}$$

In the expressions above $\mathbf{s}$, $\mathbf{c}$, and $\mathbf{N}$ have components as follows:

$$\begin{aligned} s_i &= B_1 \int_0^T \sin(\omega_C t)\phi_i(t)\,dt, \qquad i = 1, 2, \ldots \\ c_i &= B_2 \int_0^T \cos(\omega_C t)\phi_i(t)\,dt, \qquad i = 1, 2, \ldots \\ N_i &= \int_0^T N_W(t)\phi_i(t)\,dt, \qquad i = 1, 2, \ldots \end{aligned} \tag{10.165}$$

Using the orthonormal properties of the Fourier basis, we find that the **s** and **c** vectors have all zeros except for the first two components. So **s**, **c**, and **N** become

$$\mathbf{s} = \begin{bmatrix} B_1\sqrt{\frac{T}{2}} \\ 0 \\ 0 \\ \vdots \end{bmatrix}, \quad \mathbf{c} = \begin{bmatrix} 0 \\ B_2\sqrt{\frac{T}{2}} \\ 0 \\ \vdots \end{bmatrix}, \quad \mathbf{N} = \begin{bmatrix} N_1 \\ N_1 \\ N_3 \\ \vdots \end{bmatrix} \tag{10.166}$$

Under C_1, we see that the observation vector **x** is the sum of **s**, **c**, and **N**. Since the components of N are integrals of the Gaussian process $N_W(t)$, the N is a Gaussian random vector, and since B_1 and B_2 are Gaussian, the vector **x** being the sum will also be a Gaussian random vector. Similarly, under C_2, the integrals of Gaussian random processes give Gaussian random variables. So the equivalent vector formulation is Gaussian, and we can express the problem as

$$\begin{aligned} C_1&: \quad \mathbf{x} \sim N(\mathbf{m}_1, \mathbf{K}_1) \\ C_2&: \quad \mathbf{x} \sim N(\mathbf{m}_2, \mathbf{K}_2) \end{aligned} \tag{10.167}$$

The mean vectors, $\mathbf{m}_1 = E[\mathbf{x} \mid C_1]$ and $\mathbf{m}_2 = E[\mathbf{x} \mid C_2]$, are the expected values of **x** under each class. Since it is known that $E[B_1] = E[B_2] = E[N_i] = 0$, the means are determined as follows:

$$\begin{aligned} \mathbf{m}_1 &= E[\mathbf{x} \mid C_1] = E[\mathbf{s} \mid C_1] + E[\mathbf{c} \mid C_1] + E[\mathbf{N} \mid C_1] = \mathbf{0} \\ \mathbf{m}_2 &= E[\mathbf{x} \mid C_2] = E[\mathbf{N} \mid C_1] = \mathbf{0} \end{aligned} \tag{10.168}$$

It remains now to determine the conditional covariance matrices $\mathbf{K}_1$ and $\mathbf{K}_2$. As before, since the noise is Gaussian and white, the $\mathbf{K}_2 = \frac{1}{2}N_0\mathbf{I}_\infty$. Since the vector **s** plus **c** has only the first two components as nonzero entries, its covariance matrix will be the following partitioned covariance matrix:

$$\mathbf{K}_{\mathbf{s}+\mathbf{c}} = \begin{bmatrix} E[s_1^2] & E[s_1c_1] & \mathbf{0} \\ E[s_1c_1] & E[c_1^2] & \mathbf{0} \\ \mathbf{0} & \mathbf{0} & \mathbf{0} \end{bmatrix} \tag{10.169}$$

Where its components are determined as follows:

$$\begin{aligned} E[s_1^2] &= E\left[B_1^2\frac{T}{2}\right] = \alpha^2\frac{T}{2} \\ E[s_1c_2] &= E\left[B_1B_2\frac{T}{2}\right] = 0 \\ E[c_2s_1] &= E\left[B_2B_1\frac{T}{2}\right] = 0 \\ E[c_2^2] &= E\left[B_2^2\frac{T}{2}\right] = \alpha^2\frac{T}{2} \end{aligned} \tag{10.170}$$

The covariance matrix for $\mathbf{x}$ conditioned on $C_1, \mathbf{K}_1$, then becomes the sum of $\mathbf{K}_{WW}$ and $\mathbf{K}_{s+c}$ from (10.169) and is a diagonal matrix as follows:

$$\mathbf{K}_1 = \text{diag}\left[\alpha^2\frac{T}{2}+\frac{N_0}{2}, \alpha^2\frac{T}{2}+\frac{N_0}{2}, \frac{N_0}{2}, \frac{N_0}{2}, \ldots\right] \tag{10.171}$$

Similarly it can be shown that

$$\mathbf{K}_2 = \text{diag}\left[\frac{N_0}{2}, \frac{N_0}{2}, \frac{N_0}{2}, \frac{N_0}{2}, \ldots\right] \tag{10.172}$$

The problem is now a general Gaussian problem with diagonal but unequal covariance matrices. So from (9.72) the optimal decision rule is as follows:

$$\begin{aligned}
\text{If}\quad Q(\mathbf{x}) \triangleq -(\mathbf{x}-\mathbf{m}_1)^T\mathbf{K}_1^{-1}(\mathbf{x}-\mathbf{m}_1) + (\mathbf{x}-\mathbf{m}_2)^T\mathbf{K}_2^{-1}(\mathbf{x}-\mathbf{m}_2) &> T \quad \text{decide } \omega_1\\
&< \quad\ \text{decide } \omega_2\\
&= \text{decide randomly}
\end{aligned}$$

$$\begin{aligned}
\text{where } T &= 2\ln \eta_{\text{BAYES}} + \ln\frac{|\mathbf{K}_1|}{|\mathbf{K}_2|}\\
\text{and}\quad \eta_{\text{BAYES}} &= \frac{(C_{22}-C_{12})\,P(C_2)}{(C_{11}-C_{21})\,P(C_1)}
\end{aligned} \tag{10.173}$$

After substituting the zero means $\mathbf{m}_1$ and $\mathbf{m}_2$ from (10.168) and the covariance matrices $\mathbf{K}_1$ and $\mathbf{K}_2$ from above into (10.171) and (10.172), we simplify the test statistic for the decision rule to

$$\begin{aligned}
Q(\mathbf{x}) &= -\mathbf{x}^T\mathbf{K}_1^{-1}\mathbf{x} + \mathbf{x}^T\mathbf{K}_2^{-1}\mathbf{x}\\
&= -\left(\alpha^2\frac{T}{2}+\frac{N_0}{2}\right)^{-1}(x_1^2+x_2^2) + \frac{2}{N_0}(x_1^2+x_2^2) = \frac{2\alpha^2T}{N_0(\alpha^2T+N_0)}(x_1^2+x_2^2)
\end{aligned} \tag{10.174}$$

The threshold T given in (10.173) can be simplified as follows:

$$\begin{aligned}
T &= 2\ln \eta_{\text{BAYES}} + \ln\frac{|\mathbf{K}_1|}{|\mathbf{K}_2|}\\
&= 2\ln \eta_{\text{BAYES}} + 2\ln\frac{(\alpha^2T+N_0)}{N_0}
\end{aligned} \tag{10.175}$$

Using the threshold given in (10.175) and the $Q(x)$ from (10.174), we can finally write the decision rule as

$$\begin{aligned}
\text{If}\quad (x_1^2+x_2^2) &> T_2 \quad \text{decide } C_1\\
&< T_2 \quad \text{decide } C_2\\
&= T_2 \quad \text{decide randomly}\\
T_2 \triangleq \frac{N_0(\alpha^2T+N_0)}{\alpha^2T}&\left[\ln \eta_{\text{BAYES}} + \ln\frac{(\alpha^2T+N_0)}{N_0}\right]
\end{aligned} \tag{10.176}$$

By the definitions for x_1 and x_2 given in (10.165), the optimum decision rule can further be written in terms of processing the continuous signal $x(t)$ on the interval $[0, T]$ to yield a test statistic v defined as

$$v \triangleq x_1^2 + x_2^2 = \left(\sqrt{\frac{2}{T}}\int_0^T x(t)\cos(\omega_C t)\,dt\right)^2 + \left(\sqrt{\frac{2}{T}}\int_0^T x(t)\sin(\omega_C t)\,dt\right)^2 \quad (10.177)$$

The decision rule for the detection of a sinusoid with uniform random phase and Raleigh distributed amplitude is illustrated in Figure 10.13. This result will be familiar to most communications engineers, but its development may not be.

10.6.3 Detection of Known Signal with Single-Scatter Interference in AWGN

The case of single-clutter interference in additive white Gaussian noise is modeled as

$$\begin{aligned} C_1: &\quad x(t) = s(t) + Ac(t) + N_W(t), &\quad 0 \le t \le T \\ C_2: &\quad x(t) = + Ac(t) + N_W(t), &\quad 0 \le t \le T \end{aligned} \quad (10.178)$$

where $s(t)$ is an arbitrary but known deterministic signal, A is a Gaussian random variable with zero mean and variance σ_A^2, $c(t)$ is an interfering signal and $N_W(t)$ is a realization from a stationary white Gaussian process with autocorrelation function $R_{N_W N_W}(\tau) = \frac{1}{2}N_0\,\delta(\tau)$. It will be further assumed that the $P(C_1)$ and $P(C_2)$, the a priori probabilities of the classes, are known and that we desire the optimum detector in the sense of minimizing the probability of error. It will also be assumed that the random variable A is independent of the random process $N_W(t)$ and that $s(t)$ and $c(t)$ are not orthogonal but have a correlation given by r as

$$r = \int_0^T s(t)c(t)\,dt \quad (10.179)$$

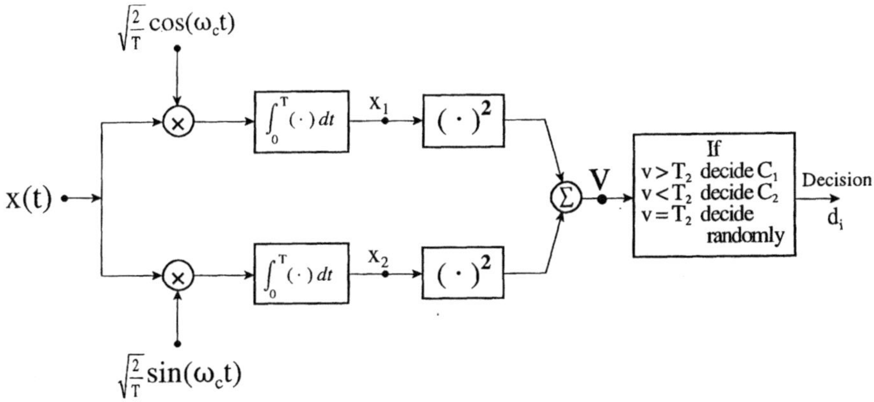

Figure 10.13 The decision rule for the detection of a sinusoid with uniform random phase and Raleigh distributed amplitude.

The general procedure for finding the optimum detector is the selection of a complete orthonormal set and an expansion of $x(t)$ using those basis functions followed by the use of previous Gaussian results for a countable observation space.

Again the choice of the CON is problem specific, and for this problem the obvious choice is to select $\phi_1(t)$ and $\phi_2(t)$ by applying the Gram-Schmidt procedure to $s(t)$ and $c(t)$. Select $\phi_1(t)$ as the normalized version of $s(t)$ as follows:

$$\phi_1(t) = \frac{1}{\sqrt{E_s}} s(t) \quad \text{where } E_s = \int_0^T s^2(t)\, dt \tag{10.180}$$

Then select $\phi_2(t)$ as the normalized version of $f_2(t) = c(t) - a_{cs}\phi_1(t)$ where a_{cs} is

$$\begin{aligned} a_{cs} &= \int_0^T c(t)\phi_1(t)\, dt \\ &= \int_0^T c(t) \frac{1}{\sqrt{E_s}} s(t)\, dt = \frac{1}{\sqrt{E_s}} r \end{aligned} \tag{10.181}$$

This gives $f_2(t)$ as

$$\begin{aligned} f_2(t) &= c(t) - \frac{1}{\sqrt{E_s}} \phi_1(t) \\ &= c(t) - \frac{1}{E_s} s(t) \end{aligned} \tag{10.182}$$

It is easily shown that $\int_0^T f_2^2(t)\, dt = \sqrt{E_c - r^2/E_s}$. So the $\phi_2(t)$ becomes

$$\phi_2(t) = \frac{c(t) - rs(t)/E_s}{\sqrt{E_c - r^2/E_s}} \tag{10.183}$$

The other basis functions $\phi_i(t)$, $i = 3, 4, \ldots,$ are arbitrarily selected to form a CON and will be shown not to effect the solution. Expanding the observation $x(t)$ over the interval $[0, T]$ in terms of the CON basis functions leads to the countable infinite vector observation as follows:

$$\begin{aligned} C_1&: \quad \mathbf{x} = \mathbf{s} + \mathbf{c} + \mathbf{N} \\ C_2&: \quad \mathbf{x} = \mathbf{c} + \mathbf{N} \end{aligned} \tag{10.184}$$

In the expression above **x**, **s**, **c**, and **N** are infinite-dimensional vectors having as elements the coefficients of the orthonormal expansion of $x(t)$, $s(t)$, $Ac(t)$ and $N_W(t)$. They are given as follows:

$$\begin{aligned}
x_i &= \int_0^T x(t)\phi_i(t)\,dt, \qquad && i = 1, 2, \ldots \\
s_i &= \int_0^T s(t)\phi_i(t)\,dt, \qquad && i = 1, 2, \ldots \\
c_i &= \int_0^T Ac(t)\phi_i(t)\,dt, \qquad && i = 1, 2, \ldots \\
N_i &= \int_0^T N_W(t)\phi_i(t)\,dt, \qquad && i = 1, 2, \ldots
\end{aligned} \tag{10.185}$$

Since $s(t)$ is $\sqrt{E_s}\phi_1(t)$, which is orthogonal with all the other basis functions, the **s** vector has all zero components except the first, which is

$$s_1 = \int_0^T s(t)\phi_1(t)\,dt = \int_0^T s(t)\,\frac{1}{\sqrt{E_s}}\,s(t)\,dt = \sqrt{E_s} \tag{10.186}$$

Thus the **s** vector can be written as $\mathbf{s} = [\sqrt{E_1}, 0, 0, \ldots]^T$.

The c_i are found by integrating the product of $Ac(t)$ and $\phi_i(t)$ over the interval $[0, T]$. Since $c(t)$ can be written from (10.183) in terms of $\phi_2(t)$ and $s(t)$, which is a scaled $\phi_1(t)$, the integrals give all zero for all $i > 2$, so we just need to find c_1 and c_2. The c_1 and c_2 are obtained as follows:

$$\begin{aligned}
c_1 &= \int_0^T Ac(t)\phi_1(t)\,dt = \int_0^T Ac(t)\,\frac{1}{\sqrt{E_s}}\,s(t)\,dt = Ar\,\frac{1}{\sqrt{E_s}} \\
c_2 &= \int_0^T Ac(t)\phi_2(t)\,dt \\
&= \int_0^T Ac(t)\,\frac{1}{\sqrt{E_C - r^2/E_s}}\left[c(t) - \frac{r}{E_s}s(t)\right]dt = A\sqrt{E_c - \frac{r^2}{E_s}}
\end{aligned} \tag{10.187}$$

Thus the **c** vector can be written in terms of A, r, and the energies of $s(t)$ and $c(t)$ as

$$\mathbf{c} = \left[Ar\frac{1}{\sqrt{E_s}},\ A\sqrt{E_c - \frac{r^2}{E_s}}, 0, 0, \ldots\right]^T \tag{10.188}$$

The **N** vector can be written in terms of the random variables $n_1, n_2, \ldots,$ as follows:

$$\mathbf{N} = [n_1, n_2, n_3, \ldots]^T \tag{10.189}$$

Using (10.184), we can write the problem in terms of the random variable A as

$$
\begin{aligned}
C_1:\quad \mathbf{x} &= \begin{bmatrix} \sqrt{E_s} + \dfrac{1}{\sqrt{E_s}} A r \\ A\sqrt{E_c - \dfrac{r^2}{E_s}} \\ 0 \\ 0 \end{bmatrix} + \begin{bmatrix} n_1 \\ n_2 \\ n_3 \\ \vdots \end{bmatrix} \\
C_2:\quad \mathbf{x} &= \begin{bmatrix} \dfrac{1}{\sqrt{E_s}} A r \\ A\sqrt{E_c - \dfrac{r^2}{E_s}} \\ 0 \\ 0 \end{bmatrix} + \begin{bmatrix} n_1 \\ n_2 \\ n_3 \\ \vdots \end{bmatrix}
\end{aligned} \tag{10.190}
$$

Since A and $\mathbf{N}$ are Gaussian, the $\mathbf{x}$ vector under each class is Gaussian as they are sums of Gaussian random vectors. To characterize the Gaussian vectors, we need to determine their mean vectors and covariance matrices. The problem can be thus written in terms of those vectors and matrices as

$$
\begin{aligned}
C_1:&\quad \mathbf{x} \sim N(\mathbf{m}_1, \mathbf{K}_1) \\
C_2:&\quad \mathbf{x} \sim N(\mathbf{m}_2, \mathbf{K}_2)
\end{aligned} \tag{10.191}
$$

where $\mathbf{m}_1$ and $\mathbf{m}_2$ are the expected value of $\mathbf{x}$ under both classes. Since $E[A]$ and all the $E[n_i]$ are zero, they simplify to

$$
\begin{aligned}
\mathbf{m}_1 &= [\sqrt{E_s}, 0, 0, \ldots]^T \\
\mathbf{m}_2 &= [0, 0, 0, \ldots]^T
\end{aligned} \tag{10.192}
$$

If the mean vector is subtracted off from $\mathbf{x}$, it is seen that the covariance matrices under each class are identical. However, because of the coupling that occurs with the component of $\mathbf{x}$, the covariance matrices are not diagonal. The A and all components of $\mathbf{N}$ are independent, since A was independent of the $N_W(t)$ and thus independent of the integrals of $N_W(t)$. So the vectors in (10.190) under C_2 are independent, and the covariance matrix is just the sum of the covariance matrix for $\mathbf{c}$ and $\mathbf{N}$. We know from the white noise assumption that $K_{NN} = \frac{1}{2}N_0\mathbf{I}_\infty$. So it remains to determine the covariance matrix for the **c** vector. The first 2 by 2 submatrix components of the $\mathbf{K}_1$ covariance matrix are determined as

$$\begin{aligned} k_{11} &= E\left[\frac{A^2r^2}{E_s}\right] = \frac{\sigma_A^2 r^2}{E_s} \\ k_{12} &= E\left[\frac{Ar}{\sqrt{E_s}} A\sqrt{E_c - r^2/E_s}\right] = \frac{\sigma_A^2 r}{\sqrt{E_s}} \sqrt{E_c - r^2/E_s} \\ k_{21} &= k_{12} \\ k_{22} &= E[A^2(E_c - r^2/E_s)] = \sigma_A^2(E_c - r^2/E_s) \end{aligned} \tag{10.193}$$

The k_{ij} for $i, j > 2$ for **c** are all zero, since there are zero components in the **c** vector past 2. Thus adding the K_{NN} to the K_{cc}, determined above, gives the observation covariance matrices $\mathbf{K}_1$ and $\mathbf{K}_2$ as

$$\mathbf{K}_1 = \mathbf{K}_2 = \begin{bmatrix} \dfrac{\sigma_A^2 r^2}{E_s} + \dfrac{1}{2}N_0 & \dfrac{1}{\sqrt{E_s}}\sigma_A^2 r\sqrt{E_c - r^2/E_s} & \mathbf{0} \\ \dfrac{1}{\sqrt{E_s}}\sigma_A^2 r\sqrt{E_c - r^2 E_s} & \sigma_A^2(E_c - r^2/E_s) + \dfrac{1}{2}N_0 & \mathbf{0} \\ \mathbf{0} & \mathbf{0} & \dfrac{1}{2}N_0\mathbf{I}_\infty \end{bmatrix} \triangleq \mathbf{K} \tag{10.194}$$

The equivalent Gaussian formulation has equal covariance matrices, but they are not diagonal. So the optimum decision rule for this case, from (9.76), is

$$\begin{aligned} &\text{If} \quad z \triangleq (\mathbf{m}_1 - \mathbf{m}_2)^T\mathbf{K}^{-1}\mathbf{x} > T_1 \quad \text{decide } C_1 \\ &\qquad\qquad\qquad\qquad\qquad < \quad \text{decide } C_2 \\ &\qquad\qquad\qquad\qquad\qquad = \quad \text{decide randomly} \\ &\text{where} \quad T_1 = \ln \eta_{\text{BAYES}} + \tfrac{1}{2}(\mathbf{m}_1^T\mathbf{K}^{-1}\mathbf{m}_1 - \mathbf{m}_2^T\mathbf{K}^{-1}\mathbf{m}_2) \end{aligned} \tag{10.195}$$

After substituting the $\mathbf{m}_1$ and $\mathbf{m}_2$ from (10.192) into (10.195), we simplify the test statistic z to

$$\begin{aligned} z &= [\sqrt{E_s}, 0, 0, \ldots] \begin{bmatrix} [\mathbf{K}^{-1}]_{11} & [K^{-1}]_{12} & \mathbf{0} \\ [\mathbf{K}^{-1}]_{21} & [K^{-1}]_{22} & \mathbf{0} \\ \mathbf{0} & \mathbf{0} & \dfrac{2}{N_0}\mathbf{I}_\infty \end{bmatrix} \begin{bmatrix} x_1 \\ x_2 \\ x_3 \\ \vdots \end{bmatrix} \\ &= \sqrt{E_s}[\mathbf{K}^{-1}]_{11}x_1 + \sqrt{E_s}[K^{-1}]_{12}x_2 \end{aligned} \tag{10.196}$$

Notice that x_3, x_4, etc., are zeroed out in the matrix multiplication and thus will not play a part in the optimal decision rule.

To simplify the expression for z further, the inverse of $\mathbf{K}$ is needed. By partitioning K into the upper 2 by 2 block the lower ∞ by ∞ scaled identity matrix and the other two off-diagonal zero matrices, we obtain the inverse by block inverses, with some medial algebra, as follows:

$$\mathbf{K}_1^{-1} = \mathbf{K}_2^{-1} = \begin{bmatrix} \dfrac{\sigma_A^2(E_c - r^2/E_s) + \frac{1}{2}N_0}{\frac{1}{2}N_0(\sigma_A^2 E_c + \frac{1}{2}N_0)} & \dfrac{-\sigma_A^2 r\sqrt{E_c - r^2/E_s}}{\frac{1}{2}N_0\sqrt{E_s}(\sigma_A^2 E_c + \frac{1}{2}N_0)} & \mathbf{0} \\ \dfrac{-\sigma_A^2 r\sqrt{E_c - r^2/E_s}}{\frac{1}{2}N_0\sqrt{E_s}(\sigma_A^2 E_c + \frac{1}{2}N_0)} & \dfrac{\sigma_A^2 r^2/E_s + \frac{1}{2}N_0}{\frac{1}{2}N_0(\sigma_A^2 E_c + \frac{1}{2}N_0)} & \mathbf{0} \\ \mathbf{0} & \mathbf{0} & \dfrac{1}{2}N_0\mathbf{I}_\infty \end{bmatrix} \triangleq \mathbf{K}^{-1} \tag{10.197}$$

After substituting the proper entries from K^{-1} into (10.196), we reduce the test statistic z to

$$z = \sqrt{E_s}\,\frac{\sigma_A^2(E_c - r^2/E_s) + \frac{1}{2}N_0}{\frac{1}{2}N_0(\sigma_A^2 E_c + \frac{1}{2}N_0)}x_1 + \frac{-\sigma_A^2 r\sqrt{E_c - r^2/E_s}}{\frac{1}{2}N_0(\sigma_A^2 E_c + \frac{1}{2}N_0)}x_2 \tag{10.198}$$

The threshold for the optimum test is determined by substituting the means from (10.192) into the T_1 of (10.195) to give

$$\begin{aligned} T_1 &= \ln \eta_{\mathrm{BAYES}} + \frac{1}{2}[\sqrt{E_s}, 0, 0, \ldots]\begin{bmatrix} [K^{-1}]_{11} & [K^{-1}]_{12} & \mathbf{0} \\ [K^{-1}]_{21} & [K^{-1}]_{22} & \mathbf{0} \\ \mathbf{0} & \mathbf{0} & \dfrac{2}{N_0}\mathbf{I}_\infty \end{bmatrix}\begin{bmatrix} \sqrt{E_s} \\ 0 \\ 0 \\ \vdots \end{bmatrix} \\ &= \ln \eta_{\mathrm{BAYES}} + \frac{1}{2}\sqrt{E_s}[K^{-1}]_{11}\sqrt{E_s} \\ &= \ln \eta_{\mathrm{BAYES}} + \frac{1}{2}E_s\frac{\sigma_A^2(E_c - r^2/E_s) + \frac{1}{2}N_0}{\frac{1}{2}N_0(\sigma_A^2 E_c + \frac{1}{2}N_0)} \end{aligned} \tag{10.199}$$

The T_1 above can be further rearranged in terms of signal to noise ratio and as noise to clutter ratio as

$$T_1 = \ln \eta_{\mathrm{BAYES}} + \frac{1}{2}\frac{E_s}{\frac{1}{2}N_0}\left(1 - \frac{r^2 E_s}{E_c - \frac{1}{2}N_0/\sigma_A^2}\right) \tag{10.200}$$

Finally, the optimum test can be put in the form of processing on the original observation $x(t)$ by determining x_1 and x_2 using the definitions for the orthonormal

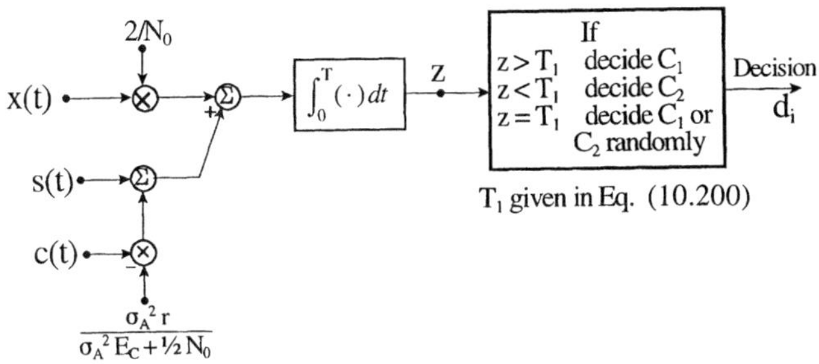

Figure 10.14 Optimal detector for a known signal $s(t)$ in additive white Gaussian noise and interference signal $c(t)$ with random amplitude $A \sim N(0, \sigma_A^2)$.

expansion. From (10.185) the x_1 and x_2 are determined as follows:

$$x_1 = \int_0^T x(t)\phi_1(t)\,dt = \int_0^T x(t)\frac{1}{\sqrt{E_s}}\,s(t)\,dt = \frac{1}{\sqrt{E_s}}\int_0^T x(t)s(t)\,dt$$

$$x_2 = \int_0^T x(t)\phi_2(t)\,dt = \int_0^T x(t)\left(c(t) - \frac{r}{E_s}s(t)\right)\frac{1}{\sqrt{E_c - r^2/E_s}}\,dt \tag{10.201}$$

$$= \frac{1}{\sqrt{E_c - r^2/E_s}}\int_0^T x(t)c(t)\,dt - \frac{r/E_s}{\sqrt{E_c - r^2/E_s}}\int_0^T x(t)s(t)\,dt$$

Substituting the x_1 and x_2 from (10.201) into the test statistic z given in (10.198) and simplifying gives z as

$$z = \frac{2}{N_0}\int_0^T x(t)s(t)\,dt - \frac{2}{N_0}\,\frac{\sigma_A^2 r}{\sigma_A^2 E_c + N_0/2}\int_0^T x(t)c(t)\,dt \quad \text{or}$$

$$= \frac{2}{N_0}\int_0^T x(t)[s(t) - k_1 c(t)]\,dt \tag{10.202}$$

$$\text{where} \quad k_1 = \frac{r}{E_c + \frac{1}{2}N_0/\sigma_A^2}$$

With all the work done, the optimal decision rule has simplified to (10.202). It can be viewed as a cross-correlation of the observed $x(t)$ with $s(t) - k_1 c(t)$, where k_1 can be interpreted as an estimate of the amplitude of the clutter pulse. The optimal decision rule is illustrated in Figure 10.14.

10.7 SUMMARY

The main topic of this chapter is the optimum detection of continuous time signals observed in additive noise over a fixed time interval. In particular, the discussion was focused on the detection of signals in additive white Gaussian noise.

The discussion began with the optimum detection of known deterministic signals observed in additive white Gaussian noise over a given time interval. The general approach to solving this continuous time detection problem was to change the problem to a countably infinite discrete detection problem by expansion in terms of a complete orthonormal (CON) set of functions. The selection of orthonormal set was motivated by the known signals we were to detect. With this infinite-dimensional Gaussian vector as a new observation vector, the general results determined in Chapter 9 were used in the limiting case. The Gram-Schmidt ortho-gonalization procedure was introduced to establish the first components of the expansion set and extend this set to form a CON set. This way the infinite-dimensional problem was reduced in dimensionality to less than or equal the number of hypotheses to be decided between. This lower-dimensional Gaussian problem was then shown to be easily solved using the results from Chapter 9 for Gaussian observation vectors.

The next generalization was the detection of known signals in nonwhite Gaussian noise. It was shown that the choice of the orthonormal set that could produce an easy solution was the Kahunen-Loeve set obtained by solving an integral equation with a kernel equal to the covariance function of the noise process. If the noise has a zero region in the spectral domain, the problem could be made singular, perfect detection, provided that the signal process has a nonzero component in that frequency band. See Van Trees [1] for a discussion of the singular problem.

The next problem explored was that of detection of known signals in nonwhite Gaussian noise plus Gaussian white noise. Again, the solution was facilitated using the Karhunen-Loeve expansion of the nonwhite component, and the possibility of singularity vanished.

The general Gaussian problem was then addressed, and the resulting optimum decision rule was left as a quadratic form involving an infinitely dimensional space that could not be simplified or reduced to a finite-dimensional space.

The chapter concluded with several examples of detection of known signal forms with random parameters. These problems illustrated the procedure to be used for such cases. The problem of detection of known signals with nonrandom parameters was not addressed. Basically this type of problem is solved by using maximum likelihood estimates of the nonrandom parameters in the likelihood functions as if they were the real parameters. Details of this method are found in Van Trees [1], Helstrom [8, 9], MacDonnaugh [4], and Srinath and Rajjasekaran [5].

Other techniques can be used to solve the continuous observation detection problem when the probability densities of the hypotheses are not known. Techniques for tests that are distribution free and information concerning this important topic are presented in the text by Srinath [5]. Of course, one way to solve this type of problem is to estimate the densities and treat them as known properties.

If the Gaussian assumption is relaxed for the continuous observation problem, the optimum detection problem is virtually impossible to solve. In general, it would require a total characterization of the signal and noise processes.

One other topic that was not included in our presentation was that of data fusion or distributed optimum detection, which has received considerable attention in the last few years. For references, consult Tenney and Sandell [14] and Therrien [13].

Sequential detection also has not been presented simply because of space limitations. The book by Wetherill and Glazebrook [12] explores this topic, but Wald [11] was the pioneer.

Another area of research not presented was the construction of decision trees from data. The interested reader will find introductory and advanced discussions of this area in Duda, Hart, and Stork [15].

PROBLEMS

10.1 Suppose that the two signals given below are equally probable and that they have been received over an additive white Gaussian noise (AWGN) channel. The noise process has a spectral density of 1/9. The observed signals under each of the classes are given by

$$C_1: \quad x(t) = 2 + N(t), \qquad 0 \le t \le 1$$
$$C_2: \quad x(t) = 1 + N(t), \qquad 0 \le t \le 1$$

(a) Find the optimum decision rule where optimum is in the minimum probability of error sense.

(b) Draw a block diagram to show how the decision rule operates on the continuous observation.

(c) Give a sufficient statistic or statistics for this problem.

10.2 Given $s_1(t)$ and $s_2(t)$ as shown in Figure P10.2 below. Suppose that the two signals are equally probable and that they have been received over an additive white Gaussian noise (AWGN) channel. The noise process has a spectral density of $N_0/2$. The observed signals under each class are given by

$$C_1: \quad x(t) = s_1(t) + N(t), \qquad 0 \le t \le T$$
$$C_2: \quad x(t) = s_2(t) + N(t), \qquad 0 \le t \le T$$

(a) Find the optimum detector where the optimum is in the minimum probability of error sense where $T = 1$.

(b) If $N_0 = 1$, calculate the probability of error for the detector determined in (a).

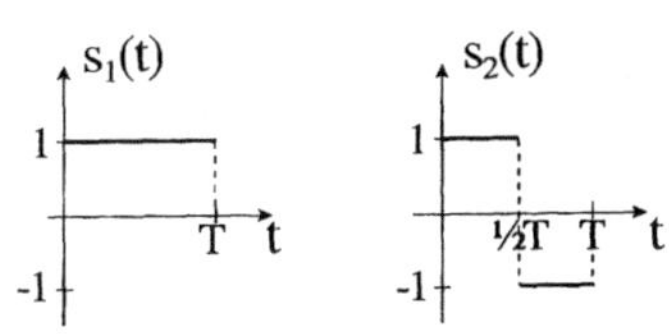

P10.2

(c) With the same signals, let $T = 2$ and repeat (a) and (b) above. Comment on your results and explain what would be gained by making T larger.

10.3 Suppose that the two signals given below are equally probable and that they have been received over an additive white Gaussian noise (AWGN) channel. The noise process has a power spectral density of $N_0/2$. The observed signals under each class are different frequencies (frequency shift keying) and given by

$$
\begin{aligned}
C_1:&\quad x(t) = 2\sin(2\pi t) + N(t), \qquad 0 \le t \le 2\\
C_2:&\quad x(t) = \sin(4\pi t) + N(t), \qquad 0 \le t \le 2
\end{aligned}
$$

(a) Show that $s_1(t)$ and $s_2(t)$ are orthogonal, and find the optimum decision rule where the optimum is in the minimum probability of error sense.

(b) Draw a block diagram to show how the decision rule operates on the continuous observation.

(c) Give a sufficient statistic or statistics for this problem.

(d) If $N_0 = 1$, for your decision rule in (b) calculate the probability of correct decision given class C_1.

10.4 Suppose that the two signals given below are equally probable and that they have been received over an additive white Gaussian noise (AWGN) channel. The noise process has a spectral density of $\frac{1}{2}$. The observed signals under each of the classes are given by

$$
\begin{aligned}
C_1:&\quad x(t) = 2\sin 2\pi t + N(t), \qquad 0 \le t \le 1\\
C_2:&\quad x(t) = 1 + N(t), \qquad 0 \le t \le 1
\end{aligned}
$$

(a) Find the optimum decision rule where the optimum is in the minimum probability of error sense.

(b) Draw a block diagram to show how the decision rule operates on the continuous observation.

(c) Give a sufficient statistic or statistics for this problem.

10.5 On–off keying (OOK). Suppose that two equally probable signals are received over an additive white Gaussian noise (AWGN) channel. The noise process has a spectral density of $N_0/2$ for all ω. The observed signals under each class are given by

$$
\begin{aligned}
C_1:&\quad x(t) = A\sin\omega_c t + N(t), \qquad 0 \le t \le T\\
C_2:&\quad x(t) = N(t), \qquad 0 \le t \le T
\end{aligned}
$$

Find the optimum detector for OOK where optimum is in the minimum probability of error sense and calculate the probability of error for the detector determined.

10.6 Coherent phase shift keying (CPSK). The observed signal $x(t)$ under two classes is given as

$$C_1: \quad x(t) = +A \sin \omega_c t + N(t), \qquad 0 \le t \le T$$
$$C_2: \quad x(t) = -A \sin \omega_c t + N(t), \qquad 0 \le t \le T$$

where A and ω_c are known constants and $N(t)$ is from a white Gaussian process with a power spectral density of $N_0/2$ for all ω. The classes are equally probable, and ω_c is an integer multiple of $2\pi/T$. Find the optimum detector for CPSK where optimum is in the minimum probability of error sense, and calculate the probability of error for the detector determined.

10.7 Coherent frequency shift keying (CFSK). Suppose that two signals are equally probable and that they have been received over an additive white Gaussian noise (AWGN) channel. The noise process has a spectral density of $N_0/2$. The observed signals, where A is a known constant, under each of the equal probable classes are given by

$$C_1: \quad x(t) = A \sin \omega_1 t + N(t), \qquad 0 \le t \le T$$
$$C_2: \quad x(t) = A \sin \omega_2 t + N(t), \qquad 0 \le t \le T$$

where ω_1 and ω_2 are integer multiples of the fundamental frequency $2\pi/T$. Find the optimum detector for CFSK where optimum is in the minimum probability of error sense, and calculate the probability of error for the detector determined.

10.8 Suppose that the two signals given below are equally probable and that they have been received over an additive white Gaussian noise (AWGN) channel. The noise process has a spectral density of $\frac{1}{2}$. The observed signals under each class are given by

$$C_1: \quad x(t) = 2e^{-t} + N(t), \qquad 0 \le t \le \frac{1}{2} \ln 2$$
$$C_2: \quad x(t) = e^{1-t} + N(t), \qquad 0 \le t \le \frac{1}{2} \ln 2$$

(a) Find the optimum decision rule where the optimum is in the minimum probability of error sense.

(b) Draw a block diagram to show how the decision rule operates on the continuous observation.

(c) Give a sufficient statistic or statistics for this problem.

(d) Calculate the probability of correct decision given class C_1 for your decision rule.

10.9 Suppose that the known deterministic signal $s_1(t)$ is received over one of two possible (AWGN) channels and that the channels are equally probable. The noise process in one channel has a spectral density of $\frac{1}{2}$ while the noise process in the second channel $N_2(t)$ has a spectral density of 1. The observed signals under each class are given by

$$\begin{aligned} C_1&: \quad x(t) = s_1(t) + N_1(t), \qquad 0 \le t \le 1 \\ C_2&: \quad x(t) = s_1(t) + N_2(t), \qquad 0 \le t \le 1 \end{aligned}$$

(a) Find the optimum decision rule where the optimum is in the minimum probability of error sense, for deciding which channel was used.

(b) Draw a block diagram to show how the decision rule operates on the continuous observation $x(t)$.

(c) Give a sufficient statistic or statistics for this problem.

(d) Calculate the probability of correct decision given class C_1 for your decision rule.

10.10 Say we observe a signal $x(t)$ that could be $s_1(t) + N_1(t)$ or $s_2(t) + N_2(t)$, with equal probability, where $s_1(t)$ and $s_2(t)$ are known different deterministic signals and $N_1(t)$ and $N_2(t)$ are realizations of two different Gaussian random processes. The observed signals under each class are thus given by

$$\begin{aligned} C_1&: \quad x(t) = s_1(t) + N_1(t), \qquad 0 \le t \le 1 \\ C_2&: \quad x(t) = s_2(t) + N_2(t), \qquad 0 \le t \le 1 \end{aligned}$$

The noise process in one channel $N_1(t)$ has a spectral density of $\frac{1}{2}$, while the noise process in the second channel $N_2(t)$ has a spectral density of 1.

(a) Find the optimum decision rule where the optimum is in the minimum probability of error sense.

(b) Draw a block diagram to show how the decision rule operates on the continuous observation $x(t)$.

(c) Give a sufficient statistic or statistics for this problem.

(d) Calculate the probability of correct decision given class C_1 for your decision rule.

10.11 Suppose that the three signals given in the equation below are equally probable and that they have been received over an additive white Gaussian noise (AWGN) channel. The noise process has a spectral density of $N_0/2$. The observed signals under each class are given by

$$\begin{aligned} C_1&: \quad x(t) = \sin(2\pi t) + N(t), \qquad 0 \le t \le T \\ C_2&: \quad x(t) = N(t), \qquad 0 \le t \le T \\ C_3&: \quad x(t) = -\sin(2\pi t) + N(t), \qquad 0 \le t \le T \end{aligned}$$

(a) Find the optimum detector where the optimum is in the minimum probability of error sense and where T is an integer.

(b) Determine the probability of error for the case where $T = 1$ and $N_0 = 1$.

(c) Repeat (b) for $T = 2$, and compare your results. Make a general statement about performance as T is increased.

(d) Show the optimal classifier for an arbitrary T not equal to an integer. How are the results changed when T is not an integer?

10.12 Suppose that $s(t)$ is a deterministic and known signal and $N(t)$ is a Gaussian white noise process with power spectral density of $\frac{1}{2}$. The observed signals $x(t)$ under each equally probable class are given by

$$\begin{aligned} C_1:&\quad X(t) = As(t) + N(t), & 0 \le t \le 1 \\ C_2:&\quad X(t) = s^2(t) + N(t), & 0 \le t \le 1 \\ C_3:&\quad X(t) = Bs(t) + N(t), & 0 \le t \le 1 \end{aligned}$$

Assume $A \sim N(0, 1), B \sim N(1, 1), A$ and B independent of $N(t)$ for all t, and

$$\int_0^1 s(t)\,dt = 0, \qquad \int_0^1 s^2(t)\,dt = 1$$

$$\int_0^1 s^3(t)\,dt = E_3, \qquad \int_0^1 s^4(t)\,dt = E_4$$

(a) Find the optimum decision rule where the optimum is in the minimum probability of error sense.

(b) Draw a block diagram to show how the decision rule operates on the continuous observation $X(t)$.

(c) Give a sufficient statistic or statistics for this problem if there are any.

10.13 Suppose that the four signals given in the equation below are equally probable and that they have been received over an additive white Gaussian noise (AWGN) channel. The noise process has a spectral density of $N_0/2$. The observed signals under each class are given by

$$\begin{aligned} C_1:&\quad x(t) = 2\sin(2\pi t) + N(t), & 0 \le t \le T \\ C_2:&\quad x(t) = \sin(2\pi t) + N(t), & 0 \le t \le T \\ C_3:&\quad x(t) = -\sin(2\pi t) + N(t), & 0 \le t \le T \\ C_4:&\quad x(t) = -2\sin(2\pi t) + N(t), & 0 \le t \le T \end{aligned}$$

(a) Find the optimum detector where the optimum is in the minimum probability of error sense and where T is an integer.

(b) Determine the probability of error for the case where $T = 1$ and $N_0 = 1$.

(c) Repeat (b) for $T = 2$, and compare your results. Make a general statement about performance as T is increased.

10.14 Suppose that the four signals given in the equation below are equally probable and that they have been received over an additive white Gaussian noise (AWGN) channel. The noise process has a spectral density of $N_0/2$. The observed signals under each class are given by

$$
\begin{aligned}
C_1:&\quad x(t) = A\cos(\omega_c t + 0) + N(t), &\quad 0 \le t \le T\\
C_2:&\quad x(t) = A\cos(\omega_c t + \pi/2) + N(t), &\quad 0 \le t \le T\\
C_3:&\quad x(t) = A\cos(\omega_c t + \pi) + N(t), &\quad 0 \le t \le T\\
C_4:&\quad x(t) = A\cos(\omega_c t + 3\pi/2) + N(t), &\quad 0 \le t \le T
\end{aligned}
$$

(a) For a known constant A find the optimum detector where the optimum is in the minimum probability of error sense and $\omega_c T$ is an integer multiple of 2π.

(b) Determine the probability of error for your detector.

(c) Consider the M-ary extension of the problem above (M-ary phase shift keying):

$$C_i:\quad x(t) = A\cos\left(\omega_c t + \frac{2\pi(i-1)}{M}\right) + N_W(t), \qquad i = 1, 2, \ldots, M$$

Find the optimum detector.

(d) Determine the probability of error for the classifier determined in (c).

10.15 Suppose that $s_1(t), s_2(t)$, and $s_3(t)$ are known orthonormal deterministic signals on the interval [0, 1] received over three different (AWGN) channels and that the channels are equally probable. The observed signals under each class are given by

$$
\begin{aligned}
C_1:&\quad x(t) = s_1(t) + N_1(t), &\quad 0 \le t \le 1\\
C_2:&\quad x(t) = s_2(t) + N_2(t), &\quad 0 \le t \le 1\\
C_3:&\quad x(t) = s_3(t) + N_3(t), &\quad 0 \le t \le 1
\end{aligned}
$$

The noise process $N_1(t)$ in the first channel has a spectral density of 1, the noise process in the second channel $N_2(t)$ has a spectral density of 2, and the noise process $N_3(t)$ in the third channel has a spectral density of 3.

(a) Find the optimum decision rule to determine which channel was observed where the optimum is in the minimum probability of error sense.

(b) Draw a block diagram to show how the decision rule operates on the continuous observation $x(t)$.

(c) Give a sufficient statistic or statistics for this problem.

(d) Calculate the probability of correct decision given class C_1 for your decision rule.

10.16 Suppose that the known deterministic signal $s_1(t)$ is received over one of three possible (AWGN) channels and that the channels are equally probable.

The noise process $N_1(t)$ in the first channel has a spectral density of $\frac{1}{2}$, the noise process $N_2(t)$ in the second channel has a spectral density of 1, and the noise process $N_3(t)$ in the third channel has a spectral density of 3/2. The observed signals under each class are given by

$$\begin{aligned} C_1:&\quad x(t) = s_1(t) + N_1(t), &\quad 0 \le t \le 1 \\ C_2:&\quad x(t) = s_1(t) + N_2(t), &\quad 0 \le t \le 1 \\ C_3:&\quad x(t) = s_1(t) + N_3(t), &\quad 0 \le t \le 1 \end{aligned}$$

(a) Find the optimum decision rule to determine which class an observation $x(t)$ is from where the optimum is in the minimum probability of error sense.

(b) Draw a block diagram to show how the decision rule operates on the continuous observation $x(t)$.

(c) Give a sufficient statistic or statistics for this problem.

(d) Calculate the probability of correct decision given class C_1 for your decision rule.

10.17 Suppose that $s_1(t)$ and $s_2(t)$ are known orthonormal deterministic signals on the interval [0, 1]. One of them (with equal probability) is received over two different (AWGN) channels with independent noise processes $N_1(t)$ and $N_2(t)$, respectively. The observed signals $x_1(t)$ and $x_2(t)$ under each class are given by

$$\begin{aligned} C_1:&\quad x_1(t) = s_1(t) + N_1(t), &\quad 0 \le t \le 1 \\ &\quad x_2(t) = s_1(t) + N_2(t), &\quad 0 \le t \le 1 \\ C_2:&\quad x_1(t) = s_2(t) + N_1(t), &\quad 0 \le t \le 1 \\ &\quad x_2(t) = s_2(t) + N_2(t), &\quad 0 \le t \le 1 \end{aligned}$$

The noise process $N_1(t)$ in the first channel has a spectral density of 1, and the noise process in the second channel $N_2(t)$ has a spectral density of 2.

(a) Find the optimum decision rule to determine from which class the observations $x_1(t)$ and $x_2(t)$ are obtained where the optimum is in the minimum probability of error sense.

(b) Draw a block diagram to show how the decision rule operates on the continuous observations $x_1(t)$ and $x_2(t)$.

(c) Give a sufficient statistic or statistics for this problem.

(d) Calculate the probability of correct decision given class C_1 for your decision rule.

10.18 Suppose that $s_1(t), s_2(t), s_3(t)$ and $s_4(t)$ are known orthonormal deterministic signals on the interval [0, 1]. $s_1(t)$ and $s_2(t)$ are sent over independent identical (AWGN) channels with noise processes $N_1(t)$ and $N_2(t)$, respectively, and $s_3(t)$ and $s_4(t)$ are sent over independent identical (AWGN) channels with noise processes $N_1(t)$ and $N_2(t)$. Assume equally likely classes and that $N_1(t)$ and $N_2(t)$ are realizations from independent Gaussian random processes with power spectral density equal to $N_0/2$. The observed signals $x_1(t)$ and $x_2(t)$ under each of the two classes are given by

$$\begin{aligned}
C_1:\ & x_1(t) = s_1(t) + N_1(t), && 0 \le t \le 1 \\
& x_2(t) = s_2(t) + N_2(t), && 0 \le t \le 1 \\
C_2:\ & x_1(t) = s_3(t) + N_1(t), && 0 \le t \le 1 \\
& x_2(t) = s_4(t) + N_2(t), && 0 \le t \le 1
\end{aligned}$$

(a) Find the optimum decision rule to determine from which class the observations $x_1(t)$ and $x_2(t)$ are obtained where the optimum is in the minimum probability of error sense.

(b) Draw a block diagram to show how the decision rule operates on the continuous observation $x(t)$.

(c) Give a sufficient statistic or statistics for this problem.

(d) Calculate the probability of correct decision given class C_1 for your decision rule.

10.19 Assume that $s_1(t)$ and $s_2(t)$ are known orthonormal deterministic signals on the interval [0, 1]. One of them (with equal probability) is received over M identical (AWGN) independent channels with noise processes $N_W(t)$ having a known power spectral density of $N_0/2$, respectively. The observed signals $x_1(t), x_2(t), \ldots, x_M(t)$ under each class are given by

$$\begin{aligned}
C_1:\ & x_1(t) = s_1(t) + N_1(t), && 0 \le t \le 1 \\
& x_2(t) = s_1(t) + N_2(t), && 0 \le t \le 1 \\
& \qquad \vdots && \quad \vdots \\
& x_M(t) = s_1(t) + N_M(t), && 0 \le t \le 1 \\
C_2:\ & x_1(t) = s_2(t) + N_1(t), && 0 \le t \le 1 \\
& x_2(t) = s_2(t) + N_2(t), && 0 \le t \le 1 \\
& \qquad \vdots && \quad \vdots \\
& x_M(t) = s_2(t) + N_M(t), && 0 \le t \le 1
\end{aligned}$$

(a) Find the optimum decision rule to determine from which class the observations $x_1(t), x_2(t), \ldots, x_M(t)$ are obtained where the optimum is in the minimum probability of error sense.

(b) Draw a block diagram to show how the decision rule operates on the continuous observations $x_i(t)$.

(c) Give a sufficient statistic or statistics for this problem.

(d) Calculate the probability of correct decision given class C_1 for your decision rule.

10.20 A zero mean wide sense stationary random process $X(t)$ has an autocorrelation function $R_{XX}(\tau)$ given by $R_{XX}(\tau) = \cos(2\pi\tau)$. Find the differential equation that the Karhunen-Loeve basis functions must satisfy for expanding the process on the interval $[-1, 1]$.

10.21 A wide sense stationary random process $x(t)$ has a zero mean and an autocorrelation function $R_{XX}(\tau) = (1 - |\tau|)[\mu(\tau + 1) - \mu(\tau - 1)]$. Find the differential equation that Karhunen-Loeve basis functions must satisfy for expanding the process on the interval $[-1, 1]$.

10.22 Find the eigen functions and values for the Karhunen-Loeve expansion of $X(t)$, a zero mean random process, that has the following autocorrelation function $R_{XX}(t, u)$ and interval of expansion. Only part (a) is easy, the others are very tedious to obtain.

(a) $R_{XX}(t, u) = \delta(t - u)$ on the interval $\{-1, 1]$.

(b) $R_{XX}(t, u) = e^{-|t-u|}\cos[\Omega_c(t - u)]$ on the interval $[0, 2\pi/\Omega_c]$.

(c) $R_{XX}(t, u) = A$ if $|t - u| < 1$, otherwise $= 0$, on the interval $[0, 1]$.

(d) $R_{XX}(t, u) = \sin\Omega_c(t - u)/(t - u)$ for a fixed Ω_c on the interval $[0, 1]$.

(e) $R_{XX}(t, u) = (1 - |t - u|/T)[\mu(\tau + T) - \mu(\tau - T)]$ on the interval $[-T, T]$.

10.23 We are given two deterministic signals $s_1(t)$ and $s_2(t)$ that are not orthogonal on a given interval $[0, T]$. Assume that

$$\int_0^T s_1^2(t)\,dt = 1, \quad \int_0^T s_2^2(t)\,dt = 1, \quad \text{and} \quad \int_0^T s_1(t)s_2(t)\,dt = E_{12}$$

We desire two new signals, call them $\phi_1(t)$ and $\phi_2(t)$, that are linear combinations of $s_1(t)$ and $s_2(t)$ given by

$$\phi_1(t) = \frac{s_1(t) + C_1 s_2(t)}{(1 + 2C_1E_{12} + C_1^2)^{1/2}}$$

$$\phi_1(t) = \frac{s_2(t) + C_2 s_1(t)}{(1 + 2C_2E_{12} + C_2^2)^{1/2}}$$

These new signals are orthogonal on the interval $[0, T]$ given by

$$\int_0^T \phi_1(t)\phi_2(t)\,dt = 0$$

Since the $\phi_1(t)$ and $\phi_2(t)$ above are normalized and orthogonal, they are orthonormal.

Substituting $\phi_1(t)$ and $\phi_2(t)$ into the integral above and dropping the nonzero normalization constants yields

$$\int_0^T [s_1(t) + C_1 s_2(t)][s_2(t) + C_2 s_1(t)]\,dt = 0$$

$$E_{12} + C_2 + C_1 + C_1 C_2 E_{12} = 0$$

The last equation gives us the conditions on C_1 and C_2 for a solution, and since there is one equation in two unknowns, there are as expected an infinite number of solutions.

(a) If $C_1 = 0$, find C_2 and verify that it results in the Gram-Schmidt set using $s_1(t)$ as the first function.

(b) If $C_2 = 0$, find C_1 and verify that it results is the Gram-Schmidt set using $s_2(t)$ as the first function.

(c) Is there a solution for $C_1 = C_2$? If so, give the solution; if not, tell why not.

(d) Suppose that $C_2 = \frac{1}{2}$. Find the solution for C_1 and give the orthonormal set.

(e) Are the KL expansion basis functions for an autocorrelation function given by

$$R_{XX}(t, u) = s_1(t)s_1(u) + s_2(t)s_2(u)$$

among those determined from above. Explain.

10.24 Say we are trying to detect one of two possible signals, both of which contain random parameters, in additive white Gaussian noise. The two classes can be modeled as

$$C_1:\quad x(t) = As_1(t) + N_W(t), \qquad 0 \le t \le T$$
$$C_2:\quad x(t) = Bs_2(t) + N_W(t), \qquad 0 \le t \le T$$

A and B are Gaussian random variables with zero mean and variances σ_A^2 and σ_B^2, respectively. Both A and B are independent of the random process $N_W(t)$ for all t in $[0, T]$. Let $P(C_1)$ and $P(C_2)$ be the known a priori probabilities of the two classes.

(a) Find the optimum classifier, in the MPE sense, for the case where $s_1(t)$ and $s_2(t)$ are orthogonal on the interval $[0, T]$.

(b) Find the optimum classifier, in the MPE sense, for the case where $s_1(t)$ and $s_2(t)$ are not orthogonal on the interval $[0, T]$.

10.25 Repeat 10.24(a) for the case where A is a uniform random variable on $[0, 1]$ and B is a uniform random variable on $[-1, 0]$.

10.26 Repeat 10.24(b) for the case where A is a uniform random variable on $[0, 1]$ and B is a uniform random variable on $[-1, 0]$.

10.27 Say we are trying to detect one of two possible signals, both of which contain the same random parameters, in additive white Gaussian noise. The two classes can be modeled as

$$C_1:\quad x(t) = As_1(t) + N_W(t), \qquad 0 \le t \le T$$
$$C_2:\quad x(t) = -As_1(t) + N_W(t), \qquad 0 \le t \le T$$

A is a Gaussian random variable with zero mean and variances σ_A^2, respectively and A is independent of the random process $N_W(t)$ for all t in $[0, T]$. Let $P(C_1)$ and $P(C_2)$ be the known a priori probabilities of the two classes.

(a) Find the optimum decision rule for deciding from which class the observation $x(t)$ is obtained where the optimality is in the MPE sense.

(b) Determine the probability of error for your decision rule found in (a).

10.28 Say we are trying to detect a sum of random amplitude signals, which contain the random parameters A_1 and A_2, in additive white Gaussian noise. The two classes can be modeled as

$$C_1:\quad x(t) = A_1s_1(t) + A_2s_2(t) + N_W(t), \qquad 0 \le t \le T$$
$$C_2:\quad x(t) = N_W(t), \qquad 0 \le t \le T$$

Let A_1 and A_2 be Gaussian random variables with zero means and variances σ_A^2 and σ_B^2, respectively. A_1 and A_2 are independent of the random process $N_W(t)$ for all t in $[0, T]$. Let $P(C_1)$ and $P(C_2)$ be the known a priori probabilities of the two classes.

(a) Find the optimum classifier, in the MPE sense, for the case where $s_1(t)$ and $s_2(t)$ are orthogonal on the interval $[0, T]$.

(b) Find the optimum classifier, in the MPE sense, for the case where $s_1(t)$ and $s_2(t)$ are not orthogonal on the interval $[0, T]$.

10.29 Assume that $s_1(t), s_2(t)$, and $s_3(t)$ are known orthonormal deterministic signals on the interval $[0, 1]$. The observed signals under each class are given by

$$C_1:\quad x(t) = s_1(t) + s_2(t) + N_W(t), \qquad 0 \le t \le 1$$
$$C_2:\quad x(t) = s_2(t) + s_3(t) + N_W(t), \qquad 0 \le t \le 1$$
$$C_3:\quad x(t) = s_1(t) + s_3(t) + N_W(t), \qquad 0 \le t \le 1$$

The Gaussian noise process $N_W(t)$ has a spectral density of $\frac{1}{5}$. The classes are equally probable.

(a) Find the optimum decision rule where the optimum is in the minimum probability of error sense.

(b) Draw a block diagram to show how the decision rule operates on the continuous observation $x(t)$.

(c) Give a sufficient statistic or statistics for this problem.

(d) Calculate a probability of correct decision given class C_1 for your decision rule.

10.30 Suppose that the following two classes are equally likely. Let A be a Gaussian random variable distributed as follows: $A \sim N(1, 1)$. The observed signal $x(t)$ under each of two classes are given by

$$C_1: \quad x(t) = A\cos(2\pi t) + N_W(t), \qquad 0 \le t \le 4$$
$$C_2: \quad x(t) = N_W(t), \qquad 0 \le t \le 4$$

$N_W(t)$ represents a white Gaussian noise process with power spectral density $\frac{1}{2}N_0$.

(a) Find the optimum decision rule where the optimum is in the minimum probability of error sense, for deciding on which hypothesis to use.

(b) Draw a block diagram to show how the decision rule operates on the continuous observation $x(t)$.

(c) Give a sufficient statistic or statistics for this problem.

(d) Calculate the probability of error for your decision rule.

10.31 Suppose that the following two classes are equally likely. The observed signal $x(t)$ under each of the two classes are given by

$$C_1: \quad x(t) = s_1(t) \cdot N_W(t), \qquad 0 \le t \le 1$$
$$C_2: \quad x(t) = s_2(t) \cdot N_W(t), \qquad 0 \le t \le 1$$

Let $s_1(t)$ and $s_2(t)$ be two known orthonormal deterministic signals and $N_W(t)$ be a white Gaussian noise process with autocorrelation function $R_{WW}(\tau) = (N_0/2)\delta(\tau)$.

(a) Find the optimum decision rule to determine from which class the observation $x(t)$ is obtained where the optimality is in the minimum probability of error sense for deciding the class.

(b) Draw a block diagram to show how the decision rule operates on the continuous observation $x(t)$.

(c) Give a sufficient statistic or statistics for this problem.

(d) Calculate the probability of error for your decision rule.

10.32 Suppose that the following two hypotheses are equally likely. Let A and B be Gaussian random variables distributed as follows: $A \sim N(1, 1)$ and $B \sim N(-1, 4)$. Assume that $N_W(t)$ is a realization from a white Gaussian noise process with power spectral density $\frac{1}{2}N_0$. The observed signal $x(t)$ under each of the two hypotheses are given by

$$C_1: \quad x(t) = A\cos(2\pi t) + N_W(t), \qquad 0 \le t \le 4$$
$$C_2: \quad x(t) = B\cos(4\pi t) + N_W(t), \qquad 0 \le t \le 4$$

(a) Find the optimum decision rule to determine from which class the observations $x(t)$ is obtained where the optimum is in the minimum probability of error sense for deciding the class.

(b) Draw a block diagram to show how the decision rule operates on the continuous observation $x(t)$.

(c) Give a sufficient statistic or statistics for this problem.

(d) Calculate the probability of correct decision given class C_1 for your decision rule.

10.33 Rework Problem (10.32) for the following equal probable classes:

$$C_1: \quad x(t) = A\cos(2\pi t) + N_W(t), \qquad 0 \le t \le 4$$
$$C_2: \quad x(t) = B\cos(2\pi t) + N_W(t), \qquad 0 \le t \le 4$$

10.34 Rework Problem (10.32) for the following equal probable classes:

$$C_1: \quad x(t) = A\cos(2\pi t), \qquad 0 \le t \le 4$$
$$C_2: \quad x(t) = B\cos(4\pi t), \qquad 0 \le t \le 4$$

10.35 Rework Problem (10.32) for the following equal probable classes:

$$C_1: \quad x(t) = A\cos(2\pi t) + N_W(t), \qquad 0 \le t \le 4$$
$$C_2: \quad x(t) = B\sin(4\pi t) + N_W(t), \qquad 0 \le t \le 4$$

REFERENCES

1. Van Trees, H. L., *Detection, Estimation and Modulation Theory*, Vol. 1, Wiley, 1968.
2. Selin, Ivan, *Detection Theory*, Princeton University Press, 1965.
3. Young, Tzay Y., and Thomas W. Calvert, *Classification, Estimation, and Pattern Recognition*, Elsevier, 1974.
4. McDonough, Robert N., and Anthony D. Whalen, *Detection of Signal in Noise*, Academic Press, 1995.
5. Srinath, M. D., P. K. Rajasekaran, and R. Viswanathan, *Statistical Signal Processing with Applications*, Prentice-Hall, 1996.
6. Papoulis, Athanasios, *Probability, Random Variables, and Stochastic Processes*, McGraw-Hill, 1965.
7. Davenport, Jr., Wilbur B., and William L. Root, *An Introduction to the Theory of Random Signals in Noise*, McGraw-Hill, 1958.
8. Helstrom, Carl W., *Statistical Theory of Signal Detection*, 2nd ed., Pergamon, 1968.
9. Helstrom, Carl W., *Elements of Signal Detection and Estimation*, Prentice-Hall, 1995.
10. Poor, H. Vincent, *An Introduction to Signal Detection and Estimation*, 2nd Ed., Springer, 1994.
11. Wald, Abraham, *Sequential Analysis*, Wiley, 1947.

12. Wetherill, G. Barrie, and Kevin D. Glazebrook, *Sequential Methods in Statistics*, 3rd ed., Chapman and Hall, 1986.
13. Therrien, Charles W., *Decision, Estimation, and Classification*, Wiley, 1989.
14. Tenney, R. R., and N. R. Sandell, Detection with distributed sensors, IEEE Transactions on Aerospace and Electronic Systems, Vol. 17, No. 4, July, 1981, p 501–509.
15. Duda, Richard O., Peter E. Hart, and David Stork, *Pattern Classification*, 2nd ed., Wiley, 2001.
16. Hancock, John C., and Paul A. Wintz, *Signal Detection Theory*, McGraw-Hill, 1966.

APPENDIX A
The Bilateral Laplace Transform

The **unilateral Laplace transform**, $F(p)$, of a given positive time function, $f(t)$, is defined as

$$F(p) \triangleq \mathscr{L}_u[f(t)] = \int_0^{\infty} f(t)e^{-pt}\, dt \tag{A.1}$$

The integral has meaning only if the integral converges. The set of values of p where the integral converges is called the **region of convergence** (ROC) for the transform of $f(t)$. The unilateral inverse transform of $F(p)$ gives the $f(t)$ and can be written as the following line integral:

$$f(t) = \mathscr{L}_u^{-1}(F(p)) = \frac{1}{2\pi j}\int_{\sigma-j\infty}^{\sigma+j\infty} F(p)e^{pt}\, dp \tag{A.2}$$

where the $\sigma - j\infty$ to $\sigma + j\infty$ line lies within the region of convergence for the transform $F(p)$. The unilateral Laplace transform works only with the positive time part of a given $f(t)$, and thus the inverse is unique for positive time functions only.

When $f(t)$ contains both a positive and negative time part, the bilateral Laplace transform is useful. The **bilateral Laplace transform** of $f(t)$ is defined as

$$F(p) \triangleq \mathscr{L}_\beta[f(t)] = \int_{\infty}^{\infty} f(t)e^{-pt}\, dt \tag{A.3}$$

and it exists for certain p, whose set forms the region of convergence for the transform $F(p)$. This region of convergence is usually a strip in the complex p-plane as shown in Figure A.1.

To obtain $f(t)$ from the $F(p)$, it can be shown that inverse bilateral Laplace transform is

$$f(t) = \mathscr{L}_\beta^{-1}[F(p)] = \frac{1}{2\pi j}\int_{\sigma-j\infty}^{\sigma+j\infty} F(p)e^{pt}\, dp \tag{A.4}$$

As this is an integral in the complex p-plane, it is referred to as the complex inversion integral and the σ must be selected so that the line $\sigma - j\infty$ to $\sigma + j\infty$ is within the region of convergence. This would require $\sigma_2 < \sigma < \sigma_1$ for the region shown in Figure A.1. This complex line integral is usually evaluated by using the Cauchy-Gorsat theorem for complex contour integration. As this is a contour inte-

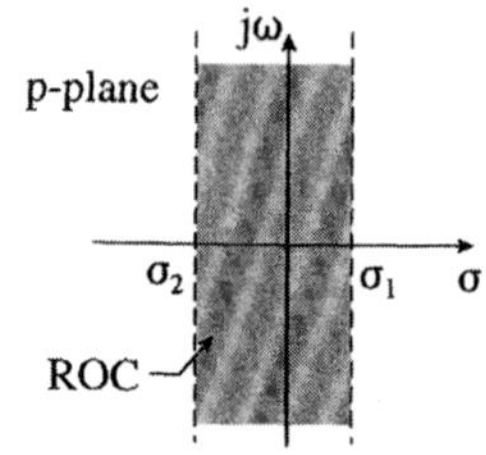

Figure A.1 Region of convergence for the bilateral Laplace transform.

gral rather than a line integral, we see that there are two ways to close the contour as shown in Figure A.2.

Closing it to the right can be shown to give the corresponding negative time answer, while closing it to the left gives the positive time answer. The contributions of the portion of the contours on the infinite semicircles can be shown to be zero for the respective positive and negative time parts of the inverse. The complex line integral can thus be broken up into two contour integrals, one for positive time and the other for negative time, as follows:

$$f(t) = \left(\frac{1}{2\pi j}\oint_{C+} F(p)e^{pt}\,dp\right)u(t) + \left(\frac{1}{2\pi j}\oint_{C-} F(p)e^{pt}\,dp\right)u(-t) \tag{A.5}$$

Using the residue theorem to evaluate the contour integrals, we can write

$$f(t) = \left(\sum_{\substack{+\text{time}\\ \text{poles}}} \text{residues of } F(p)e^{pt}\right)u(t) + \left(\sum_{\substack{-\text{time}\\ \text{poles}}} \text{residues of } F(p)e^{pt}\right)u(-t) \tag{A.6}$$

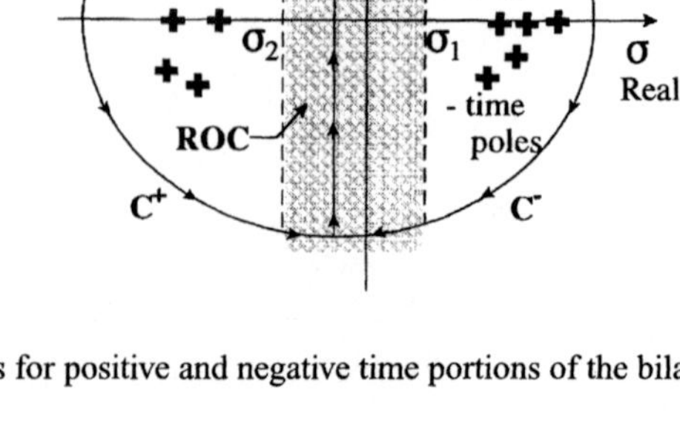

Figure A.2 Contours for positive and negative time portions of the bilateral Laplace inverse.

For rational $F(p)$, meaning that $F(p)$ is a ratio of polynomials in p, the inverse can also be found using partial fraction expansions.

The bilateral Laplace transform can be obtained entirely by using only unilateral Laplace tables for many problems. This procedure is now presented.

The integral of Eq. (A.3) can be broken up into two integrals, one for positive time and the other for negative time, as

$$\mathscr{L}_\beta[f(t)] = \int_{-\infty}^{0} f(t)e^{-pt}\,dt + \int_{0}^{\infty} f(t)e^{-pt}\,dt \tag{A.7}$$

If t is replaced by $-t$ in the first integral, the resulting $\mathscr{L}[f(t)]$ is

$$\mathscr{L}_\beta[f(t)] = \int_{\infty}^{0} f(-t)e^{+pt}(-dt) + \int_{0}^{\infty} f(t)e^{-pt}\,dt \tag{A.8}$$

Flipping the limits on the first integral cancels the minus sign on the dt, so the bilateral Laplace transform becomes

$$\mathscr{L}_\beta[f(t)] = \int_{0}^{\infty} f(-t)e^{+pt}\,dt + \int_{0}^{\infty} f(t)e^{-pt}\,dt \tag{A.9}$$

The first integral represents the unilateral Laplace transform of $f(-t)$ with p replaced by $-p$, while the second integral is the unilateral Laplace transform of the positive time part of $f(t)$.

Thus the bilateral Laplace transform can be written as

$$\mathscr{L}_\beta[f(t)] = (\mathscr{L}_u[f(-t)])_{p\to -p} + \mathscr{L}_u[f(t)] \tag{A.10}$$

The region of convergence for the bilateral Laplace transform is obtained as the intersection of the regions of convergence for each of the transforms above with the region reversed for the unilateral Laplace transform of the $f(-t)$. Thus, since only unilateral transforms are needed to obtain bilateral transforms including the regions of convergence, the standard tables of Laplace transform can be advantageously used to obtain transforms of an $f(t)$ that has negative time components.

EXAMPLE A.1

Find the bilateral Laplace transform of $f(t) = 2e^{-3t}\mu(t) + 5e^{2t}\mu(-t)$ including the region of convergence.

SOLUTION

The solution will be obtained by combining the two unilateral Laplace transforms given in Eq. (A.10). First the unilateral Laplace transform of $f(t)$ is found as

$$\begin{aligned}\mathscr{L}_u[f(t)] &= \int_0^\infty f(t)e^{-pt}\,dt \\ &= \int_0^\infty 2e^{-3t}\mu(t)e^{-pt}\,dt + \int_0^\infty 5e^{2t}\mu(-t)e^{-pt}\,dt\end{aligned} \tag{A.11}$$

The second integral is zero because the $\mu(-t)$ makes the integrand zero for $t > 0$, and therefore evaluating the first integral gives the second part of Eq. (A.10) as

$$\begin{aligned}\mathscr{L}_u[f(t)] &= \frac{2}{-(p+3)}\,e^{-(p+3)t}\Big]_0^\infty \\ &= \frac{2}{p+3} \\ &\text{ROC:}\quad \text{real }(p) > 0\end{aligned} \tag{A.12}$$

The unilateral Laplace transform of $f(-t)$ is as follows:

$$\begin{aligned}\mathscr{L}_u[f(-t)] &= \int_0^\infty f(-t)e^{-pt}\,dt \\ &= \int_0^\infty 2e^{3t}\mu(-t)e^{-pt}\,dt + \int_0^\infty 5e^{-2t}\mu(t)e^{-pt}\,dt\end{aligned} \tag{A.13}$$

Because of the $\mu(-t)$ the first integral is zero, leaving just the second integral as

$$\begin{aligned}\mathscr{L}_u[f(-t)] &= \frac{5}{p+2}\,e^{-(p+2)t}\Big]_0^\infty \\ &= \frac{5}{p+2} \\ &\text{ROC:}\quad \text{real }(p) > -2\end{aligned} \tag{A.14}$$

Then, on replacing the p by $-p$ in (A.14) and inverting the region of convergence, we have the other part of the total transform as

$$\begin{aligned}(\mathscr{L}_u[f(-t)])_{p\to -p} &= \frac{5}{-p+2} \\ \text{ROC:}\quad \text{real }(p) &< 2\end{aligned} \tag{A.15}$$

We now combine the two unilateral Laplace transforms, Eq. (A.12) and (A.15) and obtain the bilateral Laplace transform of $f(t)$ as

$$\begin{aligned}\mathscr{L}_\beta[f(t)] &= (\mathscr{L}_u[f(t)])_{p\to -p} + \mathscr{L}_u[f(t)] \\ &= \frac{2}{p+3} + \frac{5}{-p+2} \qquad \text{(A.16)} \\ &\text{ROC:} \quad -3 < \text{real}\,(p) < 2\end{aligned}$$

■

An example is now presented that shows how a partial fraction expansion and standard unilateral Laplace transform tables can be used to obtain the inverse bilateral Laplace transform whenever $F(p)$ is a rational function of p. If $F(p)$ is not rational in p, then the inverse must be evaluated using the contour integrals given in (A.5), or alternatively the residues given in (A.6).

EXAMPLE A.2

Find $\mathscr{L}_\beta^{-1}[F(p)]$ where $F(p)$ and the region of convergence are given by

$$\begin{aligned}F(p) &= \frac{2}{(-p+2)(p+1)} \qquad \text{(A.17)} \\ &\text{ROC:} \quad -1 < \text{real}\,(p) < 2\end{aligned}$$

SOLUTION

The solution can be obtained by performing a partial fraction expansion of $F(p)$, since $F(p)$ is a rational expression in p. Even though $F(p)$ contains both right and left half-plane poles, the partial fraction expansion is as usual and determined as follows:

$$F(p) = \frac{2}{(-p+2)(p+1)} = \frac{A}{-p+2} + \frac{B}{p+1}$$

where A and B are obtained in the standard way as

$$A = (-p+2)F(p)|_{p=2} = \left.\frac{2}{p+1}\right|_{p=2} = \frac{2}{3}$$

$$B = (p+1)F(p)|_{p=-1} = \left.\frac{2}{-p+2}\right|_{p=-1} = \frac{2}{3}$$

Thus the partial fraction becomes

$$F(p) = \frac{\frac{2}{3}}{-p+2} + \frac{\frac{2}{3}}{p+1}$$

Recognizing the first term as having a pole at $p = 2$, which gives a negative time answer, and the second as having a pole at $p = -1$, which gives a positive time answer, we can write the resulting $f(t)$ as

$$\mathscr{L}_\beta^{-1}[F(p)] = \mathscr{L}_\beta^{-1}\left[\frac{\frac{2}{3}}{-p+2} + \frac{\frac{2}{3}}{p+1}\right]$$
$$= \tfrac{2}{3}e^{2t}\mu(-t) + \tfrac{2}{3}e^{-t}\mu(t)$$

In using the partial fraction method, remember that the numerator of $F(p)$ must be of a lower degree than that of the denominator. If this is not the case, then the numerator must be divided by the denominator until it attains this form before a partial fraction can be performed. The next example illustrates this procedure.

EXAMPLE A.3

Find $\mathscr{L}_\beta^{-1}[F(p)]$, where $F(p)$ and the region of convergence are given by

$$F(p) = \frac{-p^2+3}{(-p+2)(p+1)}$$

$$\text{ROC:} \quad -1 < \text{real } (p) < 2$$

SOLUTION

As the degree of the numerator of $F(p)$ is the same as the degree of the denominator, namely 2, we must divide out before using the partial coefficient expansion as follows:

$$\frac{-p^2+3}{(-p+2)(p+1)} = \frac{-p^2+3}{-p^2+p+2} = 1 + \frac{-p+1}{-p^2+p+2}$$
$$= 1 + \frac{-p+1}{(-p+2)(p+1)}$$

The partial fraction expansion is easily obtained for the second term using the standard procedure as

$$\frac{-p+1}{(-p+2)(p+1)} = \frac{A}{-p+2} + \frac{B}{p+1}$$
$$= \frac{-\frac{1}{3}}{-p+2} + \frac{\frac{2}{3}}{p+1}$$

From this partial fraction the inverse bilateral Laplace transform can now be found as

$$\mathscr{L}_\beta^{-1}\left[\frac{-p^2+3}{(-p+2)(p+1)}\right] = \mathscr{L}_\beta^{-1}\left[1 + \frac{-\frac{1}{3}}{-p+2} + \frac{\frac{2}{3}}{p+1}\right]$$

The inverse bilateral Laplace of 1 is a delta function at the origin. On recognizing the inverse of the second term as a negative time part and the last term as the positive time part, we can write the final inverse as

$$f(t) = \mathscr{L}_\beta^{-1}\left[\frac{-p^2+3}{(-p+2)(p+1)}\right] = \delta(t) - \frac{1}{3}\,e^{2t}\mu(-t) + \frac{2}{3}\,e^{-t}\mu(t) \qquad \blacksquare$$

APPENDIX B
Table of Binomial Probabilities

$$b(n, k, p) \triangleq \binom{n}{k} p^k (1-p)^{n-k} = \frac{n!}{(n-k)!k!} p^k (1-p)^{n-k}, \qquad 0 < p \leq 0.5$$

$$b(n, k, p) = b(n, k, 1-p), \qquad p > 0.5$$

n	*k*	**0.05**	**0.1**	**0.15**	**0.2**	**0.25**	**0.3**	**0.35**	**0.4**	**0.45**	**0.5**
2	0	0.9025	0.81	0.7225	0.64	0.5625	0.49	0.4225	0.36	0.3025	0.25
2	1	0.095	0.18	0.255	0.32	0.375	0.42	0.455	0.48	0.495	0.5
2	2	0.0025	0.01	0.0225	0.04	0.0625	0.09	0.1225	0.16	0.2025	0.25
3	0	0.857375	0.729	0.614125	0.512	0.421875	0.343	0.274625	0.216	0.166375	0.125
3	1	0.135375	0.243	0.325125	0.384	0.421875	0.441	0.443625	0.432	0.408375	0.375
3	2	0.007125	0.027	0.057375	0.096	0.140625	0.189	0.238875	0.288	0.334125	0.375
3	3	0.000125	0.001	0.003375	0.008	0.015625	0.027	0.042875	0.064	0.091125	0.125
4	0	0.8145063	0.6561	0.5220063	0.4096	0.3164063	0.2401	0.1785063	0.1296	0.0915063	0.0625
4	1	0.171475	0.2916	0.368475	0.4096	0.421875	0.4116	0.384475	0.3456	0.299475	0.25
4	2	0.0135375	0.0486	0.0975375	0.1536	0.2109375	0.2646	0.3105375	0.3456	0.3675375	0.375
4	3	0.000475	0.0036	0.011475	0.0256	0.046875	0.0756	0.111475	0.1536	0.200475	0.25
4	4	6.25E-06	0.0001	0.0005063	0.0016	0.0039063	0.0081	0.0150063	0.0256	0.0410063	0.0625
5	0	0.7737809	0.59049	0.4437053	0.32768	0.2373047	0.16807	0.1160291	0.07776	0.0503284	0.03125
5	1	0.2036266	0.32805	0.3915047	0.4096	0.3955078	0.36015	0.3123859	0.2592	0.2058891	0.15625
5	2	0.0214344	0.0729	0.1381781	0.2048	0.2636719	0.3087	0.3364156	0.3456	0.3369094	0.3125
5	3	0.0011281	0.0081	0.0243844	0.0512	0.0878906	0.1323	0.1811469	0.2304	0.2756531	0.3125
5	4	2.969E-05	0.00045	0.0021516	0.0064	0.0146484	0.02835	0.0487703	0.0768	0.1127672	0.15625
5	5	3.125E-07	0.00001	7.594E-05	0.00032	0.0009766	0.00243	0.0052522	0.01024	0.0184528	0.03125
6	0	0.7350919	0.531441	0.3771495	0.262144	0.1779785	0.117649	0.0754189	0.046656	0.0276806	0.015625
6	1	0.2321343	0.354294	0.3993348	0.393216	0.355957	0.302526	0.243661	0.186624	0.1358868	0.09375
6	2	0.030544	0.098415	0.1761771	0.24576	0.2966309	0.324135	0.3280052	0.31104	0.2779502	0.234375
6	3	0.0021434	0.01458	0.0414534	0.08192	0.1318359	0.18522	0.2354909	0.27648	0.3032184	0.3125
6	4	8.461E-05	0.001215	0.0054865	0.01536	0.032959	0.059535	0.0951021	0.13824	0.1860659	0.234375
6	5	1.781E-06	5.4E-05	0.0003873	0.001536	0.0043945	0.010206	0.0204835	0.036864	0.0608943	0.09375
6	6	1.563E-08	0.000001	1.139E-05	6.4E-05	0.0002441	0.000729	0.0018383	0.004096	0.0083038	0.015625

n	k	0.05	0.1	0.15	0.2	0.25	0.3	0.35	0.4	0.45	0.5
7	0	0.6983373	0.4782969	0.3205771	0.2097152	0.1334839	0.0823543	0.0490223	0.0279936	0.0152244	0.007813
7	1	0.2572822	0.3720087	0.396007	0.3670016	0.3114624	0.2470629	0.1847763	0.1306368	0.087194	0.0546875
7	2	0.0406235	0.1240029	0.2096508	0.2752512	0.3114624	0.3176523	0.2984848	0.2612736	0.2140217	0.1640625
7	3	0.0035635	0.0229635	0.061662	0.114688	0.1730347	0.2268945	0.2678709	0.290304	0.2918477	0.2734375
7	4	0.0001876	0.0025515	0.0108815	0.028672	0.0576782	0.0972405	0.1442382	0.193536	0.2387845	0.2734375
7	5	5.923E-06	0.0001701	0.0011522	0.0043008	0.0115356	0.0250047	0.0466	0.0774144	0.1172215	0.1640625
7	6	1.039E-07	0.0000063	6.777E-05	0.0003584	0.0012817	0.0035721	0.0083641	0.0172032	0.0319695	0.0546875
7	7	7.813E-10	0.0000001	1.709E-06	0.0000128	6.104E-05	0.0002187	0.0006434	0.0016384	0.0037367	0.007813
8	0	0.6634204	0.4304672	0.2724905	0.1677722	0.1001129	0.057648	0.0318645	0.0167962	0.0083734	0.0039063
8	1	0.2793349	0.3826375	0.3846925	0.3355443	0.2669678	0.1976503	0.1372624	0.0895795	0.0548077	0.03125
8	2	0.0514564	0.1488035	0.2376042	0.2936013	0.3114624	0.2964755	0.2586868	0.2090189	0.1569492	0.109375
8	3	0.0054165	0.0330674	0.0838603	0.1468006	0.2076416	0.2541218	0.2785858	0.2786918	0.256826	0.21875
8	4	0.0003563	0.0045927	0.0184986	0.0458752	0.0865173	0.1361367	0.1875097	0.2322432	0.262663	0.2734375
8	5	1.5E-05	0.0004082	0.0026116	0.009175	0.0230713	0.0466754	0.0807734	0.123863	0.1719249	0.21875
8	6	3.948E-07	2.268E-05	0.0002304	0.0011469	0.0038452	0.0100019	0.0217467	0.0412877	0.0703329	0.109375
8	7	5.938E-09	7.2E-07	1.162E-05	8.192E-05	0.0003662	0.0012247	0.0033456	0.0078643	0.0164415	0.03125
8	8	3.906E-11	1E-08	2.563E-07	2.56E-06	1.526E-05	6.561E-05	0.0002252	0.0006554	0.0016815	0.0039063
9	0	0.6302494	0.3874205	0.2316169	0.1342177	0.0750847	0.0403536	0.0207119	0.0100777	0.0046054	0.0019531
9	1	0.2985392	0.3874205	0.3678622	0.3019899	0.2252541	0.1556496	0.1003731	0.0604662	0.0339122	0.0175781
9	2	0.0628504	0.1721869	0.2596674	0.3019899	0.3003387	0.2668279	0.2161882	0.1612431	0.1109855	0.0703125
9	3	0.0077185	0.044641	0.1069219	0.1761608	0.2335968	0.2668279	0.2716211	0.2508227	0.2118815	0.1640625
9	4	0.0006094	0.0074402	0.0283029	0.0660603	0.1167984	0.1715322	0.2193863	0.2508227	0.2600363	0.2460938
9	5	3.207E-05	0.0008267	0.0049946	0.0165151	0.0389328	0.0735138	0.1181311	0.1672151	0.212757	0.2460938
9	6	1.125E-06	6.124E-05	0.0005876	0.0027525	0.0086517	0.0210039	0.042406	0.0743178	0.1160493	0.1640625
9	7	2.538E-08	2.916E-06	4.444E-05	0.0002949	0.001236	0.0038579	0.009786	0.0212337	0.0406926	0.0703125
9	8	3.34E-10	8.1E-08	1.961E-06	1.843E-05	0.000103	0.0004133	0.0013173	0.0035389	0.0083235	0.0175781
9	9	1.953E-12	1E-09	3.844E-08	5.12E-07	3.815E-06	1.968E-05	7.882E-05	0.0002621	0.0007567	0.0019531

(continued)

n	*k*	**0.05**	**0.1**	**0.15**	**0.2**	**0.25**	**0.3**	**0.35**	**0.4**	**0.45**	**0.5**
10	0	0.5987369	0.3486784	0.1968744	0.1073742	0.0563135	0.0282475	0.0134627	0.0060466	0.002533	0.0009766
10	1	0.3151247	0.3874205	0.3474254	0.2684355	0.1877117	0.1210608	0.0724917	0.0403108	0.0207241	0.0097656
10	2	0.0746348	0.1937102	0.2758967	0.3019899	0.2815676	0.2334744	0.175653	0.1209324	0.0763026	0.0439453
10	3	0.0104751	0.0573956	0.1298337	0.2013266	0.2502823	0.2668279	0.2522196	0.2149908	0.1664783	0.1171875
10	4	0.0009648	0.0111603	0.0400957	0.0880804	0.145998	0.2001209	0.2376685	0.2508227	0.2383666	0.2050781
10	5	6.094E-05	0.001488	0.0084909	0.0264241	0.0583992	0.1029193	0.1535704	0.2006581	0.2340327	0.2460938
10	6	2.673E-06	0.0001378	0.0012487	0.005505	0.016222	0.0367569	0.0689098	0.1114767	0.1595678	0.2050781
10	7	8.038E-08	8.748E-06	0.0001259	0.0007864	0.0030899	0.0090017	0.021203	0.0424673	0.0746031	0.1171875
10	8	1.586E-09	3.645E-07	8.333E-06	7.373E-05	0.0003862	0.0014467	0.0042814	0.0106168	0.0228896	0.0439453
10	9	1.855E-11	9E-09	3.268E-07	4.096E-06	2.861E-05	0.0001378	0.0005123	0.0015729	0.0041617	0.0097656
10	10	9.766E-14	1E-10	5.767E-09	1.024E-07	9.537E-07	5.905E-06	2.759E-05	0.0001049	0.00034	0.0009766

APPENDIX C
Table of Discrete Random Variables and Properties

Probability Density Function	Mean and Variances
Discrete uniform random variable $(b > a)$ $f_X(x) = \sum_{i=0}^{n} \frac{1}{n+1} \delta(x - (a + i(b-a)/n))$ $\Phi_X(\omega) = \sum_{i=0}^{n} \frac{1}{n+1} e^{j\omega(a+i(b-a)/n)}$	$\eta_X = (a+b)/2$ $\sigma_X^2 = \frac{(n+2)(b-a)^2}{12n}$ $E[X^2] = ab + \frac{(2n+1)(b-a)^2}{6n}$
Bernoulli random variable $f_X(x) = (1-p)\delta(x) + p\delta(x-1)$ $\Phi_X(\omega) = (1-p) + pe^{j\omega}$	$\eta_X = p$ $\sigma_X^2 = p - p^2$ $E[X^2] = p$
Binomial random variable $f_X(x) = \sum_{k=0}^{n} \binom{n}{k} p^k (1-p)^{n-k} \delta(x-k)$ $\Phi_X(\omega) = ((1-p) + pe^{j\omega})^n$	$\eta_X = np$ $\sigma_X^2 = np(1-p)$ $E[X^2] = np(1-p) + n^2p^2$
Poisson random variable $(\lambda > 0)$ $f_X(x) = \sum_{k=0}^{\infty} \lambda^k \frac{e^{-\lambda}}{k!} \delta(x-k)$ $\Phi_X(\omega) = e^{\lambda(e^{j\omega}-1)}$	$\eta_X = \lambda$ $\sigma_X^2 = \lambda$ $E[X^2] = \lambda + \lambda^2$
Geometric random variable $f_X(x) = \sum_{k=1}^{\infty} p(1-p)^{k-1} \delta(x-k)$ $\Phi_X(\omega) = \frac{pe^{j\omega}}{1-(1-p)e^{j\omega}}$	$\eta_X = \frac{1}{p}$ $\sigma_X^2 = \frac{1-p}{p^2}$ $E[X^2] = \frac{2-p}{p^2}$

(*continued*)

Negative binomial random variable

$$f_X(x) = \sum_{j=k}^{\infty} \binom{j-1}{k-1} p^k (1-p)^{j-k} \delta(x-j)$$

$$\Phi_X(\omega) = p^k e^{j\omega k}(1-(1-p)e^{j\omega})^{-k}$$

$$\eta_X = \frac{k}{p}$$

$$\sigma_X^2 = \frac{k(1-p)}{p^2}$$

$$E[X^2] = \frac{k - kp + k^2}{p^2}$$

APPENDIX D
Table of Continuous Random Variables and Properties

Probability Density Function Characteristic Function	Mean Variance Second Moment
Uniform: $b > a$ $f_X(x) = \frac{1}{b-a}\,[\mu(x-a) - \mu(x-b)]$ $\Phi_X(\omega) = \frac{e^{j\omega b} - e^{j\omega a}}{(b-a)j\omega}$	$\eta_X = (a+b)/2$ $\sigma_X^2 = (b-a)^2/12$ $E[X^2] = (b^2 + ab + b^2)/3$
Gaussian (normal): $\sigma > 0$ $f_X(x) = \frac{1}{\sqrt{2\pi}\sigma} \exp\left\{-\frac{(x-\eta)^2}{2\sigma^2}\right\}$ $\Phi_X(\omega) = \exp(\eta\, j\omega - 1/2\sigma^2\omega^2)$	$\eta_X = \eta$ $\sigma_X^2 = \sigma^2$ $E[X^2] = \sigma^2 + \eta^2$
Exponential: $a > 0$ $f_X(x) = \begin{cases} ae^{-ax}, & x \geq 0 \\ 0, & \text{otherwise} \end{cases}$ $\Phi_X(\omega) = \frac{a}{a - j\omega}$	$\eta_X = 1/a$ $\sigma_X^2 = 1/a^2$ $E[X^2] = 2/a^2$
Rayleigh: $\alpha > 0$ $f_X(x) = \frac{x}{\alpha^2} \exp\left(-\frac{x^2}{2\alpha^2}\right)\mu(x)$ $\Phi_X(\omega) = 1 + \sqrt{\frac{\pi}{2}}\alpha j\omega \exp\left(\frac{-w^2\alpha^2}{2}\right) \operatorname{erfc}\left(-\frac{j\omega\alpha}{\sqrt{2}}\right)$	$\eta_X = \sqrt{\frac{\pi}{2}}\alpha$ $\sigma_X^2 = \left(2 - \frac{\pi}{2}\right)\alpha^2$ $E[X^2] = 2\alpha^2$

(continued)

Probability Density Function Characteristic Function	Mean Variance Second Moment
Gamma: $\alpha > 0 \quad \beta > 0$ $f_X(x) = \dfrac{1}{\beta^\alpha \Gamma(\alpha)} x^{\alpha-1} e^{-x/\beta} \mu(x)$ where $\Gamma(\alpha) = \int_0^\infty y^{\alpha-1} e^{-y}\, dy$ $\Phi_X(\omega) = (1 - j\omega\beta)^{-\alpha}$	$\eta_X = \beta\alpha$ $\sigma_X^2 = \beta^2 \alpha$ $E[X^2] = \beta^2 \alpha(1+\alpha)$
Chi square: $N > 0$ $f_X(x) = \dfrac{x^{(N/2)-1}}{2^{N/2}\Gamma(N/2)} \exp(-\frac{1}{2}x)\mu(x)$ $\Phi_X(\omega) = (1 - 2j\omega)^{-N/2}$	$\eta_X = N$ $\sigma_X^2 = 2N$ $E[X^2] = 2N + N^2$
Beta: $\alpha > 0 \; \beta > 0$ $f_X(x) = \begin{cases} kx^{\alpha-1}(1-x)^{\beta-1}, & 0 < x < 1 \\ 0, & \text{otherwise} \end{cases}$ $k = \dfrac{\Gamma(\alpha+\beta)}{\Gamma(\alpha)\Gamma(\beta)}$ and $\Gamma(\alpha) = \int_0^\infty y^{\alpha-1} e^{-y}\, dy$	$\eta_X = \alpha/(\alpha+\beta)$ $\sigma_X^2 = \dfrac{\alpha\beta}{(\alpha+\beta)^2(\alpha+\beta+1)}$ $E[X^2] = \dfrac{\alpha\beta + \alpha^2(\alpha+\beta+1)}{(\alpha+\beta)^2(\alpha+\beta+1)}$
Cauchy: $a > 0$ $f_X(x) = \dfrac{a}{\pi(a^2 + x^2)}$ $\Phi_X(\omega)$ is not useful	$\eta_X = 0$ (P.V.) σ_X^2 and $E[X^2]$ do not exist
Log normal: $\sigma > 0$ $f_X(x) = \dfrac{1}{x\sqrt{2\pi}\sigma} \exp\left\{-\dfrac{(\ln x - \eta)^2}{2\sigma^2}\right\}\mu(x)$	$\eta_X = \exp(\eta + \frac{1}{2}\sigma^2)$ $\sigma_X^2 = \exp(2\eta + 2\sigma^2)$ $\quad - \exp(2\eta + \sigma^2)$ $E[X^2] = \exp(2\alpha + 2\sigma^2)$
Double exponential: $\beta > 0$ $f_X(x) = \frac{1}{2}\beta \exp(-\beta\lvert x - \alpha \rvert)$ $\Phi_X(\omega) = \dfrac{\beta^2 e^{j\omega\alpha}}{\beta^2 + \omega^2}$	$\eta_X = \alpha$ $\sigma_X^2 = 2\beta^{-2}$ $E[X^2] = 2\beta^{-2} + \alpha^2$

APPENDIX E
Table for Gaussian Cumulative Distribution Function

$$\Phi(x) = \int_{-\infty}^{x} \frac{1}{\sqrt{2\pi}} e^{-x^2/2}\, dx, \quad \Phi(-x) = 1 - \Phi(x), \quad 0 \le x < 1.99$$

X\0.0	**0.00**	**0.01**	**0.02**	**0.03**	**0.04**	**0.05**	**0.06**	**0.07**	**0.08**	**0.09**
0	0.50000	0.50399	0.50798	0.51197	0.51595	0.51994	0.52392	0.52790	0.53188	0.53586
0.1	0.53983	0.54380	0.54776	0.55172	0.55567	0.55962	0.56356	0.56749	0.57142	0.57535
0.2	0.57926	0.58317	0.58706	0.59095	0.59483	0.59871	0.60257	0.60642	0.61026	0.61409
0.3	0.61791	0.62172	0.62552	0.62930	0.63307	0.63683	0.64058	0.64431	0.64803	0.65173
0.4	0.65542	0.65910	0.66276	0.66640	0.67003	0.67364	0.67724	0.68082	0.68439	0.68793
0.5	0.69146	0.69497	0.69847	0.70194	0.70540	0.70884	0.71226	0.71566	0.71904	0.72240
0.6	0.72575	0.72907	0.73237	0.73565	0.73891	0.74215	0.74537	0.74857	0.75175	0.75490
0.7	0.75804	0.76115	0.76424	0.76730	0.77035	0.77337	0.77637	0.77935	0.78230	0.78524
0.8	0.78814	0.79103	0.79389	0.79673	0.79955	0.80234	0.80511	0.80785	0.81057	0.81327
0.9	0.81594	0.81859	0.82121	0.82381	0.82639	0.82894	0.83147	0.83398	0.83646	0.83891
1	0.84134	0.84375	0.84614	0.84849	0.85083	0.85314	0.85543	0.85769	0.85993	0.86214
1.1	0.86433	0.86650	0.86864	0.87076	0.87286	0.87493	0.87698	0.87900	0.88100	0.88298
1.2	0.88493	0.88686	0.88877	0.89065	0.89251	0.89435	0.89617	0.89796	0.89973	0.90147
1.3	0.90320	0.90490	0.90658	0.90824	0.90988	0.91149	0.91308	0.91466	0.91621	0.91774
1.4	0.91924	0.92073	0.92220	0.92364	0.92507	0.92647	0.92785	0.92922	0.93056	0.93189
1.5	0.93319	0.93448	0.93574	0.93699	0.93822	0.93943	0.94062	0.94179	0.94295	0.94408
1.6	0.94520	0.94630	0.94738	0.94845	0.94950	0.95053	0.95154	0.95254	0.95352	0.95449
1.7	0.95543	0.95637	0.95728	0.95818	0.95907	0.95994	0.96080	0.96164	0.96246	0.96327
1.8	0.96407	0.96485	0.96562	0.96638	0.96712	0.96784	0.96856	0.96926	0.96995	0.97062
1.9	0.97128	0.97193	0.97257	0.97320	0.97381	0.97441	0.97500	0.97558	0.97615	0.97670

$$\Phi(x) = \int_{-\infty}^{x} \frac{1}{\sqrt{2\pi}} e^{-x^2/2}\,dx, \quad \Phi(-x) = 1 - \Phi(x), \quad 2 \le x < 3.59$$

X\0.0	**0.00**	**0.01**	**0.02**	**0.03**	**0.04**	**0.05**	**0.06**	**0.07**	**0.08**	**0.09**
2	0.97725	0.97778	0.97831	0.97882	0.97932	0.97982	0.98030	0.98077	0.98124	0.98169
2.1	0.98214	0.98257	0.98300	0.98341	0.98382	0.98422	0.98461	0.98500	0.98537	0.98574
2.2	0.98610	0.98645	0.98679	0.98713	0.98745	0.98778	0.98809	0.98840	0.98870	0.98899
2.3	0.98928	0.98956	0.98983	0.99010	0.99036	0.99061	0.99086	0.99111	0.99134	0.99158
2.4	0.99180	0.99202	0.99224	0.99245	0.99266	0.99286	0.99305	0.99324	0.99343	0.99361
2.5	0.99379	0.99396	0.99413	0.99430	0.99446	0.99461	0.99477	0.99492	0.99506	0.99520
2.6	0.99534	0.99547	0.99560	0.99573	0.99585	0.99598	0.99609	0.99621	0.99632	0.99643
2.7	0.99653	0.99664	0.99674	0.99683	0.99693	0.99702	0.99711	0.99720	0.99728	0.99736
2.8	0.99744	0.99752	0.99760	0.99767	0.99774	0.99781	0.99788	0.99795	0.99801	0.99807
2.9	0.99813	0.99819	0.99825	0.99831	0.99836	0.99841	0.99846	0.99851	0.99856	0.99861
3	0.99865	0.99869	0.99874	0.99878	0.99882	0.99886	0.99889	0.99893	0.99896	0.99900
3.1	0.99903	0.99906	0.99910	0.99913	0.99916	0.99918	0.99921	0.99924	0.99926	0.99929
3.2	0.99931	0.99934	0.99936	0.99938	0.99940	0.99942	0.99944	0.99946	0.99948	0.99950
3.3	0.99952	0.99953	0.99955	0.99957	0.99958	0.99960	0.99961	0.99962	0.99964	0.99965
3.4	0.99966	0.99968	0.99969	0.99970	0.99971	0.99972	0.99973	0.99974	0.99975	0.99976
3.5	0.99977	0.99978	0.99978	0.99979	0.99980	0.99981	0.99981	0.99982	0.99983	0.99983

INDEX

Printed in the United Kingdom
by Lightning Source UK Ltd.
134033UK00001B/36/A

9 780471 259756